LINEAR PARTIAL DIFFERENTIAL EQUATIONS AND FOURIER THEORY

Do you want a rigorous book that remembers where PDEs come from and what they look like? This highly visual introduction to linear PDEs and initial/boundary value problems connects the theory to physical reality, all the time providing a rigorous mathematical foundation for all solution methods.

Readers are gradually introduced to abstraction – the most powerful tool for solving problems – rather than simply drilled in the practice of imitating solutions to given examples. The book is therefore ideal for students in mathematics and physics who require a more theoretical treatment than is given in most introductory texts.

Also designed with lecturers in mind, the fully modular presentation is easily adapted to a course of one-hour lectures, and a suggested 12-week syllabus is included to aid planning. Downloadable files for the hundreds of figures, hundreds of challenging exercises, and practice problems that appear in the book are available online, as are solutions.

MARCUS PIVATO is Associate Professor in the Department of Mathematics at Trent University in Peterborough, Ontario.

LINEAR PARTIAL DIFFERENTIAL EQUATIONS AND FOURIER THEORY

MARCUS PIVATO

Trent University

CAMBRIDGE
UNIVERSITY PRESS

CAMBRIDGE UNIVERSITY PRESS
Cambridge, New York, Melbourne, Madrid, Cape Town, Singapore, São Paulo, Delhi

Cambridge University Press
The Edinburgh Building, Cambridge CB2 8RU, UK

Published in the United States of America by Cambridge University Press, New York

www.cambridge.org
Information on this title: www.cambridge.org/9780521199704

First published 2010

Printed in the United Kingdom at the University Press, Cambridge

A catalogue record for this publication is available from the British Library

Library of Congress Cataloguing in Publication data
Pivato, Marcus, 1974–
Linear partial differential equations and Fourier theory / Marcus Pivato.
p. cm.
Includes bibliographical references and index.
ISBN 978-0-521-19970-4 (hardback)
1. Differential equations, Partial. 2. Differential equations, Linear.
3. Fourier transformations. I. Title.
QA374.P58 2010
515′.353 – dc22 2009037745

ISBN 978-0-521-19970-4 Hardback
ISBN 978-0-521-13659-4 Paperback

Additional resources for this publication at http://xaravve.trentu.ca/pde

To Joseph and Emma Pivato
for their support, encouragement,
and inspiring example.

Contents

Preface   *page* xv
What's good about this book?   xviii
Suggested 12-week syllabus   xxv

Part I Motivating examples and major applications   1

1 Heat and diffusion   5
   1A   Fourier's law   5
         1A(i)   . . . in one dimension   5
         1A(ii)   . . . in many dimensions   5
   1B   The heat equation   7
         1B(i)   . . . in one dimension   7
         1B(ii)   . . . in many dimensions   8
   1C   The Laplace equation   11
   1D   The Poisson equation   14
   1E   Properties of harmonic functions   18
   1F*   Transport and diffusion   21
   1G*   Reaction and diffusion   21
   1H   Further reading   23
   1I   Practice problems   23

2 Waves and signals   25
   2A   The Laplacian and spherical means   25
   2B   The wave equation   29
         2B(i)   . . . in one dimension: the string   29
         2B(ii)   . . . in two dimensions: the drum   33
         2B(iii)   . . . in higher dimensions   35
   2C   The telegraph equation   36
   2D   Practice problems   37

3	Quantum mechanics		39
	3A	Basic framework	39
	3B	The Schrödinger equation	43
	3C	Stationary Schrödinger equation	47
	3D	Further reading	56
	3E	Practice problems	56
Part II	**General theory**		**59**
4	Linear partial differential equations		61
	4A	Functions and vectors	61
	4B	Linear operators	63
	4B(i)	. . . on finite-dimensional vector spaces	63
	4B(ii)	. . . on $\mathcal{C}^\infty$	65
	4B(iii)	Kernels	67
	4B(iv)	Eigenvalues, eigenvectors, and eigenfunctions	67
	4C	Homogeneous vs. nonhomogeneous	68
	4D	Practice problems	70
5	Classification of PDEs and problem types		73
	5A	Evolution vs. nonevolution equations	73
	5B	Initial value problems	74
	5C	Boundary value problems	75
	5C(i)	Dirichlet boundary conditions	76
	5C(ii)	Neumann boundary conditions	80
	5C(iii)	Mixed (or Robin) boundary conditions	83
	5C(iv)	Periodic boundary conditions	85
	5D	Uniqueness of solutions	87
	5D(i)	Uniqueness for the Laplace and Poisson equations	88
	5D(ii)	Uniqueness for the heat equation	91
	5D(iii)	Uniqueness for the wave equation	94
	5E*	Classification of second-order linear PDEs	97
	5E(i)	. . . in two dimensions, with constant coefficients	97
	5E(ii)	. . . in general	99
	5F	Practice problems	100
Part III	**Fourier series on bounded domains**		**105**
6	Some functional analysis		107
	6A	Inner products	107
	6B	L^2-space	109
	6C*	More about L^2-space	113
	6C(i)	Complex L^2-space	113
	6C(ii)	Riemann vs. Lebesgue integrals	114

6D	Orthogonality		116
6E	Convergence concepts		121
	6E(i)	L^2-convergence	122
	6E(ii)	Pointwise convergence	126
	6E(iii)	Uniform convergence	129
	6E(iv)	Convergence of function series	134
6F	Orthogonal and orthonormal bases		136
6G	Further reading		137
6H	Practice problems		138
7	Fourier sine series and cosine series		141
7A	Fourier (co)sine series on $[0, \pi]$		141
	7A(i)	Sine series on $[0, \pi]$	141
	7A(ii)	Cosine series on $[0, \pi]$	145
7B	Fourier (co)sine series on $[0, L]$		148
	7B(i)	Sine series on $[0, L]$	148
	7B(ii)	Cosine series on $[0, L]$	149
7C	Computing Fourier (co)sine coefficients		150
	7C(i)	Integration by parts	150
	7C(ii)	Polynomials	151
	7C(iii)	Step functions	155
	7C(iv)	Piecewise linear functions	159
	7C(v)	Differentiating Fourier (co)sine series	162
7D	Practice problems		163
8	Real Fourier series and complex Fourier series		165
8A	Real Fourier series on $[-\pi, \pi]$		165
8B	Computing real Fourier coefficients		167
	8B(i)	Polynomials	167
	8B(ii)	Step functions	168
	8B(iii)	Piecewise linear functions	170
	8B(iv)	Differentiating real Fourier series	172
8C	Relation between (co)sine series and real series		172
8D	Complex Fourier series		175
9	Multidimensional Fourier series		181
9A	. . . in two dimensions		181
9B	. . . in many dimensions		187
9C	Practice problems		195
10	Proofs of the Fourier convergence theorems		197
10A	Bessel, Riemann, and Lebesgue		197
10B	Pointwise convergence		199
10C	Uniform convergence		206

10D L^2-convergence   209
   10D(i)   Integrable functions and step functions in
              $\mathbf{L}^2[-\pi, \pi]$ 210
   10D(ii)   Convolutions and mollifiers   215
   10D(iii) Proofs of Theorems 8A.1(a) and 10D.1   223

Part IV BVP solutions via eigenfunction expansions   227

11   Boundary value problems on a line segment   229
  11A   The heat equation on a line segment   229
  11B   The wave equation on a line (the vibrating string)   233
  11C   The Poisson problem on a line segment   238
  11D   Practice problems   240

12   Boundary value problems on a square   243
  12A   The Dirichlet problem on a square   244
  12B   The heat equation on a square   251
   12B(i)   Homogeneous boundary conditions   251
   12B(ii)   Nonhomogeneous boundary conditions   255
  12C   The Poisson problem on a square   259
   12C(i)   Homogeneous boundary conditions   259
   12C(ii)   Nonhomogeneous boundary conditions   262
  12D   The wave equation on a square (the square drum)   263
  12E   Practice problems   266

13   Boundary value problems on a cube   269
  13A   The heat equation on a cube   270
  13B   The Dirichlet problem on a cube   273
  13C   The Poisson problem on a cube   276

14   Boundary value problems in polar coordinates   279
  14A   Introduction   279
  14B   The Laplace equation in polar coordinates   280
   14B(i)   Polar harmonic functions   280
   14B(ii)   Boundary value problems on a disk   284
   14B(iii) Boundary value problems on a codisk   289
   14B(iv)   Boundary value problems on an annulus   293
   14B(v)   Poisson's solution to the Dirichlet problem on
           the disk   296
  14C   Bessel functions   298
   14C(i)   Bessel's equation; eigenfunctions of $\triangle$ in polar
           coordinates   298
   14C(ii)   Boundary conditions; the roots of the Bessel
           function   301
   14C(iii) Initial conditions; Fourier–Bessel
           expansions   304

14D The Poisson equation in polar coordinates   306
14E The heat equation in polar coordinates   308
14F The wave equation in polar coordinates   309
14G The power series for a Bessel function   313
14H Properties of Bessel functions   317
14I Practice problems   322
15 Eigenfunction methods on arbitrary domains   325
15A General Solution to Poisson, heat, and wave equation
BVPs   325
15B General solution to Laplace equation BVPs   331
15C Eigenbases on Cartesian products   337
15D General method for solving I/BVPs   343
15E Eigenfunctions of self-adjoint operators   346
15E(i)   Self-adjoint operators   346
15E(ii)   Eigenfunctions and eigenbases   351
15E(iii) Symmetric elliptic operators   354
15F Further reading   355
Part V **Miscellaneous solution methods**   357
16 Separation of variables   359
16A Separation of variables in Cartesian coordinates on $\mathbb{R}^2$ 359
16B Separation of variables in Cartesian coordinates on $\mathbb{R}^D$ 361
16C Separation of variables in polar coordinates: Bessel's
equation   363
16D Separation of variables in spherical coordinates:
Legendre's equation   365
16E Separated vs. quasiseparated   375
16F The polynomial formalism   376
16G Constraints   378
16G(i)   Boundedness   378
16G(ii)   Boundary conditions   379
17 Impulse-response methods   381
17A Introduction   381
17B Approximations of identity   385
17B(i)   . . . in one dimension   385
17B(ii)   . . . in many dimensions   389
17C The Gaussian convolution solution (heat equation)   391
17C(i)   . . . in one dimension   391
17C(ii)   . . . in many dimensions   397
17D d'Alembert's solution (one-dimensional wave equation)   398
17D(i)   Unbounded domain   399
17D(ii)   Bounded domain   404

17E   Poisson's solution (Dirichlet problem on half-plane)   408
17F   Poisson's solution (Dirichlet problem on the disk)   411
17G* Properties of convolution   414
17H   Practice problems   416
18   Applications of complex analysis   421
18A   Holomorphic functions   421
18B   Conformal maps   428
18C   Contour integrals and Cauchy's theorem   440
18D   Analyticity of holomorphic maps   455
18E   Fourier series as Laurent series   459
18F* Abel means and Poisson kernels   466
18G   Poles and the residue theorem   469
18H   Improper integrals and Fourier transforms   477
18I* Homological extension of Cauchy's theorem   486
Part VI **Fourier transforms on unbounded domains**   489
19   Fourier transforms   491
19A   One-dimensional Fourier transforms   491
19B   Properties of the (one-dimensional) Fourier transform   497
19C* Parseval and Plancherel   506
19D   Two-dimensional Fourier transforms   508
19E   Three-dimensional Fourier transforms   511
19F   Fourier (co)sine transforms on the half-line   514
19G* Momentum representation and Heisenberg
      uncertainty   515
19H* Laplace transforms   520
19I   Further reading   527
19J   Practice problems   527
20   Fourier transform solutions to PDEs   531
20A   The heat equation   531
      20A(i)   Fourier transform solution   531
      20A(ii) The Gaussian convolution formula, revisited   534
20B   The wave equation   535
      20B(i)   Fourier transform solution   535
      20B(ii) Poisson's spherical mean solution; Huygens'
               principle   537
20C   The Dirichlet problem on a half-plane   540
      20C(i)   Fourier solution   541
      20C(ii) Impulse-response solution   542

20D	PDEs on the half-line	543
20E	General solution to PDEs using Fourier transforms	543
20F	Practice problems	545
Appendix A	Sets and functions	547
	A(i) Sets	547
	A(ii) Functions	548
Appendix B	Derivatives – notation	551
Appendix C	Complex numbers	553
Appendix D	Coordinate systems and domains	557
	D(i) Rectangular coordinates	557
	D(ii) Polar coordinates on $\mathbb{R}^2$	557
	D(iii) Cylindrical coordinates on $\mathbb{R}^3$	558
	D(iv) Spherical coordinates on $\mathbb{R}^3$	559
	D(v) What is a 'domain'?	559
Appendix E	Vector calculus	561
	E(i) Gradient	561
	E(ii) Divergence	562
	E(iii) The divergence theorem	564
Appendix F	Differentiation of function series	569
Appendix G	Differentiation of integrals	571
Appendix H	Taylor polynomials	573
	H(i) Taylor polynomials in one dimension	573
	H(ii) Taylor series and analytic functions	574
	H(iii) Using the Taylor series to solve ordinary differential equations	575
	H(iv) Taylor polynomials in two dimensions	578
	H(v) Taylor polynomials in many dimensions	579
References		581
Subject index		585
Notation index		597

Chapter dependency lattice

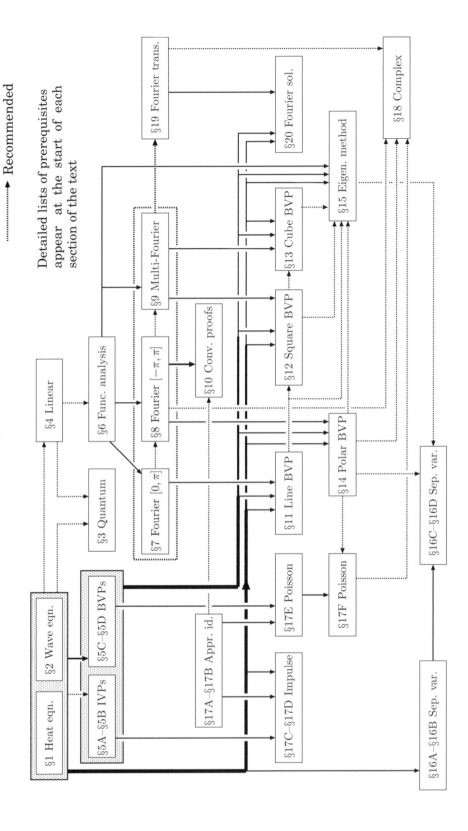

→ Required

┈┈▸ Recommended

Detailed lists of prerequisites appear at the start of each section of the text

Preface

This is a textbook for an introductory course on linear partial differential equations (PDEs) and initial/boundary value problems (I/BVPs). It also provides a mathematically rigorous introduction to Fourier analysis (Chapters 7, 8, 9, 10, and 19), which is the main tool used to solve linear PDEs in Cartesian coordinates. Finally, it introduces basic functional analysis (Chapter 6) and complex analysis (Chapter 18). The first is necessary to characterize rigorously the convergence of Fourier series, and also to discuss eigenfunctions for linear differential operators. The second provides powerful techniques to transform domains and compute integrals, and also offers additional insight into Fourier series.

This book is not intended to be comprehensive or encyclopaedic. It is designed for a one-semester course (i.e. 36–40 hours of lectures), and it is therefore strictly limited in scope. First, it deals mainly with *linear* PDEs with constant coefficients. Thus, there is no discussion of characteristics, conservation laws, shocks, variational techniques, or perturbation methods, which would be germane to other types of PDEs. Second, the book focuses mainly on concrete solution methods to specific PDEs (e.g. the Laplace, Poisson, heat, wave, and Schrödinger equations) on specific domains (e.g. line segments, boxes, disks, annuli, spheres), and spends rather little time on qualitative results about entire classes of PDEs (e.g. elliptic, parabolic, hyperbolic) on general domains. Only after a thorough exposition of these special cases does the book sketch the general theory; experience shows that this is far more pedagogically effective than presenting the general theory first. Finally, the book does not deal at all with numerical solutions or Galerkin methods.

One slightly unusual feature of this book is that, from the very beginning, it emphasizes the central role of eigenfunctions (of the Laplacian) in the solution methods for linear PDEs. Fourier series and Fourier–Bessel expansions are introduced as the orthogonal eigenfunction expansions which are most suitable in certain domains or coordinate systems. Separation of variables appears relatively late in the exposition (Chapter 16) as a convenient device to obtain such eigenfunctions.

The only techniques in the book which are not either implicitly or explicitly based on eigenfunction expansions are impulse-response functions and Green's functions (Chapter 17) and complex-analytic methods (Chapter 18).

Prerequisites and intended audience

This book is written for third-year undergraduate students in mathematics, physics, engineering, and other mathematical sciences. The only prerequisites are (1) *multivariate calculus* (i.e. partial derivatives, multivariate integration, changes of coordinate system) and (2) *linear algebra* (i.e. linear operators and their eigenvectors).

It might also be helpful for students to be familiar with the following: (1) the basic theory of ordinary differential equations (specifically, Laplace transforms, Frobenius method); (2) some elementary vector calculus (specifically, divergence and gradient operators); and (3) elementary physics (to understand the physical motivation behind many of the problems). However, none of these three things are really required.

In addition to this background knowledge, the book requires some ability at abstract mathematical reasoning. Unlike some 'applied math' texts, we do not suppress or handwave the mathematical theory behind the solution methods. At suitable moments, the exposition introduces concepts such as 'orthogonal basis', 'uniform convergence' vs. 'L_2-convergence', 'eigenfunction expansion', and 'self-adjoint operator'; thus, students must be intellectually capable of understanding abstract mathematical concepts of this nature. Likewise, the exposition is mainly organized in a 'definition → theorem → proof → example' format, rather than a 'problem → solution' format. Students must be able to understand abstract descriptions of general solution techniques, rather than simply learn by imitating worked solutions to special cases.

Conventions in the text

* in the title of a chapter or section indicates 'optional' material which is not part of the core syllabus.

(Optional) in the margin indicates that a particular theorem or statement is 'optional' in the sense that it is not required later in the text.

Ⓔ in the margin indicates the location of an exercise. (Shorter exercises are sometimes embedded within the exposition.)

♦ indicates the ends of more lengthy exercises.

□ ends the proof of a theorem.

◇ indicates the end of an example.

◇ ends the proof of a 'claim' within the proof of a theorem.

△ ends the proof of a 'subclaim' within the proof of a claim.

Acknowledgements

I would like to thank Xiaorang Li of Trent University, who read through an early draft of this book and made many helpful suggestions and corrections, and who also provided Problems 5.18, 5.19, and 6.8. I also thank Peter Nalitolela, who proofread a penultimate draft and spotted many mistakes. I especially thank Irene Pizzie of Cambridge University Press, whose very detailed and thorough copy-editing of the final manuscript resulted in many dozens of corrections and clarifications. I would like to thank several anonymous reviewers who made many useful suggestions, and I would also like to thank Peter Thompson of Cambridge University Press for recruiting these referees. I also thank Diana Gillooly of Cambridge University Press, who was very supportive and helpful throughout the entire publication process, especially concerning my desire to provide a free online version of the book, and to release the figures and problem sets under a Creative Commons License. I also thank the many students who used the early versions of this book, especially those who found mistakes or made good suggestions. Finally, I thank George Peschke of the University of Alberta, for being an inspiring example of good mathematical pedagogy.

None of these people are responsible for any remaining errors, omissions, or other flaws in the book (of which there are no doubt many). If you find an error or some other deficiency in the book, please contact me at marcuspivato@trentu.ca.

This book would not have been possible without open source software. The book was prepared entirely on the LINUX operating system (initially REDHAT,[1] and later UBUNTU[2]). All the text is written in Leslie Lamport's LaTeX2e typesetting language,[3] and was authored using Richard Stallman's EMACS editor.[4] The illustrations were hand-drawn using William Chia-Wei Cheng's excellent TGIF object-oriented drawing program.[5] Additional image manipulation and post-processing was done with GNU IMAGE MANIPULATION PROGRAM (GIMP).[6] Many of the plots were created using GNUPLOT.[7,8] I would like to take this opportunity to thank the many people in the open source community who have developed this software.

Finally, and most importantly, I would like to thank my beloved wife and partner, Reem Yassawi, and our wonderful children, Leila and Aziza, for their support and for their patience with my many long absences.

---

[1] http://www.redhat.com.   [2] http://www.ubuntu.com.   [3] http://www.latex-project.org.
[4] http://www.gnu.org/software/emacs/emacs.html.   [5] http://bourbon.usc.edu:8001/tgif.
[6] http://www.gimp.org.   [7] http://www.gnuplot.info.
[8] Many other plots were generated using Waterloo MAPLE (http://www.maplesoft.com), which unfortunately is *not* open-source.

What's good about this book?

This text has many advantages over most other introductions to partial differential equations.

Illustrations

PDEs are physically motivated and geometrical objects; they describe curves, surfaces, and scalar fields with special geometric properties, and the way these entities evolve over time under endogenous dynamics. To understand PDEs and their solutions, it is necessary to visualize them. Algebraic formulae are just a language used to communicate such visual ideas in lieu of pictures, and they generally make a poor substitute. This book has over 300 high-quality illustrations, many of which are rendered in three dimensions. In the online version of the book, most of these illustrations appear in full colour. Also, the website contains many animations which do not appear in the printed book.

Most importantly, on the book website, all illustrations are *freely available* under a Creative Commons Attribution Noncommercial Share-Alike License.[1] This means that you are free to download, modify, and utilize the illustrations to prepare your own course materials (e.g. printed lecture notes or beamer presentations), as long as you attribute the original author. Please visit <http://xaravve.trentu.ca/pde>.

Physical motivation

Connecting the math to physical reality is critical: it keeps students motivated, and helps them interpret the mathematical formalism in terms of their physical intuitions about diffusion, vibration, electrostatics, etc. Chapter 1 of this book discusses the

---

[1] See http://creativecommons.org/licenses/by-nc-sa/3.0.

physics behind the heat, Laplace, and Poisson equations, and Chapter 2 discusses the wave equation. An unusual addition to this text is Chapter 3, which discusses quantum mechanics and the Schrödinger Equation (one of the major applications of PDE theory in modern physics).

Detailed syllabus

Difficult choices must be made when turning a 600+ page textbook into a feasible 12-week lesson plan. It is easy to run out of time or inadvertently miss something important. To facilitate this task, this book provides a lecture-by-lecture breakdown of how the author covers the material (see p. xxv). Of course, each instructor can diverge from this syllabus to suit the interests/background of their students, a longer/shorter teaching semester, or their personal taste. But the prefabricated syllabus provides a base to work from, and will save most instructors a lot of time and aggravation.

Explicit prerequisites for each chapter and section

To save time, an instructor might want to skip a certain chapter or section, but worries that it may end up being important later. We resolve this problem in two ways. First, p. xiv provides a '*chapter dependency lattice*', which summarizes the large-scale structure of logical dependencies between the chapters of the book. Second, every section of every chapter begins with an explicit list of 'required' and 'recommended' prerequisite sections; this provides more detailed information about the small-scale structure of logical dependencies between sections. By tracing backward through this 'lattice of dependencies', you can figure out exactly what background material you must cover to reach a particular goal. This makes the book especially suitable for self-study.

Flat dependency lattice

There are many 'paths' through the 20-chapter dependency lattice on p. xiv every one of which is only *seven* chapters or less in length. Thus, an instructor (or an autodidact) can design many possible syllabi, depending on their interests, and can quickly move to advanced material. The 'Suggested 12-week syllabus' on p. xxv describes a gentle progression through the material, covering most of the 'core' topics in a 12-week semester, emphasizing concrete examples and gradually escalating the abstraction level. The Chapter Dependency Lattice suggests some other possibilities for 'accelerated' syllabi focusing on different themes.

- *Solving PDEs with impulse response functions.* Chapters 1, 2, 5, and 17 only.
- *Solving PDEs with Fourier transforms.* Chapters 1, 2, 5, 19, and 20 only. (Pedagogically speaking, Chapters 8 and 9 will help the student understand Chapter 19, and Chapters 11–13 will help the student understand Chapter 20. Also, it is interesting to see how the 'impulse response' methods of Chapter 17 yield the same solutions as the 'Fourier methods' of Chapter 20, using a totally different approach. However, strictly speaking, none of Chapters 8–13 or 17 is logically necessary.)
- *Solving PDEs with separation of variables.* Chapters 1, 2, and, 16 only. (However, without at least Chapters 12, and 14, the ideas of Chapter 16 will seem somewhat artificial and pointless.)
- *Solving I/BVPs using eigenfunction expansions.* Chapters 1, 2, 4, 5, 6, and 15. (It would be pedagogically better to also cover Chapters 9 and 12, and probably Chapter 14. But, strictly speaking, none of these is logically necessary.)
- *Tools for quantum mechanics.* Section 1B, then Chapters 3, 4, 6, 9, 13, 19, and 20 (skipping material on Laplace, Poisson, and wave equations in Chapters 13 and 20, and adapting the solutions to the heat equation into solutions to the Schrödinger Equation).
- *Fourier theory.* Section 4A, then Chapters 6, 7, 8, 9, 10, and 19. Finally, Sections 18A, 18C, 18E, and 18F provide a 'complex' perspective. (Section 18H also contains some useful computational tools.)
- *Crash course in complex analysis.* Chapter 18 is logically independent of the rest of the book, and rigorously develops the main ideas in complex analysis from first principles. (However, the emphasis is on applications to PDEs and Fourier theory, so some of the material may seem esoteric or unmotivated if read in isolation from other chapters.)

Highly structured exposition, with clear motivation up front

The exposition is broken into small, semi-independent logical units, each of which is clearly labelled, and which has a clear purpose or meaning which is made immediately apparent. This simplifies the instructor's task; it is not necessary to spend time restructuring and summarizing the text material because it is already structured in a manner which self-summarizes. Instead, instructors can focus more on explanation, motivation, and clarification.

Many 'practice problems' (with complete solutions and source code available online)

Frequent evaluation is critical to reinforce material taught in class. This book provides an extensive supply of (generally simple) computational 'practice problems' at the end of each chapter. Completely worked solutions to virtually all of these problems are available on the book website. Also on the book website, the LaTeX source code for all problems and solutions is *freely available* under a Creative

Commons Attribution Noncommercial Share-Alike License.[2] Thus, an instructor can download and edit this source code, and easily create quizzes, assignments, and matching solutions for students.

Challenging exercises without solutions

Complex theoretical concepts cannot really be tested in quizzes, and do not lend themselves to canned 'practice problems'. For a more theoretical course with more mathematically sophisticated students, the instructor will want to assign some proof-related exercises for homework. This book has more than 420 such exercises scattered throughout the exposition; these are flagged by an 'Ⓔ' symbol in the margin, as shown here. Many of these exercises ask the student to prove a major result from the text (or a component thereof). This is the best kind of exercise, because it reinforces the material taught in class, and gives students a sense of ownership of the mathematics. Also, students find it more fun and exciting to prove important theorems, rather than solving esoteric make-work problems.

Ⓔ

Appropriate rigour

The solutions of PDEs unfortunately involve many technicalities (e.g. different forms of convergence; derivatives of infinite function series, etc.). It is tempting to handwave and gloss over these technicalities, to avoid confusing students. But this kind of pedagogical dishonesty actually makes students *more* confused; they know something is fishy, but they can't tell quite what. Smarter students know they are being misled, and may lose respect for the book, the instructor, or even the whole subject.

In contrast, this book provides a rigorous mathematical foundation for all its solution methods. For example, Chapter 6 contains a careful explanation of L^2-spaces, the various forms of convergence for Fourier series, and the differences between them – including the 'pathologies' which can arise when one is careless about these issues. I adopt a 'triage' approach to proofs: the simplest proofs are left as exercises for the motivated student (often with a step-by-step breakdown of the best strategy). The most complex proofs I have omitted, but I provide multiple references to other recent texts. In between are those proofs which are challenging but still accessible; I provide detailed expositions of these proofs. Often, when the text contains several variants of the same theorem, I prove one variant in detail, and leave the other proofs as exercises.

---

[2] See http://creativecommons.org/licenses/by-nc-sa/3.0.

Appropriate abstraction

It is tempting to avoid abstractions (e.g. linear differential operators, eigenfunctions), and simply present *ad hoc* solutions to special cases. This cheats the student. The right abstractions provide simple, yet powerful, tools that help students understand a myriad of seemingly disparate special cases within a single unifying framework. This book provides students with the opportunity to learn an abstract perspective once they are ready for it. Some abstractions are introduced in the main exposition, others are in optional sections, or in the philosophical preambles which begin each major part of the book.

Gradual abstraction

Learning proceeds from the concrete to the abstract. Thus, the book begins each topic with a specific example or a low-dimensional formulation, and only later proceeds to a more general/abstract idea. This introduces a lot of 'redundancy' into the text, in the sense that later formulations subsume the earlier ones. So the exposition is not as 'efficient' as it could be. This is a good thing. Efficiency makes for good reference books, but lousy texts.

For example, when introducing the heat equation, Laplace equation, and wave equation in Chapters 1 and 2, I first derive and explain the one-dimensional version of each equation, then the two-dimensional version, and, finally, the general, D-dimensional version. Likewise, when developing the solution methods for BVPs in Cartesian coordinates (Chapters 11–13), I confine the exposition to the interval $[0, \pi]$, the square $[0, \pi]^2$, and the cube $[0, \pi]^3$, and assume all the coefficients in the differential equations are unity. Then the exercises ask the student to state and prove the appropriate generalization of each solution method for an interval/rectangle/box of arbitrary dimensions, and for equations with arbitrary coefficients. The general method for solving I/BVPs using eigenfunction expansions only appears in Chapter 15, after many special cases of this method have been thoroughly exposited in Cartesian and polar coordinates (Chapters 11–14).

Likewise, the development of Fourier theory proceeds in gradually escalating levels of abstraction. First we encounter Fourier (co)sine series on the interval $[0, \pi]$ (§7A); then on the interval $[0, L]$ for arbitrary $L > 0$ (§7B). Then Chapter 8 introduces 'real' Fourier series (i.e. with both sine and cosine terms), and then complex Fourier series (§8D). Then, Chapter 9 introduces two-dimensional (co)sine series and, finally, D-dimensional (co)sine series.

Expositional clarity

Computer scientists have long known that it is easy to write software that *works*, but it is much more difficult (and important) to write working software that *other people* can understand. Similarly, it is relatively easy to write formally correct mathematics; the real challenge is to make the mathematics easy to read. To achieve this, I use several techniques. I divide proofs into semi-independent modules ('claims'), each of which performs a simple, clearly defined task. I integrate these modules together in an explicit hierarchical structure (with 'subclaims' inside of 'claims'), so that their functional interdependence is clear from visual inspection. I also explain formal steps with parenthetical heuristic remarks. For example, in a long string of (in)equalities, I often attach footnotes to each step, as follows:

' $A \underset{(*)}{=} B \underset{(\dagger)}{\leq} C \underset{(\ddagger)}{<} D$. Here, $(*)$ is because [...]; $(\dagger)$ follows from [...], and $(\ddagger)$ is because [...].'

Finally, I use letters from the same 'lexicographical family' to denote objects which 'belong' together. For example: If $\mathcal{S}$ and $\mathcal{T}$ are sets, then elements of $\mathcal{S}$ should be $s_1, s_2, s_3, \ldots$, while elements of $\mathcal{T}$ are $t_1, t_2, t_3, \ldots$. If $\mathbf{v}$ is a vector, then its entries should be $v_1, \ldots, v_N$. If $\mathbf{A}$ is a matrix, then its entries should be $a_{11}, \ldots, a_{NM}$. I reserve upper-case letters (e.g. $J, K, L, M, N, \ldots$) for the bounds of intervals or indexing sets, and then use the corresponding lower-case letters (e.g. $j, k, l, m, n, \ldots$) as indexes. For example, $\forall n \in \{1, 2, \ldots, N\}$, $A_n :=$ $\sum_{j=1}^{J} \sum_{k=1}^{K} a_{jk}^n$.

Clear and explicit statements of solution techniques

Many PDE texts contain very few theorems; instead they try to develop the theory through a long sequence of worked examples, hoping that students will 'learn by imitation', and somehow absorb the important ideas 'by osmosis'. However, less gifted students often just imitate these worked examples in a slavish and uncomprehending way. Meanwhile, the more gifted students do not want to learn 'by osmosis'; they want clear and precise statements of the main ideas.

The problem is that most solution methods in PDEs, if stated as theorems in full generality, are incomprehensible to many students (especially the non-math majors). My solution is this: I provide explicit and precise statements of the solution method for almost every possible combination of (1) several major PDEs, (2) several kinds of boundary conditions, and (3) several different domains. I state these solutions as *theorems*, not as 'worked examples'. I follow each of these theorems with several completely worked examples. Some theorems I prove, but most of the proofs are left as exercises (often with step-by-step hints).

Of course, this approach is highly redundant, because I end up stating more than 20 theorems, which really are all special cases of three or four general results (for example, the general method for solving the heat equation on a compact domain with Dirichlet boundary conditions, using an eigenfunction expansion). However, this sort of redundancy is *good* in an elementary exposition. Highly 'efficient' expositions are pleasing to our aesthetic sensibilities, but they are dreadful for pedagogical purposes.

However, one must not leave the students with the impression that the theory of PDEs is a disjointed collection of special cases. To link together all the 'homogeneous Dirichlet heat equation' theorems, for example, I explicitly point out that they all utilize the same underlying strategy. Also, when a proof of one variant is left as an exercise, I encourage students to imitate the (provided) proofs of previous variants. When the students understand the underlying similarity between the various special cases, *then* it is appropriate to state the general solution. The students will almost feel they have figured it out for themselves, which is the best way to learn something.

Suggested 12-week syllabus

Week 1: *Heat and diffusion-related PDEs*

    Lecture 1: Appendix A–Appendix E   *Review of multivariate calculus; introduction to complex numbers.*

    Lecture 2: §1A–§1B   *Fourier's law; the heat equation.*

    Lecture 3: §1C–§1D   *Laplace equation; Poisson equation.*

Week 2: *Wave-related PDEs; quantum mechanics*

    Lecture 1: §1E; §2A   *Properties of harmonic functions; spherical means.*

    Lecture 2: §2B–§2C   *Wave equation; telegraph equation.*

    Lecture 3: Chapter 3   *The Schrödinger equation in quantum mechanics.*

Week 3: *General theory*

    Lecture 1: §4A–§4C   *Linear PDEs: homogeneous vs. nonhomogeneous.*

    Lecture 2: §5A; §5B   *Evolution equations and initial value problems.*

    Lecture 3: §5C   *Boundary conditions and boundary value problems.*

Week 4: *Background to Fourier theory*

    Lecture 1: §5D   *Uniqueness of solutions to BVPs;* §6A *inner products.*

    Lecture 2: §6B–§6D   *L^2-space; orthogonality.*

    Lecture 3: §6E(i)–(iii)   *L^2-convergence; pointwise convergence; uniform convergence.*

Week 5: *One-dimensional Fourier series*

    Lecture 1: §6E(iv)   *Infinite series;* §6F *orthogonal bases;* §7A *Fourier (co/sine) series: definition and examples.*

    Lecture 2: §7C(i)–(v)   *Computing Fourier series of polynomials, piecewise linear and step functions.*

    Lecture 3: §11A–§11C   *Solution to heat equation and Poisson equation on a line segment.*

Week 6: *Fourier solutions for BVPs in one and two dimensions*

    Lecture 1: §11B–§12A   *Wave equation on line segment and Laplace equation on a square.*

    Lecture 2: §9A–§9B   *Multidimensional Fourier series.*

    Lecture 3: §12B–§12C(i)   *Solution to heat equation and Poisson equation on a square.*

Week 7: *Fourier solutions for two-dimensional BVPs in Cartesian and polar coordinates*

    Lecture 1: §12C(ii), §12D   *Solution to Poisson equation and wave equation on a square.*

    Lecture 2: §5C(iv); §8A–§8B   *Periodic boundary conditions; real Fourier series.*

    Lecture 3: §14A; §14B(i)–(iv)   *Laplacian in polar coordinates; Laplace equation on (co)disk.*

Week 8: *BVPs in polar coordinates; Bessel functions*

    Lecture 1: §14C   *Bessel functions.*

    Lecture 2: §14D–§14F   *Heat, Poisson, and wave equations in polar coordinates.*

    Lecture 3: §14G   *Solving Bessel's equation with the method of Frobenius.*

Week 9: *Eigenbases; separation of variables*

    Lecture 1: §15A–§15B   *Eigenfunction solutions to BVPs.*

    Lecture 2: §15B; §16A–§16B   *Harmonic bases; separation of variables in Cartesian coordinates.*

    Lecture 3: §16C–§16D   *Separation of variables in polar and spherical coordinates; Legendre polynomials.*

Week 10: *Impulse response methods*

    Lecture 1: §17A–§17C   *Impulse response functions; convolution; approximations of identity; Gaussian convolution solution for heat equation.*

    Lecture 2: §17C–§17F *Gaussian convolutions continued; Poisson's solutions to Dirichlet problem on a half-plane and a disk.*

    Lecture 3: §14B(v); §17D   *Poisson solution on disk via polar coordinates; d'Alembert solution to wave equation.*

Week 11: *Fourier transforms*

    Lecture 1: §19A   *One-dimensional Fourier transforms.*

    Lecture 2: §19B   *Properties of one-dimensional Fourier transform.*

    Lecture 3: §20A; §20C   *Fourier transform solution to heat equation; Dirchlet problem on half-plane.*

Week 12: *Fourier transform solutions to PDEs*

    Lecture 1: §19D, §20B(i)   *Multidimensional Fourier transforms; solution to wave equation.*

Lecture 2: §20B(ii); §20E   *Poisson's spherical mean solution; Huygen's principle; the general method.*

Lecture 3: (Time permitting) §19G or §19H   *Heisenberg uncertainty or Laplace transforms.*

In a longer semester or a faster paced course, one could also cover parts of Chapter 10 (*Proofs of Fourier convergence*) and/or Chapter 18 (*Applications of complex analysis*).

Part I

Motivating examples and major applications

A *dynamical system* is a mathematical model of a system evolving in time. Most models in mathematical physics are dynamical systems. If the system has only a finite number of 'state variables', then its dynamics can be encoded in an *ordinary differential equation* (ODE), which expresses the *time derivative* of each state variable (i.e. its rate of change over time) as a function of the other state variables. For example, *celestial mechanics* concerns the evolution of a system of gravitationally interacting objects (e.g. stars and planets). In this case, the 'state variables' are vectors encoding the position and momentum of each object, and the ODE describes how the objects move and accelerate as they interact gravitationally.

However, if the system has a very large number of state variables, then it is no longer feasible to represent it with an ODE. For example, consider the flow of heat or the propagation of compression waves through a steel bar containing 10^{24} iron atoms. We *could* model this using a 10^{24}-dimensional ODE, where we explicitly track the vibrational motion of each iron atom. However, such a 'microscopic' model would be totally intractable. Furthermore, it is not necessary. The iron atoms are (mostly) immobile, and interact only with their immediate neighbours. Furthermore, nearby atoms generally have roughly the same temperature, and move in synchrony. Thus, it suffices to consider the macroscopic *temperature distribution* of the steel bar, or to study the fluctuation of a macroscopic *density field*. This temperature distribution or density field can be mathematically represented as a smooth, real-valued function over some three-dimensional domain. The flow of heat or the propagation of sound can then be described as the *evolution* of this function over time.

We now have a dynamical system where the 'state variable' is not a finite system of vectors (as in celestial mechanics), but is instead a multivariate *function*. The evolution of this function is determined by its spatial geometry – e.g. the local 'steepness' and variation of the temperature gradients between warmer and cooler regions in the bar. In other words, the *time derivative* of the function (its rate of change over time) is determined by its *spatial derivatives* (which describe its slope and curvature at each point in space). An equation that relates the different derivatives of a multivariate function in this way is a *partial differential equation* (PDE). In particular, a PDE which describes a dynamical system is called an *evolution equation*. For example, the evolution equation which describes the flow of heat through a solid is called the *heat equation*. The equation which describes compression waves is the *wave equation*.

An *equilibrium* of a dynamical system is a state which is unchanging over time; mathematically, this means that the time-derivative is equal to zero. An equilibrium of an N-dimensional evolution equation satisfies an $(N-1)$-dimensional PDE called an *equilibrium equation*. For example, the equilibrium equations corresponding to the heat equation are the *Laplace equation* and the *Poisson equation* (depending on whether or not the system is subjected to external heat input).

PDEs are thus of central importance in the thermodynamics and acoustics of continuous media (e.g. steel bars). The heat equation also describes chemical diffusion in fluids, and also the evolving probability distribution of a particle performing a random walk called *Brownian motion*. It thus finds applications everywhere from mathematical biology to mathematical finance. When diffusion or Brownian motion is combined with deterministic drift (e.g. due to prevailing wind or ocean currents) it becomes a PDE called the *Fokker–Planck equation*.

The Laplace and Poisson equations describe the equilibria of such diffusion processes. They also arise in electrostatics, where they describe the shape of an electric field in a vacuum. Finally, solutions of the two-dimensional Laplace equation are good approximations of surfaces trying to minimize their elastic potential energy – that is, soap films.

The wave equation describes the resonance of a musical instrument, the spread of ripples on a pond, seismic waves propagating through the earth's crust, and shockwaves in solar plasma. (The motion of fluids themselves is described by yet another PDE, the *Navier–Stokes equation*.) A version of the wave equation arises as a special case of Maxwell's equations of electrodynamics; this led to Maxwell's prediction of *electromagnetic waves*, which include radio, microwaves, X-rays, and visible light. When combined with a 'diffusion' term reminiscent of the heat equation, the wave equation becomes the *telegraph equation*, which describes the propagation and degradation of electrical signals travelling through a wire.

Finally, an odd-looking 'complex' version of the heat equation induces wave-like evolution in the complex-valued probability fields which describe the position and momentum of subatomic particles. This *Schrödinger equation* is the starting point of quantum mechanics, one of the two most revolutionary developments in physics in the twentieth century. The other revolutionary development was relativity theory. General relativity represents spacetime as a four-dimensional manifold, whose curvature interacts with the spatiotemporal flow of mass/energy through yet another PDE: the *Einstein equation*.

Except for the Einstein and Navier–Stokes equations, all the equations we have mentioned are *linear* PDEs. This means that a sum of two or more solutions to the PDE will also be a solution. This allows us to solve linear PDEs through the *method of superposition*: we build complex solutions by adding together many simple solutions. A particularly convenient class of simple solutions are *eigenfunctions*. Thus, an enormously powerful and general method for linear PDEs is to represent the solutions using *eigenfunction expansions*. The most natural eigenfunction expansion (in Cartesian coordinates) is the *Fourier series*.

1

Heat and diffusion

The differential equations of the propagation of heat express the most general conditions, and reduce the physical questions to problems of pure analysis, and this is the proper object of theory.

Jean Joseph Fourier

1A Fourier's law

Prerequisites: Appendix A.   **Recommended:** Appendix E.

1A(i) ... in one dimension

Figure 1A.1 depicts a material diffusing through a one-dimensional domain $\mathbb{X}$ (for example, $\mathbb{X} = \mathbb{R}$ or $\mathbb{X} = [0, L]$). Let $u(x, t)$ be the density of the material at the point $x \in \mathbb{X}$ at time $t > 0$. Intuitively, we expect the material to flow from regions of *greater* to *lesser* concentration. In other words, we expect the *flow* of the material at any point in space to be proportional to the *slope* of the curve $u(x, t)$ at that point. Thus, if $F(x, t)$ is the flow at the point x at time t, then we expect the following:

$$F(x, t) = -\kappa \cdot \partial_x u(x, t),$$

where $\kappa > 0$ is a constant measuring the rate of diffusion. This is an example of *Fourier's law*.

1A(ii) ... in many dimensions

Prerequisites: Appendix E.

Figure 1A.2 depicts a material diffusing through a two-dimensional domain $\mathbb{X} \subset \mathbb{R}^2$ (e.g. heat spreading through a region, ink diffusing in a bucket of water, etc.) We

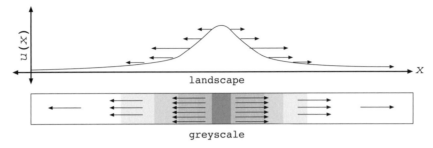

Figure 1A.1. Fourier's law of heat flow in one dimension.

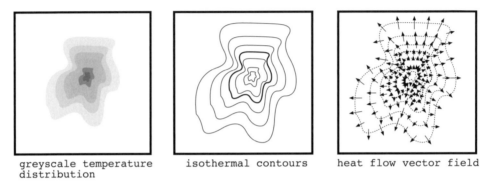

Figure 1A.2. Fourier's law of heat flow in two dimensions.

could just as easily suppose that $\mathbb{X} \subset \mathbb{R}^D$ is a D-dimensional domain. If $\mathbf{x} \in \mathbb{X}$ is a point in space, and $t \geq 0$ is a moment in time, let $u(\mathbf{x}, t)$ denote the concentration at $\mathbf{x}$ at time t. (This determines a function $u : \mathbb{X} \times \mathbb{R}_+ \longrightarrow \mathbb{R}$, called a *time-varying scalar field*.)

Now let $\vec{\mathbf{F}}(\mathbf{x}, t)$ be a D-dimensional vector describing the *flow* of the material at the point $\mathbf{x} \in \mathbb{X}$. (This determines a *time-varying vector field* $\vec{\mathbf{F}} : \mathbb{R}^D \times \mathbb{R}_+ \longrightarrow \mathbb{R}^D$.)

Again, we expect the material to flow from regions of high concentration to low concentration. In other words, material should flow *down the concentration gradient*. This is expressed by *Fourier's law of heat flow*:

$$\vec{\mathbf{F}} = -\kappa \cdot \nabla u,$$

where $\kappa > 0$ is a constant measuring the rate of diffusion.

One can imagine u as describing a distribution of highly antisocial people; each person is always fleeing everyone around them and moving in the direction with the fewest people. The constant κ measures the average walking speed of these misanthropes.

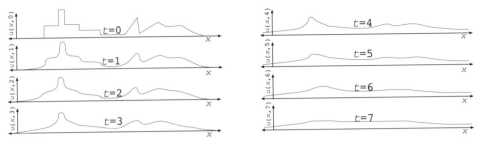

Figure 1B.1. The heat equation as 'erosion'.

1B The heat equation

Recommended: §1A.

1B(i) ... in one dimension

Prerequisites: §1A(i).

Consider a material diffusing through a one-dimensional domain $\mathbb{X}$ (for example, $\mathbb{X} = \mathbb{R}$ or $\mathbb{X} = [0, L]$). Let $u(x, t)$ be the density of the material at the location $x \in \mathbb{X}$ at time $t \in \mathbb{R}_{\not{\;}}$, and let $F(x, t)$ be the flux of the material at the location x and time t. Consider the derivative $\partial_x F(x, t)$. If $\partial_x F(x, t) > 0$, this means that the flow is *diverging*[1] at this point in space, so the material there is spreading farther apart. Hence, we expect the concentration at this point to *decrease*. Conversely, if $\partial_x F(x, t) < 0$, then the flow is *converging* at this point in space, so the material there is crowding closer together, and we expect the concentration to *increase*. To be succinct: the concentration of material will *increase* in regions where F converges and *decrease* in regions where F diverges. The equation describing this is given by

$$\partial_t u(x, t) = -\partial_x F(x, t).$$

If we combine this with Fourier's law, however, we get:

$$\partial_t u(x, t) = \kappa \cdot \partial_x \partial_x u(x, t),$$

which yields the *one-dimensional heat equation*:

$$\boxed{\partial_t u(x, t) = \kappa \cdot \partial_x^2 u(x, t).}$$

Heuristically speaking, if we imagine $u(x, t)$ as the height of some one-dimensional 'landscape', then the heat equation causes this landscape to be 'eroded', as if it were subjected to thousands of years of wind and rain (see Figure 1B.1).

---

[1] See Appendix E(ii), p. 562, for an explanation of why we say the flow is 'diverging' here.

Heat and diffusion

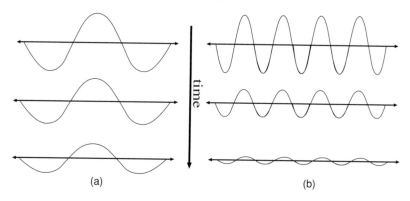

(a)   (b)

Figure 1B.2. Under the heat equation, the exponential decay of a periodic function is proportional to the square of its frequency. (a) Low frequency, slow decay; (b) high frequency, fast decay.

Example 1B.1 For simplicity we suppose $\kappa = 1$.

(a) Let $u(x, t) = e^{-9t} \cdot \sin(3x)$. Thus, u describes a spatially sinusoidal function (with spatial frequency 3) whose magnitude decays exponentially over time.

(b) *The dissipating wave.* More generally, let $u(x, t) = e^{-\omega^2 \cdot t} \cdot \sin(\omega \cdot x)$. Then u is a solution to the one-dimensional heat equation, and it looks like a standing wave whose amplitude decays exponentially over time (see Figure 1B.2). Note that the decay rate of the function u is proportional to the square of its frequency.

(c) *The (one-dimensional) Gauss–Weierstrass kernel.* Let

$$\mathcal{G}(x; t) := \frac{1}{2\sqrt{\pi t}} \exp\left(\frac{-x^2}{4t}\right).$$

Then $\mathcal{G}$ is a solution to the one-dimensional heat equation, and looks like a 'bell curve', which starts out tall and narrow, and, over time, becomes broader and flatter (Figure 1B.3). ◇

Ⓔ **Exercise 1B.1** Verify that all the functions in Examples 1B.1(a)–(c) satisfy the heat equation. ◆

All three functions in Example 1B.1 start out very tall, narrow, and pointy, and gradually become shorter, broader, and flatter. This is generally what the heat equation does; it tends to flatten things out. If u describes a physical landscape, then the heat equation describes 'erosion'.

1B(ii) . . . in many dimensions

Prerequisites: §1A(ii).

More generally, if $u : \mathbb{R}^D \times \mathbb{R}_+ \longrightarrow \mathbb{R}$ is the time-varying density of some mate-rial, and $\vec{F} : \mathbb{R}^D \times \mathbb{R}_+ \longrightarrow \mathbb{R}$ is the flux of this material, then we would expect the

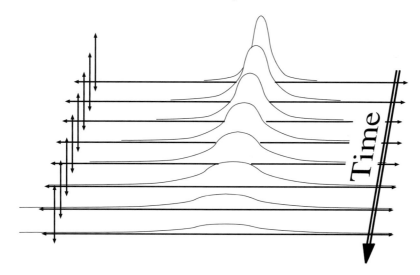

Figure 1B.3. The Gauss–Weierstrass kernel under the heat equation.

material to *increase* in regions where $\vec{\mathbf{F}}$ converges and to *decrease* in regions where $\vec{\mathbf{F}}$ diverges.[2] In other words, we have

$$\partial_t\, u = -\text{div}\,\vec{\mathbf{F}}.$$

If u is the density of some diffusing material (or heat), then $\vec{\mathbf{F}}$ is determined by Fourier's law, so we get the heat equation

$$\partial_t u = \kappa \cdot \text{div}\,\nabla u = \kappa\, \triangle u.$$

Here, $\triangle$ is the *Laplacian* operator,[3] defined as follows:

$$\boxed{\triangle u = \partial_1^2\, u + \partial_2^2\, u + \cdots + \partial_D^2\, u}$$

Exercise 1B.2 (a) If $D = 1$ and $u : \mathbb{R} \longrightarrow \mathbb{R}$, verify that div $\nabla u(x) = u''(x) = \triangle u(x)$, for all $x \in \mathbb{R}$. Ⓔ

(b) If $D = 2$ and $u : \mathbb{R}^2 \longrightarrow \mathbb{R}$, verify that div $\nabla u(x, y) = \partial_x^2 u(x, y) + \partial_y^2 u(x, y) = \triangle u(x, y)$, for all $(x, y) \in \mathbb{R}^2$.

(c) For any $D \geq 2$ and $u : \mathbb{R}^D \longrightarrow \mathbb{R}$, verify that div $\nabla u(\mathbf{x}) = \triangle u(\mathbf{x})$, for all $\mathbf{x} \in \mathbb{R}^D$. ◆

By changing to the appropriate time units, we can assume $\kappa = 1$, so the heat equation becomes

$$\boxed{\partial_t\, u = \triangle u.}$$

---

[2] See Appendix E(ii), p. 562, for a review of the 'divergence' of a vector field.
[3] Sometimes the Laplacian is written as ∇^2.

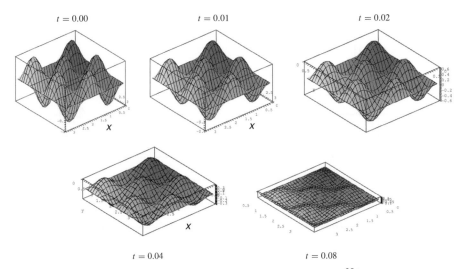

$t = 0.00$ $t = 0.01$ $t = 0.02$

$t = 0.04$ $t = 0.08$

Figure 1B.4. Five snapshots of the function $u(x, y; t) = e^{-25t} \cdot \sin(3x)\sin(4y)$ from Example 1B.2.

For example,

- if $\mathbb{X} \subset \mathbb{R}$, and $x \in \mathbb{X}$, then $\triangle u(x; t) = \partial_x^2 u(x; t)$;
- if $\mathbb{X} \subset \mathbb{R}^2$, and $(x, y) \in \mathbb{X}$, then $\triangle u(x, y; t) = \partial_x^2 u(x, y; t) + \partial_y^2 u(x, y; t)$.

Thus, as we have already seen, the one-dimensional heat equation is given by

$$\partial_t u = \partial_x^2 u,$$

and the the two-dimensional heat equation is given by

$$\boxed{\partial_t u(x, y; t) = \partial_x^2 u(x, y; t) + \partial_y^2 u(x, y; t).}$$

Example 1B.2

(a) Let $u(x, y; t) = e^{-25t} \cdot \sin(3x)\sin(4y)$. Then u is a solution to the two-dimensional heat equation, and looks like a two-dimensional 'grid' of sinusoidal hills and valleys with horizontal spacing $1/3$ and vertical spacing $1/4$. As shown in Figure 1B.4, these hills rapidly subside into a gently undulating meadow and then gradually sink into a perfectly flat landscape.

(b) The (two-dimensional) Gauss–Weierstrass kernel. Let

$$\mathcal{G}(x, y; t) := \frac{1}{4\pi t} \exp\left(\frac{-x^2 - y^2}{4t}\right).$$

Then $\mathcal{G}$ is a solution to the two-dimensional heat equation, and looks like a mountain, which begins steep and pointy and gradually 'erodes' into a broad, flat, hill.

(c) The *D*-dimensional Gauss–Weierstrass kernel is the function $\mathcal{G} : \mathbb{R}^D \times \mathbb{R}_+ \longrightarrow \mathbb{R}$ defined by

$$\mathcal{G}(\mathbf{x};t) = \frac{1}{(4\pi t)^{D/2}} \exp\left(\frac{-\|\mathbf{x}\|^2}{4t}\right).$$

Technically speaking, $\mathcal{G}(\mathbf{x};t)$ is a *D*-dimensional *symmetric normal probability distribution* with variance $\sigma = 2t$.   ◇

Exercise 1B.3 Verify that all the functions in Examples 1B.2(a)–(c) satisfy the Ⓔ
heat equation.   ◆

Exercise 1B.4 Prove the *Leibniz rule* for Laplacians: if $f, g : \mathbb{R}^D \longrightarrow \mathbb{R}$ are two Ⓔ
scalar fields, and $(f \cdot g) : \mathbb{R}^D \longrightarrow \mathbb{R}$ is their product, then, for all $\mathbf{x} \in \mathbb{R}^D$,

$$\Delta(f \cdot g)(\mathbf{x}) = g(\mathbf{x}) \cdot \left(\Delta f(\mathbf{x})\right) + 2\left(\nabla f(\mathbf{x})\right) \bullet \left(\nabla g(\mathbf{x})\right) + f(\mathbf{x}) \cdot \left(\Delta g(\mathbf{x})\right).$$

Hint: Combine the Leibniz rules for gradients and divergences (see Propositions
E.1(b) and E.2(b) in Appendix E, pp. 562 and 564).   ◆

1C The Laplace equation

Prerequisites: §1B.

If the heat equation describes the erosion/diffusion of some system, then an *equilib-
rium* or *steady-state* of the heat equation is a scalar field $h : \mathbb{R}^D \longrightarrow \mathbb{R}$ satisfying
the Laplace equation:

$$\Delta h \equiv 0.$$

A scalar field satisfying the Laplace equation is called a *harmonic function*.

Example 1C.1
(a) If $D = 1$, then $\Delta h(x) = \partial_x^2 h(x) = h''(x)$; thus, the *one-dimensional Laplace equation*
is just

$$h''(x) = 0.$$

Suppose $h(x) = 3x + 4$. Then $h'(x) = 3$ and $h''(x) = 0$, so h is harmonic. More generally, the one-dimensional harmonic functions are just the *linear* functions of the form
$h(x) = ax + b$ for some constants $a, b \in \mathbb{R}$.
(b) If $D = 2$, then $\Delta h(x, y) = \partial_x^2 h(x, y) + \partial_y^2 h(x, y)$, so the two-dimensional Laplace
equation is given by

$$\partial_x^2 h + \partial_y^2 h = 0,$$

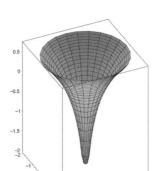
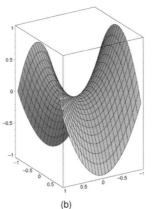
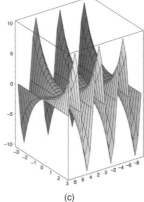

(a)   (b)   (c)

Figure 1C.1. Three harmonic functions. (a) $h(x, y) = \log(x^2 + y^2)$; (b) $h(x, y) = x^2 - y^2$; (c) $h(x, y) = \sin(x) \cdot \sinh(y)$. In all cases, note the telltale 'saddle' shape.

or, equivalently, $\partial_x^2 h = -\partial_y^2 h$. For example,

- Figure 1C.1(b) shows the harmonic function $h(x, y) = x^2 - y^2$;
- Figure 1C.1(c) shows the harmonic function $h(x, y) = \sin(x) \cdot \sinh(y)$. $\diamondsuit$

(E) **Exercise 1C.1**   Verify that these two functions are harmonic.   ♦

The surfaces in Figure 1C.1 have a 'saddle' shape, and this is typical of harmonic functions; in a sense, a harmonic function is one which is 'saddle-shaped' at every point in space. In particular, note that $h(x, y)$ has no maxima or minima anywhere; this is a universal property of harmonic functions (see Corollary 1E.2). The next example seems to contradict this assertion, but in fact it does not.

Example 1C.2   Figure 1C.1(a) shows the harmonic function $h(x, y) = \log(x^2 + y^2)$ for all $(x, y) \neq (0, 0)$. This function is well-defined everywhere except at $(0, 0)$; hence, contrary to appearances, $(0, 0)$ is *not* an extremal point. (Verification that h is harmonic is Problem 1.3, p. 23.) $\diamondsuit$

When $D \geq 3$, harmonic functions no longer define nice saddle-shaped *surfaces*, but they still have similar mathematical properties.

Example 1C.3
(a) If $D = 3$, then $\triangle h(x, y, z) = \partial_x^2 h(x, y, z) + \partial_y^2 h(x, y, z) + \partial_z^2 h(x, y, z)$. Thus, the three-dimensional Laplace equation reads as follows:

$$\boxed{\partial_x^2 h + \partial_y^2 h + \partial_z^2 h = 0.}$$

For example, let

$$h(x, y, z) = \frac{1}{\|(x, y, z)\|} = \frac{1}{\sqrt{x^2 + y^2 + z^2}}$$

for all $(x, y, z) \neq (0, 0, 0)$. Then h is harmonic everywhere except at $(0, 0, 0)$. (Verification that h is harmonic is Problem 1.4, p. 23.)

(b) For any $D \geq 3$, the D-dimensional Laplace equation reads as follows:

$$\boxed{\partial_1^2 h + \cdots + \partial_D^2 h = 0.}$$

For example, let

$$h(\mathbf{x}) = \frac{1}{\|\mathbf{x}\|^{D-2}} = \frac{1}{\left(x_1^2 + \cdots + x_D^2\right)^{\frac{D-2}{2}}}$$

for all $\mathbf{x} \neq \mathbf{0}$. Then h is harmonic everywhere in $\mathbb{R}^D \setminus \{\mathbf{0}\}$. ◇

Exercise 1C.2 Verify that h is harmonic on $\mathbb{R}^D \setminus \{\mathbf{0}\}$. ♦ Ⓔ

Harmonic functions have the convenient property that we can multiply together two lower-dimensional harmonic functions to get a higher dimensional one. For example,

- $h(x, y) = x \cdot y$ is a two-dimensional harmonic function. (**Exercise 1C.3** Verify this.) Ⓔ
- $h(x, y, z) = x \cdot (y^2 - z^2)$ is a three-dimensional harmonic function. (**Exercise 1C.4** Verify this.) Ⓔ

In general, we have the following:

Proposition 1C.4 *Suppose $u : \mathbb{R}^n \longrightarrow \mathbb{R}$ is harmonic and $v : \mathbb{R}^m \longrightarrow \mathbb{R}$ is harmonic, and define $w : \mathbb{R}^{n+m} \longrightarrow \mathbb{R}$ by $w(\mathbf{x}, \mathbf{y}) = u(\mathbf{x}) \cdot v(\mathbf{y})$ for $\mathbf{x} \in \mathbb{R}^n$ and $\mathbf{y} \in \mathbb{R}^m$. Then w is also harmonic.*

Proof **Exercise 1C.5** *Hint:* First prove that w obeys a kind of Leibniz rule: Ⓔ
$\triangle w(\mathbf{x}, \mathbf{y}) = v(\mathbf{y}) \cdot \triangle u(\mathbf{x}) + u(\mathbf{x}) \cdot \triangle v(\mathbf{y})$. □

The function $w(\mathbf{x}, \mathbf{y}) = u(\mathbf{x}) \cdot v(\mathbf{y})$ is called a *separated solution*, and this proposition illustrates a technique called *separation of variables*. Exercise 1C.6 also explores separation of variables. A full exposition of this technique appears in Chapter 16.

Exercise 1C.6 Ⓔ
(a) Let $\mu, \nu \in \mathbb{R}$ be constants, and let $f(x, y) = e^{\mu x} \cdot e^{\nu y}$. Suppose f is harmonic; what can you conclude about the relationship between μ and ν? (Justify your assertion.)

(b) Suppose $f(x, y) = X(x) \cdot Y(y)$, where $X : \mathbb{R} \longrightarrow \mathbb{R}$ and $Y : \mathbb{R} \longrightarrow \mathbb{R}$ are two smooth functions. Suppose $f(x, y)$ is harmonic.

  (i) Prove that $X''(x)/X(x) = -Y''(y)/Y(y)$ for all $x, y \in \mathbb{R}$.
 (ii) Conclude that the function $X''(x)/X(x)$ must equal a constant c independent of x. Hence $X(x)$ satisfies the ordinary differential equation $X''(x) = c \cdot X(x)$.

     Likewise, the function $Y''(y)/Y(y)$ must equal $-c$, independent of y. Hence $Y(y)$ satisfies the ordinary differential equation $Y''(y) = -c \cdot Y(y)$.
(iii) Using this information, deduce the general form for the functions $X(x)$ and $Y(y)$, and use this to obtain a general form for $f(x, y)$.   ◆

1D The Poisson equation

Prerequisites: §1C.

Imagine $p(\mathbf{x})$ is the concentration of a chemical at the point $\mathbf{x}$ in space. Suppose this chemical is being *generated* (or *depleted*) at different rates at different regions in space. Thus, in the absence of diffusion, we would have the *generation equation*:

$$\partial_t \, p(\mathbf{x}, t) = q(\mathbf{x}),$$

where $q(\mathbf{x})$ is the rate at which the chemical is being created/destroyed at $\mathbf{x}$ (we assume that q is constant in time).

If we now include the effects of diffusion, we get the *generation–diffusion equation*:

$$\partial_t \, p = \kappa \triangle p + q.$$

A *steady state* of this equation is a scalar field p satisfying the *Poisson equation*:

$$\boxed{\triangle p = Q,}$$

where $Q(\mathbf{x}) = -q(\mathbf{x})/\kappa$.

Example 1D.1 One-dimensional Poisson equation
If $D = 1$, then $\triangle p(x) = \partial_x^2 \, p(x) = p''(x)$; thus, the *one-dimensional Poisson equation* is just

$$\boxed{p''(x) = Q(x).}$$

We can solve this equation by twice-integrating the function $Q(x)$. If $p(x) = \int \int Q(x)$ is some double-antiderivative of G, then p clearly satisfies the Poisson equation. For example:

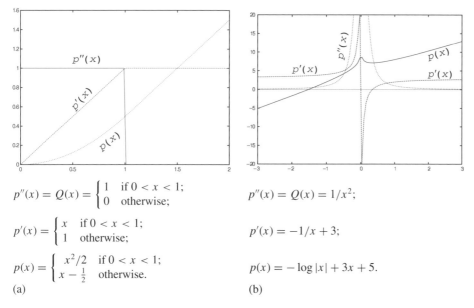

$$p''(x) = Q(x) = \begin{cases} 1 & \text{if } 0 < x < 1; \\ 0 & \text{otherwise;} \end{cases}$$

$$p''(x) = Q(x) = 1/x^2;$$

$$p'(x) = \begin{cases} x & \text{if } 0 < x < 1; \\ 1 & \text{otherwise;} \end{cases}$$

$$p'(x) = -1/x + 3;$$

$$p(x) = \begin{cases} x^2/2 & \text{if } 0 < x < 1; \\ x - \frac{1}{2} & \text{otherwise.} \end{cases}$$

$$p(x) = -\log|x| + 3x + 5.$$

(a) (b)

Figure 1D.1. Two one-dimensional potentials.

(a) Suppose

$$Q(x) = \begin{cases} 1 & \text{if } 0 < x < 1; \\ 0 & \text{otherwise.} \end{cases}$$

Then define

$$p(x) = \int_0^x \int_0^y q(z)dz\,dy = \begin{cases} 0 & \text{if } x < 0; \\ x^2/2 & \text{if } 0 < x < 1; \\ x - \frac{1}{2} & \text{if } 1 < x \end{cases}$$

(Figure 1D.1(a));

(b) If $Q(x) = 1/x^2$ (for $x \neq 0$), then $p(x) = \int \int Q(x) = -\log|x| + ax + b$ (for $x \neq 0$), where $a, b \in \mathbb{R}$ are arbitrary constants (see Figure 1D.1(b)). ◇

Exercise 1D.1 Verify that the functions $p(x)$ in Examples 1D.1 (a) and (b) are Ⓔ both solutions to their respective Poisson equations. ◆

Example 1D.2 Electrical/gravitational fields
The Poisson equation also arises in classical field theory.[4] Suppose, for any point $\mathbf{x} = (x_1, x_2, x_3)$ in three-dimensional space, that $q(\mathbf{x})$ is the charge density at $\mathbf{x}$, and that $p(\mathbf{x})$ is the electric potential field at $\mathbf{x}$. Then we have the

---

[4] For a quick yet lucid introduction to electrostatics, see Stevens (1995), chap. 3.

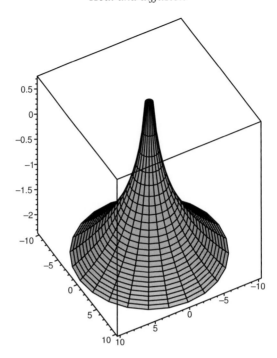

Figure 1D.2. The two-dimensional potential field generated by a concentration of charge at the origin.

following:

$$\triangle \, p(\mathbf{x}) = \kappa \, q(\mathbf{x}) \qquad (\kappa \text{ some constant}). \qquad (1\text{D}.1)$$

If $q(\mathbf{x})$ were the *mass* density at $\mathbf{x}$ and $p(\mathbf{x})$ were the *gravitational* potential energy, then we would get the same equation. (See Figure 1D.2 for an example of such a potential in two dimensions.)

In particular, in a region where there is no charge/mass (i.e. where $q \equiv 0$), equation (1D.1) reduces to the Laplace equation $\triangle p \equiv 0$. Because of this, solutions to the Poisson equation (and especially the Laplace equation) are sometimes called *potentials*.   ◇

Example 1D.3 The Coulomb potential
Let $D = 3$, and let

$$p(x, y, z) = \frac{1}{\|(x, y, z)\|} = \frac{1}{\sqrt{x^2 + y^2 + z^2}}.$$

In Example 1C.3(a), we asserted that $p(x, y, z)$ was harmonic everywhere except at $(0, 0, 0)$, where it is not well-defined. For physical reasons, it is 'reasonable' to write the equation

$$\triangle p(0, 0, 0) = \delta_0, \tag{1D.2}$$

where δ_0 is the 'Dirac delta function' (representing an infinite concentration of charge at zero).[5] Then $p(x, y, z)$ describes the electric potential generated by a *point charge*.   ◇

Ⓔ

Exercise 1D.2 Check that

$$\nabla p(x, y, z) = \frac{-(x, y, z)}{\|(x, y, z)\|^3}.$$

This is the electric field generated by a point charge, as given by *Coulomb's law* from classical electrostatics.   ♦

Ⓔ

Exercise 1D.3

(a) Let $q : \mathbb{R}^3 \longrightarrow \mathbb{R}$ be a scalar field describing a charge density distribution. If $\vec{E} : \mathbb{R}^3 \longrightarrow \mathbb{R}^3$ is the electric field generated by q, then *Gauss's law* says $\operatorname{div} \vec{E} = \kappa q$, where κ is a constant. If $p : \mathbb{R}^3 \longrightarrow \mathbb{R}$ is the electric potential field associated with $\vec{E}$, then, by definition, $\vec{E} = \nabla p$. Use these facts to derive equation (1D.1).

(b) Suppose q is independent of the x_3 coordinate; that is, $q(x_1, x_2, x_3) = Q(x_1, x_2)$ for some function $Q : \mathbb{R}^2 \longrightarrow \mathbb{R}$. Show that p is also independent of the x_3 coordinate; that is, $p(x_1, x_2, x_3) = P(x_1, x_2)$ for some function $P : \mathbb{R}^2 \longrightarrow \mathbb{R}$. Show that P and Q satisfy the *two*-dimensional version of the Poisson equation, that is that $\triangle P = \kappa Q$.

     (This is significant because many physical problems have (approximate) translational symmetry along one dimension (e.g. an electric field generated by a long, uniformly charged wire or plate). Thus, we can reduce the problem to two dimensions, where powerful methods from complex analysis can be applied; see Section 18B, p. 428.)   ♦

     Note that the electric/gravitational potential field is *not uniquely defined* by equation (1D.1). If $p(\mathbf{x})$ solves the Poisson equation (1D.1), then so does $\widetilde{p}(\mathbf{x}) = p(\mathbf{x}) + a$ for any constant $a \in \mathbb{R}$. Thus we say that the potential field is well-defined *up to addition of a constant*; this is similar to the way in which the antiderivative

---

[5] Equation (1D.2) seems mathematically nonsensical but it *can* be made mathematically meaningful using *distribution theory*. However, this is far beyond the scope of this book, so, for our purposes, we will interpret equation (1D.2) as purely metaphorical.

$\int Q(x)$ of a function is only well-defined up to some constant.[6] This is an example of a more general phenomenon.

Proposition 1D.4   *Let $X \subset \mathbb{R}^D$ be some domain, and let $p : X \longrightarrow \mathbb{R}$ and $h : X \longrightarrow \mathbb{R}$ be two functions on X. Let $\widetilde{p}(x) := p(x) + h(x)$ for all $x \in X$. Suppose that h is harmonic, i.e. $\triangle h \equiv 0$. If p satisfies the Poisson equation $\triangle p \equiv q$, then $\widetilde{p}$ also satisfies this Poisson equation.*

Ⓔ   *Proof* **Exercise 1D.4.** *Hint:* Note that $\triangle \widetilde{p}(x) = \triangle p(x) + \triangle h(x)$.   □

For example, if $Q(x) = 1/x^2$, as in Example 1D.1(b), then $p(x) = -\log(x)$ is a solution to the Poisson equation $p''(x) = 1/x^2$. If $h(x)$ is a one-dimensional harmonic function, then $h(x) = ax + b$ for some constants a and b (see Example 1C.1(a)). Thus $\widetilde{p}(x) = -\log(x) + ax + b$, and we have already seen that these are also valid solutions to this Poisson equation.

1E Properties of harmonic functions

Prerequisites: §1C, Appendix H(ii).   **Prerequisites** (for proofs): §2A, §17G, Appendix E(iii).

Recall that a function $h : \mathbb{R}^D \longrightarrow \mathbb{R}$ is *harmonic* if $\triangle h \equiv 0$. Harmonic functions have nice geometric properties, which can be loosely summarized as 'smooth and gently curving'.

Theorem 1E.1 Mean value theorem

Let $f : \mathbb{R}^D \longrightarrow \mathbb{R}$ be a scalar field. Then f is harmonic if and only if f is integrable, and

$$\text{for any } x \in \mathbb{R}^D, \text{ and any } R > 0, f(x) = \frac{1}{A(R)} \int_{\mathbb{S}(x;R)} f(s) ds. \qquad (1E.1)$$

Here, $\mathbb{S}(x; R) := \{ s \in \mathbb{R}^D; \ \|s - x\| = R \}$ is the $(D-1)$-dimensional sphere of radius R around x, and $A(R)$ is the $(D-1)$-dimensional surface area of $\mathbb{S}(x; R)$.

Ⓔ   *Proof* **Exercise 1E.1.**

   (a)   Suppose f is integrable and statement (1E.1) is true. Use the *spherical means formula* for the Laplacian (Theorem 2A.1) to show that f is harmonic.

---

[6] For the purposes of the physical theory, this constant *does not matter*, because the field p is physically interpreted only by computing the *potential difference* between two points, and the constant a will always cancel out in this computation. Thus, the two potential fields $p(x)$ and $\widetilde{p}(x) = p(x) + a$ will generate identical physical predictions.

(b) Now suppose f is harmonic. Define $\phi : \mathbb{R}_{\neq} \longrightarrow \mathbb{R}$ by

$$\phi(R) := \frac{1}{A(R)} \int_{\mathbb{S}(\mathbf{x};R)} f(\mathbf{s})d\mathbf{s}.$$

Show that

$$\phi'(R) = \frac{K}{A(R)} \int_{\mathbb{S}(\mathbf{x};R)} \partial_{\perp} f(\mathbf{s})d\mathbf{s},$$

for some constant $K > 0$. Here, $\partial_{\perp} f(\mathbf{s})$ is the *outward normal derivative* of f at the point $\mathbf{s}$ on the sphere (see p. 567 for an abstract definition; see §5C(ii), p. 80, for more information).

(c) Let $\mathbb{B}(\mathbf{x}; R) := \{\mathbf{b} \in \mathbb{R}^D; \|\mathbf{b} - \mathbf{x}\| \le R\}$ be the *ball* of radius R around $\mathbf{x}$. Apply *Green's Formula* (Theorem E.5(a), p. 567) to show that

$$\phi'(R) = \frac{K}{A(R)} \int_{\mathbb{B}(\mathbf{x};R)} \triangle f(\mathbf{b})d\mathbf{b}.$$

(d) Deduce that, if f is harmonic, then ϕ must be constant.

(e) Use the fact that f is continuous to show that $\lim_{r \to 0} \phi(r) = f(\mathbf{x})$. Deduce that $\phi(r) = f(\mathbf{x})$ for all $r \ge 0$. Conclude that, if f is harmonic, then statement (1E.1) must be true.   □

Corollary 1E.2 Maximum principle for harmonic functions

Let $\mathbb{X} \subset \mathbb{R}^D$ be a domain, and let $u : \mathbb{X} \longrightarrow \mathbb{R}$ be a nonconstant harmonic function. Then u has no local maximal or minimal points anywhere in the interior of $\mathbb{X}$.

If $\mathbb{X}$ is bounded (hence compact), then u does obtain a maximum and minimum, but only on the boundary of $\mathbb{X}$.

Proof (By contradiction.) Suppose $\mathbf{x}$ was a local maximum of u somewhere in the interior of $\mathbb{X}$. Let $R > 0$ be small enough that $\mathbb{S}(\mathbf{x}; R) \subset \mathbb{X}$, and such that

$$u(\mathbf{x}) \ge u(\mathbf{s}) \quad \text{for all } \mathbf{s} \in \mathbb{S}(\mathbf{x}; R), \tag{1E.2}$$

where this inequality is strict for at least one $\mathbf{s}_0 \in \mathbb{S}(\mathbf{x}; R)$.

Claim 1 *There is a nonempty open subset $\mathbb{Y} \subset \mathbb{S}(\mathbf{x}; R)$ such that $u(\mathbf{x}) > u(\mathbf{y})$ for all $\mathbf{y}$ in $\mathbb{Y}$.*

Proof We know that $u(\mathbf{x}) > u(\mathbf{s}_0)$. But u is continuous, so there must be some open neighbourhood $\mathbb{Y}$ around $\mathbf{s}_0$ such that $u(\mathbf{x}) > u(\mathbf{y})$ for all $\mathbf{y}$ in $\mathbb{Y}$.   ◇Claim 1

Equation (1E.2) and Claim 1 imply that

$$f(\mathbf{x}) > \frac{1}{A(R)} \int_{\mathbb{S}(\mathbf{x};R)} f(\mathbf{s})d\mathbf{s}.$$

But this contradicts the mean value theorem. By contradiction, $\mathbf{x}$ cannot be a local maximum. (The proof for local minima is analogous.) $\square$

A function $F : \mathbb{R}^D \longrightarrow \mathbb{R}$ is *spherically symmetric* if $F(\mathbf{x})$ depends only on the norm $\|\mathbf{x}\|$ (i.e. $F(\mathbf{x}) = f(\|\mathbf{x}\|)$ for some function $f : \mathbb{R}_+ \longrightarrow \mathbb{R}$). For example, the function $F(\mathbf{x}) := \exp(-\|\mathbf{x}\|^2)$ is spherically symmetric.

If $h, F : \mathbb{R}^D \longrightarrow \mathbb{R}$ are two integrable functions, then their *convolution* is the function $h * F : \mathbb{R}^D \longrightarrow \mathbb{R}$ defined by

$$h * F(\mathbf{x}) := \int_{\mathbb{R}^D} h(\mathbf{y}) \cdot F(\mathbf{x} - \mathbf{y}) d\mathbf{y}, \quad \text{for all } \mathbf{x} \in \mathbb{R}^D$$

(if this integral converges). We will encounter convolutions in §10D(ii), p. 215 (where they will be used to prove the L^2-convergence of a Fourier series), and again in Chapter 17 (where they will be used to construct 'impulse-response' solutions for PDEs). For now, we state the following simple consequence of the mean value theorem.

Lemma 1E.3 *If $h : \mathbb{R}^D \longrightarrow \mathbb{R}$ is harmonic and $F : \mathbb{R}^D \longrightarrow \mathbb{R}$ is integrable and spherically symmetric, then $h * F = K \cdot h$, where $K \in \mathbb{R}$ is some constant.*

Ⓔ *Proof* **Exercise 1E.2.** $\square$

Proposition 1E.4 Smoothness of harmonic functions
If $h : \mathbb{R}^D \longrightarrow \mathbb{R}$ is a harmonic function, then h is infinitely differentiable.

Proof Let $F : \mathbb{R}^D \longrightarrow \mathbb{R}$ be some infinitely differentiable, spherically symmetric, integrable function. For example, we could take $F(\mathbf{x}) := \exp(-\|\mathbf{x}\|^2)$. Then Proposition 17G.2(f), p. 415, says that $h * F$ is infinitely differentiable. But Lemma 1E.3 implies that $h * F = Kh$ for some constant $K \in \mathbb{R}$; thus, h is also infinitely differentiable.

(For another proof, see Evans (1991), §2.2, Theorem 6.) $\square$

Actually, we can go even further than this. A function $h : \mathbb{X} \longrightarrow \mathbb{R}$ is *analytic* if, for every $\mathbf{x} \in \mathbb{X}$, there is a multivariate Taylor series expansion for h around $\mathbf{x}$ with a nonzero radius of convergence.[7]

Proposition 1E.5 Harmonic functions are analytic
Let $\mathbb{X} \subseteq \mathbb{R}^D$ be an open set. If $h : \mathbb{X} \longrightarrow \mathbb{R}$ is a harmonic function, then h is analytic on $\mathbb{X}$.

Proof For the case $D = 2$, see Corollary 18D.2, p. 456. For the general case $D \geq 2$, see Evans (1991), §2.2, Theorem 10. $\square$

---

[7] See Appendices H(ii) and H(v), pp. 574, 579.

1F* Transport and diffusion

Prerequisites: §1B, §6A.

If $u : \mathbb{R}^D \longrightarrow \mathbb{R}$ is a 'mountain', then recall that $\nabla u(\mathbf{x})$ points in the direction of *most rapid ascent* at $\mathbf{x}$. If $\vec{v} \in \mathbb{R}^D$ is a vector, then the dot product $\vec{v} \bullet \nabla u(\mathbf{x})$ measures how rapidly you would be ascending if you walked in direction $\vec{v}$.

Suppose $u : \mathbb{R}^D \longrightarrow \mathbb{R}$ describes a pile of leafs, and there is a steady wind blowing in the direction $\vec{v} \in \mathbb{R}^D$. We would expect the pile to move slowly in the direction $\vec{v}$. Suppose you were an observer fixed at location $\mathbf{x}$. The pile is moving past you in direction $\vec{v}$, which is the same as you walking along the pile in direction $-\vec{v}$; thus, you would expect the height of the pile at your location to increase/decrease at a rate $-\vec{v} \bullet \nabla u(\mathbf{x})$. The pile thus satisfies the *transport equation*:

$$\partial_t u = -\vec{v} \bullet \nabla u.$$

Now suppose that the wind does not blow in a *constant* direction, but instead has some complex spatial pattern. The wind velocity is therefore determined by a *vector field* $\vec{\mathbf{V}} : \mathbb{R}^D \longrightarrow \mathbb{R}^D$. As the wind picks up leaves and carries them around, the *flux* of leaves at a point $\mathbf{x} \in \mathbb{X}$ is then given by the vector $\vec{\mathbf{F}}(\mathbf{x}) = u(\mathbf{x}) \cdot \vec{\mathbf{V}}(\mathbf{x})$. But the rate at which leaves are piling up at each location is the *divergence* of the flux. This results in *Liouville's equation*:

$$\partial_t u = -\operatorname{div} \vec{\mathbf{F}} = -\operatorname{div}(u \cdot \vec{\mathbf{V}}) \underset{(*)}{=} -\vec{\mathbf{V}} \bullet \nabla u - u \cdot \operatorname{div} \vec{\mathbf{V}}.$$

Here, $(*)$ is by the Leibniz rule for divergence (Proposition E.2(b), p. 564).

Liouville's equation describes the rate at which u-material accumulates when it is being pushed around by the $\vec{\mathbf{V}}$-vector field. Another example: $\vec{\mathbf{V}}(\mathbf{x})$ describes the flow of water at $\mathbf{x}$, and $u(\mathbf{x})$ is the buildup of some sediment at $\mathbf{x}$.

Now suppose that, in addition to the deterministic force $\vec{\mathbf{V}}$ acting on the leaves, there is also a 'random' component. In other words, while being blown around by the wind, the leaves are also subject to some random diffusion. To describe this, we combine *Liouville's equation* with the *heat equation*, to obtain the *Fokker–Planck equation*:

$$\partial_t u = \kappa \, \triangle u - \vec{\mathbf{V}} \bullet \nabla u - u \cdot \operatorname{div} \vec{\mathbf{V}}.$$

1G* Reaction and diffusion

Prerequisites: §1B.

Suppose A, B and C are three chemicals, satisfying the following chemical reaction:

$$2A + B \Longrightarrow C.$$

As this reaction proceeds, the A and B species are consumed and C is produced. Thus, if a, b, and c are the concentrations of the three chemicals, we have

$$\partial_t\, c = R(t) = -\partial_t\, b = -\frac{1}{2}\partial_t\, a,$$

where $R(t)$ is the rate of the reaction at time t. The rate $R(t)$ is determined by the concentrations of A and B, and by a rate constant ρ. Each chemical reaction requires two molecules of A and one of B; thus, the reaction rate is given by

$$R(t) = \rho \cdot a(t)^2 \cdot b(t).$$

Hence, we get three ordinary differential equations, called the *reaction kinetic equations* of the system:

$$\left. \begin{array}{l} \partial_t\, a(t) = -2\rho \cdot a(t)^2 \cdot b(t), \\ \partial_t\, b(t) = -\rho \cdot a(t)^2 \cdot b(t), \\ \partial_t\, c(t) = \rho \cdot a(t)^2 \cdot b(t). \end{array} \right\} \qquad (1\text{G}.1)$$

Now suppose that the chemicals A, B, and C are in solution, but are not uniformly mixed. At any location $\mathbf{x} \in \mathbb{X}$ and time $t > 0$, let $a(\mathbf{x}, t)$ be the concentration of chemical A at location $\mathbf{x}$ at time t; likewise, let $b(\mathbf{x}, t)$ be the concentration of B and $c(\mathbf{x}, t)$ be the concentration of C. (This determines three *time-varying scalar fields*, $a, b, c : \mathbb{R}^3 \times \mathbb{R} \longrightarrow \mathbb{R}$.) As the chemicals react, their concentrations at each point in space may change. Thus, the functions a, b, c will obey equations (1G.1) at each point in space. That is, for every $\mathbf{x} \in \mathbb{R}^3$ and $t \in \mathbb{R}$, we have

$$\partial_t\, a(\mathbf{x}; t) \approx -2\rho \cdot a(\mathbf{x}; t)^2 \cdot b(\mathbf{x}; t),$$

etc. However, the dissolved chemicals are also subject to *diffusion* forces. In other words, each of the functions a, b, and c will also be obeying the heat equation. Thus, we obtain the following system:

$$\partial_t\, a = \kappa_a \cdot \triangle a(\mathbf{x}; t) - 2\rho \cdot a(\mathbf{x}; t)^2 \cdot b(\mathbf{x}; t),$$

$$\partial_t\, b = \kappa_b \cdot \triangle b(\mathbf{x}; t) - \rho \cdot a(\mathbf{x}; t)^2 \cdot b(\mathbf{x}; t),$$

$$\partial_t\, c = \kappa_c \cdot \triangle c(\mathbf{x}; t) + \rho \cdot a(\mathbf{x}; t)^2 \cdot b(\mathbf{x}; t),$$

where $\kappa_a, \kappa_b, \kappa_c > 0$ are three different diffusivity constants.

   This is an example of a *reaction–diffusion system*. In general, in a reaction–diffusion system involving N distinct chemicals, the concentrations of the different species are described by a *concentration vector field* $\mathbf{u} : \mathbb{R}^3 \times \mathbb{R} \longrightarrow \mathbb{R}^N$, and the chemical reaction is described by a *rate function* $F : \mathbb{R}^N \longrightarrow \mathbb{R}^N$. For example, in the previous example, $\mathbf{u}(\mathbf{x}, t) = (a(\mathbf{x}, t), b(\mathbf{x}, t), c(\mathbf{x}, t))$, and

$$F(a, b, c) = \left[-2\rho a^2 b, -\rho a^2 b, \rho a^2 b \right].$$

The *reaction–diffusion equations* for the system then take the form

$$\partial_t u_n = \kappa_n \triangle u_n + F_n(\mathbf{u}),$$

for $n = 1, \ldots, N$.

1H Further reading

An analogy of the Laplacian can be defined on any Riemannian manifold, where it is sometimes called the *Laplace–Beltrami operator*. The study of harmonic functions on manifolds yields important geometric insights (Warner (1983), Chavel (1993)).

The reaction–diffusion systems from §1G play an important role in modern mathematical biology (Murray (1993)).

The heat equation also arises frequently in the theory of Brownian motion and other Gaussian stochastic processes on $\mathbb{R}^D$ (Strook (1993)).

1I Practice problems

1.1 Let $f : \mathbb{R}^4 \longrightarrow \mathbb{R}$ be a differentiable scalar field. Show that $\operatorname{div} \nabla f(x_1, x_2, x_3, x_4) = \triangle f(x_1, x_2, x_3, x_4)$.

1.2 Let $f(x, y; t) = \exp(-34t) \cdot \sin(3x + 5y)$. Show that $f(x, y; t)$ satisfies the two-dimensional heat equation $\partial_t f(x, y; t) = \triangle f(x, y; t)$.

1.3 Let $u(x, y) = \log(x^2 + y^2)$. Show that $u(x, y)$ satisfies the (two-dimensional) Laplace equation, everywhere except at $(x, y) = (0, 0)$. *Remark:* If $(x, y) \in \mathbb{R}^2$, recall that $\|(x, y)\| := \sqrt{x^2 + y^2}$. Thus, $\log(x^2 + y^2) = 2 \log \|(x, y)\|$. This function is sometimes called the *logarithmic potential*.

1.4 If $(x, y, z) \in \mathbb{R}^3$, recall that $\|(x, y, z)\| := \sqrt{x^2 + y^2 + z^2}$. Define

$$u(x, y, z) = \frac{1}{\|(x, y, z)\|} = \frac{1}{\sqrt{x^2 + y^2 + z^2}}.$$

Show that u satisfies the (three-dimensional) Laplace equation, everywhere except at $(x, y, z) = (0, 0, 0)$. *Remark:* Observe that

$$\nabla u(x, y, z) = \frac{-(x, y, z)}{\|(x, y, z)\|^3}.$$

What force field does this remind you of? *Hint:* $u(x, y, z)$ is sometimes called the *Coulomb potential*.

1.5 Let

$$u(x, y; t) = \frac{1}{4\pi t} \exp\left(\frac{-\|(x, y)\|^2}{4t}\right) = \frac{1}{4\pi t} \exp\left(\frac{-x^2 - y^2}{4t}\right)$$

be the (two-dimensional) *Gauss–Weierstrass kernel*. Show that u satisfies the (two-dimensional) heat equation $\partial_t u = \triangle u$.

1.6 Let α and β be real numbers, and let $h(x, y) = \sinh(\alpha x) \cdot \sin(\beta y)$.

(a) Compute $\triangle h(x, y)$.

(b) Suppose h is *harmonic*. Write an equation describing the relationship between α and β.

2

Waves and signals

There is geometry in the humming of the strings.

Pythagoras

2A The Laplacian and spherical means

Prerequisites: Appendix A, Appendix B, Appendix H(v).   **Recommended:** §1B.

Let $u : \mathbb{R}^D \longrightarrow \mathbb{R}$ be a function of D variables. Recall that the Laplacian of u is defined as follows:

$$\Delta u = \partial_1^2 u + \partial_2^2 u + \cdots + \partial_D^2 u.$$

In this section, we will show that $\Delta u(\mathbf{x})$ measures the discrepancy between $u(\mathbf{x})$ and the 'average' of u in a small neighbourhood around $\mathbf{x}$.

Let $\mathbb{S}(\epsilon)$ be the D-dimensional 'sphere' of radius ϵ around 0.

- If $D = 1$, then $\mathbb{S}(\epsilon)$ is just a set with two points: $\mathbb{S}(\epsilon) = \{-\epsilon, +\epsilon\}$.
- If $D = 2$, then $\mathbb{S}(\epsilon)$ is the *circle* of radius ϵ: $\mathbb{S}(\epsilon) = \{(x, y) \in \mathbb{R}^2; x^2 + y^2 = \epsilon^2\}$.
- If $D = 3$, then $\mathbb{S}(\epsilon)$ is the three-dimensional spherical shell of radius ϵ:

$$\mathbb{S}(\epsilon) = \left\{(x, y, z) \in \mathbb{R}^3; \ x^2 + y^2 + z^2 = \epsilon^2\right\}.$$

- More generally, for any dimension D,

$$\mathbb{S}(\epsilon) = \left\{(x_1, x_2, \ldots, x_D) \in \mathbb{R}^D; \ x_1^2 + x_2^2 + \cdots + x_D^2 = \epsilon^2\right\}.$$

Let A_ϵ be the 'surface area' of the sphere.

- If $D = 1$, then $\mathbb{S}(\epsilon) = \{-\epsilon, +\epsilon\}$ is a finite set, with two points, so we say $A_\epsilon = 2$.
- If $D = 2$, then $\mathbb{S}(\epsilon)$ is the circle of radius ϵ; the *perimeter* of this circle is $2\pi\epsilon$, so we say $A_\epsilon = 2\pi\epsilon$.
- If $D = 3$, then $\mathbb{S}(\epsilon)$ is the sphere of radius ϵ, which has *surface area* $4\pi\epsilon^2$.

25

Let

$$\mathbf{M}_\epsilon \, f(0) := \frac{1}{A_\epsilon} \int_{\mathbb{S}(\epsilon)} f(\mathbf{s}) ds;$$

then $\mathbf{M}_\epsilon \, f(0)$ is the *average value* of $f(\mathbf{s})$ over all $\mathbf{s}$ on the surface of the ϵ-radius sphere around 0, which is called the *spherical mean* of f at 0. The interpretation of the integral sign '$\int$' depends on the dimension D of the space. For example, '$\int$' represents a *surface integral* if $D = 3$, a *line integral* if $D = 2$, and a simple two-point sum if $D = 1$. Thus we have the following.

- If $D = 1$, then $\mathbb{S}(\epsilon) = \{-\epsilon, +\epsilon\}$, so that $\int_{\mathbb{S}(\epsilon)} f(\mathbf{s}) ds = f(\epsilon) + f(-\epsilon)$; thus,

$$\mathbf{M}_\epsilon \, f = \frac{f(\epsilon) + f(-\epsilon)}{2}.$$

- If $D = 2$, then any point on the circle has the form $(\epsilon \cos(\theta), \epsilon \sin(\theta))$ for some angle $\theta \in [0, 2\pi)$. Thus,

$$\int_{\mathbb{S}(\epsilon)} f(\mathbf{s}) ds = \int_0^{2\pi} f\left(\epsilon \cos(\theta), \epsilon \sin(\theta)\right) \epsilon \, d\theta,$$

so that

$$\mathbf{M}_\epsilon \, f = \frac{1}{2\pi\epsilon} \int_0^{2\pi} f\left(\epsilon \cos(\theta), \epsilon \sin(\theta)\right) \epsilon d\theta = \frac{1}{2\pi} \int_0^{2\pi} f\left(\epsilon \cos(\theta), \epsilon \sin(\theta)\right) d\theta.$$

Likewise, for any $\mathbf{x} \in \mathbb{R}^D$, we define

$$\mathbf{M}_\epsilon \, f(\mathbf{x}) := \frac{1}{A_\epsilon} \int_{\mathbb{S}(\epsilon)} f(\mathbf{x} + \mathbf{s}) ds$$

to be the average value of f over an ϵ-radius sphere around $\mathbf{x}$. Suppose $f : \mathbb{R}^D \longrightarrow \mathbb{R}$ is a smooth scalar field, and $\mathbf{x} \in \mathbb{R}^D$. One interpretation of the Laplacian is as follows: $\triangle f(\mathbf{x})$ measures the disparity between $f(\mathbf{x})$ and the *average value* of f in the immediate vicinity of $\mathbf{x}$. This is the meaning of the following theorem.

Theorem 2A.1

(a) *If $f : \mathbb{R} \longrightarrow \mathbb{R}$ is a smooth scalar field, then (as shown in Figure 2A.1), for any $x \in \mathbb{R}$,*

$$\triangle f(x) = \lim_{\epsilon \to 0} \frac{2}{\epsilon^2} \left[\mathbf{M}_\epsilon \, f(x) - f(x)\right] = \lim_{\epsilon \to 0} \frac{2}{\epsilon^2} \left[\frac{f(x - \epsilon) + f(x + \epsilon)}{2} - f(x)\right].$$

(b)[1] *If $f : \mathbb{R}^D \longrightarrow \mathbb{R}$ is a smooth scalar field, then, for any $\mathbf{x} \in \mathbb{R}^D$,*

$$\triangle f(\mathbf{x}) = \lim_{\epsilon \to 0} \frac{C}{\epsilon^2} \left[\mathbf{M}_\epsilon \, f(\mathbf{x}) - f(\mathbf{x})\right] = \lim_{\epsilon \to 0} \frac{C}{\epsilon^2} \left[\frac{1}{A_\epsilon} \int_{\mathbb{S}(\epsilon)} f(\mathbf{x} + \mathbf{s}) ds - f(\mathbf{x})\right].$$

(Here C is a constant determined by the dimension D.)

---

[1] Part (b) of Theorem 2A.1 is not necessary for the physical derivation of the wave equation which appears later in this chapter. However, part (b) *is* required to prove the Mean Value Theorem for harmonic functions (Theorem 1E.1, p. 18).

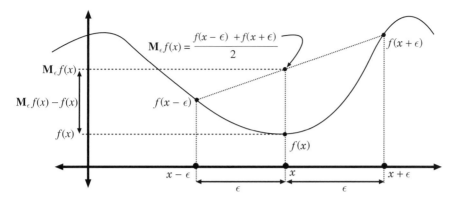

Figure 2A.1. Local averages: $f(x)$ vs. $\mathbf{M}_\epsilon\,f(x) := \frac{f(x-\epsilon)+f(x+\epsilon)}{2}$.

Proof (a) Using *Taylor's theorem* (see Appendix H(i), p. 573), we have

$$f(x+\epsilon) = f(x) + \epsilon f'(x) + \frac{\epsilon^2}{2} f''(x) + \mathcal{O}(\epsilon^3)$$

and

$$f(x-\epsilon) = f(x) - \epsilon f'(x) + \frac{\epsilon^2}{2} f''(x) + \mathcal{O}(\epsilon^3).$$

Here, $f'(x) = \partial_x f(x)$ and $f''(x) = \partial_x^2 f(x)$. The expression '$\mathcal{O}(\epsilon)$' means 'some function (we don't care which one) such that $\lim_{\epsilon \to 0} \mathcal{O}(\epsilon) = 0$'.[2] Likewise, '$\mathcal{O}(\epsilon^3)$' means 'some function (we don't care which one) such that $\lim_{\epsilon \to 0} \frac{\mathcal{O}(\epsilon^3)}{\epsilon^2} = 0$.' Summing these two equations, we obtain

$$f(x+\epsilon) + f(x-\epsilon) = 2f(x) + \epsilon^2 \cdot f''(x) + \mathcal{O}(\epsilon^3).$$

Thus,

$$\frac{f(x-\epsilon) - 2f(x) + f(x+\epsilon)}{\epsilon^2} = f''(x) + \mathcal{O}(\epsilon)$$

(because $\mathcal{O}(\epsilon^3)/\epsilon^2 = \mathcal{O}(\epsilon)$). Now take the limit as $\epsilon \to 0$ to obtain

$$\lim_{\epsilon \to 0} \frac{f(x-\epsilon) - 2f(x) + f(x+\epsilon)}{\epsilon^2} = \lim_{\epsilon \to 0} f''(x) + \mathcal{O}(\epsilon) = f''(x) = \triangle f(x),$$

as desired.

---

[2] Actually, '$\mathcal{O}(\epsilon)$' means slightly more than this – see Appendix H(i). However, for our purposes, this will be sufficient.

(b) Define the *Hessian derivative matrix* of f at $\mathbf{x}$:

$$\mathsf{D}^2 f(\mathbf{x}) = \begin{bmatrix} \partial_1^2 f & \partial_1 \partial_2 f & \cdots & \partial_1 \partial_D f \\ \partial_2 \partial_1 f & \partial_2^2 f & \cdots & \partial_2 \partial_D f \\ \vdots & \vdots & \ddots & \vdots \\ \partial_D \partial_1 f & \partial_D \partial_2 f & \cdots & \partial_D^2 f \end{bmatrix}.$$

Then, for any $\mathbf{s} \in \mathbb{S}(\epsilon)$, the *multivariate Taylor theorem* (see Appendix H(v), p. 579) says:

$$f(\mathbf{x} + \mathbf{s}) = f(\mathbf{x}) + \mathbf{s} \bullet \nabla f(\mathbf{x}) + \frac{1}{2} \mathbf{s} \bullet \mathsf{D}^2 f(\mathbf{x}) \cdot \mathbf{s} + \mathcal{O}(\epsilon^3).$$

Now, if $\mathbf{s} = (s_1, s_2, \ldots, s_D)$, then $\mathbf{s} \bullet \mathsf{D}^2 f(\mathbf{x}) \cdot \mathbf{s} = \sum_{c,d=1}^{D} s_c \cdot s_d \cdot \partial_c \partial_d f(\mathbf{x})$. Thus, for any $\epsilon > 0$, we have

$$A_\epsilon \cdot \mathbf{M}_\epsilon f(\mathbf{x}) = \int_{\mathbb{S}(\epsilon)} f(\mathbf{x} + \mathbf{s}) d\mathbf{s}$$

$$= \int_{\mathbb{S}(\epsilon)} f(\mathbf{x}) d\mathbf{s} + \int_{\mathbb{S}(\epsilon)} \mathbf{s} \bullet \nabla f(\mathbf{x}) d\mathbf{s}$$

$$+ \frac{1}{2} \int_{\mathbb{S}(\epsilon)} \mathbf{s} \bullet \mathsf{D}^2 f(\mathbf{x}) \cdot \mathbf{s} \, d\mathbf{s} + \int_{\mathbb{S}(\epsilon)} \mathcal{O}(\epsilon^3) d\mathbf{s}$$

$$= A_\epsilon f(\mathbf{x}) + \nabla f(\mathbf{x}) \bullet \int_{\mathbb{S}(\epsilon)} \mathbf{s} \, d\mathbf{s}$$

$$+ \frac{1}{2} \int_{\mathbb{S}(\epsilon)} \left(\sum_{c,d=1}^{D} s_c s_d \cdot \partial_c \partial_d f(\mathbf{x}) \right) d\mathbf{s} + \mathcal{O}(\epsilon^{D+2})$$

$$= A_\epsilon f(\mathbf{x}) + \underbrace{\nabla f(\mathbf{x}) \bullet \mathbf{0}}_{(*)}$$

$$+ \frac{1}{2} \sum_{c,d=1}^{D} \left(\partial_c \partial_d f(\mathbf{x}) \cdot \left(\int_{\mathbb{S}(\epsilon)} s_c s_d \, d\mathbf{s} \right) \right) + \mathcal{O}(\epsilon^{D+2})$$

$$= A_\epsilon f(\mathbf{x}) + \underbrace{\frac{1}{2} \sum_{d=1}^{D} \left(\partial_d^2 f(\mathbf{x}) \cdot \left(\int_{\mathbb{S}(\epsilon)} s_d^2 \, d\mathbf{s} \right) \right)}_{(\dagger)} + \mathcal{O}(\epsilon^{D+2})$$

$$= A_\epsilon f(\mathbf{x}) + \frac{1}{2} \triangle f(\mathbf{x}) \cdot \epsilon^{D+1} K + \mathcal{O}(\epsilon^{D+2}),$$

where $K := \int_{\mathbb{S}(1)} s_1^2 \, d\mathbf{s}$. Here, $(*)$ is because $\int_{\mathbb{S}(\epsilon)} \mathbf{s} \, d\mathbf{s} = \mathbf{0}$, because the centre-of-mass of a sphere is at its centre, namely $\mathbf{0}$; $(\dagger)$ is because, if $c, d \in [1 \ldots D]$, and
Ⓔ $c \neq d$, then $\int_{\mathbb{S}(\epsilon)} s_c s_d d\mathbf{s} = 0$ (**Exercise 2A.1** *Hint:* Use symmetry.)

Thus,

$$A_\epsilon \cdot \mathbf{M}_\epsilon \, f(\mathbf{x}) - A_\epsilon \, f(\mathbf{x}) = \frac{\epsilon^{D+1} K}{2} \triangle f(\mathbf{x}) + \mathcal{O}(\epsilon^{D+2}),$$

so

$$\begin{aligned}
\mathbf{M}_\epsilon \, f(\mathbf{x}) - f(\mathbf{x}) &= \frac{\epsilon^{D+1} K}{2A_\epsilon} \triangle f(\mathbf{x}) + \frac{1}{A_\epsilon} \mathcal{O}(\epsilon^{D+2}) \\
&\underset{(*)}{=} \frac{\epsilon^{D+1} K}{2A_1 \cdot \epsilon^{D-1}} \triangle f(\mathbf{x}) + \mathcal{O}\left(\frac{\epsilon^{D+2}}{\epsilon^{D-1}}\right) \\
&= \frac{\epsilon^2 K}{2A_1} \triangle f(\mathbf{x}) + \mathcal{O}(\epsilon^3),
\end{aligned}$$

where $(*)$ is because $A_\epsilon = A_1 \cdot \epsilon^{D-1}$. Thus,

$$\frac{2A_1}{K \, \epsilon^2} \left(\mathbf{M}_\epsilon \, f(\mathbf{x}) - f(\mathbf{x}) \right) = \triangle f(\mathbf{x}) + \mathcal{O}(\epsilon).$$

Now take the limit as $\epsilon \to 0$ and set $C := 2A_1/K$ to prove part (b). $\qquad \square$

Exercise 2A.2 Let $f : \mathbb{R}^D \longrightarrow \mathbb{R}$ be a smooth scalar field, such that $\mathbf{M}_\epsilon \, f(\mathbf{x}) =$ ⒠ $f(\mathbf{x})$ for all $\mathbf{x} \in \mathbb{R}^D$. Show that f is harmonic. ◆

2B The wave equation

Prerequisites: §2A.

2B(i) ... in one dimension: the string

We want to describe mathematically vibrations propagating through a stretched elastic cord. We will represent the cord with a one-dimensional domain $\mathbb{X}$; either $\mathbb{X} = [0, L]$ or $\mathbb{X} = \mathbb{R}$. We will make several simplifying assumptions as follows.

(W1) The cord has uniform thickness and density. Thus, there is a constant *linear density* $\rho > 0$, so that a cord segment of length ℓ has mass $\rho\ell$.

(W2) The cord is *perfectly elastic*, meaning that it is infinitely flexible and does not resist bending in any way. Likewise, there is no air friction to resist the motion of the cord.

(W3) The only force acting on the cord is *tension*, which is a force of magnitude T pulling the cord to the right, balanced by an equal but opposite force of magnitude $-T$ pulling the cord to the left. These two forces are in balance, so the cord exhibits no horizontal motion. The tension T must be constant along the whole length of the cord. Thus, the

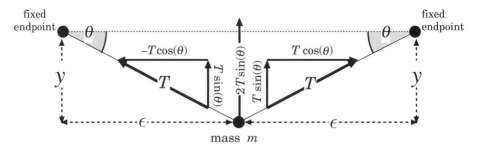

Figure 2B.1. Bead on a string.

equilibrium state for the cord is to be perfectly straight. Vibrations are deviations from this straight position.[3]

(W4) The vibrational motion of the cord is entirely *vertical*; there is no horizontal component to the vibration. Thus, we can describe the motion using a scalar-valued function $u(x, t)$, where $u(x, t)$ is the vertical displacement of the cord (from its flat equilibrium) at point x at time t. We assume that $u(x, t)$ is relatively small relative to the length of the cord, so that the cord is not significantly stretched by the vibrations.[4]

For simplicity, imagine first a single bead of mass m suspended at the midpoint of a (massless) elastic cord of length 2ϵ, stretched between two endpoints. Suppose we displace the bead by a distance y from its equilibrium, as shown in Figure 2B.1. The tension force T now pulls the bead diagonally towards each endpoint with force T. The horizontal components of the two tension forces are equal and opposite, so they cancel and the bead experiences no net horizontal force. Suppose the cord makes an angle θ with the horizontal; then the vertical component of each tension force is $T \sin(\theta)$, so the total vertical force acting on the bead is $2T \sin(\theta)$. But $\theta = \arctan(\epsilon/y)$ by the geometry of the triangles in Figure 2B.1, so $\sin(\theta) = \frac{y}{\sqrt{y^2+\epsilon^2}}$. Thus, the vertical force acting on the bead is given by

$$F = 2T \sin(\theta) = \frac{2Ty}{\sqrt{y^2 + \epsilon^2}}. \tag{2B.1}$$

Now we return to our original problem of the vibrating string. Imagine that we replace the string with a 'necklace' made up of small beads of mass m separated by massless elastic strings of length ϵ as shown in Figure 2B.2. Each of these beads, in isolation, behaves like the 'bead on a string' in Figure 2B.1. However, now

---

[3] We could also incorporate the force of gravity as a constant downward force. In this case, the equilibrium position for the cord is to sag downwards in a 'catenary' curve. Vibrations are then deviations from this curve. This does not change the mathematics of this derivation, so we will assume for simplicity that gravity is absent and the cord is straight.

[4] If $u(x, t)$ were large, then the vibrations would stretch the cord and a *restoring force* would act against this stretching, as described by *Hooke's law*. By assuming that the vibrations are small, we are assuming we can neglect Hooke's law.

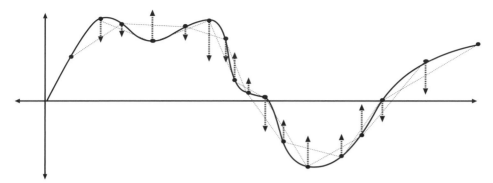

Figure 2B.2. Each bead feels a negative force proportional to its deviation from the local average.

the vertical displacement of each bead is not computed relative to the horizontal, but instead relative to the *average height* of the two neighbouring beads. Thus, in equation (2B.1), we set $y := u(x) - \mathbf{M}_\epsilon\, u(x)$, where $u(x)$ is the height of bead x, and where $\mathbf{M}_\epsilon\, u := \frac{1}{2}[u(x - \epsilon) + u(x + \epsilon)]$ is the average of its neighbours. Substituting this into equation (2B.1), we obtain

$$F_\epsilon(x) = \frac{2T[u(x) - \mathbf{M}_\epsilon\, u(x)]}{\sqrt{[u(x) - \mathbf{M}_\epsilon\, u(x)]^2 + \epsilon^2}}. \tag{2B.2}$$

(The 'ϵ' subscript in 'F_ϵ' is to remind us that this is just an ϵ-bead approximation of the real string.) Each bead represents a length-ϵ segment of the original string, so if the string has linear density ρ, then each bead must have mass $m_\epsilon := \rho\epsilon$. Thus, by Newton's law, the vertical acceleration of bead x must be as follows:

$$a_\epsilon(x) = \frac{F_\epsilon(x)}{m_\epsilon} = \frac{2T[u(x) - \mathbf{M}_\epsilon\, u(x)]}{\rho\,\epsilon\,\sqrt{[u(x) - \mathbf{M}_\epsilon\, u(x)]^2 + \epsilon^2}}$$

$$= \frac{2T[u(x) - \mathbf{M}_\epsilon\, u(x)]}{\rho\,\epsilon^2\,\sqrt{[u(x) - \mathbf{M}_\epsilon\, u(x)]^2/\epsilon^2 + 1}}. \tag{2B.3}$$

Now we take the limit as $\epsilon \to 0$ to calculate the vertical acceleration of the string at x:

$$a(x) = \lim_{\epsilon \to 0} a_\epsilon(x) = \frac{T}{\rho}\lim_{\epsilon \to 0}\frac{2}{\epsilon^2}\left[u(x) - \mathbf{M}_\epsilon\, u(x)\right]\cdot\lim_{\epsilon \to 0}\frac{1}{\sqrt{[u(x) - \mathbf{M}_\epsilon\, u(x)]^2/\epsilon^2 + 1}}$$

$$\underset{(*)}{=} \frac{T}{\rho}\,\partial_x^2 u(x)\frac{1}{\lim_{\epsilon \to 0}\sqrt{\epsilon^2 \cdot \partial_x^2 u(x)^2 + 1}} \underset{(\dagger)}{=} \frac{T}{\rho}\,\partial_x^2 u(x). \tag{2B.4}$$

Here, $(*)$ is because Theorem 2A.1(a), p. 26, says that $\lim_{\epsilon \to 0}\frac{2}{\epsilon^2}[u(x) - \mathbf{M}_\epsilon\, u(x)] = \partial_x^2 u(x)$. Finally, $(\dagger)$ is because, for any value of $u'' \in \mathbb{R}$, we have

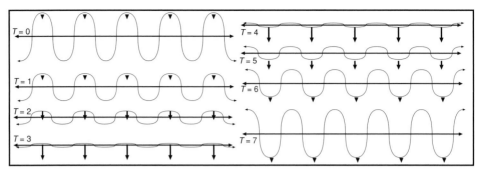

Figure 2B.3. One-dimensional standing wave.

$\lim_{\epsilon \to 0} \sqrt{\epsilon^2 u'' + 1} = 1$. We conclude that

$$a(x) = \frac{T}{\rho} \partial_x^2 u(x) = \lambda^2 \partial_x^2 u(x),$$

where $\lambda := \sqrt{T/\rho}$. Now, the position (and hence velocity and acceleration) of the cord is changing in time. Thus, a and u are functions of t as well as x. And of course the acceleration $a(x, t)$ is nothing more than the second derivative of u with respect to t. Hence we have the *one-dimensional* wave equation:

$$\boxed{\partial_t^2 u(x, t) = \lambda^2 \cdot \partial_x^2 u(x, t).}$$

This equation describes the propagation of a transverse wave along an idealized string, or electrical pulses propagating in a wire.

Example 2B.1 Standing waves
(a) Suppose $\lambda^2 = 4$, and let $u(x; t) = \sin(3x) \cdot \cos(6t)$. Then u satisfies the wave equation and describes a *standing wave* with a *temporal frequency* of 6 and a *wave number* (or *spatial frequency*) of 3 (see Figure 2B.3).
(b) More generally, fix $\omega > 0$; if $u(x; t) = \sin(\omega \cdot x) \cdot \cos(\lambda \cdot \omega \cdot t)$, then u satisfies the wave equation and describes a *standing wave* of *temporal frequency* $\lambda \cdot \omega$ and *wave number* ω.   ◇

Ⓔ **Exercise 2B.1** Verify Examples 2B.1(a) and (b).   ◆

Example 2B.2 Travelling waves
(a) Suppose $\lambda^2 = 4$, and let $u(x; t) = \sin(3x - 6t)$. Then u satisfies the wave equation and describes a *sinusoidal travelling wave* with *temporal frequency* 6 and *wave number* 3. The wave crests move rightwards along the cord with velocity 2 (see Figure 2B.4(a)).
(b) More generally, fix $\omega \in \mathbb{R}$ and let $u(x; t) = \sin(\omega \cdot x - \lambda \cdot \omega \cdot t)$. Then u satisfies the wave equation and describes a *sinusoidal travelling wave* of *wave number* ω. The wave crests move rightwards along the cord with velocity λ.

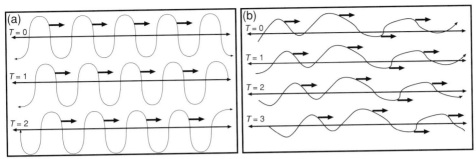

Figure 2B.4. (a) One-dimensional sinusoidal travelling wave. (b) General one-dimensional travelling wave.

(c) More generally, suppose that f is any function of one variable, and define $u(x;t) = f(x - \lambda \cdot t)$. Then u satisfies the wave equation and describes a *travelling wave*, whose shape is given by f, and which moves rightwards along the cord with velocity λ (see Figure 2B.4(b)). ◇

Exercise 2B.2 Verify Examples 2B.2(a)–(c). ♦ Ⓔ

Exercise 2B.3 According to Example 2B.2(c), you can turn any function into a Ⓔ travelling wave. Can you turn any function into a standing wave? Why or why not? ♦

2B(ii) ... in two dimensions: the drum

Now suppose $D = 2$, and imagine a two-dimensional 'rubber sheet'. Suppose $u(x, y; t)$ is the vertical displacement of the rubber sheet at the point $(x, y) \in \mathbb{R}^2$ at time t. To derive the two-dimensional wave equation, we approximate this rubber sheet as a two-dimensional 'mesh' of tiny beads connected by massless, tense elastic strings of length ϵ. Each bead (x, y) feels a net vertical force $F = F_x + F_y$, where F_x is the vertical force arising from the tension in the x direction and F_y is the vertical force from the tension in the y direction. Both of these are expressed by a formula similar to equation (2B.2). Thus, if bead (x, y) has mass m_ϵ, then it experiences an acceleration $a = F/m_\epsilon = F_x/m_\epsilon + F_y/m_\epsilon = a_x + a_y$, where $a_x := F_x/m_\epsilon$ and $a_y := F_y/m_\epsilon$, and each of these is expressed by a formula similar to equation (2B.3). Taking the limit as $\epsilon \to 0$, as in equation (2B.4), we deduce that

$$a(x, y) = \lim_{\epsilon \to 0} a_{x,\epsilon}(x, y) + \lim_{\epsilon \to 0} a_{y,\epsilon}(x, y) = \lambda^2 \, \partial_x^2 u(x, y) + \lambda^2 \, \partial_y^2 u(x, y),$$

where λ is a constant determined by the density and tension of the rubber membrane. Again, we recall that u and a are also functions of time, and that

$a(x, y; t) = \partial_t^2 u(x, y; t)$. Thus, we have the *two-dimensional* wave equation:

$$\boxed{\partial_t^2 u(x, y; t) = \lambda^2 \cdot \partial_x^2 u(x, y; t) + \lambda^2 \cdot \partial_y^2 u(x, y; t),}$$ (2B.5)

or, more abstractly,

$$\boxed{\partial_t^2 u = \lambda^2 \cdot \Delta u.}$$

This equation describes the propagation of wave energy through any medium with a linear restoring force. For example:

- transverse waves on an idealized rubber sheet;
- ripples on the surface of a pool of water;
- acoustic vibrations on a drumskin.

Example 2B.3 Two-dimensional standing waves
(a) Suppose $\lambda^2 = 9$, and let $u(x, y; t) = \sin(3x) \cdot \sin(4y) \cdot \cos(15t)$. This describes a two-dimensional standing wave with temporal frequency 15.
(b) More generally, fix $\boldsymbol{\omega} = (\omega_1, \omega_2) \in \mathbb{R}^2$ and let $\Omega = \|\boldsymbol{\omega}\|_2 = \sqrt{\omega_1^2 + \omega_2^2}$. Then the function

$$u(\mathbf{x}; t) := \sin(\omega_1 x) \cdot \sin(\omega_2 y) \cdot \cos(\lambda \cdot \Omega t)$$

satisfies the two-dimensional wave equation and describes a standing wave with temporal frequency $\lambda \cdot \Omega$.
(c) Even more generally, fix $\boldsymbol{\omega} = (\omega_1, \omega_2) \in \mathbb{R}^2$ and let $\Omega = \|\boldsymbol{\omega}\|_2 = \sqrt{\omega_1^2 + \omega_2^2}$, as before. Let

$$SC_1(x) = \text{either } \sin(x) \text{ or } \cos(x);$$

let

$$SC_2(y) = \text{either } \sin(y) \text{ or } \cos(y);$$

and let

$$SC_t(t) = \text{either } \sin(t) \text{ or } \cos(t).$$

Then

$$u(\mathbf{x}; t) = SC_1(\omega_1 x) \cdot SC_2(\omega_2 y) \cdot SC_t(\lambda \cdot \Omega t)$$

satisfies the two-dimensional wave equation and describes a standing wave with temporal frequency $\lambda \cdot \Omega$.   ◇

ⓔ   **Exercise 2B.4**   Check Examples 2B.3(a)–(c).   ◆

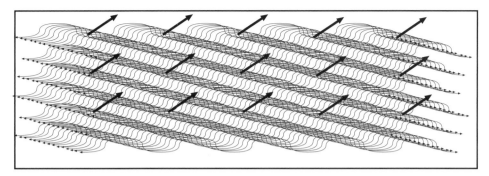

Figure 2B.5. Two-dimensional travelling wave.

Example 2B.4 Two-dimensional travelling waves

(a) Suppose $\lambda^2 = 9$, and let $u(x, y; t) = \sin(3x + 4y + 15t)$. Then u satisfies the two-dimensional wave equation, and describes a sinusoidal travelling wave with *wave vector* $\omega = (3, 4)$ and temporal frequency 15 (see Figure 2B.5).

(b) More generally, fix $\omega = (\omega_1, \omega_2) \in \mathbb{R}^2$ and let $\Omega = \|\omega\|_2 = \sqrt{\omega_1^2 + \omega_2^2}$. Then

$$u(\mathbf{x}; t) = \sin\left(\omega_1 x + \omega_2 y + \lambda \cdot \Omega t\right) \quad \text{and} \quad v(\mathbf{x}; t) = \cos\left(\omega_1 x + \omega_2 y + \lambda \cdot \Omega t\right)$$

both satisfy the two-dimensional wave equation and describe sinusoidal travelling waves with wave vector ω and temporal frequency $\lambda \cdot \Omega$.   ◇

Exercise 2B.5 Check Examples 2B.4(a) and (b).   ◆   Ⓔ

2B(iii) ... in higher dimensions

The same reasoning applies for $D \geq 3$. For example, the three-dimensional wave equation describes the propagation of (small-amplitude[5]) sound-waves in air or water. In general, the wave equation takes the form

$$\partial_t^2 u = \lambda^2 \triangle u,$$

where λ is some constant (determined by the density, elasticity, pressure, etc. of the medium) which describes the speed-of-propagation of the waves.

By a suitable choice of time units, we can always assume that $\lambda = 1$. Hence, from now on we will consider the simplest form of the wave equation:

$$\boxed{\partial_t^2 u = \triangle u.}$$

---

[5] At large amplitudes, nonlinear effects become important and invalidate the physical argument used here.

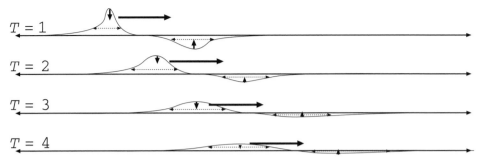

Figure 2C.1. A solution to the telegraph equation propagates like a wave, but it also diffuses over time due to noise, and decays exponentially in magnitude due to 'leakage'.

For example, fix $\omega = (\omega_1, \ldots, \omega_D) \in \mathbb{R}^D$ and let $\Omega = \|\omega\|_2 = \sqrt{\omega_1^2 + \cdots + \omega_D^2}$. Then

$$u(\mathbf{x}; t) = \sin\left(\omega_1 x_1 + \omega_2 x_2 + \cdots + \omega_D x_D + \Omega t\right) = \sin\left(\omega \bullet \mathbf{x} + \lambda \cdot \Omega \cdot t\right)$$

satisfies the D-dimensional wave equation and describes a transverse wave of with wave vector ω propagating across D-dimensional space.

Ⓔ **Exercise 2B.6** Check this.   ◆

2C The telegraph equation

Recommended: §2B(i), §1B(i).

Imagine a signal propagating through a medium with a linear restoring force (e.g. an electrical pulse in a wire, a vibration on a string). In an ideal universe, the signal obeys the wave equation. However, in the real universe, *damping effects* interfere. First, energy might 'leak' out of the system. For example, if a wire is imperfectly insulated, then current can leak out into surrounding space. Also, the signal may get blurred by noise or frictional effects. For example, an electric wire will pick up radio waves ('crosstalk') from other nearby wires, while losing energy to electrical resistance. A guitar string will pick up vibrations from the air, while losing energy to friction.

Thus, intuitively, we expect the signal to propagate like a wave, but to be gradually smeared out and attenuated by noise and leakage (see Figure 2C.1). The model for such a system is the *telegraph equation*:

$$\kappa_2 \partial_t^2 u + \kappa_1 \partial_t u + \kappa_0 u = \lambda \triangle u$$

(where $\kappa_2, \kappa_1, \kappa_0, \lambda > 0$ are constants).

Heuristically speaking, this equation is a 'sum' of two equations. The first,

$$\kappa_2 \partial_t^2 u = \lambda_1 \triangle u,$$

is a version of the wave equation, and describes the 'ideal' signal, while the second,

$$\kappa_1 \partial_t u = -\kappa_0 u + \lambda_2 \triangle u,$$

describes energy lost due to leakage and frictional forces.

2D Practice problems

2.1 By explicitly computing derivatives, show that the following functions satisfy the (one-dimensional) wave equation $\partial_t^2 u = \partial_x^2 u$:

(a) $u(x, t) = \sin(7x) \cos(7t)$;
(b) $u(x, t) = \sin(3x) \cos(3t)$;
(c) $u(x, t) = \frac{1}{(x-t)^2}$ (for $x \neq t$);
(d) $u(x, t) = (x - t)^2 - 3(x - t) + 2$;
(e) $v(x, t) = (x - t)^2$.

2.2 Let $f : \mathbb{R} \longrightarrow \mathbb{R}$ be any twice-differentiable function. Define $u : \mathbb{R} \times \mathbb{R} \longrightarrow \mathbb{R}$ by $u(x, t) := f(x - t)$, for all $(x, t) \in \mathbb{R} \times \mathbb{R}$. Does u satisfy the (one-dimensional) wave equation $\partial_t^2 u = \triangle u$? Justify your answer.

2.3 Let $u(x, t)$ be as in Problem 2.1(a) and let $v(x, t)$ be as in Problem 2.1(e) and suppose $w(x, t) = 3u(x, t) - 2v(x, t)$. Conclude that w also satisfies the wave equation, without explicitly computing any derivatives of w.

2.4 Suppose $u(x, t)$ and $v(x, t)$ are both solutions to the wave equation, and $w(x, t) = 5u(x, t) + 2v(x, t)$. Conclude that w also satisfies the wave equation.

2.5 Let

$$u(x, t) = \int_{x-t}^{x+t} \cos(y) dy = \sin(x + t) - \sin(x - t).$$

Show that u satisfies the (one-dimensional) wave equation $\partial_t^2 u = \triangle u$.

2.6 By explicitly computing derivatives, show that the following functions satisfy the (two-dimensional) wave equation $\partial_t^2 u = \triangle u$:

(a) $u(x, y; t) = \sinh(3x) \cdot \cos(5y) \cdot \cos(4t)$;
(b) $u(x, y; t) = \sin(x) \cos(2y) \sin(\sqrt{5}t)$;
(c) $u(x, y; t) = \sin(3x - 4y) \cos(5t)$.

3

Quantum mechanics

[M]odern physics has definitely decided in favor of Plato. In fact the smallest units of matter are not physical objects in the ordinary sense; they are forms, ideas which can be expressed unambiguously only in mathematical language.

Werner Heisenberg

3A Basic framework

Prerequisites: Appendix C, §1B(ii).

Near the beginning of the twentieth century, physicists realized that electromagnetic waves sometimes exhibited particle-like properties, as if light were composed of discrete 'photons'. In 1923, Louis de Broglie proposed that, conversely, particles of matter might have wave-like properties. This was confirmed in 1927 by C. J. Davisson and L. H. Germer and, independently, by G. P. Thompson, who showed that an electron beam exhibited an unmistakable diffraction pattern when scattered off a metal plate, as if the beam were composed of 'electron waves'. Systems with many interacting particles exhibit even more curious phenomena. *Quantum mechanics* is a theory which explains these phenomena.

We will not attempt here to provide a physical justification for quantum mechanics. Historically, quantum theory developed through a combination of vaguely implausible physical analogies and wild guesses motivated by inexplicable empirical phenomena. By now, these analogies and guesses have been overwhelmingly vindicated by experimental evidence. The best justification for quantum mechanics is that it 'works', by which we mean that its theoretical predictions match all available empirical data with astonishing accuracy.

Unlike the heat equation in §1B and the wave equation in §2B, we cannot derive quantum theory from 'first principles', because the postulates of quantum mechanics *are* the first principles. Instead, we will simply state the main assumptions of the theory, which are far from self-evident, but which we hope you will accept because of the weight of empirical evidence in their favour. Quantum theory describes any physical system via a *probability distribution* on a certain *statespace*. This probability distribution evolves over time; the evolution is driven by a potential energy function, as described by a partial differential equation called the *Schrödinger equation*. We will now examine each of these concepts in turn.

Statespace A system of N interacting particles moving in three-dimensional space can be completely described using the $3N$-dimensional *statespace* $\mathbb{X} := \mathbb{R}^{3N}$. An element of $\mathbb{X}$ consists of a list of N ordered triples:

$$\mathbf{x} = (x_{11}, x_{12}, x_{13}; x_{21}, x_{22}, x_{23}; \cdots; x_{N1}, x_{N2}, x_{N3}) \in \mathbb{R}^{3N},$$

where (x_{11}, x_{12}, x_{13}) is the spatial position of particle #1, (x_{21}, x_{22}, x_{23}) is the spatial position of particle #2, and so on.

Example 3A.1
(a) *Single electron* A single electron is a one-particle system, so it would be represented using a three-dimensional statespace $\mathbb{X} = \mathbb{R}^3$. If the electron was confined to a two-dimensional space (e.g. a conducting plate), we would use $\mathbb{X} = \mathbb{R}^2$. If the electron was confined to a one-dimensional space (e.g. a conducting wire), we would use $\mathbb{X} = \mathbb{R}$.
(b) *Hydrogen atom* The common isotope of hydrogen contains a single proton and a single electron, so it is a two-particle system, and would be represented using a six-dimensional statespace $\mathbb{X} = \mathbb{R}^6$. An element of $\mathbb{X}$ has the form $\mathbf{x} = (x_1^p, x_2^p, x_3^p; x_1^e, x_2^e, x_3^e)$, where (x_1^p, x_2^p, x_3^p) are the coordinates of the proton and (x_1^e, x_2^e, x_3^e) are those of the electron.   ◇

Readers familiar with classical mechanics may be wondering how *momentum* is represented in this statespace. Why isn't the statespace $6N$-dimensional, with three 'position' and three *momentum* coordinates for each particle? The answer, as we will see later, is that the *momentum* of a quantum system is implicitly encoded in the wavefunction which describes its position (see §19G, p. 515).

Potential energy We define a *potential energy* (or *voltage*) function $V : \mathbb{X} \longrightarrow \mathbb{R}$, which describes which states are 'preferred' by the quantum system. Loosely speaking, the system will 'avoid' states of high potential energy and 'seek' states of low energy. The voltage function is usually defined using reasoning familiar from 'classical' physics.

Example 3A.2 Electron in an ambient field
Imagine a single electron moving through an ambient electric field $\vec{E}$. The statespace for this system is $\mathbb{X} = \mathbb{R}^3$, as in Example 3A.1(a). The potential function V is just

the voltage of the electric field; in other words, V is any scalar function such that $-q_e \cdot \vec{\mathbf{E}} = \nabla V$, where q_e is the charge of the electron.

(a) *Null field* If $\vec{\mathbf{E}} \equiv 0$, then V will be a constant, which we can assume is zero: $V \equiv 0$.

(b) *Constant field* If $\vec{\mathbf{E}} \equiv (E, 0, 0)$, for some constant $E \in \mathbb{R}$, then $V(x, y, z) = -q_e E x + c$, where c is an arbitrary constant, which we normally set to zero.

(c) *Coulomb field* Suppose the electric field $\vec{\mathbf{E}}$ is generated by a (stationary) point charge Q at the origin. Let ϵ_0 be the 'permittivity of free space'. Then Coulomb's law says that the electric voltage is given by

$$V(\mathbf{x}) := \frac{q_e \cdot Q}{4\pi \epsilon_0 \cdot |\mathbf{x}|}, \quad \text{for all } \mathbf{x} \in \mathbb{R}^3.$$

In SI units, $q_e \approx 1.60 \times 10^{-19}$ C and $\epsilon_0 \approx 8.85 \times 10^{-12}$ C/N m². However, for simplicity, we will normally adopt 'atomic units' of charge and field strength, where $q_e = 1$ and $4\pi \epsilon_0 = 1$. Then the above expression becomes $V(\mathbf{x}) = Q/|\mathbf{x}|$.

(d) *Potential well* Sometimes we confine the electron to some bounded region $\mathbb{B} \subset \mathbb{R}^3$, by setting the voltage equal to 'positive infinity' outside $\mathbb{B}$. For example, a low-energy electron in a cube made of conducting metal can move freely about the cube, but cannot leave[1] the cube. If the subset $\mathbb{B}$ represents the cube, then we define $V : \mathbb{X} \longrightarrow [0, \infty]$ by

$$V(\mathbf{x}) = \begin{cases} 0 & \text{if } \mathbf{x} \in \mathbb{B}; \\ +\infty & \text{if } \mathbf{x} \notin \mathbb{B}. \end{cases}$$

(If '$+\infty$' makes you uncomfortable, then replace it with some 'really big' number.) ◇

Example 3A.3 Hydrogen atom

The system is an electron and a proton; the statespace of this system is $\mathbb{X} = \mathbb{R}^6$ as in Example 3A.1(b). Assuming there is no external electric field, the voltage function is defined as follows:

$$V(\mathbf{x}^p, \mathbf{x}^e) := \frac{q_e^2}{4\pi \epsilon_0 \cdot |\mathbf{x}^p - \mathbf{x}^e|}, \quad \text{for all } (\mathbf{x}^p, \mathbf{x}^e) \in \mathbb{R}^6,$$

where $\mathbf{x}^p$ is the position of the proton, $\mathbf{x}^e$ is the position of the electron, and q_e is the charge of the electron (which is also the charge of the proton, with reversed sign). If we adopt 'atomic' units, where $q_e := 1$ and $4\pi \epsilon_0 = 1$. then this expression simplifies to

$$V(\mathbf{x}^p, \mathbf{x}^e) := \frac{1}{|\mathbf{x}^p - \mathbf{x}^e|}, \quad \text{for all } (\mathbf{x}^p, \mathbf{x}^e) \in \mathbb{R}^6.$$ ◇

---

[1] 'Cannot leave' of course really means 'is very highly unlikely to leave'.

Probability and wavefunctions   Our knowledge of the classical properties of a quantum system is inherently incomplete. All we have is a time-varying probability distribution $\rho : \mathbb{X} \times \mathbb{R} \longrightarrow \mathbb{R}_{\not\,}$ which describes where the particles are likely or unlikely to be at a given moment in time.

As time passes, the probability distribution ρ evolves. However, ρ itself cannot exhibit the 'wave-like' properties of a quantum system (e.g. destructive interference), because ρ is a nonnegative function (and we need to add negative to positive values to get destructive interference). So, we introduce a complex-valued *wavefunction* $\omega : \mathbb{X} \times \mathbb{R} \longrightarrow \mathbb{C}$. The wavefunction ω determines ρ via the following equation:

$$\rho_t(\mathbf{x}) := |\omega_t(\mathbf{x})|^2, \quad \text{for all } \mathbf{x} \in \mathbb{X} \text{ and } t \in \mathbb{R}.$$

(Here, as always in this book, we define $\rho_t(\mathbf{x}) := \rho(\mathbf{x}; t)$ and $\omega_t(\mathbf{x}) := \omega(\mathbf{x}; t)$; subscripts do *not* indicate derivatives.) Now, ρ_t is supposed to be a probability density function, so ω_t must satisfy the condition

$$\int_{\mathbb{X}} |\omega_t(\mathbf{x})|^2 \, d\mathbf{x} = 1, \quad \text{for all } t \in \mathbb{R}. \tag{3A.1}$$

It is acceptable (and convenient) to relax condition (3A.1), and instead simply require

$$\int_{\mathbb{X}} |\omega_t(\mathbf{x})|^2 d\mathbf{x} = W < \infty, \quad \text{for all } t \in \mathbb{R}, \tag{3A.2}$$

where W is some finite constant, independent of t. In this case, we define $\rho_t(\mathbf{x}) := \frac{1}{W}|\omega_t(\mathbf{x})|^2$ for all $\mathbf{x} \in \mathbb{X}$. It follows that any physically meaningful solution to the Schrödinger equation must satisfy condition (3A.2). This excludes, for example, solutions where the magnitude of the wavefunction grows exponentially in the $\mathbf{x}$ or t variables.

For any fixed $t \in \mathbb{R}$, condition (3A.2) is usually expressed by saying that ω_t is *square-integrable*. Let $\mathbf{L}^2(\mathbb{X})$ denote the set of all square-integrable functions on $\mathbb{X}$. If $\omega_t \in \mathbf{L}^2(\mathbb{X})$, then the L^2-*norm* of ω is defined by

$$\|\omega_t\|_2 := \sqrt{\int_{\mathbb{X}} |\omega_t(\mathbf{x})|^2 \, d\mathbf{x}}.$$

Thus, a fundamental postulate of quantum theory is as follows.

Let $\omega : \mathbb{X} \times \mathbb{R} \longrightarrow \mathbb{C}$ be a wavefunction. To be physically meaningful, we must have $\omega_t \in \mathbf{L}^2(\mathbb{X})$ for all $t \in \mathbb{R}$. Furthermore, $\|\omega_t\|_2$ must be constant in time.

We refer the reader to §6B, p. 109, for more information on L^2-norms and L^2-spaces.

3B The Schrödinger equation

Prerequisites: §3A.   **Recommended:** §4B.

The wavefunction ω evolves over time in response to the potential field V. Let $\hbar$ be the 'rationalized' Planck constant given by

$$\hbar := \frac{h}{2\pi} \approx \frac{1}{2\pi} \times 6.6256 \times 10^{-34} \text{ Js} \approx 1.0545 \times 10^{-34} \text{ Js}.$$

Then the wavefunction's evolution is described by the *Schrödinger equation*:

$$i\hbar \, \partial_t \, \omega = H\omega, \tag{3B.1}$$

where H is a linear differential operator called the *Hamiltonian operator*, defined by

$$H\omega_t(\mathbf{x}) := \frac{-\hbar^2}{2} \blacktriangle \omega_t(\mathbf{x}) + V(\mathbf{x}) \cdot \omega_t(\mathbf{x}), \quad \text{for all } \mathbf{x} \in \mathbb{X} \text{ and } t \in \mathbb{R}. \tag{3B.2}$$

Here, $\blacktriangle \omega_t$ is like the Laplacian of ω_t, except that the components for each particle are divided by the *rest mass* of that particle. The *potential function* $V : \mathbb{X} \longrightarrow \mathbb{R}$ encodes all the exogenous aspects of the system we are modelling (e.g. the presence of ambient electric fields). Substituting equation (3B.2) into equation (3B.1), we get

$$i\hbar \, \partial_t \, \omega = \frac{-\hbar^2}{2} \blacktriangle \omega + V \cdot \omega. \tag{3B.3}$$

In 'atomic units', $\hbar = 1$, so the Schrödinger equation (3B.3) becomes

$$i\partial_t \, \omega_t(\mathbf{x}) = \frac{-1}{2} \blacktriangle \omega_t(\mathbf{x}) + V(\mathbf{x}) \cdot \omega_t(\mathbf{x}), \quad \text{for all } \mathbf{x} \in \mathbb{X} \text{ and } t \in \mathbb{R}.$$

Example 3B.1

(a) *Free electron*   Let $m_e \approx 9.11 \times 10^{-31}$ kg be the rest mass of an electron. A solitary electron in a null electric field (as in Example 3A.2(a)) satisfies the *free Schrödinger equation*:

$$i\hbar \, \partial_t \, \omega_t(\mathbf{x}) = \frac{-\hbar^2}{2 m_e} \triangle \omega_t(\mathbf{x}). \tag{3B.4}$$

(In this case $\blacktriangle = \frac{1}{m_e}\triangle$ and $V \equiv 0$ because the ambient field is null.) In atomic units, we set $m_e := 1$ and $\hbar := 1$, so equation (3B.4) becomes

$$i\partial_t \, \omega = \frac{-1}{2} \triangle \omega = \frac{-1}{2} \left(\partial_1^2 \omega + \partial_2^2 \omega + \partial_3^2 \omega \right). \tag{3B.5}$$

(b) *Electron vs. point charge*   Consider the *Coulomb* electric field, generated by a (stationary) point charge Q at the origin, as in Example 3A.2(c). A solitary electron in this electric field satisfies the Schrödinger equation:

$$i\hbar \, \partial_t \, \omega_t(\mathbf{x}) = \frac{-\hbar^2}{2\,m_e} \, \triangle \, \omega_t(\mathbf{x}) + \frac{q_e \cdot Q}{4\pi \epsilon_0 \cdot |\mathbf{x}|} \, \omega_t(\mathbf{x}).$$

In atomic units, we have $m_e := 1$, $q_e := 1$, etc. Let $\widetilde{Q} = Q/q_e$ be the charge Q converted in units of electron charge. Then the previous expression simplifies to

$$i\,\partial_t \, \omega_t(\mathbf{x}) = \frac{-1}{2} \, \triangle \, \omega_t(\mathbf{x}) + \frac{\widetilde{Q}}{|\mathbf{x}|} \, \omega_t(\mathbf{x}).$$

(c) *Hydrogen atom* (See Example 3A.3.) An interacting proton–electron pair (in the absence of an ambient field) satisfies the two-particle Schrödinger equation:

$$i\hbar \, \partial_t \, \omega_t(\mathbf{x}^p, \mathbf{x}^e) = \frac{-\hbar^2}{2\,m_p} \, \triangle_p \, \omega_t(\mathbf{x}^p, \mathbf{x}^e) + \frac{-\hbar^2}{2\,m_e} \, \triangle_e \, \omega_t(\mathbf{x}^p, \mathbf{x}^e) + \frac{q_e^2 \cdot \omega_t(\mathbf{x}^p, \mathbf{x}^e)}{4\pi \epsilon_0 \cdot |\mathbf{x}^p - \mathbf{x}^e|},$$

(3B.6)

where $\triangle_p \omega := \partial_{x_1^p}^2 \omega + \partial_{x_2^p}^2 \omega + \partial_{x_3^p}^2 \omega$ is the Laplacian in the 'proton' position coordinates, and $m_p \approx 1.6727 \times 10^{-27}$ kg is the rest mass of a proton. Likewise, $\triangle_e \omega := \partial_{x_1^e}^2 \omega + \partial_{x_2^e}^2 \omega + \partial_{x_3^e}^2 \omega$ is the Laplacian in the 'electron' position coordinates, and m_e is the rest mass of the electron. In atomic units, we have $4\pi \epsilon_0 = 1$, $q_e = 1$, and $m_e = 1$. If $\widetilde{m}_p \approx 1864$ is the ratio of proton mass to electron mass, then $2\widetilde{m}_p \approx 3728$, and equation (3B.6) becomes

$$i\partial_t \, \omega_t(\mathbf{x}^p, \mathbf{x}^e) = \frac{-1}{3728} \, \triangle_p \, \omega_t(\mathbf{x}^p, \mathbf{x}^e) + \frac{-1}{2} \, \triangle_e \, \omega_t(\mathbf{x}^p, \mathbf{x}^e) + \frac{\omega_t(\mathbf{x}^p, \mathbf{x}^e)}{|\mathbf{x}^p - \mathbf{x}^e|}. \qquad \diamond$$

The major mathematical problems of quantum mechanics come down to finding solutions to the Schrödinger equation for various physical systems. In general it is very difficult to solve the Schrödinger equation for most 'realistic' potential functions. We will confine ourselves to a few 'toy models' to illustrate the essential ideas.

Example 3B.2 Free electron with known velocity (null field)
Consider a single electron in a null electromagnetic field. Suppose an experiment has precisely measured the 'classical' velocity of the electron and determined it to be $\mathbf{v} = (v_1, 0, 0)$. Then the wavefunction of the electron is given by[2]

$$\omega_t(\mathbf{x}) = \exp\left(\frac{-i}{\hbar} \frac{m_e v_1^2}{2} t\right) \cdot \exp\left(\frac{i}{\hbar} m_e v_1 \cdot x_1\right)$$

(3B.7)

(see Figure 3B.1). This ω satisfies the free Schrödinger equation (3B.4). (See Problem 3.1.)

---

[2] We will not attempt here to justify *why* this is the correct wavefunction for a particle with this velocity. It is not obvious.

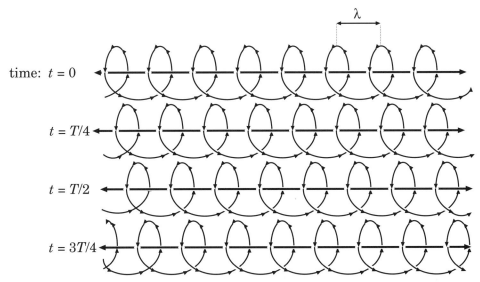

Figure 3B.1. Four successive 'snapshots' of the wavefunction of a single electron in a zero potential, with a precisely known velocity. Only one spatial dimension is shown. The angle of the spiral indicates complex phase.

Exercise 3B.1
     Ⓔ
(a) Check that the spatial wavelength λ of the function ω is given by $\lambda = 2\pi\hbar/p_1 = h/m_e v$. This is the so-called *de Broglie wavelength* of an electron with velocity v.
(b) Check that the temporal period of ω is given by $T := 2h/m_e v^2$.
(c) Conclude that the *phase velocity* of ω (i.e. the speed at which the wavefronts propagate through space) is equal to v.   ♦

More generally, suppose the electron has a precisely known velocity $\mathbf{v} = (v_1, v_2, v_3)$, with corresponding momentum vector $\mathbf{p} := m_e \mathbf{v}$. Then the wavefunction of the electron is given by

$$\omega_t(\mathbf{x}) = \exp\left(\frac{-\mathbf{i}}{\hbar} E_k t\right) \cdot \exp\left(\frac{\mathbf{i}}{\hbar} \mathbf{p} \bullet \mathbf{x}\right), \tag{3B.8}$$

where $E_k := \frac{1}{2}m_e|\mathbf{v}|^2$ is kinetic energy, and $\mathbf{p} \bullet \mathbf{x} := p_1 x_1 + p_2 x_2 + p_3 x_3$. If we convert to atomic units, then $E_k = \frac{1}{2}|\mathbf{v}|^2$ and $\mathbf{p} = \mathbf{v}$, and this function takes the simpler form

$$\omega_t(\mathbf{x}) = \exp\left(\frac{-\mathbf{i}|\mathbf{v}|^2 t}{2}\right) \cdot \exp\left(\mathbf{i}\,\mathbf{v} \bullet \mathbf{x}\right).$$

This ω satisfies the free Schrödinger equation (3B.5). (See Problem 3.2.)

     The wavefunction, equation (3B.8), represents a state of maximal uncertainty about the position of the electron. This is an extreme manifestation of the famous

Heisenberg uncertainty principle; by assuming that the electron's *velocity* was 'precisely determined', we have forced its *position* to be entirely undetermined (see §19G for more information).

Indeed, the wavefunction (3B.8) violates our 'fundamental postulate' – the function ω_t is *not* square-integrable because $|\omega_t(\mathbf{x})| = 1$ for all $\mathbf{x} \in \mathbb{R}$, so $\int_{\mathbb{R}^3} |\omega_t(\mathbf{x})|^2 \, d\mathbf{x} = \infty$. Thus, wavefunction (3B.8) cannot be translated into a probability distribution, so it is not physically meaningful. This is not too surprising because wavefunction (3B.8) seems to suggest that the electron is equally likely to be anywhere in the (infinite) 'universe' $\mathbb{R}^3$! It is well known that the location of a quantum particle can be 'dispersed' over some region of space, but this seems a bit extreme. There are two solutions to this problem.

- Let $\mathbb{B}(R) \subset \mathbb{R}^3$ be a ball of radius R, where R is much larger than the physical system or laboratory apparatus we are modelling (e.g. $R = 1$ lightyear). Define the wavefunction $\omega_t^{(R)}(\mathbf{x})$ by equation (3B.8) for all $\mathbf{x} \in \mathbb{B}(R)$, and set $\omega_t^{(R)}(\mathbf{x}) = 0$ for all $\mathbf{x} \notin \mathbb{B}(R)$. This means that the position of the electron is still extremely dispersed (indeed, 'infinitely' dispersed for the purposes of any laboratory experiment), but the function $\omega_t^{(R)}$ is still square-integrable. Note that the function $\omega_t^{(R)}$ violates the Schrödinger equation at the boundary of $\mathbb{B}(R)$, but this boundary occurs very far from the physical system we are studying, so it does not matter. In a sense, the solution (3B.8) can be seen as the 'limit' of $\omega^{(R)}$ as $R \to \infty$.
- Reject the wavefunction (3B.8) as 'physically meaningless'. Our starting assumption – an electron with a precisely known velocity – has led to a contradiction. Our conclusion: a free quantum particle can *never* have a precisely known classical velocity. Any physically meaningful wavefunction in a vacuum must contain a 'mixture' of several velocities.   ◇

Remark *The meaning of phase*

At any point $\mathbf{x}$ in space and moment t in time, the wavefunction $\omega_t(\mathbf{x})$ can be described by its *amplitude* $A_t(\mathbf{x}) := |\omega_t(\mathbf{x})|$ and its *phase* $\phi_t(\mathbf{x}) := \omega_t(\mathbf{x})/A_t(\mathbf{x})$. We have already discussed the physical meaning of the amplitude: $|A_t(\mathbf{x})|^2$ is the *probability* that a classical measurement will produce the outcome $\mathbf{x}$ at time t. What is the meaning of phase?

The phase $\phi_t(\mathbf{x})$ is a complex number of modulus one – an element of the unit circle in the complex plane (hence $\phi_t(\mathbf{x})$ is sometimes called the *phase angle*). The 'oscillation' of the wavefunction ω over time can be imagined in terms of the 'rotation' of $\phi_t(\mathbf{x})$ around the circle. The 'wave-like' properties of quantum systems (e.g. interference patterns) occur because wavefunctions with different phases will partially cancel one another when they are superposed. In other words, it is because of *phase* that the Schrödinger equation yields 'wave-like' phenomena, instead of yielding 'diffusive' phenomena like the heat equation.

However, like potential energy, phase is *not directly physically observable*. We can observe the phase *difference* between wavefunction α and wavefunction β (by observing the cancellation between α and β), just as we can observe the potential energy *difference* between point A and point B (by measuring the energy released by a particle moving from point A to point B). However, it is not physically meaningful to speak of the 'absolute phase' of wavefunction α, just as it is not physically meaningful to speak of the 'absolute potential energy' of point A.

Indeed, inspection of the Schrödinger equation (3B.3) on p. 43 will reveal that the speed of phase rotation of a wavefunction ω at point $\mathbf{x}$ is determined by the magnitude of the potential function V at $\mathbf{x}$. But we can arbitrarily increase V by a constant, without changing its physical meaning. Thus, we can arbitrarily 'accelerate' the phase rotation of the wavefunction without changing the physical meaning of the solution.

3C Stationary Schrödinger equation

Prerequisites: §3B.   **Recommended:** §4B(iv).

A 'stationary' state of a quantum system is one where the probability density does not change with time. This represents a physical system which is in some kind of long-term equilibrium. Note that a stationary quantum state does *not* mean that the particles are 'not moving' (whatever 'moving' means for quanta). It instead means that they are moving in some kind of regular, confined pattern (i.e. an 'orbit') which remains qualitatively the same over time. For example, the orbital of an electron in a hydrogen atom should be a stationary state because (unless the electron absorbs or emits energy) the orbital should stay the same over time.

Mathematically speaking, a stationary wavefunction ω yields a time-invariant probability density function $\rho : \mathbb{X} \longrightarrow \mathbb{R}$ such that, for any $t \in \mathbb{R}$,

$$|\omega_t(\mathbf{x})|^2 = \rho(\mathbf{x}), \quad \text{for all } \mathbf{x} \in \mathbb{X}.$$

The simplest way to achieve this is to assume that ω has the *separated* form

$$\omega_t(\mathbf{x}) = \phi(t) \cdot \omega_0(\mathbf{x}), \tag{3C.1}$$

where $\omega_0 : \mathbb{X} \longrightarrow \mathbb{C}$ and $\phi : \mathbb{R} \longrightarrow \mathbb{C}$ satisfy the following conditions:

$$|\phi(t)| = 1, \text{ for all } t \in \mathbb{R}, \quad \text{and} \quad |\omega_0(\mathbf{x})| = \sqrt{\rho(\mathbf{x})}, \text{ for all } x \in \mathbb{X}. \tag{3C.2}$$

Lemma 3C.1 *Suppose* $\omega_t(\mathbf{x}) = \phi(t) \cdot \omega_0(\mathbf{x})$ *is a separated solution to the Schrödinger Equation, as in equations (3C.1) and (3C.2). Then there is some constant* $E \in \mathbb{R}$ *such that*

- $\phi(t) = \exp(-\mathbf{i}Et/\hbar)$, *for all* $t \in \mathbb{R}$;
- $H\omega_0 = E \cdot \omega_0$ *(in other words ω_0 is an* eigenfunction[3] *of the Hamiltonian operator* H, *with* eigenvalue E);
- *thus,* $\omega_t(\mathbf{x}) = e^{-\mathbf{i}Et/\hbar} \cdot \omega_0(\mathbf{x})$, *for all* $\mathbf{x} \in \mathbb{X}$ *and* $t \in \mathbb{R}$.

Ⓔ *Proof* **Exercise 3C.1.** *Hint:* Use separation of variables.[4]   □

Physically speaking, E corresponds to the *total energy* (potential + kinetic) of the quantum system.[5] Thus, Lemma 3C.1 yields one of the key concepts of quantum theory.

Eigenfunctions of the Hamiltonian correspond to stationary quantum states. The eigenvalues of these eigenfunctions correspond to the energy level of these states.

Thus, to get stationary states, we must solve the *stationary Schrödinger equation*:

$$H\omega_0 = E \cdot \omega_0,$$

where $E \in \mathbb{R}$ is an unknown constant (the energy eigenvalue) and $\omega_0 : \mathbb{X} \longrightarrow \mathbb{C}$ is an unknown wavefunction.

Example 3C.2 The free electron
Recall the 'free electron' of Example 3B.2. If the electron has velocity v, then the function ω in equation (3B.7) yields a solution to the stationary Schrödinger equation, with eigenvalue $E = \frac{1}{2}m_e v^2$ (see Problem 3.3). Observe that E corresponds to the classical *kinetic energy* of an electron with velocity v.   ◇

Example 3C.3 One-dimensional square potential well: finite voltage
Consider an electron confined to a one-dimensional environment (e.g. a long conducting wire). Thus, $\mathbb{X} := \mathbb{R}$, and the wavefunction $\omega_0 : \mathbb{R} \times \mathbb{R} \longrightarrow \mathbb{C}$ obeys the one-dimensional Schrödinger equation:

$$\mathbf{i}\partial_t \,\omega_0 = \frac{-1}{2}\partial_x^2 \,\omega_0 + V \cdot \omega_0,$$

where $V : \mathbb{R} \longrightarrow \mathbb{R}$ is the potential energy function, and we have adopted atomic units. Let $V_0 > 0$ be some constant, and suppose that

$$V(x) = \begin{cases} 0 & \text{if } 0 \le x \le L; \\ V_0 & \text{if } x < 0 \text{ or } L < x. \end{cases}$$

---

[3] See §4B(iv), p. 67.   [4] See Chapter 16.
[5] This is not obvious, but it is a consequence of the fact that the Hamiltonian $H\omega$ measures the total energy of the wavefunction ω. Loosely speaking, the term $\frac{\hbar^2}{2}\blacktriangle\omega$ represents the 'kinetic energy' of ω, while the term $V \cdot \omega$ represents the 'potential energy'.

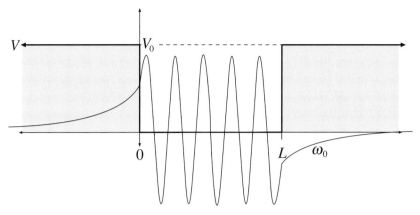

Figure 3C.1. The (stationary) wavefunction of an electron in a one-dimensional 'square' potential well, with finite voltage gaps.

Physically, this means that V defines a 'potential energy well', which tries to confine the electron in the interval $[0, L]$, between two 'walls', which are voltage gaps of height V_0 (see Figure 3C.1). The corresponding stationary Schrödinger equation is:

$$\frac{-1}{2}\partial_x^2 \omega_0 + V \cdot \omega_0 = E \cdot \omega_0, \tag{3C.3}$$

where $E > 0$ is an (unknown) eigenvalue which corresponds to the energy of the electron. The function V only takes two values, so we can split equation (3C.3) into two equations, one inside the interval $[0, L]$, and one outside it:

$$\begin{aligned} -\tfrac{1}{2}\partial_x^2 \omega_0(x) &= E \cdot \omega_0(x), && \text{for } x \in [0, L]; \\ -\tfrac{1}{2}\partial_x^2 \omega_0(x) &= (E - V_0) \cdot \omega_0(x), && \text{for } x \notin [0, L]. \end{aligned} \tag{3C.4}$$

Assume that $E < V_0$. This means that the electron's energy is less than the voltage gap, so the electron has insufficient energy to 'escape' the interval (at least in classical theory). The (physically meaningful) solutions to equation (3C.4) have the following form:

$$\omega_0(x) = \begin{cases} C \exp(\epsilon' x), & \text{if } x \in (-\infty, 0]; \\ A \sin(\epsilon x) + B \cos(\epsilon x), & \text{if } x \in [0, L]; \\ D \exp(-\epsilon' x), & \text{if } L \in [L, \infty). \end{cases} \tag{3C.5}$$

(See Figure 3C.1.) Here, $\epsilon := \sqrt{2E}$ and $\epsilon' := \sqrt{2E - 2V_0}$, and $A, B, C, D \in \mathbb{C}$ are constants. The corresponding solution to the full Schrödinger equation is given by

$$\omega_t(x) = \begin{cases} Ce^{-i(E-V_0)t} \cdot \exp(\epsilon'x), & \text{if } x \in (-\infty, 0]; \\ e^{-iEt} \cdot (A\sin(\epsilon x) + B\cos(\epsilon x)), & \text{if } x \in [0, L]; \\ De^{-i(E-V_0)t} \cdot \exp(-\epsilon'x), & \text{if } L \in [L, \infty); \end{cases}$$

for all $t \in \mathbb{R}$. This has two consequences.

(a) With nonzero probability, the electron might be found *outside* the interval $[0, L]$. In other words, it is quantumly possible for the electron to 'escape' from the potential well, something which is classically impossible.[6] This phenomenon is called *quantum tunnelling* (because the electron can 'tunnel' through the wall of the well).

(b) The system has a physically meaningful solution only for certain values of E. In other words, the electron is only 'allowed' to reside at certain discrete *energy levels*; this phenomenon is called *quantization of energy*.

To see (a), recall that the electron has a probability distribution given by

$$\rho(x) := \frac{1}{W} |\omega_0(x)|^2, \quad \text{where } W := \int_{-\infty}^{\infty} |\omega_0(x)|^2 \, dx.$$

Thus, if $C \neq 0$, then $\rho(x) \neq 0$ for $x < 0$, while, if $D \neq 0$, then $\rho(x) \neq 0$ for $x > L$. Either way, the electron has nonzero probability of 'tunnelling' out of the well.

To see (b), note that we must choose A, B, C, D so that ω_0 is continuously differentiable at the boundary points $x = 0$ and $x = L$. This means we must have

$$\begin{aligned} B = A\sin(0) + B\cos(0) &= \omega_0(0) = C\exp(0) = C; \\ \epsilon A = A\epsilon\cos(0) - B\epsilon\sin(0) &= \omega_0'(0) = \epsilon'C\exp(0) = \epsilon'C; \\ A\sin(\epsilon L) + B\cos(\epsilon L) &= \omega_0(L) = D\exp(-\epsilon'L); \\ A\epsilon\cos(\epsilon L) - B\epsilon\sin(\epsilon L) &= \omega_0'(L) = -\epsilon'D\exp(-\epsilon'L). \end{aligned} \qquad (3\text{C}.6)$$

Clearly, we can satisfy the first two equations in (3C.6) by setting $B := C := \frac{\epsilon}{\epsilon'}A$. The third and fourth equations in (3C.6) then become

$$e^{\epsilon'L} \cdot \left(\sin(\epsilon L) + \frac{\epsilon}{\epsilon'}\cos(\epsilon L)\right) \cdot A = D = \frac{-\epsilon}{\epsilon'}e^{\epsilon'L} \cdot \left(\cos(\epsilon L) - \frac{\epsilon}{\epsilon'}\sin(\epsilon L)\right)A.$$

$$(3\text{C}.7)$$

---

[6] Many older texts observe that the electron 'can penetrate the classically forbidden region', which has caused mirth to generations of physics students.

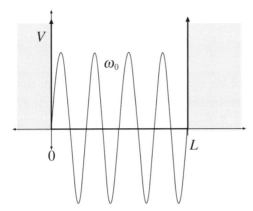

Figure 3C.2. The (stationary) wavefunction of an electron in an infinite potential well.

Cancelling the factors $e^{\epsilon' L}$ and A from both sides and substituting $\epsilon := \sqrt{2E}$ and $\epsilon' := \sqrt{2E - 2V_0}$, we see that equation (3C.7) is satisfiable if and only if

$$\sin\left(\sqrt{2E} \cdot L\right) + \frac{\sqrt{E} \cdot \cos(\sqrt{2E} \cdot L)}{\sqrt{E - V_0}} = \frac{-\sqrt{E} \cdot \cos\left(\sqrt{2E} \cdot L\right)}{\sqrt{E - V_0}}$$

$$+ \frac{E \cdot \sin(\sqrt{2E} \cdot L)}{E - V_0}. \qquad (3C.8)$$

Hence, equation (3C.4) has a physically meaningful solution only for those values of E which satisfy the transcendental equation (3C.8). The set of solutions to equation (3C.8) is an infinite discrete subset of $\mathbb{R}$; each solution for equation (3C.8) corresponds to an allowed 'energy level' for the physical system.   ◇

Example 3C.4 One-dimensional square potential well: infinite voltage
We can further simplify the model of Example 3C.3 by setting $V_0 := +\infty$, which physically represents a 'huge' voltage gap that totally confines the electron within the interval $[0, L]$ (see Figure 3C.2). In this case, $\epsilon' = \infty$, so $\exp(\epsilon' x) = 0$ for all $x < 0$ and $\exp(-\epsilon' x) = 0$ for all $x > L$. Hence, if ω_0 is as in equation (3C.5), then $\omega_0(x) \equiv 0$ for all $x \notin [0, L]$, and the constants C and D are no longer physically meaningful; we set $C = 0 = D$ for simplicity. Also, we must have $\omega_0(0) = 0 = \omega_0(L)$ to get a continuous solution; thus, we must set $B := 0$ in equation (3C.5). Thus, the stationary solution in equation (3C.5) becomes

$$\omega_0(x) = \begin{cases} 0 & \text{if } x \notin [0, L]; \\ A \cdot \sin(\sqrt{2E}\, x) & \text{if } x \in [0, L], \end{cases}$$

where A is a constant, and E satisfies the equation

$$\sin(\sqrt{2E}\,L) = 0 \tag{3C.9}$$

(Figure 3C.2). Assume for simplicity that $L := \pi$. Then equation (3C.9) is true if and only if $\sqrt{2E}$ is an integer, which means $2E \in \{0, 1, 4, 9, 16, 25, \ldots\}$, which means $E \in \{0, \frac{1}{2}, 2, \frac{9}{2}, 8, \frac{25}{2}, \ldots\}$. Here we see the phenomenon of *quantization of energy* in its simplest form.   ◇

The set of eigenvalues of a linear operator is called the *spectrum* of that operator. For example, in Example 3C.4 the spectrum of the Hamiltonian operator H is the set $\{0, \frac{1}{2}, 2, \frac{9}{2}, 8, \frac{25}{2}, \ldots\}$. In quantum theory, the spectrum of the Hamiltonian is the set of allowed energy levels of the system.

Example 3C.5 Three-dimensional square potential well: infinite voltage
We can easily generalize Example 3C.4 to three dimensions. Let $\mathbb{X} := \mathbb{R}^3$, and let $\mathbb{B} := [0, \pi]^3$ be a cube with one corner at the origin, having sidelength $L = \pi$. We use the potential function $V : \mathbb{X} \longrightarrow \mathbb{R}$ defined as

$$V(\mathbf{x}) = \begin{cases} 0 & \text{if } \mathbf{x} \in \mathbb{B}; \\ +\infty & \text{if } \mathbf{x} \notin \mathbb{B}. \end{cases}$$

Physically, this represents an electron confined within a cube of perfectly conducting material with perfectly insulating boundaries.[7] Suppose the electron has energy E. The corresponding stationary Schrödinger Equation is given by

$$\begin{aligned} -\tfrac{1}{2} \triangle \omega_0(\mathbf{x}) &= E \cdot \omega_0(\mathbf{x}) & \text{for } \mathbf{x} \in \mathbb{B}; \\ -\tfrac{1}{2} \triangle \omega_0(\mathbf{x}) &= -\infty \cdot \omega_0(\mathbf{x}) & \text{for } \mathbf{x} \notin \mathbb{B} \end{aligned} \tag{3C.10}$$

(in atomic units). By reasoning similar to that in Example 3C.4, we find that the physically meaningful solutions to equation (3C.10) have the following form:

$$\omega_0(\mathbf{x}) = \begin{cases} \dfrac{\sqrt{2}}{\pi^{3/2}} \sin(n_1 x_1) \cdot \sin(n_2 x_2) \cdot \sin(n_3 x_3) & \text{if } \mathbf{x} = (x_1, x_2, x_3) \in \mathbb{B}; \\ 0 & \text{if } \mathbf{x} \notin \mathbb{B}, \end{cases} \tag{3C.11}$$

where n_1, n_2, and n_3 are arbitrary integers (called the *quantum numbers* of the solution), and $E = \frac{1}{2}(n_1^2 + n_2^2 + n_3^2)$ is the associated energy eigenvalue.

---

[7] Alternatively, it could be any kind of particle, confined in a cubical cavity with impenetrable boundaries.

The corresponding solution to the full Schrödinger Equation for all $t \in \mathbb{R}$ is given by

$$\omega_t(\mathbf{x}) = \begin{cases} \dfrac{\sqrt{2}}{\pi^{3/2}} e^{-\mathrm{i}(n_1^2 + n_2^2 + n_3^2)t/2} \cdot \sin(n_1 x_1) \cdot \sin(n_2 x_2) \cdot \sin(n_3 x_3) & \text{if } \mathbf{x} \in \mathbb{B}; \\ 0 & \text{if } \mathbf{x} \notin \mathbb{B}. \end{cases}$$ ◇

Exercise 3C.2 Ⓔ
(a) Check that equation (3C.11) is a solution for equation (3C.10).
(b) Check that $\rho := |\omega|^2$ is a probability density, by confirming that

$$\int_{\mathbb{X}} |\omega_0(\mathbf{x})|^2 \, d\mathbf{x} = \frac{2}{\pi^3} \int_0^\pi \int_0^\pi \int_0^\pi \sin(n_1 x_1)^2 \cdot \sin(n_2 x_2)^2 \cdot \sin(n_3 x_3)^2 \, dx_1 \, dx_2 \, dx_3 = 1$$

(this is the reason for using the constant $\sqrt{2}/\pi^{3/2}$).   ◆

Example 3C.6 Hydrogen atom
In Example 3A.3, p. 41, we described the hydrogen atom as a two-particle system, with a six-dimensional statespace. However, the corresponding Schrödinger equation (Example 3B.1(c)) is already too complicated for us to solve it here, so we will work with a simplified model.

Because the proton is 1864 times as massive as the electron, we can treat the proton as remaining effectively immobile while the electron moves around it. Thus, we can model the hydrogen atom as a *one*-particle system: a single electron moving in a Coulomb potential well, as described in Example 3B.1(b). The electron then satisfies the Schrödinger equation:

$$\mathrm{i}\hbar \, \partial_t \, \omega_t(\mathbf{x}) = \frac{-\hbar^2}{2m_e} \triangle \omega_t(\mathbf{x}) + \frac{q_e^2}{4\pi\epsilon_0 \cdot |\mathbf{x}|} \cdot \omega_t(\mathbf{x}), \quad \forall \mathbf{x} \in \mathbb{R}^3. \quad (3C.12)$$

(Recall that m_e is the mass of the electron, q_e is the charge of both electron and proton, ϵ_0 is the 'permittivity of free space', and $\hbar$ is the rationalized Planck constant.) Assuming the electron is in a stable orbital, we can replace equation (3C.12) with the *stationary* Schrödinger equation:

$$\frac{-\hbar^2}{2m_e} \triangle \omega_0(\mathbf{x}) + \frac{q_e^2}{4\pi\epsilon_0 \cdot |\mathbf{x}|} \cdot \omega_0(\mathbf{x}) = E \cdot \omega_0(\mathbf{x}), \quad \forall \mathbf{x} \in \mathbb{R}^3, \quad (3C.13)$$

where E is the 'energy level' of the electron. One solution to this equation is

$$\omega(\mathbf{x}) = \frac{b^{3/2}}{\sqrt{\pi}} \exp(-b|\mathbf{x}|), \quad \text{where } b := \frac{m \, q_e^2}{4\pi\epsilon_0 \, \hbar^2}, \quad (3C.14)$$

with corresponding energy eigenvalue

$$E = \frac{-\hbar^2}{2m} \cdot b^2 = \frac{-m \, q_e^4}{32\pi^2 \epsilon_0^2 \, \hbar^2}. \quad (3C.15)$$

(See Figure 3C.3.)

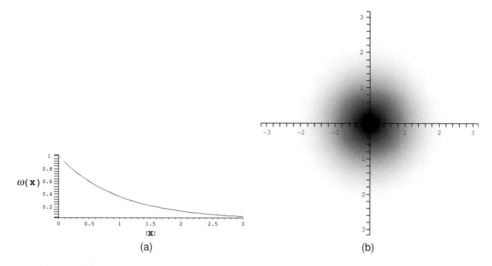

Figure 3C.3. The groundstate wavefunction for the hydrogen atom in Example 3C.6. (a) Probability density as a function of distance from the nucleus. (b) Probability density visualized in three dimensions.

Ⓔ **Exercise 3C.3**
  (a) Verify that the function ω_0 in equation (3C.14) is a solution to equation (3C.13), with E given by equation (3C.15).
  (b) Verify that the function ω_0 defines a probability density, by checking that $\int_{\mathbb{X}} |\omega|^2 = 1$. ◆

There are many other, more complicated, solutions to equation (3C.13). However, equation (3C.14) is the simplest solution, and has the *lowest* energy eigenvalue E of any solution. In other words, the solution (3C.13) describes an electron in the *ground state*: the orbital of lowest potential energy, where the electron is 'closest' to the nucleus. This solution immediately yields two experimentally testable predictions.

  (a) The *ionization potential* for the hydrogen atom, which is the energy required to 'ionize' the atom by stripping off the electron and removing it to an infinite distance from the nucleus.
  (b) The *Bohr radius* of the hydrogen atom; that is, the 'most probable' distance of the electron from the nucleus.

To see (a), recall that E is the sum of potential and kinetic energy for the electron. We assert (without proof) that there exist solutions to the stationary Schrödinger equation (3C.13) with energy eigenvalues arbitrarily close to zero (note that E is negative). These zero-energy solutions represent orbitals where the electron has been removed to some very large distance from the nucleus and the atom is essentially ionized. Thus, the energy difference between these 'ionized' states and ω_0 is $E - 0 = E$, and this is the energy necessary to 'ionize' the atom when the electron is in the orbital described by ω_0.

By substituting in the numerical values $q_e \approx 1.60 \times 10^{-19}$ C, $\epsilon_0 \approx 8.85 \times 10^{-12}$ C/Nm^2, $m_e \approx 9.11 \times 10^{-31}$ kg, and $\hbar \approx 1.0545 \times 10^{-34}$ J s, the reader can verify that, in fact, $E \approx -2.1796 \times 10^{-18}$ J ≈ -13.605 eV, which is very close to -13.595 eV, the experimentally determined ionization potential for a hydrogen atom.[8]

To see (b), observe that the probability density function for the distance r of the electron from the nucleus is given by

$$P(r) = 4\pi r^2 |\omega(r)|^2 = 4b^3 r^2 \exp(-2b|\mathbf{x}|)$$

(**Exercise 3C.4**). The *mode* of the radial probability distribution is the maximal Ⓔ point of $P(r)$; if we solve the equation $P'(r) = 0$, we find that the mode occurs at

$$r := \frac{1}{b} = \frac{4\pi \epsilon_0 \hbar^2}{m_e q_e^2} \approx 5.29172 \times 10^{-11} \text{ m.} \qquad \diamond$$

The Balmer lines Recall that the *spectrum* of the Hamiltonian operator H is the set of all eigenvalues of H. Let $\mathcal{E} = \{E_0 < E_1 < E_2 < \cdots\}$ be the spectrum of the Hamiltonian of the hydrogen atom from Example 3C.6, with the elements listed in increasing order. Thus, the smallest eigenvalue is $E_0 \approx -13.605$, the energy eigenvalue of the aforementioned ground state ω_0. The other, larger eigenvalues correspond to electron orbitals with higher potential energy.

When the electron 'falls' from a high-energy orbital (with eigenvalue E_n, for some $n \in \mathbb{N}$) to a low-energy orbital (with eigenvalue E_m, where $m < n$), it releases the energy difference and emits a photon with energy $(E_n - E_m)$. Conversely, to 'jump' from a low E_m-energy orbital to a higher E_n-energy orbital, the electron must *absorb* a photon, and this photon must have exactly energy $(E_n - E_m)$.

Thus, the hydrogen atom can only emit or absorb photons of energy $|E_n - E_m|$, for some $n, m \in \mathbb{N}$. Let $\mathcal{E}' := \{|E_n - E_m|; \ n, m \in \mathbb{N}\}$. We call $\mathcal{E}'$ the *energy spectrum* of the hydrogen atom.

Planck's law says that a photon with energy E has frequency $f = E/h$, where $h \approx 6.626 \times 10^{-34}$ J s is Planck's constant. Thus, if $\mathcal{F} = \{E/h; E \in \mathcal{E}'\}$, then a hydrogen atom can only emit/absorb a photon whose frequency is in $\mathcal{F}$; we say that $\mathcal{F}$ is the *frequency spectrum* of the hydrogen atom.

Here lies the explanation for the empirical observations of nineteenth century physicists such as Balmer, Lyman, Rydberg, and Paschen, who found that an energized hydrogen gas has a distinct *emission spectrum* of frequencies at which it

---

[8] The error of 0.01 eV is mainly due to our simplifying assumption of an 'immobile' proton.

emits light, and an identical *absorption spectrum* of frequencies which the gas can absorb. Indeed, every chemical element has its own distinct spectrum; astronomers use these 'spectral signatures' to measure the concentrations of chemical elements in the stars of distant galaxies. Now we see that

The (frequency) spectrum of an atom is determined by the (eigenvalue) spectrum of the corresponding Hamiltonian.

3D Further reading

Unfortunately, most other texts on partial differential equations do not discuss the Schrödinger equation; one of the few exceptions is the excellent text by Asmar (2005). For a lucid, fast, yet precise introduction to quantum mechanics in general, see McWeeny (1972). For a more comprehensive textbook on quantum theory, see Bohm (1979). A completely different approach to quantum theory uses Feynman's *path integrals*; for a good introduction to this approach, see Stevens (1995), which also contains excellent introductions to classical mechanics, electromagnetism, statistical physics, and special relativity. For a rigorous mathematical approach to quantum theory, an excellent introduction is Prugovecki (1981); another source is Blank, Exner, and Havlicek (1994).

3E Practice problems

3.1 Let $v_1 \in \mathbb{R}$ be a constant. Consider the function $\omega : \mathbb{R}^3 \times \mathbb{R} \longrightarrow \mathbb{C}$, defined as

$$\omega_t(x_1, x_2, x_3) = \exp\left(\frac{-\mathbf{i}}{\hbar} \frac{m_e v_1^2}{2} t\right) \cdot \exp\left(\frac{\mathbf{i}}{\hbar} m_e v_1 \cdot x_1\right).$$

Show that ω satisfies the (free) *Schrödinger equation:*

$$\mathbf{i}\hbar \, \partial_t \, \omega_t(\mathbf{x}) = \frac{-\hbar^2}{2 m_e} \Delta \, \omega_t(\mathbf{x}).$$

3.2 Let $\mathbf{v} := (v_1, v_2, v_3)$ be a three-dimensional velocity vector, and let $|\mathbf{v}|^2 = v_1^2 + v_2^2 + v_3^2$. Consider the function $\omega : \mathbb{R}^3 \times \mathbb{R} \longrightarrow \mathbb{C}$ defined as

$$\omega_t(x_1, x_2, x_3) = \exp\left(-\mathbf{i}\,|\mathbf{v}|^2\,t/2\right) \cdot \exp\left(\mathbf{i}\,\mathbf{v} \bullet \mathbf{x}\right).$$

Show that ω satisfies the (free) *Schrödinger equation:* $\mathbf{i}\partial_t\, \omega = \dfrac{-1}{2} \Delta\, \omega.$

3.3 Consider the stationary Schrödinger equation for a null potential:

$$\mathsf{H}\,\omega_0 = E \cdot \omega_0, \quad \text{where } \mathsf{H} = \frac{-\hbar^2}{2m_e} \Delta.$$

Let $v \in \mathbb{R}$ be a constant. Consider the function $\omega_0 : \mathbb{R}^3 \longrightarrow \mathbb{C}$ defined as

$$\omega_0(x_1, x_2, x_3) = \exp\left(\frac{\mathbf{i}}{\hbar} m_e v_1 \cdot x_1\right).$$

Show that ω_0 is a solution to the above stationary Schrödinger equation, with eigenvalue $E = \frac{1}{2} m_e v^2$.

3.4 Complete Exercise 3C.2(a) (p. 53).

3.5 Complete Exercise 3C.3(a) (p. 54).

Part II

General theory

Differential equations encode the underlying dynamical laws which govern a physical system. But a physical system is more than an abstract collection of laws; it is a specific configuration of matter and energy, which begins in a specific initial state (mathematically encoded as *initial conditions*) and which is embedded in a specific environment (encoded by certain *boundary conditions*). Thus, to model this physical system, we must find functions that satisfy the underlying differential equations, while simultaneously satisfying these initial conditions and boundary conditions; this is called an *initial/boundary value problem* (I/BVP).

Before we can develop solution methods for PDEs in general, and I/BVPs in particular, we must answer some qualitative questions. Under what conditions does a solution even *exist*? If it exists, is it unique? If the solution is not unique, then how can we parameterize the set of all possible solutions? Does this set have some kind of order or structure?

The beauty of *linear* differential equations is that linearity makes these questions much easier to answer. A linear PDE can be seen as a linear equation in an infinite-dimensional vector space, where the 'vectors' are *functions*. Most of the methods and concepts of linear algebra can be translated almost verbatim into this context. In particular, the set of solutions to a homogeneous linear PDE (or homogeneous linear I/BVP) forms a *linear subspace*. If we can find a convenient basis for this subspace, we thereby obtain a powerful and general solution technique for solving that I/BVP.

In Chapter 4, we will briefly review linear algebra in the setting of infinite-dimensional spaces of functions. In §5A–§5C, we will formally define the common types of initial and boundary value problems, and provide applications, interpretations, and examples. Finally, in §5D, we will establish the existence and uniqueness of solutions to the Laplace, Poisson, heat, and wave equations for a very broad class of initial/boundary conditions. The proofs of these existence/uniqueness theorems use methods of vector calculus, and provide important insights into the geometry of the solutions.

4

Linear partial differential equations

The Universe is a grand book which cannot be read until one first learns the language in which it is composed. It is written in the language of mathematics.

Galileo Galilei

4A Functions and vectors

Prerequisites: Appendix A.

Vectors If $\mathbf{v} = \begin{bmatrix} 2 \\ 7 \\ -3 \end{bmatrix}$ and $\mathbf{w} = \begin{bmatrix} -1.5 \\ 3 \\ 1 \end{bmatrix}$, then we can add these two vectors *componentwise* as follows:

$$\mathbf{v} + \mathbf{w} = \begin{bmatrix} 2 - 1.5 \\ 7 + 3 \\ -3 + 1 \end{bmatrix} = \begin{bmatrix} 0.5 \\ 10 \\ -2 \end{bmatrix}.$$

In general, if $\mathbf{v}, \mathbf{w} \in \mathbb{R}^3$, then $\mathbf{u} = \mathbf{v} + \mathbf{w}$ is defined by

$$u_n = v_n + w_n, \quad \text{for } n = 1, 2, 3 \tag{4A.1}$$

(see Figure 4A.1(a)). Think of $\mathbf{v}$ as a function $v : \{1, 2, 3\} \longrightarrow \mathbb{R}$, where $v(1) = 2$, $v(2) = 7$, and $v(3) = -3$. If we likewise represent $\mathbf{w}$ with $w : \{1, 2, 3\} \longrightarrow \mathbb{R}$ and $\mathbf{u}$ with $u : \{1, 2, 3\} \longrightarrow \mathbb{R}$, then we can rewrite equation (4A.1) as '$u(n) = v(n) + w(n)$ for $n = 1, 2, 3$'. In a similar fashion, any N-dimensional vector $\mathbf{u} = (u_1, u_2, \ldots, u_N)$ can be thought of as a function $u : [1 \ldots N] \longrightarrow \mathbb{R}$.

Functions as vectors Letting N go to infinity, we can imagine any function $f : \mathbb{R} \longrightarrow \mathbb{R}$ as a sort of 'infinite-dimensional vector' (see Figure 4A.2). Indeed, if f and g are two functions, we can add them *pointwise*, to get a new function

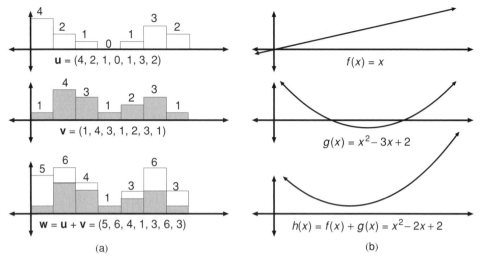

Figure 4A.1. (a) We add vectors *componentwise*: if $\mathbf{u} = (4, 2, 1, 0, 1, 3, 2)$ and $\mathbf{v} = (1, 4, 3, 1, 2, 3, 1)$, then the equation '$\mathbf{w} = \mathbf{v} + \mathbf{w}$' means that $\mathbf{w} = (5, 6, 4, 1, 3, 6, 3)$. (b) We add two functions *pointwise*: if $f(x) = x$, and $g(x) = x^2 - 3x + 2$, then the equation '$h = f + g$' means that $h(x) = f(x) + g(x) = x^2 - 2x + 2$ for every x.

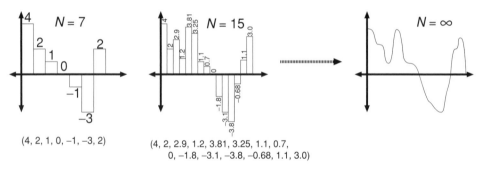

Figure 4A.2. We can think of a function as an 'infinite-dimensional vector'.

$h = f + g$, where

$$h(x) = f(x) + g(x), \quad \text{for all } x \in \mathbb{R} \tag{4A.2}$$

(see Figure 4A.1(b)). Note the similarity between equations (4A.2) and (4A.1), and the similarity between Figures 4A.1(a) and (b).

One of the most important ideas in the theory of PDEs is that *functions are infinite-dimensional vectors*. Just as with finite vectors, we can add them together, act on them with linear operators, or represent them in different *coordinate systems* on infinite-dimensional space. Also, the vector space $\mathbb{R}^D$ has a natural geometric structure; we can identify a similar geometry in infinite dimensions.

Let $\mathbb{X} \subseteq \mathbb{R}^D$ be some domain. The vector space of all continuous functions from $\mathbb{X}$ into $\mathbb{R}^m$ is denoted $\mathcal{C}(\mathbb{X}; \mathbb{R}^m)$. That is:

$$\mathcal{C}(\mathbb{X}; \mathbb{R}^m) := \{f : \mathbb{X} \longrightarrow \mathbb{R}^m; \ f \text{ is continuous}\}.$$

When $\mathbb{X}$ and $\mathbb{R}^m$ are obvious from the context, we may just write '$\mathcal{C}$'.

Exercise 4A.1 Show that $\mathcal{C}(\mathbb{X}; \mathbb{R}^m)$ is a vector space. ♦ Ⓔ

A scalar field $f : \mathbb{X} \longrightarrow \mathbb{R}$ is *infinitely differentiable* (or *smooth*) if, for every $N > 0$ and every $i_1, i_2, \ldots, i_N \in [1 \ldots D]$, the Nth derivative $\partial_{i_1} \partial_{i_2} \ldots \partial_{i_N} f(\mathbf{x})$ exists at each $\mathbf{x} \in \mathbb{X}$. A vector field $f : \mathbb{X} \longrightarrow \mathbb{R}^m$ is *infinitely differentiable* (or *smooth*) if $f(\mathbf{x}) := (f_1(\mathbf{x}), \ldots, f_m(\mathbf{x}))$, where each of the scalar fields $f_1, \ldots, f_m : \mathbb{X} \longrightarrow \mathbb{R}$ is infinitely differentiable. The vector space of all smooth functions from $\mathbb{X}$ into $\mathbb{R}^m$ is denoted $\mathcal{C}^\infty(\mathbb{X}; \mathbb{R}^m)$. That is:

$$\mathcal{C}^\infty(\mathbb{X}; \mathbb{R}^m) := \{f : \mathbb{X} \longrightarrow \mathbb{R}^m; \ f \text{ is infinitely differentiable}\}.$$

When $\mathbb{X}$ and $\mathbb{R}^m$ are obvious from the context, we may just write '$\mathcal{C}^\infty$'.

Example 4A.1
(a) $\mathcal{C}^\infty(\mathbb{R}^2; \mathbb{R})$ is the space of all smooth *scalar fields* on the *plane* (i.e. all functions $u : \mathbb{R}^2 \longrightarrow \mathbb{R}$).
(b) $\mathcal{C}^\infty(\mathbb{R}; \mathbb{R}^3)$ is the space of all smooth *curves* in *three-dimensional space*. ◇

Exercise 4A.2 Show that $\mathcal{C}^\infty(\mathbb{X}; \mathbb{R}^m)$ is a vector space and thus a linear subspace Ⓔ of $\mathcal{C}(\mathbb{X}; \mathbb{R}^m)$. ♦

4B Linear operators

Prerequisites: §4A.

4B(i) ... on finite-dimensional vector spaces

Let $\mathbf{v} := \begin{bmatrix} 2 \\ 7 \end{bmatrix}$ and $\mathbf{w} := \begin{bmatrix} -1.5 \\ 3 \end{bmatrix}$, and let $\mathbf{u} := \mathbf{v} + \mathbf{w} = \begin{bmatrix} 0.5 \\ 10 \end{bmatrix}$. If $\mathbf{A} := \begin{bmatrix} 1 & -1 \\ 4 & 0 \end{bmatrix}$, then $\mathbf{A} \cdot \mathbf{u} = \mathbf{A} \cdot \mathbf{v} + \mathbf{A} \cdot \mathbf{w}$. That is:

$$\begin{bmatrix} 1 & -1 \\ 4 & 0 \end{bmatrix} \cdot \begin{bmatrix} 0.5 \\ 10 \end{bmatrix} = \begin{bmatrix} -9.5 \\ 2 \end{bmatrix} = \begin{bmatrix} -5 \\ 8 \end{bmatrix} + \begin{bmatrix} -4.5 \\ -6 \end{bmatrix}$$

$$= \begin{bmatrix} 1 & -1 \\ 4 & 0 \end{bmatrix} \cdot \begin{bmatrix} 2 \\ 7 \end{bmatrix} + \begin{bmatrix} 1 & -1 \\ 4 & 0 \end{bmatrix} \cdot \begin{bmatrix} -1.5 \\ 3 \end{bmatrix}.$$

Also, if $\mathbf{x} = 3\mathbf{v} = \begin{bmatrix} 6 \\ 21 \end{bmatrix}$, then $\mathbf{Ax} = 3\mathbf{Av}$. That is:

$$\begin{bmatrix} 1 & -1 \\ 4 & 0 \end{bmatrix} \cdot \begin{bmatrix} 6 \\ 21 \end{bmatrix} = \begin{bmatrix} -15 \\ 24 \end{bmatrix} = 3 \begin{bmatrix} -5 \\ 8 \end{bmatrix} = 3 \begin{bmatrix} 1 & -1 \\ 4 & 0 \end{bmatrix} \cdot \begin{bmatrix} 2 \\ 7 \end{bmatrix}.$$

In other words, multiplication by the matrix $\mathbf{A}$ is a *linear operator* on the vector space $\mathbb{R}^2$. In general, a function $L : \mathbb{R}^N \longrightarrow \mathbb{R}^M$ is *linear* if

- for all $\mathbf{v}, \mathbf{w} \in \mathbb{R}^N$, we have $L(\mathbf{v} + \mathbf{w}) = L(\mathbf{v}) + L(\mathbf{w})$;
- for all $\mathbf{v} \in \mathbb{R}^N$ and $r \in \mathbb{R}$, we have $L(r \cdot \mathbf{v}) = r \cdot L(\mathbf{v})$.

Every linear function from $\mathbb{R}^N$ to $\mathbb{R}^M$ corresponds to multiplication by some $N \times M$ matrix.

Example 4B.1

(a) Difference operator. Suppose $D : \mathbb{R}^5 \longrightarrow \mathbb{R}^4$ is the following function:

$$D \begin{bmatrix} x_0 \\ x_1 \\ x_2 \\ x_3 \\ x_4 \end{bmatrix} = \begin{bmatrix} x_1 - x_0 \\ x_2 - x_1 \\ x_3 - x_2 \\ x_4 - x_3 \end{bmatrix}.$$

Then D corresponds to multiplication by the matrix

$$\begin{bmatrix} -1 & 1 & & & \\ & -1 & 1 & & \\ & & -1 & 1 & \\ & & & -1 & 1 \end{bmatrix}.$$

(b) Summation operator. Suppose $S : \mathbb{R}^4 \longrightarrow \mathbb{R}^5$ is the following function:

$$S \begin{bmatrix} x_1 \\ x_2 \\ x_3 \\ x_4 \end{bmatrix} = \begin{bmatrix} 0 \\ x_1 \\ x_1 + x_2 \\ x_1 + x_2 + x_3 \\ x_1 + x_2 + x_3 + x_4 \end{bmatrix}.$$

Then S corresponds to multiplication by the matrix

$$\begin{bmatrix} 0 & 0 & 0 & 0 \\ 1 & 0 & 0 & 0 \\ 1 & 1 & 0 & 0 \\ 1 & 1 & 1 & 0 \\ 1 & 1 & 1 & 1 \end{bmatrix}.$$

(c) Multiplication operator. Suppose $M : \mathbb{R}^5 \longrightarrow \mathbb{R}^5$ is the following function.

$$
M \begin{bmatrix} x_1 \\ x_2 \\ x_3 \\ x_4 \\ x_5 \end{bmatrix} = \begin{bmatrix} 3 \cdot x_1 \\ 2 \cdot x_2 \\ -5 \cdot x_3 \\ \frac{3}{4} \cdot x_4 \\ \sqrt{2} \cdot x_5 \end{bmatrix}.
$$

Then M corresponds to multiplication by the matrix

$$
\begin{bmatrix} 3 & & & & \\ & 2 & & & \\ & & -5 & & \\ & & & \frac{3}{4} & \\ & & & & \sqrt{2} \end{bmatrix}.
$$ ◇

Remark Note that the transformation D is a *left-inverse* to the transformation S. That is, $D \circ S = \mathbf{Id}$. However, D is *not* a *right-inverse* to S, because if $\mathbf{x} = (x_0, x_1, \dots, x_4)$, then $S \circ D(\mathbf{x}) = \mathbf{x} - (x_0, x_0, \dots, x_0)$.

4B(ii) ... on C^∞

Recommended: §1B, §1C, §2B.

A transformation $\mathsf{L} : C^\infty \longrightarrow C^\infty$ is called a *linear operator* if, for any two differentiable functions $f, g \in C^\infty$, we have $\mathsf{L}(f + g) = \mathsf{L}(f) + \mathsf{L}(g)$, and, for any real number $r \in \mathbb{R}$, we have $\mathsf{L}(r \cdot f) = r \cdot \mathsf{L}(f)$.

Example 4B.2
(a) Differentiation. If $f, g : \mathbb{R} \longrightarrow \mathbb{R}$ are differentiable functions, and $h = f + g$, then we know that, for any $x \in \mathbb{R}$,

$$ h'(x) = f'(x) + g'(x). $$

Also, if $h = r \cdot f$, then $h'(x) = r \cdot f'(x)$. Thus, if we define the operation $\mathsf{D} : C^\infty(\mathbb{R}; \mathbb{R}) \longrightarrow C^\infty(\mathbb{R}; \mathbb{R})$ by $\mathsf{D}[f] = f'$, then D is a linear transformation of $C^\infty(\mathbb{R}; \mathbb{R})$. For example, sin and cos are elements of $C^\infty(\mathbb{R}; \mathbb{R})$, and we have

$$ \mathsf{D}[\sin] = \cos \quad \text{and} \quad \mathsf{D}[\cos] = -\sin. $$

More generally, if $f, g : \mathbb{R}^D \longrightarrow \mathbb{R}$ and $h = f + g$, then, for any $i \in [1 \dots D]$,

$$ \partial_j h = \partial_j f + \partial_j g. $$

Also, if $h = r \cdot f$, then $\partial_j h = r \cdot \partial_j f$. In other words, the transformation $\partial_j : C^\infty(\mathbb{R}^D; \mathbb{R}) \longrightarrow C^\infty(\mathbb{R}^D; \mathbb{R})$ is a linear operator.

(b) Integration. If $f, g : \mathbb{R} \longrightarrow \mathbb{R}$ are integrable functions, and $h = f + g$, then we know that, for any $x \in \mathbb{R}$,

$$\int_0^x h(y)dy = \int_0^x f(y)dy + \int_0^x g(y)dy.$$

Also, if $h = r \cdot f$, then

$$\int_0^x h(y)dy = r \cdot \int_0^x f(y)dy.$$

Thus, if we define the operation $\mathsf{S} : \mathcal{C}^\infty(\mathbb{R}; \mathbb{R}) \longrightarrow \mathcal{C}^\infty(\mathbb{R}; \mathbb{R})$ by

$$\mathsf{S}[f](x) = \int_0^x f(y)dy, \quad \text{for all } x \in \mathbb{R},$$

then S is a linear transformation. For example, $\sin$ and $\cos$ are elements of $\mathcal{C}^\infty(\mathbb{R}; \mathbb{R})$, and we have

$$\mathsf{S}[\sin] = 1 - \cos \quad \text{and} \quad \mathsf{S}[\cos] = \sin.$$

(c) Multiplication. If $\gamma : \mathbb{R}^D \longrightarrow \mathbb{R}$ is a scalar field, then define the operator $\Gamma : \mathcal{C}^\infty \longrightarrow \mathcal{C}^\infty$ by $\Gamma[f] = \gamma \cdot f$. In other words, for all $\mathbf{x} \in \mathbb{R}^D$, $\Gamma[f](\mathbf{x}) = \gamma(\mathbf{x}) \cdot f(\mathbf{x})$. Then Γ is a linear function, because, for any $f, g \in \mathcal{C}^\infty$, $\Gamma[f + g] = \gamma \cdot [f + g] = \gamma \cdot f + \gamma \cdot g = \Gamma[f] + \Gamma[g]$.   ◇

Remark Note that the transformation D is a *left-inverse* for the transformation S, because the fundamental theorem of calculus says that $\mathsf{D} \circ \mathsf{S}(f) = f$ for any $f \in \mathcal{C}^\infty(\mathbb{R})$. However, D is *not* a *right-inverse* for S, because in general $\mathsf{S} \circ \mathsf{D}(f) = f - c$, where $c = f(0)$ is a constant.

Ⓔ **Exercise 4B.1** Compare the three linear transformations in Example 4B.2 with those from Example 4B.1. Do you notice any similarities?   ♦

Remark Unlike linear transformations on $\mathbb{R}^N$, there is, in general, no way to express a linear transformation on $\mathcal{C}^\infty$ in terms of multiplication by some matrix. To convince yourself of this, try to express the three transformations from Example 4B.2 in terms of 'matrix multiplication'.

Any combination of linear operations is also a linear operation. In particular, any combination of differentiation and multiplication operations is linear. Thus, for example, the second-derivative operator $\mathsf{D}^2[f] = \partial_x^2 f$ is linear, and the Laplacian operator

$$\triangle f = \partial_1^2 f + \cdots + \partial_D^2 f$$

is also linear; in other words, $\triangle[f + g] = \triangle f + \triangle g$.

A linear transformation that is formed by adding and/or composing multiplications and differentiations is called a *linear differential operator*. For example, the Laplacian $\triangle$ is a linear differential operator.

4B(iii) Kernels

If L is a linear function, then the *kernel* of L is the set of all vectors **v** such that $L(\mathbf{v}) = 0$.

Example 4B.3
(a) Consider the differentiation operator ∂_x on the space $C^\infty(\mathbb{R}; \mathbb{R})$. The kernel of ∂_x is the set of all functions $u : \mathbb{R} \longrightarrow \mathbb{R}$ such that $\partial_x u \equiv 0$; in other words, the set of all *constant* functions.
(b) The kernel of ∂_x^2 is the set of all functions $u : \mathbb{R} \longrightarrow \mathbb{R}$ such that $\partial_x^2 u \equiv 0$; in other words, the set of all *flat* functions of the form $u(x) = ax + b$. ◇

Many partial differential equations are really equations for the kernel of some differential operator.

Example 4B.4
(a) *Laplace's equation '$\Delta u \equiv 0$'* really just says: 'u is in the kernel of Δ'.
(b) The *heat equation '$\partial_t u = \Delta u$'* really just says: 'u is in the kernel of the operator $\mathsf{L} = \partial_t - \Delta$'. ◇

4B(iv) Eigenvalues, eigenvectors, and eigenfunctions

If L is a linear operator on some vector space, then an *eigenvector* of L is a vector **v** such that

$$\mathsf{L}(\mathbf{v}) = \lambda \cdot \mathbf{v},$$

for some constant $\lambda \in \mathbb{C}$, called the associated *eigenvalue*.

Example 4B.5 If $L : \mathbb{R}^2 \longrightarrow \mathbb{R}^2$ is defined by the matrix $\begin{bmatrix} 0 & 1 \\ 1 & 0 \end{bmatrix}$ and $\mathbf{v} = \begin{bmatrix} -1 \\ 1 \end{bmatrix}$, then $L(\mathbf{v}) = \begin{bmatrix} 1 \\ -1 \end{bmatrix} = -\mathbf{v}$, so **v** is an eigenvector for L, with eigenvalue $\lambda = -1$. ◇

If L is a linear operator on C^∞, then an eigenvector of L is sometimes called an *eigenfunction*.

Example 4B.6 Let $n, m \in \mathbb{N}$. Define $u(x, y) = \sin(n \cdot x) \cdot \sin(m \cdot y)$. Then

$$\Delta u(x, y) = -(n^2 + m^2) \cdot \sin(n \cdot x) \cdot \sin(m \cdot y) = \lambda \cdot u(x, y),$$

where $\lambda = -(n^2 + m^2)$. Thus, u is an eigenfunction of the linear operator Δ, with eigenvalue λ. ◇

Exercise 4B.2 Verify the claims in Example 4B.6. ♦ Ⓔ

Eigenfunctions of linear differential operators (particularly eigenfunctions of $\triangle$) play a central role in the solution of linear PDEs. This is implicit in Chapters 11–14 and 20, and is made explicit in Chapter 15.

4C Homogeneous vs. nonhomogeneous

Prerequisites: §4B.

If $\mathsf{L} : C^\infty \longrightarrow C^\infty$ is a linear differential operator, then the equation '$\mathsf{L}u \equiv 0$' is called a *homogeneous linear* partial differential equation.

Example 4C.1 The following are linear homogeneous PDEs. Here $\mathbb{X} \subset \mathbb{R}^D$ is some domain.

(a) Laplace's equation.[1] Here, $C^\infty = C^\infty(\mathbb{X}; \mathbb{R})$, and $\mathsf{L} = \triangle$.
(b) Heat equation.[2] $C^\infty = C^\infty(\mathbb{X} \times \mathbb{R}; \mathbb{R})$, and $\mathsf{L} = \partial_t - \triangle$.
(c) Wave equation.[3] $C^\infty = C^\infty(\mathbb{X} \times \mathbb{R}; \mathbb{R})$, and $\mathsf{L} = \partial_t^2 - \triangle$.
(d) Schrödinger equation.[4] $C^\infty = C^\infty(\mathbb{R}^{3N} \times \mathbb{R}; \mathbb{C})$, and, for any $\omega \in C^\infty$ and $(\mathbf{x}; t) \in \mathbb{R}^{3N} \times \mathbb{R}$,

$$\mathsf{L}\omega(\mathbf{x}; t) := \frac{-\hbar^2}{2} \blacktriangle \omega(\mathbf{x}; t) + V(\mathbf{x}) \cdot \omega(\mathbf{x}; t) - \mathbf{i}\hbar\, \partial_t\, \omega(\mathbf{x}; t).$$

(Here, $V : \mathbb{R}^{3N} \longrightarrow \mathbb{R}$ is some *potential function*, and $\blacktriangle$ is like a Laplacian operator, except that the components for each particle are divided by the rest mass of that particle.)
(e) Fokker–Planck equation.[5] $C^\infty = C^\infty(\mathbb{X} \times \mathbb{R}; \mathbb{R})$, and, for any $u \in C^\infty$,

$$\mathsf{L}(u) = \partial_t\, u - \triangle u + \vec{\mathbf{V}} \bullet \nabla u + u \cdot \operatorname{div} \vec{\mathbf{V}}. \qquad\qquad \diamond$$

Linear homogeneous PDEs are nice because we can combine two solutions together to obtain a third solution.

Example 4C.2
(a) Let $u(x; t) = 7 \sin [2t + 2x]$ and $v(x; t) = 3 \sin [17t + 17x]$ be two travelling wave solutions to the wave equation. Then $w(x; t) = u(x; t) + v(x; t) = 7 \sin(2t + 2x) + 3 \sin(17t + 17x)$ is also a solution (see Figure 4C.1). To use a musical analogy: if we think of u and v as two 'pure tones', then we can think of w as a 'chord'.
(b) Let

$$f(x; t) = \frac{1}{2\sqrt{\pi t}} \exp\left[\frac{-x^2}{4t}\right],$$

$$g(x; t) = \frac{1}{2\sqrt{\pi t}} \exp\left[\frac{-(x-3)^2}{4t}\right],$$

---

[1] See §1C, p. 11.   [2] See §1B, p. 7.   [3] See §2B, p. 29.   [4] See §3B, p. 43.   [5] See §1F, p. 21.

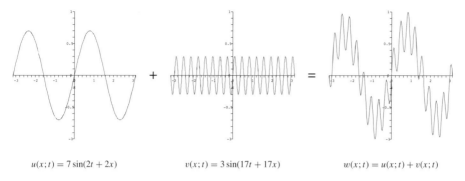

$u(x;t) = 7\sin(2t + 2x)$ $v(x;t) = 3\sin(17t + 17x)$ $w(x;t) = u(x;t) + v(x;t)$

Figure 4C.1. Two travelling waves $u(x;t)$ and $v(x;t)$ add to produce a solution $w(x;t)$; see Example 4C.2(a).

and

$$h(x;t) = \frac{1}{2\sqrt{\pi t}} \exp\left[\frac{-(x-5)^2}{4t}\right]$$

be one-dimensional Gauss–Weierstrass kernels, centered at 0, 3, and 5, respectively. Thus, f, g, and h are all solutions to the heat equation. Then, $F(x) = f(x) + 7 \cdot g(x) + h(x)$ is also a solution to the heat equation. If a Gauss–Weierstrass kernel models the erosion of a single 'mountain', then the function F models the erosion of a little 'mountain range', with peaks at 0, 3, and 5, and where the middle peak is seven times higher than the other two. ◇

These examples illustrate a general principle, outlined in Theorem 4C.3.

Theorem 4C.3 Superposition principle for homogeneous linear PDEs
Suppose L *is a linear differential operator and* $u_1, u_2 \in C^\infty$ *are solutions to the homogeneous linear PDE* 'Lu = 0'. *Then, for any* $c_1, c_2 \in \mathbb{R}$, $u = c_1 \cdot u_1 + c_2 \cdot u_2$ *is also a solution.*

Proof **Exercise 4C.1.** □ Ⓔ

If $q \in C^\infty$ is some fixed nonzero function, then the equation 'L$p \equiv q$' is called a *nonhomogeneous* linear partial differential equation.

Example 4C.4 The following are linear *non*homogeneous PDEs.
(a) The *antidifferentiation equation* $p' = q$ is familiar from first-year calculus. The *fundamental theorem of calculus* says that one solution to this equation is the integral function $p(x) = \int_0^x q(y)dy$.
(b) The *Poisson equation*,[6] '$\triangle p = q$', is a *nonhomogeneous* linear PDE. ◇

Recall Examples 1D.1 and 1D.2, p. 15, where we obtained *new* solutions to a nonhomogeneous equation by taking a single solution and adding solutions of

---

[6] See §1D, p. 14.

the *homogeneous* equation to this solution. These examples illustrates a general principle, outlined in Theorem 4C.5.

Theorem 4C.5 Subtraction principle for nonhomogeneous linear PDEs

*Suppose L is a linear differential operator, and $q \in C^\infty$. Let $p_1 \in C^\infty$ be a solution to the nonhomogeneous linear PDE '$\mathsf{L}p_1 = q$'. If $h \in C^\infty$ is any solution to the homogeneous equation (i.e. $\mathsf{L}h = 0$), then $p_2 = p_1 + h$ is another solution to the non*homogeneous *equation. In summary:*

$$\left(\mathsf{L}p_1 = q; \quad \mathsf{L}h = 0; \quad and \ p_2 = p_1 + h. \right) \Longrightarrow \left(\mathsf{L}p_2 = q \right).$$

Ⓔ   *Proof* **Exercise 4C.2.**   □

If $\mathsf{P} : C^\infty \longrightarrow C^\infty$ is *not* a linear operator, then a PDE of the form '$\mathsf{P}u \equiv 0$' or '$\mathsf{P}u \equiv g$' is called a *nonlinear* PDE. For example, if $F : \mathbb{R}^D \longrightarrow \mathbb{R}^D$ is some nonlinear 'rate function' describing chemical reactions, then the *reaction–diffusion equation*[7]

$$\partial_t \mathbf{u} = \triangle \mathbf{u} + F(\mathbf{u}),$$

is a *nonlinear* PDE, corresponding to the nonlinear differential operator $\mathsf{P}(\mathbf{u}) := \partial_t \mathbf{u} - \triangle \mathbf{u} - F(\mathbf{u})$.

   The theory of linear partial differential equations is relatively simple, because solutions to linear PDEs interact in very nice ways, as shown by Theorems 4C.3 and 4C.5. The theory of *non*linear PDEs is much more complicated; furthermore, many of the methods which *do* exist for solving nonlinear PDEs involve somehow 'approximating' them with linear ones. In this book we shall concern ourselves only with linear PDEs.

4D Practice problems

4.1 For each of the following equations: u is an unknown function; q is always some fixed, predetermined function; and λ is always a constant. In each case, is the equation linear? If it is linear, is it homogeneous? Justify your answers.

(a) Heat equation: $\partial_t u(\mathbf{x}) = \triangle u(\mathbf{x})$;
(b) Poisson equation: $\triangle u(\mathbf{x}) = q(\mathbf{x})$;
(c) Laplace equation: $\triangle u(\mathbf{x}) = 0$;
(d) Monge–Ampère equation: $q(x, y) = \det \begin{bmatrix} \partial_x^2 u(x, y) & \partial_x \partial_y u(x, y) \\ \partial_x \partial_y u(x, y) & \partial_y^2 u(x, y) \end{bmatrix}$;
(e) reaction-diffusion equation: $\partial_t u(\mathbf{x}; t) = \triangle u(\mathbf{x}; t) + q(u(\mathbf{x}; t))$;
(f) scalar conservation law: $\partial_t u(x; t) = -\partial_x (q \circ u)(x; t)$;

---

[7] See §1G, p. 21.

(g) Helmholtz equation: $\triangle u(\mathbf{x}) = \lambda \cdot u(\mathbf{x})$;

(h) Airy equation: $\partial_t u(x;t) = -\partial_x^3 u(x;t)$;

(i) Beam equation: $\partial_t u(x;t) = -\partial_x^4 u(x;t)$;

(j) Schrödinger equation: $\partial_t u(\mathbf{x};t) = \mathbf{i}\,\triangle u(\mathbf{x};t) + q(\mathbf{x};t) \cdot u(\mathbf{x};t)$;

(k) Burger equation: $\partial_t u(x;t) = -u(x;t) \cdot \partial_x u(x;t)$;

(l) Eikonal equation: $|\partial_x u(x)| = 1$.

4.2 Which of the following are eigenfunctions for the two-dimensional Laplacian $\triangle = \partial_x^2 + \partial_y^2$? In each case, if u is an eigenfunction, what is the eigenvalue?

(a) $u(x, y) = \sin(x)\sin(y)$ (see Figure 5F.1(a), p. 102);

(b) $u(x, y) = \sin(x) + \sin(y)$ (see Figure 5F.1(b), p. 102);

(c) $u(x, y) = \cos(2x) + \cos(y)$ (see Figure 5F.1(c), p. 102);

(d) $u(x, y) = \sin(3x) \cdot \cos(4y)$;

(e) $u(x, y) = \sin(3x) + \cos(4y)$;

(f) $u(x, y) = \sin(3x) + \cos(3y)$;

(g) $u(x, y) = \sin(3x) \cdot \cosh(4y)$;

(h) $u(x, y) = \sinh(3x) \cdot \cosh(4y)$;

(i) $u(x, y) = \sinh(3x) + \cosh(4y)$;

(j) $u(x, y) = \sinh(3x) + \cosh(3y)$;

(k) $u(x, y) = \sin(3x + 4y)$;

(l) $u(x, y) = \sinh(3x + 4y)$;

(m) $u(x, y) = \sin^3(x) \cdot \cos^4(y)$;

(n) $u(x, y) = e^{3x} \cdot e^{4y}$;

(o) $u(x, y) = e^{3x} + e^{4y}$;

(p) $u(x, y) = e^{3x} + e^{3y}$.

5

Classification of PDEs and problem types

If one looks at the different problems of the integral calculus which
arise naturally when one wishes to go deep into the different parts of
physics, it is impossible not to be struck by the analogies existing.
Whether it be electrostatics or electrodynamics, the propogation of heat,
optics, elasticity, or hydrodynamics, we are led always to differential
equations of the same family.

Henri Poincaré

5A Evolution vs. nonevolution equations

Recommended: §1B, §1C, §2B, §4B.

An *evolution equation* is a PDE with a distinguished 'time' coordinate, t. In other
words, it describes functions of the form $u(\mathbf{x}; t)$, and the equation has the following
form:

$$\mathsf{D}_t\, u = \mathsf{D}_{\mathbf{x}}\, u,$$

where D_t is some differential operator involving only derivatives in the t variable
(e.g. ∂_t, ∂_t^2, etc.), while $\mathsf{D}_{\mathbf{x}}$ is some differential operator involving only derivatives
in the $\mathbf{x}$ variables (e.g. ∂_x, ∂_y^2, $\triangle$, etc.).

Example 5A.1 The following are evolution equations:

(a) the heat equation '$\partial_t\, u = \triangle u$' of §1B;
(b) the wave equation '$\partial_t^2\, u = \triangle u$' of §2B;
(c) the telegraph equation '$\kappa_2 \partial_t^2\, u + \kappa_1 \partial_t\, u = -\kappa_0 u + \triangle u$' of §2C;
(d) the Schrödinger equation '$\partial_t\, \omega = \frac{1}{i\hbar} \mathsf{H}\omega$' of §3B (here H is a Hamiltonian
    operator);
(e) Liouville's equation, the Fokker–Planck equation, and reaction–diffusion
    equations.   ◇

73

Non-Example 5A.2 The following are *not* evolution equations:

(a) the Laplace equation '$\triangle u = 0$' of §1C;
(b) the Poisson equation '$\triangle u = q$' of §1D;
(c) the Helmholtz equation '$\triangle u = \lambda u$' (where $\lambda \in \mathbb{C}$ is a constant, i.e. an eigenvalue of $\triangle$);
(d) the stationary Schrödinger equation $H\omega_0 = E \cdot \omega_0$ (where $E \in \mathbb{C}$ is a constant eigenvalue). ◇

In mathematical models of physical phenomena, most PDEs are evolution equations. Nonevolutionary PDEs generally arise as *stationary state* equations for evolution PDEs (e.g. the Laplace equation) or as *resonance states* (e.g. Sturm–Liouville, Helmholtz).

Order The *order* of the differential operator $\partial_x^2 \partial_y^3$ is $2 + 3 = 5$. More generally, the *order* of the differential operator $\partial_1^{k_1} \partial_2^{k_2} \ldots \partial_D^{k_D}$ is the sum $k_1 + \cdots + k_D$. The *order* of a general differential operator is the highest order of any of its terms. For example, the Laplacian is second order. The *order* of a PDE is the highest order of the differential operator that appears in it. Thus, the transport equation, Liouville's equation, and the (nondiffusive) reaction equation are *first order*, but all the other equations we have looked at (the heat equation, the wave equation, etc.) are of *second order*.

5B Initial value problems

Prerequisites: §5A.

Let $\mathbb{X} \subset \mathbb{R}^D$ be some domain, and let L be a differential operator on $\mathcal{C}^\infty(\mathbb{X}; \mathbb{R})$. Consider the evolution equation

$$\partial_t u = L u \tag{5B.1}$$

for an unknown function $u : \mathbb{X} \times \mathbb{R}_+ \longrightarrow \mathbb{R}$. An *initial value problem* (IVP) for equation (5B.1) is the following problem:

Given some function $f_0 : \mathbb{X} \longrightarrow \mathbb{R}$ (the initial conditions), find a continuous function $u : \mathbb{X} \times \mathbb{R}_+ \longrightarrow \mathbb{R}$ which satisfies equation (5B.1) and also satisfies $u(\mathbf{x}, 0) = f_0(\mathbf{x})$, for all $\mathbf{x} \in \mathbb{X}$.

For example, suppose the domain $\mathbb{X}$ is an iron pan being heated on a gas flame stove. You turn off the flame (so there is no further heat entering the system) and then throw some vegetables into the pan. Thus, equation (5B.1) is the heat equation, and f_0 describes the initial distribution of heat: cold vegetables in a hot pan. The initial value problem asks: 'How fast do the vegetables cook? How fast does the pan cool?'

Next, consider the second-order evolution equation,

$$\partial_t^2 u = L u, \tag{5B.2}$$

for an unknown function $u : \mathbb{X} \times \mathbb{R}_{\not{+}} \longrightarrow \mathbb{R}$. An *initial value problem* (or *IVP*, or *Cauchy problem*) for equation (5B.2) is as follows:

Given a function $f_0 : \mathbb{X} \longrightarrow \mathbb{R}$ (the initial position), and/or another function $f_1 : \mathbb{X} \longrightarrow \mathbb{R}$ (the initial velocity), find a continuously differentiable function $u : \mathbb{X} \times \mathbb{R}_{\not{+}} \longrightarrow \mathbb{R}$ which satisfies equation (5B.2) and also satisfies equation $u(\mathbf{x}, 0) = f_0(\mathbf{x})$ and $\partial_t u(\mathbf{x}, 0) = f_1(\mathbf{x})$, for all $\mathbf{x} \in \mathbb{X}$.

For example, suppose equation (5B.1) is the wave equation on $\mathbb{X} = [0, L]$. Imagine $[0, L]$ as a vibrating string. Thus, f_0 describes the initial displacement of the string and f_1 describes its initial momentum.

If $f_0 \neq 0$, and $f_1 \equiv 0$, then the string is initially at rest, but is released from a displaced state; in other words, it is *plucked* (e.g. in a guitar or a harp). Hence, the initial value problem asks: 'How does a guitar string sound when it is plucked?'

On the other hand, if $f_0 \equiv 0$, and $f_1 \neq 0$, then the string is initially flat, but is imparted with nonzero momentum; in other words, it is *struck* (e.g. by the hammer in the piano). Hence, the initial value problem asks: 'How does a piano string sound when it is struck?'

5C Boundary value problems

Prerequisites: Appendix D, §1C.   **Recommended:** §5B.

If $\mathbb{X} \subset \mathbb{R}^D$ is a finite domain, then $\partial\mathbb{X}$ denotes its *boundary*. The *interior* of $\mathbb{X}$ is the set int $(\mathbb{X})$ of all points in $\mathbb{X}$ *not* on the boundary.

Example 5C.1
(a) If $\mathbb{I} = [0, 1] \subset \mathbb{R}$ is the *unit interval*, then $\partial\mathbb{I} = \{0, 1\}$ is a two-point set, and int $(\mathbb{I}) = (0, 1)$.
(b) If $\mathbb{X} = [0, 1]^2 \subset \mathbb{R}^2$ is the *unit square*, then int $(\mathbb{X}) = (0, 1)^2$, and

$$\partial\mathbb{X} = \{(x, y) \in \mathbb{X}; \ x = 0 \text{ or } x = 1 \text{ or } y = 0 \text{ or } y = 1\}.$$

(c) In polar coordinates on $\mathbb{R}^2$, let $\mathbb{D} = \{(r, \theta); \ r \leq 1, \theta \in [-\pi, \pi)\}$ be the *unit disk*. Then $\partial\mathbb{D} = \{(1, \theta); \ \theta \in [-\pi, \pi)\}$ is the *unit circle*, and int $(\mathbb{D}) = \{(r, \theta); \ r < 1, \ \theta \in [-\pi, \pi)\}$.
(d) In spherical coordinates on $\mathbb{R}^3$, let $\mathbb{B} = \{\mathbf{x} \in \mathbb{R}^3; \|\mathbf{x}\| \leq 1\}$ be the three-dimensional *unit ball* in $\mathbb{R}^3$. Then $\partial\mathbb{B} = \mathbb{S} := \{\{\mathbf{x} \in \mathbb{R}^D; \|\mathbf{x}\| = 1\}$ is the *unit sphere*, and int $(\mathbb{B}) = \{\mathbf{x} \in \mathbb{R}^D; \|\mathbf{x}\| < 1\}$.
(e) In cylindrical coordinates on $\mathbb{R}^3$, let $\mathbb{X} = \{(r, \theta, z); \ r \leq R, -\pi \leq \theta \leq \pi, 0 \leq z \leq L\}$ be the *finite cylinder* in $\mathbb{R}^3$. Then $\partial\mathbb{X} = \{(r, \theta, z); \ r = R \text{ or } z = 0 \text{ or } z = L\}$.   ◇

A *boundary value problem* (or *BVP*) is a problem of the following kind.

Find a continuous function u : $\mathbb{X} \longrightarrow \mathbb{R}$ *such that*

(1) *u satisfies some PDE at all* **x** *in the interior of* $\mathbb{X}$;
(2) *u also satisfies some other equation* (maybe a differential equation) *for all* **s** *on the boundary of* $\mathbb{X}$.

The condition *u* must satisfy on the boundary of $\mathbb{X}$ is called a *boundary condition*. Note that there is no 'time variable' in our formulation of a BVP; thus, typically the PDE in question is an 'equilibrium' equation, like the Laplace equation or the Poisson equation.

If we try to solve an evolution equation with specified initial conditions *and* specified boundary conditions, then we are confronted with an 'initial/boundary value problem'. Formally, an *initial/boundary value problem* (I/BVP) is a problem of the following kind.

Find a continuous function u : $\mathbb{X} \times \mathbb{R}_+ \longrightarrow \mathbb{R}$ *such that*

(1) *u satisfies some* (evolution) *PDE at all* **x** *in the interior of* $\mathbb{X} \times \mathbb{R}_+$;
(2) *u satisfies some boundary condition for all* $(\mathbf{s}; t)$ *in* $(\partial\mathbb{X}) \times \mathbb{R}_+$;
(3) $u(\mathbf{x}; 0)$ *also satisfies some initial condition* (as described in §5B) *for all* $\mathbf{x} \in \mathbb{X}$.

We will consider four kinds of boundary conditions: *Dirichlet, Neumann, mixed,* and *periodic*. Each of these boundary conditions has a particular physical inter-pretation and yields particular kinds of solutions for a partial differential equation.

5C(i) Dirichlet boundary conditions

Let $\mathbb{X}$ be a domain, and let $u : \mathbb{X} \longrightarrow \mathbb{R}$ be a function. We say that u satisfies *homogeneous Dirichlet boundary conditions (HDBC)* on $\mathbb{X}$ if

$$\text{for all } \mathbf{s} \in \partial\mathbb{X}, \quad u(\mathbf{s}) \equiv 0.$$

Physical interpretation

 Thermodynamic (heat equation, Laplace equation, or Poisson equation). In this case, *u* represents a temperature distribution. We imagine that the domain $\mathbb{X}$ represents some physical object, whose boundary $\partial\mathbb{X}$ is made out of metal or some other material which conducts heat almost perfectly. Hence, we can assume that *the temperature on the boundary is always equal to the temperature of the surrounding environment.*

  We further assume that this environment has a constant temperature T_E (for exam-ple, $\mathbb{X}$ is immersed in a 'bath' of some uniformly mixed fluid), which remains constant during the experiment (for example, the fluid is present in large enough quantities that the heat flowing into/out of $\mathbb{X}$ does not measurably change it). We can then assume that the ambient temperature is $T_E \equiv 0$, by simply subtracting a constant temperature

of T_E off the inside and the outside. (This is like changing from measuring temperature in Kelvin to measuring in Celsius; you're just adding 273 to both sides, which
makes no mathematical difference.)

Electrostatic (Laplace equation or Poisson equation). In this case, u represents an electrostatic potential. The domain $\mathbb{X}$ represents some compartment or region in space
whose boundary $\partial\mathbb{X}$ is made out of metal or some other perfect electrical conductor.
Thus, the electrostatic potential within the metal boundary is a constant, which we
can normalize to be zero.

Acoustic (wave equation). In this case, u represents the vibrations of some vibrating
medium (e.g. a violin string or a drum skin). Homogeneous Dirichlet boundary
conditions mean that the medium is *fixed* on the boundary $\partial\mathbb{X}$ (e.g. a violin string
is clamped at its endpoints; a drumskin is pulled down tightly around the rim of the
drum).

The set of *infinitely differentiable* functions from $\mathbb{X}$ to $\mathbb{R}$ which satisfy homogeneous
Dirichlet boundary conditions will be denoted $\mathcal{C}_0^\infty(\mathbb{X};\mathbb{R})$ or $\mathcal{C}_0^\infty(\mathbb{X})$. Thus, for
example,

$$\mathcal{C}_0^\infty[0, L] = \left\{ f : [0, L] \longrightarrow \mathbb{R}; \ f \text{ is smooth, and } f(0) = 0 = f(L)\right\}.$$

The set of *continuous* functions from $\mathbb{X}$ to $\mathbb{R}$ which satisfy homogeneous Dirichlet
boundary conditions will be denoted $\mathcal{C}_0(\mathbb{X};\mathbb{R})$ or $\mathcal{C}_0(\mathbb{X})$.

Example 5C.2

(a) Suppose $\mathbb{X} = [0, 1]$, and $f : \mathbb{X} \longrightarrow \mathbb{R}$ is defined by $f(x) = x(1 - x)$. Then $f(0) =$
$0 = f(1)$, and f is smooth, so $f \in \mathcal{C}_0^\infty[0, 1]$ (see Figure 5C.1).

(b) Let $\mathbb{X} = [0, \pi]$.

    (1) For any $n \in \mathbb{N}$, let $\mathbf{S}_n(x) = \sin(n \cdot x)$ (see Figure 6D.1, p. 118). Then $\mathbf{S}_n \in$
$\mathcal{C}_0^\infty[0, \pi]$.

    (2) If $f(x) = 5 \sin(x) - 3 \sin(2x) + 7 \sin(3x)$, then $f \in \mathcal{C}_0^\infty[0, \pi]$. More generally,
any finite sum $\sum_{n=1}^{N} B_n \mathbf{S}_n(x)$ (for some constants B_n) is in $\mathcal{C}_0^\infty[0, \pi]$.

    (3) If $f(x) = \sum_{n=1}^{\infty} B_n \mathbf{S}_n(x)$ is a *uniformly convergent* Fourier sine series,[1] then $f \in$
$\mathcal{C}_0^\infty[0, \pi]$.

(c) Let $\mathbb{D} = \{(r, \theta); \ r \leq 1\}$ be the unit disk. Let $f : \mathbb{D} \longrightarrow \mathbb{R}$ be the 'cone' in Figure
5C.2(a), defined: $f(r, \theta) = (1 - r)$. Then f is continuous, and $f \equiv 0$ on the boundary
of the disk, so f satisfies Dirichlet boundary conditions. Thus, $f \in \mathcal{C}_0(\mathbb{D})$. However, f
is not smooth (it is nondifferentiable at zero), so $f \notin \mathcal{C}_0^\infty(\mathbb{D})$.

(d) Let $f : \mathbb{D} \longrightarrow \mathbb{R}$ be the 'dome' in Figure 5C.2(b), defined by $f(r, \theta) = 1 - r^2$. Then
$f \in \mathcal{C}_0^\infty(\mathbb{D})$.

(e) Let $\mathbb{X} = [0, \pi] \times [0, \pi]$ be the square of sidelength π.

---

[1] See §7B, p. 148.

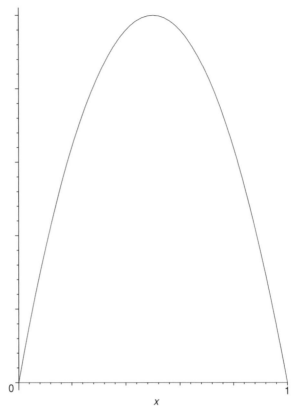

0

1

x

Figure 5C.1. $f(x) = x(1 - x)$ satisfies homogeneous Dirichlet boundary conditions on the interval $[0, 1]$.

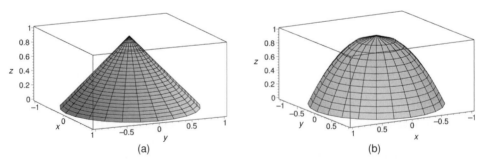

(a)   (b)

Figure 5C.2. (a) $f(r, \theta) = 1 - r$ satisfies homogeneous Dirichlet boundary conditions on the disk $\mathbb{D} = \{(r, \theta); r \leq 1\}$, but is not smooth at zero. (b) $f(r, \theta) = 1 - r^2$ satisfies homogeneous Dirichlet boundary conditions on the disk $\mathbb{D} = \{(r, \theta); r \leq 1\}$, and is smooth everywhere.

(1) For any $(n, m) \in \mathbb{N}^2$, let $\mathbf{S}_{n,m}(x, y) = \sin(n \cdot x) \cdot \sin(m \cdot y)$. Then $\mathbf{S}_{n,m} \in C_0^\infty(\mathbb{X})$
(see Figure 9A.2, p. 185).

(2) If $f(x) = 5 \sin(x) \sin(2y) - 3 \sin(2x) \sin(7y) + 7 \sin(3x) \sin(y)$, then $f \in C_0^\infty$
$(\mathbb{X})$. More generally, any finite sum $\sum_{n=1}^N \sum_{m=1}^M B_{n,m}\mathbf{S}_{n,m}(x)$ is in $C_0^\infty(\mathbb{X})$.

(3) If $f = \sum_{n,m=1}^\infty B_{n,m}\mathbf{S}_{n,m}$ is a *uniformly convergent* two-dimensional Fourier sine
series,[2] then $f \in C_0^\infty(\mathbb{X})$.   ◇

(E)

Exercise 5C.1

(i) Verify Examples 5C.2(b)–(e).

(ii) Show that $C_0^\infty(\mathbb{X})$ is a vector space.

(iii) Show that $C_0(\mathbb{X})$ is a vector space.   ♦

Arbitrary *nonhomogeneous Dirichlet boundary conditions* are imposed by fixing
some function $b : \partial\mathbb{X} \longrightarrow \mathbb{R}$, and then requiring

$$u(\mathbf{s}) = b(\mathbf{s}), \quad \text{for all } \mathbf{s} \in \partial\mathbb{X}. \tag{5C.1}$$

For example, the *classical Dirichlet problem* is to find a continuous function
$u : \mathbb{X} \longrightarrow \mathbb{R}$ satisfying the Dirichlet condition (5C.1), such that u also satisfies
Laplace's Equation: $\triangle u(\mathbf{x}) = 0$ for all $\mathbf{x} \in \text{int}(\mathbb{X})$.

Physical interpretations

Thermodynamic. Here u describes a stationary temperature distribution on $\mathbb{X}$, where
the temperature is *fixed* on the boundary. Different parts of the boundary may have
different temperatures, so heat may be flowing *through* the region $\mathbb{X}$ from warmer
boundary regions to cooler boundary regions, but the actual temperature distribution
within $\mathbb{X}$ is in equilibrium.

Electrostatic. Here u describes an electrostatic potential field within the region $\mathbb{X}$. The
voltage level on the boundaries is fixed (e.g. boundaries of $\mathbb{X}$ are wired up to batteries
which maintain a constant voltage). However, different parts of the boundary may
have different voltages (the boundary is not a perfect conductor).

Minimal surface. Here u describes a minimal-energy surface (e.g. a soap film). The
boundary of the surface is clamped in some position (e.g. the wire frame around
the soap film); the interior of the surface must adapt to find the minimal energy
configuration compatible with these boundary conditions. Minimal surfaces of low
curvature are well-approximated by harmonic functions.

For example, if $\mathbb{X} = [0, L]$, and $b(0)$ and $b(L)$ are two constants, then the
Dirichlet problem is to find $u : [0, L] \longrightarrow \mathbb{R}$ such that

$$u(0) = b(0), \ u(L) = b(L), \quad \text{and} \ \ \partial_x^2 u(x) = 0, \quad \text{for } 0 < x < L. \tag{5C.2}$$

---

[2] See §9A, p. 181.

That is, the temperature at the left-hand endpoint is fixed at $b(0)$, and at the right-hand endpoint it is fixed at $b(L)$. The unique solution to this problem is the function $u(x) = (b(L) - b(0))x/L + b(0)$.

Ⓔ **Exercise 5C.2**   Verify the unique solution to the Dirichlet problem given above.   ◆

5C(ii) Neumann boundary conditions

Suppose $\mathbb{X}$ is a domain with boundary $\partial\mathbb{X}$, and $u : \mathbb{X} \longrightarrow \mathbb{R}$ is some function. Then for any boundary point $s \in \partial\mathbb{X}$, we use '$\partial_\perp u(s)$' to denote the *outward normal derivative*[3] of u on the boundary. Physically, $\partial_\perp u(s)$ is the *rate of change in u as you leave $\mathbb{X}$ by passing through $\partial\mathbb{X}$ in a perpendicular direction*.

Example 5C.3
(a) If $\mathbb{X} = [0, 1]$, then $\partial_\perp u(0) = -\partial_x u(0)$ and $\partial_\perp u(1) = \partial_x u(1)$.
(b) Suppose $\mathbb{X} = [0, 1]^2 \subset \mathbb{R}^2$ is the unit square, and $(x, y) \in \partial\mathbb{X}$. There are four cases:

  - if $x = 0$ (left edge), then $\partial_\perp u(0, y) = -\partial_x u(0, y)$;
  - if $x = 1$ (right edge), then $\partial_\perp u(1, y) = \partial_x u(1, y)$;
  - if $y = 0$ (top edge), then $\partial_\perp u(x, 0) = -\partial_y u(x, 0)$;
  - if $y = 1$ (bottom edge), then $\partial_\perp u(x, 1) = \partial_y u(x, 1)$.

  (If more than one of these conditions is true – for example at $(0, 0)$ – then (x, y) is a corner, and $\partial_\perp u(x, y)$ is not well-defined.)
(c) Let $\mathbb{D} = \{(r, \theta); r < 1\}$ be the unit disk in the plane. Then $\partial\mathbb{D}$ is the set $\{(1, \theta); \theta \in [-\pi, \pi)\}$, and for any $(1, \theta) \in \partial\mathbb{D}$, $\partial_\perp u(1, \theta) = \partial_r u(1, \theta)$.
(d) Let $\mathbb{D} = \{(r, \theta); r < R\}$ be the disk of radius R. Then $\partial\mathbb{D} = \{(R, \theta); \theta \in [-\pi, \pi)\}$, and for any $(R, \theta) \in \partial\mathbb{D}$, $\partial_\perp u(R, \theta) = \partial_r u(R, \theta)$.
(e) Let $\mathbb{B} = \{(r, \phi, \theta); r < 1\}$ be the unit ball in $\mathbb{R}^3$. Then $\partial\mathbb{B} = \{(r, \phi, \theta); r = 1\}$ is the unit sphere. If $u(r, \phi, \theta)$ is a function in polar coordinates, then, for any boundary point $s = (1, \phi, \theta)$, $\partial_\perp u(s) = \partial_r u(s)$.
(f) Suppose $\mathbb{X} = \{(r, \theta, z); r \le R, 0 \le z \le L, -\pi \le \theta < \pi\}$, is the *finite cylinder*, and $(r, \theta, z) \in \partial\mathbb{X}$. There are three cases:

  - if $r = R$ (sides), then $\partial_\perp u(R, \theta, z) = \partial_r u(R, \theta, z)$;
  - if $z = 0$ (bottom disk), then $\partial_\perp u(r, \theta, 0) = -\partial_z u(r, \theta, 0)$;
  - if $z = L$ (top disk), then $\partial_\perp u(r, \theta, L) = \partial_z u(r, \theta, L)$.   ◇

We say that u satisfies *homogeneous Neumann boundary conditions* if

$$\partial_\perp u(s) = 0 \quad \text{for all } s \in \partial\mathbb{X}. \tag{5C.3}$$

---

[3] This is sometimes indicated as $\frac{\partial u}{\partial \mathbf{n}}$ or $\frac{\partial u}{\partial v}$, or as $\nabla u \bullet \vec{\mathbf{N}}$, or as $\nabla u \bullet \vec{\mathbf{n}}$.

Physical interpretations

Thermodynamic (heat, Laplace, or Poisson equations). Suppose u represents a temperature distribution. Recall that Fourier's Law of Heat Flow (§1A, p. 5) says that $\nabla u(\mathbf{s})$ is the speed and direction in which heat is flowing at $\mathbf{s}$. Recall that $\partial_\perp u(\mathbf{s})$ is the component of $\nabla u(\mathbf{s})$ which is perpendicular to $\partial\mathbb{X}$. Thus, homogeneous Neumann BC means that $\nabla u(\mathbf{s})$ is *parallel* to the boundary for all $\mathbf{s} \in \partial\mathbb{X}$. In other words *no heat is crossing the boundary*. This means that the boundary is a *perfect insulator*.

If u represents the concentration of a diffusing substance, then $\nabla u(\mathbf{s})$ is the flux of this substance at $\mathbf{s}$. Homogeneous Neumann boundary conditions mean that the boundary is an *impermeable barrier* to this substance.

Electrostatic (Laplace or Poisson equations). Suppose u represents an electric potential. Thus $\nabla u(\mathbf{s})$ is the *electric field* at $\mathbf{s}$. Homogeneous Neumann BC means that $\nabla u(\mathbf{s})$ is *parallel* to the boundary for all $\mathbf{s} \in \partial\mathbb{X}$; i.e. no field lines penetrate the boundary.

The set of *continuous* functions from $\mathbb{X}$ to $\mathbb{R}$ which satisfy homogeneous Neumann boundary conditions will be denoted $C_\perp(\mathbb{X})$. The set of *infinitely differentiable* functions from $\mathbb{X}$ to $\mathbb{R}$ which satisfy homogeneous Neumann boundary conditions will be denoted $C_\perp^\infty(\mathbb{X})$. Thus, for example,

$$C_\perp^\infty[0, L] = \left\{ f : [0, L] \longrightarrow \mathbb{R}; \ f \text{ is smooth, and } f'(0) = 0 = f'(L) \right\}.$$

Example 5C.4

(a) Let $\mathbb{X} = [0, 1]$, and let $f : [0, 1] \longrightarrow \mathbb{R}$ be defined by $f(x) = \frac{1}{2}x^2 - \frac{1}{3}x^3$ (see Figure 5C.3(a)). Then $f'(0) = 0 = f'(1)$, and f is smooth, so $f \in C_\perp^\infty[0, 1]$.

(b) Let $\mathbb{X} = [0, \pi]$.

  (1) For any $n \in \mathbb{N}$, let $\mathbf{C}_n(x) = \cos(n \cdot x)$ (see Figure 6D.1, p. 118). Then $\mathbf{C}_n \in C_\perp^\infty[0, \pi]$.

  (2) If $f(x) = 5\cos(x) - 3\cos(2x) + 7\cos(3x)$, then $f \in C_\perp^\infty[0, \pi]$. More generally, any finite sum $\sum_{n=1}^{N} A_n \mathbf{C}_n(x)$ (for some constants A_n) is in $C_\perp^\infty[0, \pi]$.

  (3) If $f(x) = \sum_{n=1}^{\infty} A_n \mathbf{C}_n(x)$ is a *uniformly convergent* Fourier cosine series,[4] and the derivative series $f'(x) = -\sum_{n=1}^{\infty} n A_n \mathbf{S}_n(x)$ is *also* uniformly convergent, then $f \in C_\perp^\infty[0, \pi]$.

(c) Let $\mathbb{D} = \{(r, \theta); \ r \leq 1\}$ be the unit disk.

  (1) Let $f : \mathbb{D} \longrightarrow \mathbb{R}$ be the 'witch's hat' of Figure 5C.3(b), defined: $f(r, \theta) := (1 - r)^2$. Then $\partial_\perp f \equiv 0$ on the boundary of the disk, so f satisfies Neumann boundary conditions. Also, f is continuous on $\mathbb{D}$; hence $f \in C_\perp(\mathbb{D})$. However, f is not smooth (it is nondifferentiable at zero), so $f \notin C_\perp^\infty(\mathbb{D})$.

---

[4] See §7B, p. 148.

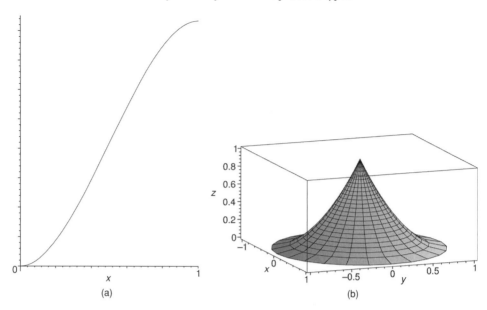

(a)   (b)

Figure 5C.3. (a) $f(x) = \frac{1}{2}x^2 - \frac{1}{3}x^3$ satsfies homogeneous Neumann boundary conditions on the interval $[0, 1]$. (b) $f(r, \theta) = (1 - r)^2$ satisfies homogeneous Neumann boundary conditions on the disk $\mathbb{D} = \{(r, \theta); r \leq 1\}$, but is not differentiable at zero.

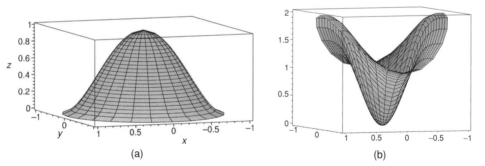

(a)   (b)

Figure 5C.4. (a) $f(r, \theta) = (1 - r^2)^2$ satisfies homogeneous Neumann boundary conditions on the disk, and is smooth everywhere. (b) $f(r, \theta) = (1 + \cos(\theta)^2) \cdot (1 - (1 - r^2)^4)$ does *not* satisfy homogeneous Neumann boundary conditions on the disk, and is not constant on the boundary.

(2) Let $f : \mathbb{D} \longrightarrow \mathbb{R}$ be the 'bell' of Figure 5C.4(a), defined: $f(r, \theta) := (1 - r^2)^2$. Then $\partial_\perp f \equiv 0$ on the boundary of the disk, and f is smooth everywhere on $\mathbb{D}$, so $f \in C_\perp^\infty(\mathbb{D})$.

(3) Let $f : \mathbb{D} \longrightarrow \mathbb{R}$ be the 'flower vase' of Figure 5C.4(b), defined: $f(r, \theta) := (1 + \cos(\theta)^2) \cdot (1 - (1 - r^2)^4)$. Then $\partial_\perp f \equiv 0$ on the boundary of the disk, and f is smooth everywhere on $\mathbb{D}$, so $f \in C_\perp^\infty(\mathbb{D})$. Note that, in this case, the angular derivative is nonzero, so f is not constant on the boundary of the disk.

(d) Let $\mathbb{X} = [0, \pi] \times [0, \pi]$ be the square of sidelength π.

   (1) For any $(n, m) \in \mathbb{N}^2$, let $\mathbf{C}_{n,m}(x, y) = \cos(nx) \cdot \cos(my)$ (see Figure 9A.2, p. 185). Then $\mathbf{C}_{n,m} \in \mathcal{C}^\infty_\perp(\mathbb{X})$.
   (2) If $f(x) = 5 \cos(x) \cos(2y) - 3 \cos(2x) \cos(7y) + 7 \cos(3x) \cos(y)$, then $f \in \mathcal{C}^\infty_\perp(\mathbb{X})$. More generally, any finite sum $\sum_{n=1}^{N} \sum_{m=1}^{M} A_{n,m} \mathbf{C}_{n,m}(x)$ (for some constants $A_{n,m}$) is in $\mathcal{C}^\infty_\perp(\mathbb{X})$.
   (3) More generally, if $f = \sum_{n,m=0}^{\infty} A_{n,m} \mathbf{C}_{n,m}$ is a *uniformly convergent* two-dimensional Fourier cosine series,[5] and the derivative series

$$\partial_x f(x, y) = - \sum_{n,m=0}^{\infty} n A_{n,m} \sin(nx) \cdot \cos(my),$$

$$\partial_y f(x, y) = - \sum_{n,m=0}^{\infty} m A_{n,m} \cos(nx) \cdot \sin(my),$$

are *also* uniformly convergent, then $f \in \mathcal{C}^\infty_\perp(\mathbb{X})$. $\diamond$

Exercise 5C.3   Verify Examples 5C.4(b)–(d).   ◆ Ⓔ

Arbitrary *nonhomogeneous Neumann boundary conditions* are imposed by fixing a function $b : \partial\mathbb{X} \longrightarrow \mathbb{R}$, and then requiring

$$\partial_\perp u(\mathbf{s}) = b(\mathbf{s}) \quad \text{for all } \mathbf{s} \in \partial\mathbb{X}. \tag{5C.4}$$

For example, the *classical Neumann problem* is to find a continuously differentiable function $u : \mathbb{X} \longrightarrow \mathbb{R}$ satisfying the Neumann condition (5C.4), such that u also satisfies Laplace's equation: $\Delta u(\mathbf{x}) = 0$ for all $\mathbf{x} \in \text{int}(\mathbb{X})$.

Physical interpretations

   Thermodynamic. Here u represents a temperature distribution, or the concentration of some diffusing material. Recall that Fourier's law (§1A, p. 5) says that $\nabla u(\mathbf{s})$ is the flux of heat (or material) at $\mathbf{s}$. Thus, for any $\mathbf{s} \in \partial\mathbb{X}$, the derivative $\partial_\perp u(\mathbf{s})$ is the flux of heat/material *across the boundary* at $\mathbf{s}$. The nonhomogeneous Neumann boundary condition $\partial_\perp u(\mathbf{s}) = b(\mathbf{s})$ means that heat (or material) is being 'pumped' across the boundary at a constant rate described by the function $b(\mathbf{s})$.

   Electrostatic. Here, u represents an electric potential. Thus $\nabla u(\mathbf{s})$ is the *electric field* at $\mathbf{s}$. Nonhomogeneous Neumann boundary conditions mean that the field vector perpendicular to the boundary is determined by the function $b(\mathbf{s})$.

5C(iii) *Mixed (or Robin) boundary conditions*

These are a combination of Dirichlet and Neumann-type conditions obtained as follows. Fix functions $b : \partial\mathbb{X} \longrightarrow \mathbb{R}$, and $h, h_\perp : \partial\mathbb{X} \longrightarrow \mathbb{R}$. Then $(h, h_\perp, b)$-*mixed*

---

[5] See §9A, p. 181.

boundary conditions are given as follows:

$$h(\mathbf{s}) \cdot u(\mathbf{s}) + h_\perp(\mathbf{s}) \cdot \partial_\perp u(\mathbf{s}) = b(x) \quad \text{for all } \mathbf{s} \in \partial \mathbb{X}. \tag{5C.5}$$

For example:

- *Dirichlet conditions* corresponds to $h \equiv 1$ and $h_\perp \equiv 0$.
- *Neumann conditions* corresponds to $h \equiv 0$ and $h_\perp \equiv 1$.
- *No* boundary conditions corresponds to $h \equiv h_\perp \equiv 0$.
- *Newton's law of cooling* reads as follows:

$$\partial_\perp u = c \cdot (u - T_E). \tag{5C.6}$$

This describes a situation where the boundary is an *imperfect conductor* (with conductivity constant c), and is immersed in a bath with ambient temperature T_E. Thus, heat leaks in or out of the boundary at a rate proportional to c times the difference between the internal temperature u and the external temperature T_E. Equation (5C.6) can be rewritten as follows:

$$c \cdot u - \partial_\perp u = b,$$

where $b = c \cdot T_E$. This is the mixed boundary equation (5C.5), with $h \equiv c$ and $h_\perp \equiv -1$.

- *Homogeneous* mixed boundary conditions take the following form:

$$h \cdot u + h_\perp \cdot \partial_\perp u \equiv 0.$$

The set of functions in $\mathcal{C}^\infty(\mathbb{X})$ satisfying this property will be denoted $\mathcal{C}^\infty_{h,h_\perp}(\mathbb{X})$. Thus, for example, if $\mathbb{X} = [0, L]$, and $h(0)$, $h_\perp(0)$, $h(L)$, and $h_\perp(L)$ are four constants, then

$$\mathcal{C}^\infty_{h,h_\perp}[0, L] = \left\{ f : [0, L] \longrightarrow \mathbb{R}; \quad \begin{array}{l} f \text{ is differentiable, } \ h(0)f(0) - h_\perp(0)f'(0) = 0 \\ \text{and } h(L)f(L) + h_\perp(L)f'(L) = 0. \end{array} \right\}.$$

Remarks

(a) Note that there is some redundancy in this formulation. Equation (5C.5) is equivalent to

$$k \cdot h(\mathbf{s}) \cdot u(\mathbf{s}) + k \cdot h_\perp(\mathbf{s}) \cdot \partial_\perp u(\mathbf{s}) = k \cdot b(\mathbf{s}),$$

for any constant $k \neq 0$. Normally we choose k so that at least one of the coefficients h or $h_\perp$ is equal to 1.

(b) Some authors (e.g. Pinsky (1998)) call this *general boundary conditions*, and, for mathematical convenience, write this as

$$\cos(\alpha)u + L \cdot \sin(\alpha)\partial_\perp u = T, \tag{5C.7}$$

where α and T are parameters. Here, the '$\cos(\alpha)$, $\sin(\alpha)$' coefficients of equation (5C.7) are just a mathematical 'gadget' to express concisely any weighted combination of Dirichlet and Neumann conditions. An expression of type (5C.5) can be transformed into one of type (5C.7) as follows. Let

$$\alpha := \arctan\left(\frac{h_\perp}{L \cdot h}\right)$$

(if $h = 0$, then set $\alpha = \pi/2$) and let

$$T := b\frac{\cos(\alpha) + L\sin(\alpha)}{h + h_\perp}.$$

Going the other way is easier; simply define $h := \cos(\alpha)$, $h_\perp := L \cdot \sin(\alpha)$, and $T := b$.

5C(iv) Periodic boundary conditions

Periodic boundary conditions means that function u 'looks the same' on opposite edges of the domain. For example, if we are solving a PDE on the *interval* $[-\pi, \pi]$, then periodic boundary conditions are imposed by requiring

$$u(-\pi) = u(\pi) \quad \text{and} \quad u'(-\pi) = u'(\pi).$$

Interpretation #1 Pretend that u is actually a small piece of an infinitely extended, periodic function $\tilde{u} : \mathbb{R} \longrightarrow \mathbb{R}$, where, for any $x \in \mathbb{R}$ and $n \in \mathbb{Z}$, we have:

$$\tilde{u}(x + 2n\pi) = u(x).$$

Thus u must have the same value – and the same derivative – at x and $x + 2n\pi$, for any $x \in \mathbb{R}$. In particular, u must have the same value and derivative at $-\pi$ and π. This explains the name 'periodic boundary conditions'.

Interpretation #2 Suppose you 'glue together' the left and right ends of the interval $[-\pi, \pi]$ (i.e. glue $-\pi$ to π). Then the interval looks like a circle (where $-\pi$ and π actually become the 'same' point). Thus u must have the same value – and the same derivative – at $-\pi$ and π.

Example 5C.5
(a) $u(x) = \sin(x)$ and $v(x) = \cos(x)$ have periodic boundary conditions.
(b) For any $n \in \mathbb{N}$, the functions $S_n(x) = \sin(nx)$ and $C_n(x) = \cos(nx)$ have periodic boundary conditions. (See Figure 6D.1, p. 118.)
(c) $\sin(3x) + 2\cos(4x)$ has periodic boundary conditions.
(d) If $u_1(x)$ and $u_2(x)$ have periodic boundary conditions, and c_1, c_2 are any constants, then $u(x) = c_1 u_1(x) + c_2 u_2(x)$ also has periodic boundary conditions. ◇

Exercise 5C.4 Verify these examples. ♦ Ⓔ

On the *square* $[-\pi, \pi] \times [-\pi, \pi]$, periodic boundary conditions are imposed by requiring:

(P1) $u(x, -\pi) = u(x, \pi)$ and $\partial_y u(x, -\pi) = \partial_y u(x, \pi)$, for all $x \in [-\pi, \pi]$.
(P2) $u(-\pi, y) = u(\pi, y)$ and $\partial_x u(-\pi, y) = \partial_x u(\pi, y)$ for all $y \in [-\pi, \pi]$.

Figure 5C.5. If we 'glue' the opposite edges of a square together, we get a torus.

Interpretation #1 Pretend that u is actually a small piece of an infinitely extended, doubly periodic function $\tilde{u} : \mathbb{R}^2 \longrightarrow \mathbb{R}$, where, for every $(x, y) \in \mathbb{R}^2$ and every $n, m \in \mathbb{Z}$, we have

$$\tilde{u}(x + 2n\pi, y + 2m\pi) = u(x, y).$$

Ⓔ **Exercise 5C.5** Explain how conditions (P1) and (P2) arise naturally from this interpretation.   ◆

Interpretation #2 Glue the top edge of the square to the bottom edge, and the right edge to the left edge. In other words, pretend that the square is really a *torus* (Figure 5C.5).

Example 5C.6
(a) The functions $u(x, y) = \sin(x)\sin(y)$ and $v(x, y) = \cos(x)\cos(y)$ have periodic boundary conditions. So do the functions $w(x, y) = \sin(x)\cos(y)$ and $w(x, y) = \cos(x)\sin(y)$.
(b) For any $(n, m) \in \mathbb{N}^2$, the functions $\mathbf{S}_{n,m}(x) = \sin(nx)\sin(my)$ and $\mathbf{C}_{n,m}(x) = \cos(nx)\cos(mx)$ have periodic boundary conditions. (See Figure 9A.2, p. 185.)
(c) $\sin(3x)\sin(2y) + 2\cos(4x)\cos(7y)$ has periodic boundary conditions.
(d) If $u_1(x, y)$ and $u_2(x, y)$ have periodic boundary conditions, and c_1, c_2 are any constants, then $u(x, y) = c_1 u_1(x, y) + c_2 u_2(x, y)$ also has periodic boundary conditions.   ◇

Ⓔ **Exercise 5C.6** Verify these examples.   ◆

On the D-dimensional *cube* $[-\pi, \pi]^D$, we require, for $d = 1, 2, \ldots, D$ and all $x_1, \ldots, x_D \in [-\pi, \pi]$, that

$$u(x_1, \ldots, x_{d-1}, -\pi, x_{d+1}, \ldots, x_D) = u(x_1, \ldots, x_{d-1}, \pi, x_{d+1}, \ldots, x_D)$$

and

$$\partial_d u(x_1, \ldots, x_{d-1}, -\pi, x_{d+1}, \ldots, x_D) = \partial_d u(x_1, \ldots, x_{d-1}, \pi, x_{d+1}, \ldots, x_D).$$

Again, the idea is that we are identifying $[-\pi, \pi]^D$ with the D-dimensional *torus*. The space of all functions satisfying these conditions will be denoted $\mathcal{C}^\infty_{\text{per}}[-\pi, \pi]^D$. Thus, for example,

$$\mathcal{C}^\infty_{\text{per}}[-\pi, \pi] = \left\{ f : [-\pi, \pi] \longrightarrow \mathbb{R}; \quad f \text{ is differentiable,} \right.$$

$$\left. f(-\pi) = f(\pi) \text{ and } f'(-\pi) = f'(\pi) \right\}$$

$$\mathcal{C}^\infty_{\text{per}}[-\pi, \pi]^2 = \left\{ f : [-\pi, \pi] \times [-\pi, \pi] \longrightarrow \mathbb{R}; \quad f \text{ is differentiable,} \right.$$

$$\left. \text{and satisfies (P1) and (P2) above} \right\}.$$

5D Uniqueness of solutions

Prerequisites: §1B, §2B, §1C, §5B, §5C.
Prerequisites (for proofs): §1E, Appendix E(iii), Appendix G.

Differential equations are interesting primarily because they can be used to express the laws governing physical phenomena (e.g. heat flow, wave motion, electrostatics, etc.). By specifying particular initial conditions and boundary conditions, we try to encode mathematically the physical conditions, constraints, and external influences which are present in a particular situation. A solution to the differential equation which satisfies these initial/boundary conditions thus constitutes a *prediction* about what will occur under these physical conditions.

However, this strategy can only succeed if there is a *unique* solution to the differential equation with particular initial/boundary conditions. If there are many mathematically correct solutions, then we cannot make a clear prediction about *which* of them will really occur. Sometimes we can reject some solutions as being 'unphysical' (e.g. they are nondifferentiable, or discontinuous, or contain unacceptable infinities, or predict negative values for a necessarily positive quantity like density). However, these notions of 'unphysicality' really just represent further mathematical constraints which we are implicitly imposing on the solution. If multiple solutions still exist, we should try to impose further constraints (i.e. construct a more detailed or well-specified model) until we get a unique solution. Thus, the question of *uniqueness of solutions* is extremely important in the general theory of differential equations (both ordinary and partial). In this section, we will establish sufficient conditions for the uniqueness of solutions to I/BVPs for the Laplace, Poisson, heat, and wave equations.

Let $\mathcal{S} \subset \mathbb{R}^D$. We say that $\mathcal{S}$ is a *smooth graph* if there is an open subset $\mathbb{U} \subset \mathbb{R}^{D-1}$, a function $f : \mathbb{U} \longrightarrow \mathbb{R}$, and some $d \in [1 \dots D]$, such that $\mathcal{S}$ 'looks like' the graph of the function f, plotted over the domain $\mathbb{U}$, with the value of f plotted in the dth coordinate. In other words,

$$S = \{(u_1, \ldots, u_{d-1}, y, u_d, \ldots, u_{D-1}); (u_1, \ldots, u_{D-1}) \in \mathbb{U},$$

$$y = f(u_1, \ldots, u_{D-1})\}.$$

Intuitively, this means that S looks like a smooth surface (oriented 'roughly perpendicular' to the dth dimension). More generally, if $S \subset \mathbb{R}^D$, we say that S is a *smooth hypersurface* if, for each $\mathbf{s} \in S$, there exists some $\epsilon > 0$ such that $\mathbb{B}(\mathbf{s}, \epsilon) \cap S$ is a smooth graph.

Example 5D.1

(a) Let $\mathbb{P} \subset \mathbb{R}^D$ be any $(D-1)$-dimensional hyperplane; then $\mathbb{P}$ is a smooth hypersurface.

(b) Let $\mathbb{S}^1 := \{\mathbf{s} \in \mathbb{R}^2; |\mathbf{s}| = 1\}$ be the unit circle in $\mathbb{R}^2$. Then $\mathbb{S}^1$ is a smooth hypersurface in $\mathbb{R}^2$.

(c) Let $\mathbb{S}^2 := \{\mathbf{s} \in \mathbb{R}^3; |\mathbf{s}| = 1\}$ be the unit sphere in $\mathbb{R}^3$. Then $\mathbb{S}^2$ is a smooth hypersurface in $\mathbb{R}^3$.

(d) Let $\mathbb{S}^{D-1} := \{\mathbf{s} \in \mathbb{R}^D; |\mathbf{s}| = 1\}$ be the unit hypersphere in $\mathbb{R}^D$. Then $\mathbb{S}^{D-1}$ is a smooth hypersurface in $\mathbb{R}^D$.

(e) Let $S \subset \mathbb{R}^D$ be any smooth hypersurface, and let $\mathbb{U} \subset \mathbb{R}^D$ be an open set. Then $S \cap \mathbb{U}$ is also a smooth hypersurface (if it is nonempty).   ◇

ⓔ **Exercise 5D.1**   Verify these examples.   ♦

A domain $\mathbb{X} \subset \mathbb{R}^D$ has a *piecewise smooth boundary* if $\partial\mathbb{X}$ is a finite union of smooth hypersurfaces. If $u : \mathbb{X} \longrightarrow \mathbb{R}$ is some differentiable function, then this implies that the normal derivative $\partial_\perp u(\mathbf{s})$ is well-defined for $\mathbf{s} \in \partial\mathbb{X}$, except for those $\mathbf{s}$ on the (negligible) regions where two or more of these smooth hypersurfaces intersect. This means that it is meaningful to impose Neumann boundary conditions on u. It also means that certain methods from vector calculus can be applied to u (see Appendix E(iii), p. 564).

Example 5D.2   Every domain in Example 5C.1, p. 75, has a piecewise smooth boundary.   ◇

ⓔ **Exercise 5D.2**   Verify Example 5D.2.   ♦

Indeed, every domain we will consider in this book will have a piecewise smooth boundary, as does any domain which is likely to arise in any physically realistic model. Hence, it suffices to obtain uniqueness results for such domains.

5D(i) Uniqueness for the Laplace and Poisson equations

Let $\mathbb{X} \subset \mathbb{R}^D$ be a domain and let $u : \mathbb{X} \longrightarrow \mathbb{R}$. We say that u is *continuous and harmonic* on $\mathbb{X}$ if u is continuous on $\mathbb{X}$ and $\triangle u(\mathbf{x}) = 0$ for all $\mathbf{x} \in \text{int}(\mathbb{X})$.

Lemma 5D.3 Solution uniqueness for Laplace equation; homogeneous BC

Let $\mathbb{X} \subset \mathbb{R}^D$ be a bounded domain, and suppose $u : \mathbb{X} \longrightarrow \mathbb{R}$ is continuous and harmonic on $\mathbb{X}$. Then various homogeneous boundary conditions constrain the solution as follows.

(a) (Homogeneous Dirichlet BC) *If $u(\mathbf{s}) = 0$ for all $\mathbf{s} \in \partial\mathbb{X}$, then u must be the constant 0 function; i.e. $u(\mathbf{x}) = 0$, for all $\mathbf{x} \in \mathbb{X}$.*

(b) (Homogeneous Neumann BC) *Suppose $\mathbb{X}$ has a piecewise smooth boundary. If $\partial_\perp u(\mathbf{s}) = 0$ for all $\mathbf{s} \in \partial\mathbb{X}$, then u must be a constant; i.e. $u(\mathbf{x}) = C$, for all $\mathbf{x} \in \mathbb{X}$.*

(c) (Homogeneous Robin BC) *Suppose $\mathbb{X}$ has a piecewise smooth boundary, and let $h, h_\perp : \partial\mathbb{X} \longrightarrow \mathbb{R}_{\neq}$ be two other continuous non-negative functions such that $h(\mathbf{s}) + h_\perp(\mathbf{s}) > 0$ for all $\mathbf{s} \in \partial\mathbb{X}$. If $h(\mathbf{s})u(\mathbf{s}) + h_\perp(\mathbf{s})\partial_\perp u(\mathbf{s}) = 0$ for all $\mathbf{s} \in \partial\mathbb{X}$, then u must be a constant function.*

> *Furthermore, if h is nonzero somewhere on $\partial\mathbb{X}$, then $u(\mathbf{x}) = 0$, for all $\mathbf{x} \in \mathbb{X}$.*

Proof (a) If $u : \mathbb{X} \longrightarrow \mathbb{R}$ is harmonic, then the maximum principle (Corollary 1E.2, p. 19) says that any maximum/minimum of u occurs somewhere on $\partial\mathbb{X}$. But $u(\mathbf{s}) = 0$ for all $\mathbf{s} \in \partial\mathbb{X}$; thus, $\max_{\mathbb{X}}(u) = 0 = \min_{\mathbb{X}}(u)$; thus, $u \equiv 0$.

(If $\mathbb{X}$ has a piecewise smooth boundary, then another proof of (a) arises by setting $h \equiv 1$ and $h_\perp \equiv 0$ in part (c).)

To prove (b), set $h \equiv 0$ and $h_\perp \equiv 1$ in part (c).

To prove (c), we will use Green's formula. We begin with the following claim.

Claim 1 *For all $\mathbf{s} \in \partial\mathbb{X}$, we have $u(\mathbf{s}) \cdot \partial_\perp u(\mathbf{s}) \leq 0$.*

Proof The homogeneous Robin boundary conditions say $h(\mathbf{s})u(\mathbf{s}) + h_\perp(\mathbf{s})\partial_\perp u(\mathbf{s}) = 0$. Multiplying by $u(\mathbf{s})$, we get

$$h(\mathbf{s})u^2(\mathbf{s}) + u(\mathbf{s})h_\perp(\mathbf{s})\partial_\perp u(\mathbf{s}) = 0. \tag{5D.1}$$

If $h_\perp(\mathbf{s}) = 0$, then $h(\mathbf{s})$ must be nonzero, and equation (5D.1) reduces to $h(\mathbf{s})u^2(\mathbf{s}) = 0$, which means $u(\mathbf{s}) = 0$, which means $u(\mathbf{s}) \cdot \partial_\perp u(\mathbf{s}) \leq 0$, as desired.

If $h_\perp(\mathbf{s}) \neq 0$, then we can rearrange equation (5D.1) to get

$$u(\mathbf{s}) \cdot \partial_\perp u(\mathbf{s}) = \underset{(*)}{\underbrace{\frac{-h(\mathbf{s})u^2(\mathbf{s})}{h_\perp(\mathbf{s})}}} \leq 0,$$

where $(*)$ is because because $h(\mathbf{s}), h_\perp(\mathbf{s}) \geq 0$ by hypothesis, and of course $u^2(\mathbf{s}) \geq 0$. The claim follows. $\diamond$ Claim 1

Now, if u is harmonic, then u is infinitely differentiable, by Proposition 1E.4, p. 20. Thus, we can apply vector calculus techniques from Appendix E(iii). We have

$$0 \underset{(*)}{\geq} \int_{\partial\mathbb{X}} u(\mathbf{s}) \cdot \partial_\perp u(\mathbf{s}) d\mathbf{s} \underset{(\dagger)}{=} \int_{\mathbb{X}} u(\mathbf{x}) \bigtriangleup u(\mathbf{x}) + |\nabla u(\mathbf{x})|^2 \, d\mathbf{x}$$

$$\underset{(\ddagger)}{=} \int_{\mathbb{X}} |\nabla u(\mathbf{x})|^2 \, d\mathbf{x} \underset{(\diamond)}{\geq} 0. \tag{5D.2}$$

Here, $(*)$ is by Claim 1, $(\dagger)$ is by Green's formula (Theorem E.5(b), p. 567), $(\ddagger)$ is because $\triangle u \equiv 0$, and $(\diamond)$ is because $|\nabla u(\mathbf{x})|^2 \geq 0$ for all $\mathbf{x} \in \mathbb{X}$.

The inequalities (5D.2) imply that

$$\int_{\mathbb{X}} |\nabla u(\mathbf{x})|^2 \, d\mathbf{x} = 0.$$

This implies, however, that $|\nabla u(\mathbf{x})| = 0$ for all $\mathbf{x} \in \mathbb{X}$, which means $\nabla u \equiv 0$, which means u is a constant on $\mathbb{X}$, as desired.

Now, if $\nabla u \equiv 0$, then clearly $\partial_\perp u(\mathbf{s}) = 0$ for all $\mathbf{s} \in \partial \mathbb{X}$. Thus, the Robin boundary conditions reduce to $h(\mathbf{s})u(\mathbf{s}) = 0$. If $h(\mathbf{s}) \neq 0$ for some $\mathbf{s} \in \partial \mathbb{X}$, then we get $u(\mathbf{s}) = 0$. But since u is a constant, this means that $u \equiv 0$. $\qquad \square$

One of the nice things about *linear* differential equations is that linearity enormously simplifies the problem of solution uniqueness. First we show that the only solution satisfying *homogeneous* boundary conditions (and, if applicable, *zero* initial conditions) is the constant zero function (as in Lemma 5D.3 above). Then it is easy to deduce uniqueness for arbitrary initial/boundary conditions.

Corollary 5D.4 Solution uniqueness: Laplace equation, nonhomogeneous BC

Let $\mathbb{X} \subset \mathbb{R}^D$ be a bounded domain, and let $b : \partial \mathbb{X} \longrightarrow \mathbb{R}$ be continuous.

(a) *There exists at most one continuous, harmonic function $u : \mathbb{X} \longrightarrow \mathbb{R}$ which satisfies the* nonhomogeneous Dirichlet BC $u(\mathbf{s}) = b(\mathbf{s})$ *for all $\mathbf{s} \in \partial \mathbb{X}$.*

(b) *Suppose $\mathbb{X}$ has a piecewise smooth boundary.*

    (i) *If $\int_{\partial \mathbb{X}} b(\mathbf{s}) d\mathbf{s} \neq 0$, then there is no continuous harmonic function $u : \mathbb{X} \longrightarrow \mathbb{R}$ which satisfies the* nonhomogeneous Neumann BC $\partial_\perp u(\mathbf{s}) = b(\mathbf{s})$ *for all $\mathbf{s} \in \partial \mathbb{X}$.*

    (ii) *Suppose $\int_{\partial \mathbb{X}} b(\mathbf{s}) d\mathbf{s} = 0$. If $u_1, u_2 : \mathbb{X} \longrightarrow \mathbb{R}$ are two continuous harmonic functions which both satisfy the* nonhomogeneous Neumann BC $\partial_\perp u(\mathbf{s}) = b(\mathbf{s})$ *for all $\mathbf{s} \in \partial \mathbb{X}$, then $u_1 = u_2 + C$ for some constant C.*

(c) *Suppose $\mathbb{X}$ has a piecewise smooth boundary, and let $h, h_\perp : \partial \mathbb{X} \longrightarrow \mathbb{R}_+$ be two other continuous nonnegative functions such that $h(\mathbf{s}) + h_\perp(\mathbf{s}) > 0$ for all $\mathbf{s} \in \partial \mathbb{X}$. If $u_1, u_2 : \mathbb{X} \longrightarrow \mathbb{R}$ are two continuous harmonic functions which both satisfy the* nonhomogeneous Robin BC $h(\mathbf{s})u(\mathbf{s}) + h_\perp(\mathbf{s})\partial_\perp u(\mathbf{s}) = b(\mathbf{s})$ *for all $\mathbf{s} \in \partial \mathbb{X}$, then $u_1 = u_2 + C$ for some constant C. Furthermore, if h is nonzero somewhere on $\partial \mathbb{X}$, then $u_1 = u_2$.*

  Ⓔ *Proof* **Exercise 5D.3.** *Hint:* For (a), (c), and (b)(ii), suppose that $u_1, u_2 : \mathbb{X} \longrightarrow \mathbb{R}$ are two continuous harmonic functions with the desired nonhomogeneous boundary conditions. Then $(u_1 - u_2)$ is a continuous harmonic function satisfying *homogeneous* boundary conditions of the same kind; now apply the appropriate part of Lemma 5D.3 to conclude that $(u_1 - u_2)$ is zero or a constant.

For (b)(i), use Green's formula (Theorem E.5(a), p. 567). $\qquad \square$

Exercise 5D.4 Let $\mathbb{X} = \mathbb{D} = \{(r, \theta); \ \theta \in [-\pi, \pi), \ r \leq 1\}$ be the closed unit disk (E)
(in polar coordinates). Consider the function $h : \mathbb{D} \longrightarrow \mathbb{R}$ defined by $h(r, \theta) = \log(r)$. In Cartesian coordinates, h has the form $h(x, y) = \log(x^2 + y^2)$ (see Figure 1C.1(a), p. 12). In Example 1C.2 we observed that h is harmonic. But h satisfies homogeneous Dirichlet BC on $\partial\mathbb{D}$, so it seems to be a counterexample to Lemma 5D.3(a). Also, $\partial_\perp h(x) = 1$ for all $x \in \partial\mathbb{D}$, so h seems to be a counterexample to Corollary 5D.4(b)(i).

Why is this function *not* a counterexample to Lemma 5D.3 or Corollary 5D.4(b)(i)? ◆

Theorem 5D.5 Solution uniqueness: Poisson equation, nonhomogeneous BC

Let $\mathbb{X} \subset \mathbb{R}^D$ *be a bounded domain with a piecewise smooth boundary. Let* $q : \mathbb{X} \longrightarrow \mathbb{R}$ *be a continuous function* (e.g. describing an electric charge or heat source), *and let* $b : \partial\mathbb{X} \longrightarrow \mathbb{R}$ *be another continuous function* (a boundary condition). *Then there is at most one continuous function* $u : \mathbb{X} \longrightarrow \mathbb{R}$ *satisfying the Poisson equation* $\Delta u = q$, *and satisfying either of the following nonhomogeneous boundary conditions:*

(a) (Nonhomogeneous Dirichlet BC) $u(\mathbf{s}) = b(\mathbf{s})$ *for all* $\mathbf{s} \in \partial\mathbb{X}$.
(b) (Nonhomogeneous Robin BC) $h(\mathbf{s})u(\mathbf{s}) + h_\perp(\mathbf{s})\partial_\perp u(\mathbf{s}) = b(\mathbf{s})$ *for all* $\mathbf{s} \in \partial\mathbb{X}$, *where* $h, h_\perp : \partial\mathbb{X} \longrightarrow \mathbb{R}_{\not\perp}$ *are two other nonnegative functions, and* h *is nontrivial.*

Furthermore, if u_1 *and* u_2 *are two functions satisfying* $\Delta u = q$, *and also satisfying*

(c) (nonhomogeneous Neumann BC) $\partial_\perp u(\mathbf{s}) = b(\mathbf{s})$ *for all* $\mathbf{s} \in \partial\mathbb{X}$,

then $u_1 = u_2 + C$, *where* C *is a constant.*

Proof Suppose u_1 and u_2 were two continuous functions satisfying one of (a) or (b), and such that $\Delta u_1 = q = \Delta u_2$. Let $u = u_1 - u_2$. Then u is continuous, harmonic, and satisfies one of (a) or (c) in Lemma 5D.3. Thus, $u \equiv 0$. But this means that $u_1 \equiv u_2$. Hence, there can be at most one solution. The proof for (c) is **Exercise 5D.5**. □ (E)

5D(ii) *Uniqueness for the heat equation*

Throughout this section, if $u : \mathbb{X} \times \mathbb{R}_{\not\perp} \longrightarrow \mathbb{R}$ is a time-varying scalar field, and $t \in \mathbb{R}_{\not\perp}$, then define the function $u_t : \mathbb{X} \longrightarrow \mathbb{R}$ by $u_t(\mathbf{x}) := u(\mathbf{x}; t)$, for all $\mathbf{x} \in \mathbb{X}$. (*Note:* u_t does *not* denote the time-derivative.)

If $f : \mathbb{X} \longrightarrow \mathbb{R}$ is any integrable function, then the L^2-*norm* of f is defined

$$\|f\|_2 := \left(\int_\mathbb{X} |f(\mathbf{x})|^2 \, d\mathbf{x} \right)^{1/2}.$$

(See §6B for more information.) We begin with a result which reinforces our intuition that the heat equation resembles 'melting' or 'erosion'.

Lemma 5D.6 L^2-norm decay for heat equation

Let $\mathbb{X} \subset \mathbb{R}^D$ be a bounded domain with a piecewise smooth boundary. Suppose that $u : \mathbb{X} \times \mathbb{R}_{\not+} \longrightarrow \mathbb{R}$ satisfies the following three conditions:

(a) *(Regularity) u is continuous on $\mathbb{X} \times \mathbb{R}_{\not+}$, and $\partial_t u$ and $\partial_1^2 u, \dots, \partial_D^2 u$ are continuous on $\mathrm{int}\,(\mathbb{X}) \times \mathbb{R}_+$;*

(b) *(Heat equation) $\partial_t u = \Delta u$;*

(c) *(Homogeneous Dirichlet/Neumann BC)[6] For all $s \in \partial\mathbb{X}$ and $t \in \mathbb{R}_{\not+}$, either $u_t(s) = 0$ or $\partial_\perp u_t(s) = 0$.*

Define the function $E : \mathbb{R}_{\not+} \longrightarrow \mathbb{R}_{\not+}$ by

$$E(t) := \|u_t\|_2^2 = \int_{\mathbb{X}} |u_t(\mathbf{x})|^2 \, d\mathbf{x}, \quad \text{for all } t \in \mathbb{R}_{\not+}. \tag{5D.3}$$

Then E is differentiable and nonincreasing; that is, $E'(t) \leq 0$ for all $t \in \mathbb{R}_{\not+}$.

Proof For any $\mathbf{x} \in \mathbb{X}$ and $t \in \mathbb{R}_{\not+}$, we have

$$\partial_t |u_t(\mathbf{x})|^2 \underset{(*)}{=} 2u_t(\mathbf{x}) \cdot \partial_t u_t(\mathbf{x}) \underset{(\dagger)}{=} 2u_t(\mathbf{x}) \cdot \Delta u_t(\mathbf{x}), \tag{5D.4}$$

where $(*)$ is the Leibniz rule, and $(\dagger)$ is because u satisfies the heat equation by hypothesis (b). Thus,

$$E'(t) \underset{(*)}{=} \int_{\mathbb{X}} \partial_t |u_t(\mathbf{x})|^2 \, d\mathbf{x} \underset{(\dagger)}{=} 2 \int_{\mathbb{X}} u_t(\mathbf{x}) \cdot \Delta u_t(\mathbf{x}) d\mathbf{x}. \tag{5D.5}$$

Here $(*)$ comes from differentiating the integral (5D.3) using Proposition G.1, p. 571. Meanwhile, $(\dagger)$ is, by equation (5D.4).

Claim 1 *For all $t \in \mathbb{R}_{\not+}$,*

$$\int_{\mathbb{X}} u_t(\mathbf{x}) \cdot \Delta u_t(\mathbf{x}) d\mathbf{x} = -\int_{\mathbb{X}} \|\nabla u_t(\mathbf{x})\|^2 \, d\mathbf{x}.$$

Proof For all $s \in \partial\mathbb{X}$, either $u_t(s) = 0$ or $\partial_\perp u_t(s) = 0$ by hypothesis (c). But $\partial_\perp u_t(s) = \nabla u_t(s) \bullet \vec{\mathbf{N}}(s)$ (where $\vec{\mathbf{N}}(s)$ is the unit normal vector at s), so this implies that $u_t(s) \cdot \nabla u_t(s) \bullet \vec{\mathbf{N}}(s) = 0$ for all $s \in \partial\mathbb{X}$. Thus,

$$0 = \int_{\partial\mathbb{X}} u_t(s) \cdot \nabla u_t(s) \bullet \vec{\mathbf{N}}(s) ds \underset{(*)}{=} \int_{\mathbb{X}} \mathrm{div}\,(u_t \cdot \nabla u_t)(\mathbf{x}) d\mathbf{x}$$

$$\underset{(\dagger)}{=} \int_{\mathbb{X}} \left(u_t \cdot \mathrm{div}\,\nabla u_t + \nabla u_t \bullet \nabla u_t \right) (\mathbf{x}) d\mathbf{x}$$

$$\underset{(\ddagger)}{=} \int_{\mathbb{X}} u_t(\mathbf{x}) \cdot \Delta u_t(\mathbf{x}) d\mathbf{x} + \int_{\mathbb{X}} \|\nabla u_t(\mathbf{x})\|^2 \, d\mathbf{x}.$$

---

[6] Note that this allows different boundary points to satisfy different homogeneous boundary conditions at different times.

Here, (∗) is the divergence theorem, Theorem E.4, p. 566, (†) is by the Leibniz rule for divergences (Proposition E.2(b), p. 564), and (‡) is because $\mathrm{div}\, \nabla u = \triangle u$, while $\nabla u_t \bullet \nabla u_t = \|\nabla u_t(\mathbf{x})\|^2$. We thus have

$$\int_{\mathbb{X}} u_t \cdot \triangle u_t + \int_{\mathbb{X}} \|\nabla u_t\|^2 = 0.$$

Rearranging this equation yields the claim. $\diamondsuit_{\text{Claim 1}}$

Applying Claim 1 to equation (5D.5), we get

$$E'(t) = -2 \int_{\mathbb{X}} \|\nabla u_t(\mathbf{x})\|^2 \, d\mathbf{x} \leq 0$$

because $\|\nabla u_t(\mathbf{x})\|^2 \geq 0$ for all $\mathbf{x} \in \mathbb{X}$.   □

Lemma 5D.7 Solution uniqueness for heat equation; homogeneous I/BC
Let $\mathbb{X} \subset \mathbb{R}^D$ be a bounded domain with a piecewise smooth boundary. Suppose that $u : \mathbb{X} \times \mathbb{R}_{\not\!+} \longrightarrow \mathbb{R}$ satisfies the following four conditions:

(a) *(Regularity) u is continuous on $\mathbb{X} \times \mathbb{R}_{\not\!+}$, and $\partial_t u$ and $\partial_1^2 u, \ldots, \partial_D^2 u$ are continuous on $\mathrm{int}\,(\mathbb{X}) \times \mathbb{R}_+$;*
(b) *(Heat equation) $\partial_t u = \triangle u$;*
(c) *(Zero initial condition) $u_0(\mathbf{x}) = 0$ for all $\mathbf{x} \in \mathbb{X}$;*
(d) *(Homogeneous Dirichlet/Neumann BC)[7] For all $\mathbf{s} \in \partial\mathbb{X}$ and $t \in \mathbb{R}_{\not\!+}$, either $u_t(\mathbf{s}) = 0$ or $\partial_\perp u_t(\mathbf{s}) = 0$.*

Then u must be the constant 0 function: $u \equiv 0$.

Proof   Define $E : \mathbb{R}_{\not\!+} \longrightarrow \mathbb{R}_{\not\!+}$ as in Lemma 5D.6. Then E is a nonincreasing function. But $E(0) = 0$, because $u_0 \equiv 0$ by hypothesis (c). Thus, $E(t) = 0$ for all $t \in \mathbb{R}_{\not\!+}$. Thus, we must have $u_t \equiv 0$ for all $t \in \mathbb{R}_{\not\!+}$.   □

Theorem 5D.8 Uniqueness: forced heat equation, nonhomogeneous I/BC
Let $\mathbb{X} \subset \mathbb{R}^D$ be a bounded domain with a piecewise smooth boundary. Let $\mathcal{I} : \mathbb{X} \longrightarrow \mathbb{R}$ be a continuous function (describing an initial condition), and let $b : \partial\mathbb{X} \times \mathbb{R}_{\not\!+} \longrightarrow \mathbb{R}$, and $h, h_\perp : \partial\mathbb{X} \times \mathbb{R}_{\not\!+} \longrightarrow \mathbb{R}$ be three other continuous functions (describing time-varying boundary conditions). Let $f : \mathrm{int}\,(\mathbb{X}) \times \mathbb{R}_{\not\!+} \longrightarrow \mathbb{R}$ be another continuous function (describing exogenous heat being 'forced' into or out of the system). Then there is at most one solution function $u : \mathbb{X} \times \mathbb{R}_{\not\!+} \longrightarrow \mathbb{R}$ satisfying the following four conditions:

(a) *(Regularity) u is continuous on $\mathbb{X} \times \mathbb{R}_{\not\!+}$, and $\partial_t u$ and $\partial_1^2 u, \ldots, \partial_D^2 u$ are continuous on $\mathrm{int}\,(\mathbb{X}) \times \mathbb{R}_+$;*
(b) *(Heat equation with forcing) $\partial_t u = \triangle u + f$;*

---

[7] Note that this allows different boundary points to satisfy different homogeneous boundary conditions at different times.

(c) (Initial condition) $u(\mathbf{x}, 0) = \mathcal{I}(\mathbf{x})$ *for all* $\mathbf{x} \in \mathbb{X}$;

(d) (Nonhomogeneous mixed BC)[8] $h(\mathbf{s}, t) \cdot u_t(\mathbf{s}) + h_\perp(\mathbf{s}, t) \cdot \partial_\perp u_t(\mathbf{s}) = b(x, t)$, *for all* $\mathbf{s} \in \partial \mathbb{X}$ *and* $t \in \mathbb{R}_{\not{}}$.

Proof Suppose u_1 and u_2 were two functions satisfying all of (a)–(d). Let $u = u_1 - u_2$. Then u satisfies all of (a)–(d) in Lemma 5D.7. Thus, $u \equiv 0$. But this means that $u_1 \equiv u_2$. Hence, there can be at most one solution.   □

5D(iii) Uniqueness for the wave equation

Throughout this section, if $u : \mathbb{X} \times \mathbb{R}_{\not{}} \longrightarrow \mathbb{R}$ is a time-varying scalar field, and $t \in \mathbb{R}_{\not{}}$, then define the function $u_t : \mathbb{X} \longrightarrow \mathbb{R}$ by $u_t(\mathbf{x}) := u(\mathbf{x}; t)$, for all $\mathbf{x} \in \mathbb{X}$. (*Note:* u_t does *not* denote the time-derivative.) For all $t \geq 0$, the *energy* of u is defined:

$$E(t) := \frac{1}{2} \int_{\mathbb{X}} |\partial_t u_t(\mathbf{x})|^2 + \|\nabla u_t(\mathbf{x})\|^2 \ d\mathbf{x}. \tag{5D.6}$$

We begin with a result that has an appealing physical interpretation.

Lemma 5D.9 Conservation of energy for wave equation
Let $\mathbb{X} \subset \mathbb{R}^D$ *be a bounded domain with a piecewise smooth boundary. Suppose* $u : \mathbb{X} \times \mathbb{R}_{\not{}} \longrightarrow \mathbb{R}$ *satisfies the following three conditions:*

(a) (*Regularity*) u *is continuous on* $\mathbb{X} \times \mathbb{R}_{\not{}}$, *and* $u \in \mathcal{C}^2$ (int $(\mathbb{X}) \times \mathbb{R}_+$);

(b) (*Wave equation*) $\partial_t^2 u = \Delta u$;

(c) (*Homogeneous Dirichlet/Neumann BC*)[9] *For all* $\mathbf{s} \in \partial \mathbb{X}$, *either* $u_t(\mathbf{s}) = 0$ *for all* $t \geq 0$, *or* $\partial_\perp u_t(\mathbf{s}) = 0$ *for all* $t \geq 0$.

Then E is constant in time; that is, $\partial_t E(t) = 0$ *for all* $t > 0$.

Proof The Leibniz rule says that

$$\partial_t |\partial_t u|^2 = (\partial_t^2 u) \cdot (\partial_t u) + (\partial_t u) \cdot (\partial_t^2 u);$$

$$= 2 \cdot (\partial_t u) \cdot (\partial_t^2 u); \tag{5D.7}$$

$$\partial_t \|\nabla u\|^2 = (\partial_t \nabla u) \bullet (\nabla u) + (\nabla u) \bullet (\partial_t \nabla u)$$

$$= 2 \cdot (\nabla u) \bullet (\partial_t \nabla u)$$

$$= 2 \cdot (\nabla u) \bullet (\nabla \partial_t u). \tag{5D.8}$$

---

[8] Note that this includes nonhomogeneous Dirichlet BC (set $h_\perp \equiv 0$) and nonhomogeneous Neumann BC (set $h \equiv 0$) as special cases. Also note that by varying h and $h_\perp$, we can allow different boundary points to satisfy different nonhomogeneous boundary conditions at different times.

[9] This allows different boundary points to satisfy different homogeneous boundary conditions; but each particular boundary point must satisfy the *same* homogeneous boundary condition at all times.

Thus,

$$\partial_t E \underset{(*)}{=} \frac{1}{2} \int_{\mathbb{X}} \left(\partial_t |\partial_t u|^2 + \partial_t \|\nabla u\|^2 \right)$$

$$\underset{(\dagger)}{=} \int_{\mathbb{X}} \left(\partial_t u \cdot \partial_t^2 u + (\nabla u) \bullet (\nabla \partial_t u) \right). \tag{5D.9}$$

Here (*) comes from differentiating the integral (5D.6) using Proposition G.1, p. 571). Meanwhile, (†) comes from substituting equations (5D.7) and (5D.8).

Claim 1 *Fix* $\mathbf{s} \in \partial \mathbb{X}$ *and let* $\vec{\mathbf{N}}(\mathbf{s})$ *be the outward unit normal vector to* $\partial \mathbb{X}$ *at* $\mathbf{s}$. *Then* $\partial_t u_t(\mathbf{s}) \cdot \nabla u_t(\mathbf{s}) \bullet \vec{\mathbf{N}}(\mathbf{s}) = 0$, *for all* $t > 0$.

Proof By hypothesis (c), either $\partial_\perp u_t(\mathbf{s}) = 0$ for all $t > 0$, or $u_t(\mathbf{s}) = 0$ for all $t > 0$. Thus, either $\nabla u_t(\mathbf{s}) \bullet \vec{\mathbf{N}}(\mathbf{s}) = 0$ for all $t > 0$, or $\partial_t u_t(\mathbf{s}) = 0$ for all $t > 0$. In either case, $\partial_t u_t(\mathbf{s}) \cdot \nabla u_t(\mathbf{s}) \bullet \vec{\mathbf{N}}(\mathbf{s}) = 0$ for all $t > 0$. $\diamond_{\text{Claim 1}}$

Claim 2 *For any* $t \in \mathbb{R}_+$,

$$\int_{\mathbb{X}} \nabla u_t \bullet \nabla \partial_t u_t = - \int_{\mathbb{X}} \partial_t u_t \cdot \Delta u_t.$$

Proof Integrating Claim 1 over $\partial \mathbb{X}$, we get

$$0 = \int_{\partial \mathbb{X}} \partial_t u_t(\mathbf{s}) \cdot \nabla u_t(\mathbf{s}) \bullet \vec{\mathbf{N}}(\mathbf{s}) d\mathbf{s} \underset{(*)}{=} \int_{\mathbb{X}} \text{div} \left(\partial_t u_t \cdot \nabla u_t \right)(\mathbf{x}) d\mathbf{x}$$

$$\underset{(\dagger)}{=} \int_{\mathbb{X}} \left(\partial_t u_t \cdot \text{div} \, \nabla u_t + \nabla \partial_t u_t \bullet \nabla u_t \right)(\mathbf{x}) d\mathbf{x}$$

$$\underset{(\ddagger)}{=} \int_{\mathbb{X}} (\partial_t u_t \cdot \Delta u_t)(\mathbf{x}) d\mathbf{x} + \int_{\mathbb{X}} (\nabla \partial_t u_t \bullet \nabla u_t)(\mathbf{x}) d\mathbf{x}.$$

Here, (*) is the divergence theorem, Theorem E.4, p. 566, (†) is by the Leibniz Rule for divergences (Proposition E.2(b), p. 564), and (‡) is because $\text{div} \, \nabla u_t = \Delta u_t$. We thus have

$$\int_{\mathbb{X}} \nabla u_t \bullet \nabla \partial_t u_t + \int_{\mathbb{X}} \partial_t u_t \cdot \Delta u_t = 0.$$

Rearranging this equation yields the claim. $\diamond_{\text{Claim 2}}$

Putting it all together, we get:

$$\partial_t E \underset{(\dagger)}{=} \int_{\mathbb{X}} \partial_t u \cdot \partial_t^2 u + \int_{\mathbb{X}} (\nabla u) \bullet (\nabla \partial_t u)$$

$$\underset{(\ddagger)}{=} \int_{\mathbb{X}} \partial_t u \cdot \partial_t^2 u - \int_{\mathbb{X}} \partial_t u \cdot \Delta u = \int_{\mathbb{X}} \partial_t u \cdot \left(\partial_t^2 u - \Delta u \right)$$

$$\underset{(*)}{=} \int_{\mathbb{X}} \partial_t u \cdot 0 = 0,$$

as desired. Here, (†) is by equation (5D.9), (‡) is by Claim 2, and (∗) is because $\partial_t^2 u - \Delta u \equiv 0$ because u satisfies the wave equation by hypothesis (b). ☐

Physical interpretation $E(t)$ can be interpreted as the *total energy* in the system at time t. The first term in the integrand of equation (5D.6) measures the *kinetic* energy of the wave motion, while the second term measures the *potential* energy stored in the deformation of the medium. With this physical interpretation, Lemma 5D.9 simply asserts the principle of *Conservation of Energy*: E must be constant in time, because no energy enters or leaves the system, by hypotheses (b) and (c).

Lemma 5D.10 Solution uniqueness for wave equation; homogeneous I/BC
Let $\mathbb{X} \subset \mathbb{R}^D$ be a bounded domain with a piecewise smooth boundary. Suppose $u : \mathbb{X} \times \mathbb{R}_{\not{}} \longrightarrow \mathbb{R}$ satisfies all five of the following conditions:

(a) *(Regularity)* u *is continuous on* $\mathbb{X} \times \mathbb{R}_{\not{}}$, *and* $u \in C^2 (\text{int}(\mathbb{X}) \times \mathbb{R}_+)$;
(b) *(Wave equation)* $\partial_t^2 u = \Delta u$;
(c) *(Zero initial position)* $u_0(\mathbf{x}) = 0$, *for all* $\mathbf{x} \in \mathbb{X}$;
(d) *(Zero initial velocity)* $\partial_t u_0(\mathbf{x}) = 0$ *for all* $\mathbf{x} \in \mathbb{X}$;
(e) *(Homogeneous Dirichlet/Neumann BC)*[10] *For all* $\mathbf{s} \in \partial\mathbb{X}$, *either* $u_t(\mathbf{s}) = 0$ *for all* $t \geq 0$, *or* $\partial_\perp u_t(\mathbf{s}) = 0$ *for all* $t \geq 0$.

Then u *must be the constant* 0 *function:* $u \equiv 0$.

Proof Let $E : \mathbb{R}_{\not{}} \longrightarrow \mathbb{R}_{\not{}}$ be the energy function from Lemma 5D.9. Then E is a constant. But $E(0) = 0$ because $u_0 \equiv 0$ and $\partial_t u_0 \equiv 0$, by hypotheses (c) and (d). Thus, $E(t) = 0$ for all $t \geq 0$. But this implies that $|\partial_t u_t(\mathbf{x})|^2 = 0$, and hence $\partial_t u_t(\mathbf{x}) = 0$, for all $\mathbf{x} \in \mathbb{X}$ and $t > 0$. Thus, u is constant in time. Since $u_0 \equiv 0$, we conclude that $u_t \equiv 0$ for all $t \geq 0$, as desired. ☐

Theorem 5D.11 Uniqueness: forced wave equation, nonhomogeneous I/BC
Let $\mathbb{X} \subset \mathbb{R}^D$ be a bounded domain with a piecewise smooth boundary. Let $\mathcal{I}_0, \mathcal{I}_1 : \mathbb{X} \longrightarrow \mathbb{R}$ be continuous functions (describing initial position and velocity). Let $b : \partial\mathbb{X} \times \mathbb{R}_{\not{}} \longrightarrow \mathbb{R}$ be another continuous function (describing a time-varying boundary condition). Let $f : \text{int}(\mathbb{X}) \times \mathbb{R}_{\not{}} \longrightarrow \mathbb{R}$ be another continuous function (describing exogenous vibrations being 'forced' into the system). Then there is at most one solution function $u : \mathbb{X} \times \mathbb{R}_{\not{}} \longrightarrow \mathbb{R}$ satisfying all five of the following conditions:

(a) *(Regularity)* u *is continuous on* $\mathbb{X} \times \mathbb{R}_{\not{}}$, *and* $u \in C^2 (\text{int}(\mathbb{X}) \times \mathbb{R}_+)$;
(b) *(Wave equation with forcing)* $\partial_t^2 u = \Delta u + f$;
(c) *(Initial position)* $u(\mathbf{x}, 0) = \mathcal{I}_0(\mathbf{x})$; *for all* $\mathbf{x} \in \mathbb{X}$;

---

[10] This allows different boundary points to satisfy different homogeneous boundary conditions; but each particular boundary point must satisfy the *same* homogeneous boundary condition at all times.

(d) (Initial velocity) $\partial_t u(\mathbf{x}, 0) = \mathcal{I}_1(\mathbf{x})$; *for all* $\mathbf{x} \in \mathbb{X}$;
(e) (Nonhomogeneous Dirichlet/Neumann BC) *For all* $\mathbf{s} \in \partial\mathbb{X}$, *either* $u(\mathbf{s}, t) = b(\mathbf{s}, t)$ *for all* $t \geq 0$, *or* $\partial_\perp u(\mathbf{s}, t) = b(\mathbf{s}, t)$ *for all* $t \geq 0$.

Proof Suppose u_1 and u_2 were two functions satisfying all of (a)–(e). Let $u = u_1 - u_2$. Then u satisfies all of (a)–(e), in Lemma 5D.10. Thus, $u \equiv 0$. But this means that $u_1 \equiv u_2$. Hence, there can be at most one solution. $\qquad\square$

Remarks

(a) Earlier, we observed that the *initial position* problem for the (unforced) wave equation represents a 'plucked string' (e.g. in a guitar), while the *initial velocity* problem represents a 'struck string' (e.g. in a piano). Continuing the musical analogy, the *forced* wave equation represents a *rubbed* string (e.g. in a violin or cello), as well as any other musical instrument driven by an exogenous vibration (e.g. any wind instrument).

(b) Note that Theorems 5D.5, 5D.8, and 5D.11 apply under much more general conditions than any of the solution methods we will actually develop in this book (i.e. they work for almost any 'reasonable' domain, we allow for possible 'forcing', and we even allow the boundary conditions to vary in time). This is a recurring theme in differential equation theory; it is generally possible to prove 'qualitative' results (e.g. about existence, uniqueness, or general properties of solutions) in much more general settings than it is possible to get 'quantitative' results (i.e. explicit formulae for solutions). Indeed, for most *nonlinear* differential equations, qualitative results are pretty much all you can ever get.

5E* Classification of second-order linear PDEs

Prerequisites: §5A. **Recommended:** §1B, §1C, §1F, §2B.

5E(i) ... in two dimensions, with constant coefficients

Recall that $\mathcal{C}^\infty(\mathbb{R}^2; \mathbb{R})$ is the space of all differentiable scalar fields on the plane $\mathbb{R}^2$. In general, a second-order linear differential operator L on $\mathcal{C}^\infty(\mathbb{R}^2; \mathbb{R})$ with constant coefficients looks like the following:

$$\mathsf{L}u = a \cdot \partial_x^2 u + b \cdot \partial_x \partial_y u + c \cdot \partial_y^2 u + d \cdot \partial_x u + e \cdot \partial_y u + f \cdot u, \quad (5\text{E}.1)$$

where a, b, c, d, e, f are constants. Define:

$$\alpha = f, \quad \beta = \begin{bmatrix} d \\ e \end{bmatrix}, \quad \text{and} \quad \Gamma = \begin{bmatrix} a & \frac{1}{2}b \\ \frac{1}{2}b & c \end{bmatrix} = \begin{bmatrix} \gamma_{11} & \gamma_{12} \\ \gamma_{21} & \gamma_{22} \end{bmatrix}.$$

Then we can rewrite equation (5E.1) as follows:

$$\mathsf{L}u = \sum_{c,d=1}^{2} \gamma_{c,d} \cdot \partial_c \, \partial_d \, u + \sum_{d=1}^{2} \beta_d \cdot \partial_d \, u + \alpha \cdot u.$$

Any 2×2 symmetric matrix Γ defines a *quadratic form* $G : \mathbb{R}^2 \longrightarrow \mathbb{R}$ by

$$G(x, y) = [x \ y] \cdot \begin{bmatrix} \gamma_{11} & \gamma_{12} \\ \gamma_{21} & \gamma_{22} \end{bmatrix} \cdot \begin{bmatrix} x \\ y \end{bmatrix} = \gamma_{11} \cdot x^2 + \left(\gamma_{12} + \gamma_{21} \right) \cdot xy + \gamma_{22} \cdot y^2.$$

We say Γ is *positive definite* if, for all $x, y \in \mathbb{R}$, we have

- $G(x, y) \geq 0$;
- $G(x, y) = 0$ if and only if $x = 0 = y$.

Geometrically, this means that the graph of G defines an *elliptic paraboloid* in $\mathbb{R}^2 \times \mathbb{R}$, which curves upwards in every direction. Equivalently, Γ is positive definite if there is a constant $K > 0$ such that

$$G(x, y) \geq K \cdot (x^2 + y^2)$$

for every $(x, y) \in \mathbb{R}^2$. We say Γ is *negative definite* if $-\Gamma$ is positive definite.

The differential operator L from equation (5E.1) is called *elliptic* if the matrix Γ is either positive definite or negative definite.

Example 5E.1 If $\mathsf{L} = \triangle$, then $\Gamma = \begin{bmatrix} 1 & \\ & 1 \end{bmatrix}$ is just the identity matrix, while $\beta = 0$ and $\alpha = 0$. The identity matrix is clearly positive definite; thus, $\triangle$ is an elliptic differential operator. $\diamondsuit$

Suppose that L is an elliptic differential operator. Then we have the following.

- An *elliptic* PDE is one of the form $\mathsf{L}u = 0$ (or $\mathsf{L}u = g$); for example, the *Laplace equation* is elliptic.
- A *parabolic* PDE is one of the form $\partial_t = \mathsf{L}u$; for example, the two-dimensional *heat equation* is parabolic.
- A *hyperbolic* PDE is one of the form $\partial_t^2 = \mathsf{L}u$; for example, the two-dimensional *wave equation* is hyperbolic.

(See Remark 16F.4, p. 378, for a partial justification of this terminology.)

Ⓔ **Exercise 5E.1** Show that Γ is positive definite if and only if $0 < \det(\Gamma) = ac - \frac{1}{4}b^2$. In other words, L is elliptic if and only if $4ac - b^2 > 0$.   ♦

5E(ii) ... in general

Recall that $C^\infty(\mathbb{R}^D; \mathbb{R})$ is the space of all differentiable scalar fields on D-dimensional space. The general second-order linear differential operator on $C^\infty(\mathbb{R}^D; \mathbb{R})$ has the form

$$\mathsf{L}u = \sum_{c,d=1}^{D} \gamma_{c,d} \cdot \partial_c \, \partial_d \, u + \sum_{d=1}^{D} \beta_d \cdot \partial_d \, u + \alpha \cdot u, \qquad (5\text{E}.2)$$

where $\alpha : \mathbb{R}^D \times \mathbb{R} \longrightarrow \mathbb{R}$ is some time-varying scalar field, $(\beta_1, \ldots, \beta_D) = \boldsymbol{\beta} : \mathbb{R}^D \times \mathbb{R} \longrightarrow \mathbb{R}^D$ is a time-varying vector field, and $\gamma_{c,d} : \mathbb{R}^D \times \mathbb{R} \longrightarrow \mathbb{R}$ are functions such that, for any $\mathbf{x} \in \mathbb{R}^D$ and $t \in \mathbb{R}$, the matrix

$$\Gamma(\mathbf{x}; t) = \begin{bmatrix} \gamma_{11}(\mathbf{x}; t) & \cdots & \gamma_{1D}(\mathbf{x}; t) \\ \vdots & \ddots & \vdots \\ \gamma_{D1}(\mathbf{x}; t) & \cdots & \gamma_{DD}(\mathbf{x}; t) \end{bmatrix}$$

is *symmetric* (i.e. $\gamma_{cd} = \gamma_{dc}$).

Example 5E.2
(a) If $\mathsf{L} = \triangle$, then $\boldsymbol{\beta} \equiv 0$, $\alpha = 0$, and

$$\Gamma \equiv \mathbf{Id} = \begin{bmatrix} 1 & 0 & \cdots & 0 \\ 0 & 1 & \cdots & 0 \\ \vdots & \vdots & \ddots & \vdots \\ 0 & 0 & \cdots & 1 \end{bmatrix}.$$

(b) The Fokker–Planck equation (see §1F, p. 21) has the form $\partial_t \, u = \mathsf{L}u$, where $\alpha = -\operatorname{div} \vec{\mathbf{V}}(\mathbf{x})$, $\boldsymbol{\beta}(\mathbf{x}) = -\nabla \vec{\mathbf{V}}(\mathbf{x})$, and $\Gamma \equiv \mathbf{Id}$.   ◇

Exercise 5E.2   Verify Example 5E.2(b).   ♦   Ⓔ

If the functions $\gamma_{c,d}$, β_d and α are independent of $\mathbf{x}$, then we say L is *spatially homogeneous*. If they are also independent of t, we say that L has *constant coefficients*.

Any symmetric matrix Γ defines a *quadratic form* $G : \mathbb{R}^D \longrightarrow \mathbb{R}$ by

$$G(\mathbf{x}) = [x_1 \ldots x_D] \begin{bmatrix} \gamma_{11} & \cdots & \gamma_{1D} \\ \vdots & \ddots & \vdots \\ \gamma_{D1} & \cdots & \gamma_{DD} \end{bmatrix} \begin{bmatrix} x_1 \\ \vdots \\ x_D \end{bmatrix} = \sum_{c,d=1}^{D} \gamma_{c,d} \cdot x_c \cdot x_d.$$

Γ is called *positive definite* if, for all $\mathbf{x} \in \mathbb{R}^D$, we have:

• $G(\mathbf{x}) \geq 0$;
• $G(\mathbf{x}) = 0$ if and only if $\mathbf{x} = 0$.

Equivalently, Γ is positive definite if there is a constant $K_\Gamma > 0$ such that $G(\mathbf{x}) \geq K_\Gamma \cdot \|\mathbf{x}\|^2$ for every $\mathbf{x} \in \mathbb{R}^D$. On the other hand, Γ is *negative definite* if $-\Gamma$ is positive definite.

The differential operator L from equation (5E.2) is *elliptic* if the matrix $\Gamma(\mathbf{x}; t)$ is either positive definite or negative definite for every $(\mathbf{x}; t) \in \mathbb{R}^D \times \mathbb{R}_{\not{}}$, and, furthermore, there is some $K > 0$ such that $K_{\Gamma(\mathbf{x};t)} \geq K$ for all $(\mathbf{x}; t) \in \mathbb{R}^D \times \mathbb{R}_{\not{}}$.

Ⓔ **Exercise 5E.3** Show that the Laplacian and the Fokker–Planck operator are both elliptic. ♦

Suppose that L is an elliptic differential operator. Then we have the following.

- An *elliptic* PDE is one of the form $\mathsf{L}u = 0$ (or $\mathsf{L}u = g$).
- A *parabolic* PDE is one of the form $\partial_t = \mathsf{L}u$.
- A *hyperbolic* PDE is one of the form $\partial_t^2 = \mathsf{L}u$.

Example 5E.3

(a) Laplace's equation and Poisson's equation are *elliptic* PDEs.
(b) The heat equation and the Fokker–Planck equation are *parabolic*.
(c) The wave equation is *hyperbolic*.   ◇

Parabolic equations are 'generalized heat equations', describing *diffusion through an inhomogeneous,*[11] *anisotropic*[12] *medium with drift.* The terms in $\Gamma(\mathbf{x}; t)$ describe the inhomogeneity and anisotropy of the diffusion,[13] while the vector field $\boldsymbol{\beta}$ describes the drift.

Hyperbolic equations are 'generalized wave equations', describing *wave propagation* through an inhomogeneous, anisotropic medium with drift – for example, sound waves propagating through an air mass with variable temperature and pressure and wind blowing.

5F Practice problems

Evolution equations and initial value problems For each of the following equations: u is an unknown function; q is always some fixed, predetermined function; λ is always a constant. In each case, determine the *order* of the equation and decide: Is this an *evolution equation*? Why or why not?

5.1 Heat equation: $\partial_t u(\mathbf{x}) = \Delta u(\mathbf{x})$.

5.2 Poisson equation: $\Delta u(\mathbf{x}) = q(\mathbf{x})$.

---

[11] *Homogeneous* means 'looks the same everywhere in space'; *inhomogeneous* means the opposite.
[12] *Isotropic* means 'looks the same in every direction'; *anisotropic* means the opposite.
[13] If the medium is homogeneous, then Γ is constant. If the medium is isotropic, then $\Gamma = \mathbf{Id}$.

5.3 Laplace equation: $\triangle u(\mathbf{x}) = 0$.

5.4 Monge–Ampère equation: $q(x, y) = \det \begin{bmatrix} \partial_x^2 u(x, y) & \partial_x \partial_y u(x, y) \\ \partial_x \partial_y u(x, y) & \partial_y^2 u(x, y) \end{bmatrix}$.

5.5 Reaction–diffusion equation: $\partial_t u(\mathbf{x}; t) = \triangle u(\mathbf{x}; t) + q(u(\mathbf{x}; t))$.

5.6 Scalar conservation law: $\partial_t u(x; t) = -\partial_x (q \circ u)(x; t)$.

5.7 Helmholtz equation: $\triangle u(\mathbf{x}) = \lambda \cdot u(\mathbf{x})$.

5.8 Airy equation: $\partial_t u(x; t) = -\partial_x^3 u(x; t)$.

5.9 Beam equation: $\partial_t u(x; t) = -\partial_x^4 u(x; t)$.

5.10 Schrödinger equation: $\partial_t u(\mathbf{x}; t) = \mathbf{i} \triangle u(\mathbf{x}; t) + q(\mathbf{x}; t) \cdot u(\mathbf{x}; t)$.

5.11 Burger equation: $\partial_t u(x; t) = -u(x; t) \cdot \partial_x u(x; t)$.

5.12 Eikonal equation: $|\partial_x u(x)| = 1$.

Boundary value problems

5.13 Each of the following functions is defined on the interval $[0, \pi]$, in Cartesian coordinates. For each function, decide whether it satisfies: homogeneous *Dirichlet* BC; homogeneous *Neumann* BC; homogeneous *Robin*[14] BC; periodic BC. Justify your answers.

(a) $u(x) = \sin(3x)$;
(b) $u(x) = \sin(x) + 3\sin(2x) - 4\sin(7x)$;
(c) $u(x) = \cos(x) + 3\sin(3x) - 2\cos(6x)$;
(d) $u(x) = 3 + \cos(2x) - 4\cos(6x)$;
(e) $u(x) = 5 + \cos(2x) - 4\cos(6x)$.

5.14 Each of the following functions is defined on the interval $[-\pi, \pi]$, in Cartesian coordinates. For each function, decide whether it satisfies: homogeneous *Dirichlet* BC; homogeneous *Neumann* BC; homogeneous *Robin* BC; periodic BC. Justify your answers.

(a) $u(x) = \sin(x) + 5\sin(2x) - 2\sin(3x)$;
(b) $u(x) = 3\cos(x) - 3\sin(2x) - 4\cos(2x)$;
(c) $u(x) = 6 + \cos(x) - 3\cos(2x)$.

5.15 Each of the following functions is defined on the box $[0, \pi]^2$, in Cartesian coordinates. For each function, decide whether it satisfies: homogeneous *Dirichlet* BC;

---

[14] Here, 'Robin' BC means *nontrivial* Robin BC – i.e. *not* just homogenous Dirichlet or Neumann.

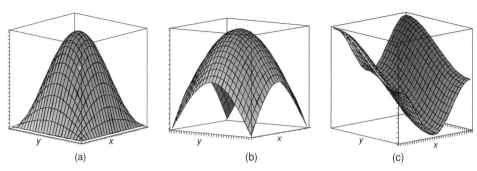

(a)   (b)   (c)

Figure 5F.1. (a) $f(x, y) = \sin(x)\sin(y)$; (b) $g(x, y) = \sin(x) + \sin(y)$; (c) $h(x, y) = \cos(2x) + \cos(y)$. See Problems 5.15(a)–(c).

homogeneous *Neumann* BC; homogeneous *Robin* BC; periodic BC. Justify your answers.

(a) $f(x, y) = \sin(x)\sin(y)$ (see Figure 5F.1(a));
(b) $g(x, y) = \sin(x) + \sin(y)$ (see Figure 5F.1(b));
(c) $h(x, y) = \cos(2x) + \cos(y)$ (see Figure 5F.1(c));
(d) $u(x, y) = \sin(5x)\sin(3y)$;
(e) $u(x, y) = \cos(-2x)\cos(7y)$.

5.16 Each of the following functions is defined on the unit disk

$$\mathbb{D} = \{(r, \theta);\ 0 \leq r \leq 1,\ \text{and}\ \theta \in [0, 2\pi)\},$$

in polar coordinates. For each function, decide whether it satisfies: homogeneous *Dirichlet* BC; homogeneous *Neumann* BC; homogeneous *Robin* BC. Justify your answers.

(a) $u(r, \theta) = (1 - r^2)$;
(b) $u(r, \theta) = 1 - r^3$;
(c) $u(r, \theta) = 3 + (1 - r^2)^2$;
(d) $u(r, \theta) = \sin(\theta)(1 - r^2)^2$;
(e) $u(r, \theta) = \cos(2\theta)(e - e^r)$.

5.17 Each of the following functions is defined on the three-dimensional unit ball

$$\mathbb{B} = \left\{(r, \theta, \varphi);\ 0 \leq r \leq 1, \theta \in [0, 2\pi),\ \text{and}\ \varphi \in \left[\frac{-\pi}{2}, \frac{\pi}{2}\right]\right\},$$

in spherical coordinates. For each function, decide whether it satisfies: homogeneous *Dirichlet* BC; homogeneous *Neumann* BC; homogeneous *Robin* BC. Justify your answers.

(a) $u(r, \theta, \varphi) = (1 - r)^2$;
(b) $u(r, \theta, \varphi) = (1 - r)^3 + 5$.

5.18 Which Neumann BVP has solution(s) on the domain $\mathbb{X} = [0, 1]$?

    (a) $u''(x) = 0, u'(0) = 1, u'(1) = 1$;

    (b) $u''(x) = 0, u'(0) = 1, u'(1) = 2$;

    (c) $u''(x) = 0, u'(0) = 1, u'(1) = -1$;

    (d) $u''(x) = 0, u'(0) = 1, u'(1) = -2$.

5.19 Which BVP of the Laplace equation on the unit disk $\mathbb{D}$ has a solution? Which BVP has more than one solution?

    (a) $\Delta u = 0, u(1, \theta) = 0$, for all $\theta \in [-\pi, \pi)$;

    (b) $\Delta u = 0, u(1, \theta) = \sin\theta$, for all $\theta \in [-\pi, \pi)$;

    (c) $\Delta u = 0, \partial_\perp u(1, \theta) = \sin(\theta)$, for all $\theta \in [-\pi, \pi)$;

    (d) $\Delta u = 0, \partial_\perp u(1, \theta) = 1 + \cos(\theta)$, for all $\theta \in [-\pi, \pi)$.

Part III

Fourier series on bounded domains

Any complex sound is a combination of simple 'pure tones' of different frequencies. For example, a musical *chord* is a superposition of three (or more) musical notes, each with a different frequency. In fact, a musical note itself is not really a single frequency at all; a note consists of a 'fundamental' frequency, plus a cascade of higher-frequency 'harmonics'. The energy distribution of these harmonics is part of what gives each musical instrument its distinctive sound. The decomposition of a sound into separate frequencies is sometimes called its *power spectrum*. A crude graphical representation of this power spectrum is visible on most modern stereo systems (the little jiggling red bars).

Fourier theory is based on the idea that a real-valued function is like a sound, which can be represented as a superposition of 'pure tones' (i.e. sine waves and/or cosine waves) of distinct frequencies. This provides a 'coordinate system' for expressing functions, and within this coordinate system we can express the solutions for many partial differential equations in a simple and elegant way. Fourier theory is also an essential tool in probability theory and signal analysis (although we will not discuss these applications in this book).

The idea of Fourier theory is simple, but to make this idea rigorous enough to be useful, we must deploy some formidable mathematical machinery. So we will begin by developing the necessary background concerning inner products, orthogonality, and the convergence of functions.

6

Some functional analysis

Mathematical science is in my opinion an indivisible whole, an
organism whose vitality is conditioned upon the connection of its parts.

David Hilbert

6A Inner products

Prerequisites: §4A.

Let $\mathbf{x}, \mathbf{y} \in \mathbb{R}^D$, with $\mathbf{x} = (x_1, \ldots, x_D)$ and $\mathbf{y} = (y_1, \ldots, y_D)$. The *inner product*[1] of $\mathbf{x}, \mathbf{y}$ is defined:

$$\langle \mathbf{x}, \mathbf{y} \rangle := x_1 y_1 + x_2 y_2 + \cdots + x_D y_D.$$

The inner product describes the geometric relationship between $\mathbf{x}$ and $\mathbf{y}$, via the following formula:

$$\langle \mathbf{x}, \mathbf{y} \rangle := \|\mathbf{x}\| \cdot \|\mathbf{y}\| \cdot \cos(\theta),$$

where $\|\mathbf{x}\|$ and $\|\mathbf{y}\|$ are the lengths of vectors $\mathbf{x}$ and $\mathbf{y}$, and θ is the angle between them. (**Exercise 6A.1** Verify this.) In particular, if $\mathbf{x}$ and $\mathbf{y}$ are *perpendicular*, then $\theta = \pm\frac{\pi}{2}$, and then $\langle \mathbf{x}, \mathbf{y} \rangle = 0$; we then say that $\mathbf{x}$ and $\mathbf{y}$ are *orthogonal*. For example, $\mathbf{x} = \begin{bmatrix} 1 \\ 1 \end{bmatrix}$ and $\mathbf{y} = \begin{bmatrix} 1 \\ -1 \end{bmatrix}$ are orthogonal in $\mathbb{R}^2$, while ⒠

$$\mathbf{u} = \begin{bmatrix} 1 \\ 0 \\ 0 \\ 0 \end{bmatrix}, \quad \mathbf{v} = \begin{bmatrix} 0 \\ 0 \\ \frac{1}{\sqrt{2}} \\ \frac{1}{\sqrt{2}} \end{bmatrix}, \quad \text{and} \quad \mathbf{w} = \begin{bmatrix} 0 \\ 1 \\ 0 \\ 0 \end{bmatrix}$$

are all orthogonal to one another in $\mathbb{R}^4$. Indeed, $\mathbf{u}$, $\mathbf{v}$, and $\mathbf{w}$ also have unit norm; we call any such collection an *orthonormal set* of vectors. Thus, $\{\mathbf{u}, \mathbf{v}, \mathbf{w}\}$ is an

---

[1] This is sometimes called the *dot product*, and denoted '$\mathbf{x} \bullet \mathbf{y}$'.

orthonormal set. However, $\{\mathbf{x}, \mathbf{y}\}$ is orthogonal but *not* orthonormal (because $\|\mathbf{x}\| = \|\mathbf{y}\| = \sqrt{2} \neq 1$).

The *norm* of a vector satisfies the following equation:

$$\|\mathbf{x}\| = \left(x_1^2 + x_2^2 + \cdots + x_D^2\right)^{1/2} = \langle \mathbf{x}, \mathbf{x} \rangle^{1/2}.$$

If $\mathbf{x}_1, \ldots, \mathbf{x}_N$ are a collection of mutually orthogonal vectors, and $\mathbf{x} = \mathbf{x}_1 + \cdots + \mathbf{x}_N$, then we have the generalized *Pythagorean formula*:

$$\|\mathbf{x}\|^2 = \|\mathbf{x}_1\|^2 + \|\mathbf{x}_2\|^2 + \cdots + \|\mathbf{x}_N\|^2.$$

Ⓔ **Exercise 6A.2** Verify the Pythagorean formula.   ◆

An *orthonormal basis* of $\mathbb{R}^D$ is any collection of mutually orthogonal vectors $\{\mathbf{v}_1, \mathbf{v}_2, \ldots, \mathbf{v}_D\}$, all of norm 1, such that, for any $\mathbf{w} \in \mathbb{R}^D$, if we define $\omega_d = \langle \mathbf{w}, \mathbf{v}_d \rangle$ for all $d \in [1 \ldots D]$, then

$$\mathbf{w} = \omega_1 \mathbf{v}_1 + \omega_2 \mathbf{v}_2 + \cdots + \omega_D \mathbf{v}_D.$$

In other words, the set $\{\mathbf{v}_1, \mathbf{v}_2, \ldots, \mathbf{v}_D\}$ defines a *coordinate system* for $\mathbb{R}^D$, and in this coordinate system the vector $\mathbf{w}$ has coordinates $(\omega_1, \omega_2, \ldots, \omega_D)$. If $\mathbf{x} \in \mathbb{R}^D$ is another vector, and $\xi_d = \langle \mathbf{x}, \mathbf{v}_d \rangle$ all $d \in [1 \ldots D]$, then we also have

$$\mathbf{x} = \xi_1 \mathbf{v}_1 + \xi_2 \mathbf{v}_2 + \cdots + \xi_D \mathbf{v}_D.$$

We can then compute $\langle \mathbf{w}, \mathbf{x} \rangle$ using *Parseval's equality*:

$$\langle \mathbf{w}, \mathbf{x} \rangle = \omega_1 \xi_1 + \omega_2 \xi_2 + \cdots + \omega_D \xi_D.$$

Ⓔ **Exercise 6A.3** Prove Parseval's equality.   ◆

If $\mathbf{x} = \mathbf{w}$, we get the following version of the generalized Pythagorean formula:

$$\|\mathbf{w}\|^2 = \omega_1^2 + \omega_2^2 + \cdots + \omega_D^2.$$

Example 6A.1

(a) $\left\{ \begin{bmatrix} 1 \\ 0 \\ \vdots \\ 0 \end{bmatrix}, \begin{bmatrix} 0 \\ 1 \\ \vdots \\ 0 \end{bmatrix}, \ldots, \begin{bmatrix} 0 \\ 0 \\ \vdots \\ 1 \end{bmatrix} \right\}$ is an orthonormal basis for $\mathbb{R}^D$.

(b) If $\mathbf{v}_1 = \begin{bmatrix} \sqrt{3}/2 \\ 1/2 \end{bmatrix}$ and $\mathbf{v}_2 = \begin{bmatrix} -1/2 \\ \sqrt{3}/2 \end{bmatrix}$, then $\{\mathbf{v}_1, \mathbf{v}_2\}$ is an orthonormal basis of $\mathbb{R}^2$. If $\mathbf{w} = \begin{bmatrix} 2 \\ 4 \end{bmatrix}$, then $\omega_1 = \sqrt{3} + 2$ and $\omega_2 = 2\sqrt{3} - 1$, so that

$$\begin{bmatrix} 2 \\ 4 \end{bmatrix} = \omega_1 \mathbf{v}_1 + \omega_2 \mathbf{v}_2 = \left(\sqrt{3} + 2\right) \cdot \begin{bmatrix} \sqrt{3}/2 \\ 1/2 \end{bmatrix} + \left(2\sqrt{3} - 1\right) \cdot \begin{bmatrix} -1/2 \\ \sqrt{3}/2 \end{bmatrix}.$$

Thus, $\|\mathbf{w}\|_2^2 = 2^2 + 4^2 = 20$, and also, by Parseval's equality, $20 = \omega_1^2 + \omega_2^2 = (\sqrt{3} + 2)^2 + (1 - 2\sqrt{3})^2.$ ◇

Exercise 6A.4   Verify these claims.   ◆ Ⓔ

6B   L²-space

The ideas of §6A generalize to spaces of functions. Suppose $\mathbb{X} \subset \mathbb{R}^D$ is some bounded domain, and let $M := \int_{\mathbb{X}} 1 \, d\mathbf{x}$ be the *volume*[2] of the domain $\mathbb{X}$. (The second column of Table 6.1 provides examples of M for various domains.)

If $f, g : \mathbb{X} \longrightarrow \mathbb{R}$ are integrable functions, then the *inner product* of f and g is defined:

$$\langle f, g \rangle := \frac{1}{M} \int_{\mathbb{X}} f(\mathbf{x}) \cdot g(\mathbf{x}) d\mathbf{x}. \tag{6B.1}$$

Example 6B.1
(a) Suppose $\mathbb{X} = [0, 3] = \{x \in \mathbb{R}; \; 0 \le x \le 3\}$. Then $M = 3$. If $f(x) = x^2 + 1$ and $g(x) = x$ for all $x \in [0, 3]$, then

$$\langle f, g \rangle = \frac{1}{3} \int_0^3 f(x)g(x) dx = \frac{1}{3} \int_0^3 (x^3 + x) dx = \frac{27}{4} + \frac{3}{2}.$$

(b) The third column of Table 6.1 provides examples of $\langle f, g \rangle$ for various other domains.   ◇

The *L²-norm* of an integrable function $f : \mathbb{X} \longrightarrow \mathbb{R}$ is defined:

$$\|f\|_2 := \langle f, f \rangle^{1/2} = \left(\frac{1}{M} \int_{\mathbb{X}} f^2(\mathbf{x}) d\mathbf{x} \right)^{1/2}. \tag{6B.2}$$

(See Figure 6B.1; of course, this integral may not converge.) The set of all integrable functions on $\mathbb{X}$ with finite L^2-norm is denoted $\mathbf{L}^2(\mathbb{X})$, and is called *L²-space*. For example, any bounded, continuous function $f : \mathbb{X} \longrightarrow \mathbb{R}$ is in $\mathbf{L}^2(\mathbb{X})$.

Example 6B.2
(a) Suppose $\mathbb{X} = [0, 3]$, as in Example 6B.1, and let $f(x) = x + 1$. Then $f \in \mathbf{L}^2[0, 3]$ because

$$\|f\|_2^2 = \langle f, f \rangle = \frac{1}{3} \int_0^3 (x + 1)^2 dx$$

$$= \frac{1}{3} \int_0^3 x^2 + 2x + 1 \, dx = \frac{1}{3} \left(\frac{x^3}{3} + x^2 + x \right) \Big|_{x=0}^{x=3} = 7;$$

hence $\|f\|_2 = \sqrt{7} < \infty.$

────────

[2] Or *length*, if $D = 1$, or *area* if $D = 2, \ldots$

Table 6.1. *Inner products on various domains*

	Domain		M	Inner product
Unit interval	$\mathbb{X} = [0, 1] \subset \mathbb{R}$	length	$M = 1$	$\langle f, g \rangle = \displaystyle\int_0^1 f(x) \cdot g(x) \, dx$
π interval	$\mathbb{X} = [0, \pi] \subset \mathbb{R}$	length	$M = \pi$	$\langle f, g \rangle = \dfrac{1}{\pi} \displaystyle\int_0^\pi f(x) \cdot g(x) \, dx$
Unit square	$\mathbb{X} = [0, 1] \times [0, 1] \subset \mathbb{R}^2$	area	$M = 1$	$\langle f, g \rangle = \displaystyle\int_0^1 \int_0^1 f(x, y) \cdot g(x, y) \, dx \, dy$
$\pi \times \pi$ square	$\mathbb{X} = [0, \pi] \times [0, \pi] \subset \mathbb{R}^2$	area	$M = \pi^2$	$\langle f, g \rangle = \dfrac{1}{\pi^2} \displaystyle\int_0^\pi \int_0^\pi f(x, y) \cdot g(x, y) \, dx \, dy$
Unit disk (polar coordinates)	$\mathbb{X} = \{(r, \theta); \ r \le 1\} \subset \mathbb{R}^2$	area	$M = \pi$	$\langle f, g \rangle = \dfrac{1}{\pi} \displaystyle\int_0^1 \int_0^\pi \int_{-\pi}^\pi f(r, \theta) \cdot g(r, \theta) \, r \cdot d\theta \, dr$
Unit cube	$\mathbb{X} = [0, 1] \times [0, 1] \times [0, 1] \subset \mathbb{R}^3$	volume	$M = 1$	$\langle f, g \rangle = \displaystyle\int_0^1 \int_0^1 \int_0^1 f(x, y, z) \cdot g(x, y, z) \, dx \, dy \, dz$

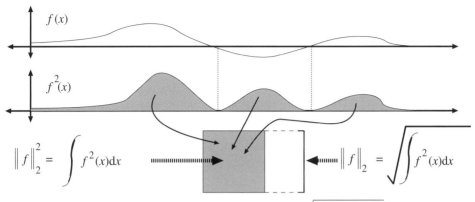

Figure 6B.1. The L^2-norm of f: $\|f\|_2 = \sqrt{\int_{\mathbb{X}} |f(x)|^2 \, dx}$.

(b) Let $\mathbb{X} = (0, 1]$, and suppose $f \in \mathcal{C}^\infty(0, 1]$ is defined as $f(x) := 1/x$. Then $\|f\|_2 = \infty$, so $f \notin \mathbf{L}^2(0, 1]$. ◇

Remark Some authors define the inner product as $\langle f, g \rangle := \int_{\mathbb{X}} f(\mathbf{x}) \cdot g(\mathbf{x}) d\mathbf{x}$ and define the L^2-norm as $\|f\|_2 := (\int_{\mathbb{X}} f^2(\mathbf{x})d\mathbf{x})^{1/2}$. In other words, these authors do *not* divide by the volume M of the domain. This yields a mathematically equivalent theory. The advantage of our definition is greater computational convenience in some situations. (For example, if $\mathbb{1}_{\mathbb{X}}$ is the constant 1-valued function, then, in our definition, $\|\mathbb{1}_{\mathbb{X}}\|_2 = 1$.) When comparing formulae from different books, you should always check their respective definitions of L^2-norm.

L^2-space on an infinite domain Suppose $\mathbb{X} \subset \mathbb{R}^D$ is a region of *infinite* volume (or length, area, etc.). For example, maybe $\mathbb{X} = \mathbb{R}_+$ is the *positive half-line*, or perhaps $\mathbb{X} = \mathbb{R}^D$. In this case, $M = \infty$, so it doesn't make any sense to divide by M. If $f, g : \mathbb{X} \longrightarrow \mathbb{R}$ are integrable functions, then the *inner product* of f and g is defined:

$$\langle f, g \rangle := \int_{\mathbb{X}} f(\mathbf{x}) \cdot g(\mathbf{x})d\mathbf{x}. \tag{6B.3}$$

Example 6B.3 Suppose $\mathbb{X} = \mathbb{R}$. If

$$f(x) = e^{-|x|} \quad \text{and} \quad g(x) = \begin{cases} 1 & \text{If } 0 < x < 7; \\ 0 & \text{otherwise,} \end{cases}$$

then

$$\langle f, g \rangle = \int_{-\infty}^{\infty} f(x)g(x)dx = \int_0^7 e^{-x}dx$$

$$= -(e^{-7} - e^0) = 1 - \frac{1}{e^7}. \qquad ◇$$

The L^2-*norm* of an integrable function $f : \mathbb{X} \longrightarrow \mathbb{R}$ is defined:

$$\|f\|_2 = \langle f, f \rangle^{1/2} = \left(\int_\mathbb{X} f^2(\mathbf{x}) d\mathbf{x} \right)^{1/2}. \tag{6B.4}$$

Again, this integral may not converge. Indeed, even if f is bounded and continuous everywhere, this integral may still equal infinity. The set of all integrable functions on $\mathbb{X}$ with finite L^2-norm is denoted $\mathbf{L}^2(\mathbb{X})$, and is called L^2-*space*. (You may recall that on p. 42 of §3A we discussed how L^2-space arises naturally in quantum mechanics as the space of 'physically meaningful' wavefunctions.)

Proposition 6B.4 Properties of the inner product
Whether it is defined using equation (6B.1) *or* (6B.3), *the inner product has the following properties.*

(a) Bilinearity *For any $f_1, f_2, g_1, g_2 \in \mathbf{L}^2(\mathbb{X})$, and any constants $r_1, r_2, s_1, s_2 \in \mathbb{R}$,*

$$\langle r_1 f_1 + r_2 f_2, s_1 g_1 + s_2 g_2 \rangle = r_1 s_1 \langle f_1, g_1 \rangle + r_1 s_2 \langle f_1, g_2 \rangle$$
$$+ r_2 s_1 \langle f_2, g_1 \rangle + r_2 s_2 \langle f_2, g_2 \rangle .$$

(b) Symmetry *For any $f, g \in \mathbf{L}^2(\mathbb{X})$, $\langle f, g \rangle = \langle g, f \rangle$.*
(c) Positive-definite *For any $f \in \mathbf{L}^2(\mathbb{X})$, $\langle f, f \rangle \geq 0$. Also, $\langle f, f \rangle = 0$ if and only if $f = 0$.*

Ⓔ *Proof* **Exercise 6B.1.**   □

If $\mathbf{v}, \mathbf{w} \in \mathbb{R}^D$, recall that $\langle \mathbf{v}, \mathbf{w} \rangle = \|\mathbf{v}\| \cdot \|\mathbf{w}\| \cdot \cos(\theta)$, where θ is the angle between $\mathbf{v}$ to $\mathbf{w}$. In particular, this implies that

$$\left| \langle \mathbf{v}, \mathbf{w} \rangle \right| \leq \|\mathbf{v}\| \cdot \|\mathbf{w}\|. \tag{6B.5}$$

If $f, g \in \mathbf{L}^2(\mathbb{X})$ are two functions, then it doesn't make sense to talk about the 'angle' between f and g (as 'vectors' in $\mathbf{L}^2(\mathbb{X})$). But an inequality analogous to equation (6B.5) is still true.

Theorem 6B.5 Cauchy–Bunyakowski–Schwarz inequality
Let $f, g \in \mathbf{L}^2(\mathbb{X})$. Then $|\langle f, g \rangle| \leq \|f\|_2 \cdot \|g\|_2$.

Proof Let $A = \|g\|_2^2$, $B := \langle f, g \rangle$, and $C := \|f\|_2^2$; thus, we are trying to show that $B \leq \sqrt{A} \cdot \sqrt{C}$. Define $q : \mathbb{R} \longrightarrow \mathbb{R}$ by $q(t) := \|f - t \cdot g\|_2^2$. Then

$$q(t) = \langle f - t \cdot g, f - t \cdot g \rangle \underset{(b)}{=} \langle f, f \rangle - t \langle f, g \rangle - t \langle g, f \rangle + t^2 \langle g, g \rangle$$
$$= \|f\|_2^2 + 2 \langle f, g \rangle t + \|g\|_2^2 t^2 = C + 2Bt + At^2, \tag{6B.6}$$

a quadratic polynomial in t. (Here, step (b) is by Proposition 6B.4(a).)

Now, $q(t) = \|f - t \cdot g\|_2^2 \geq 0$ for all $t \in \mathbb{R}$; thus, $q(t)$ has at most one root, so the discriminant of the quadratic polynomial (6B.6) is not positive. That is $4B^2 - 4AC \leq 0$. Thus, $B^2 \leq AC$, and thus, $B \leq \sqrt{A} \cdot \sqrt{C}$, as desired. $\qquad\square$

Note The CBS inequality involves three integrals: $\langle f, g \rangle$, $\|f\|_2$, and $\|g\|_2$. But the proof of Theorem 6B.5 *does not involve any integrals at all.* Instead, it just uses simple algebraic manipulations of the inner product operator. In particular, this means that the same proof works whether we define the inner product using equation (6B.1) or equation (6B.3). Indeed, the CBS inequality is not really about L^2-spaces, *per se* – it is actually a theorem about a much broader class of abstract geometric structures, called *inner product spaces*. An enormous amount of knowledge about $\mathbf{L}^2(\mathbb{X})$ can be obtained from this abstract geometric approach, usually through simple algebraic arguments like the proof of Theorem 6B.5 (i.e. without lots of messy integration technicalities). This is the beginning of a beautiful area of mathematics called *Hilbert space theory* (see Conway (1990) for an excellent introduction).

6C* More about L^2-space

Prerequisites: §6B, Appendix C.

This section contains some material which is not directly germane to the solution methods we present later in the book, but may be interesting to some students who want a broader perspective.

6C(i) Complex L^2-space

In §6B we introduced the inner product for real-valued functions. The inner product for complex-valued functions is slightly different. For any $z = x + y\mathbf{i} \in \mathbb{C}$, let $\bar{z} := x - y\mathbf{i}$ denote the *complex conjugate* of z. Let $\mathbb{X} \subset \mathbb{R}^D$ be some domain, and let $f, g : \mathbb{X} \longrightarrow \mathbb{C}$ be complex-valued functions. We define:

$$\langle f, g \rangle := \int_{\mathbb{X}} f(\mathbf{x}) \cdot \overline{g(\mathbf{x})} d\mathbf{x}. \tag{6C.1}$$

If g is real-valued, then $\bar{g} = g$, and then equation (6C.1) is equivalent to equation (6B.4).

For any $z \in \mathbb{C}$, recall that $z \cdot \bar{z} = |z|^2$. Thus, if f is a complex-valued function, then $f(x)\overline{f}(x) = |f(x)|^2$. It follows that we can define the L^2-*norm* of an integrable function $f : \mathbb{X} \longrightarrow \mathbb{C}$ just as before:

$$\|f\|_2 = \langle f, f \rangle^{1/2} = \left(\int_{\mathbb{X}} |f|^2(\mathbf{x}) d\mathbf{x} \right)^{1/2},$$

and this quantity will always be a real number (when the integral converges). We define $\mathbf{L}^2(\mathbb{X}; \mathbb{C})$ to be the set of all integrable functions $f : \mathbb{X} \longrightarrow \mathbb{C}$ such that $\|f\|_2 < \infty$.[3]

Proposition 6C.1 Properties of the complex inner product
The inner product on $\mathbf{L}^2(\mathbb{X}; \mathbb{C})$ has the following properties.

(a) *Sesquilinearity*　*For any $f_1, f_2, g_1, g_2 \in \mathbf{L}^2(\mathbb{X}; \mathbb{C})$, and any constants $b_1, b_2, c_1,$ $c_2 \in \mathbb{C}$,*

$$\langle b_1 f_1 + b_2 f_2, c_1 g_1 + c_2 g_2 \rangle = b_1 \bar{c}_1 \langle f_1, g_1 \rangle + b_1 \bar{c}_2 \langle f_1, g_2 \rangle$$
$$+ b_2 \bar{c}_1 \langle f_2, g_1 \rangle + b_2 \bar{c}_2 \langle f_2, g_2 \rangle.$$

(b) *Hermitian*　*For any $f, g \in \mathbf{L}^2(\mathbb{X}; \mathbb{C})$, $\langle f, g \rangle = \overline{\langle g, f \rangle}$.*
(c) *Positive-definite*　*For any $f \in \mathbf{L}^2(\mathbb{X}; \mathbb{C})$, $\langle f, f \rangle$ is a real number and $\langle f, f \rangle \geq 0$. Also, $\langle f, f \rangle = 0$ if and only if $f = 0$.*
(e) *CBS inequality*　*For any $f, g \in \mathbf{L}^2(\mathbb{X}; \mathbb{C})$, $|\langle f, g \rangle| \leq \|f\|_2 \cdot \|g\|_2$.*

Ⓔ　*Proof* **Exercise 6C.1.** *Hint:* Imitate the proofs of Proposition 6B.4 and Theorem 6B.5. In your proof of the CBS inequality, don't forget that $\langle f, g \rangle + \overline{\langle f, g \rangle} = 2\mathrm{Re}\,[\langle f, g \rangle]$. ☐

6C(ii)　*Riemann vs. Lebesgue integrals*

We have defined $\mathbf{L}^2(\mathbb{X})$ to be the set of all 'integrable' functions on $\mathbb{X}$ with finite L^2-norm, but we have been somewhat vague about what we mean by 'integrable'. The most familiar and elementary integral is the *Riemann integral*. For example, if $\mathbb{X} = [a, b]$, and $f : \mathbb{X} \longrightarrow \mathbb{R}$, then the Riemann integral of f is defined:

$$\int_a^b f(x)\mathrm{d}x := \lim_{N \to \infty} \frac{b-a}{N} \sum_{n=1}^{N} f\left(a + \frac{n(b-a)}{N}\right). \tag{6C.2}$$

A similar (but more complicated) definition can be given if $\mathbb{X}$ is an arbitrary domain in $\mathbb{R}^D$. We say f is *Riemann integrable* if the limit (6C.2) exists and is finite.

However, this is not what we mean here by 'integrable'. The problem is that the limit (6C.2) only exists if the function f is reasonably 'nice' (e.g. piecewise continuous). We need an integral which works even for extremely 'nasty' functions (e.g. functions which are discontinuous everywhere; functions which have a 'fractal' structure, etc.). This object is called the *Lebesgue integral*; its definition is similar to that in equation (6C.2) but much more complicated.

Loosely speaking, the 'Riemann sum' in equation (6C.2) chops the interval $[a, b]$ up into N equal subintervals. The corresponding sum in the Lebesgue integral

---

[3] We are using $\mathbf{L}^2(\mathbb{X})$ to refer only to real-valued functions. In more advanced books, the notation $L^2(\mathbb{X})$ denotes the set of *complex*-valued L^2-functions; if one wants to refer only to *real*-valued L^2-functions, one must use the notation $L^2(\mathbb{X}; \mathbb{R})$.

allows us to chop $[a, b]$ into any number of 'Borel-measurable subsets'. A 'Borel-measurable subset' is any open set, any closed set, any (countably infinite) union or intersection of open or closed sets, any (countably infinite) union or intersection of *these* sets, etc. Clearly 'measurable subsets' can become very complex. The Lebesgue integral is obtained by taking a limit over all possible 'Riemann sums' obtained using such 'measurable partitions' of $[a, b]$. This is a very versatile and powerful construction, which can integrate incredibly bizarre and pathological functions. (See Remark 10D.3, p. 213, for further discussion of Riemann vs. Lebesgue integration.)

You might ask, Why would I want to integrate bizarre and pathological functions? Indeed, the sorts of functions which arise in applied mathematics are almost always piecewise continuous, and, for them, the Riemann integral works just fine. To answer this, consider the difference between the following two equations:

$$x^2 = \frac{16}{9};$$
(6C.3a)

$$x^2 = 2.$$
(6C.3b)

Both equations have solutions, but they are different. The solutions to equation (6C.3a) are rational numbers, for which we have an *exact* expression $x = \pm 4/3$. The solutions to equation (6C.3b) are irrational numbers, for which we have only approximate expressions: $x = \pm\sqrt{2} \approx \pm 1.414213562\ldots$

Irrational numbers are 'pathological': they do not admit nice, simple, exact expressions like $4/3$. We might be inclined to ignore such pathological objects in our mathematics – to pretend they don't exist. Indeed, this was precisely the attitude of the ancient Greeks, whose mathematics was based entirely on rational numbers. The problem is that, in this 'ancient Greek' mathematical universe, equation (6C.3b) *has no solution*. This is not only inconvenient, it is profoundly counterintuitive; after all, $\sqrt{2}$ is simply the length of the hypotenuse of a right angle triangle whose other sides both have length 1. And surely the *sidelength* of a triangle should be a number.

Furthermore we can find rational numbers which seem to be arbitrarily good *approximations* to a solution of equation (6C.3b). For example,

$$\left(\frac{1414}{100}\right)^2 = 1.999\,396;$$

$$\left(\frac{1\,414\,213}{100\,000}\right)^2 = 1.999\,998\,409;$$

$$\left(\frac{141\,421\,356}{10\,000\,000}\right)^2 = 1.999\,999\,993;$$

$$\vdots \quad \vdots \quad \vdots$$

It certainly seems like this sequence of rational numbers is converging to 'something'. Our name for that 'something' is $\sqrt{2}$. In fact, this is the only way we can *ever* specify $\sqrt{2}$. Since we cannot express $\sqrt{2}$ as a fraction or some simple decimal expansion, we can only say, '$\sqrt{2}$ is the number to which the above sequence of rational numbers seems to be converging'.

But how do we know that any such number exists? Couldn't there just be a 'hole' in the real number line where we think $\sqrt{2}$ is supposed to be? The answer is that the set $\mathbb{R}$ is *complete* – that is, any sequence in $\mathbb{R}$ which 'looks like it is converging'[4] does, in fact, converge to some limit point in $\mathbb{R}$. Because $\mathbb{R}$ is complete, we are confident that $\sqrt{2}$ exists, even though we can never precisely specify its value.

Now let's return to $\mathbf{L}^2(\mathbb{X})$. Like the real line $\mathbb{R}$, the space $\mathbf{L}^2(\mathbb{X})$ has a *geometry*: a notion of 'distance' defined by the L^2-norm $\|\bullet\|_2$. This geometry provides us with a notion of *convergence* in $\mathbf{L}^2(\mathbb{X})$ (see §6E(i), p. 122). Like $\mathbb{R}$, we would like $\mathbf{L}^2(\mathbb{X})$ to be *complete*, so that any sequence of functions which 'looks like it is converging' does, in fact, converge to some limit point in $\mathbf{L}^2(\mathbb{X})$.

Unfortunately, a sequence of perfectly 'nice' functions in $\mathbf{L}^2(\mathbb{X})$ can converge to a totally 'pathological' limit function, the same way that a sequence of 'nice' rational numbers can converge to an irrational number. If we exclude the pathological functions from $\mathbf{L}^2(\mathbb{X})$, we will be like the ancient Greeks, who excluded irrational numbers from their mathematics. We will encounter situations where a certain equation 'should' have a solution, but *doesn't*, just as the Greeks discovered that the equation $x^2 = 2$ had no solution in their mathematics.

Thus our definition of $\mathbf{L}^2(\mathbb{X})$ *must* include some pathological functions. But if these pathological functions are in $\mathbf{L}^2(\mathbb{X})$, and $\mathbf{L}^2(\mathbb{X})$ is defined as the set of elements with finite norm, and the norm $\|f\|_2$ is defined using an integral like equation (6B.2), then we must have a way of integrating these pathological functions. Hence the necessity of the Lebesgue integral.

Fortunately, all the functions we will encounter in this book are Riemann integrable. For the purposes of solving PDEs, you do not need to know how to compute the Lebesgue integral. But it is important to know that it exists, and that somewhere, in the background, its presence is making all the mathematics work properly.

6D Orthogonality

Prerequisites: §6A.

Two functions $f, g \in \mathbf{L}^2(\mathbb{X})$ are *orthogonal* if $\langle f, g \rangle = 0$. Intuitively, this means that f and g are 'perpendicular' vectors in the infinite-dimensional vector space $\mathbf{L}^2(\mathbb{X})$.

----

[4] Technically, any *Cauchy sequence*.

Example 6D.1 Treat sin and cos as elements of $\mathbf{L}^2[-\pi, \pi]$. Then they are orthogonal:

$$\langle \sin, \cos \rangle = \frac{1}{2\pi} \int_{-\pi}^{\pi} \sin(x) \cos(x) dx = 0. \qquad \diamond$$

Exercise 6D.1 Verify Example 6D.1. ◆ Ⓔ

An *orthogonal set* of functions is a set $\{f_1, f_2, f_3, \ldots\}$ of elements in $\mathbf{L}^2(\mathbb{X})$ such that $\langle f_j, f_k \rangle = 0$ whenever $j \neq k$. If, in addition, $\|f_j\|_2 = 1$ for all j, then we say this is an *orthonormal set* of functions. Fourier analysis is based on the orthogonality of certain families of trigonometric functions. Example 6D.1 was an example of this, which generalizes as follows.

Proposition 6D.2 Trigonometric orthogonality on $[-\pi, \pi]$

For every $n \in \mathbb{N}$, define the functions $\mathbf{S}_n, \mathbf{C}_n : [-\pi, \pi] \longrightarrow \mathbb{R}$ by $\mathbf{S}_n(x) := \sin(nx)$ and $\mathbf{C}_n(x) := \cos(nx)$, for all $x \in [-\pi, \pi]$ (see Figure 6D.1). Then the set $\{\mathbf{C}_0, \mathbf{C}_1, \mathbf{C}_2, \ldots; \mathbf{S}_1, \mathbf{S}_2, \mathbf{S}_3, \ldots\}$ is an orthogonal set of functions for $\mathbf{L}^2[-\pi, \pi]$. In other words,

(a) $\langle \mathbf{S}_n, \mathbf{S}_m \rangle = \frac{1}{2\pi} \int_{-\pi}^{\pi} \sin(nx) \sin(mx) dx = 0$, *whenever $n \neq m$;*
(b) $\langle \mathbf{C}_n, \mathbf{C}_m \rangle = \frac{1}{2\pi} \int_{-\pi}^{\pi} \cos(nx) \cos(mx) dx = 0$, *whenever $n \neq m$;*
(c) $\langle \mathbf{S}_n, \mathbf{C}_m \rangle = \frac{1}{2\pi} \int_{-\pi}^{\pi} \sin(nx) \cos(mx) dx = 0$, *for any n and m.*
(d) *However, these functions are not orthonormal, because they do not have unit norm. Instead, for any $n \neq 0$,*

$$\|\mathbf{C}_n\|_2 = \sqrt{\frac{1}{2\pi} \int_{-\pi}^{\pi} \cos(nx)^2 dx} = \frac{1}{\sqrt{2}},$$

and

$$\|\mathbf{S}_n\|_2 = \sqrt{\frac{1}{2\pi} \int_{-\pi}^{\pi} \sin(nx)^2 dx} = \frac{1}{\sqrt{2}}.$$

Proof **Exercise 6D.2.** *Hint:* Use the following trigonometric identities: $2 \sin(\alpha)$ Ⓔ $\cos(\beta) = \sin(\alpha + \beta) + \sin(\alpha - \beta)$, $2 \sin(\alpha) \sin(\beta) = \cos(\alpha - \beta) - \cos(\alpha + \beta)$, and $2 \cos(\alpha) \cos(\beta) = \cos(\alpha + \beta) + \cos(\alpha - \beta)$. □

Remark Note that $\mathbf{C}_0(x) = 1$ is just the *constant* function.

It is important to remember that the statement 'f and g are orthogonal' depends upon the *domain* $\mathbb{X}$ which we are considering. For example, compare the following result to Proposition 6D.2.

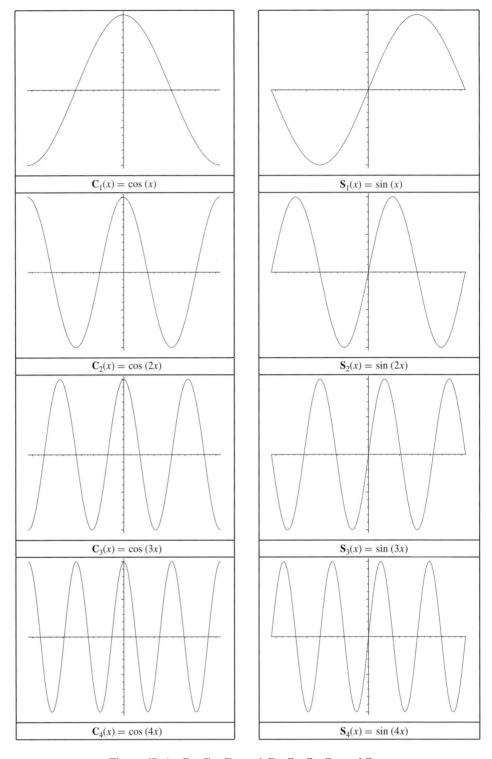

Figure 6D.1. $\mathbf{C}_1$, $\mathbf{C}_2$, $\mathbf{C}_3$, and $\mathbf{C}_4$; $\mathbf{S}_1$, $\mathbf{S}_2$, $\mathbf{S}_3$, and $\mathbf{S}_4$.

Proposition 6D.3 Trigonometric orthogonality on $[0, L]$

Let $L > 0$, and, for every $n \in \mathbb{N}$, define the functions $\mathbf{S}_n, \mathbf{C}_n : [0, L] \longrightarrow \mathbb{R}$ by $\mathbf{S}_n(x) := \sin\left(\frac{n\pi x}{L}\right)$ and $\mathbf{C}_n(x) := \cos\left(\frac{n\pi x}{L}\right)$, for all $x \in [0, L]$.

(a) *The set $\{\mathbf{C}_0, \mathbf{C}_1, \mathbf{C}_2, \ldots\}$ is an* orthogonal set *of functions for $\mathbf{L}^2[0, L]$. In other words,*

$$\langle \mathbf{C}_n, \mathbf{C}_m \rangle = \frac{1}{L} \int_0^L \cos\left(\frac{n\pi}{L}x\right) \cos\left(\frac{m\pi}{L}x\right) dx = 0,$$

whenever $n \neq m$. However, these functions are not orthonormal, *because they do not have unit norm. Instead, for any $n \neq 0$,*

$$\|\mathbf{C}_n\|_2 = \sqrt{\frac{1}{L} \int_0^L \cos\left(\frac{n\pi}{L}x\right)^2 dx} = \frac{1}{\sqrt{2}}.$$

(b) *The set $\{\mathbf{S}_1, \mathbf{S}_2, \mathbf{S}_3, \ldots\}$ is an* orthogonal set *of functions for $\mathbf{L}^2[0, L]$. In other words,*

$$\langle \mathbf{S}_n, \mathbf{S}_m \rangle = \frac{1}{L} \int_0^L \sin\left(\frac{n\pi}{L}x\right) \sin\left(\frac{m\pi}{L}x\right) dx = 0,$$

whenever $n \neq m$. However, these functions are not orthonormal, *because they do not have unit norm. Instead, for any $n \neq 0$,*

$$\|\mathbf{S}_n\|_2 = \sqrt{\frac{1}{L} \int_0^L \sin\left(\frac{n\pi}{L}x\right)^2 dx} = \frac{1}{\sqrt{2}}.$$

(c) *The functions $\mathbf{C}_n$ and $\mathbf{S}_m$ are* not *orthogonal to one another on $[0, L]$. Instead*

$$\langle \mathbf{S}_n, \mathbf{C}_m \rangle = \frac{1}{L} \int_0^L \sin\left(\frac{n\pi}{L}x\right) \cos\left(\frac{m\pi}{L}x\right) dx$$

$$= \begin{cases} 0 & \text{if } n + m \text{ is even}; \\[2mm] \dfrac{2n}{\pi(n^2 - m^2)} & \text{if } n + m \text{ is odd.} \end{cases}$$

Proof **Exercise 6D.3.** □ Ⓔ

The trigonometric functions are just one of several important orthogonal sets of functions. Different orthogonal sets are useful for different domains or different applications. For example, in some cases, it is convenient to use a collection of orthogonal *polynomial* functions. Several orthogonal polynomial families exist, including the *Legendre polynomials* (see §16D, p. 365), the *Chebyshev polynomials* (see Exercise 14B.1(e), p. 282), the *Hermite polynomials*, and the *Laguerre polynomials*. See Broman (1989), chap. 3, for a good introduction.

In the study of partial differential equations, the following fact is particularly important:

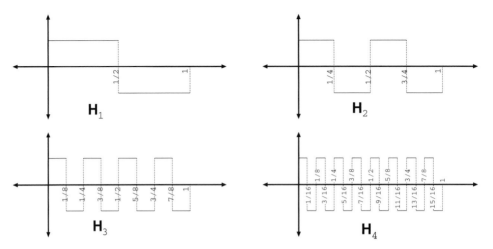

Figure 6D.2. Four Haar Basis elements: H_1, H_2, H_3, H_4.

Let $X \subset \mathbb{R}^D$ be any domain. If f, $g : X \longrightarrow \mathbb{C}$ are two eigenfunctions of the Laplacian with different eigenvalues, then f and g are orthogonal in $L^2(X)$.

(See Proposition 15E.9, p. 351, for a precise statement of this.) Because of this, we can have orthogonal sets whose members are eigenfunctions of the Laplacian (see Theorem 15E.12, p. 353). These orthogonal sets are the 'building blocks' with which we can construct solutions to a PDE satisfying prescribed initial conditions or boundary conditions. This is the basic strategy behind the solution methods of Chapters 11–14.

Ⓔ **Exercise 6D.4** Figure 6D.2 portrays the *the Haar basis*. We define $H_0 \equiv 1$, and, for any natural number $N \in \mathbb{N}$, we define the Nth *Haar function* $H_N : [0, 1] \longrightarrow \mathbb{R}$ as follows:

$$H_N(x) = \begin{cases} 1 & \text{if } \dfrac{2n}{2^N} \leq x < \dfrac{2n+1}{2^N}, & \text{for some } n \in [0 \ldots 2^{N-1}); \\[3mm] -1 & \text{if } \dfrac{2n+1}{2^N} \leq x < \dfrac{2n+2}{2^N}, & \text{for some } n \in [0 \ldots 2^{N-1}). \end{cases}$$

(a) Show that the set $\{H_0, H_1, H_2, H_3, \ldots\}$ is an orthonormal set in $L^2[0, 1]$.
(b) There is another way to define the Haar basis. First recall that any number $x \in [0, 1]$ has a unique *binary expansion* of the form

$$x = \frac{x_1}{2} + \frac{x_2}{4} + \frac{x_3}{8} + \frac{x_4}{16} + \cdots + \frac{x_n}{2^n} + \cdots,$$

where $x_1, x_2, x_3, x_4, \ldots$ are all either 0 or 1. Show that, for any $n \geq 1$,

$$H_n(x) = (-1)^{x_n} = \begin{cases} 1 & \text{if } x_n = 0; \\ -1 & \text{if } x_n = 1. \end{cases} \qquad \blacklozenge$$

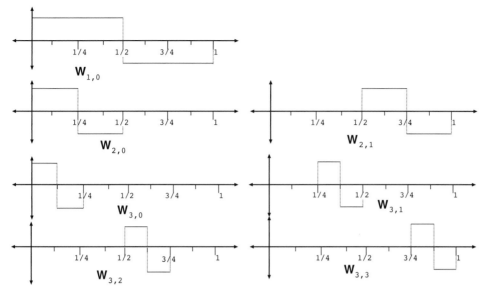

Figure 6D.3. Seven wavelet basis elements: $\mathbf{W}_{1,0}$; $\mathbf{W}_{2,0}$, $\mathbf{W}_{2,1}$; $\mathbf{W}_{3,0}$, $\mathbf{W}_{3,1}$, $\mathbf{W}_{3,2}$, $\mathbf{W}_{3,3}$.

Exercise 6D.5   Figure 6D.3 portrays a *wavelet basis*. We define $\mathbf{W}_0 \equiv 1$, and, for any $N \in \mathbb{N}$ and $n \in [0 \ldots 2^{N-1})$, we define: ⓔ

$$
\mathbf{W}_{n;N}(x) = \begin{cases} 1 & \text{if } \dfrac{2n}{2^N} \leq x < \dfrac{2n+1}{2^N}; \\[2mm] -1 & \text{if } \dfrac{2n+1}{2^N} \leq x < \dfrac{2n+2}{2^N}; \\[2mm] 0 & \text{otherwise.} \end{cases}
$$

Show that the set

$$\{\mathbf{W}_0; \mathbf{W}_{1,0}; \mathbf{W}_{2,0}, \mathbf{W}_{2,1}; \mathbf{W}_{3,0}, \mathbf{W}_{3,1}, \mathbf{W}_{3,2}, \mathbf{W}_{3,3}; \mathbf{W}_{4,0}, \ldots, \mathbf{W}_{4,7};$$
$$\mathbf{W}_{5,0}, \ldots, \mathbf{W}_{5,15}; \ldots\}$$

is an *orthogonal* set in $\mathbf{L}^2[0, 1]$, but is *not* orthonormal: for any N and n, we have $\|\mathbf{W}_{n;N}\|_2 = \frac{1}{2^{(N-1)/2}}$. ◆

6E Convergence concepts

Prerequisites: §4A.

If $\{x_1, x_2, x_3, \ldots\}$ is a sequence of numbers, we know what it means to say '$\lim_{n\to\infty} x_n = x$'. We can think of convergence as a kind of 'approximation'.

Heuristically speaking, if the sequence $\{x_n\}_{n=1}^{\infty}$ converges to x, then, for very large n, the number x_n is a *good approximation* of x.

If $\{f_1, f_2, f_3, \dots\}$ is a sequence of functions, and f is some other function, then we might want to say that '$\lim_{n\to\infty} f_n = f$'. We again imagine convergence as a kind of 'approximation'. Heuristically speaking, if the sequence $\{f_n\}_{n=1}^{\infty}$ converges to f, then, for very large n, the function f_n is a *good approximation* of f.

There are several ways we can interpret 'good approximation', and these in turn lead to several different notions of 'convergence'. Thus, convergence of *functions* is a much more subtle concept that convergence of *numbers*. We will deal with *three* kinds of convergence here: L^2-convergence, *pointwise* convergence, and *uniform* convergence.

6E(i) L^2-convergence

Let $\mathbb{X} \subset \mathbb{R}^D$ be some domain, and define

$$
M := \begin{cases} \displaystyle\int_{\mathbb{X}} 1 \, d\mathbf{x} & \text{if } \mathbb{X} \text{ is a finite domain;} \\ 1 & \text{if } \mathbb{X} \text{ is an infinite domain.} \end{cases}
$$

If $f, g \in \mathbf{L}^2(\mathbb{X})$, then the L^2-*distance* between f and g is just

$$
\| f - g \|_2 := \left(\frac{1}{M} \int_{\mathbb{X}} |f(\mathbf{x}) - g(\mathbf{x})|^2 \, d\mathbf{x} \right)^{1/2},
$$

If we think of f as an 'approximation' of g, then $\|f - g\|_2$ measures the *root-mean-squared error* of this approximation.

Lemma 6E.1 $\|\bullet\|_2$ *is a norm. That is:*

(a) *for any $f : \mathbb{X} \longrightarrow \mathbb{R}$ and $r \in \mathbb{R}$, $\|r \cdot f\|_2 = |r| \cdot \|f\|_2$;*
(b) *(triangle inequality) For any $f, g : \mathbb{X} \longrightarrow \mathbb{R}$, $\|f + g\|_2 \leq \|f\|_2 + \|g\|_2$;*
(c) *for any $f : \mathbb{X} \longrightarrow \mathbb{R}$, $\|f\|_2 = 0$ if and only if $f \equiv 0$.*

ⓔ   *Proof* **Exercise 6E.1.**   □

If $\{f_1, f_2, f_3, \dots\}$ is a sequence of successive approximations of f, then we say the sequence *converges to f in L^2* if $\lim_{n\to\infty} \|f_n - f\|_2 = 0$ (sometimes this is called *convergence in mean square*); see Figure 6E.1. We then write $f = L^2\text{-}\lim_{n\to\infty} f_n$.

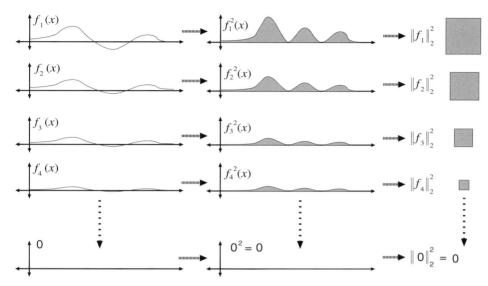

Figure 6E.1. The sequence $\{f_1, f_2, f_3, \ldots\}$ converges to the constant 0 function in $\mathbf{L}^2(\mathbb{X})$.

Example 6E.2 In each of the following examples, let $\mathbb{X} = [0, 1]$.

(a) Suppose

$$f_n(x) = \begin{cases} 1 & \text{if } 1/n < x < 2/n; \\ 0 & \text{otherwise.} \end{cases}$$

(See Figure 6E.2(a).) Then $\|f_n\|_2 = \frac{1}{\sqrt{n}}$ (**Exercise 6E.2**). Hence,   Ⓔ

$$\lim_{n\to\infty} \|f_n\|_2 = \lim_{n\to\infty} \frac{1}{\sqrt{n}} = 0,$$

so the sequence $\{f_1, f_2, f_3, \ldots\}$ converges to the constant 0 function in $\mathbf{L}^2[0, 1]$.

(b) For all $n \in \mathbb{N}$, let

$$f_n(x) = \begin{cases} n & \text{if } 1/n < x < 2/n; \\ 0 & \text{otherwise.} \end{cases}$$

(See Figure 6E.2(b).) Then $\|f_n\|_2 = \sqrt{n}$ (**Exercise 6E.3**). Hence,   Ⓔ

$$\lim_{n\to\infty} \|f_n\|_2 = \lim_{n\to\infty} \sqrt{n} = \infty,$$

so the sequence $\{f_1, f_2, f_3, \ldots\}$ does *not* converge to zero in $\mathbf{L}^2[0, 1]$.

(c) For each $n \in \mathbb{N}$, let

$$f_n(x) = \begin{cases} 1 & \text{if } |\frac{1}{2} - x| < \frac{1}{n}; \\ 0 & \text{otherwise.} \end{cases}$$

Then the sequence $\{f_n\}_{n=1}^{\infty}$ converges to 0 in L^2 (**Exercise 6E.4**).   Ⓔ

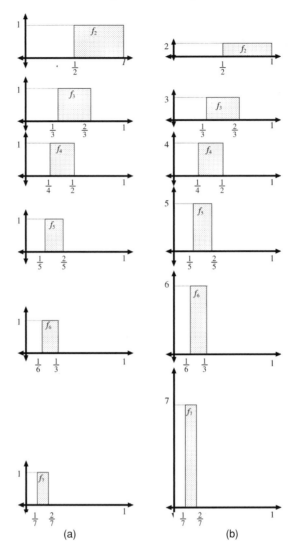

Figure 6E.2. (a) Examples 6E.2(a), 6E.5(a), and 6E.9(a); (b) Examples 6E.2(b) and 6E.5(b).

(d) For all $n \in \mathbb{N}$, let

$$f_n(x) = \frac{1}{1 + n \cdot \left| x - \frac{1}{2} \right|}.$$

Figure 6E.3 portrays elements f_1, f_{10}, f_{100}, and f_{1000}; these strongly suggest that the sequence is converging to the constant 0 function in $\mathbf{L}^2[0, 1]$. The proof of this is **Exercise 6E.5**.

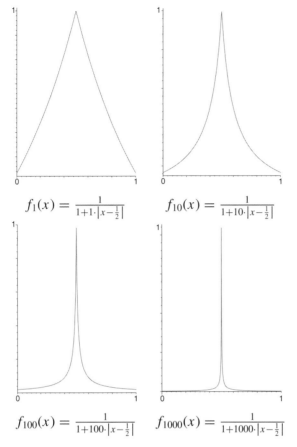

$$f_1(x) = \frac{1}{1+1\cdot|x-\frac{1}{2}|} \qquad f_{10}(x) = \frac{1}{1+10\cdot|x-\frac{1}{2}|}$$

$$f_{100}(x) = \frac{1}{1+100\cdot|x-\frac{1}{2}|} \qquad f_{1000}(x) = \frac{1}{1+1000\cdot|x-\frac{1}{2}|}$$

Figure 6E.3. Examples 6E.2(c) and 6E.5(c). If $f_n(x) = \frac{1}{1+n\cdot|x-\frac{1}{2}|}$, then the sequence $\{f_1, f_2, f_3, \ldots\}$ converges to the constant 0 function in $\mathbf{L}^2[0, 1]$.

(e) Recall the *wavelet* functions from Example 6D.4(b). For any $N \in \mathbb{N}$ and $n \in [0 \ldots 2^{N-1})$, we had

$$\|\mathbf{W}_{N,n}\|_2 = \frac{1}{2^{(N-1)/2}}.$$

Thus, the sequence of wavelet basis elements converges to the constant 0 function in $\mathbf{L}^2[0, 1]$. ◇

Note that, if we define $g_n = f - f_n$ for all $n \in \mathbb{N}$, then

$$\left(f_n \xrightarrow[n\to\infty]{} f \text{ in } L^2 \right) \iff \left(g_n \xrightarrow[n\to\infty]{} 0 \text{ in } L^2 \right).$$

Hence, to understand L^2-convergence in general, it is sufficient to understand L^2-convergence to the constant 0 function.

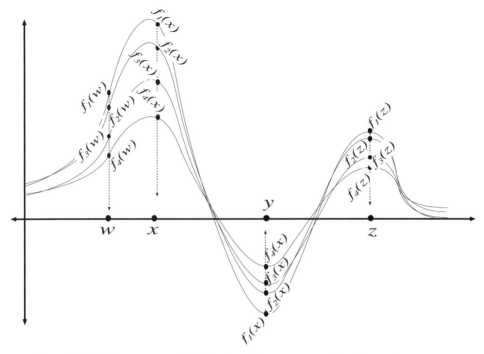

Figure 6E.4. The sequence $\{f_1, f_2, f_3, \ldots\}$ converges *pointwise* to the constant
0 function. Thus, if we pick some random points $w, x, y, z \in \mathbb{X}$, we see that
$\lim_{n\to\infty} f_n(w) = 0$, $\lim_{n\to\infty} f_n(x) = 0$, $\lim_{n\to\infty} f_n(y) = 0$, and $\lim_{n\to\infty} f_n(z) =$
0.

Lemma 6E.3 *The inner product function $\langle \bullet, \bullet \rangle$ is continuous with respect to L^2-convergence. That is, if $\{f_1, f_2, f_3, \ldots\}$ and $\{g_1, g_2, g_3, \ldots\}$ are two sequences of functions in $\mathbf{L}^2(\mathbb{X})$, and $\mathbf{L}^2\text{-}\lim_{n\to\infty} f_n = f$ and $\mathbf{L}^2\text{-}\lim_{n\to\infty} g_n = g$, then $\lim_{n\to\infty} \langle f_n, g_n \rangle = \langle f, g \rangle$.*

Ⓔ *Proof* **Exercise 6E.6**.   □

6E(ii) Pointwise convergence

Convergence in L^2 only means that the *average* approximation error gets small.
It does *not* mean that $\lim_{n\to\infty} f_n(\mathbf{x}) = f(\mathbf{x})$ for every $\mathbf{x} \in \mathbb{X}$. If *this* equation is
true, then we say that the sequence $\{f_1, f_2, \ldots\}$ converges *pointwise* to f (see
Figure 6E.4). We then write $f \equiv \lim_{n\to\infty} f_n$. Pointwise convergence is generally
considered stronger than L^2 convergence because of the following result.

Theorem 6E.4 *Let $\mathbb{X} \subset \mathbb{R}^D$ be a bounded domain, and let $\{f_1, f_2, \ldots\}$ be a
sequence of functions in $\mathbf{L}^2(\mathbb{X})$. Suppose the following.*

(a) *All the functions are uniformly bounded; that is, there is some M > 0 such that* $|f_n(x)| <$ *M for all n* $\in \mathbb{N}$ *and all x* $\in \mathbb{X}$.

(b) *The sequence* $\{f_1, f_2, \ldots\}$ *converges* pointwise *to some function f* $\in L^2(\mathbb{X})$. *Then the sequence* $\{f_1, f_2, \ldots\}$ *also converges to f in* L^2.

Proof **Exercise 6E.7.** *Hint:* You may use the following special case of Lebesgue's ⓔ dominated convergence theorem.[5]

Let $\{g_1, g_2, \ldots\}$ *be a sequence of integrable functions on the domain* $\mathbb{X}$. *Let g* : $\mathbb{X} \longrightarrow \mathbb{R}$ *be another such function. Suppose that*

(a) *There is some some L > 0 such that* $|g_n(x)| < L$ *for all n* $\in \mathbb{N}$ *and all x* $\in \mathbb{X}$.

(b) *For all x* $\in \mathbb{X}$, $\lim_{n\to\infty} g_n(x) = g(x)$.

Then $\lim_{n\to\infty} \int_{\mathbb{X}} g_n(x) dx = \int_{\mathbb{X}} g(x) dx$.

Let $g_n := |f - f_n|^2$ for all $n \in \mathbb{N}$, and let $g = 0$. Apply the dominated convergence theorem. ☐

Example 6E.5 In each of the following examples, let $\mathbb{X} = [0, 1]$.

(a) As in Example 6E.2(a), for each $n \in \mathbb{N}$, let

$$f_n(x) = \begin{cases} 1 & \text{if } 1/n < x < 2/n; \\ 0 & \text{otherwise.} \end{cases}$$

(See Figure 6E.2(a).) The sequence $\{f_n\}_{n=1}^{\infty}$ converges pointwise to the constant 0 function on [0, 1]. Also, as predicted by Theorem 6E.4, the sequence $\{f_n\}_{n=1}^{\infty}$ converges to the constant 0 function in L^2 (see Example 6E.2(a)).

(b) As in Example 6E.2(b), for each $n \in \mathbb{N}$, let

$$f_n(x) = \begin{cases} n & \text{if } 1/n < x < 2/n; \\ 0 & \text{otherwise.} \end{cases}$$

(See Figure 6E.2(b).) Then this sequence converges *pointwise* to the constant 0 function, but does *not* converge to zero in $\mathbf{L}^2[0, 1]$. This illustrates the importance of the *boundedness* hypothesis in Theorem 6E.4.

(c) As in Example 6E.2(c), for each $n \in \mathbb{N}$, let

$$f_n(x) = \begin{cases} 1 & \text{if } |\frac{1}{2} - x| < \frac{1}{n}; \\ 0 & \text{otherwise.} \end{cases}$$

Then the sequence $\{f_n\}_{n=1}^{\infty}$ does *not* converge to 0 in pointwise, although it *does* converge in L^2.

(d) Recall the functions

$$f_n(x) = \frac{1}{1 + n \cdot |x - \frac{1}{2}|}$$

[5] See Folland (1984), Theorem 2.24, p. 53, or Kolmogorov and Fomīn (1975), §30.1, p. 303.

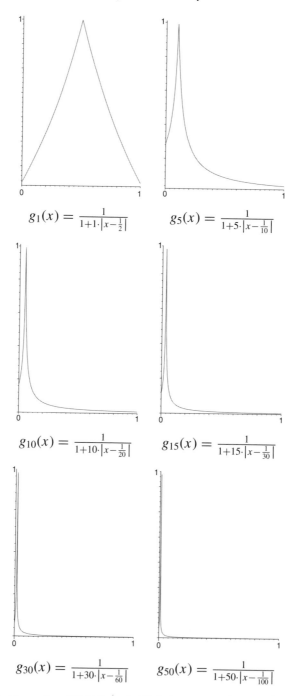

$$g_1(x) = \frac{1}{1+1\cdot\left|x-\frac{1}{2}\right|} \qquad g_5(x) = \frac{1}{1+5\cdot\left|x-\frac{1}{10}\right|}$$

$$g_{10}(x) = \frac{1}{1+10\cdot\left|x-\frac{1}{20}\right|} \qquad g_{15}(x) = \frac{1}{1+15\cdot\left|x-\frac{1}{30}\right|}$$

$$g_{30}(x) = \frac{1}{1+30\cdot\left|x-\frac{1}{60}\right|} \qquad g_{50}(x) = \frac{1}{1+50\cdot\left|x-\frac{1}{100}\right|}$$

Figure 6E.5. Examples 6E.5(d) and 6E.9(d). If $g_n(x) = \frac{1}{1+n\cdot\left|x-\frac{1}{2n}\right|}$, then the sequence $\{g_1, g_2, g_3, \ldots\}$ converges pointwise to the constant 0 function on $[0, 1]$.

from Example 6E.2(d). This sequence of functions converges to zero in $\mathbf{L}^2[0, 1]$; however, it does *not* converge to zero pointwise (**Exercise 6E.8**).   (E)

(e) For all $n \in \mathbb{N}$, let

$$g_n(x) = \frac{1}{1 + n \cdot \left| x - \frac{1}{2n} \right|}.$$

Figure 6E.5 portrays elements $g_1, g_5, g_{10}, g_{15}, g_{30}$, and g_{50}; these strongly suggest that the sequence is converging pointwise to the constant 0 function on $[0, 1]$. The proof of this is **Exercise 6E.9**.   (E)

(f) Recall from Example 6E.2(e) that the sequence of wavelet basis elements $\{\mathbf{W}_{N;n}\}$ converges to zero in $\mathbf{L}^2[0, 1]$. Note, however, that it does *not* converge to zero pointwise (**Exercise 6E.10**).   ◇ (E)

Note that, if we define $g_n = f - f_n$ for all $n \in \mathbb{N}$, then

$$\left(f_n \xrightarrow[n \to \infty]{} f \text{ pointwise} \right) \iff \left(g_n \xrightarrow[n \to \infty]{} 0 \text{ pointwise} \right).$$

Hence, to understand pointwise convergence in general, it is sufficient to understand pointwise convergence to the constant 0 function.

6E(iii) Uniform convergence

There is an even stronger form of convergence. If $f : \mathbb{X} \longrightarrow \mathbb{R}$ is a function, then the *uniform norm* of f is defined:

$$\| f \|_\infty := \sup_{\mathbf{x} \in \mathbb{X}} \left| f(\mathbf{x}) \right|.$$

This measures the farthest deviation of the function f from zero (see Figure 6E.6).

Example 6E.6 Suppose $\mathbb{X} = [0, 1]$, and $f(x) = \frac{1}{3}x^3 - \frac{1}{4}x$ (as in Figure 6E.7(a)). The minimal point of f is $x = \frac{1}{2}$, where $f'\left(\frac{1}{2}\right) = 0$ and $f\left(\frac{1}{2}\right) = \frac{-1}{12}$. The maximal point of f is $x = 1$, where $f(1) = \frac{1}{12}$. Thus, $|f(x)|$ takes a maximum value of $\frac{1}{12}$ at either point, so that

$$\| f \|_\infty = \sup_{0 \le x \le 1} \left| \frac{1}{3}x^3 - \frac{1}{4}x \right| = \frac{1}{12}.$$ ◇

Lemma 6E.7 $\| \bullet \|_\infty$ *is a norm. That is*

(a) *for any $f : \mathbb{X} \longrightarrow \mathbb{R}$ and $r \in \mathbb{R}$, $\| r \cdot f \|_\infty = |r| \cdot \| f \|_\infty$;*

(b) *(triangle inequality) for any $f, g : \mathbb{X} \longrightarrow \mathbb{R}$, $\| f + g \|_\infty \le \| f \|_\infty + \| g \|_\infty$;*

(c) *for any $f : \mathbb{X} \longrightarrow \mathbb{R}$, $\| f \|_\infty = 0$ if and only if $f \equiv 0$.*

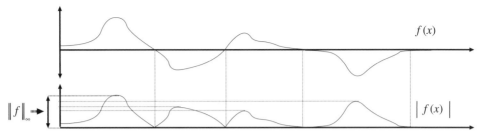

Figure 6E.6. The uniform norm of f is defined: $\|f\|_\infty := \sup_{x\in\mathbb{X}} |f(x)|$.

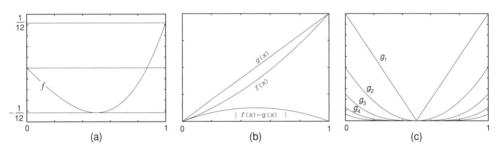

Figure 6E.7. (a) The uniform norm of $f(x) = \frac{1}{3}x^3 - \frac{1}{4}x$ (Example 6E.6). (b)
The uniform distance between $f(x) = x(x+1)$ and $g(x) = 2x$ (Example 6E.8).
(c) $g_n(x) = \left|x - \frac{1}{2}\right|^n$, for $n = 1, 2, 3, 4, 5$ (Example 6E.9(b))

Ⓔ *Proof* **Exercise 6E.11.**   □

The *uniform distance* between two functions f and g is then given by

$$\|f - g\|_\infty = \sup_{\mathbf{x}\in\mathbb{X}} \left|f(\mathbf{x}) - g(\mathbf{x})\right|.$$

One way to interpret this is portrayed in Figure 6E.8. Define a 'tube' of width ϵ
around the function f. If $\|f - g\|_\infty < \epsilon$, this means that $g(x)$ is confined within
this tube for all x.

Example 6E.8 Let $\mathbb{X} = [0, 1]$, and suppose $f(x) = x(x+1)$ and $g(x) = 2x$ (as
in Figure 6E.7(b)). For any $x \in [0, 1]$,

$$|f(x) - g(x)| = \left|x^2 + x - 2x\right| = \left|x^2 - x\right| = x - x^2.$$

(because it is non-negative). This expression takes its maximum at $x = 1/2$ (to see
this, solve for $f'(x) = 0$), and its value at $x = 1/2$ is $1/4$. Thus,

$$\|f - g\|_\infty = \sup_{x\in\mathbb{X}} \left|x(x - 1)\right| = \frac{1}{4}.$$ ◇

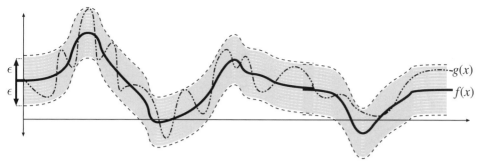

Figure 6E.8. If $\|f - g\|_\infty < \epsilon$, this means that $g(x)$ is confined within an ϵ-tube around f for all x.

Let $\{g_1, g_2, g_3, \ldots\}$ be functions from $\mathbb{X}$ to $\mathbb{R}$, and let $f : \mathbb{X} \longrightarrow \mathbb{R}$ be some other function. The sequence $\{g_1, g_2, g_3, \ldots\}$ *converges uniformly* to f if $\lim_{n\to\infty} \|g_n - f\|_\infty = 0$. We then write $f = \text{unif-}\lim_{n\to\infty} g_n$. This means, not only that $\lim_{n\to\infty} g_n(\mathbf{x}) = f(\mathbf{x})$ for every $\mathbf{x} \in \mathbb{X}$, but also that the functions g_n converge to f everywhere at the same 'speed'. This is portrayed in Figure 6E.9. For any $\epsilon > 0$, we can define a 'tube' of width ϵ around f, and, no matter how small we make this tube, the sequence $\{g_1, g_2, g_3, \ldots\}$ will eventually enter this tube and remain there. To be precise: there is some N such that, for all $n > N$, the function g_n is confined within the ϵ-tube around f; i.e., $\|f - g_n\|_\infty < \epsilon$.

Example 6E.9 In each of the following examples, let $\mathbb{X} = [0, 1]$.

(a) Suppose, as in Example 6E.5(a), p. 127, and Figure 6E.2(b), p. 124, that

$$g_n(x) = \begin{cases} 1 & \text{if } \frac{1}{n} < x < \frac{2}{n}; \\ 0 & \text{otherwise.} \end{cases}$$

Then the sequence $\{g_1, g_2, \ldots\}$ converges *pointwise* to the constant 0 function, but does *not* converge to 0 uniformly on $[0, 1]$. (**Exercise 6E.12** Verify these claims.) Ⓔ

(b) If $g_n(x) = \left|x - \frac{1}{2}\right|^n$ (see Figure 6E.7(c)), then $\|g_n\|_\infty = \frac{1}{2^n}$ (**Exercise 6E.13**). Thus, Ⓔ the sequence $\{g_1, g_2, \ldots\}$ converges to zero uniformly on $[0, 1]$, because

$$\lim_{n\to\infty} \|g_n\|_\infty = \lim_{n\to\infty} \frac{1}{2^n} = 0.$$

(c) If $g_n(x) = 1/n$ for all $x \in [0, 1]$, then the sequence $\{g_1, g_2, \ldots\}$ converges to zero uniformly on $[0, 1]$ (**Exercise 6E.14**). Ⓔ

(d) Recall the functions

$$g_n(x) = \frac{1}{1 + n \cdot \left|x - \frac{1}{2n}\right|}$$

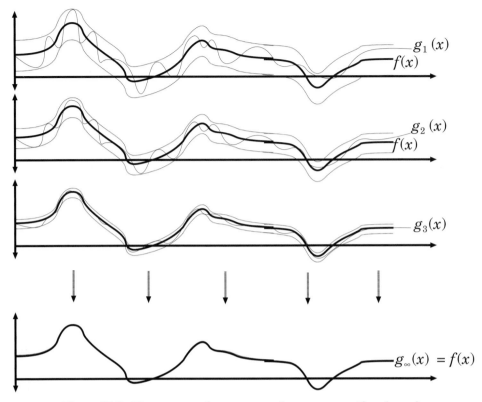

Figure 6E.9. The sequence $\{g_1, g_2, g_3, \ldots\}$ converges *uniformly* to f.

from Example 6E.5(e) (see Figure 6E.5, p. 128). The sequence $\{g_1, g_2, \ldots\}$ converges
pointwise to the constant 0 function, but does *not* converge to zero uniformly on $[0, 1]$.
(**Exercise 6E.15** Verify these claims.)   ◇

Note that, if we define $g_n = f - f_n$ for all $n \in \mathbb{N}$, then

$$\left(f_n \xrightarrow[n\to\infty]{} f \text{ uniformly} \right) \Longleftrightarrow \left(g_n \xrightarrow[n\to\infty]{} 0 \text{ uniformly} \right).$$

Hence, to understand uniform convergence in general, it is sufficient to understand
uniform convergence to the constant 0 function.

Uniform convergence is the 'best' kind of convergence. It has the most useful
consequences, but it is also the most difficult to achieve. (In many cases, we
must settle for pointwise or L^2-convergence instead.) For example, the following
consequences of uniform convergence are extremely useful.

Proposition 6E.10 *Let $\mathbb{X} \subset \mathbb{R}^D$ be some domain. Let $\{f_1, f_2, f_3, \ldots\}$ be functions
from $\mathbb{X}$ to $\mathbb{R}$, and let $f : \mathbb{X} \longrightarrow \mathbb{R}$ be some other function. Suppose $f_n \xrightarrow[n\to\infty]{} f$
uniformly.*

(a) *If $\{f_n\}_{n=1}^{\infty}$ are all continuous on $\mathbb{X}$, then f is also continuous on $\mathbb{X}$.*
(b) *If $\mathbb{X}$ is compact (that is, closed and bounded), then*

$$\lim_{n\to\infty} \int_{\mathbb{X}} f_n(x)dx = \int_{\mathbb{X}} f(x)dx.$$

(c) *Suppose the functions $\{f_n\}_{n=1}^{\infty}$ are all differentiable on $\mathbb{X}$, and suppose $f_n' \xrightarrow[n\to\infty]{} F$ uniformly. Then f is also differentiable, and $f' = F$.*

Proof (a) **Exercise 6E.16.** (Slightly challenging; for students with some analysis Ⓔ background.)

For (b) and (c) see, e.g., Asmar (2005), §2.9, Theorems 4 and 5. □

Note that Proposition 6E.10(a) and (c) are *false* if we replace 'uniformly' with 'pointwise' or 'in L^2'. (Proposition 6E.10(b) is sometimes true under these conditions, but only if we also add additional hypotheses.) Indeed, the next result says that uniform convergence is logically stronger than either pointwise or L^2-convergence.

Corollary 6E.11 *Let $\{f_1, f_2, f_3, \ldots\}$ be functions from $\mathbb{X}$ to $\mathbb{R}$, and let $f : \mathbb{X} \longrightarrow \mathbb{R}$ be some other function.*

(a) *If $f_n \xrightarrow[n\to\infty]{} f$ uniformly, then $f_n \xrightarrow[n\to\infty]{} f$ pointwise.*
(b) *Suppose $\mathbb{X}$ is compact (that is, closed and bounded). If $f_n \xrightarrow[n\to\infty]{} f$ uniformly, then*

  (i) *$f_n \xrightarrow[n\to\infty]{} f$ in L^2;*
  (ii) *for any $g \in \mathbf{L}^2(\mathbb{X})$, we have $\lim_{n\to\infty} \langle f_n, g \rangle = \langle f, g \rangle$.*

Proof **Exercise 6E.17.** (a) is easy. For (b), use Proposition 6E.10(b). □ Ⓔ

Sometimes, uniform convergence is a little too much to ask for, and we must settle for a slightly weaker form of convergence. Let $\mathbb{X} \subset \mathbb{R}^D$ be some domain. Let $\{g_1, g_2, g_3, \ldots\}$ be functions from $\mathbb{X}$ to $\mathbb{R}$, and let $f : \mathbb{X} \longrightarrow \mathbb{R}$ be some other function. The sequence $\{g_1, g_2, g_3, \ldots\}$ *converges semiuniformly* to f if

(a) $\{g_1, g_2, g_3, \ldots\}$ converges *pointwise* to f on $\mathbb{X}$, i.e. $f(x) = \lim_{n\to\infty} g_n(x)$ for all $x \in \mathbb{X}$;
(b) $\{g_1, g_2, g_3, \ldots\}$ converges *uniformly* to f on any *closed subset* of int$(\mathbb{X})$; in other words, if $\mathbb{Y} \subset$ int$(\mathbb{X})$ is any closed set, then

$$\lim_{n\to\infty} \left(\sup_{y\in\mathbb{Y}} |f(y) - g_n(y)| \right) = 0.$$

Heuristically speaking, this means that the sequence $\{g_n\}_{n=1}^{\infty}$ is 'trying' to converge to f uniformly on $\mathbb{X}$, but it is maybe getting 'stuck' at some of the boundary points of $\mathbb{X}$.

Example 6E.12 Let $\mathbb{X} := (0, 1)$. Recall the functions

$$g_n(x) = \frac{1}{1 + n \cdot \left| x - \frac{1}{2n} \right|}$$

from Figure 6E.5, p. 128. By Example 6E.9(d), p. 131, we know that this sequence *doesn't* converge uniformly to 0 on $(0, 1)$. However, it does converge *semi*uniformly to 0. First, we know it converges pointwise on $(0, 1)$, by Example 6E.5(e), p. 129. Second, if $0 < a < b < 1$, it is easy to check that $\{g_n\}_{n=1}^{\infty}$ converges to f uniformly on the closed interval $[a, b]$ (**Exercise 6E.18**). It follows that $\{g_n\}_{n=1}^{\infty}$ converges to f uniformly on any closed subset of $(0, 1)$.   ◇

Summary The various forms of convergence are logically related as follows:

(uniform convergence) $\Rightarrow$ (semiuniform convergence) $\Rightarrow$ (pointwise convergence).

Also, if $\mathbb{X}$ is compact, then

$$\text{(uniform convergence)} \Longrightarrow \text{(convergence in } L^2\text{)}.$$

Finally, if the sequence of functions is uniformly bounded and $\mathbb{X}$ is compact, then

$$\text{(pointwise convergence)} \Longrightarrow \text{(convergence in } L^2\text{)}.$$

However, the opposite implications are *not* true. In general:

(convergence in L^2) $\nRightarrow$ (pointwise convergence) $\nRightarrow$ (uniform convergence).

6E(iv) *Convergence of function series*

Let $\{f_1, f_2, f_3, \ldots\}$ be functions from $\mathbb{X}$ to $\mathbb{R}$. The *function series* $\sum_{n=1}^{\infty} f_n$ is the formal infinite summation of these functions; we would like to think of this series as defining another function from $\mathbb{X}$ to $\mathbb{R}$. Intuitively, the symbol $\sum_{n=1}^{\infty} f_n$ should represent the function which arises as the limit $\lim_{N \to \infty} F_N$, where, for each $N \in \mathbb{N}$, $F_N(x) := \sum_{n=1}^{N} f_n(x) = f_1(x) + f_2(x) + \cdots + f_N(x)$ is the Nth *partial sum*. To make this precise, we must specify the sense in which the partial sums $\{F_1, F_2, \ldots\}$ converge. If $F : \mathbb{X} \longrightarrow \mathbb{R}$ is this putative limit function, then we say that the series $\sum_{n=1}^{\infty} f_n$

- converges *in* L^2 to F if $F = L^2\text{-}\lim_{N \to \infty} \sum_{n=1}^{N} f_n$ (we then write $F \underset{L^2}{\approx} \sum_{n=1}^{\infty} f_n$);
- converges *pointwise* to F if, for each $x \in \mathbb{X}$, $F(x) = \lim_{N \to \infty} \sum_{n=1}^{N} f_n(x)$ (we then write $F \equiv \sum_{n=1}^{\infty} f_n$);
- converges *uniformly* to F if $F = \text{unif-}\lim_{N \to \infty} \sum_{n=1}^{N} f_n$ (we then write $F \underset{\text{unif}}{\equiv} \sum_{n=1}^{\infty} f_n$).

The following result provides a useful condition for the uniform convergence of an infinite summation of functions; we will use this result often in our study of Fourier series and other eigenfunction expansions in Chapters 7–9.

Proposition 6E.13 Weierstrass M-test

Let $\{f_1, f_2, f_3, \ldots\}$ be functions from $\mathbb{X}$ to $\mathbb{R}$. For every $n \in \mathbb{N}$, let $M_n := \|f_n\|_\infty$. If $\sum_{n=1}^\infty M_n < \infty$, then the series $\sum_{n=1}^\infty f_n$ converges uniformly on $\mathbb{X}$.

Proof **Exercise 6E.19.**   Ⓔ

(a) Show that the series converges *pointwise* to some limit function $f : \mathbb{X} \longrightarrow \mathbb{R}$.

(b) For any $N \in \mathbb{N}$, show that

$$\left\| F - \sum_{n=1}^N f_n \right\|_\infty \leq \sum_{n=N+1}^\infty M_n.$$

(c) Show that $\lim_{N \to \infty} \sum_{n=N+1}^\infty M_n = 0$.   □

The next three sufficient conditions for convergence are also sometimes useful (but they are not used later in this book).

Proposition 6E.14 Dirichlet test

Let $\{f_1, f_2, f_3, \ldots\}$ be functions from $\mathbb{X}$ to $\mathbb{R}$. Let $\{c_k\}_{k=1}^\infty$ be a sequence of positive real numbers. Then the series $\sum_{n=1}^\infty c_n f_n$ converges uniformly on $\mathbb{X}$ if   (Optional)

- $\lim_{n \to \infty} c_k = 0$; *and*
- *there is some $M > 0$ such that, for all $N \in \mathbb{N}$, we have $\left\| \sum_{n=1}^N f_n \right\|_\infty < M$.*

Proof See Asmar (2005), appendix to §2.10.   □

Proposition 6E.15 Cauchy's criterion

Let $\{f_1, f_2, f_3, \ldots\}$ be functions from $\mathbb{X}$ to $\mathbb{R}$. For every $N \in \mathbb{N}$, let $C_N := \sup_{M>N} \left\| \sum_{n=N}^M f_n \right\|_\infty$. Then   (Optional)

$$\left(\text{the series } \sum_{n=1}^\infty f_n \text{ converges uniformly on } \mathbb{X} \right) \iff \left(\lim_{N \to \infty} C_N = 0 \right).$$

Proof See Churchill and Brown (1987), §88.   □

Proposition 6E.16 Abel's test

Let $\mathbb{X} \subset \mathbb{R}^N$ and $\mathbb{Y} \subset \mathbb{R}^M$ be two domains. Let $\{f_1, f_2, f_3, \ldots\}$ be a sequence of functions from $\mathbb{X}$ to $\mathbb{R}$, such that the series $\sum_{n=1}^\infty f_n$ converges uniformly on $\mathbb{X}$. Let $\{g_1, g_2, g_3, \ldots\}$ be another sequence of functions from $\mathbb{Y}$ to $\mathbb{R}$, and consider the sequence $\{h_1, h_2, \ldots\}$ of functions from $\mathbb{X} \times \mathbb{Y}$ to $\mathbb{R}$, defined by $h_n(x, y) := f_n(x)g_n(y)$. Suppose the following.   (Optional)

(a) *The sequence* $\{g_n\}_{n=1}^{\infty}$ *is uniformly bounded; i.e., there is some $M > 0$ such that $|g_n(y)| < M$ for all $n \in \mathbb{N}$ and $y \in \mathbb{Y}$.*

(b) *The sequence* $\{g_n\}_{n=1}^{\infty}$ *is monotonic; i.e., either $g_1(y) \leq g_2(y) \leq g_3(y) \leq \cdots$ for all $y \in \mathbb{Y}$, or $g_1(y) \geq g_2(y) \geq g_3(y) \geq \cdots$ for all $y \in \mathbb{Y}$.*

Then the series $\sum_{n=1}^{\infty} h_n$ converges uniformly on $\mathbb{X} \times \mathbb{Y}$.

Proof See Churchill and Brown (1987), §88. □

6F Orthogonal and orthonormal bases

Prerequisites: §6A, §6E(i). Recommended: §6E(iv).

An *orthogonal set* in $\mathbf{L}^2(\mathbb{X})$ is a (finite or infinite) collection of functions $\{\mathbf{b}_1, \mathbf{b}_2, \mathbf{b}_3, \ldots\}$ such that $\langle \mathbf{b}_k, \mathbf{b}_j \rangle = 0$ whenever $k \neq j$. Intuitively, the vectors $\{\mathbf{b}_1, \mathbf{b}_2, \mathbf{b}_3, \ldots\}$ are all 'perpendicular' to one another in the infinite-dimensional geometry of $\mathbf{L}^2(\mathbb{X})$. One consequence is an L^2-version of the *Pythagorean formula*: For any $N \in \mathbb{N}$ and any real numbers $r_1, r_2, \ldots, r_N \in \mathbb{R}$, we have

$$\|r_1\mathbf{b}_1 + r_2\mathbf{b}_2 + \cdots + r_N\mathbf{b}_n\|_2^2 = r_1^2 \|\mathbf{b}_1\|_2^2 + r_2^2 \|\mathbf{b}_N\|_2^2 + \cdots + r_N^2 \|\mathbf{b}_N\|_2^2. \tag{6F.1}$$

ⓔ **Exercise 6F.1** Verify the L^2 Pythagorean formula. ◆

An *orthogonal basis* for $\mathbf{L}^2(\mathbb{X})$ is an infinite collection of functions $\{\mathbf{b}_1, \mathbf{b}_2, \mathbf{b}_3, \ldots\}$ such that

- $\{\mathbf{b}_1, \mathbf{b}_2, \mathbf{b}_3, \ldots\}$ form an orthogonal set (i.e. $\langle \mathbf{b}_k, \mathbf{b}_j \rangle = 0$ whenever $k \neq j$);
- for any $\mathbf{g} \in \mathbf{L}^2(\mathbb{X})$, if we define

$$\gamma_n = \frac{\langle \mathbf{g}, \mathbf{b}_n \rangle}{\|\mathbf{b}_n\|_2^2},$$

for all $n \in \mathbb{N}$, then $\mathbf{g} \underset{\mathit{L2}}{\approx} \sum_{n=1}^{\infty} \gamma_n \mathbf{b}_n$.

Recall that this means that

$$\lim_{N \to \infty} \left\| \mathbf{g} - \sum_{n=1}^{N} \gamma_n \mathbf{b}_n \right\|_2 = 0.$$

In other words, we can approximate $\mathbf{g}$ as closely as we want in L^2-norm with a partial sum $\sum_{n=1}^{N} \gamma_n \mathbf{b}_n$, if we make N large enough.

An *orthonormal basis* for $\mathbf{L}^2(\mathbb{X})$ is an infinite collection of functions $\{\mathbf{b}_1, \mathbf{b}_2, \mathbf{b}_3, \ldots\}$ such that

- $\|\mathbf{b}_k\|_2 = 1$ for every k;
- $\{\mathbf{b}_1, \mathbf{b}_2, \mathbf{b}_3, \ldots\}$ is an orthogonal basis for $\mathbf{L}^2(\mathbb{X})$. In other words, $\langle \mathbf{b}_k, \mathbf{b}_j \rangle = 0$ whenever $k \neq j$, and, for any $\mathbf{g} \in \mathbf{L}^2(\mathbb{X})$, if we define $\gamma_n = \langle \mathbf{g}, \mathbf{b}_n \rangle$ for all $n \in \mathbb{N}$, then $\mathbf{g} \underset{L2}{\approx} \sum_{n=1}^{\infty} \gamma_n \mathbf{b}_n$.

One consequence of this is Theorem 6F.1.

Theorem 6F.1 Parseval's equality

Let $\{\mathbf{b}_1, \mathbf{b}_2, \mathbf{b}_3, \ldots\}$ be an orthonormal basis for $\mathbf{L}^2(\mathbb{X})$, and let $\mathbf{f}, \mathbf{g} \in \mathbf{L}^2(\mathbb{X})$. Let $\varphi_n := \langle \mathbf{f}, \mathbf{b}_n \rangle$ and $\gamma_n := \langle \mathbf{g}, \mathbf{b}_n \rangle$ for all $n \in \mathbb{N}$. Then

(a) $\langle \mathbf{f}, \mathbf{g} \rangle = \sum_{n=1}^{\infty} \varphi_n \gamma_n$;

(b) $\|\mathbf{g}\|_2^2 = \sum_{n=1}^{\infty} |\gamma_n|^2$.

Proof **Exercise 6F.2.** *Hint:* For all $N \in \mathbb{N}$, let $\mathbf{F}_N := \sum_{n=1}^{N} \varphi_n \mathbf{b}_n$ and $\mathbf{G}_N := \sum_{n=1}^{N} \gamma_n \mathbf{b}_n$. Ⓔ

(i) Show that $\langle \mathbf{F}_N, \mathbf{G}_N \rangle = \sum_{n=1}^{N} \varphi_n \gamma_n$. (*Hint:* the functions $\{\mathbf{b}_1, \ldots, \mathbf{b}_N\}$ are orthonormal.)

(ii) To prove (a), show that $\langle \mathbf{f}, \mathbf{g} \rangle = \lim_{N \to \infty} \langle \mathbf{F}_N, \mathbf{G}_N \rangle$. (*Hint:* Use Lemma 6E.3.)

(iii) To prove (b), set $\mathbf{f} = \mathbf{g}$ in (a). □

The idea of Fourier analysis is to find an *orthogonal basis* for an L^2-space, using familiar trigonometric functions. We will return to this in Chapter 7.

6G Further reading

Most of the mathematically rigorous texts on partial differential equations (such as Churchill and Brown (1987), Evans (1991), appendix D, or Asmar (2005)) contain detailed and thorough discussions of L^2-space, orthogonal basis, and the various convergence concepts discussed in this chapter. This is because almost all solutions to partial differential equations arise through some sort of infinite series or approximating sequence; hence it is essential to understand completely the various forms of function convergence and their relationships.

The convergence of sequences of functions is part of a subject called *real analysis*, and any advanced textbook on real analysis will contain extensive material on convergence. There are many other forms of function convergence we have not even mentioned in this chapter, including $\mathbf{L}^p$-*convergence* (for any value of p between 1 and ∞), convergence *in measure*, convergence *almost everywhere*, and *weak** convergence. Different convergence modes are useful in different contexts, and the logical relationships between them are fairly subtle. See Folland (1984), §2.4, for a good summary. Other standard references are: Kolmogorov and Fomīn (1975), §28.4–28.5, §37; Wheeden and Zygmund (1977), chap. 8; Rudin (1987); or Royden (1988).

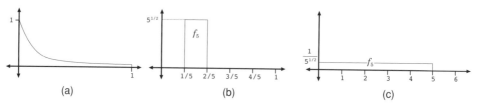

Figure 6H.1. Functions for (a) Problem 6.1; (b) Problem (6.2); (c) Problem 6.3.

The geometry of infinite-dimensional vector spaces is called *functional analysis*, and is logically distinct from the convergence theory for functions (although, of course, most of the important infinite-dimensional spaces are spaces of functions). Infinite-dimensional vector spaces fall into several broad classes, depending upon the richness of the geometric and topological structure, which include *Hilbert spaces* (such as $\mathbf{L}^2(\mathbb{X})$), *Banach spaces* (such as $\mathcal{C}(\mathbb{X})$ or $\mathbf{L}^1(\mathbb{X})$) and *locally convex spaces*. An excellent introduction to functional analysis is given in Conway (1990). Other standard references are Kolmogorov and Fomīn (1975), chap. 4, and Folland (1984), chap. 5. Hilbert spaces are the mathematical foundation of quantum mechanics; see Prugovecki (1981) and Blank *et al.* (1994).

6H Practice problems

6.1 Let $\mathbb{X} = (0, 1]$. For any $n \in \mathbb{N}$, define the function $f_n : (0, 1] \longrightarrow \mathbb{R}$ by $f_n(x) = \exp(-nx)$ (Figure 6H.1(a)).

  (a) Compute $\| f_n \|_2$ for all $n \in \mathbb{N}$.
  (b) Does the sequence $\{ f_n \}_{n=1}^{\infty}$ converge to the constant 0 function in $\mathbf{L}^2(0, 1]$? Explain.
  (c) Compute $\| f_n \|_\infty$ for all $n \in \mathbb{N}$.
  (d) Does the sequence $\{ f_n \}_{n=1}^{\infty}$ converge to the constant 0 function uniformly on $(0, 1]$? Explain.
  (e) Does the sequence $\{ f_n \}_{n=1}^{\infty}$ converge to the constant 0 function pointwise on $(0, 1]$? Explain.

6.2 Let $\mathbb{X} = [0, 1]$. For any $n \in \mathbb{N}$, define $f_n : [0, 1] \longrightarrow \mathbb{R}$ by

$$f_n(x) = \begin{cases} \sqrt{n} & \text{if } \frac{1}{n} \leq x < \frac{2}{n}; \\ 0 & \text{otherwise.} \end{cases}$$

(See Figure 6H.1(b).)

  (a) Does the sequence $\{ f_n \}_{n=1}^{\infty}$ converge to the constant 0 function pointwise on $[0, 1]$? Explain.
  (b) Compute $\| f_n \|_2$ for all $n \in \mathbb{N}$.
  (c) Does the sequence $\{ f_n \}_{n=1}^{\infty}$ converge to the constant 0 function in $\mathbf{L}^2[0, 1]$? Explain.
  (d) Compute $\| f_n \|_\infty$ for all $n \in \mathbb{N}$.
  (e) Does the sequence $\{ f_n \}_{n=1}^{\infty}$ converge to the constant 0 function uniformly on $[0, 1]$? Explain.

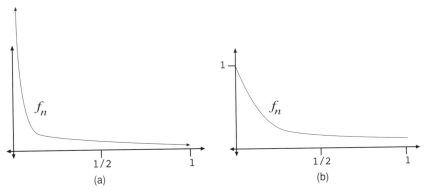

Figure 6H.2. Functions for (a) Problem 6.4; (b) Problem 6.5.

6.3 Let $\mathbb{X} = \mathbb{R}$. For any $n \in \mathbb{N}$, define $f_n : \mathbb{R} \longrightarrow \mathbb{R}$ by

$$f_n(x) = \begin{cases} \frac{1}{\sqrt{n}} & \text{if } 0 \le x < n; \\ 0 & \text{otherwise.} \end{cases}$$

(See Figure 6H.1(c).)

(a) Compute $\|f_n\|_\infty$ for all $n \in \mathbb{N}$.
(b) Does the sequence $\{f_n\}_{n=1}^\infty$ converge to the constant 0 function uniformly on $\mathbb{R}$? Explain.
(c) Does the sequence $\{f_n\}_{n=1}^\infty$ converge to the constant 0 function pointwise on $\mathbb{R}$? Explain.
(d) Compute $\|f_n\|_2$ for all $n \in \mathbb{N}$.
(e) Does the sequence $\{f_n\}_{n=1}^\infty$ converge to the constant 0 function in $\mathbf{L}^2(\mathbb{R})$? Explain.

6.4 Let $\mathbb{X} = (0, 1]$. For all $n \in \mathbb{N}$, define $f_n : (0, 1] \longrightarrow \mathbb{R}$ by $f_n(x) = \frac{1}{\sqrt[3]{nx}}$ (for all $x \in (0, 1]$) (Figure 6H.2(a)).

(a) Does the sequence $\{f_n\}_{n=1}^\infty$ converge to the constant 0 function *pointwise* on $(0, 1]$? Why or why not?
(b) Compute $\|f_n\|_2$ for all $n \in \mathbb{N}$.
(c) Does the sequence $\{f_n\}_{n=1}^\infty$ converge to the constant 0 function in $\mathbf{L}^2(0, 1]$? Why or why not?
(d) Compute $\|f_n\|_\infty$ for all $n \in \mathbb{N}$.
(e) Does the sequence $\{f_n\}_{n=1}^\infty$ converge to the constant 0 function uniformly on $(0, 1]$? Explain.

6.5 Let $\mathbb{X} = [0, 1]$. For all $n \in \mathbb{N}$, define $f_n : [0, 1] \longrightarrow \mathbb{R}$ by $f_n(x) = \frac{1}{(nx+1)^2}$ (for all $x \in [0, 1]$) (Figure 6H.2(b)).

(a) Does the sequence $\{f_n\}_{n=1}^\infty$ converge to the constant 0 function pointwise on $[0, 1]$? Explain.
(b) Compute $\|f_n\|_2$ for all $n \in \mathbb{N}$.

(c) Does the sequence $\{f_n\}_{n=1}^{\infty}$ converge to the constant 0 function in $\mathbf{L}^2[0, 1]$? Explain.

(d) Compute $\|f_n\|_{\infty}$ for all $n \in \mathbb{N}$. (*Hint:* Look at Figure 6H.2(b). Where is the value of $f_n(x)$ largest?)

(e) Does the sequence $\{f_n\}_{n=1}^{\infty}$ converge to the constant 0 function uniformly on $[0, 1]$? Explain.

6.6 In each of the following cases, you are given two functions $f, g : [0, \pi] \longrightarrow \mathbb{R}$. Compute the inner product $\langle f, g \rangle$.

(a) $f(x) = \sin(3x)$, $g(x) = \sin(2x)$.
(b) $f(x) = \sin(nx)$, $g(x) = \sin(mx)$, with $n \neq m$.
(c) $f(x) = \sin(nx) = g(x)$ for some $n \in \mathbb{N}$.   *Question:* What is $\|f\|_2$?
(d) $f(x) = \cos(3x)$, $g(x) = \cos(2x)$.
(e) $f(x) = \cos(nx)$, $g(x) = \cos(mx)$, with $n \neq m$.
(f) $f(x) = \sin(3x)$, $g(x) = \cos(2x)$.

6.7 In each of the following cases, you are given two functions $f, g : [-\pi, \pi] \longrightarrow \mathbb{R}$. Compute the inner product $\langle f, g \rangle$.

(a) $f(x) = \sin(nx)$, $g(x) = \sin(mx)$, with $n \neq m$.
(b) $f(x) = \sin(nx) = g(x)$ for some $n \in \mathbb{N}$.   *Question:* What is $\|f\|_2$?
(c) $f(x) = \cos(nx)$, $g(x) = \cos(mx)$, with $n \neq m$.
(d) $f(x) = \sin(3x)$, $g(x) = \cos(2x)$.

6.8 Determine if f_n converges to f pointwise, in $L^2(\mathbb{X})$, or uniformly for the following cases.

(a) $f_n(x) = e^{-nx^2}$, $f(x) = 0$, $\mathbb{X} = [-1, 1]$.
(b) $f_n(x) = n \sin(x/n)$, $f(x) = x$, $\mathbb{X} = [-\pi, \pi]$.

7

Fourier sine series and cosine series

The art of doing mathematics consists in finding that special case which
contains all the germs of generality.

David Hilbert

7A Fourier (co)sine series on $[0, \pi]$

Prerequisites: §6E(iv), §6F.

Throughout this section, for all $n \in \mathbb{N}$, we define the functions $\mathbf{S}_n : [0, \pi] \longrightarrow \mathbb{R}$ and $\mathbf{C}_n : [0, \pi] \longrightarrow \mathbb{R}$ by $\mathbf{S}_n(x) := \sin(nx)$ and $\mathbf{C}_n(x) := \cos(nx)$, for all $x \in [0, \pi]$ (see Figure 6D.1, p. 118).

7A(i) Sine series on $[0, \pi]$

Recommended: §5C(i).

Suppose $f \in \mathbf{L}^2[0, \pi]$ (i.e. $f : [0, \pi] \longrightarrow \mathbb{R}$ is a function with $\|f\|_2 < \infty$). We define the *Fourier sine coefficients* of f:

$$B_n := \frac{\langle f, \mathbf{S}_n \rangle}{\|\mathbf{S}_n\|_2^2} = \boxed{\frac{2}{\pi} \int_0^\pi f(x) \sin(nx) \mathrm{d}x,} \quad \text{for all } n \geq 1. \quad (7\text{A}.1)$$

The *Fourier sine series* of f is then the infinite summation of functions:

$$\boxed{\sum_{n=1}^\infty B_n \mathbf{S}_n(x).} \quad (7\text{A}.2)$$

A function $f : [0, \pi] \longrightarrow \mathbb{R}$ is *continuously differentiable* on $[0, \pi]$ if f is continuous on $[0, \pi]$ (hence, bounded), $f'(x)$ exists for all $x \in (0, \pi)$, and furthermore, the function $f' : (0, \pi) \longrightarrow \mathbb{R}$ is itself bounded and continuous on $(0, \pi)$. Let $\mathcal{C}^1[0, \pi]$ be the space of all continuously differentiable functions.

141

We say f is *piecewise continuously differentiable* (or *piecewise C^1*, or *section-ally smooth*) if there exist points $0 = j_0 < j_1 < j_2 < \cdots < j_{M+1} = \pi$ (for some $M \in \mathbb{N}$) such that f is bounded and continuously differentiable on each of the open intervals (j_m, j_{m+1}); these are called *C^1 intervals* for f. In particular, any continuously differentiable function on $[0, \pi]$ is piecewise continuously differentiable (in this case, $M = 0$ and the set $\{j_1, \ldots, j_M\}$ is empty, so all of $(0, \pi)$ is a C^1 interval).

Ⓔ **Exercise 7A.1**

   (a) Show that any continuously differentiable function has finite L^2-norm. In other words, $C^1[0, \pi] \subset L^2[0, \pi]$.
   (b) Show that any piecewise C^1 function on $[0, \pi]$ is in $L^2[0, \pi]$. ◆

Theorem 7A.1 Fourier sine series convergence on $[0, \pi]$

   (a) *The set $\{S_1, S_2, S_3, \ldots\}$ is an orthogonal basis for $L^2[0, \pi]$. Thus, if $f \in L^2[0, \pi]$, then the sine series (7A.2) converges to f in L^2-norm, i.e. $f \underset{L^2}{\approx} \sum_{n=1}^{\infty} B_n S_n$. Furthermore, the coefficient sequence $\{B_n\}_{n=1}^{\infty}$ is the unique sequence of coefficients with this property. In other words, if $\{B_n'\}_{n=1}^{\infty}$ is some other sequence of coefficients such that $f \underset{L^2}{\approx} \sum_{n=1}^{\infty} B_n' S_n$, then we must have $B_n' = B_n$ for all $n \in \mathbb{N}$.*
   (b) *If $f \in C^1[0, \pi]$, then the sine series (7A.2) converges pointwise on $(0, \pi)$. More generally, if f is piecewise C^1, then the sine series (7A.2) converges to f pointwise on each C^1 interval for f. In other words, if $\{j_1, \ldots, j_m\}$ is the set of discontinuity points of f and/or f', and $j_m < x < j_{m+1}$, then $f(x) = \lim_{N \to \infty} \sum_{n=1}^{N} B_n \sin(nx)$.*
   (c) *If $\sum_{n=1}^{\infty} |B_n| < \infty$, then the sine series (7A.2) converges to f uniformly on $[0, \pi]$.*
   (d) (i) *If f is continuous and piecewise differentiable on $[0, \pi]$, and $f' \in L^2[0, \pi]$, and f satisfies homogeneous Dirichlet boundary conditions (i.e. $f(0) = f(\pi) = 0$), then the sine series (7A.2) converges to f uniformly on $[0, \pi]$.*
       (ii) *Conversely, if the sine series (7A.2) converges to f uniformly on $[0, \pi]$, then f is continuous on $[0, \pi]$, and satisfies homogeneous Dirichlet boundary conditions.*
   (e) *If f is piecewise C^1, and $\mathbb{K} \subset (j_m, j_{m+1})$ is any closed subset of a C^1 interval of f, then the series (7A.2) converges uniformly to f on $\mathbb{K}$.*
   (f) *Suppose $\{B_n\}_{n=1}^{\infty}$ is a non-negative sequence decreasing to zero. (That is, $B_1 \geq B_2 \geq \cdots \geq 0$ and $\lim_{n \to \infty} B_n = 0$.) If $0 < a < b < \pi$, then the series (7A.2) converges uniformly to f on $[a, b]$.*

Ⓔ *Proof* (c) **Exercise 7A.2.** (*Hint:* Use the Weierstrass M-test, Proposition 6E.13, p. 135.)

Ⓔ   (a), (b) (d)(i), (e) **Exercise 7A.3.** (*Hint:* Use Theorem 8A.1(a), (b), (d), (e), p. 166, and Proposition 8C.5(a) and Lemma 8C.6(a), p. 174).)

Ⓔ   (d)(ii) **Exercise 7A.4.**
       (f) See Asmar (2005), Theorem 2 of §2.10.   □

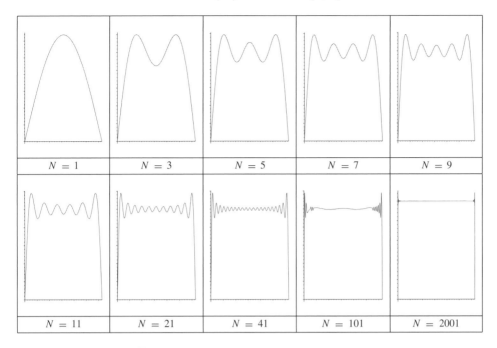

Figure 7A.1. $\frac{4}{\pi} \sum_{\substack{n=1 \\ n\,\text{odd}}}^{N} \frac{1}{n} \sin(nx)$, for $N = 1, 3, 5, 7, 9, 11, 21, 41, 101$, and 2001. Note the Gibbs phenomenon in the plots for large N.

Example 7A.2

(a) If $f(x) = \sin(5x) - 2\sin(3x)$, then the Fourier sine series of f is just $\sin(5x) - 2\sin(3x)$. In other words, the Fourier coefficients B_n are all zero, except that $B_3 = -2$ and $B_5 = 1$.

(b) Suppose $f(x) \equiv 1$. For all $n \in \mathbb{N}$,

$$B_n = \frac{2}{\pi} \int_0^\pi \sin(nx)\mathrm{d}x = \frac{-2}{n\pi} \cos(nx) \Big|_{x=0}^{x=\pi} = \frac{2}{n\pi}\left[1 - (-1)^n\right]$$

$$= \begin{cases} \frac{4}{n\pi} & \text{if } n \text{ is } odd; \\ 0 & \text{if } n \text{ is } even. \end{cases}$$

Thus, the Fourier sine series is as follows:

$$\frac{4}{\pi} \sum_{\substack{n=1 \\ n\,\text{odd}}}^{\infty} \frac{1}{n} \sin(nx) = \frac{4}{\pi}\left(\sin(x) + \frac{\sin(3x)}{3} + \frac{\sin(5x)}{5} + \cdots\right). \tag{7A.3}$$

Theorem 7A.1(a) says that

$$1 \underset{12}{\approx} \frac{4}{\pi} \sum_{\substack{n=1 \\ n\,\text{odd}}}^{\infty} \frac{1}{n} \sin(nx).$$

Figure 7A.1 displays some partial sums of the series (7A.3). The function $f \equiv 1$ is clearly continuously differentiable, so, by Theorem 7A.1(b), the Fourier sine series converges pointwise to 1 on the interior of the interval $[0, \pi]$. However, the series does *not* converge to f at the points 0 or π. This is betrayed by the violent oscillations of the partial sums near these points; this is an example of the *Gibbs phenomenon*.

Since the Fourier sine series does not converge at the endpoints 0 and π, we know automatically that it does not converge to f uniformly on $[0, \pi]$. However, we could have also deduced this fact by noticing that f does *not* have homogeneous Dirichlet boundary conditions (because $f(0) = 1 = f(\pi)$), whereas every finite sum of $\sin(nx)$-type functions *does* have homogeneous Dirichlet BC. Thus, the series (7A.3) is 'trying' to converge to f, but it is 'stuck' at the endpoints 0 and π. (This is the idea behind Theorem 7A.1(d).)

(c) If $f(x) = \cos(mx)$, then the Fourier sine series of f is

$$\frac{4}{\pi} \sum_{\substack{n=1 \\ n+m \text{ odd}}}^{\infty} \frac{n}{n^2 - m^2} \sin(nx). \qquad \diamondsuit$$

Ⓔ **Exercise 7A.5** Verify Example 7A.2(c). (*Hint:* Use Theorem 6D.3, p. 119.) ◆

Example 7A.3 $\sinh(\alpha x)$

If $\alpha > 0$, and $f(x) = \sinh(\alpha x)$, then its Fourier sine series is given by

$$\sinh(\alpha x) \underset{\overline{12}}{\approx} \frac{2\sinh(\alpha\pi)}{\pi} \sum_{n=1}^{\infty} \frac{n(-1)^{n+1}}{\alpha^2 + n^2} \cdot \sin(nx).$$

To prove this, we must show that, for all $n > 0$,

$$B_n = \frac{2}{\pi} \int_0^\pi \sinh(\alpha x) \cdot \sin(nx)dx = \frac{2\sinh(\alpha\pi)}{\pi} \frac{n(-1)^{n+1}}{\alpha^2 + n^2}.$$

To begin with, let $I := \int_0^\pi \sinh(\alpha x) \cdot \sin(nx)dx$. Then, applying integration by parts, we obtain

$$I = \frac{-1}{n}\left[\sinh(\alpha x) \cdot \cos(nx) \Big|_{x=0}^{x=\pi} - \alpha \cdot \int_0^\pi \cosh(\alpha x) \cdot \cos(nx)dx \right]$$

$$= \frac{-1}{n}\left[\sinh(\alpha\pi) \cdot (-1)^n - \frac{\alpha}{n} \cdot \left(\cosh(\alpha x) \cdot \sin(nx) \Big|_{x=0}^{x=\pi} \right.\right.$$

$$\left.\left. - \alpha \int_0^\pi \sinh(\alpha x) \cdot \sin(nx)dx \right) \right]$$

$$= \frac{-1}{n}\left[\sinh(\alpha\pi) \cdot (-1)^n - \frac{\alpha}{n} \cdot (0 - \alpha \cdot I) \right]$$

$$= \frac{-\sinh(\alpha\pi) \cdot (-1)^n}{n} - \frac{\alpha^2}{n^2} I.$$

Hence

$$I = \frac{-\sinh(\alpha\pi) \cdot (-1)^n}{n} - \frac{\alpha^2}{n^2} I;$$

thus

$$\left(1 + \frac{\alpha^2}{n^2}\right) I = \frac{-\sinh(\alpha\pi) \cdot (-1)^n}{n};$$

i.e.

$$\left(\frac{n^2 + \alpha^2}{n^2}\right) I = \frac{\sinh(\alpha\pi) \cdot (-1)^{n+1}}{n};$$

so that

$$I = \frac{n \cdot \sinh(\alpha\pi) \cdot (-1)^{n+1}}{n^2 + \alpha^2}.$$

Thus,

$$B_n = \frac{2}{\pi} I = \frac{2}{\pi} \frac{n \cdot \sinh(\alpha\pi) \cdot (-1)^{n+1}}{n^2 + \alpha^2}.$$

The function sinh is clearly continuously differentiable, so Theorem 7A.1(b) implies that the Fourier sine series converges to $\sinh(\alpha x)$ pointwise on the open interval $(0, \pi)$.   ◇

Exercise 7A.6   Verify that the Fourier sine series in Example 7A.3 does Ⓔ
not converge uniformly to $\sinh(\alpha x)$ on $[0, \pi]$. (*Hint:* What is the value of $\sinh(\alpha\pi)$?).   ♦

7A(ii) Cosine series on $[0, \pi]$

Recommended: §5C(ii).

If $f \in \mathbf{L}^2[0, \pi]$, we define the *Fourier cosine coefficients* of f:

$$A_0 := \langle f, \mathbb{1} \rangle = \boxed{\frac{1}{\pi} \int_0^\pi f(x)dx,}$$

$$A_n := \frac{\langle f, \mathbf{C}_n \rangle}{\|\mathbf{C}_n\|_2^2} = \boxed{\frac{2}{\pi} \int_0^\pi f(x)\cos(nx)dx,} \quad \text{for all } n \in \mathbb{N}. \qquad (7A.4)$$

The *Fourier cosine series* of f is then the infinite summation of functions:

$$\boxed{\sum_{n=0}^\infty A_n \mathbf{C}_n(x).} \qquad (7A.5)$$

Theorem 7A.4 Fourier cosine series convergence on $[0, \pi]$
(a) *The set* $\{\mathbf{C}_0, \mathbf{C}_1, \mathbf{C}_2, \ldots\}$ *is an orthogonal basis for* $\mathbf{L}^2[0, \pi]$. *Thus, if* $f \in \mathbf{L}^2[0, \pi]$, *then the cosine series* (7A.5) *converges to* f *in* $\mathbf{L}^2$-*norm, i.e.* $f \underset{12}{\approx} \sum_{n=0}^\infty A_n \mathbf{C}_n$.

Furthermore, the coefficient sequence $\{A_n\}_{n=0}^{\infty}$ *is the unique sequence of coefficients with this property. In other words, if* $\{A'_n\}_{n=0}^{\infty}$ *is some other sequence of coefficients such that* $f \underset{\overline{L^2}}{\approx} \sum_{n=0}^{\infty} A'_n C_n$, *then we must have* $A'_n = A_n$ *for all* $n \in \mathbb{N}$.

(b) *If* $f \in C^1[0, \pi]$, *then the cosine series (7A.5) converges pointwise on* $(0, \pi)$. *If* f *is piecewise* C^1 *on* $[0, \pi]$, *then the cosine series (7A.5) converges to* f *pointwise on each* C^1 *interval for* f. *In other words, if* $\{j_1, \ldots, j_m\}$ *is the set of discontinuity points of* f *and/or* f', *and* $j_m < x < j_{m+1}$, *then* $f(x) = \lim_{N \to \infty} \sum_{n=0}^{N} A_n \cos(nx)$.

(c) *If* $\sum_{n=0}^{\infty} |A_n| < \infty$, *then the cosine series (7A.5) converges to* f *uniformly on* $[0, \pi]$.

(d) (i) *If* f *is continuous and piecewise differentiable on* $[0, \pi]$, *and* $f' \in \mathbf{L}^2[0, \pi]$, *then the cosine series (7A.5) converges to* f *uniformly on* $[0, \pi]$.

    (ii) *Conversely, if* $\sum_{n=0}^{\infty} n |A_n| < \infty$, *then* $f \in C^1[0, \pi]$, *and* f *satisfies homogeneous Neumann boundary conditions (i.e.* $f'(0) = f'(\pi) = 0$).

(e) *If* f *is piecewise* C^1, *and* $\mathbb{K} \subset (j_m, j_{m+1})$ *is any closed subset of a* C^1 *interval of* f, *then the series (7A.5) converges uniformly to* f *on* $\mathbb{K}$.

(f) *Suppose* $\{A_n\}_{n=0}^{\infty}$ *is a non-negative sequence decreasing to zero. (That is,* $A_0 \geq A_1 \geq A_2 \geq \cdots \geq 0$ *and* $\lim_{n \to \infty} A_n = 0$.) *If* $0 < a < b < \pi$, *then the series (7A.5) converges uniformly to* f *on* $[a, b]$.

(Ⓔ) *Proof* (c) **Exercise 7A.7.** (*Hint:* Use the Weierstrass M-test, Proposition 6E.13, p. 135.)

(Ⓔ)   (a), (b), (d)(i), (e) **Exercise 7A.8.** (*Hint:* Use Theorem 8A.1(a), (b), (d), (e), p. 166, and Proposition 8C.5(b) and Lemma 8C.6(b), p. 174.)

(Ⓔ)   (d)(ii) **Exercise 7A.9.** (*Hint:* Use Theorem 7C.10(b), p. 162.)

    (f) Asmar (2005), Theorem 2 of §2.10.   □

Example 7A.5

(a) *If* $f(x) = \cos(13x)$, *then the Fourier cosine series of* f *is just* $\cos(13x)$. *In other words, the Fourier coefficients* A_n *are all zero, except that* $A_{13} = 1$.

(b) *Suppose* $f(x) \equiv 1$. *Then* $f = C_0$, *so the Fourier cosine coefficients are* $A_0 = 1$, *while* $A_1 = A_2 = A_3 = \ldots 0$.

(c) *Let* $f(x) = \sin(mx)$. *If* m *is even, then the Fourier cosine series of* f *is*

$$\frac{4}{\pi} \sum_{\substack{n=1 \\ n \text{ odd}}}^{\infty} \frac{n}{n^2 - m^2} \cos(nx).$$

If m *is odd, then the Fourier cosine series of* f *is*

$$\frac{2}{\pi m} + \frac{4}{\pi} \sum_{\substack{n=2 \\ n \text{ even}}}^{\infty} \frac{n}{n^2 - m^2} \cos(nx). \qquad \diamondsuit$$

(Ⓔ) **Exercise 7A.10**   Verify Example 7A.5(c). (*Hint:* Use Theorem 6D.3, p. 119.) ◆

Example 7A.6 $\cosh(x)$

Suppose $f(x) = \cosh(x)$. Then the Fourier cosine series of f is given by

$$\cosh(x) \underset{12}{\approx} \frac{\sinh(\pi)}{\pi} + \frac{2 \sinh(\pi)}{\pi} \sum_{n=1}^{\infty} \frac{(-1)^n \cdot \cos(nx)}{n^2 + 1}.$$

To see this, first note that

$$A_0 = \frac{1}{\pi} \int_0^\pi \cosh(x) dx = \frac{1}{\pi} \sinh(x) \Big|_{x=0}^{x=\pi} = \frac{\sinh(\pi)}{\pi}$$

(because $\sinh(0) = 0$).

Next, let $I := \int_0^\pi \cosh(x) \cdot \cos(nx) dx$. Then applying integration by parts yields

$$I = \frac{1}{n} \left(\cosh(x) \cdot \sin(nx) \Big|_{x=0}^{x=\pi} - \int_0^\pi \sinh(x) \cdot \sin(nx) dx \right)$$

$$= \frac{-1}{n} \int_0^\pi \sinh(x) \cdot \sin(nx) dx$$

$$= \frac{1}{n^2} \left(\sinh(x) \cdot \cos(nx) \Big|_{x=0}^{x=\pi} - \int_0^\pi \cosh(x) \cdot \cos(nx) dx \right)$$

$$= \frac{1}{n^2} (\sinh(\pi) \cdot \cos(n\pi) - I) = \frac{1}{n^2} \left((-1)^n \sinh(\pi) - I \right).$$

Thus,

$$I = \frac{1}{n^2} \left((-1)^n \cdot \sinh(\pi) - I \right).$$

Hence,

$$(n^2 + 1)I = (-1)^n \cdot \sinh(\pi).$$

Hence,

$$I = \frac{(-1)^n \cdot \sinh(\pi)}{n^2 + 1}.$$

Thus,

$$A_n = \frac{2}{\pi} I = \frac{2 (-1)^n \cdot \sinh(\pi)}{n^2 + 1}. \qquad \Diamond$$

Remark (a) Almost any introduction to the theory of partial differential equations will contain a discussion of the Fourier convergence theorems. For example, see: duChateau and Zachmann (1986), Theorem 6.1; Haberman (1987), §3.2; and Powers (1999), §1.3–§1.7.

(b) Please see Remark 8D.3, p. 177, for further technical remarks about the (non)convergence of Fourier (co)sine series, in situations where the hypotheses of Theorems 7A.1 and 7A.4 are not satisfied.

7B Fourier (co)sine series on $[0, L]$

Prerequisites: §6E, §6F.   **Recommended:** §7A.

Throughout this section, let $L > 0$ be some positive real number. For all $n \in$ $\mathbb{N}$, we define the functions $\mathbf{S}_n : [0, L] \longrightarrow \mathbb{R}$ and $\mathbf{C}_n : [0, L] \longrightarrow \mathbb{R}$ by $\mathbf{S}_n(x) :=$ $\sin\left(\frac{n\pi x}{L}\right)$ and $\mathbf{C}_n(x) := \cos\left(\frac{n\pi x}{L}\right)$, for all $x \in [0, L]$ (see Figure 6D.1, p. 118). Note that, if $L = \pi$, then $\mathbf{S}_n(x) = \sin(nx)$ and $\mathbf{C}_n(x) = \cos(nx)$, as in §7A. The results in this section exactly parallel those in §7A, except that we replace π with L to obtain slightly greater generality. In principle, every statement in this section is equivalent to the corresponding statement in §7A, through the change of variables $y = x/\pi$ (it is a useful exercise to reflect on this as you read this section).

7B(i) Sine series on $[0, L]$

Recommended: §5C(i), §7A(i).

Fix $L > 0$, and let $[0, L]$ be an interval of length L. If $f \in \mathbf{L}^2[0, L]$, we define the *Fourier sine coefficients* of f:

$$
B_n := \frac{\langle f, \mathbf{S}_n \rangle}{\|\mathbf{S}_n\|_2^2} = \boxed{\frac{2}{L} \int_0^L f(x) \sin\left(\frac{n\pi x}{L}\right) dx,} \quad \text{for all } n \geq 1.
$$

The *Fourier sine series* of f is then the infinite summation of functions:

$$
\boxed{\sum_{n=1}^{\infty} B_n \mathbf{S}_n(x).} \tag{7B.1}
$$

A function $f : [0, L] \longrightarrow \mathbb{R}$ is *continuously differentiable* on $[0, L]$ if f is continuous on $[0, L]$ (hence, bounded), and $f'(x)$ exists for all $x \in (0, L)$, and furthermore, the function $f' : (0, L) \longrightarrow \mathbb{R}$ is itself bounded and continuous on $(0, L)$. Let $\mathcal{C}^1[0, L]$ be the space of all continuously differentiable functions.

We say $f : [0, L] \longrightarrow \mathbb{R}$ is *piecewise continuously differentiable* (or *piecewise C^1*, or *sectionally smooth*) if there exist points $0 = j_0 < j_1 < j_2 < \cdots < j_{M+1} = L$ such that f is bounded and continuously differentiable on each of the open intervals (j_m, j_{m+1}); these are called $\mathcal{C}^1$ *intervals* for f. In particular, any continuously differentiable function on $[0, L]$ is piecewise continuously differentiable (in this case, all of $(0, L)$ is a $\mathcal{C}^1$ interval).

Theorem 7B.1 Fourier sine series convergence on $[0, L]$
Parts (a)–(f) of Theorem 7A.1, p. 142, are all still true if you replace 'π' with 'L' everywhere.

Proof **Exercise 7B.1.** (*Hint:* Use the change-of-variables $y = \frac{\pi}{L} x$ to pass from ⓔ
$y \in [0, L]$ to $x \in [0, \pi]$.) □

Example 7B.2

(a) If $f(x) = \sin\left(\frac{5\pi}{L} x\right)$, then the Fourier sine series of f is just $\sin\left(\frac{5\pi}{L} x\right)$. In other words,
the Fourier coefficients B_n are all zero, except that $B_5 = 1$.

(b) Suppose $f(x) \equiv 1$. For all $n \in \mathbb{N}$,

$$B_n = \frac{2}{L} \int_0^L \sin\left(\frac{n\pi x}{L}\right) dx = \frac{-2}{n\pi} \cos\left(\frac{n\pi x}{L}\right) \Big|_{x=0}^{x=L}$$

$$= \frac{2}{n\pi}\left[1 - (-1)^n\right] = \begin{cases} \frac{4}{n\pi} & \text{if } n \text{ is odd;} \\ 0 & \text{if } n \text{ is even.} \end{cases}$$

Thus, the Fourier sine series is given by

$$\frac{4}{\pi} \sum_{\substack{n=1 \\ n \text{ odd}}}^{\infty} \frac{1}{n} \sin\left(\frac{n\pi}{L} x\right).$$

Figure 7A.1 displays some partial sums of this series (in the case $L = \pi$). The *Gibbs
phenomenon* is clearly evident, just as in Example 7A.2(b), p. 143.

(c) If $f(x) = \cos\left(\frac{m\pi}{L} x\right)$, then the Fourier sine series of f is:

$$\frac{4}{\pi} \sum_{\substack{n=1 \\ n+m \text{ odd}}}^{\infty} \frac{n}{n^2 - m^2} \sin\left(\frac{n\pi}{L} x\right).$$

(**Exercise 7B.2** *Hint:* Use Theorem 6D.3, p. 119.) ⓔ

(d) If $\alpha > 0$, and $f(x) = \sinh\left(\frac{\alpha\pi x}{L}\right)$, then its Fourier sine coefficients are

$$B_n = \frac{2}{L} \int_0^L \sinh\left(\frac{\alpha\pi x}{L}\right) \cdot \sin\left(\frac{n\pi x}{L}\right) dx = \frac{2\sinh(\alpha\pi)}{\pi} \frac{n(-1)^{n+1}}{\alpha^2 + n^2}.$$

(**Exercise 7B.3**). ◇ ⓔ

7B(ii) *Cosine series on* $[0, L]$

Recommended: §5C(ii), §7A(ii).

If $f \in \mathbf{L}^2[0, L]$, we define the *Fourier cosine coefficients* of f:

$$A_0 := \langle f, \mathbb{1} \rangle = \boxed{\frac{1}{L} \int_0^L f(x) dx,}$$

$$A_n := \frac{\langle f, \mathbf{C}_n \rangle}{\|\mathbf{C}_n\|_2^2} = \boxed{\frac{2}{L} \int_0^L f(x) \cos\left(\frac{n\pi x}{L}\right) dx,} \quad \text{for all } n > 0.$$

The *Fourier cosine series* of f is then the infinite summation of functions:

$$\sum_{n=0}^{\infty} A_n C_n(x).$$ (7B.2)

Theorem 7B.3 Fourier cosine series convergence on $[0, L]$

Parts (a)–(f) of Theorem 7A.4, p. 145, are all still true if you replace 'π' with 'L' everywhere.

Ⓔ *Proof* **Exercise 7B.4.** (*Hint:* Use the change-of-variables $y := \frac{\pi}{L}x$ to pass from $x \in [0, L]$ to $y \in [0, \pi]$.)   □

Example 7B.4

(a) If $f(x) = \cos\left(\frac{13\pi}{L}x\right)$, then the Fourier cosine series of f is just $\cos\left(\frac{13\pi}{L}x\right)$. In other words, the Fourier coefficients A_n are all zero, except that $A_{13} = 1$.

(b) Suppose $f(x) \equiv 1$. Then $f = C_0$, so the Fourier cosine coefficients are $A_0 = 1$, while $A_1 = A_2 = A_3 = \ldots 0$.

(c) Let $f(x) = \sin\left(\frac{m\pi}{L}x\right)$. If m is *even*, then the Fourier cosine series of f is

$$\frac{4}{\pi} \sum_{\substack{n=1 \\ n \text{ odd}}}^{\infty} \frac{n}{n^2 - m^2} \cos\left(\frac{n\pi}{L}x\right).$$

If m is *odd*, then the Fourier cosine series of f is

$$\frac{2}{\pi m} + \frac{4}{\pi} \sum_{\substack{n=2 \\ n \text{ even}}}^{\infty} \frac{n}{n^2 - m^2} \cos(nx).$$ ◇

Ⓔ **Exercise 7B.5** Verify Example 7B.4(c). (*Hint:* Use Theorem 6D.3, p. 119.) ◆

7C Computing Fourier (co)sine coefficients

Prerequisites: §7B.

When computing the Fourier sine coefficient $B_n = \frac{2}{L} \int_0^L f(x) \cdot \sin\left(\frac{n\pi}{L}x\right) dx$, it is simpler to first compute the integral $\int_0^L f(x) \cdot \sin\left(\frac{n\pi}{L}x\right) dx$, and then multiply the result by $2/L$. Likewise, to compute a Fourier cosine coefficients, first compute the integral $\int_0^L f(x) \cdot \cos\left(\frac{n\pi}{L}x\right) dx$, and then multiply the result by $2/L$. In this section, we review some useful techniques to compute these integrals.

7C(i) Integration by parts

Computing Fourier coefficients almost always involves integration by parts. Generally, if you can't compute it with integration by parts, you can't compute it. When evaluating a Fourier integral by parts, one almost always ends up with boundary terms of the form 'cos($n\pi$)' or 'sin$\left(\frac{n}{2}\pi\right)$', etc. The following formulae are useful in this regard.

$$\boxed{\sin(n\pi) = 0 \quad \text{for any } n \in \mathbb{Z};} \tag{7C.1}$$

for example, $\sin(-\pi) = \sin(0) = \sin(\pi) = \sin(2\pi) = \sin(3\pi) = 0$.

$$\boxed{\cos(n\pi) = (-1)^n \quad \text{for any } n \in \mathbb{Z};} \tag{7C.2}$$

for example, $\cos(-\pi) = -1$, $\cos(0) = 1$, $\cos(\pi) = -1$, $\cos(2\pi) = 1$, $\cos(3\pi) = -1$, etc.

$$\boxed{\sin\left(\frac{n}{2}\pi\right) = \begin{cases} 0 & \text{if } n \text{ is } even; \\ (-1)^k & \text{if } n \text{ is } odd, \text{ and } n = 2k + 1; \end{cases}} \tag{7C.3}$$

for example, $\sin(0) = 0$, $\sin\left(\frac{1}{2}\pi\right) = 1$, $\sin(\pi) = 0$, $\sin\left(\frac{3}{2}\pi\right) = -1$, etc.

$$\boxed{\cos\left(\frac{n}{2}\pi\right) = \begin{cases} 0 & \text{if } n \text{ is } odd; \\ (-1)^k & \text{if } n \text{ is } even, \text{ and } n = 2k; \end{cases}} \tag{7C.4}$$

for example, $\cos(0) = 1$, $\cos\left(\frac{1}{2}\pi\right) = 0$, $\cos(\pi) = -1$, $\cos\left(\frac{3}{2}\pi\right) = 0$, etc.

Exercise 7C.1 Verify equations (7C.1), (7C.2), (7C.3), and (7C.4). ♦ Ⓔ

7C(ii) Polynomials

Theorem 7C.1 *Let $n \in \mathbb{N}$. Then*

(a)
$$\int_0^L \sin\left(\frac{n\pi}{L}x\right)dx = \begin{cases} \dfrac{2L}{n\pi} & \text{if } n \text{ is } odd; \\ 0 & \text{if } n \text{ is } even. \end{cases} \tag{7C.5}$$

(b)
$$\int_0^L \cos\left(\frac{n\pi}{L}x\right)dx = \begin{cases} L & \text{if } n = 0; \\ 0 & \text{if } n > 0. \end{cases} \tag{7C.6}$$

For any $k \in \{1, 2, 3, \ldots\}$, we have the following recurrence relations:

(c)
$$\int_0^L x^k \cdot \sin\left(\frac{n\pi}{L}x\right)dx = \frac{(-1)^{n+1}}{n} \cdot \frac{L^{k+1}}{\pi} + \frac{k}{n} \cdot \frac{L}{\pi} \int_0^L x^{k-1} \cdot \cos\left(\frac{n\pi}{L}x\right), \tag{7C.7}$$

(d)
$$\int_0^L x^k \cdot \cos\left(\frac{n\pi}{L}x\right)dx = \frac{-k}{n} \cdot \frac{L}{\pi} \int_0^L x^{k-1} \cdot \sin\left(\frac{n\pi}{L}x\right). \tag{7C.8}$$

Proof **Exercise 7C.2.** (*Hint:* For (c) and (d), use integration by parts.) □ Ⓔ

Example 7C.2 In all of the following examples, let $L = \pi$.

(a) $\dfrac{2}{\pi} \displaystyle\int_0^\pi \sin(nx)dx = \dfrac{2}{\pi} \dfrac{1-(-1)^n}{n}.$

(b) $\dfrac{2}{\pi} \displaystyle\int_0^\pi x \cdot \sin(nx)dx = (-1)^{n+1} \dfrac{2}{n}.$

(c) $\dfrac{2}{\pi} \displaystyle\int_0^\pi x^2 \cdot \sin(nx)dx = (-1)^{n+1} \dfrac{2\pi}{n} + \dfrac{4}{\pi n^3} \left((-1)^n - 1\right).$

(d) $\dfrac{2}{\pi} \displaystyle\int_0^\pi x^3 \cdot \sin(nx)dx = (-1)^n \left(\dfrac{12}{n^3} - \dfrac{2\pi^2}{n}\right).$

(e) $\dfrac{2}{\pi} \displaystyle\int_0^\pi \cos(nx)dx = \begin{cases} 2 & \text{if } n = 0; \\ 0 & \text{if } n > 0. \end{cases}$

(f) $\dfrac{2}{\pi} \displaystyle\int_0^\pi x \cdot \cos(nx)dx = \dfrac{2}{\pi n^2} \left((-1)^n - 1\right),$ if $n > 0.$

(g) $\dfrac{2}{\pi} \displaystyle\int_0^\pi x^2 \cdot \cos(nx)dx = (-1)^n \dfrac{4}{n^2},$ if $n > 0.$

(h) $\dfrac{2}{\pi} \displaystyle\int_0^\pi x^3 \cdot \cos(nx)dx = (-1)^n \dfrac{6\pi}{n^2} - \dfrac{12}{\pi n^4} \left((-1)^n - 1\right),$ if $n > 0.$ ◇

Proof (b) We will show this in two ways. First, by direct computation:

$$\int_0^\pi x \cdot \sin(nx)dx = \dfrac{-1}{n} \left(x \cdot \cos(nx) \Big|_{x=0}^{x=\pi} - \int_0^\pi \cos(nx)dx \right)$$

$$= \dfrac{-1}{n} \left(\pi \cdot \cos(n\pi) - \dfrac{1}{n} \sin(nx) \Big|_{x=0}^{x=\pi} \right)$$

$$= \dfrac{-1}{n}(-1)^n \pi = \dfrac{(-1)^{n+1}\pi}{n}.$$

Thus,

$$\dfrac{2}{\pi} \int_0^\pi x \cdot \sin(nx)dx = \dfrac{2(-1)^{n+1}}{n},$$

as desired.

Next, we verify (b) using Theorem 7C.1. Setting $L = \pi$ and $k = 1$ in equation (7C.7), we have:

$$\int_0^\pi x \cdot \sin(nx)dx = \dfrac{(-1)^{n+1}}{n} \cdot \dfrac{\pi^{1+1}}{\pi} + \dfrac{1}{n} \cdot \dfrac{\pi}{\pi} \int_0^\pi x^{k-1} \cdot \cos(nx)dx$$

$$= \dfrac{(-1)^{n+1}}{n} \cdot \pi + \dfrac{1}{n} \int_0^\pi \cos(nx)dx = \dfrac{(-1)^{n+1}}{n} \cdot \pi,$$

because $\int_0^\pi \cos(nx)dx = 0$ by equation (7C.6). Thus,

$$\frac{2}{\pi} \int_0^\pi x \cdot \sin(nx)dx = \frac{2(-1)^{n+1}}{n},$$

as desired.

(c)

$$\int_0^\pi x^2 \cdot \sin(nx)dx = \frac{-1}{n} \left(x^2 \cdot \cos(nx) \Big|_{x=0}^{x=\pi} - 2 \int_0^\pi x \cos(nx)dx \right)$$

$$= \frac{-1}{n} \left[\pi^2 \cdot \cos(n\pi) - \frac{2}{n} \left(x \cdot \sin(nx) \Big|_{x=0}^{x=\pi} - \int_0^\pi \sin(nx)dx \right) \right]$$

$$= \frac{-1}{n} \left[\pi^2 \cdot (-1)^n + \frac{2}{n} \left(\frac{-1}{n} \cos(nx) \Big|_{x=0}^{x=\pi} \right) \right]$$

$$= \frac{-1}{n} \left[\pi^2 \cdot (-1)^n - \frac{2}{n^2} \left((-1)^n - 1 \right) \right]$$

$$= \frac{2}{n^3} \left((-1)^n - 1 \right) + \frac{(-1)^{n+1}\pi^2}{n}.$$

The result follows. (**Exercise 7C.3** Verify (c) using Theorem 7C.1.)   Ⓔ
   (g) We will show this in two ways. First, by direct computation:

$$\int_0^\pi x^2 \cdot \cos(nx)dx = \frac{1}{n} \left[x^2 \cdot \sin(nx) \Big|_{x=0}^{x=\pi} - 2 \int_0^\pi x \cdot \sin(nx)dx \right]$$

$$= \frac{-2}{n} \int_0^\pi x \cdot \sin(nx)dx$$

(because $\sin(nx) = \sin(0) = 0$)

$$= \frac{2}{n^2} \left[x \cdot \cos(nx) \Big|_{x=0}^{x=\pi} - \int_0^\pi \cos(nx)dx \right]$$

$$= \frac{2}{n^2} \left[\pi \cdot (-1)^n - \frac{1}{n} \sin(nx) \Big|_{x=0}^{x=\pi} \right]$$

$$= \frac{2\pi \cdot (-1)^n}{n^2}.$$

Thus,

$$\frac{2}{\pi} \int_0^\pi x^2 \cdot \cos(nx)dx = \frac{4 \cdot (-1)^n}{n^2},$$

as desired.

Next, we verify (g) using Theorem 7C.1. Setting $L = \pi$ and $k = 2$ in equation (7C.8), we have:

$$\int_0^\pi x^2 \cdot \cos(nx)dx = \frac{-k}{n} \cdot \frac{L}{\pi} \int_0^\pi x^{k-1} \cdot \sin(nx) = \frac{-2}{n} \cdot \int_0^\pi x \cdot \sin(nx).$$
(7C.9)

Next, applying equation (7C.7) with $k = 1$, we get

$$\int_0^\pi x \cdot \sin(nx) = \frac{(-1)^{n+1}}{n} \cdot \frac{\pi^2}{\pi} + \frac{1}{n} \cdot \frac{\pi}{\pi} \int_0^\pi \cos(nx)$$

$$= \frac{(-1)^{n+1}\pi}{n} + \frac{1}{n} \int_0^\pi \cos(nx).$$

Substituting this into equation (7C.9), we get

$$\int_0^\pi x^2 \cdot \cos(nx)dx = \frac{-2}{n} \cdot \left[\frac{(-1)^{n+1}\pi}{n} + \frac{1}{n} \int_0^\pi \cos(nx) \right]. \quad (7C.10)$$

We are assuming $n > 0$, but equation (7C.6) says $\int_0^\pi \cos(nx) = 0$. Thus, we can simplify equation (7C.10) to conclude:

$$\frac{2}{\pi} \int_0^\pi x^2 \cdot \cos(nx)dx = \frac{2}{\pi} \cdot \frac{-2}{n} \cdot \frac{(-1)^{n+1}\pi}{n} = \frac{4(-1)^n}{n^2},$$

as desired. ☐

Ⓔ **Exercise 7C.4** Verify all of the other parts of Example 7C.2, both using Theorem 7C.1 and through direct integration. ◆

To compute the Fourier series of an arbitrary polynomial, we integrate one term at a time.

Example 7C.3 Let $L = \pi$ and let $f(x) = x^2 - \pi \cdot x$. Then the Fourier sine series of f is given by

$$\frac{-8}{\pi} \sum_{\substack{n=1 \\ n \text{ odd}}}^\infty \frac{1}{n^3} \sin(nx) = \frac{-8}{\pi} \left(\sin(x) + \frac{\sin(3x)}{27} + \frac{\sin(5x)}{125} + \frac{\sin(7x)}{343} + \cdots \right).$$

To see this, first note that, by Example 7C.2(b),

$$\int_0^\pi x \cdot \sin(nx)dx = \frac{-1}{n}(-1)^n \pi = \frac{(-1)^{n+1}\pi}{n}.$$

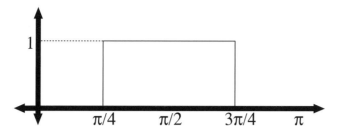

Figure 7C.1. Step function (see Example 7C.4).

Next, by Example 7C.2(c),

$$\int_0^\pi x^2 \cdot \sin(nx)\,dx = \frac{2}{n^3}\left((-1)^n - 1\right) + \frac{(-1)^{n+1}\pi^2}{n}.$$

Thus,

$$\int_0^\pi \left(x^2 - \pi x\right) \cdot \sin(nx)\,dx = \int_0^\pi x^2 \cdot \sin(nx)\,dx - \pi \cdot \int_0^\pi x \cdot \sin(nx)\,dx$$

$$= \frac{2}{n^3}\left((-1)^n - 1\right) + \frac{(-1)^{n+1}\pi^2}{n} - \pi \cdot \frac{(-1)^{n+1}\pi}{n}$$

$$= \frac{2}{n^3}\left((-1)^n - 1\right).$$

Thus,

$$B_n = \frac{2}{\pi}\int_0^\pi \left(x^2 - \pi x\right) \cdot \sin(nx)\,dx = \frac{4}{\pi n^3}\left((-1)^n - 1\right)$$

$$= \begin{cases} -8/\pi n^3 & \text{if } n \text{ is odd}; \\ 0 & \text{if } n \text{ is even}. \end{cases}$$

◇

7C(iii) *Step functions*

Example 7C.4 Let $L = \pi$, and suppose

$$f(x) = \begin{cases} 1 & \text{if } \frac{\pi}{4} \le x \le \frac{3\pi}{4}; \\ 0 & \text{otherwise}. \end{cases}$$

(See Figure 7C.1.) Then the Fourier sine coefficients of f are given by

$$B_n = \begin{cases} 0 & \text{if } n \text{ is } even; \\ \frac{2\sqrt{2}(-1)^k}{n\pi} & \text{if } n \text{ is } odd, \text{ and } n = 4k \pm 1 \text{ for some } k \in \mathbb{N}. \end{cases}$$

To see this, observe that

$$\int_0^\pi f(x)\sin(nx)dx = \int_{\pi/4}^{3\pi/4} \sin(nx)dx = \left.\frac{-1}{n}\cos(nx)\right|_{x=\pi/4}^{x=3\pi/4}$$

$$= \frac{-1}{n}\left(\cos\left(\frac{3n\pi}{4}\right) - \cos\left(\frac{n\pi}{4}\right)\right)$$

$$= \begin{cases} 0 & \text{if } n \text{ is } even; \\ \frac{\sqrt{2}(-1)^{k+1}}{n} & \text{if } n \text{ is } odd, \text{ and } n = 4k \pm 1 \text{ for some } k \in \mathbb{N} \end{cases}$$

Ⓔ (**Exercise 7C.5**). Thus, the Fourier sine series for f is given by

$$\frac{2\sqrt{2}}{\pi}\left(\sin(x) + \sum_{k=1}^N (-1)^k \left(\frac{\sin\left((4k-1)x\right)}{4k-1} + \frac{\sin\left((4k+1)x\right)}{4k+1}\right)\right)$$

Ⓔ (**Exercise 7C.6**).

Figure 7C.2 shows some of the partial sums of this series. The series converges *pointwise* to $f(x)$ in the interior of the intervals $[0, \frac{\pi}{4})$, $(\frac{\pi}{4}, \frac{3\pi}{4})$, and $(\frac{3\pi}{4}, \pi]$. However, it does not converge to f at the discontinuity points $\frac{\pi}{4}$ and $\frac{3\pi}{4}$. In the plots, this is betrayed by the violent oscillations of the partial sums near these discontinuity points – this is an example of the *Gibbs phenomenon*.   ◇

Example 7C.4 is an example of a *step function*. A function $F : [0, L] \longrightarrow \mathbb{R}$ is a *step function* (see Figure 7C.3(a)) if there are numbers $0 = x_0 < x_1 < x_2 < x_3 < \cdots < x_{M-1} < x_M = L$ and constants $a_1, a_2, \ldots, a_M \in \mathbb{R}$ such that

$$F(x) = \begin{cases} a_1 & \text{if } 0 \leq x \leq x_1; \\ a_2 & \text{if } x_1 < x \leq x_2; \\ \vdots & \vdots \\ a_m & \text{if } x_{m-1} < x \leq x_m; \\ \vdots & \vdots \\ a_M & \text{if } x_{M-1} < x \leq L. \end{cases} \tag{7C.11}$$

For instance, in Example 7C.4, $M = 3$; $x_0 = 0$, $x_1 = \frac{\pi}{4}$, $x_2 = \frac{3\pi}{4}$, and $x_3 = \pi$; $a_1 = 0 = a_3$, and $a_2 = 1$.

To compute the Fourier coefficients of a step function, we simply break the integral into 'pieces', as in Example 7C.4. The general formula is given by the following theorem, but it is really not worth memorizing the formula. Instead, understand the idea.

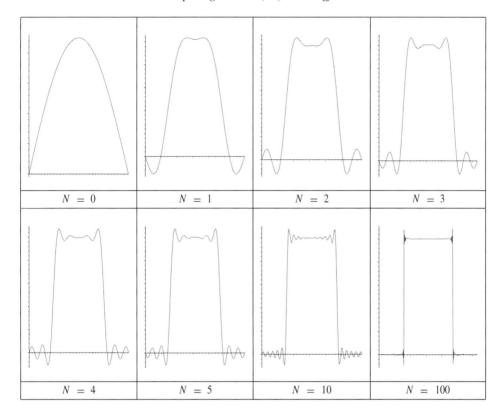

Figure 7C.2. Partial Fourier sine series for Example 7C.4, for $N = 0, 1, 2,$ $3, 4, 5, 10,$ and 100. Note the Gibbs phenomenon in the plots for large N.

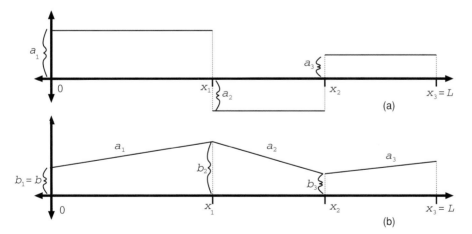

Figure 7C.3. (a) Step function. (b) Piecewise linear function.

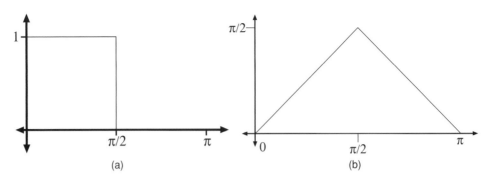

Figure 7C.4. (a) The step function $g(x)$ in Example 7C.6. (b) The tent function $f(x)$ in Example 7C.7.

Theorem 7C.5 *Suppose $F : [0, L] \longrightarrow \mathbb{R}$ is a step function like that in equation (7C.11). Then the Fourier coefficients of F are given by*

$$\frac{1}{L} \int_0^L F(x) = \frac{1}{L} \sum_{m=1}^{M} a_m \cdot (x_m - x_{m-1});$$

$$\frac{2}{L} \int_0^L F(x) \cdot \cos\left(\frac{n\pi}{L}x\right) dx = \frac{-2}{\pi n} \sum_{m=1}^{M-1} \sin\left(\frac{n\pi}{L} \cdot x_m\right) \cdot \left(a_{m+1} - a_m\right);$$

$$\frac{2}{L} \int_0^L F(x) \cdot \sin\left(\frac{n\pi}{L}x\right) dx = \frac{2}{\pi n} \left(a_1 + (-1)^{n+1} a_M\right)$$

$$+ \frac{2}{\pi n} \sum_{m=1}^{M-1} \cos\left(\frac{n\pi}{L} \cdot x_m\right) \cdot \left(a_{m+1} - a_m\right).$$

Ⓔ   *Proof* **Exercise 7C.7.** (*Hint:* Integrate the function piecewise.)   □

Remark Note that the Fourier series of a step function f will converge uniformly to f on the *interior* of each 'step', but will *not* converge to f at any of the step boundaries, because f is not continuous at these points.

Example 7C.6 Suppose $L = \pi$, and

$$g(x) = \begin{cases} 1 & \text{if } 0 \le x < \frac{\pi}{2}; \\ 0 & \text{if } \frac{\pi}{2} \le x. \end{cases}$$

(See Figure 7C.4(a).) Then the Fourier cosine series of $g(x)$ is given by

$$\frac{1}{2} + \frac{2}{\pi} \sum_{k=0}^{\infty} \frac{(-1)^k}{2k+1} \cos\left((2k+1)x\right).$$

In other words, $A_0 = 1/2$ and, for all $n > 0$,

$$A_n = \begin{cases} \frac{2}{\pi} \frac{(-1)^k}{2k+1} & \text{if } n \text{ is odd and } n = 2k+1; \\ 0 & \text{if } n \text{ is even.} \end{cases} \qquad \Diamond$$

Exercise 7C.8 Show this in two ways: first by direct integration, and then by applying the formula from Theorem 7C.5. ⓔ ♦

7C(iv) Piecewise linear functions

Example 7C.7 The tent function
Let $\mathbb{X} = [0, \pi]$ and let

$$f(x) = \begin{cases} x & \text{if } 0 \le x \le \frac{\pi}{2}; \\ \pi - x & \text{if } \frac{\pi}{2} < x \le \pi. \end{cases}$$

(See Figure 7C.4(b).) The Fourier sine series of f is:

$$\frac{4}{\pi} \sum_{\substack{n=1 \\ n \text{ odd}; \\ n=2k+1}}^{\infty} \frac{(-1)^k}{n^2} \sin(nx).$$

To prove this, we must show that, for all $n > 0$,

$$B_n = \frac{2}{\pi} \int_0^\pi f(x) \sin(nx) dx = \begin{cases} \frac{4}{n^2\pi}(-1)^k & \text{if } n \text{ is odd, } n = 2k+1; \\ 0 & \text{if } n \text{ is even.} \end{cases}$$

To verify this, we observe that

$$\int_0^\pi f(x) \sin(nx) dx = \int_0^{\pi/2} x \sin(nx) dx + \int_{\pi/2}^\pi (\pi - x) \sin(nx) dx. \qquad \Diamond$$

Exercise 7C.9 Complete the computation of B_n. ♦ ⓔ

The tent function in Example 7C.7 is *piecewise linear*. A function $F : [0, L] \longrightarrow \mathbb{R}$ is *piecewise linear* (see Figure 7C.3(b), p. 157) if there are numbers $0 = x_0 < x_1 < x_1 < x_2 < \cdots < x_{M-1} < x_M = L$ and constants $a_1, a_2, \ldots, a_M \in \mathbb{R}$ and $b \in \mathbb{R}$ such that

$$F(x) = \begin{cases} a_1(x - L) + b_1 & \text{if } 0 \le x \le x_1; \\ a_2(x - x_1) + b_2 & \text{if } x_1 < x \le x_2; \\ \vdots & \vdots \\ a_m(x - x_m) + b_{m+1} & \text{if } x_m < x \le x_{m+1}; \\ \vdots & \vdots \\ a_M(x - x_{M-1}) + b_M & \text{if } x_{M-1} < x \le L \end{cases} \qquad (7C.12)$$

where $b_1 = b$, and, for all $m > 1$, $b_m = a_m(x_m - x_{m-1}) + b_{m-1}$.

For instance, in Example 7C.7, $M = 2$, $x_1 = \frac{\pi}{2}$ and $x_2 = \pi$; $a_1 = 1$ and $a_2 = -1$.

To compute the Fourier coefficients of a piecewise linear function, we can break the integral into 'pieces', as in Example 7C.7. The general formula is given by the following theorem, but it is really not worth memorizing the formula. Instead, understand the *idea*.

Theorem 7C.8 *Suppose* $F : [0, L] \longrightarrow \mathbb{R}$ *is a piecewise linear function like that in equation (7C.12). Then the Fourier coefficients of* F *are given by*

$$\frac{1}{L} \int_0^L F(x) = \frac{1}{L} \sum_{m=1}^{M} \frac{a_m}{2} (x_m - x_{m-1})^2 + b_m \cdot (x_m - x_{m-1});$$

$$\frac{2}{L} \int_0^L F(x) \cdot \cos \left(\frac{n\pi}{L} x \right) \mathrm{d}x = \frac{2L}{(\pi n)^2} \sum_{m=1}^{M} \cos \left(\frac{n\pi}{L} \cdot x_m \right) \cdot \left(a_m - a_{m+1} \right);$$

$$\frac{2}{L} \int_0^L F(x) \cdot \sin \left(\frac{n\pi}{L} x \right) \mathrm{d}x = \frac{2L}{(\pi n)^2} \sum_{m=1}^{M-1} \sin \left(\frac{n\pi}{L} \cdot x_m \right) \cdot \left(a_m - a_{m+1} \right);$$

where we define $a_{M+1} := a_1$ *for convenience.*

Ⓔ *Proof* **Exercise 7C.10.** (*Hint:* Invoke Theorem 7C.5 and integration by parts.) □

Note that the summands in this theorem read '$a_m - a_{m+1}$', not the other way around.

Example 7C.9 Cosine series of the tent function
Let $\mathbb{X} = [0, \pi]$ and let

$$f(x) = \begin{cases} x & \text{if } 0 \leq x \leq \frac{\pi}{2}; \\ \pi - x & \text{if } \frac{\pi}{2} < x \leq \pi, \end{cases}$$

as in Example 7C.7. The Fourier cosine series of f is given by

$$\frac{\pi}{4} - \frac{8}{\pi} \sum_{\substack{n=1, \\ n=4j+2, \\ \text{for some } j}}^{\infty} \frac{1}{n^2} \cos(nx).$$

In other words,

$$f(x) = \frac{\pi}{4} - \frac{8}{\pi} \left(\frac{\cos(2x)}{4} + \frac{\cos(6x)}{36} + \frac{\cos(10x)}{100} + \frac{\cos(14x)}{196} \right.$$

$$\left. + \frac{\cos(18x)}{324} + \cdots \right).$$

To see this, first observe that

$$A_0 = \frac{1}{\pi} \int_0^\pi f(x)\mathrm{d}x = \frac{1}{\pi} \left(\int_0^{\pi/2} x\,\mathrm{d}x + \int_{\pi/2}^\pi (\pi - x)\mathrm{d}x \right)$$

$$= \frac{1}{\pi} \left(\left[\frac{x^2}{2} \right]_0^{\pi/2} + \frac{\pi^2}{2} - \left[\frac{x^2}{2} \right]_{\pi/2}^\pi \right) = \frac{1}{\pi} \left(\frac{\pi^2}{8} + \frac{\pi^2}{2} - \left(\frac{\pi^2}{2} - \frac{\pi^2}{8} \right) \right)$$

$$= \frac{\pi^2}{4\pi} = \frac{\pi}{4}.$$

Now compute A_n for $n > 0$. First,

$$\int_0^{\pi/2} x\cos(nx)\mathrm{d}x = \frac{1}{n} \left[x\sin(nx) \Big|_0^{\pi/2} - \int_0^{\pi/2} \sin(nx)\mathrm{d}x \right]$$

$$= \frac{1}{n} \left[\frac{\pi}{2} \sin\left(\frac{n\pi}{2}\right) + \frac{1}{n}\cos(nx) \Big|_0^{\pi/2} \right]$$

$$= \frac{\pi}{2n} \sin\left(\frac{n\pi}{2}\right) + \frac{1}{n^2}\cos\left(\frac{n\pi}{2}\right) - \frac{1}{n^2}.$$

Next,

$$\int_{\pi/2}^\pi x\cos(nx)\mathrm{d}x = \frac{1}{n} \left[x\sin(nx) \Big|_{\pi/2}^\pi - \int_{\pi/2}^\pi \sin(nx)\mathrm{d}x \right]$$

$$= \frac{1}{n} \left[\frac{-\pi}{2} \sin\left(\frac{n\pi}{2}\right) + \frac{1}{n}\cos(nx) \Big|_{\pi/2}^\pi \right]$$

$$= \frac{-\pi}{2n} \sin\left(\frac{n\pi}{2}\right) + \frac{(-1)^n}{n^2} - \frac{1}{n^2}\cos\left(\frac{n\pi}{2}\right).$$

Finally,

$$\int_{\pi/2}^\pi \pi\cos(nx)\mathrm{d}x = \frac{\pi}{n} \sin(nx) \Big|_{\pi/2}^\pi$$

$$= \frac{-\pi}{n} \sin\left(\frac{n\pi}{2}\right).$$

Putting it all together, we have:

$$\int_0^\pi f(x)\cos(nx)\mathrm{d}x = \int_0^{\pi/2} x\cos(nx)\mathrm{d}x + \int_{\pi/2}^\pi \pi\cos(nx)\mathrm{d}x - \int_{\pi/2}^\pi x\cos(nx)\mathrm{d}x$$

$$= \frac{\pi}{2n} \sin\left(\frac{n\pi}{2}\right) + \frac{1}{n^2}\cos\left(\frac{n\pi}{2}\right) - \frac{1}{n^2} - \frac{\pi}{n}\sin\left(\frac{n\pi}{2}\right)$$

$$+ \frac{\pi}{2n} \sin\left(\frac{n\pi}{2}\right) - \frac{(-1)^n}{n^2} + \frac{1}{n^2}\cos\left(\frac{n\pi}{2}\right)$$

$$= \frac{2}{n^2}\cos\left(\frac{n\pi}{2}\right) - \frac{1 + (-1)^n}{n^2}.$$

Now,

$$\cos\left(\frac{n\pi}{2}\right) = \begin{cases} (-1)^k & \text{if } n \text{ is even and } n = 2k; \\ 0 & \text{if } n \text{ is odd.} \end{cases}$$

while

$$1 + (-1)^n = \begin{cases} 2 & \text{if } n \text{ is even;} \\ 0 & \text{if } n \text{ is odd.} \end{cases}$$

Thus,

$$2\cos\left(\frac{n\pi}{2}\right) - \left(1 + (-1)^n\right)$$

$$= \begin{cases} -4 & \text{if } n \text{ is even, } n = 2k \text{ and } k = 2j+1 \text{ for some } j; \\ 0 & \text{otherwise} \end{cases}$$

$$= \begin{cases} -4 & \text{if } n = 4j+2 \text{ for some } j; \\ 0 & \text{otherwise.} \end{cases}$$

(For example, $n = 2, 6, 10, 14, 18, \ldots$) Thus

$$A_n = \frac{2}{\pi}\int_0^\pi f(x)\cos(nx)\mathrm{d}x = \begin{cases} \dfrac{-8}{n^2\pi} & \text{if } n = 4j+2 \text{ for some } j; \\ 0 & \text{otherwise.} \end{cases} \qquad \diamondsuit$$

7C(v) Differentiating Fourier (co)sine series

Prerequisites: §7B, Appendix F.

Suppose $f(x) = 3\sin(x) - 5\sin(2x) + 7\sin(3x)$. Then $f'(x) = 3\cos(x) - 10\cos(2x) + 21\cos(3x)$. Likewise, if $f(x) = 3 + 2\cos(x) - 6\cos(2x) + 11\cos(3x)$, then $f'(x) = -2\sin(x) + 12\sin(2x) - 33\sin(3x)$. This illustrates a general pattern.

Theorem 7C.10 *Suppose $f \in \mathcal{C}^\infty[0, L]$.*

(a) *Suppose f has Fourier sine series $\sum_{n=1}^\infty B_n S_n(x)$. If $\sum_{n=1}^\infty n|B_n| < \infty$, then f' has Fourier cosine series: $f'(x) = \frac{\pi}{L}\sum_{n=1}^\infty n B_n C_n(x)$, and this series converges uniformly.*

(b) *Suppose f has Fourier cosine series $\sum_{n=0}^\infty A_n C_n(x)$. If $\sum_{n=1}^\infty n|A_n| < \infty$, then f' has Fourier sine series: $f'(x) = \frac{-\pi}{L}\sum_{n=1}^\infty n A_n S_n(x)$, and this series converges uniformly.*

ⓔ *Proof* **Exercise 7C.11.** (*Hint:* Apply Proposition F.1, p. 569.)   □

Consequence If $f(x) = A \cos\left(\frac{n\pi x}{L}\right) + B \sin\left(\frac{n\pi x}{L}\right)$ for some $A, B \in \mathbb{R}$, then $f''(x) = -\left(\frac{n\pi}{L}\right)^2 \cdot f(x)$. In other words, f is an *eigenfunction*[1] for the differentiation operator ∂_x^2, with eigenvalue $\lambda = -\left(\frac{n\pi}{L}\right)^2$. More generally, for any $k \in \mathbb{N}$, we have $\partial_x^{2k} f = \lambda^k \cdot f$.

7D Practice problems

In all of these problems, the domain is $\mathbb{X} = [0, \pi]$.

7.1 Let $\alpha > 0$ be a constant. Compute the Fourier *sine* series of $f(x) = \exp(\alpha \cdot x)$. At which points does the series converge pointwise? Why? Does the series converge uniformly? Why or why not?

7.2 Compute the Fourier *cosine* series of $f(x) = \sinh(x)$. At which points does the series converge pointwise? Why? Does the series converge uniformly? Why or why not?

7.3 Let $\alpha > 0$ be a constant. Compute the Fourier *sine* series of $f(x) = \cosh(\alpha x)$. At which points does the series converge pointwise? Why? Does the series converge uniformly? Why or why not?

7.4 Compute the Fourier *cosine* series of $f(x) = x$. At which points does the series converge pointwise? Why? Does the series converge uniformly? Why or why not?

7.5 Let

$$g(x) = \begin{cases} 1 & \text{if } 0 \le x < \frac{\pi}{2}; \\ 0 & \text{if } \frac{\pi}{2} \le x \end{cases}$$

(Figure 7C.4(a), p. 158).

(a) Compute the Fourier *cosine* series of $g(x)$. At which points does the series converge pointwise? Why? Does the series converge uniformly? Why or why not?
(b) Compute the Fourier *sine* series of $g(x)$. At which points does the series converge pointwise? Why? Does the series converge uniformly? Why or why not?

7.6 Compute the Fourier *cosine* series of

$$g(x) = \begin{cases} 3 & \text{if } 0 \le x < \frac{\pi}{2}; \\ 1 & \text{if } \frac{\pi}{2} \le x. \end{cases}$$

At which points does the series converge pointwise? Why? Does the series converge uniformly? Why or why not?

7.7 Compute the Fourier *sine* series of

$$f(x) = \begin{cases} x & \text{if } 0 \le x \le \frac{\pi}{2}; \\ \pi - x & \text{if } \frac{\pi}{2} < x \le \pi \end{cases}$$

---

[1] See §4B(iv), p. 67.

(Figure 7C.4(b), p. 158) At which points does the series converge pointwise? Why?
Does the series converge uniformly? Why or why not?

Hint: Note that

$$\int_0^\pi f(x)\sin(nx)dx = \int_0^{\pi/2} x\sin(nx)dx + \int_{\pi/2}^\pi (\pi - x)\sin(nx)dx.$$

7.8 Let $f : [0, \pi] \longrightarrow \mathbb{R}$ be defined:

$$f(x) = \begin{cases} x & \text{if } 0 \le x \le \frac{\pi}{2}; \\ 0 & \text{if } \frac{\pi}{2} < x \le \pi. \end{cases}$$

Compute the Fourier *sine* series for $f(x)$. At which points does the series converge
pointwise? Why? Does the series converge uniformly? Why or why not?

8

Real Fourier series and complex Fourier series

Ordinary language is totally unsuited for expressing what physics really
asserts, since the words of everyday life are not sufficiently abstract.
Only mathematics and mathematical logic can say as little as the
physicist means to say.

Bertrand Russell

8A Real Fourier series on $[-\pi, \pi]$

Prerequisites: §6E, §6F.   **Recommended:** §7A, §5C(iv).

Throughout this section, for all $n \in \mathbb{N}$, we define the functions $\mathbf{S}_n : [-\pi, \pi] \longrightarrow \mathbb{R}$
and $\mathbf{C}_n : [-\pi, \pi] \longrightarrow \mathbb{R}$ by $\mathbf{S}_n(x) := \sin(nx)$ and $\mathbf{C}_n(x) := \cos(nx)$, for all $x \in$
$[-\pi, \pi]$ (see Figure 6D.1, p. 118). If $f : [-\pi, \pi] \longrightarrow \mathbb{R}$ is any function with
$\|f\|_2 < \infty$, we define the *(real) Fourier coefficients*:

$$A_0 := \langle f, \mathbf{C}_0 \rangle = \langle f, 1\!\!1 \rangle = \boxed{\frac{1}{2\pi} \int_{-\pi}^{\pi} f(x)\mathrm{d}x;}$$

$$A_n := \frac{\langle f, \mathbf{C}_n \rangle}{\|\mathbf{C}_n\|_2^2} = \boxed{\frac{1}{\pi} \int_{-\pi}^{\pi} f(x) \cos(nx)\mathrm{d}x;}$$

and

$$B_n := \frac{\langle f, \mathbf{S}_n \rangle}{\|\mathbf{S}_n\|_2^2} = \boxed{\frac{1}{\pi} \int_{-\pi}^{\pi} f(x) \sin(nx)\mathrm{d}x,} \quad \text{for all } n \geq 1.$$

The *(real) Fourier series* of f is then the infinite summation of functions:

$$\boxed{A_0 + \sum_{n=1}^{\infty} A_n \mathbf{C}_n(x) + \sum_{n=1}^{\infty} B_n \mathbf{S}_n(x).} \tag{8A.1}$$

165

We define *continuously differentiable* and *piecewise continuously differentiable* functions on $[-\pi, \pi]$ in a manner exactly analogous to the definitions on $[0, \pi]$ (p. 142). Let $\mathcal{C}^1[-\pi, \pi]$ be the set of all continuously differentiable functions $f : [-\pi, \pi] \longrightarrow \mathbb{R}$.

Ⓔ **Exercise 8A.1**

(a) Show that any continuously differentiable function has finite L^2-norm. In other words, $\mathcal{C}^1[-\pi, \pi] \subset \mathbf{L}^2[\pi, \pi]$.

(b) Show that any piecewise $\mathcal{C}^1$ function on $[-\pi, \pi]$ is in $\mathbf{L}^2[-\pi, \pi]$. ◆

Theorem 8A.1 Fourier convergence on $[-\pi, \pi]$

(a) *The set $\{\mathbb{1}, \mathbf{S}_1, \mathbf{C}_1, \mathbf{S}_2, \mathbf{C}_2, \ldots\}$ is an orthogonal basis for $\mathbf{L}^2[-\pi, \pi]$. Thus, if $f \in \mathbf{L}^2[-\pi, \pi]$, then the Fourier series (8A.1) converges to f in L^2-norm. Furthermore, the coefficient sequences $\{A_n\}_{n=0}^{\infty}$ and $\{B_n\}_{n=1}^{\infty}$ are the unique sequences of coefficients with this property. In other words, if $\{A_n'\}_{n=0}^{\infty}$ and $\{B_n'\}_{n=1}^{\infty}$ are two other sequences of coefficients such that $f \underset{L^2}{\approx} \sum_{n=0}^{\infty} A_n' \mathbf{C}_n + \sum_{n=1}^{\infty} B_n' \mathbf{S}_n$, then we must have $A_n' = A_n$ and $B_n' = B_n$ for all $n \in \mathbb{N}$.*

(b) *If $f \in \mathcal{C}^1[-\pi, \pi]$ then the Fourier series (8A.1) converges pointwise on $(-\pi, \pi)$. More generally, if f is piecewise $\mathcal{C}^1$, then the real Fourier series (8A.1) converges to f pointwise on each $\mathcal{C}^1$ interval for f. In other words, if $\{j_1, \ldots, j_m\}$ is the set of discontinuity points of f and/or f' in $[-\pi, \pi]$, and $j_m < x < j_{m+1}$, then*

$$f(x) = A_0 + \lim_{N \to \infty} \sum_{n=1}^{N} \Big(A_n \cos(nx) + B_n \sin(nx) \Big).$$

(c) *If $\sum_{n=0}^{\infty} |A_n| + \sum_{n=1}^{\infty} |B_n| < \infty$, then the series (8A.1) converges to f uniformly on $[-\pi, \pi]$.*

(d) *Suppose $f : [-\pi, \pi] \longrightarrow \mathbb{R}$ is continuous and piecewise differentiable, $f' \in \mathbf{L}^2[-\pi, \pi]$, and $f(-\pi) = f(\pi)$. Then the series (8A.1) converges to f uniformly on $[-\pi, \pi]$.*

(e) *If f is piecewise $\mathcal{C}^1$, and $\mathbb{K} \subset (j_m, j_{m+1})$ is any closed subset of a $\mathcal{C}^1$ interval of f, then the series (8A.1) converges uniformly to f on $\mathbb{K}$.*

Proof For a proof of (a) see §10D, p. 209. For a proof of (b), see §10B, p. 199. (Alternatively, (b) follows immediately from (e).) For a proof of (d) see §10C, p. 206.

Ⓔ (c) **Exercise 8A.2.** (*Hint:* Use the Weierstrass M-test, Proposition 6E.13, p. 135).

Ⓔ (e) **Exercise 8A.3.** (*Hint:* Use Theorem 8D.1(e) and Proposition 8D.2, p. 176). □

There is nothing special about the interval $[-\pi, \pi]$. Real Fourier series can be defined for functions on an interval $[-L, L]$ for any $L > 0$. We chose $L = \pi$ because it makes the computations simpler. If $L \neq \pi$, then we can define a Fourier series analogous to equation (8A.1) using the functions $\mathbf{S}_n(x) = \sin\left(\frac{n\pi x}{L}\right)$ and $\mathbf{C}_n(x) = \cos\left(\frac{n\pi x}{L}\right)$.

Exercise 8A.4 Let $L > 0$, and let $f : [-L, L] \longrightarrow \mathbb{R}$. Generalize all parts of ⓔ Theorem 8A.1 to characterize the convergence of the real Fourier series of f. ◆

Remark Please see Remark 8D.3, p. 177, for further technical remarks about the (non)convergence of real Fourier series, in situations where the hypotheses of Theorem 8A.1 are not satisfied.

8B Computing real Fourier coefficients

Prerequisites: §8A.   Recommended: §7C.

When computing the real Fourier coefficient $A_n = \frac{1}{\pi} \int_{-\pi}^{\pi} f(x) \cdot \cos(nx) \mathrm{d}x$ (or $B_n = \frac{1}{\pi} \int_{-\pi}^{\pi} f(x) \cdot \sin(nx) \mathrm{d}x$), it is simpler to compute the integral $\int_{-\pi}^{\pi} f(x) \cdot \cos(nx) \mathrm{d}x$ (or $\int_{-\pi}^{\pi} f(x) \cdot \sin(nx) \mathrm{d}x$) first, and then multiply the result by $1/\pi$. In this section, we review some useful techniques to compute this integral.

8B(i) Polynomials

Recommended: §7C(ii).

Theorem 8B.1

$$\int_{-\pi}^{\pi} \sin(nx)\mathrm{d}x = 0 = \int_{-\pi}^{\pi} \cos(nx)\mathrm{d}x.$$

For any $k \in \{1, 2, 3, \ldots\}$, we have the following recurrence relations.

- *If k is even,*

$$\int_{-\pi}^{\pi} x^k \cdot \sin(nx)\mathrm{d}x = 0;$$

$$\int_{-\pi}^{\pi} x^k \cdot \cos(nx)\mathrm{d}x = \frac{-k}{n} \int_{-\pi}^{\pi} x^{k-1} \cdot \sin(nx)\mathrm{d}x.$$

- *If $k > 0$ is odd,*

$$\int_{-\pi}^{\pi} x^k \cdot \sin(nx)\mathrm{d}x = \frac{2(-1)^{n+1}\pi^k}{n} + \frac{k}{n} \int_{-\pi}^{\pi} x^{k-1} \cdot \cos(nx)\mathrm{d}x;$$

$$\int_{-\pi}^{\pi} x^k \cdot \cos(nx)\mathrm{d}x = 0.$$

ⓔ *Proof* **Exercise 8B.1.** (*Hint:* Use integration by parts.)   □

Example 8B.2

(a) $p(x) = x$. Since $k = 1$ is *odd*, we have

$$\frac{1}{\pi} \int_{-\pi}^{\pi} x \cdot \cos(nx)dx = 0;$$

$$\frac{1}{\pi} \int_{-\pi}^{\pi} x \cdot \sin(nx)dx = \frac{2(-1)^{n+1}\pi^0}{n} + \frac{1}{n\pi} \int_{-\pi}^{\pi} \cos(nx)dx$$

$$\underset{(*)}{=} \frac{2(-1)^{n+1}}{n},$$

where equality $(*)$ follows from case $k = 0$ in Theorem 8B.1.

(b) $p(x) = x^2$. Since $k = 2$ is *even*, we have, for all n,

$$\frac{1}{\pi} \int_{-\pi}^{\pi} x^2 \sin(nx)dx = 0;$$

$$\frac{1}{\pi} \int_{-\pi}^{\pi} x^2 \cos(nx)dx = \frac{-2}{n\pi} \int_{-\pi}^{\pi} x^1 \cdot \sin(nx)dx$$

$$\underset{(*)}{=} \frac{-2}{n} \left(\frac{2(-1)^{n+1}}{n} \right) = \frac{4(-1)^n}{n^2},$$

where equality $(*)$ follows from the previous example.   ◇

8B(ii) Step functions

Recommended: §7C(iii).

A function $F : [-\pi, \pi] \longrightarrow \mathbb{R}$ is a *step function* (see Figure 8B.1(a)) if there are numbers $-\pi = x_0 < x_1 < x_2 < x_3 < \cdots < x_{M-1} < x_M = \pi$ and constants $a_1, a_2, \ldots, a_M \in \mathbb{R}$ such that

$$F(x) = \begin{cases} a_1 & \text{if } -\pi \le x \le x_1; \\ a_2 & \text{if } x_1 < x \le x_2; \\ \vdots & \vdots \\ a_m & \text{if } x_{m-1} < x \le x_m; \\ \vdots & \vdots \\ a_M & \text{if } x_{M-1} < x \le \pi. \end{cases} \tag{8B.1}$$

To compute the Fourier coefficients of a step function, we break the integral into 'pieces'. The general formula is given by the following theorem, but it is really not worth memorizing the formula. Instead, understand the *idea*.

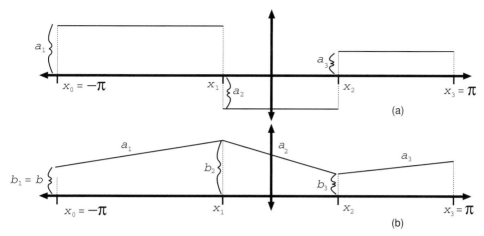

Figure 8B.1. (a) Step function. (b) Piecewise linear function.

Theorem 8B.3 *Suppose* $F : [-\pi, \pi] \longrightarrow \mathbb{R}$ *is a step function like that in equation* (8B.1). *Then the Fourier coefficients of F are given by*

$$\frac{1}{2\pi} \int_{-\pi}^{\pi} F(x) dx = \frac{1}{2\pi} \sum_{m=1}^{M} a_m \cdot (x_m - x_{m-1});$$

$$\frac{1}{\pi} \int_{-\pi}^{\pi} F(x) \cdot \cos(nx) dx = \frac{-1}{\pi n} \sum_{m=1}^{M-1} \sin(n \cdot x_m) \cdot \left(a_{m+1} - a_m\right);$$

$$\frac{1}{\pi} \int_{-\pi}^{\pi} F(x) \cdot \sin(nx) dx = \frac{(-1)^n}{\pi n} \left(a_1 - a_M\right) + \frac{1}{\pi n} \sum_{m=1}^{M-1} \cos(n \cdot x_m) \cdot \left(a_{m+1} - a_m\right).$$

Proof **Exercise 8B.2.** *Hint:* Integrate the function piecewise. Use the fact that Ⓔ

$$\int_{x_{m-1}}^{x_m} f(x) \sin(nx) = \frac{a_m}{n} \left(\cos(n \cdot x_{m-1}) - \cos(n \cdot x_m)\right)$$

and

$$\int_{x_{m-1}}^{x_m} f(x) \cos(nx) = \frac{a_m}{n} \left(\cos(n \cdot x_m) - \cos(n \cdot x_{m-1})\right).$$ □

Remark Note that the Fourier series of a step function f will converge uniformly to f on the *interior* of each 'step', but will *not* converge to f at any of the step boundaries, because f is not continuous at these points.

Example 8B.4 Suppose

$$f(x) = \begin{cases} -3 & \text{if} \quad -\pi \le x < \frac{-\pi}{2}; \\ 5 & \text{if} \quad \frac{-\pi}{2} \le x < \frac{\pi}{2}; \\ 2 & \text{if} \quad \frac{\pi}{2} \le x \le \pi \end{cases}$$

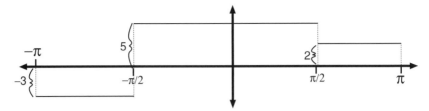

Figure 8B.2. The step function in Example 8B.4.

(see Figure 8B.2). In the notation of Theorem 8B.3, we have $M = 3$, and

$$x_0 = -\pi; \quad x_1 = \frac{-\pi}{2}; \quad x_2 = \frac{\pi}{2}; \quad x_3 = \pi;$$

$$a_1 = -3; \quad a_2 = 5; \quad a_3 = 2.$$

Thus,

$$A_n = \frac{-1}{\pi n}\left[8 \cdot \sin\left(n \cdot \frac{-\pi}{2}\right) - 3 \cdot \sin\left(n \cdot \frac{\pi}{2}\right)\right]$$

$$= \begin{cases} 0 & \text{if } n \text{ is even;} \\ (-1)^k \cdot \dfrac{11}{\pi n} & \text{if } n = 2k + 1 \text{ is odd,} \end{cases}$$

and

$$B_n = \frac{1}{\pi n}\left[8 \cdot \cos\left(n \cdot \frac{-\pi}{2}\right) - 3 \cdot \cos\left(n \cdot \frac{\pi}{2}\right) - 5 \cdot \cos\left(n \cdot \pi\right)\right]$$

$$= \begin{cases} \dfrac{5}{\pi n} & \text{if } n \text{ is odd;} \\ \dfrac{5}{\pi n}\left((-1)^k - 1\right) & \text{if } n = 2k \text{ is even} \end{cases}$$

$$= \begin{cases} \dfrac{5}{\pi n} & \text{if } n \text{ is odd;} \\ 0 & \text{if } n \text{ is divisible by 4;} \\ \dfrac{-10}{\pi n} & \text{if } n \text{ is even but } not \text{ divisible by 4.} \end{cases} \qquad \diamond$$

8B(iii) *Piecewise linear functions*

Recommended: §7C(iv).

A continuous function $F : [-\pi, \pi] \longrightarrow \mathbb{R}$ is *piecewise linear* (see Figure 8B.1(b))
if there are numbers $-\pi = x_0 < x_1 < x_1 < x_2 < \cdots < x_{M-1} < x_M = \pi$ and

constants $a_1, a_2, \ldots, a_M \in \mathbb{R}$ and $b \in \mathbb{R}$ such that

$$F(x) = \begin{cases} a_1(x - \pi) + b_1 & \text{if } -\pi < x < x_1; \\ a_2(x - x_1) + b_2 & \text{if } x_1 < x < x_2; \\ \quad \vdots & \vdots \\ a_m(x - x_m) + b_{m+1} & \text{if } x_m < x < x_{m+1}; \\ \quad \vdots & \vdots \\ a_M(x - x_{M-1}) + b_M & \text{if } x_{M-1} < x < \pi, \end{cases}$$ (8B.2)

where $b_1 = b$, and, for all $m > 1$, $b_m = a_m(x_m - x_{m-1}) + b_{m-1}$.

Example 8B.5 If $f(x) = |x|$, then f is piecewise linear, with: $x_0 = -\pi$, $x_1 = 0$, and $x_2 = \pi$; $a_1 = -1$ and $a_2 = 1$; $b_1 = \pi$, and $b_2 = 0$. ◇

To compute the Fourier coefficients of a piecewise linear function, we break the integral into 'pieces'. The general formula is given by the following theorem, but it is really not worth memorizing the formula. Instead, understand the *idea*.

Theorem 8B.6 *Suppose* $F : [-\pi, \pi] \longrightarrow \mathbb{R}$ *is a piecewise linear function like that in equation* (8B.2). *Then the Fourier coefficients of* F *are given by*

$$\frac{1}{2\pi} \int_{-\pi}^{\pi} F(x)dx = \frac{1}{2\pi} \sum_{m=1}^{M} \frac{a_m}{2}(x_m - x_{m-1})^2 + b_m \cdot (x_m - x_{m-1});$$

$$\frac{1}{\pi} \int_{-\pi}^{\pi} F(x) \cdot \cos(nx)dx = \frac{1}{\pi n^2} \sum_{m=1}^{M} \cos(nx_m) \cdot \left(a_m - a_{m+1}\right);$$

$$\frac{1}{\pi} \int_{-\pi}^{\pi} F(x) \cdot \sin(nx)dx = \frac{1}{\pi n^2} \sum_{m=1}^{M-1} \sin(nx_m) \cdot \left(a_m - a_{m+1}\right).$$

(Here, we define $a_{M+1} := a_1$ *for convenience.)*

Proof **Exercise 8B.3.** (*Hint:* Invoke Theorem 8B.3 and use integration by parts.) $\qquad\qquad\square$ Ⓔ

Note that the summands in this theorem read '$a_m - a_{m+1}$', not the other way around.

Example 8B.7 Recall $f(x) = |x|$ from Example 8B.5. Applying Theorem 8B.6, we have

$$A_0 = \frac{1}{2\pi} \left[\frac{-1}{2}(0 + \pi)^2 + \pi \cdot (0 + \pi) + \frac{1}{2}(\pi - 0)^2 + 0 \cdot (\pi - 0) \right] = \frac{\pi}{2};$$

$$A_n = \frac{\pi}{\pi n^2} \left[(-1 - 1) \cdot \cos(n0)(1 + 1) \cdot \cos(n\pi) \right]$$

$$= \frac{1}{\pi n^2} \left[-2 + 2(-1)^n \right] = \frac{-2}{\pi n^2} \left[1 - (-1)^n \right],$$

while $B_n = 0$ for all $n \in \mathbb{N}$, because f is an even function. ◇

8B(iv) Differentiating real Fourier series

Prerequisites: §8A, Appendix F.

Suppose $f(x) = 3 + 2\cos(x) - 6\cos(2x) + 11\cos(3x) + 3\sin(x) - 5\sin(2x) + 7\sin(3x)$. Then $f'(x) = -2\sin(x) + 12\sin(2x) - 33\sin(3x) + 3\cos(x) - 10\cos(2x) + 21\cos(3x)$. This illustrates a general pattern.

Theorem 8B.8 *Let* $f \in \mathcal{C}^\infty[-\pi, \pi]$, *and suppose* f *has Fourier series* $\sum_{n=0}^\infty A_n \mathbf{C}_n + \sum_{n=1}^\infty B_n \mathbf{S}_n$. *If* $\sum_{n=1}^\infty n|A_n| < \infty$ *and* $\sum_{n=1}^\infty n|B_n| < \infty$, *then* f' *has Fourier series:* $\sum_{n=1}^\infty n(B_n\mathbf{C}_n - A_n\mathbf{S}_n)$.

ⓔ *Proof* **Exercise 8B.4.** (*Hint:* Apply Proposition F.1, p. 569.)   □

Consequence If $f(x) = A\cos(nx) + B\sin(nx)$ for some $A, B \in \mathbb{R}$, then $f''(x) = -n^2 f(x)$. In other words, f is an *eigenfunction* for the differentation operator ∂_x^2, with eigenvalue $-n^2$. Hence, for any $k \in \mathbb{N}$, we have $\partial_x^{2k} f = (-1)^k (n)^{2k} \cdot f$.

8C Relation between (co)sine series and real series

Prerequisites: §7A, §8A.

We have seen in §8A how the collection $\{\mathbf{C}_n\}_{n=0}^\infty \cup \{\mathbf{S}_n\}_{n=1}^\infty$ forms an orthogonal basis for $\mathbf{L}^2[-\pi, \pi]$. However, if we confine our attention to *half* this interval – that is, to $\mathbf{L}^2[0, \pi]$ – then the results of §7A imply that we only need half as many basis elements; either the collection $\{\mathbf{C}_n\}_{n=0}^\infty$ or the collection $\{\mathbf{S}_n\}_{n=1}^\infty$ will suffice. Why is this? And what is the relationship between the Fourier (co)sine series of §7A and the Fourier series of §8A?

A function $f : [-L, L] \longrightarrow \mathbb{R}$ is *even* if $f(-x) = f(x)$ for all $x \in [0, L]$. For example, the following functions are even:

- $f(x) = 1$;
- $f(x) = |x|$;
- $f(x) = x^2$;
- $f(x) = x^k$ for any even $k \in \mathbb{N}$;
- $f(x) = \cos(x)$.

A function $f : [-L, L] \longrightarrow \mathbb{R}$ is *odd* if $f(-x) = -f(x)$ for all $x \in [0, L]$. For example, the following functions are odd:

- $f(x) = x$;
- $f(x) = x^3$;
- $f(x) = x^k$ for any odd $k \in \mathbb{N}$;
- $f(x) = \sin(x)$;

Every function can be 'split' into an 'even part' and an 'odd part'.

Proposition 8C.1

(a) *For any* $f : [-L, L] \longrightarrow \mathbb{R}$, *there is a unique even function* $\check{f}$ *and a unique odd function* $\acute{f}$ *such that* $f = \check{f} + \acute{f}$. *To be specific:*

$$\check{f}(x) = \frac{f(x) + f(-x)}{2} \quad and \quad \acute{f}(x) = \frac{f(x) - f(-x)}{2}.$$

(b) *If* f *is even, then* $f = \check{f}$ *and* $\acute{f} = 0$.
(c) *If* f *is odd, then* $\check{f} = 0$ *and* $f = \acute{f}$.

Proof **Exercise 8C.1.** □ Ⓔ

The equation $f = \check{f} + \acute{f}$ is called the *even–odd decomposition* of f. Next, we define the vector spaces:

$$\mathbf{L}^2_{\text{even}}[-\pi, \pi] := \{\text{all } even \text{ elements in } \mathbf{L}^2[-\pi, \pi]\};$$

$$\mathbf{L}^2_{\text{odd}}[-\pi, \pi] := \{\text{all } odd \text{ elements in } \mathbf{L}^2[-\pi, \pi]\}.$$

Proposition 8C.1 implies that any $f \in \mathbf{L}^2[-\pi, \pi]$ can be written (in a unique way) as $f = \check{f} + \acute{f}$ for some $\check{f} \in \mathbf{L}^2_{\text{even}}[-\pi, \pi]$ and $\acute{f} \in \mathbf{L}^2_{\text{odd}}[-\pi, \pi]$. (This is sometimes indicated by writing $\mathbf{L}^2[-\pi, \pi] = \mathbf{L}^2_{\text{even}}[-\pi, \pi] \oplus \mathbf{L}^2_{\text{odd}}[-\pi, \pi]$.)

Lemma 8C.2 *Let* $n \in \mathbb{N}$.

(a) *The function* $\mathbf{C}_n(x) = \cos(nx)$ *is even.*
(b) *The function* $\mathbf{S}_n(x) = \sin(nx)$ *is odd.*

Let $f : [-\pi, \pi] \longrightarrow \mathbb{R}$ *be any function.*

(c) *If* $f(x) = \sum_{n=0}^{\infty} A_n \mathbf{C}_n(x)$, *then* f *is even.*
(d) *If* $f(x) = \sum_{n=1}^{\infty} B_n \mathbf{S}_n(x)$, *then* f *is odd.*

Proof **Exercise 8C.2.** □ Ⓔ

In other words, cosine series are even and sine series are odd. The converse is also true. To be precise, consider the following.

Proposition 8C.3 *Let* $f : [-\pi, \pi] \longrightarrow \mathbb{R}$ *be any function, and suppose* f *has real Fourier series* $f(x) = A_0 + \sum_{n=1}^{\infty} A_n \mathbf{C}_n(x) + \sum_{n=1}^{\infty} B_n \mathbf{S}_n(x)$. *Then*

(a) *if* f *is odd, then* $A_n = 0$ *for every* $n \in \mathbb{N}$;
(b) *if* f *is even, then* $B_n = 0$ *for every* $n \in \mathbb{N}$.

Proof **Exercise 8C.3.** □ Ⓔ

From this, Proposition 8C.4 follows immediately.

Proposition 8C.4

(a) *The set* $\{\mathbf{C}_0, \mathbf{C}_1, \mathbf{C}_2, \ldots\}$ *is an orthogonal basis for* $\mathbf{L}^2_{\text{even}}[-\pi, \pi]$ *(where* $\mathbf{C}_0 = 1\!1$*).*
(b) *The set* $\{\mathbf{S}_1, \mathbf{S}_2, \mathbf{S}_3, \ldots\}$ *is an orthogonal basis for* $\mathbf{L}^2_{\text{odd}}[-\pi, \pi]$.

(c) *Suppose f has even–odd decomposition $f = \check{f} + \acute{f}$, and f has real Fourier series $f(x) = A_0 + \sum_{n=1}^{\infty} A_n C_n(x) + \sum_{n=1}^{\infty} B_n S_n(x)$. Then $\check{f}(x) = \sum_{n=0}^{\infty} A_n C_n(x)$ and $\acute{f}(x) = \sum_{n=1}^{\infty} B_n S_n(x)$.*

Ⓔ *Proof* **Exercise 8C.4.**   □

If $f : [0, \pi] \longrightarrow \mathbb{R}$, then we can 'extend' f to a function on $[-\pi, \pi]$ in two ways.

- The *even* extension of f is defined: $f_{\text{even}}(x) = f(|x|)$ for all $x \in [-\pi, \pi]$.
- The *odd* extension of f is defined:

$$f_{\text{odd}}(x) = \begin{cases} f(x) & \text{if } x > 0; \\ 0 & \text{if } x = 0; \\ -f(-x) & \text{if } x < 0. \end{cases}$$

Ⓔ **Exercise 8C.5**
(a) Show that f_{even} is even and f_{odd} is odd.
(b) For all $x \in [0, \pi]$, show that $f_{\text{even}}(x) = f(x) = f_{\text{odd}}(x)$.   ◆

Proposition 8C.5 *Let $f : [0, \pi] \longrightarrow \mathbb{R}$ have even extension $f_{\text{even}} : [-\pi, \pi] \longrightarrow \mathbb{R}$ and odd extension $f_{\text{odd}} : [-\pi, \pi] \longrightarrow \mathbb{R}$.*

(a) *The Fourier sine series for f is the same as the real Fourier series for f_{odd}. In other words, the nth Fourier sine coefficient is given: $B_n = \frac{1}{\pi} \int_{-\pi}^{\pi} f_{\text{odd}}(x) S_n(x) dx$.*
(b) *The Fourier cosine series for f is the same as the real Fourier series for f_{even}. In other words, the nth Fourier cosine coefficient is given: $A_n = \frac{1}{\pi} \int_{-\pi}^{\pi} f_{\text{even}}(x) C_n(x) dx$.*

Ⓔ *Proof* **Exercise 8C.6.**   □

Let $f \in C^1[0, \pi]$. Recall that Theorem 7A.1(d) (p. 142) says that the Fourier sine series of f converges to f *uniformly* on $[0, \pi]$ if and only if f satisfies homogeneous Dirichlet boundary conditions on $[0, \pi]$ (i.e. $f(0) = f(\pi) = 0$). On the other hand, Theorem 7A.4(d) (p. 145) says that the Fourier cosine series of f *always* converges to f uniformly on $[0, \pi]$ if $f \in C^1[0, \pi]$; furthermore, if the formal derivative of this cosine series converges to f' uniformly on $[0, \pi]$, then f satisfies homogeneous *Neumann* boundary conditions on $[0, \pi]$ (i.e. $f'(0) = f'(\pi) = 0$). Meanwhile, if $F \in C^1[-\pi, \pi]$, then Theorem 8A.1(d) (p. 166) says that the (real) Fourier series of F converges to F uniformly on $[-\pi, \pi]$ if F satisfies *periodic* boundary conditions on $[-\pi, \pi]$ (i.e. $F(-\pi) = F(\pi)$). The next result explains the logical relationship between these three statements.

Lemma 8C.6 *Let $f : [0, \pi] \longrightarrow \mathbb{R}$ have even extension $f_{\text{even}} : [-\pi, \pi] \longrightarrow \mathbb{R}$ and odd extension $f_{\text{odd}} : [-\pi, \pi] \longrightarrow \mathbb{R}$. Suppose f is right-continuous at 0 and left-continuous at π.*

(a) f_{odd} *is continuous at zero and satisfies periodic boundary conditions on* $[-\pi, \pi]$, *if and only if* f *satisfies homogeneous Dirichlet boundary conditions on* $[0, \pi]$.

(b) f_{even} *is **always** continuous at zero and **always** satisfies periodic boundary conditions on* $[-\pi, \pi]$. *However, the derivative* f'_{even} *is continuous at zero and satisfies periodic boundary conditions on* $[-\pi, \pi]$ *if and only if* f *satisfies homogeneous Neumann boundary conditions on* $[0, \pi]$.

Proof **Exercise 8C.7.**   □   Ⓔ

8D Complex Fourier series

Prerequisites: §6C(i), §6E, §6F, Appendix C.   **Recommended:** §8A.

Let $f, g : \mathbb{X} \longrightarrow \mathbb{C}$ be complex-valued functions. Recall from §6C(i) that we define their *inner product*:

$$\langle f, g \rangle := \frac{1}{M} \int_{\mathbb{X}} f(\mathbf{x}) \cdot \overline{g(\mathbf{x})} d\mathbf{x},$$

where M is the length/area/volume of domain $\mathbb{X}$. Once again,

$$\|f\|_2 := \langle f, f \rangle^{1/2} = \left(\frac{1}{M} \int_{\mathbb{X}} f(\mathbf{x}) \overline{f(\mathbf{x})} d\mathbf{x} \right)^{1/2} = \left(\frac{1}{M} \int_{\mathbb{X}} |f(\mathbf{x})|^2 \, d\mathbf{x} \right)^{1/2}.$$

The concepts of orthogonality, L^2-distance, and L^2-convergence are exactly the same as before. Let $\mathbf{L}^2([-L, L]; \, \mathbb{C})$ be the set of all complex-valued functions $f : [-L, L] \longrightarrow \mathbb{C}$ with $\|f\|_2 < \infty$. For all $n \in \mathbb{Z}$, let

$$\mathbf{E}_n(x) := \exp\left(\frac{\pi i n x}{L} \right).$$

(Thus, $\mathbf{E}_0 = \mathbb{1}$ is the constant unit function.) For all $n > 0$, note that Euler's Formula (see p. 553) implies that

$$\mathbf{E}_n(x) = \mathbf{C}_n(x) + \mathbf{i} \cdot \mathbf{S}_n(x);$$
$$\text{and } \mathbf{E}_{-n}(x) = \mathbf{C}_n(x) - \mathbf{i} \cdot \mathbf{S}_n(x). \tag{8D.1}$$

Also, note that $\langle \mathbf{E}_n, \mathbf{E}_m \rangle = 0$ if $n \neq m$, and $\|\mathbf{E}_n\|_2 = 1$ (**Exercise 8D.1**), so these   Ⓔ functions form an *orthonormal set*.

If $f : [-L, L] \longrightarrow \mathbb{C}$ is any function with $\|f\|_2 < \infty$, then we define the *(complex) Fourier coefficients* of f:

$$\widehat{f}_n := \langle f, \mathbf{E}_n \rangle = \frac{1}{2L} \int_{-L}^{L} f(x) \cdot \exp\left(\frac{-\pi i n x}{L} \right) dx. \tag{8D.2}$$

The *(complex) Fourier series* of f is then the infinite summation of functions:

$$\sum_{n=-\infty}^{\infty} \widehat{f_n} \cdot \mathbf{E}_n. \qquad (8D.3)$$

(Note that, in this sum, n ranges from $-\infty$ to ∞.)

Theorem 8D.1 Complex Fourier convergence

(a) *The set* $\{\ldots, \mathbf{E}_{-1}, \mathbf{E}_0, \mathbf{E}_1, \ldots\}$ *is an orthonormal basis for* $\mathbf{L}^2([-L, L]; \mathbb{C})$. *Thus, if* $f \in \mathbf{L}^2([-L, L]; \mathbb{C})$, *then the complex Fourier series* (8D.3) *converges to* f *in* $\mathbf{L}^2$-*norm. Furthermore,* $\{\widehat{f_n}\}_{n=-\infty}^{\infty}$ *is the unique sequence of coefficients with this property.*

(b) *If* f *is continuously differentiable*[1] *on* $[-\pi, \pi]$, *then the Fourier series* (8D.3) *converges pointwise on* $(-\pi, \pi)$. *More generally, if* f *is piecewise* $\mathcal{C}^1$, *then the complex Fourier series* (8D.3) *converges to* f *pointwise on each* $\mathcal{C}^1$ *interval for* f. *In other words, if* $\{j_1, \ldots, j_m\}$ *is the set of discontinuity points of* f *and/or* f' *in* $[-L, L]$, *and* $j_m < x < j_{m+1}$, *then* $f(x) = \lim_{N \to \infty} \sum_{n=-N}^{N} \widehat{f_n} \mathbf{E}_n(x)$.

(c) *If* $\sum_{n=-\infty}^{\infty} |\widehat{f_n}| < \infty$, *then the series* (8D.3) *converges to* f *uniformly on* $[-\pi, \pi]$.

(d) *Suppose* $f : [-\pi, \pi] \longrightarrow \mathbb{R}$ *is continuous and piecewise differentiable,* $f' \in \mathbf{L}^2[-\pi, \pi]$, *and* $f(-\pi) = f(\pi)$. *Then the series* (8D.3) *converges to* f *uniformly on* $[-\pi, \pi]$.

(e) *If* f *is piecewise* $\mathcal{C}^1$, *and* $\mathbb{K} \subset (j_m, j_{m+1})$ *is any closed subset of a* $\mathcal{C}^1$ *interval of* f, *then the series* (8D.3) *converges uniformly to* f *on* $\mathbb{K}$.

Ⓔ *Proof* (a) **Exercise 8D.2.** (*Hint:* Use Theorem 8A.1(a), p. 166, and Proposition 8D.2 below.)

For a direct proof of (a), see Katznelson (1976), §I.5.5.

Ⓔ   (b) **Exercise 8D.3.** (*Hint:* (i) Use Theorem 8A.1(b) on p. 166 and Proposition 8D.2 below. (ii) For a second proof, derive (b) from (e).)

Ⓔ   (c) **Exercise 8D.4.** (*Hint:* Use the Weierstrass M-test, Proposition 6E.13, p. 135.)

Ⓔ   (d) is **Exercise 8D.5.** (*Hint:* Use Theorem 8A.1(d), p. 166, and Proposition 8D.2 below.)

For a direct proof of (d) see Wheeden and Zygmund (1977), Theorem 12.20.

For (e) see Katznelson (1976), p. 53, or Folland (1984), Theorem 8.4.3.   □

Proposition 8D.2 Relation between real and complex Fourier series

Let $f : [-\pi, \pi] \longrightarrow \mathbb{R}$ *be a real-valued function, and let* $\{A_n\}_{n=0}^{\infty}$ *and* $\{B_n\}_{n=1}^{\infty}$ *be its real Fourier coefficients, as defined on p. 165. We can also regard* f *as a*

---

[1] This means that $f(x) = f_r(x) + \mathbf{i} f_i(x)$, where $f_r : [-L, L] \longrightarrow \mathbb{R}$ and $f_i : [-L, L] \longrightarrow \mathbb{R}$ are both continuously differentiable, real-valued functions.

complex-valued function; let $\{\widehat{f}_n\}_{n=-\infty}^{\infty}$ *be the complex Fourier coefficients of* f, *as defined by equation* (8D.2), *p. 175. Let* $n \in \mathbb{N}_+$. *Then*

(a) $\widehat{f}_n = \frac{1}{2}(A_n - \mathbf{i}B_n)$, *and* $\widehat{f}_{-n} = \overline{\widehat{f}_n} = \frac{1}{2}(A_n + \mathbf{i}B_n)$;

(b) *thus,* $A_n = \widehat{f}_n + \widehat{f}_{-n}$, *and* $B_n = \mathbf{i}(\widehat{f}_n - \widehat{f}_{-n})$;

(c) $\widehat{f}_0 = A_0$.

Proof **Exercise 8D.6.** (*Hint:* Use equations (8D.1).) □ Ⓔ

Exercise 8D.7 Show that Theorem 8D.1(a) and Theorem 8A.1(a) are equivalent, Ⓔ using Proposition 8D.2. ♦

Remark 8D.3 Further remarks on Fourier convergence

(a) In Theorems 7A.1(b), 7A.4(b), 8A.1(b), and 8D.1(b), if x is a discontinuity point of f, then the Fourier (co)sine series converges to the average of the 'left-hand' and 'right-hand' limits of f at x, namely:

$$\frac{f(x-) + f(x+)}{2}, \quad \text{where} \quad f(x-) := \lim_{y \nearrow x} f(y) \quad \text{and} \quad f(x+) := \lim_{y \searrow x} f(y).$$

(b) If the hypothesis of Theorems 7A.1(c), 7A.4(c), 8A.1(c), or 8D.1(c) is satisfied, then we say that the Fourier series (real, complex, sine or cosine) converges *absolutely*. (In fact, Theorems 7A.1(d)(i), 7A.4(d)(i), 8A.1(d), or 8D.1(d) can be strengthened to yield absolute convergence.) Absolute convergence is stronger than uniform convergence, and functions with absolutely convergent Fourier series form a special class; see Katznelson (1976), §I.6 for more information.

(c) In Theorems 7A.1(e), 7A.4(e), 8A.1(e), and 8D.1(e), we don't quite need f to be *differentiable* to guarantee uniform convergence of the Fourier (co)sine series. Let $\alpha > 0$ be a constant; we say that f is α-*Hölder continuous* on $[-\pi, \pi]$ if there is some $M < \infty$ such that

$$\text{for all } x, y \in [0, \pi], \quad \frac{|f(x) - f(y)|}{|x - y|^\alpha} \leq M.$$

Bernstein's theorem says: if f is α-Hölder continuous for some $\alpha > 1/2$, then the Fourier series (real, complex, sine or cosine) of f will converge uniformly (indeed, absolutely) to f; see Katznelson (1976), Theorem 6.3, or Folland (1984), Theorem 8.39. (If f is differentiable, then f would be α-Hölder continuous with $\alpha = 1$, so Bernstein's theorem immediately implies Theorems 7A.1(e) and 7A.4(e).)

(d) The *total variation* of f is defined:

$$\text{var}(f) := \sup_{N \in \mathbb{N}} \ \sup_{-\pi \leq x_0 < \cdots < x_N \leq \pi} \ \sum_{n=1}^{N} \left| f(x_n) - f(x_{n-1}) \right| \underset{(*)}{=} \int_{-\pi}^{\pi} |f'(x)| \, dx.$$

Here, the supremum is taken over all finite increasing sequences $\{-\pi \leq x_0 < x_1 < \cdots < x_N \leq \pi\}$ (for any $N \in \mathbb{N}$), and equality $(*)$ is true if and only if f is continuously differentiable. *Zygmund's theorem* says: if $\text{var}(f) < \infty$ (i.e. f has *bounded variation*) and f is α-Hölder continuous for some $\alpha > 0$, then the Fourier series of f

will converge uniformly (indeed, absolutely) to f on $[-\pi, \pi]$; see Katznelson (1976), Theorem 6.4.

(e) However, merely being *continuous* is *not* sufficient for uniform Fourier convergence, or even pointwise convergence. There exists a continuous function $f : [0, \pi] \longrightarrow \mathbb{R}$ whose Fourier series does *not* converge pointwise on $(0, \pi)$; i.e. the series diverges at some points in $(0, \pi)$; see Katznelson (1976), Theorem 2.1, or Wheeden and Zygmund (1977), Theorem 12.35. Thus, Theorems 7A.1(b), 7A.4(b), 8A.1(b), and 8D.1(b) are *false* if we replace 'differentiable' with 'continuous'.

(f) Fix $p \in [1, \infty)$. For any $f : [-\pi, \pi] \longrightarrow \mathbb{C}$, we define the L^p-*norm* of f:

$$\|f\|_p = \left(\int_{-\pi}^{\pi} |f(x)|^p \, dx \right)^{1/p}.$$

(Thus, if $p = 2$, we get the familiar L^2-norm $\|f\|_2$.) Let $\mathbf{L}^p[-\pi, \pi]$ be the set of all integrable functions $f : [-\pi, \pi] \longrightarrow \mathbb{C}$ such that $\|f\|_p < \infty$. Theorem 8D.1(a) says that, if $f \in \mathbf{L}^2[-\pi, \pi]$, then the complex Fourier series of f converges to f in L^2-norm. The Fourier series of f also converges in L^p-norm for any other $p \in (1, \infty)$. That is, for any $p \in (1, \infty)$ and any $f \in \mathbf{L}^p[-\pi, \pi]$, we have

$$\lim_{N \to \infty} \left\| f - \sum_{n=-N}^{N} \widehat{f}_n \mathbf{E}_n \right\|_p = 0$$

See Katznelson (1976), Theorem 1.5. If $f \in \mathbf{L}^p[-\pi, \pi]$ is purely real-valued, then the same statement holds for the real Fourier series:

$$\lim_{N \to \infty} \left\| f - \left(A_0 + \sum_{n=1}^{N} A_n \mathbf{C}_n + \sum_{n=1}^{N} B_n \mathbf{S}_n \right) \right\|_p = 0.$$

To understand the significance of L^p-convergence, we remark that if p is very large, then L^p convergence is 'almost' the same as uniform convergence. Also note the following.

Ⓔ
- If $p > q$, then $\mathbf{L}^p[-\pi, \pi] \subset \mathbf{L}^q[-\pi, \pi]$ (**Exercise 8D.8**). For example, if $f \in \mathbf{L}^3[-\pi, \pi]$, then it follows that $f \in \mathbf{L}^2[-\pi, \pi]$ (but not vice versa). If $f \in \mathbf{L}^2[-\pi, \pi]$, then it follows that $f \in \mathbf{L}^{3/2}[-\pi, \pi]$ (but not vice versa).
- If $p > q$, and the Fourier series of f converges to f in L^p-norm, then it also converges to f in L^q-norm; see, e.g., Folland (1984), Proposition 6.12. For example, if $f \in \mathbf{L}^2[\pi, \pi]$, then Theorem 8D.1(a) implies that the Fourier series of f converges to f in L^q-norm for all $q \in [1, 2]$. (However, if $q < 2$, then there are functions in $\mathbf{L}^q[-\pi, \pi]$ to which Theorem 8D.1(a) does not apply.)

Finally, similar L^p-convergence statements hold for the Fourier (co)sine series of real-valued functions in $\mathbf{L}^p[0, \pi]$.

(g) The pointwise convergence of a Fourier series is a somewhat subtle and complicated business, once you depart from the realm of C^1 functions. In particular, the Fourier series of continuous (but nondifferentiable) functions can be badly behaved. This is perplexing, because we know that Fourier series converge in L^2-norm for any function in $\mathbf{L}^2[-\pi, \pi]$ (which includes all sorts of strange functions which are not differentiable anywhere). To bridge the gap between L^2- and pointwise convergence, a variety of other 'summation schemes' have been introduced for Fourier coefficients. These include

- the *Cesáro mean*, $\lim_{N\to\infty} \frac{1}{N} \sum_{n=1}^{N} S_N(f)$, where $S_N(f) := \sum_{n=-N}^{N} \widehat{f_n} \mathbf{E}_n$ is the Nth partial sum of the complex Fourier series (8D.3);
- the *Abel mean*, $\lim_{r \nearrow 1} \sum_{n=-\infty}^{\infty} r^{|n|} \widehat{f_n} \mathbf{E}_n$.

These sums have somewhat nicer convergence properties than the 'standard' Fourier series (8D.3). (See §18F, p. 466, for further discussion of the Abel mean.)

(h) There is a close relationship between the Fourier series of complex-valued functions on $[-\pi, \pi]$ and the Laurent series of complex-analytic functions defined near the unit circle; see §18E, p. 459.

(i) Remark (h) and the periodic boundary conditions required for Theorem 8D.1(d) both suggest that the Fourier series 'wants' us to identify the interval $(-\pi, \pi]$ with the unit circle $\mathbb{S}$ in the complex plane, via the bijection $\phi : (-\pi, \pi] \longrightarrow \mathbb{S}$ defined by $\phi(x) = e^{\mathrm{i}x}$. Now, $\mathbb{S}$ is an *abelian group* under the complex multiplication operator. That is: if $s, t \in \mathbb{S}$, then their product $s \cdot t$ is also in $\mathbb{S}$, the multiplicative inverse s^{-1} is in $\mathbb{S}$, and the identity element 1 is an element of $\mathbb{S}$. Furthermore, $\mathbb{S}$ is a compact subset of $\mathbb{C}$, and the multiplication operation is continuous with respect to the topology of $\mathbb{S}$. In summary, $\mathbb{S}$ is a *compact abelian topological group*. The functions $\{\mathbf{E}_n\}_{n=-\infty}^{\infty}$ are then *continuous homomorphisms* from $\mathbb{S}$ into $\mathbb{S}$ (these are called the *characters* of the group).

The existence of the Fourier series (8D.3) and the convergence properties enumerated in Theorem 8D.1 are actually a *consequence* of these facts. In fact, if $\mathbb{G}$ is *any* compact abelian topological group, then one can develop a version of Fourier analysis on $\mathbb{G}$. The *characters* of $\mathbb{G}$ are the continuous homomorphisms from $\mathbb{G}$ into the unit circle group $\mathbb{S}$. The set of all characters of $\mathbb{G}$ forms an orthonormal basis for $\mathbf{L}^2(\mathbb{G})$, so that almost any 'reasonable' function $f : \mathbb{G} \longrightarrow \mathbb{C}$ can be expressed as a complex-linear combination of these characters.

The study of Fourier series, their summability, and their generalizations to other compact abelian groups is called *harmonic analysis*, and is a crucial tool in many areas of mathematics, including the ergodic theory of dynamical systems and the representation theory of Lie groups. See Katznelson (1976), Wheeden and Zygmund (1977), chap. 12, or Folland (1984), chap. 8, to learn more about this vast and fascinating area of mathematics.

9

Multidimensional Fourier series

The scientist does not study nature because it is useful; he studies it because he delights in it, and he delights in it because it is beautiful. If nature were not beautiful, it would not be worth knowing, and if nature were not worth knowing, life would not be worth living.

Henri Poincaré

9A ...in two dimensions

Prerequisites: §6E, §6F.   Recommended: §7B.

Let $X, Y > 0$, and let $\mathbb{X} := [0, X] \times [0, Y]$ be an $X \times Y$ rectangle in the plane. Suppose $f : \mathbb{X} \longrightarrow \mathbb{R}$ is a real-valued function of two variables. For all $n, m \in \mathbb{N}_+ := \{1, 2, 3, \ldots\}$, we define the *two-dimensional Fourier sine coefficients*:

$$B_{n,m} := \boxed{\frac{4}{XY} \int_0^X \int_0^Y f(x, y) \sin\left(\frac{\pi n x}{X}\right) \sin\left(\frac{\pi m y}{Y}\right) dx\, dy.}$$

The *two-dimensional Fourier sine series* of f is the doubly infinite summation:

$$\boxed{\sum_{n,m=1}^{\infty} B_{n,m} \sin\left(\frac{\pi n x}{X}\right) \sin\left(\frac{\pi m y}{Y}\right).} \tag{9A.1}$$

Note that we are now summing over *two* independent indices, n and m.

Example 9A.1   Let $X = \pi = Y$, so that $\mathbb{X} = [0, \pi] \times [0, \pi]$, and let $f(x, y) = x \cdot y$. Then f has the two-dimensional Fourier sine series:

$$4 \sum_{n,m=1}^{\infty} \frac{(-1)^{n+m}}{nm} \sin(nx) \sin(my).$$

181

To see this, recall from Example 7C.2(c), p. 152, that we know that

$$\frac{2}{\pi} \int_0^\pi x \sin(x) dx = \frac{2(-1)^{n+1}}{n}.$$

Thus,

$$B_{n,m} = \frac{4}{\pi^2} \int_0^\pi \int_0^\pi xy \cdot \sin(nx) \sin(my) dx dy$$

$$= \left(\frac{2}{\pi} \int_0^\pi x \sin(nx) dx \right) \cdot \left(\frac{2}{\pi} \int_0^\pi y \sin(my) dy \right)$$

$$= \left(\frac{2(-1)^{n+1}}{n} \right) \cdot \left(\frac{2(-1)^{m+1}}{m} \right) = \frac{4(-1)^{m+n}}{nm}. \qquad \diamond$$

Example 9A.2 Let $X = \pi = Y$, so that $\mathbb{X} = [0, \pi] \times [0, \pi]$, and let $f(x, y) = 1$ be the constant 1 function. Then f has the two-dimensional Fourier sine series:

$$\frac{4}{\pi^2} \sum_{n,m=1}^\infty \frac{[1 - (-1)^n][1 - (-1)^m]}{n \quad m} \sin(nx) \sin(my)$$

$$= \frac{16}{\pi^2} \sum_{\substack{n,m=1 \\ \text{both odd}}}^\infty \frac{1}{n \cdot m} \sin(nx) \sin(my). \qquad \diamond$$

Ⓔ **Exercise 9A.1** Verify this. ◆

For all $n, m \in \mathbb{N} := \{0, 1, 2, 3, \ldots\}$, we define the *two-dimensional Fourier cosine coefficients* of f

$$\boxed{A_0 := \frac{1}{XY} \int_0^X \int_0^Y f(x, y) dx\, dy;}$$

$$\boxed{A_{n,0} := \frac{2}{XY} \int_0^X \int_0^Y f(x, y) \cos\left(\frac{\pi nx}{X}\right) dx\, dy} \quad \text{for } n > 0;$$

$$\boxed{A_{0,m} := \frac{2}{XY} \int_0^X \int_0^Y f(x, y) \cos\left(\frac{\pi my}{X}\right) dx\, dy} \quad \text{for } m > 0;$$

$$\boxed{A_{n,m} := \frac{4}{XY} \int_0^X \int_0^Y f(x, y) \cos\left(\frac{\pi nx}{X}\right) \cos\left(\frac{\pi my}{Y}\right) dx\, dy} \quad \text{for } n, m > 0.$$

The *two-dimensional Fourier cosine series* of f is the doubly infinite summation:

$$\sum_{n,m=0}^{\infty} A_{n,m} \cos\left(\frac{\pi n x}{X}\right) \cos\left(\frac{\pi m y}{Y}\right). \qquad (9A.2)$$

In what sense do these series converge to f? For any $n, m \in \mathbb{N}$, define the functions $\mathbf{C}_{n,m}, \mathbf{S}_{n,m} : [0, X] \times [0, Y] \longrightarrow \mathbb{R}$ by

$$\mathbf{C}_{n,m}(x, y) := \cos\left(\frac{\pi n x}{X}\right) \cdot \cos\left(\frac{\pi m y}{Y}\right),$$

and

$$\mathbf{S}_{n,m}(x, y) := \sin\left(\frac{\pi n x}{X}\right) \cdot \sin\left(\frac{\pi m y}{Y}\right),$$

for all $(x, y) \in [0, X] \times [0, Y]$ (see Figures 9A.1 and 9A.2).

Theorem 9A.3 Two-dimensional co/sine series convergence
Let $X, Y > 0$, and let $\mathbb{X} := [0, X] \times [0, Y]$.

(a) (i) *The set $\{\mathbf{S}_{n,m}; n, m \in \mathbb{N}_+\}$ is an orthogonal basis for $\mathbf{L}^2(\mathbb{X})$.*
   (ii) *The set $\{\mathbf{C}_{n,m}; n, m \in \mathbb{N}\}$ is also an orthogonal basis for $\mathbf{L}^2(\mathbb{X})$.*
   (iii) *Thus, if $f \in \mathbf{L}^2(\mathbb{X})$, then the series (9A.1) and (9A.2) both converge to f in L^2-norm. Furthermore, the coefficient sequences $\{A_{n,m}\}_{n,m=0}^{\infty}$ and $\{B_{n,m}\}_{n,m=1}^{\infty}$ are the unique sequences of coefficients with this property.*

(b) *If $f \in C^1(\mathbb{X})$ (i.e. f is continuously differentiable on $\mathbb{X}$), then the series (9A.1) and (9A.2) both converge to f pointwise on $(0, X) \times (0, Y)$.*

(c) (i) *If $\sum_{n,m=1}^{\infty} |B_{n,m}| < \infty$, then the two-dimensional Fourier sine series (9A.1) converges to f uniformly on $\mathbb{X}$.*
   (ii) *If $\sum_{n,m=0}^{\infty} |A_{n,m}| < \infty$, then the two-dimensional Fourier cosine series (9A.2) converges to f uniformly on $\mathbb{X}$.*

(d) (i) *If $f \in C^1(\mathbb{X})$, and the derivative functions $\partial_x f$ and $\partial_y f$ are both in $\mathbf{L}^2(\mathbb{X})$, and f satisfies homogeneous Dirichlet boundary conditions[1] on $\mathbb{X}$, then the two-dimensional Fourier sine series (9A.1) converges to f uniformly on $\mathbb{X}$.*
   (ii) *Conversely, if the series (9A.1) converges to f uniformly on $\mathbb{X}$, then f is continuous and satisfies homogeneous Dirichlet boundary conditions.*

(e) (i) *If $f \in C^1(\mathbb{X})$, and the derivative functions $\partial_x f$ and $\partial_y f$ are both in $\mathbf{L}^2(\mathbb{X})$, then the two-dimensional Fourier cosine series (9A.2) converges to f uniformly on $\mathbb{X}$.*
   (ii) *Conversely, if $\sum_{n,m=1}^{\infty} n |A_{nm}| < \infty$ and $\sum_{n,m=1}^{\infty} m |A_{nm}| < \infty$, then $f \in C^1(\mathbb{X})$, and f satisfies homogeneous Neumann boundary conditions[2] on $\mathbb{X}$.*

Proof This is just the case $D = 2$ of Theorem 9B.1, p. 189.   ☐

---

[1] That is, $f(0, y) = 0 = f(X, y)$ for all $y \in [0, Y]$, and $f(x, 0) = 0 = f(x, Y)$ for all $x \in [0, X]$.
[2] That is, $\partial_x f(0, y) = 0 = \partial_x f(X, y)$ for all $y \in [0, Y]$, and $\partial_y f(x, 0) = 0 = \partial_y f(x, Y)$ for all $x \in [0, X]$.

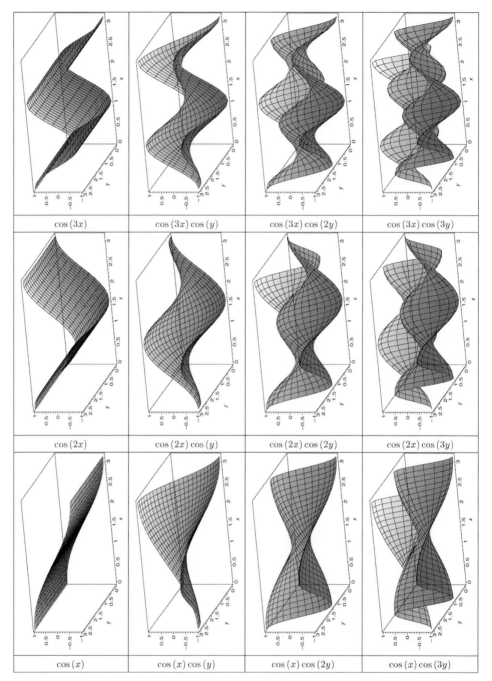

Figure 9A.1. $\mathbf{C}_{n,m}$ for $n = 1, \ldots, 3$ and $m = 0, \ldots, 3$ (rotate page).

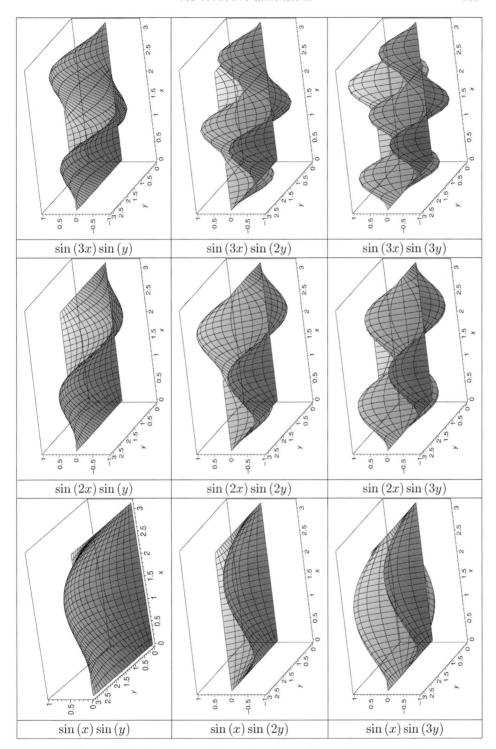

Figure 9A.2. $\mathbf{S}_{n,m}$ for $n = 1, \ldots, 3$ and $m = 1, \ldots, 3$ (rotate page).

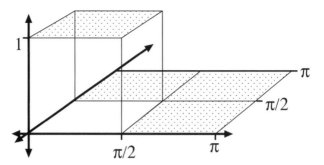

Figure 9A.3. The box function $f(x, y)$ in Example 9A.4.

Example 9A.4 Suppose $X = \pi = Y$, and

$$f(x, y) = \begin{cases} 1 & \text{if } 0 \le x < \frac{\pi}{2} \text{ and } 0 \le y < \frac{\pi}{2}; \\ 0 & \text{if } \frac{\pi}{2} \le x \text{ or } \frac{\pi}{2} \le y \end{cases}$$

(see Figure 9A.3). Then the two-dimensional Fourier cosine series of f is given by

$$\frac{1}{4} + \frac{1}{\pi} \sum_{k=0}^{\infty} \frac{(-1)^k}{2k+1} \cos\left((2k+1)x\right) + \frac{1}{\pi} \sum_{j=0}^{\infty} \frac{(-1)^j}{2j+1} \cos\left((2j+1)y\right)$$

$$+ \frac{4}{\pi^2} \sum_{k,j=0}^{\infty} \frac{(-1)^{k+j}}{(2k+1)(2j+1)} \cos\left((2k+1)x\right) \cdot \cos\left((2j+1)y\right).$$

To see this, note that $f(x, y) = g(x) \cdot g(y)$, where

$$g(x) = \begin{cases} 1 & \text{if } 0 \le x < \frac{\pi}{2}; \\ 0 & \text{if } \frac{\pi}{2} \le x. \end{cases}$$

Recall from Example 7C.6, p. 158, that the (one-dimensional) Fourier cosine series of $g(x)$ is

$$g(x) \underset{\text{\tiny L2}}{\approx} \frac{1}{2} + \frac{2}{\pi} \sum_{k=0}^{\infty} \frac{(-1)^k}{2k+1} \cos\left((2k+1)x\right).$$

Thus, the cosine series for $f(x, y)$ is given by

$$f(x, y) = g(x) \cdot g(y)$$

$$\underset{\text{\tiny L2}}{\approx} \left[\frac{1}{2} + \frac{2}{\pi} \sum_{k=0}^{\infty} \frac{(-1)^k}{2k+1} \cos\left((2k+1)x\right) \right]$$

$$\times \left[\frac{1}{2} + \frac{2}{\pi} \sum_{j=0}^{\infty} \frac{(-1)^j}{2j+1} \cos\left((2j+1)y\right) \right]. \qquad \diamond$$

Mixed Fourier series (optional)

We can also define the *mixed Fourier sine/cosine coefficients*:

$$C_{n,0}^{[sc]} := \frac{2}{XY} \int_0^X \int_0^Y f(x, y) \sin\left(\frac{\pi n x}{X}\right) dx\, dy, \quad \text{for } n > 0;$$

$$C_{n,m}^{[sc]} := \frac{4}{XY} \int_0^X \int_0^Y f(x, y) \sin\left(\frac{\pi n x}{X}\right) \cos\left(\frac{\pi m y}{Y}\right) dx\, dy, \quad \text{for } n, m > 0;$$

$$C_{0,m}^{[cs]} := \frac{2}{XY} \int_0^X \int_0^Y f(x, y) \sin\left(\frac{\pi m y}{Y}\right) dx\, dy, \quad \text{for } m > 0;$$

$$C_{n,m}^{[cs]} := \frac{4}{XY} \int_0^X \int_0^Y f(x, y) \cos\left(\frac{\pi n x}{X}\right) \sin\left(\frac{\pi m y}{Y}\right) dx\, dy, \quad \text{for } n, m > 0.$$

The *mixed Fourier sine/cosine series* of f are then given by

$$\sum_{n=1,m=0}^{\infty} C_{n,m}^{[sc]} \sin\left(\frac{\pi n x}{X}\right) \cos\left(\frac{\pi m y}{Y}\right) \tag{9A.3}$$

and

$$\sum_{n=0,m=1}^{\infty} C_{n,m}^{[cs]} \cos\left(\frac{\pi n x}{X}\right) \sin\left(\frac{\pi m y}{Y}\right).$$

For any $n, m \in \mathbb{N}$, define the functions $\mathbf{M}_{n,m}^{[sc]}, \mathbf{M}_{n,m}^{[cs]} : [0, X] \times [0, Y] \longrightarrow \mathbb{R}$ by

$$\mathbf{M}_{n,m}^{[sc]}(x, y) := \sin\left(\frac{\pi n_1 x}{X}\right) \cos\left(\frac{\pi n_2 y}{Y}\right)$$

and

$$\mathbf{M}_{n,m}^{[cs]}(x, y) := \cos\left(\frac{\pi n_1 x}{X}\right) \sin\left(\frac{\pi n_2 y}{Y}\right),$$

for all $(x, y) \in [0, X] \times [0, Y]$.

Proposition 9A.5 Two-dimensional mixed co/sine series convergence

Let $\mathbb{X} := [0, X] \times [0, Y]$. *The sets of 'mixed' functions,* $\{\mathbf{M}_{n,m}^{[sc]}; n \in \mathbb{N}_+, m \in \mathbb{N}\}$ *and* $\{\mathbf{M}_{n,m}^{[cs]}; n \in \mathbb{N}, m \in \mathbb{N}_+\}$ *are both* **orthogonal basis** *for* $\mathbf{L}^2(\mathbb{X})$. *In other words, if* $f \in \mathbf{L}^2(\mathbb{X})$, *then the series* (9A.3) *both converge to* f *in* L^2.

Exercise 9A.2 Formulate conditions for pointwise and uniform convergence of ⒺΕ
the mixed series. ◆

9B ... in many dimensions

Prerequisites: §6E, §6F.   **Recommended:** §9A.

Let $X_1, \ldots, X_D > 0$, and let $\mathbb{X} := [0, X_1] \times \cdots \times [0, X_D]$ be an $X_1 \times \cdots \times X_D$ box in D-dimensional space. For any $\mathbf{n} \in \mathbb{N}^D$, define the functions $\mathbf{C_n} : \mathbb{X} \longrightarrow \mathbb{R}$

and $\mathbf{S_n} : \mathbb{X} \longrightarrow \mathbb{R}$ as follows:

$$\mathbf{C_n}(x_1, \ldots, x_D) := \boxed{\cos\left(\frac{\pi n_1 x_1}{X_1}\right) \cos\left(\frac{\pi n_2 x_2}{X_2}\right) \cdots \cos\left(\frac{\pi n_D x_D}{X_D}\right),} \quad (9\text{B}.1)$$

$$\mathbf{S_n}(x_1, \ldots, x_D) := \boxed{\sin\left(\frac{\pi n_1 x_1}{X_1}\right) \sin\left(\frac{\pi n_2 x_2}{X_2}\right) \cdots \sin\left(\frac{\pi n_D x_D}{X_D}\right),} \quad (9\text{B}.2)$$

for any $\mathbf{x} = (x_1, x_2, \ldots, x_D) \in \mathbb{X}$. Also, for any sequence $\boldsymbol{\omega} = (\omega_1, \ldots, \omega_D)$ of D symbols 's' and 'c', we can define the 'mixed' functions, $\mathbf{M_n^\omega} : \mathbb{X} \longrightarrow \mathbb{R}$. For example, if $D = 3$, then define

$$\mathbf{M_n^{[scs]}}(x, y, z) := \sin\left(\frac{\pi n_1 x}{X_x}\right) \cos\left(\frac{\pi n_2 y}{X_y}\right) \sin\left(\frac{\pi n_3 z}{X_z}\right).$$

If $f : \mathbb{X} \longrightarrow \mathbb{R}$ is any function with $\|f\|_2 < \infty$, then, for all $\mathbf{n} \in \mathbb{N}_+^D$, we define the *multiple Fourier sine coefficients*:

$$B_\mathbf{n} := \frac{\langle f, \mathbf{S_n} \rangle}{\|\mathbf{S_n}\|_2^2} = \boxed{\frac{2^D}{X_1 \cdots X_D} \int_\mathbb{X} f(\mathbf{x}) \cdot \mathbf{S_n}(\mathbf{x}) d\mathbf{x}.}$$

The *multiple Fourier sine series* of f is then given by

$$\boxed{\sum_{\mathbf{n} \in \mathbb{N}_+^D} B_\mathbf{n} \mathbf{S_n}.} \quad (9\text{B}.3)$$

For all $\mathbf{n} \in \mathbb{N}^D$, we define the *multiple Fourier cosine coefficients*:

$$A_0 := \langle f, \mathbb{1} \rangle = \boxed{\frac{1}{X_1 \cdots X_D} \int_\mathbb{X} f(\mathbf{x}) d\mathbf{x}}$$

and

$$A_\mathbf{n} := \frac{\langle f, \mathbf{C_n} \rangle}{\|\mathbf{C_n}\|_2^2} = \boxed{\frac{2^{d_\mathbf{n}}}{X_1 \cdots X_D} \int_\mathbb{X} f(\mathbf{x}) \cdot \mathbf{C_n}(\mathbf{x}) d\mathbf{x},}$$

where, for each $\mathbf{n} \in \mathbb{N}^D$, the number $d_\mathbf{n}$ is the number of nonzero entries in $\mathbf{n} = (n_1, n_2, \ldots, n_D)$. The *multiple Fourier cosine series* of f is then given by

$$\boxed{\sum_{\mathbf{n} \in \mathbb{N}^D} A_\mathbf{n} \mathbf{C_n},} \quad \text{where } \mathbb{N} := \{0, 1, 2, 3, \ldots\}. \quad (9\text{B}.4)$$

Finally, we define the *mixed Fourier sine/cosine coefficients*:

$$C_{\mathbf{n}}^{\omega} := \frac{\langle f, \mathbf{M}_{\mathbf{n}}^{\omega} \rangle}{\left\| \mathbf{M}_{\mathbf{n}}^{\omega} \right\|_2^2} = \frac{2^{d_{\mathbf{n}}}}{X_1 \cdots X_D} \int_{\mathbb{X}} f(\mathbf{x}) \cdot \mathbf{M}_n^{\omega}(\mathbf{x}) d\mathbf{x},$$

where, for each $\mathbf{n} \in \mathbb{N}^D$, the number $d_{\mathbf{n}}$ is the number of nonzero entries n_i in $\mathbf{n} = (n_1, \ldots, n_D)$. The *mixed Fourier sine/cosine series* of f is then given by

$$\boxed{\sum_{\mathbf{n} \in \mathbb{N}^D} C_{\mathbf{n}}^{\omega} \mathbf{M}_{\mathbf{n}}^{\omega}.} \tag{9B.5}$$

Theorem 9B.1 Multidimensional co/sine series convergence on $\mathbb{X}$

Let $\mathbb{X} := [0, X_1] \times \cdots \times [0, X_D]$ be a D-dimensional box.

(a) (i) *The set $\left\{ \mathbf{S}_{\mathbf{n}} ; \mathbf{n} \in \mathbb{N}_+^D \right\}$ is an orthogonal basis for $\mathbf{L}^2(\mathbb{X})$.*

    (ii) *The set $\left\{ \mathbf{C}_{\mathbf{n}} ; \mathbf{n} \in \mathbb{N}^D \right\}$ is an orthogonal basis for $\mathbf{L}^2(\mathbb{X})$.*

    (iii) *For any sequence ω of D symbols 's' and 'c', the set of 'mixed' functions, $\left\{ \mathbf{M}_{\mathbf{n}}^{\omega} ; \mathbf{n} \in \mathbb{N}^D \right\}$, is an orthogonal basis for $\mathbf{L}^2(\mathbb{X})$.*

    (iv) *In other words, if $f \in \mathbf{L}^2(\mathbb{X})$, then the series (9B.3), (9B.4), and (9B.5) all converge to f in L^2-norm. Furthermore, the coefficient sequences $\{A_{\mathbf{n}}\}_{\mathbf{n} \in \mathbb{N}^D}$, $\{B_{\mathbf{n}}\}_{\mathbf{n} \in \mathbb{N}_+^D}$, and $\{C_{\mathbf{n}}^{\omega}\}_{\mathbf{n} \in \mathbb{N}^D}$ are the unique sequences of coefficients with these properties.*

(b) *If $f \in \mathcal{C}^1(\mathbb{X})$ (i.e. f is continuously differentiable on $\mathbb{X}$), then the series (9B.3), (9B.4), and (9B.5) converge pointwise on the interior of $\mathbb{X}$.*

(c) (i) *If $\sum_{\mathbf{n} \in \mathbb{N}_+^D} |B_{\mathbf{n}}| < \infty$, then the multidimensional Fourier sine series (9B.3) converges to f uniformly on $\mathbb{X}$.*

    (ii) *If $\sum_{\mathbf{n} \in \mathbb{N}^D} |A_{\mathbf{n}}| < \infty$, then the multidimensional Fourier cosine series (9B.4) converges to f uniformly on $\mathbb{X}$.*

(d) (i) *If $f \in \mathcal{C}^1(\mathbb{X})$, and the derivative functions $\partial_k f$ are themselves in $\mathbf{L}^2(\mathbb{X})$ for all $k \in [1...D]$, and f satisfies homogeneous Dirichlet boundary conditions on $\mathbb{X}$, then the multidimensional Fourier sine series (9B.3) converges to f uniformly on $\mathbb{X}$.*

    (ii) *Conversely, if the series (9B.3) converges to f uniformly on $\mathbb{X}$, then f is continuous and satisfies homogeneous Dirichlet boundary conditions.*

(e) (i) *If $f \in \mathcal{C}^1(\mathbb{X})$, and the derivative functions $\partial_k f$ are themselves in $\mathbf{L}^2(\mathbb{X})$ for all $k \in [1...D]$, then the multidimensional Fourier cosine series (9B.4) converges to f uniformly on $\mathbb{X}$.*

    (ii) *Conversely, if $\sum_{\mathbf{n} \in \mathbb{N}^D} (n_1 + \cdots + n_D) |A_{\mathbf{n}}| < \infty$, then $f \in \mathcal{C}^1(\mathbb{X})$, and f satisfies homogeneous Neumann boundary conditions.*

Proof The proof of (c) is **Exercise 9B.1.** (*Hint:* Use the Weierstrass M-test, Ⓔ Proposition 6E.13, p. 135.)

(E)   The proofs of (d)(ii) and (e)(ii) are **Exercise 9B.2.** (*Hint:* Generalize the solutions to Exercises 7A.4 and 7A.9, pp. 142 and 146.)

(E)   The proof of (a) is **Exercise 9B.3.** (*Hint:* Prove this by induction on the dimension D. The base case ($D = 1$) is Theorems 7A.1(a) and 7A.4(a), pp. 142 and 145. Use Lemma 15C.2(f) (p. 337) to handle the induction step.)

We will prove (b), (d)(i), and (e)(i) by induction on the dimension D. The base cases ($D = 1$) are Theorems 7A.1((b), (d)(i)) and 7A.4((b), (d)(i)), pp. 142 and 145.

For the induction step, suppose the theorem is true for D, and consider $D + 1$. Let $\mathbb{X} := [0, X_0] \times [0, X_1] \times \cdots \times [0, X_D]$ be a $(D + 1)$-dimensional box. Note that $\mathbb{X} := [0, X_0] \times \mathbb{X}^*$, where $\mathbb{X}^* := [0, X_1] \times \cdots \times [0, X_D]$ is a D-dimensional box. If $f : \mathbb{X} \longrightarrow \mathbb{R}$, then for all $y \in [0, X_0]$, let $f^y : \mathbb{X}^* \longrightarrow \mathbb{R}$ be the function defined by $f^y(\mathbf{x}) := f(y, \mathbf{x})$ for all $\mathbf{x} \in \mathbb{X}^*$.

Claim 1

(a) *If $f \in \mathcal{C}^1(\mathbb{X})$, then $f^y \in \mathcal{C}^1(\mathbb{X}^*)$ for all $y \in [0, X_0]$.*

(b) *Furthermore, if $\partial_k f \in \mathbf{L}^2(\mathbb{X})$ for all $k \in [1 \ldots D]$, then $\partial_k (f^y) \in \mathbf{L}^2(\mathbb{X}^*)$ for all $k \in [1 \ldots D]$ and all $y \in [0, X_0]$.*

(c) *If f satisfies homogenous Dirichlet BC on $\mathbb{X}$, then f^y satisfies homogenous Dirichlet BC on $\mathbb{X}^*$, for all $y \in [0, X_0]$.*

(E)   *Proof* **Exercise 9B.4.**   ◇Claim 1

For all $\mathbf{n} \in \mathbb{N}^D$, define $\mathbf{C}_\mathbf{n}^*, \mathbf{S}_\mathbf{n}^* : \mathbb{X}^* \longrightarrow \mathbb{R}$ as in equations (9B.1) and (9B.2). For every $y \in [0, X_0]$, let

$$A_\mathbf{n}^y := \frac{\langle f^y, \mathbf{C}_\mathbf{n}^* \rangle}{\left\| \mathbf{C}_\mathbf{n}^* \right\|_2^2} \quad \text{and} \quad B_\mathbf{n}^y := \frac{\langle f^y, \mathbf{S}_\mathbf{n}^* \rangle}{\left\| \mathbf{S}_\mathbf{n}^* \right\|_2^2}$$

be the D-dimensional Fourier (co)sine coefficients for f^y, so that f^y has D-dimensional Fourier (co)sine series:

$$\sum_{\mathbf{n} \in \mathbb{N}_+^D} B_\mathbf{n}^y \mathbf{S}_\mathbf{n}^* \underset{L2}{\approx} f^y \underset{L2}{\approx} \sum_{\mathbf{n} \in \mathbb{N}^D} A_\mathbf{n}^y \mathbf{C}_\mathbf{n}^*. \tag{9B.6}$$

Claim 2 *For all $y \in [0, X_0]$, the two series in equation (9B.6) converge to f^y in the desired fashion (i.e. pointwise or uniform) on $\mathbb{X}^*$.*

(E)   *Proof* **Exercise 9B.5.** (*Hint:* Use the induction hypothesis and Claim 1.)   ◇Claim 2

Fix $\mathbf{n} \in \mathbb{N}^D$. Define $\alpha_\mathbf{n}, \beta_\mathbf{n} : [0, X_0] \longrightarrow \mathbb{R}$ by $\alpha_\mathbf{n}(y) := A_\mathbf{n}^y$ and $\beta_\mathbf{n}(y) := B_\mathbf{n}^y$ for all $y \in [0, X_0]$.

Claim 3 *For all $\mathbf{n} \in \mathbb{N}^D$, $\alpha_\mathbf{n} \in \mathbf{L}^2[0, X_0]$ and $\beta_\mathbf{n} \in \mathbf{L}^2[0, X_0]$.*

Proof We have

$$
\|\alpha_{\mathbf{n}}\|_2^2 \;=\; \frac{1}{X_0} \int_0^{X_0} |\alpha_{\mathbf{n}}(y)|^2 \, \mathrm{d}y \;=\; \frac{1}{X_0} \int_0^{X_0} \left| \frac{\langle f^y, \mathbf{C}_{\mathbf{n}}^* \rangle}{\|\mathbf{C}_{\mathbf{n}}^*\|_2^2} \right|^2 \mathrm{d}y
$$

$$
= \; \frac{1}{X_0 \cdot \|\mathbf{C}_{\mathbf{n}}^*\|_2^4} \int_0^{X_0} |\langle f^y, \mathbf{C}_{\mathbf{n}}^* \rangle|^2 \, \mathrm{d}y
$$

$$
\underset{(*)}{\leq} \; \frac{1}{X_0 \cdot \|\mathbf{C}_{\mathbf{n}}^*\|_2^4} \int_0^{X_0} \|f^y\|_2^2 \cdot \|\mathbf{C}_{\mathbf{n}}^*\|_2^2 \, \mathrm{d}y
$$

$$
= \; \frac{1}{X_0 \cdot \|\mathbf{C}_{\mathbf{n}}^*\|_2^2} \int_0^{X_0} \|f^y\|_2^2 \, \mathrm{d}y
$$

$$
= \; \frac{1}{X_0 \cdot \|\mathbf{C}_{\mathbf{n}}^*\|_2^2} \left(\int_0^{X_0} \frac{1}{X_1 \cdots X_D} \int_{\mathbb{X}^*} |f^y(\mathbf{x})|^2 \, \mathrm{d}\mathbf{x} \right) \mathrm{d}y
$$

$$
= \; \frac{1}{X_0 \cdots X_D \cdot \|\mathbf{C}_{\mathbf{n}}^*\|_2^2} \int_{\mathbb{X}} |f(y, \mathbf{x})|^2 \, \mathrm{d}(y; \mathbf{x}) \;=\; \frac{1}{\|\mathbf{C}_{\mathbf{n}}^*\|_2^2} \|f\|_2^2 .
$$

Here, $(*)$ is the Cauchy–Bunyakowski–Schwarz Inequality (Theorem 6B.5, p. 112). Thus, $\|\alpha_{\mathbf{n}}\|_2^2 < \infty$ because $\|f\|_2^2 < \infty$ because $f \in \mathbf{L}^2(\mathbb{X})$ by hypothesis. Thus, $\alpha_{\mathbf{n}} \in \mathbf{L}^2[0, X_0]$. The proof that $\beta_{\mathbf{n}} \in \mathbf{L}^2[0, X_0]$ is similar. $\diamond_{\text{Claim 3}}$

For all $m \in \mathbb{N}$, define $\mathbf{S}_m, \mathbf{C}_m : [0, X_0] \longrightarrow \mathbb{R}$ by $\mathbf{S}_m(y) := \sin(\pi m y / X_0)$ and $\mathbf{C}_m(y) := \cos(\pi m y / X_0)$, for all $y \in [0, X_0]$. For all $m \in \mathbb{N}$, let

$$
A_m^{\mathbf{n}} := \frac{\langle \alpha_{\mathbf{n}}, \mathbf{C}_m \rangle}{\|\mathbf{C}_m\|_2^2} \quad \text{and} \quad B_m^{\mathbf{n}} := \frac{\langle \beta_{\mathbf{n}}, \mathbf{S}_m \rangle}{\|\mathbf{S}_m\|_2^2}
$$

be the one-dimensional Fourier (co)sine coefficients for the functions $\alpha_{\mathbf{n}}$ and $\beta_{\mathbf{n}}$, so that we get the one-dimensional Fourier (co)sine series:

$$
\alpha_{\mathbf{n}} \underset{\scriptscriptstyle L2}{\approx} \sum_{m=0}^{\infty} A_m^{\mathbf{n}} \mathbf{C}_m \quad \text{and} \quad \beta_{\mathbf{n}} \underset{\scriptscriptstyle L2}{\approx} \sum_{m=1}^{\infty} B_m^{\mathbf{n}} \mathbf{S}_m . \tag{9B.7}
$$

For all $\mathbf{n} \in \mathbb{N}^D$ and all $m \in \mathbb{N}$, define $\mathbf{S}_{m;\mathbf{n}}, \mathbf{C}_{m;\mathbf{n}} : \mathbb{X} \longrightarrow \mathbb{R}$ by $\mathbf{S}_{m;\mathbf{n}}(y; \mathbf{x}) := \mathbf{S}_m(y) \cdot \mathbf{S}_{\mathbf{n}}(\mathbf{x})$ and $\mathbf{C}_{m;\mathbf{n}}(y; \mathbf{x}) := \mathbf{C}_m(y) \cdot \mathbf{C}_{\mathbf{n}}(\mathbf{x})$, for all $y \in [0, X_0]$ and $\mathbf{x} \in \mathbb{X}^*$. Then let

$$
A_{m;\mathbf{n}} := \frac{\langle f, \mathbf{C}_{m;\mathbf{n}} \rangle}{\|\mathbf{C}_{m;\mathbf{n}}\|_2^2} \quad \text{and} \quad B_{m;\mathbf{n}} := \frac{\langle f, \mathbf{S}_{m;\mathbf{n}} \rangle}{\|\mathbf{S}_{m;\mathbf{n}}\|_2^2}
$$

be the $(D{+}1)$-dimensional Fourier (co)sine coefficients for the function f, so that we get the $(D{+}1)$-dimensional Fourier (co)sine series

$$
\sum_{m=0}^{\infty} \sum_{\mathbf{n} \in \mathbb{N}^D} A_{m;\mathbf{n}} \, \mathbf{C}_{m;\mathbf{n}} \underset{\scriptscriptstyle L2}{\approx} f \underset{\scriptscriptstyle L2}{\approx} \sum_{m=1}^{\infty} \sum_{\mathbf{n} \in \mathbb{N}_+^D} B_{m;\mathbf{n}} \, \mathbf{S}_{m;\mathbf{n}} . \tag{9B.8}
$$

Claim 4 *For all* $\mathbf{n} \in \mathbb{N}^D$ *and all* $m \in \mathbb{N}$, $A_m^{\mathbf{n}} = A_{m;\mathbf{n}}$ *and* $B_m^{\mathbf{n}} = B_{m;\mathbf{n}}$.

(E)   *Proof* **Exercise 9B.6.**

◇ Claim 4

Let $\partial_0 f$ be the derivative of f in the 0th (or 'y') coordinate, which we regard as a function $\partial_0 f : \mathbb{X} \longrightarrow \mathbb{R}$.

Claim 5

(a) *If* $f \in C^1(\mathbb{X})$, *then, for all* $\mathbf{n} \in \mathbb{N}^D$, $\alpha_{\mathbf{n}} \in C^1[0, X_0]$, *and* $\beta_{\mathbf{n}} \in C^1[0, X_0]$.

(b) *Furthermore, if* $\partial_0 f \in \mathbf{L}^2(\mathbb{X})$, *then, for all* $\mathbf{n} \in \mathbb{N}^D$, $\alpha'_{\mathbf{n}} \in \mathbf{L}^2[0, X_0]$ *and* $\beta'_{\mathbf{n}} \in \mathbf{L}^2[0, X_0]$.

(c) *If* f *satisfies homogeneous Dirichlet BC on* $\mathbb{X}$, *then* $\beta_{\mathbf{n}}$ *satisfies homogeneous Dirichlet BC on* $[0, X_0]$, *for all* $\mathbf{n} \in \mathbb{N}_+^D$.

Proof To prove (a), we proceed as follows.

(E)   ### Exercise 9B.7

(a) Show that f is *uniformly* continuous on $\mathbb{X}$. (*Hint:* Note that f is continuous on $\mathbb{X}$, and $\mathbb{X}$ is compact.)

(b) Show that, for any $y_0 \in [0, X_0]$, the functions f^y converge *uniformly* to f^{y_0} as $y \to y_0$.

(c) For any fixed $\mathbf{n} \in \mathbb{N}$, deduce that $\lim_{y \to y_0} A_{\mathbf{n}}^y = A_{\mathbf{n}}^{y_0}$ and $\lim_{y \to y_0} B_{\mathbf{n}}^y = B_{\mathbf{n}}^{y_0}$. (*Hint:* Use Corollary 6E.11(b)(ii), p. 133.)

(d) Conclude that the functions $\alpha_{\mathbf{n}}$ and $\beta_{\mathbf{n}}$ are continuous at y_0.   ◆

The conclusion of Exercise 9B.7(d) holds for all $y_0 \in [0, X_0]$ and all $\mathbf{n} \in \mathbb{N}$. Thus, the functions $\alpha_{\mathbf{n}}$ and $\beta_{\mathbf{n}}$ are continuous on $[0, X_0]$, for all $\mathbf{n} \in \mathbb{N}$. For all $y \in [0, X_0]$, let $(\partial_0 f)^y : \mathbb{X}^* \longrightarrow \mathbb{R}$ be the function defined by $(\partial_0 f)^y(\mathbf{x}) := \partial_0 f(y, \mathbf{x})$ for all $\mathbf{x} \in \mathbb{X}^*$.

(E)   **Exercise 9B.8** Suppose $f \in C^1(\mathbb{X})$. Use Proposition G.1, p. 571, to show, for all $\mathbf{n} \in \mathbb{N}^D$, that the functions $\alpha_{\mathbf{n}}$ and $\beta_{\mathbf{n}}$ are differentiable on $[0, X_0]$; furthermore, for all $y \in [0, X_0]$,

$$\alpha'_{\mathbf{n}}(y) = \frac{\langle (\partial_0 f)^y, \mathbf{C_n} \rangle}{\|\mathbf{C_n}\|_2^2} \quad \text{and} \quad \beta'_{\mathbf{n}}(y) = \frac{\langle (\partial_0 f)^y, \mathbf{S_n} \rangle}{\|\mathbf{S_n}\|_2^2}. \tag{9B.9}$$

(E)   **Exercise 9B.9** Using the same technique as in Exercise 9B.7, use equation (9B.9) to prove that the functions $\alpha'_{\mathbf{n}}$ and $\beta'_{\mathbf{n}}$ are continuous on $[0, X_0]$.   ◆

Thus, $\alpha_{\mathbf{n}} \in C^1[0, X_0]$ and $\beta_{\mathbf{n}} \in C^1[0, X_0]$; this proves part (a) of the Claim.

(E)   The proof of (b) is **Exercise 9B.10.** (*Hint:* Imitate the proof of Claim 3.)

(E)   The proof of (c) is **Exercise 9B.11.**

◇ Claim 5

Claim 6 *The one-dimensional Fourier cosine series in equation (9B.7) converges to* $\alpha_{\mathbf{n}}$, *and the one-dimensional Fourier sine series in equation (9B.7) converges to* $\beta_{\mathbf{n}}$ *in the desired fashion* (i.e. pointwise or uniform), *for all* $\mathbf{n} \in \mathbb{N}^D$.

Proof Proof **Exercise 9B.12.** (*Hint:* Use Theorems 7A.1(b) and (d)(i) and ⒠
7A.4(b) and (d)(i) and Claim 5.) $\diamond$ Claim 6

Now, Claim 4 implies that the $(D+1)$-dimensional Fourier (co)sine series in (9B.8) can be rewritten as

$$\sum_{m=0}^{\infty} \sum_{\mathbf{n} \in \mathbb{N}^D} A_m^{\mathbf{n}} \, C_m \, \mathbf{C_n} \underset{L2}{\approx} f \underset{L2}{\approx} \sum_{m=1}^{\infty} \sum_{\mathbf{n} \in \mathbb{N}_+^D} B_m^{\mathbf{n}} \, S_m \, \mathbf{S_n}. \tag{9B.10}$$

Exercise 9B.13 Suppose $f \in \mathcal{C}^1(\mathbb{X})$. Use the 'pointwise' versions of Claims 2 ⒠
and 6 to show that the two series in equation (9B.10) converge to f pointwise on the interior of $\mathbb{X}$. ◆

This proves part (b) of Theorem 9B.1.

Exercise 9B.14 (Hard) Suppose $f \in \mathcal{C}^1(\mathbb{X})$, the derivative functions $\partial_0 f$, ⒠
$\partial_1 f, \dots, \partial_D f$ are all in $\mathbf{L}^2(\mathbb{X})$, and (for the sine series) f satisfies homogeneous Dirichlet boundary conditions. Use the 'uniform' versions of Claims 2 and 6 to show that the two series in equation (9B.10) converge to f uniformly if $f \in \mathcal{C}^1(\mathbb{X})$. ◆

This proves parts (d)(i) and (e)(i) of Theorem 9B.1 □

Remarks
(a) If f is a *piecewise* $\mathcal{C}^1$ function on the interval $[0, \pi]$, then Theorems 7A.1 and 7A.4 also yield pointwise convergence and 'local' uniform convergence of one-dimensional Fourier (co)sine to f inside the '$\mathcal{C}^1$ intervals' of f. Likewise, if f is a 'piecewise $\mathcal{C}^1$ function' on the D-dimensional domain $\mathbb{X}$, then one can extend Theorem 9B.1 to get pointwise convergence and 'local' uniform convergence of D-dimensional Fourier (co)sine to f inside the '$\mathcal{C}^1$ regions' of f; however, it is too technically complicated to state this formally here.
(b) Remark 8D.3, p. 177, provided some technical remarks about the (non)convergence of one-dimensional Fourier (co)sine series, when the hypotheses of Theorems 7A.1 and 7A.4 are further weakened. Similar remarks apply to D-dimensional Fourier series.
(c) It is also possible to define D-dimensional complex Fourier series on the D-dimensional box $[-\pi, \pi]^D$, in a manner analogous to the results of Section 8D, and then state and prove a theorem analogous to Theorem 9B.1 for such D-dimensional complex Fourier series. (**Exercise 9B.15** (Challenging) Do this.) ⒠

In Chapters 11–14, we will often propose a multiple Fourier series (or similar object) as the solution to some PDE, perhaps with certain boundary conditions. To verify that the Fourier series really satisfies the PDE, we must be able to compute its Laplacian. If we also require the Fourier series solution to satisfy some Neumann boundary conditions, then we must be able to compute its normal derivatives on the boundary of the domain. For these purposes, the next result is crucial.

Proposition 9B.2 The derivatives of a multiple Fourier (co)sine series
Let $\mathbb{X} := [0, X_1] \times \cdots \times [0, X_D]$. *Let* $f : \mathbb{X} \longrightarrow \mathbb{R}$ *have the uniformly convergent Fourier series*

$$f \underset{\text{unif}}{\equiv} A_0 + \sum_{\mathbf{n} \in \mathbb{N}^D} A_{\mathbf{n}} \mathbf{C_n} + \sum_{\mathbf{n} \in \mathbb{N}_+^D} B_{\mathbf{n}} \mathbf{S_n}.$$

(a) *Fix* $i \in [1...D]$. *Suppose that*

$$\sum_{\mathbf{n} \in \mathbb{N}^D} n_i |A_{\mathbf{n}}| + \sum_{\mathbf{n} \in \mathbb{N}_+^D} n_i |B_{\mathbf{n}}| < \infty.$$

Then the function $\partial_i f$ *exists, and*

$$\partial_i f \underset{L2}{\approx} \sum_{\mathbf{n} \in \mathbb{N}^D} \left(\frac{\pi n_i}{X_i} \right) \cdot \left(B_{\mathbf{n}} \mathbf{S}'_{\mathbf{n}} - A_{\mathbf{n}} \mathbf{C}'_{\mathbf{n}} \right).$$

Here, for all $\mathbf{n} \in \mathbb{N}^D$, *and all* $\mathbf{x} \in \mathbb{X}$, *we define*

$$\mathbf{C}'_{\mathbf{n}}(\mathbf{x}) := \sin\left(\frac{\pi n_i x_i}{X_i} \right) \cdot \mathbf{C_n}(\mathbf{x}) / \cos\left(\frac{\pi n_i x_i}{X_i} \right)$$

and

$$\mathbf{S}'_{\mathbf{n}}(\mathbf{x}) := \cos\left(\frac{\pi n_i x_i}{X_i} \right) \cdot \mathbf{S_n}(\mathbf{x}) / \sin\left(\frac{\pi n_i x_i}{X_i} \right).$$

(b) *Fix* $i \in [1...D]$. *Suppose that*

$$\sum_{\mathbf{n} \in \mathbb{N}^D} n_i^2 |A_{\mathbf{n}}| + \sum_{\mathbf{n} \in \mathbb{N}_+^D} n_i^2 |B_{\mathbf{n}}| < \infty.$$

Then the function $\partial_i^2 f$ *exists, and*

$$\partial_i^2 f \underset{L2}{\approx} \sum_{\mathbf{n} \in \mathbb{N}^D} -\left(\frac{\pi n_i}{X_i} \right)^2 \cdot \left(A_{\mathbf{n}} \mathbf{C_n} + B_{\mathbf{n}} \mathbf{S_n} \right).$$

(c) *Suppose that*

$$\sum_{\mathbf{n} \in \mathbb{N}^D} |\mathbf{n}|^2 |A_{\mathbf{n}}| + \sum_{\mathbf{n} \in \mathbb{N}^D} |\mathbf{n}|^2 |B_{\mathbf{n}}| < \infty$$

(where we define $|\mathbf{n}|^2 := n_1^2 + \cdots + n_D^2$). *Then* f *is twice-differentiable, and*

$$\triangle f \underset{L2}{\approx} -\pi^2 \sum_{\mathbf{n} \in \mathbb{N}^D} \left[\left(\frac{n_1}{X_1} \right)^2 + \cdots + \left(\frac{n_D}{X_D} \right)^2 \right] \cdot \left(A_{\mathbf{n}} \mathbf{C_n} + B_{\mathbf{n}} \mathbf{S_n} \right).$$

Ⓔ *Proof* **Exercise 9B.16.** (*Hint:* Apply Proposition F.1, p. 569.) ☐

Example 9B.3 Fix $\mathbf{n} \in \mathbb{N}^D$. If $f = A \cdot \mathbf{C_n} + B \cdot \mathbf{S_n}$, then

$$\triangle f = -\pi^2 \left[\left(\frac{n_1}{X_1} \right)^2 + \cdots + \left(\frac{n_D}{X_D} \right)^2 \right] \cdot f.$$

In particular, if $X_1 = \cdots = X_D = \pi$, then this simplifies to $\triangle f = -|\mathbf{n}|^2 \cdot f$. In other words, f is an *eigenfunction* of the Laplacian operator, with eigenvalue $\lambda = -|\mathbf{n}|^2$. ◇

9C Practice problems

Compute the two-dimensional Fourier sine transforms of the following functions. For each question, also determine the following: at which points does the series converge pointwise? Why? Does the series converge uniformly? Why or why not?

9.1 $f(x, y) = x^2 \cdot y$.

9.2 $g(x, y) = x + y$.

9.3 $f(x, y) = \cos(Nx) \cdot \cos(My)$, for some integers $M, N > 0$.

9.4 $f(x, y) = \sin(Nx) \cdot \sinh(Ny)$, for some integer $N > 0$.

10

Proofs of the Fourier convergence theorems

The profound study of nature is the most fertile source of
mathematical discoveries.

Jean Joseph Fourier

In this chapter, we will prove Theorem 8A.1(a), (b), and (d), p. 166 (and
thus indirectly prove Theorems 7A.1(a), (b), and (d) and 7A.4(a), (b), and (d),
pp. 142 and 145, respectively). Along the way, we will introduce some ideas which
are of independent interest: Bessel's inequality, the Riemann–Lebesgue lemma,
the Dirichlet kernel, convolutions and mollifiers, and the relationship between the
smoothness of a function and the asymptotic decay of its Fourier coefficients. This
chapter assumes no prior knowledge of analysis, beyond some background from
Chapter 6. However, the presentation is slightly more abstract than most of the
book, and is intended for more 'theoretically inclined' students.

10A Bessel, Riemann, and Lebesgue

Prerequisites: §6D.   **Recommended:** §7A, §8A.

We begin with a general result, which is true for any orthonormal set in any L^2
space.

Theorem 10A.1 Bessel's inequality
*Let $\mathbb{X} \subset \mathbb{R}^D$ be any bounded domain. Let $\{\phi_n\}_{n=1}^{\infty}$ be any orthonormal set of
functions in $\mathbf{L}^2(\mathbb{X})$. Let $f \in \mathbf{L}^2(\mathbb{X})$, and, for all $n \in \mathbb{N}$, let $c_n := \langle f, \phi_n \rangle$. Then, for
all $N \in \mathbb{N}$,*

$$\sum_{n=1}^{N} |c_n|^2 \leq \| f \|_2^2 .$$

In particular, $\lim_{n \to \infty} c_n = 0$.

Proof Without loss of generality, suppose $|\mathbb{X}| = 1$, so that $\langle f, g \rangle = \int_{\mathbb{X}} f(x)\, g(x)\, dx$ for any $f, g \in \mathbf{L}^2(\mathbb{X})$. First note that

$$\left[f(x) - \sum_{n=1}^{N} c_n \phi_n(x) \right]^2$$

$$= f(x)^2 - 2f(x) \sum_{n=1}^{N} c_n \phi_n(x) + \left(\sum_{n=1}^{N} c_n \phi_n(x) \right) \cdot \left(\sum_{m=1}^{N} c_m \phi_m(x) \right)$$

$$= f(x)^2 - 2 \sum_{n=1}^{N} c_n f(x) \phi_n(x) + \sum_{n,m=1}^{N} c_n c_m \phi_n(x) \phi_m(x). \qquad (10\text{A}.1)$$

Thus,

$$0 \leq \left\| f - \sum_{n=1}^{N} c_n \phi_n \right\|_2^2 = \int_{\mathbb{X}} \left[f(x) - \sum_{n=1}^{N} c_n \phi_n(x) \right]^2 dx$$

$$\underset{(*)}{=} \int_{\mathbb{X}} \left(f(x)^2 - 2 \sum_{n=1}^{N} c_n f(x) \phi_n(x) + \sum_{n,m=1}^{N} c_n c_m \phi_n(x) \phi_m(x) \right) dx$$

$$= \int_{\mathbb{X}} f(x)^2\, dx - 2 \sum_{n=1}^{N} c_n \int_{\mathbb{X}} f(x) \phi_n(x)\, dx + \sum_{n,m=1}^{N} c_n c_m \int_{\mathbb{X}} \phi_n(x) \phi_m(x)\, dx$$

$$= \underbrace{\langle f, f \rangle}_{\|f\|_2^2} - 2 \sum_{n=1}^{N} c_n \underbrace{\langle f, \phi_n \rangle}_{c_n} + \sum_{n,m=1}^{N} c_n c_m \underbrace{\langle \phi_n, \phi_m \rangle}_{\substack{=1 \text{ if} n=m \\ =0 \text{ if } n \neq m}}$$

$$= \|f\|_2^2 - 2 \sum_{n=1}^{N} c_n^2 + \sum_{n=1}^{N} c_n^2 = \|f\|_2^2 - \sum_{n=1}^{N} c_n^2.$$

Here, $(*)$ is by equation (10A.1). Thus, $0 \leq \|f\|_2^2 - \sum_{n=1}^{N} c_n^2$. Thus $\sum_{n=1}^{N} c_n^2 \leq \|f\|_2^2$, as desired. $\qquad \square$

Example 10A.2 Suppose $f \in \mathbf{L}^2[-\pi, \pi]$ has real Fourier coefficients $\{A_n\}_{n=0}^{\infty}$ and $\{B_n\}_{n=1}^{\infty}$, as defined on p. 165. Then, for all $N \in \mathbb{N}$,

$$A_0^2 + \sum_{n=1}^{N} \frac{|A_n|^2}{2} + \sum_{n=1}^{N} \frac{|B_n|^2}{2} \leq \|f\|_2^2. \qquad \diamondsuit$$

ⓔ **Exercise 10A.1** Prove Example 10A.2. (*Hint:* Let $\mathbb{X} = [-\pi, \pi]$ and let $\{\phi_k\}_{k=1}^{\infty} = \{\sqrt{2}\mathbf{C}_n\}_{n=0}^{\infty} \sqcup \{\sqrt{2}\mathbf{S}_n\}_{n=1}^{\infty}$. Show that $\{\phi_k\}_{k=1}^{\infty}$ is an orthonormal set of functions (use Proposition 6D.2, p. 117). Now apply Bessel's inequality.) $\qquad \blacklozenge$

Corollary 10A.3 Riemann–Lebesgue lemma

(a) *Suppose $f \in \mathbf{L}^2[-\pi, \pi]$ has real Fourier coefficients $\{A_n\}_{n=0}^{\infty}$ and $\{B_n\}_{n=1}^{\infty}$, as defined on p. 165. Then $\lim_{n\to\infty} A_n = 0$ and $\lim_{n\to\infty} B_n = 0$.*

(b) *Suppose $f \in \mathbf{L}^2[0, \pi]$ has Fourier cosine coefficients $\{A_n\}_{n=0}^{\infty}$, as defined by equation (7A.4), p. 145, and Fourier sine coefficients $\{B_n\}_{n=1}^{\infty}$, as defined by equation (7A.1), p. 141. Then $\lim_{n\to\infty} A_n = 0$ and $\lim_{n\to\infty} B_n = 0$.*

Proof **Exercise 10A.2.** (*Hint:* Use Example 10A.2.) □ Ⓔ

10B Pointwise convergence

Prerequisites: §8A, §10A.   **Recommended:** §17B.

In this section we will prove Theorem 8A.1(b), through a common strategy in harmonic analysis: the use of a *summation kernel*. For all $N \in \mathbb{N}$, the Nth *Dirichlet kernel* is the function $\mathbf{D}_N : [-2\pi, 2\pi] \longrightarrow \mathbb{R}$ defined by

$$\mathbf{D}_N(x) := 1 + 2 \sum_{n=1}^{N} \cos(nx)$$

(see Figure 10B.1). Note that $\mathbf{D}_N$ is 2π-periodic (i.e. $\mathbf{D}_N(x + 2\pi) = \mathbf{D}_N(x)$ for all $x \in [-2\pi, 0]$). Thus, we could represent $\mathbf{D}_N$ as a function from $[-\pi, \pi]$ into $\mathbb{R}$. However, it is sometimes convenient to extend $\mathbf{D}_N$ to $[-2\pi, 2\pi]$. For example, for any function $f : [-\pi, \pi] \longrightarrow \mathbb{R}$, the *convolution* of $\mathbf{D}_N$ and f is the function $\mathbf{D}_N * f : [-\pi, \pi] \longrightarrow \mathbb{R}$ defined by

$$\mathbf{D}_N * f(x) := \frac{1}{2\pi} \int_{-\pi}^{\pi} f(y) \mathbf{D}_N(x - y) dy, \quad \text{for all } x \in [-\pi, \pi].$$

(Note that, to define $\mathbf{D}_N * f$, we must evaluate $\mathbf{D}_N(z)$ for all $z \in [-2\pi, 2\pi]$.) The connection between Dirichlet kernels and Fourier series is given by Lemma 10B.1.

Lemma 10B.1 *Let $f \in \mathbf{L}^2[-\pi, \pi]$, and, for all $n \in \mathbb{N}$, let*

$$A_n := \frac{1}{\pi} \int_{-\pi}^{\pi} \cos(ny) f(y) dy \quad and \quad B_n := \frac{1}{\pi} \int_{-\pi}^{\pi} \sin(ny) f(y) dy$$

be the real Fourier coefficients of f. Then, for any $N \in \mathbb{N}$, and every $x \in [-\pi, \pi]$, we have

$$A_0 + \sum_{n=1}^{N} A_n \mathbf{C}_n(x) + \sum_{n=1}^{N} B_n \mathbf{S}_n(x) = \mathbf{D}_N * f(x).$$

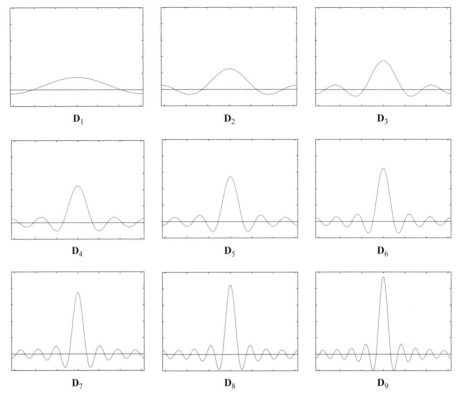

Figure 10B.1. The Dirichlet kernels $\mathbf{D}_1, \mathbf{D}_2, \ldots, \mathbf{D}_9$ plotted on the interval $[-\pi, \pi]$. Note the increasing concentration of the function near $x = 0$. (In the terminology of Sections 10D(ii) and 17B, the sequence $\{\mathbf{D}_1, \mathbf{D}_2, \ldots\}$ is like an *approximation of the identity*.)

Proof For any $x \in [-\pi, \pi]$, we have

$$A_0 + \sum_{n=1}^{N} A_n \mathbf{C}_n(x) + \sum_{n=1}^{N} B_n \mathbf{S}_n(x)$$

$$= A_0 + \sum_{n=1}^{N} \cos(nx) \left(\frac{1}{\pi} \int_{-\pi}^{\pi} \cos(ny) f(y) dy \right)$$

$$+ \sum_{n=1}^{N} \sin(nx) \left(\frac{1}{\pi} \int_{-\pi}^{\pi} \sin(ny) f(y) dy \right)$$

$$= A_0 + \sum_{n=1}^{N} \frac{1}{\pi} \left(\int_{-\pi}^{\pi} \cos(nx) \cos(ny) f(y) dy + \int_{-\pi}^{\pi} \sin(nx) \sin(ny) f(y) dy \right)$$

$$= A_0 + \sum_{n=1}^{N} \frac{1}{\pi} \int_{-\pi}^{\pi} \Big(\cos(nx) \cos(ny) + \sin(nx) \sin(ny) \Big) f(y) dy$$

$$\underset{(*)}{=} \frac{1}{2\pi} \int_{-\pi}^{\pi} f(y)dy + \sum_{n=1}^{N} \frac{1}{\pi} \int_{-\pi}^{\pi} \cos\left(n(x-y)\right) f(y)dy$$

$$= \frac{1}{2\pi} \int_{-\pi}^{\pi} \left(f(y) + 2\sum_{n=1}^{N} \cos\left(n(x-y)\right) f(y) \right) dy$$

$$= \frac{1}{2\pi} \int_{-\pi}^{\pi} f(y) \mathbf{D}_N(x-y)dy = \mathbf{D}_N * f(x).$$

Here, $(*)$ uses the fact that $A_0 := (1/2\pi) \int_{-\pi}^{\pi} f(y)dy$, and also the well known trigonometric identity $\cos(u-v) = \cos(u)\cos(v) + \sin(u)\sin(v)$ (with $u = nx$ and $v = ny$). □

Remark See Exercise 18F.7, p. 469, for another proof of Lemma 10B.1 for complex Fourier series.

Figure 10B.1 shows how the 'mass' of the Dirichlet kernel $\mathbf{D}_N$ becomes increasingly concentrated near $x = 0$ as $N \to \infty$. In the terminology of Sections 10D and 17B (pp. 209 and 385), the sequence $\{\mathbf{D}_1, \mathbf{D}_2, \ldots\}$ is like an *approximation of the identity*. Thus, our strategy is to show that $\mathbf{D}_N * f(x) \to f(x)$ as $N \to \infty$, whenever f is continuous at x. Indeed, we will go further: when f is *discontinuous* at x, we will show that $\mathbf{D}_N * f(x)$ converges to the average of the *left-hand* and *right-hand limits* of f at x. First we need some technical results.

Lemma 10B.2

(a) *For any $N \in \mathbb{N}$, we have $\int_0^{\pi} \mathbf{D}_N(x)dx = \pi$.*

(b) *For any $N \in \mathbb{N}$ and $x \in (-\pi, 0) \sqcup (0, \pi)$, we have*

$$\mathbf{D}_N(x) = \frac{\sin((2N+1)x/2)}{\sin(x/2)}.$$

(c) *Let $g : [0, \pi] \longrightarrow \mathbb{R}$ be a piecewise continuous function. Then*

$$\lim_{N \to \infty} \int_0^{\pi} g(x) \sin\left(\frac{(2N+1)x}{2}\right) dx = 0.$$

Proof The proof of (b) is **Exercise 10B.1**. (*Hint:* Use trigonometric identities.) ⒠
To prove (a), note that

$$\int_0^{\pi} \mathbf{D}_N(x)dx = \int_0^{\pi} 1 + 2\sum_{n=1}^{N} \cos(nx)dx = \int_0^{\pi} 1dx + 2\sum_{n=1}^{N} \int_0^{\pi} \cos(nx)dx$$

$$= \pi + 2\sum_{n=1}^{N} 0 = \pi.$$

To prove (c), first observe that

$$\sin\left(\frac{(2N+1)x}{2}\right) = \sin\left(Nx + \frac{x}{2}\right)$$

$$= \sin(Nx)\cos(x/2) + \cos(Nx)\sin(x/2), \quad (10B.1)$$

where the last step uses the well known trigonometric identity $\sin(u + v) = \sin(u)\cos(v) + \cos(u)\sin(v)$ (with $u := Nx$ and $v := x/2$). Thus,

$$\int_0^\pi g(x)\sin\left(\frac{(2N+1)x}{2}\right)dx$$

$$\underset{(\dagger)}{=} \int_0^\pi g(x)\Big(\sin(Nx)\cos(x/2) + \cos(Nx)\sin(x/2)\Big)dx$$

$$= \int_0^\pi \underbrace{g(x)\cos(x/2)}_{G_1(x)}\ \underbrace{\sin(Nx)}_{S_N(x)}dx + \int_0^\pi \underbrace{g(x)\sin(x/2)}_{G_2(x)}\ \underbrace{\cos(Nx)}_{C_N(x)}dx$$

$$\underset{(*)}{=} \frac{2\pi}{2\pi}\int_0^\pi G_1(x)\,S_N(x)dx + \frac{2\pi}{2\pi}\int_0^\pi G_2(x)\,C_N(x)dx$$

$$\underset{(\ddagger)}{=} \frac{\pi}{2}\langle G_1, S_N\rangle + \frac{\pi}{2}\langle G_2, C_N\rangle$$

$$\xrightarrow[N\to\infty]{} 0 + 0,$$

by Corollary 10A.3(b) (the Riemann–Lebesgue lemma). Here (†) is by equation (10B.1) and (‡) is by definition of the inner product on $\mathbf{L}^2[0, \pi]$. In (∗), we define the functions $G_1(x) := g(x)\cos(x/2)$ $G_2(x) := g(x)\sin(x/2)$; these functions are piecewise continuous because g is piecewise continuous; thus they are in $\mathbf{L}^2[0, \pi]$, so the Riemann–Lebesgue lemma is applicable. □

Let $f : [-\pi, \pi] \longrightarrow \mathbb{R}$ be a function. For any $x \in [-\pi, \pi)$, the *right-hand limit* of f at x is defined:

$$\lim_{y \searrow x} f(y) := \lim_{\epsilon \to 0} f(x + |\epsilon|) \quad \text{(if this limit exists)}.$$

Likewise, for any $x \in (-\pi, \pi]$, the *left-hand limit* of f at x is defined:

$$\lim_{y \nearrow x} f(y) := \lim_{\epsilon \to 0} f(x - |\epsilon|) \quad \text{(if this limit exists)};$$

see Figure 10B.2(a). Clearly, if f is continuous at x, then the left-hand and right-hand limits both exist, and $\lim_{y \searrow x} f(y) = f(x) = \lim_{y \nearrow x} f(y)$. However, the left-hand and right-hand limits may exist even when f is not continuous.

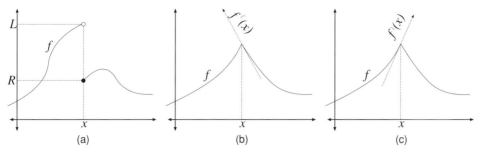

Figure 10B.2. (a) Left-hand and right-hand limits. Here, $L := \lim_{y \nearrow x} f(x)$ and $R := \lim_{y \searrow x} f(x)$. (b) The right-hand derivative $f^{\langle}(x)$. (c) The left-hand derivative $f^{\rangle}(x)$.

For any $x \in [-\pi, \pi)$, let $f(x^+) := \lim_{y \searrow x} f(y)$. The *right-hand derivative* of f at x is defined:

$$f^{\langle}(x) := \lim_{y \searrow x} \frac{f(y) - f(x^+)}{y - x} = \lim_{\epsilon \to 0} \frac{f(x + |\epsilon|) - f(x^+)}{|\epsilon|} \quad \text{(if this limit exists)};$$

see Figure 10B.2(b). Likewise, for any $x \in (-\pi, \pi]$, let $f(x^-) := \lim_{y \nearrow x} f(y)$. The *left-hand derivative* of f at x is defined:

$$f^{\rangle}(x) := \lim_{y \nearrow x} \frac{f(y) - f(x^-)}{y - x} = \lim_{\epsilon \to 0} \frac{f(x - |\epsilon|) - f(x^-)}{-|\epsilon|} \quad \text{(if this limit exists)};$$

see Figure 10B.2(c). If $f^{\langle}(x)$ and $f^{\rangle}(x)$ both exist, then we say f is *semidifferentiable* at x. Clearly, f, is differentiable at x if, and only if, f is continuous at x (so that $f(x^-) = f(x^+)$), and f is semidifferentiable at x, *and* $f^{\langle}(x) = f^{\rangle}(x)$. In this case, $f'(x) = f^{\langle}(x) = f^{\rangle}(x)$. However, f can be semidifferentiable at x even when f is not differentiable (or even continuous) at x.

Lemma 10B.3 *Let $\widetilde{f} : [-\pi, \pi] \longrightarrow \mathbb{R}$ be a piecewise continuous function which is semidifferentiable at 0. Then*

$$\lim_{N \to \infty} \int_{-\pi}^{\pi} \widetilde{f}(x) \, \mathbf{D}_N(x) \mathrm{d}x = \pi \cdot \left(\lim_{x \nearrow 0} \widetilde{f}(x) + \lim_{x \searrow 0} \widetilde{f}(x) \right).$$

Proof It suffices to show that

$$\lim_{N \to \infty} \int_{-\pi}^{0} \widetilde{f}(x) \, \mathbf{D}_N(x) \mathrm{d}x = \pi \cdot \lim_{x \nearrow 0} \widetilde{f}(x) \tag{10B.2}$$

and

$$\lim_{N \to \infty} \int_{0}^{\pi} \widetilde{f}(x) \, \mathbf{D}_N(x) \mathrm{d}x = \pi \cdot \lim_{x \searrow 0} \widetilde{f}(x). \tag{10B.3}$$

We will prove equation (10B.3). Let $\tilde{f}(0^+) := \lim_{x \searrow 0} \tilde{f}(x)$, and consider the function $g : [0, \pi] \longrightarrow \mathbb{R}$ defined by

$$g(x) := \frac{\tilde{f}(x) - \tilde{f}(0^+)}{\sin(x/2)}$$

if $x > 0$, while $g(0) := 2 \tilde{f}'(0)$.

Claim 1 *g is piecewise continuous on $[0, \pi]$.*

Proof Clearly, g is piecewise continuous on $(0, \pi]$ because $\tilde{f}$ is piecewise continuous, while $\sin(x/2)$ is nonzero on $(0, \pi]$. The only potential location of an unbounded discontinuity is at 0. But

$$\lim_{x \searrow 0} g(x) = \lim_{x \searrow 0} \frac{\tilde{f}(x) - \tilde{f}(0^+)}{\sin(x/2)} = \lim_{x \searrow 0} \left(\frac{\tilde{f}(x) - \tilde{f}(0^+)}{x} \right) \cdot \left(\frac{x}{\sin(x/2)} \right)$$

$$= \underbrace{\left(\lim_{x \searrow 0} \frac{\tilde{f}(x) - \tilde{f}(0^+)}{x - 0} \right)}_{\tilde{f}'(0)} \cdot 2 \cdot \underbrace{\left(\lim_{x \searrow 0} \frac{x/2}{\sin(x/2)} \right)}_{=1} = 2 \tilde{f}'(0) =: g(0).$$

Thus, g is (right-)continuous at 0, as desired. $\quad\Diamond_{\text{Claim 1}}$

Now,

$$\lim_{N \to \infty} \int_0^\pi \tilde{f}(x) \mathbf{D}_N(x) dx$$

$$= \lim_{N \to \infty} \int_0^\pi \left(\tilde{f}(0^+) + \tilde{f}(x) - \tilde{f}(0^+) \right) \mathbf{D}_N(x) dx$$

$$= \lim_{N \to \infty} \int_0^\pi \tilde{f}(0^+) \mathbf{D}_N(x) dx + \int_0^\pi \left(\tilde{f}(x) - \tilde{f}(0^+) \right) \mathbf{D}_N(x) dx$$

$$\underset{(a)}{=} \pi \tilde{f}(0^+) + \int_0^\pi \left(\tilde{f}(x) - \tilde{f}(0^+) \right) \mathbf{D}_N(x) dx$$

$$\underset{(b)}{=} \pi \tilde{f}(0^+) + \lim_{N \to \infty} \int_0^\pi \frac{\tilde{f}(x) - \tilde{f}(0^+)}{\sin(x/2)} \cdot \sin\left(\frac{(2N + 1)x}{2} \right) dx$$

$$= \pi \tilde{f}(0^+) + \lim_{N \to \infty} \int_0^\pi g(x) \cdot \sin\left(\frac{(2N + 1)x}{2} \right) dx$$

$$\underset{(c)}{=} \pi \tilde{f}(0^+) + 0 = \pi \tilde{f}(0^+),$$

as desired. Here, (a) is by Lemma 10B.2(a); (b) is by Lemma 10B.2(b); and (c) is by Lemma 10B.2(c), which is applicable because g is piecewise continuous by Claim 1.

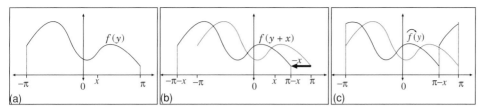

Figure 10B.3. The 2π-periodic phase-shift of a function. (a) A function $f : [-\pi, \pi] \longrightarrow \mathbb{R}$. (b) The function $y \mapsto f(x + y)$. (c) The function $\widehat{f}$: $[-\pi, \pi] \longrightarrow \mathbb{R}$.

This proves equation (10B.3). The proof of equation (10B.2) is **Exercise 10B.2**. Ⓔ Adding together equations (10B.2) and (10B.3) proves the lemma. ☐

Lemma 10B.4 *Let $f : [-\pi, \pi] \longrightarrow \mathbb{R}$ be piecewise continuous, and suppose that f is semidifferentiable at x (i.e. $f^{\triangleleft}(x)$ and $f^{\triangleright}(x)$ exist). Then*

$$\lim_{N \to \infty} \mathbf{D}_N * f(x) = \frac{1}{2} \left(\lim_{y \searrow x}(y) + \lim_{y \nearrow x} f(y) \right). \tag{10B.4}$$

In particular, if f is continuous and semidifferentiable at x, then $\lim_{N \to \infty} \mathbf{D}_N * f(x) = f(x)$.

Proof Suppose $x \in [0, \pi]$ (the case $x \in [-\pi, 0]$ is handled similarly). Define $\widehat{f} : [-\pi, \pi] \longrightarrow \mathbb{R}$:

$$\widehat{f}(y) := \begin{cases} f(y + x) & \text{if } y \in [-\pi, \pi - x]; \\ f(y + x - 2\pi) & \text{if } y \in [\pi - x, \pi]. \end{cases}$$

(Effectively, we are treating f as a 2π-periodic function, and 'phase-shifting' f by x; see Figure 10B.3.) Then

$$2\pi \cdot \mathbf{D}_N * f(x) = \int_{-\pi}^{\pi} f(y) \mathbf{D}_N(x - y) dy \underset{(*)}{=} \int_{-\pi}^{\pi} f(y) \mathbf{D}_N(y - x) dy$$

$$\underset{(c)}{=} \int_{-\pi-x}^{\pi-x} f(z + x) \mathbf{D}_N(z) dz$$

$$= \int_{-\pi-x}^{-\pi} f(z + x) \mathbf{D}_N(z) dz + \int_{-\pi}^{\pi-x} f(z + x) \mathbf{D}_N(z) dz$$

$$\underset{(@)}{=} \int_{\pi-x}^{\pi} f(w + x - 2\pi) \mathbf{D}_N(w - 2\pi) dw + \int_{-\pi}^{\pi-x} f(z + x) \mathbf{D}_N(z) dz$$

$$\underset{(\dagger)}{=} \int_{\pi-x}^{\pi} \widehat{f}(w) \mathbf{D}_N(w) dw + \int_{-\pi}^{\pi-x} \widehat{f}(z) \mathbf{D}_N(z) dz$$

$$= \int_{-\pi}^{\pi} \widehat{f}\,(z)\mathbf{D}_N(z)\mathrm{d}z \underset{(\diamond)}{=} \pi \cdot \left(\lim_{z \searrow 0} \widehat{f}\,(z) + \lim_{z \nearrow 0} \widehat{f}\,(z) \right)$$

$$\underset{(\ddagger)}{=} \pi \cdot \left(\lim_{z \searrow 0} f(z+x) + \lim_{z \nearrow 0} f(z+x) \right)$$

$$\underset{(c)}{=} \pi \cdot \left(\lim_{y \searrow x} f(y) + \lim_{y \nearrow x} f(y) \right).$$

Now divide both sides by 2π to get equation (10B.4).

Here, $(*)$ is because $\mathbf{D}_N$ is *even* (i.e. $\mathbf{D}_N(-r) = \mathbf{D}_N(r)$ for all $r \in \mathbb{R}$). Both equalities marked (c) are the change of variables $z := y - x$ (so that $y = z + x$). Likewise, equality (@) is the change of variables $w := z + 2\pi$ (so that $z = w - 2\pi$). Both $(\dagger)$ and $(\ddagger)$ use the definition of $\widehat{f}$, and $(\dagger)$ also uses the fact that $\mathbf{D}_N$ is 2π-periodic, so that $\mathbf{D}_N(w - 2\pi) = \mathbf{D}_N(w)$ for all $w \in [\pi - x, \pi]$. Finally, $(\diamond)$ is by Lemma 10B.3 applied to $\widehat{f}$ (which is continuous and semidifferentiable at 0 because f is continuous and semidifferentiable at x). $\qquad \square$

Proof of Theorem 8A.1(b) Let $x \in [-\pi, \pi]$, and suppose f is continuous and differentiable at x. Then

$$\lim_{N \to \infty} A_0 + \sum_{n=1}^{N} A_n \mathbf{C}_n(x) + \sum_{n=1}^{N} B_n \mathbf{S}_n(x) \underset{(*)}{=} \lim_{N \to \infty} \mathbf{D}_N * f(x) \underset{(\dagger)}{=} f(x),$$

as desired. Here, $(*)$ is by Lemma 10B.1 and $(\dagger)$ is by Lemma 10B.4. $\qquad \square$

Remarks

(a) Note that we have actually proved a slightly stronger result than Theorem 8A.1(b). If f is discontinuous, but semidifferentiable at x, then Lemmas 10B.1 and 10B.4 together imply that

$$\lim_{N \to \infty} A_0 + \sum_{n=1}^{N} A_n \mathbf{C}_n(x) + \sum_{n=1}^{N} B_n \mathbf{S}_n(x) = \frac{1}{2} \left(\lim_{y \searrow x} f(y) + \lim_{y \nearrow x} f(y) \right).$$

This is how the 'Pointwise Fourier Convergence Theorem' is stated in some texts.

(b) For other good expositions of this material, see Churchill and Brown (1987), §30–31, pp. 87–92; Broman (1989), Corollary 1.4.5, p. 16; Powers (1999), §1.7, p. 79; and Asmar (2005), Theorem 1, §2.2, p. 30.

10C Uniform convergence

Prerequisites: §8A, §10A.

In this section, we will prove Theorem 8A.1(d). First we state a 'discrete' version of the Cauchy–Bunyakowski–Schwarz Inequality (Theorem 6B.5, p. 112).

Lemma 10C.1 Cauchy–Bunyakowski–Schwarz Inequality in $\ell^2(\mathbb{N})$

Let $(a_n)_{n=1}^\infty$ and $(b_n)_{n=1}^\infty$ be two infinite sequences of real numbers. Then

$$\left(\sum_{n=1}^\infty a_n b_n\right)^2 \leq \left(\sum_{n=1}^\infty a_n^2\right) \cdot \left(\sum_{n=1}^\infty b_n^2\right),$$

whenever these sums are finite.

Proof **Exercise 10C.1.** (*Hint:* Imitate the proof of Theorem 6B.5, p. 112.) $\qquad\square\qquad$ Ⓔ

Remark For any infinite sequences of real numbers $\mathbf{a} := (a_n)_{n=1}^\infty$ and $\mathbf{b} := (b_n)_{n=1}^\infty$, we can define $\langle \mathbf{a}, \mathbf{b}\rangle := \sum_{n=1}^\infty a_n b_n$ and $\|\mathbf{a}\|_2 := \sqrt{\langle \mathbf{a}, \mathbf{a}\rangle} = \sqrt{\sum_{n=1}^\infty a_n^2}$. The set of all sequences $\mathbf{a}$, such that $\|\mathbf{a}\|_2 < \infty$, is denoted $\ell^2(\mathbb{N})$. Lemma 10C.1 can then be reformulated as the following statement: 'For all $\mathbf{a}, \mathbf{b} \in \ell^2(\mathbb{N})$, $|\langle \mathbf{a}, \mathbf{b}\rangle| \leq \|\mathbf{a}\|_2 \cdot \|\mathbf{b}\|_2$'.

Next, we will prove a result which relates the 'smoothness' of the function f to the 'asymptotic decay rate' of its Fourier coefficients.

Lemma 10C.2 *Let $f : [-\pi, \pi] \longrightarrow \mathbb{R}$ be continuous, with $f(-\pi) = f(\pi)$. Let $\{A_n\}_{n=0}^\infty$ and $\{B_n\}_{n=1}^\infty$ be the real Fourier coefficients of f, as defined on p. 165. If f is piecewise differentiable on $[-\pi, \pi]$, and $f' \in \mathbf{L}^2[-\pi, \pi]$, then the sequences $\{A_n\}_{n=0}^\infty$ and $\{B_n\}_{n=1}^\infty$ converge to zero fast enough that $\sum_{n=1}^\infty |A_n| < \infty$ and $\sum_{n=1}^\infty |B_n| < \infty$.*

Proof If $f' \in \mathbf{L}^2[-\pi, \pi]$, then we can compute its real Fourier coefficients $\{A_n'\}_{n=0}^\infty$ and $\{B_n'\}_{n=1}^\infty$.

Claim 1 *For all $n \in \mathbb{N}$, $A_n = -B_n'/n$ and $B_n = A_n'/n$.*

Proof By definition,

$$A_n' := \frac{1}{\pi}\int_{-\pi}^\pi f'(x)\cos(nx)\,dx$$

$$\underset{(p)}{=} \frac{1}{\pi} f(x)\cos(nx) \Big|_{x=-\pi}^{x=\pi} + \frac{1}{\pi}\int_{-\pi}^\pi f(x)\,n\sin(nx)\,dx$$

$$\underset{(c)}{=} \frac{(-1)^n}{\pi}\Big(f(\pi) - f(-\pi)\Big) + n\,B_n \underset{(*)}{=} n\,B_n.$$

Likewise,

$$B_n' := \frac{1}{\pi}\int_{-\pi}^\pi f'(x)\sin(nx)\,dx$$

$$\underset{(p)}{=} \frac{1}{\pi} f(x)\sin(nx) \Big|_{x=-\pi}^{x=\pi} - \frac{1}{\pi}\int_{-\pi}^\pi f(x)\,n\cos(nx)\,dx$$

$$\underset{(s)}{=} \Big(0 - 0\Big) - n\,A_n = -n\,A_n.$$

Here, (p) is integration by parts, (c) is because $\cos(-n\pi) = (-1)^n = \cos(n\pi)$, (∗) is because $f(-\pi) = f(\pi)$, and (s) is because $\sin(-n\pi) = 0 = \sin(n\pi)$. Thus, $B'_n = -n A_n$ and $A'_n = n B_n$; hence $A_n = -B'_n/n$ and $B_n = A'_n/n$.

$\diamond$ Claim 1

Let $K := \sum_{n=1}^{\infty} 1/n^2$ (a finite value). Then

$$\left(\sum_{n=1}^{\infty} |A_n| \right)^2 \underset{(*)}{=} \left(\sum_{n=1}^{\infty} \frac{1}{n} |B'_n| \right)^2 \underset{\text{(CBS)}}{\leq} \left(\sum_{n=1}^{\infty} \frac{1}{n^2} \right) \cdot \left(\sum_{n=1}^{\infty} |B'_n|^2 \right)$$

$$= K \cdot \sum_{n=1}^{\infty} |B'_n|^2 \underset{\text{(B)}}{\leq} K \cdot \|f'\|_2^2 \underset{(\dagger)}{<} \infty.$$

Here, (∗) is by Claim 1, (CBS) is by Lemma 10C.1, and (B) is by Bessel's inequality (Theorem 10A.1, p. 197). Finally, (†) is because $f' \in \mathbf{L}^2[-\pi, \pi]$ by hypothesis. It follows that $\sum_{n=1}^{\infty} |A_n| < \infty$. The proof that $\sum_{n=1}^{\infty} |B_n| < \infty$ is similar. $\square$

Proof of Theorem 8A.1(d)   If $f : [-\pi, \pi] \longrightarrow \mathbb{R}$ is continuous and piecewise differentiable, $f' \in \mathbf{L}^2[-\pi, \pi]$, and $f(-\pi) = f(\pi)$, then Lemma 10C.2 implies that $\sum_{n=1}^{\infty} |A_n| + \sum_{n=1}^{\infty} |B_n| < \infty$. But then Theorem 8A.1(c) says that the Fourier series of f converges uniformly. (Theorem 8A.1(c), in turn, is a direct consequence of the Weierstrass M test, Proposition 6E.13, p. 135). $\square$

Remarks

(a) For other treatments of the material in this section, see Churchill and Brown (1987), §34–35, pp. 105–109, or Asmar (2005), Theorem 3, §2.9, p. 90.

(b) The connection between the smoothness of f and the asymptotic decay of its Fourier coefficients is a recurring theme in harmonic analysis. In general, the 'smoother' a function is, the 'faster' its Fourier coefficients decay to zero. The weakest statement of this kind is the Riemann–Lebesgue lemma (Corollary 10A.3, p. 199), which says that if f is merely in L^2, then its Fourier coefficients must converge to zero – although perhaps very slowly. (In the context of Fourier *transforms* of functions on $\mathbb{R}$, the corresponding statement is Theorem 19B.1, p. 497.) If f is 'slightly smoother' – specifically, if f is *absolutely continuous* or if f has *bounded variation* – then its Fourier coefficients decay to zero with speed comparable to the sequence $\{1, \frac{1}{2}, \frac{1}{3}, \ldots, \frac{1}{n}, \ldots\}$; see Katznelson (1976), Theorems 4.3 and 4.5. If f is differentiable, then Lemma 10C.2 says that its Fourier coefficients must decay fast enough that the sums $\sum_{n=1}^{\infty} |A_n|$ and $\sum_{n=1}^{\infty} |B_n|$ converge. (For Fourier *transforms* of functions on $\mathbb{R}$, the corresponding result is Theorem 19B.7, p. 500.) More generally, if f is k times differentiable on $[-\pi, \pi]$, then its Fourier coefficients must decay fast enough that the sums $\sum_{n=1}^{\infty} n^{k-1}|A_n|$ and $\sum_{n=1}^{\infty} n^{k-1}|B_n|$

converge; see Katznelson (1976), Theorem 4.3. Finally, if f is *analytic*,[1] on $[-\pi, \pi]$, then its Fourier coefficients must decay *exponentially quickly* to zero; that is, for small enough $r > 0$, we have $\lim_{n\to\infty} r^n |A_n| = 0$ and $\lim_{n\to\infty} r^n |B_n| = 0$ (see Proposition 18E.3, p. 463).

At the other extreme, what about a sequence of Fourier coefficients which does *not* satisfy the Riemann–Lebesgue lemma – that is, which does not converge to zero? This corresponds to the Fourier series of an object which is more 'singular' than any function can be: a *Schwartz distribution* or a *measure* on $[-\pi, \pi]$, which can have 'infinitely dense' concentrations of mass at some points. See Katznelson (1976), §1.7, pp. 34–46, or Folland (1984), §8.5 and §8.8.

10D *L²*-convergence

Prerequisites: §6B.

In this section, we will prove Theorem 8A.1(a) (concerning the L^2-convergence of Fourier series). For any $k \in \mathbb{N}$, let $\mathcal{C}^k_{\text{per}}[-\pi, \pi]$ be the set of functions f which are k times continuously differentiable on $[-\pi, \pi]$, and such that $f(-\pi) = f(\pi)$, $f'(-\pi) = f'(\pi)$, $f''(-\pi) = f''(\pi)$, ..., and $f^{(k)}(-\pi) = f^{(k)}(\pi)$. If $f \in \mathcal{C}^1_{\text{per}}[-\pi, \pi]$, then Theorem 8A.1(d) (which we proved in §10C) says that the Fourier series of f converges *uniformly*. Then Corollary 6E.11(b)(i) (p. 133) immediately implies that the Fourier series of f converges in L^2-norm. Unfortunately, this argument does not work for most functions in $\mathbf{L}^2[-\pi, \pi]$, which are *not* in $\mathcal{C}^1_{\text{per}}[-\pi, \pi]$. Our strategy will be to show that $\mathcal{C}^1_{\text{per}}[-\pi, \pi]$ is *dense* in $\mathbf{L}^2[-\pi, \pi]$; thus, the L^2-convergence of Fourier series in $\mathcal{C}^1_{\text{per}}[-\pi, \pi]$ can be 'leveraged' to obtain L^2-convergence for all functions in $\mathbf{L}^2[-\pi, \pi]$.

A subset $\mathcal{G} \subset \mathbf{L}^2[-\pi, \pi]$ is *dense* in $\mathbf{L}^2[-\pi, \pi]$ if, for any $f \in \mathbf{L}^2[-\pi, \pi]$ and any $\epsilon > 0$, we can find some $g \in \mathcal{G}$ such that $\|f - g\|_2 < \epsilon$. In other words, any element of $\mathbf{L}^2[-\pi, \pi]$ can be approximated arbitrarily closely[2] by elements of $\mathcal{G}$. Aside from Theorem 8A.1(a), the major goal of this section is to prove the following result.

Theorem 10D.1 *For all $k \in \mathbb{N}$, the subset $\mathcal{C}^k_{\text{per}}[-\pi, \pi]$ is dense in $\mathbf{L}^2[-\pi, \pi]$.*

To achieve this proof, we must first examine the structure of integrable functions and develop some useful machinery involving 'convolutions' and 'mollifiers'. Then we will prove Theorem 10D.1. Once Theorem 10D.1 is established, we will prove Theorem 8A.1(a) by using Theorem 8A.1(d) and the triangle inequality.

---

[1] See Appendix H, p. 573
[2] In the same way, the set $\mathbb{Q}$ of rational numbers is *dense* in the set $\mathbb{R}$ of real numbers: any real number can be approximated arbitrarily closely by rational numbers. Indeed, we exploit this fact every time we approximate a real number using a decimal expansion – e.g. $\pi \approx 3.141592653 = (3\,141\,592\,653) \div (100\,000\,000)$.

10D(i) Integrable functions and step functions in $\mathbf{L}^2[-\pi, \pi]$

Prerequisites: §6B, §6E(i).

We have defined $\mathbf{L}^2[-\pi, \pi]$ to be the set of 'integrable' functions $f : [-\pi, \pi] \longrightarrow \mathbb{R}$ such that $\int_{-\pi}^{\pi} |f(x)|^2 dx < \infty$. But what exactly does *integrable* mean? To explain this, let $\mathsf{Step}[-\pi, \pi]$ be the set of all *step functions* on $[-\pi, \pi]$ (see §8B(ii), p. 168, for the definition of step functions). If $f : [-\pi, \pi] \longrightarrow \mathbb{R}$ is any bounded function, then we can 'approximate' f using step functions in a natural way. First, let $\mathcal{Y} := \{-\pi = y_0 < y_1 < y_2 < y_3 < \cdots < y_{M-1} < y_M = \pi\}$ be some finite 'mesh' of points in $[-\pi, \pi]$. For all $n \in \mathbb{N}$, let

$$\underline{a}_n := \inf_{y_{n-1} \leq x \leq y_n} f(x) \quad \text{and} \quad \overline{a}_n := \sup_{y_{n-1} \leq x \leq y_n} f(x).$$

Then define step functions $\underline{S}_{\mathcal{Y}} : [-\pi, \pi] \longrightarrow \mathbb{R}$ and $\overline{S}_{\mathcal{Y}} : [-\pi, \pi] \longrightarrow \mathbb{R}$ as follows:

$$\underline{S}_{\mathcal{Y}}(x) := \begin{cases} \underline{a}_1 & \text{if} \quad -\pi \leq x \leq y_1; \\ \underline{a}_2 & \text{if} \quad y_1 < x \leq y_2; \\ \quad \vdots & \\ \underline{a}_m & \text{if} \quad y_{m-1} < x \leq y_m; \\ \quad \vdots & \\ \underline{a}_M & \text{if} \quad y_{M-1} < x \leq \pi, \end{cases}$$

$$\overline{S}_{\mathcal{Y}}(x) := \begin{cases} \overline{a}_1 & \text{if} \quad -\pi \leq x \leq y_1; \\ \overline{a}_2 & \text{if} \quad y_1 < x \leq y_2; \\ \quad \vdots & \\ \overline{a}_m & \text{if} \quad y_{m-1} < x \leq y_m; \\ \quad \vdots & \\ \overline{a}_M & \text{if} \quad y_{M-1} < x \leq \pi. \end{cases}$$

It is easy to compute the integrals of $\underline{S}_{\mathcal{Y}}$ and $\overline{S}_{\mathcal{Y}}$:

$$\int_{-\pi}^{\pi} \underline{S}_{\mathcal{Y}}(x) dx = \sum_{n=1}^{N} \underline{a}_n \cdot |y_n - y_{n-1}|; \quad \int_{-\pi}^{\pi} \overline{S}_{\mathcal{Y}}(x) dx = \sum_{n=1}^{N} \overline{a}_n \cdot |y_n - y_{n-1}|.$$

(You may recognize these as upper and lower *Riemann sums* of f.) If the mesh $\{y_0, y_1, y_2, \ldots, y_M\}$ is 'dense' enough in $[-\pi, \pi]$, so that $\underline{S}_{\mathcal{Y}}$ and $\overline{S}_{\mathcal{Y}}$ are 'good approximations' of f, then we might expect $\int_{-\pi}^{\pi} \underline{S}_{\mathcal{Y}}(x) dx$ and $\int_{-\pi}^{\pi} \overline{S}_{\mathcal{Y}}(x) dx$ to be good approximations of $\int_{-\pi}^{\pi} f(x) dx$ (if the integral of f is well defined). Furthermore, it is clear from their definitions that $\underline{S}_{\mathcal{Y}}(x) \leq f(x) \leq \overline{S}_{\mathcal{Y}}(x)$ for all

$x \in [-\pi, \pi]$; thus we would expect

$$\int_{-\pi}^{\pi} \underline{S}_y(x)dx \leq \int_{-\pi}^{\pi} f(x)dx \leq \int_{-\pi}^{\pi} \overline{S}_y(x)dx,$$

whenever the integral $\int_{-\pi}^{\pi} f(x)dx$ exists. Let $\mathfrak{Y}$ be the set of all finite 'meshes' of points in $[-\pi, \pi]$. Formally:

$$\mathfrak{Y} := \left\{ \begin{array}{l} \mathcal{Y} \subset [-\pi, \pi]; \quad \mathcal{Y} = \{y_0, y_1, y_2, \dots, y_M\}, \quad \text{for some} M \in \mathbb{N} \\ \text{and } -\pi = y_0 < y_1 < y_2 < \cdots < y_M = \pi \end{array} \right\}.$$

We define the *lower* and *upper semi-integrals* of f:

$$\underline{I}(f) := \sup_{\mathcal{Y} \in \mathfrak{Y}} \int_{-\pi}^{\pi} \underline{S}_y(x)dx \qquad\qquad (10D.1)$$

$$= \sup \left\{ \sum_{n=1}^{N} \left(|y_n - y_{n-1}| \cdot \inf_{y_{n-1} \leq x \leq y_n} f(x) \right); M \in \mathbb{N}, -\pi = y_0 < y_1 < \cdots < y_M = \pi \right\};$$

$$\overline{I}(f) := \inf_{\mathcal{Y} \in \mathfrak{Y}} \int_{-\pi}^{\pi} \overline{S}_y(x)dx \qquad\qquad (10D.2)$$

$$= \inf \left\{ \sum_{n=1}^{N} \left(|y_n - y_{n-1}| \cdot \sup_{y_{n-1} \leq x \leq y_n} f(x) \right); M \in \mathbb{N}, -\pi = y_0 < y_1 < \cdots < y_M = \pi \right\}.$$

It is easy to see that $\underline{I}(f) \leq \overline{I}(f)$. (**Exercise 10D.1** Check this.) Indeed, if f is Ⓔ a sufficiently 'pathological' function, then we may have $\underline{I}(f) < \overline{I}(f)$. If $\underline{I}(f) = \overline{I}(f)$, then we say that f is *(Riemann)-integrable*, and we define the *(Riemann) integral* of f:

$$\int_{-\pi}^{\pi} f(x)dx := \underline{I}(f) = \overline{I}(f).$$

For example:

- any bounded, piecewise continuous function on $[-\pi, \pi]$ is Riemann-integrable;
- any continuous function on $[-\pi, \pi]$ is Riemann-integrable;
- any step function on $[-\pi, \pi]$ is Riemann-integrable.

If $f : [-\pi, \pi] \longrightarrow \mathbb{R}$ is *not* bounded, then the definitions of $\underline{S}_y$ and/or $\overline{S}_y$ make no sense (because at least one of them is defined as '∞' or '$-\infty$' on some interval). Thus, at least one of the expressions (10D.1) and (10D.2) is not well defined if f is unbounded. In this case, for any $N \in \mathbb{N}$, we define the 'truncated' functions $f_N^+ : [-\pi, \pi] \longrightarrow [0, N]$ and $f_N^- : [-\pi, \pi] \longrightarrow [-N, 0]$ as follows:

$$f_N^+(x) := \begin{cases} 0 & \text{if} \quad f(x) \leq 0; \\ f(x) & \text{if} \quad 0 \leq f(x) \leq N; \\ N & \text{if} \quad N \leq f(x), \end{cases}$$

and

$$f_N^-(x) := \begin{cases} -N & \text{if} & f(x) \leq -N; \\ f(x) & \text{if} & -N \leq f(x) \leq 0; \\ 0 & \text{if} & 0 \leq f(x). \end{cases}$$

The functions f_N^+ and f_N^- are clearly bounded, so their Riemann integrals are potentially well defined. If f_N^+ and f_N^- are integrable for all $N \in \mathbb{N}$, then we say that f is *(Riemann)-measurable*. We then define

$$\int_{-\pi}^{\pi} f(x)dx := \lim_{N \to \infty} \int_{-\pi}^{\pi} f_N^+(x)dx + \lim_{N \to \infty} \int_{-\pi}^{\pi} f_N^-(x)dx.$$

If both these limits are finite, then $\int_{-\pi}^{\pi} f(x)dx$ is well defined, and we say that the unbounded function f is *(Riemann)-integrable*. The set of all integrable functions (bounded or unbounded) is denoted $\mathbf{L}^1[-\pi, \pi]$, and, for any $f \in \mathbf{L}^1[-\pi, \pi]$, we define

$$\|f\|_1 = \int_{-\pi}^{\pi} |f(x)|dx.$$

We can now define $\mathbf{L}^2[-\pi, \pi]$:

$$\mathbf{L}^2[-\pi, \pi] := \left\{ \begin{array}{l} \text{all measurable functions } f : [-\pi, \pi] \longrightarrow \mathbb{R} \text{ such that} \\ f^2 : [-\pi, \pi] \longrightarrow \mathbb{R} \text{ is integrable; i.e. } \int_{-\pi}^{\pi} |f(x)|^2 dx < \infty \end{array} \right\}.$$

Proposition 10D.2 Step$[-\pi, \pi]$ *is dense in* $\mathbf{L}^2[-\pi, \pi]$.

Proof Let $f \in \mathbf{L}^2[-\pi, \pi]$ and let $\epsilon > 0$. We want to find some $S \in$ Step$[-\pi, \pi]$ such that $\|f - S\|_2 < \epsilon$.

First suppose that f is *bounded*. Since f^2 is integrable, we know that $\underline{I}(f^2) = \int_{-\pi}^{\pi} f^2(x)dx$, where $\underline{I}(f^2)$ is defined by equation (10D.1). Thus, we can find some step function $S_0 \in$ Step$[-\pi, \pi]$ such that $0 \leq S_0(x) \leq f^2(x)$ for all $x \in [-\pi, \pi]$, and such that

$$0 \leq \int_{-\pi}^{\pi} f(x)^2 dx - \int_{-\pi}^{\pi} S_0(x)dx < \epsilon. \tag{10D.3}$$

Define the step function $S \in$ Step$[-\pi, \pi]$ by $S(x) := \text{sign}[f(x)] \cdot \sqrt{S_0(x)}$. Thus, $S^2(x) = S_0(x)$, and the sign of S agrees with that of f everywhere. Observe that

$$(f - S)^2 = \frac{f - S}{f + S} \cdot (f - S)(f + S) = \frac{f - S}{f + S} \cdot (f^2 - S^2)$$

$$\underset{(*)}{\leq} f^2 - S^2 = f^2 - S_0. \tag{10D.4}$$

Here, (∗) is because $0 < (f - S)/(f + S) < 1$ (because, for all x, either $f(x) \leq S(x) \leq 0$ or $0 \leq S(x) \leq f(x)$), while $f^2 - S^2 \geq 0$ (because $f^2 \geq S_0 = S^2$). Thus,

$$0 \leq \|f - S\|_2^2 = \int_{-\pi}^{\pi} (f(x) - S(x))^2 \, dx \underset{(*)}{\leq} \int_{-\pi}^{\pi} f(x)^2 - S_0(x) dx$$

$$= \int_{-\pi}^{\pi} f(x)^2 \, dx - \int_{-\pi}^{\pi} S_0(x) dx \underset{(\dagger)}{<} \epsilon,$$

where (∗) is by equation (10D.4) and (†) is by equation (10D.3).

This works for any $\epsilon > 0$; thus the set Step$[-\pi, \pi]$ is dense in the space of bounded elements of $\mathbf{L}^2[-\pi, \pi]$.

The case when f is unbounded is **Exercise 10D.2**. (*Hint:* Approximate f with Ⓔ bounded functions.) □

Remark 10D.3 To avoid developing a considerable amount of technical background, we have defined $\mathbf{L}^2[-\pi, \pi]$ using the *Riemann* integral. The 'true' definition of $\mathbf{L}^2[-\pi, \pi]$ involves the more powerful and versatile *Lebesgue* integral. (See §6C(ii), p. 114, for an earlier discussion of Lebesgue integration.) The definition of the Lebesgue integral is similar to the Riemann integral, but, instead of approximating f using step functions, we use *simple* functions. A simple function is a piecewise constant function, like a step function, but, instead of open intervals, the 'pieces' of a simple function are *Borel-measurable subsets* of $[-\pi, \pi]$. A Borel-measurable subset is a countable union of countable intersections of countable unions of countable intersections of . . . of countable unions/intersections of open and/or closed subsets of $[-\pi, \pi]$. In particular, any interval is Borel-measurable (so any step function is a simple function), but Borel-measurable subsets can be very complicated indeed. Thus, 'simple' functions are capable of approximating even pathological, wildly discontinuous functions on $[-\pi, \pi]$, so that the Lebesgue integral can be evaluated even on such crazy functions. The set of Lebesgue-integrable functions is thus much larger than the set of Riemann-integrable functions. Every Riemann-integrable function is Lebesgue-integrable (and its Lebesgue integral is the same as its Riemann integral), but not vice versa.

The analogy of Proposition 10D.2 is still true if we define $\mathbf{L}^2[-\pi, \pi]$ using *Lebesgue*-integrable functions, and if we replace Step$[-\pi, \pi]$ with the set of all *simple* functions. The other results in this section can also be extended to the Lebesgue version of $\mathbf{L}^2[-\pi, \pi]$, but at the cost of considerable technical complexity.

Let $f : [-\pi, \pi] \longrightarrow \mathbb{R}$. Let $f^{\circ} : [-2\pi, 2\pi] \longrightarrow \mathbb{R}$ be the 2π-*periodic extension* of f, defined:

$$f^{\circ}(x) := \begin{cases} f(x + 2\pi) & \text{if} \quad -2\pi \leq x < -\pi; \\ f(x) & \text{if} \quad -\pi \leq x \leq \pi; \\ f(x - 2\pi) & \text{if} \quad \pi < x \leq 2\pi \end{cases}$$

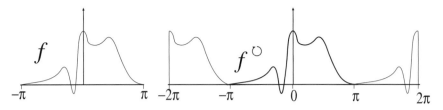

Figure 10D.1. $f^{\circlearrowleft} : [-2\pi, 2\pi]$ is the 2π-periodic extension of $f : [-\pi, \pi] \longrightarrow \mathbb{R}$.

(see Figure 10D.1). (Observe that $f^{\circlearrowleft}$ is continuous if, and only if, f is continuous and $f(-\pi) = f(\pi)$.) For any $t \in \mathbb{R}$, define the function $\widehat{f^{t}} : [-\pi, \pi] \longrightarrow \mathbb{R}$ by $\widehat{f^{t}}(x) = f^{\circlearrowleft}(x - t)$. (For example, the function $\widehat{f}$ defined on p. 205 could be written $\widehat{f} = \widehat{f^{-x}}$; see Figure 10B.3, p. 205.)

Lemma 10D.4 *Let $f \in \mathbf{L}^2[-\pi, \pi]$. Then $f = \mathbf{L}^2\text{–}\lim_{t \to 0} \widehat{f^{t}}$.*

Proof We will employ a classic strategy in real analysis: first prove the result for some 'nice' class of functions, and then prove it for all functions by approximating them with these nice functions. In this case, the nice functions are the step functions.

Claim 1 *Let $S \in \mathsf{Step}[-\pi, \pi]$. Then $S = \mathbf{L}^2\text{–}\lim_{t \to 0} \widehat{S^{t}}$.*

Ⓔ *Proof* **Exercise 10D.3.**   ◇ Claim 1

Now, let $f \in \mathbf{L}^2[-\pi, \pi]$, and let $\epsilon > 0$. Proposition 10D.2 says there is some $S \in \mathsf{Step}[-\pi, \pi]$ such that

$$\| S - f \|_2 < \frac{\epsilon}{3}. \tag{10D.5}$$

Claim 2 *For all $t \in \mathbb{R}$, $\| \widehat{S^{t}} - \widehat{f^{t}} \|_2 = \| S - f \|_2$.*

Ⓔ *Proof* **Exercise 10D.4.**   ◇ Claim 2

Now, using Claim 1, find $\delta > 0$ such that, if $|t| < \delta$, then

$$\left\| S - \widehat{S^{t}} \right\|_2 < \frac{\epsilon}{3}. \tag{10D.6}$$

Then

$$\left\| f - \widehat{f^{t}} \right\|_2 = \left\| f - S + S - \widehat{S^{t}} + \widehat{S^{t}} - \widehat{f^{t}} \right\|_2$$

$$\underset{(\triangle)}{\leq} \| f - S \|_2 + \left\| S - \widehat{S^{t}} \right\|_2 + \left\| \widehat{S^{t}} - \widehat{f^{t}} \right\|_2$$

$$\underset{(*)}{\leq} \frac{\epsilon}{3} + \frac{\epsilon}{3} + \frac{\epsilon}{3} = \epsilon.$$

Here ($\triangle$) is the triangle inequality, and ($*$) is by equations (10D.5) and (10D.6) and Claim 2. This works for all $\epsilon > 0$; thus, $f = \mathbf{L}^2\text{--}\lim_{t \to 0} \hat{f}^t$. $\qquad\square$

There is one final technical result we will need about $\mathbf{L}^2[-\pi, \pi]$. If f_1, $f_2, \ldots, f_N \in \mathbf{L}^2[-\pi, \pi]$ and $r_1, r_2, \ldots, r_N \in \mathbb{R}$ are real numbers, then the triangle inequality implies that

$$\|r_1 f_1 + r_2 f_2 + \cdots + r_N f_N\|_2 \leq |r_1| \cdot \|f_1\|_2 + |r_2| \cdot \|f_2\|_2 + \cdots + |r_N| \cdot \|f_N\|_2.$$

This is a special case of *Minkowski's inequality*. The next result says that the same inequality holds if we sum together a 'continuum' of functions.

Theorem 10D.5 Minkowski's inequality for integrals
Let $a < b$, and, for all $t \in [a, b]$, let $f_t \in \mathbf{L}^2[-\pi, \pi]$. Define $F : [a, b] \times [-\pi, \pi]$ $\longrightarrow \mathbb{R}$ by $F(t, x) = f_t(x)$ for all $(t, x) \in [a, b] \times [-\pi, \pi]$, and suppose that the family $\{f_t\}_{t \in [a,b]}$ is such that the function F is integrable on $[a, b] \times [-\pi, \pi]$. Let $R : [a, b] \longrightarrow \mathbb{R}$ be some other integrable function, and define $G : [-\pi, \pi] \longrightarrow \mathbb{R}$ by

$$G(x) := \int_a^b R(t) f_t(x) \mathrm{d}t, \quad \text{for all } x \in [-\pi, \pi].$$

Then $G \in \mathbf{L}^2[-\pi, \pi]$, and

$$\|G\|_2 \leq \int_a^b |R(t)| \cdot \|f_t\|_2 \, \mathrm{d}t.$$

In particular, if $\|f_t\|_2 < M$ for all $t \in [a, b]$, then $\|G\|_2 \leq M \cdot \|R\|_1$, where $\|R\|_1 := \int_a^b |R(t)| \mathrm{d}t$.

Proof See Folland (1984), Theorem 6.18. $\qquad\square$

10D(ii) *Convolutions and mollifiers*

Prerequisites: §10D(i).　　　**Recommended:** §17B.

Let $f, g : [-\pi, \pi] \longrightarrow \mathbb{R}$ be two integrable functions. Let $\overset{\circ}{g} : [-2\pi, 2\pi] \longrightarrow \mathbb{R}$ be the 2π-*periodic extension* of g (see Figure 10D.1, p. 214). The (2π-periodic) *convolution* of f and g is the function $f * g : [-\pi, \pi] \longrightarrow \mathbb{R}$ defined:

$$f * g(x) := \frac{1}{2\pi} \int_{-\pi}^{\pi} f(y) \overset{\circ}{g}(x - y) \mathrm{d}y, \quad \text{for all } x \in [-\pi, \pi].$$

Convolution is an important and versatile mathematical operation, which appears frequently in harmonic analysis, probability theory, and the study of partial differential equations. We will encounter it again in Chapter 17, in the context of 'impulse-response' solutions to boundary value problems. In this subsection, we will develop the theory of convolutions on $[-\pi, \pi]$. We will actually develop slightly more than we need in order to prove Theorems 10D.1 and 8A.1(a). Results which are not logically required for the proofs of Theorems 10D.1 and 8A.1(a) are marked with the margin symbol '(Optional)' and can be skipped on a first reading; however, we feel that these results are interesting enough in themselves to be worth including in the exposition.

Lemma 10D.6 Properties of convolutions

Let $f, g : [-\pi, \pi] \longrightarrow \mathbb{R}$ be integrable functions. The convolution of f and g has the following properties.

(a) *(Commutativity)* $f * g = g * f$.

(b) *(Linearity) If $h : [-\pi, \pi] \longrightarrow \mathbb{R}$ is another integrable function, then $f * (g + h) = f * g + f * h$ and $(f + g) * h = f * h + g * h$.*

(c) *(Optional) If $f, g \in \mathbf{L}^2[-\pi, \pi]$, then $f * g$ is bounded: for all $x \in [-\pi, \pi]$, we have $|f * g(x)| \leq \|f\|_2 \cdot \|g\|_2$. (In other words, $\|f * g\|_\infty \leq \|f\|_2 \cdot \|g\|_2$.)*

Ⓔ   *Proof* (a) is **Exercise 10D.5.**

To prove (b), let $x \in [-\pi, \pi]$. Then

$$f * (g + h)(x) = \frac{1}{2\pi} \int_{-\pi}^{\pi} f(y) \left(\overset{\circ}{g}(x - y) + \overset{\circ}{h}(x - y) \right) dy$$

$$= \frac{1}{2\pi} \int_{-\pi}^{\pi} f(y) \overset{\circ}{g}(x - y) dy + \frac{1}{2\pi} \int_{-\pi}^{\pi} f(y) \overset{\circ}{h}(x - y) dy$$

$$= f * g(x) + f * h(x).$$

(c) (Optional) Let $x \in [-\pi, \pi]$. Define $h \in \mathbf{L}^2[-\pi, \pi]$ by $h(y) := \overset{\circ}{g}(x - y)$. Then

$$f * g(x) = \frac{1}{2\pi} \int_{-\pi}^{\pi} f(y) \overset{\circ}{g}(x - y) dy = \frac{1}{2\pi} \int_{-\pi}^{\pi} f(y) h(y) dy = \langle f, h \rangle.$$

Thus,

$$|f * g(x)| = |\langle f, h \rangle| \underset{\text{(CBS)}}{\leq} \|f\|_2 \cdot \|h\|_2, \tag{10D.7}$$

where (CBS) is the Cauchy–Bunyakowski–Schwarz Inequality (Theorem 6B.5, p. 112). But

$$\|h\|_2^2 = \frac{1}{2\pi} \int_{-\pi}^{\pi} h(y)^2 dy = \frac{1}{2\pi} \int_{-\pi}^{\pi} \overset{\circ}{g}(x - y)^2 dy \underset{(*)}{=\!=\!=} \frac{-1}{2\pi} \int_{x+\pi}^{x-\pi} \overset{\circ}{g}(z)^2 dz$$

$$= \frac{1}{2\pi} \int_{x-\pi}^{x+\pi} \overset{\circ}{g}(z)^2 dz \underset{(\dagger)}{=\!=\!=} \frac{1}{2\pi} \int_{-\pi}^{\pi} g(z)^2 dz = \|g\|_2^2. \tag{10D.8}$$

Here, (∗) is the change of variables $z = x - y$ (so that $dz = -dy$) and (†) is by definition of the periodic extension g° of g.

Combining equations (10D.5) and (10D.8), we conclude that $|f * g(x)| < \|f\|_2 \cdot \|g\|_2$, as claimed. □

Remarks

(a) Proposition 17G.1, p. 414, provides an analog to Lemma 10D.6 for convolutions on $\mathbb{R}^D$.

(b) There is also an interesting relationship between convolution and complex Fourier coefficients; see Lemma 18F.3, p. 469.

Elements of $\mathbf{L}^1[-\pi, \pi]$ and $\mathbf{L}^2[-\pi, \pi]$ need not be differentiable, or even continuous (indeed, some of these functions are discontinuous 'almost everywhere'). But the convolution of even two highly discontinuous elements of $\mathbf{L}^2[-\pi, \pi]$ will be a continuous function. Furthermore, convolution with a smooth function has a powerful 'smoothing' effect on even the nastiest elements of $\mathbf{L}^1[-\pi, \pi]$.

Lemma 10D.7 *Let* $f, g \in \mathbf{L}^1[-\pi, \pi]$.

(a) $f * g(-\pi) = f * g(\pi)$.

(b) *If* $f \in \mathbf{L}^1[-\pi, \pi]$ *and* g *is continuous with* $g(-\pi) = g(\pi)$, *then* $f * g$ *is continuous.*

(c) *(Optional) If* $f, g \in \mathbf{L}^2[-\pi, \pi]$, *then* $f * g$ *is continuous.*

(d) *If* g *is differentiable on* $[-\pi, \pi]$, *then* $f * g$ *is also differentiable on* $[-\pi, \pi]$, *and* $(f * g)' = f * (g')$.

(e) *If* $g \in \mathcal{C}^1[-\pi, \pi]$, *then* $f * g \in \mathcal{C}^1_{\mathrm{per}}[-\pi, \pi]$.

(f) *(Optional) For any* $k \in \mathbb{N}$, *if* $g \in \mathcal{C}^k[-\pi, \pi]$, *then*[3] $f * g \in \mathcal{C}^k_{\mathrm{per}}[-\pi, \pi]$. *Furthermore,* $(f * g)' = f * g', (f * g)'' = f * g'', \ldots,$ *and* $(f * g)^{(k)} = f * g^{(k)}$.

Proof (a)

$$f * g(\pi) = \frac{1}{2\pi} \int_{-\pi}^{\pi} f(y) g^{\circ}(\pi - y) dy \underset{(*)}{=} \frac{1}{2\pi} \int_{-\pi}^{\pi} f(y) g^{\circ}(\pi - y - 2\pi) dy$$

$$= \frac{1}{2\pi} \int_{-\pi}^{\pi} f(y) g^{\circ}(-\pi - y) dy = f * g(-\pi).$$

Here, (∗) is because g° is 2π-periodic.

(b) Fix $x \in [-\pi, \pi]$ and let $\epsilon > 0$. We must find some $\delta > 0$ such that, for any $x_1 \in [-\pi, \pi]$, if $|x - x_1| < \delta$ then $|f * g(x) - f * g(x_1)| < \epsilon$. But

$$f * g(x) - f * g(x_1) = \frac{1}{2\pi} \int_{-\pi}^{\pi} f(y) g^{\circ}(x - y) dy - \frac{1}{2\pi} \int_{-\pi}^{\pi} f(y) g^{\circ}(x_1 - y) dy$$

$$= \frac{1}{2\pi} \int_{-\pi}^{\pi} f(y) g^{\circ}(x - y) - f(y) g^{\circ}(x_1 - y) dy$$

$$= \frac{1}{2\pi} \int_{-\pi}^{\pi} f(y) \left(g^{\circ}(x - y) - g^{\circ}(x_1 - y) \right) dy. \qquad (10D.9)$$

---

[3] See p. 209 for the definition of $\mathcal{C}^k_{\mathrm{per}}[-\pi, \pi]$.

Since g is continuous on $[-\pi, \pi]$ and $g(-\pi) = g(\pi)$, it follows that $\overset{\circ}{g}$ is continuous on $[-2\pi, 2\pi]$; since $[-2\pi, 2\pi]$ is a closed and bounded set, it then follows that $\overset{\circ}{g}$ is *uniformly* continuous on $[-2\pi, 2\pi]$. That is, there is some $\delta > 0$ such that, for any $z, z_1 \in [-2\pi, 2\pi]$,

$$\text{if } |z - z_1| < \delta, \text{ then } \quad |\overset{\circ}{g}(z) - \overset{\circ}{g}(z_1)| < \frac{2\pi \epsilon}{\|f\|_1}. \qquad (10\text{D}.10)$$

Now, suppose $|x - x_1| < \delta$. Then

$$|f * g(x) - f * g(x_1)| \underset{(*)}{=} \left| \frac{1}{2\pi} \int_{-\pi}^{\pi} f(y) \left(\overset{\circ}{g}(x - y) - \overset{\circ}{g}(x_1 - y) \right) dy \right|$$

$$\underset{(\triangle)}{\leq} \frac{1}{2\pi} \int_{-\pi}^{\pi} |f(y)| \cdot \left| \overset{\circ}{g}(x - y) - \overset{\circ}{g}(x_1 - y) \right| dy$$

$$\underset{(\dagger)}{\leq} \frac{1}{2\pi} \int_{-\pi}^{\pi} |f(y)| \cdot \frac{2\pi \epsilon}{\|f\|_1} dy = \frac{\epsilon}{\|f\|_1} \cdot \int_{-\pi}^{\pi} |f(y)| \, dy$$

$$= \frac{\epsilon}{\|f\|_1} \cdot \|f\|_1 = \epsilon.$$

Here, $(*)$ is by equation (10D.9), $(\triangle)$ is the triangle inequality for integrals, and $(\dagger)$ is by equation (10D.10), because $|(x - y) - (x_1 - y)| < \delta$ for all $y \in [-\pi, \pi]$, because $|x - x_1| < \delta$.

Thus, if $|x - x_1| < \delta$ then $|f * g(x) - f * g(x_1)| < \epsilon$. This argument works for any $\epsilon > 0$ and $x \in [-\pi, \pi]$. Thus, $f * g$ is continuous, as desired.

(c) (Optional) Fix $x \in [-\pi, \pi]$ and let $\epsilon > 0$. We must find some $\delta > 0$ such that, for any $t \in [-\pi, \pi]$, if $|t| < \delta$ then $|f * g(x) - f * g(x - t)| < \epsilon$. But

$$f * g(x - t) = \frac{1}{2\pi} \int_{-\pi}^{\pi} f(y) \, \overset{\circ}{g}(x - t - y) dy$$

$$= \frac{1}{2\pi} \int_{-\pi}^{\pi} f(y) \, \widehat{(g^t)} \, \overset{\circ}{} (x - y) dy = f * \widehat{g^t}(x).$$

Thus,

$$f * g(x) - f * g(x - t) = f * g(x) - f * \widehat{g^t}(x) \underset{(*)}{=} f * (g - \widehat{g^t})(x),$$

so

$$\left| f * g(x) - f * g(x - t) \right| = \left| f * (g - \widehat{g^t})(x) \right|$$

$$\underset{(\dagger)}{\leq} \|f\|_2 \cdot \left\| g - \widehat{g^t} \right\|_2. \qquad (10\text{D}.11)$$

Here, $(*)$ is by Lemma 10D.6(b) and $(\dagger)$ is by Lemma 10D.6(c). However, Lemma 10D.4, p. 214, says that $g = \mathbf{L}^2\text{-}\lim_{t \to 0} \widehat{g^t}$. Thus, there exists some $\delta > 0$

such that, if $|t| < \delta$, then $\| g - \widehat{g^t} \|_2 < \epsilon / \| f \|_2$. Thus, if $|t| < \delta$, then

$$\left| f * g(x) - f * g(x - t) \right| \underset{(*)}{\leq} \| f \|_2 \cdot \left\| g - \widehat{g^t} \right\|_2 \leq \| f \|_2 \cdot \frac{\epsilon}{\| f \|_2} = \epsilon,$$

where $(*)$ is by equation (10D.11). This argument works for any $\epsilon > 0$ and $x \in [-\pi, \pi]$. Thus, $f * g$ is continuous, as desired.

(d) We have

$$2\pi \, (f * g)'(x) = 2\pi \, \partial_x \, (f * g)(x) = \partial_x \int_{-\pi}^{\pi} f(y) \cdot g(x - y) \mathrm{d}y$$

$$\underset{(*)}{=} \int_{-\pi}^{\pi} f(y) \cdot \partial_x \, g(x - y) \mathrm{d}y = \int_{-\pi}^{\pi} f(y) \cdot g'(x - y) \mathrm{d}y$$

$$= 2\pi \, f * (g')(x).$$

Here, $(*)$ is by Proposition G.1, p. 571.

(e) Follows immediately from (a), (b) and (d).

(f) is **Exercise 10D.6.** (*Hint:* Use proof by induction, along with parts (b) and (d).) ⓔ □

Remarks

(a) Proposition 17G.2, p. 415, provides an analog to Lemma 10D.7 for convolutions on $\mathbb{R}^D$.

(b) (For algebraists) Let $\mathcal{C}_{\mathrm{per}}[-\pi, \pi]$ be the set of all continuous functions $f : [-\pi, \pi] \longrightarrow \mathbb{R}$ such that $f(-\pi) = f(\pi)$. Then Lemmas 10D.6(a) and (b) and 10D.7(a) and (b) imply that $\mathcal{C}_{\mathrm{per}}[-\pi, \pi]$ is a *commutative ring*, where functions are added pointwise, and where the convolution operator '$*$' plays the role of 'multiplication'. Furthermore, Lemma 10D.7(f) says that, for all $k \in \mathbb{N}$, the set $\mathcal{C}_{\mathrm{per}}^k[-\pi, \pi]$ is an *ideal* of the ring $\mathcal{C}_{\mathrm{per}}[-\pi, \pi]$. Note that this ring does *not* have a multiplicative identity element. However, it does have 'approximations' of identity, as we shall now see.

For all $n \in \mathbb{N}$, let $\gamma_n : [-\pi, \pi] \longrightarrow \mathbb{R}$ be a non-negative function. The sequence $\{\gamma_n\}_{n=1}^{\infty}$ is called an *approximation of identity* if it has the following properties:

(AI1) $\frac{1}{2\pi} \int_{-\pi}^{\pi} \gamma_n(y) \mathrm{d}y = 1$ for all $n \in \mathbb{N}$;

(AI2) for any $\epsilon > 0$,

$$\lim_{n \to \infty} \frac{1}{2\pi} \int_{-\epsilon}^{\epsilon} \gamma_n(x) \mathrm{d}x = 1.$$

(See Figure 10D.2.)

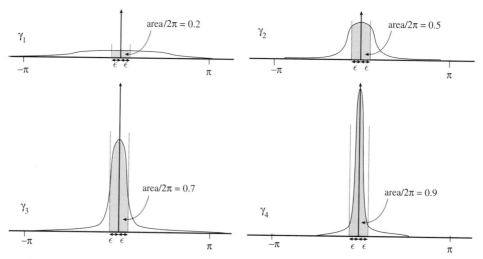

Figure 10D.2. Approximation of identity on $[-\pi, \pi]$. Here, $\epsilon > 0$ is fixed, and $\lim_{n\to\infty} \frac{1}{2\pi} \int_{-\epsilon}^{\epsilon} \gamma_n(x)dx = 1$.

Example 10D.8 Let $\Gamma : [-\pi, \pi] \longrightarrow \mathbb{R}$ be any non-negative function with $\frac{1}{2\pi} \int_{-\pi}^{\pi} \Gamma(x)dx = 1$. For all $n \in \mathbb{N}$, define $\gamma_n : [-\pi, \pi] \longrightarrow \mathbb{R}$ by

$$
\gamma_n(x) := \begin{cases} 0 & \text{if } x < -\pi/n; \\ n\Gamma(nx) & \text{if } -\pi/n \le x \le \pi/n; \\ 0 & \text{if } \pi/n < x \end{cases}
$$

(see Figure 10D.3). Then $\{\gamma_n\}_{n=1}^{\infty}$ is an approximation of identity
(E) (**Exercise 10D.7**).   ◇

The term 'approximation of identity' is due to the following result.

Proposition 10D.9 *Let* $\{\gamma_n\}_{n=1}^{\infty}$ *be an approximation of identity. Let* $f :$ $[-\pi, \pi] \longrightarrow \mathbb{R}$ *be some integrable function.*

(a) *If* $f \in \mathbf{L}^2[-\pi, \pi]$, *then* $f = \mathbf{L}^2\text{-}\lim_{n\to\infty} \gamma_n * f$.
(b) (*Optional*) *If* $x \in (-\pi, \pi)$ *and* f *is continuous at* x, *then* $f(x) = \mathbf{L}^2\text{-}\lim_{n\to\infty} \gamma_n * f(x)$.

Proof (a) Fix $\epsilon > 0$. We must find $N \in \mathbb{N}$ such that, for all $n > N$, $\|f - \gamma_n *f\|_2 < \epsilon$. First, find some $\eta > 0$ which is small enough that

$$
\left(2\|f\|_2 + 1\right) \cdot \eta < \epsilon. \tag{10D.12}
$$

Now, Lemma 10D.4, p. 214, says that there is some $\delta > 0$ such that,

$$
\text{for any } t \in (-\delta, \delta), \quad \left\|f - f^t\right\|_2 < \eta. \tag{10D.13}
$$

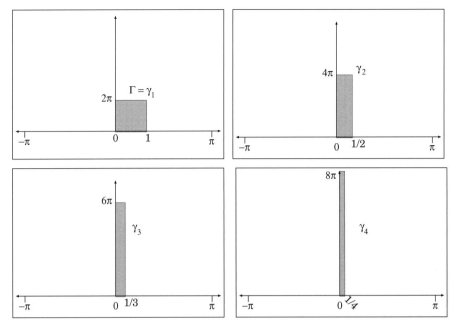

Figure 10D.3. The functions described in Example 10D.8.

Next, property (AI2) says there is some $N \in \mathbb{N}$ such that,

$$\text{for all } n > N, \qquad 1 - \eta < \frac{1}{2\pi} \int_{-\delta}^{\delta} \gamma_n(y) \mathrm{d}y \leq 1. \tag{10D.14}$$

Now, for any $x \in [-\pi, \pi]$ and $n \in \mathbb{N}$, observe that

$$f(x) = f(x) \cdot 1 \underset{(*)}{=} f(x) \frac{1}{2\pi} \int_{-\pi}^{\pi} \gamma_n(y) \mathrm{d}y = \frac{1}{2\pi} \int_{-\pi}^{\pi} f(x) \gamma_n(y) \mathrm{d}y, \tag{10D.15}$$

where $(*)$ is by property (AI1). Thus, for all $x \in [-\pi, \pi]$ and $n \in \mathbb{N}$,

$$f(x) - \gamma_n * f(x) \underset{(*)}{=} \frac{1}{2\pi} \int_{-\pi}^{\pi} f(x) \gamma_n(t) \mathrm{d}t - \frac{1}{2\pi} \int_{-\pi}^{\pi} f^{\circ}(x - t) \gamma_n(t) \mathrm{d}t$$

$$= \frac{1}{2\pi} \int_{-\pi}^{\pi} \left(f(x) - f^{\circ}(x - t) \right) \gamma_n(t) \mathrm{d}t$$

$$= \frac{1}{2\pi} \int_{-\pi}^{\pi} \left(f(x) - f^{\widehat{t}}(x) \right) \gamma_n(t) \mathrm{d}t$$

$$= \frac{1}{2\pi} \int_{-\pi}^{\pi} \gamma_n(t) \cdot F_t(x) \mathrm{d}t,$$

where $(*)$ is by equation (10D.15) and where, for all $t \in [-\pi, \pi]$, we define the function $F_t : [-\pi, \pi] \longrightarrow \mathbb{R}$ by $F_t(x) := f(x) - f^{\widehat{t}}(x)$ for all $x \in [-\pi, \pi]$. Thus

$$\|f - \gamma_n * f\|_2 = \left\| \frac{1}{2\pi} \int_{-\pi}^{\pi} \gamma_n(t) \cdot F_t dt \right\|_2 \underset{(M)}{\leq} \frac{1}{2\pi} \int_{-\pi}^{\pi} |\gamma_n(t)| \cdot \|F_t\|_2 \, dt$$

$$= \frac{1}{2\pi} \int_{-\pi}^{-\delta} \gamma_n(t) \cdot \|F_t\|_2 \, dt + \frac{1}{2\pi} \int_{-\delta}^{\delta} \gamma_n(t) \cdot \|F_t\|_2 \, dt + \frac{1}{2\pi} \int_{\delta}^{\pi} \gamma_n(t) \cdot \|F_t\|_2 \, dt$$

$$\underset{(*)}{\leq} \frac{\|f\|_2}{\pi} \int_{-\pi}^{-\delta} \gamma_n(t) dt + \frac{1}{2\pi} \int_{-\delta}^{\delta} \gamma_n(t) \cdot \|F_t\|_2 \, dt + \frac{\|f\|_2}{\pi} \int_{\delta}^{\pi} \gamma_n(t) dt$$

$$\underset{(\dagger)}{\leq} \frac{\|f\|_2}{\pi} \left(\int_{-\pi}^{-\delta} \gamma_n(t) dt + \int_{\delta}^{\pi} \gamma_n(t) dt \right) + \frac{\eta}{2\pi} \int_{-\delta}^{\delta} \gamma_n(t) dt$$

$$\underset{(\diamond)}{<} 2 \|f\|_2 \cdot \eta + \eta \cdot 1 = (2 \|f\|_2 + 1) \cdot \eta \underset{(\ddagger)}{\leq} \epsilon.$$

Here, (M) is Minkowski's inequality for integrals (Theorem 10D.5, p. 215). Next, $(*)$ is because

$$\|F_t\|_2 = \left\| f - f^{\widehat{t}} \right\|_2 \underset{(\triangle)}{\leq} \|f\|_2 + \left\| f^{\widehat{t}} \right\|_2 = \|f\|_2 + \|f\|_2 = 2 \|f\|_2.$$

Next, $(\dagger)$ is because $\|F_t\|_2 < \eta$ for all $t \in (-\delta, \delta)$, by equation (10D.13). Inequality $(\diamond)$ is because equation (10D.14) says $1 - \eta < \frac{1}{2\pi} \int_{-\delta}^{\delta} \gamma_n(t) dt \leq 1$; thus, we must have $\frac{1}{2\pi} \int_{-\pi}^{-\delta} \gamma_n(t) dt + \frac{1}{2\pi} \int_{\delta}^{\pi} \gamma_n(t) dt < \eta$. Finally, $(\ddagger)$ is by equation (10D.12).

This argument works for any $\epsilon > 0$. We conclude that $f = \mathbf{L}^2\text{–}\lim_{n\to\infty} \gamma_n * f$.

Ⓔ (b) **Exercise 10D.8.**   □

For any $k \in \mathbb{N}$, a C^k-*mollifier* is an approximation of identity $\{\gamma_n\}_{n=1}^{\infty}$ such that $\gamma_n \in C^k_{\text{per}}[-\pi, \pi]$ for all $n \in \mathbb{N}$. Lemma 10D.7(f) says that you can 'mollify' some initially pathological function f into a nice smooth approximation by convolving it with γ_n. Our final task in this section is to show how to construct such a C^k-mollifier.

Lemma 10D.10   *Let* $\Gamma \in C^k[-\pi, \pi]$ *be such that* $\frac{1}{2\pi} \int_{-\pi}^{\pi} \Gamma(x) dx = 1$, *and such that*

$$\Gamma(-\pi) = \Gamma'(-\pi) = \Gamma''(-\pi) = \cdots = \Gamma^{(k)}(-\pi) = 0;$$

$$\Gamma(\pi) = \Gamma'(\pi) = \Gamma''(\pi) = \cdots = \Gamma^{(k)}(\pi) = 0.$$

Define $\{\gamma_n\}_{n=1}^{\infty}$ *as in Example 10D.8. Then* $\{\gamma_n\}_{n=1}^{\infty}$ *is a* C^k-*mollifier.*

Ⓔ *Proof* **Exercise 10D.9.**   □

Example 10D.11 Let $g(x) = (x + \pi)^{k+1}(x - \pi)^{k+1}$, let $G = \frac{1}{2\pi} \int_{-\pi}^{\pi} g(x)dx$, and then let $\Gamma(x) := g(x)/G$. Then Γ satisfies the hypotheses of Lemma 10D.10 (**Exercise 10D.10**). ◇ Ⓔ

Remark For more information about convolutions and mollifiers, see Wheeden and Zygmund (1977), chap. 9, and Folland (1984), §8.2.

10D(iii) Proofs of Theorems 8A.1(a) and 10D.1

Prerequisites: §8A, §10A, §10D(ii).

Proof of Theorem 10D.1 Let $\{\gamma_n\}_{n=1}^{\infty} \subset \mathcal{C}_{per}^k[-\pi, \pi]$ be the $\mathcal{C}^k$-mollifier from Lemma 10D.10. Then Proposition 10D.9(a) says that $f = \mathbf{L}^2 \text{-}\lim_{n \to \infty} \gamma_n * f$. Thus, for any $\epsilon > 0$, we can find some $n \in \mathbb{N}$ such that $\| f - \gamma_n * f \|_2 < \epsilon$. Furthermore, $\gamma_n \in \mathcal{C}^k[-\pi, \pi]$, so Lemma 10D.7(f) says that $\gamma_n * f \in \mathcal{C}_{per}^k[-\pi, \pi]$, for all $n \in \mathbb{N}$. □

The proof of Theorem 8A.1(a) now follows a standard strategy in analysis: approximate the function f with a 'nice' function $\tilde{f}$; establish convergence for the Fourier series of $\tilde{f}$ first; and then use the triangle inequality to 'leverage' this into convergence for the Fourier series of f.

Proof of Theorem 8A.1(a) Let $f \in \mathbf{L}^2[-\pi, \pi]$. Fix $\epsilon > 0$. Theorem 10D.1 says there exists some $\tilde{f} \in \mathcal{C}_{per}^1[-\pi, \pi]$ such that

$$\| f - \tilde{f} \|_2 < \frac{\epsilon}{3}. \tag{10D.16}$$

Let $\{A_n\}_{n=0}^{\infty}$ and $\{B_n\}_{n=1}^{\infty}$ be the real Fourier coefficients for f, and let $\{\tilde{A}_n\}_{n=0}^{\infty}$ and $\{\tilde{B}_n\}_{n=1}^{\infty}$ be the real Fourier coefficients for $\tilde{f}$. Let $\overline{f} := f - \tilde{f}$, and let $\{\overline{A}_n\}_{n=0}^{\infty}$ and $\{\overline{B}_n\}_{n=1}^{\infty}$ be the real Fourier coefficients for $\overline{f}$. Then, for all $n \in \mathbb{N}$, we have

$$A_n = \overline{A}_n + \tilde{A}_n \quad \text{and} \quad B_n = \overline{B}_n + \tilde{B}_n. \tag{10D.17}$$

Also, for any $N \in \mathbb{N}$, we have

$$\left\| \overline{A}_0 + \sum_{n=0}^{N} \overline{A}_n \mathbf{C}_n + \sum_{n=1}^{N} \overline{B}_n \mathbf{S}_n \right\|_2^2 \underset{(\triangle)}{\leq} \overline{A}_0^2 + \sum_{n=0}^{\infty} |\overline{A}_n|^2 \cdot \| \mathbf{C}_n \|_2^2 + \sum_{n=1}^{\infty} |\overline{B}_n|^2 \cdot \| \mathbf{C}_n \|_2^2$$

$$\underset{(\dagger)}{=} \overline{A}_0^2 + \sum_{n=0}^{\infty} \frac{|\overline{A}_n|^2}{2} + \sum_{n=1}^{\infty} \frac{|\overline{B}_n|^2}{2}$$

$$\underset{(B)}{\leq} \| \overline{f} \|_2^2 = \| f - \tilde{f} \|_2^2 \underset{(*)}{<} \left(\frac{\epsilon}{3} \right)^2. \tag{10D.18}$$

Here: ($\triangle$) is by the triangle inequality,[4] and ($\dagger$) is because $\|\mathbf{C}_n\|_2^2 = \frac{1}{2} = \|\mathbf{S}_n\|_2^2$ for all $n \in \mathbb{N}$ (by Proposition 6D.2, p. 117); (B) is Bessel's Inequality (Example 10A.2, p. 198); and ($*$) is by equation (10D.16).

Now, $\tilde{f} \in \mathcal{C}^1_{\mathrm{per}}[-\pi, \pi]$, so Theorem 8A.1(d) (which we proved in Section 10C) says that

$$\mathrm{unif}\!-\!\lim_{N\to\infty} \left(\tilde{A}_0 + \sum_{n=0}^{N} \tilde{A}_n \mathbf{C}_n + \sum_{n=1}^{N} \tilde{B}_n \mathbf{S}_n \right) = \tilde{f}.$$

Thus Corollary 6E.11(b)(i), p. 133, implies that

$$\mathbf{L}^2\!-\!\lim_{N\to\infty} \left(\tilde{A}_0 + \sum_{n=0}^{N} \tilde{A}_n \mathbf{C}_n + \sum_{n=1}^{N} \tilde{B}_n \mathbf{S}_n \right) = \tilde{f}.$$

Thus, there exists some $N \in \mathbb{N}$ such that

$$\left\| \tilde{A}_0 + \sum_{n=0}^{N} \tilde{A}_n \mathbf{C}_n + \sum_{n=1}^{N} \tilde{B}_n \mathbf{S}_n - \tilde{f} \right\|_2 < \frac{\epsilon}{3}. \tag{10D.19}$$

Thus,

$$\left\| A_0 + \sum_{n=0}^{N} A_n \mathbf{C}_n + \sum_{n=1}^{N} B_n \mathbf{S}_n - f \right\|_2$$

$$\underset{(\dagger)}{=} \left\| (\overline{A}_0 + \tilde{A}_0) + \sum_{n=0}^{N} (\overline{A}_n + \tilde{A}_n) \mathbf{C}_n + \sum_{n=1}^{N} (\overline{B}_n + \tilde{B}_n) \mathbf{S}_n - \overline{f} + \tilde{f} - f \right\|_2$$

$$= \left\| \overline{A}_0 + \sum_{n=0}^{N} \overline{A}_n \mathbf{C}_n + \sum_{n=1}^{N} \overline{B}_n \mathbf{S}_n + \tilde{A}_0 + \sum_{n=0}^{N} \tilde{A}_n \mathbf{C}_n + \sum_{n=1}^{N} \tilde{B}_n \mathbf{S}_n - \overline{f} + \tilde{f} - f \right\|_2$$

$$\underset{(\triangle)}{\leq} \left\| \overline{A}_0 + \sum_{n=0}^{N} \overline{A}_n \mathbf{C}_n + \sum_{n=1}^{N} \overline{B}_n \mathbf{S}_n \right\|_2 + \left\| \tilde{A}_0 + \sum_{n=0}^{N} \tilde{A}_n \mathbf{C}_n + \sum_{n=1}^{N} \tilde{B}_n \mathbf{S}_n - \overline{f} \right\|_2 + \left\| \tilde{f} - f \right\|_2$$

$$\underset{(*)}{\leq} \frac{\epsilon}{3} + \frac{\epsilon}{3} + \frac{\epsilon}{3} = \epsilon.$$

Here, ($\dagger$) is by equation (10D.17), ($\triangle$) is the triangle inequality, and ($*$) is by inequalities (10D.16), (10D.18), and (10D.19).

This argument works for any $\epsilon > 0$. We conclude that

$$A_0 + \sum_{n=0}^{\infty} A_n \mathbf{C}_n + \sum_{n=1}^{\infty} B_n \mathbf{S}_n \underset{12}{\approx} f. \qquad \square$$

---

[4] Actually, this is an equality, because of the L^2 Pythagorean formula (equation (6F.1), p. 136).

Recall that a function $f : [-\pi, \pi] \longrightarrow \mathbb{R}$ is *analytic* if f is infinitely differen-tiable, and the Taylor expansion of f around any $x \in [-\pi, \pi]$ has a nonzero radius of convergence.[5] Let $\mathcal{C}^\omega_{\text{per}}[-\pi, \pi]$ be the set of all analytic functions f on $[-\pi, \pi]$ such that $f(-\pi) = f(\pi)$, and $f^{(k)}(-\pi) = f^{(k)}(\pi)$ for all $k \in \mathbb{N}$. For example, the functions sin and cos are in $\mathcal{C}^\omega_{\text{per}}[-\pi, \pi]$. Elements of $\mathcal{C}^\omega_{\text{per}}[-\pi, \pi]$ are some of the 'nicest' possible functions on $[-\pi, \pi]$. On the other hand, arbitrary elements of $\mathbf{L}^2[-\pi, \pi]$ can by quite 'nasty' (i.e. nondifferentiable, discontinuous). Thus, the following result is quite striking.

Corollary 10D.12 $\mathcal{C}^\omega_{\text{per}}[-\pi, \pi]$ *is dense in* $\mathbf{L}^2[-\pi, \pi]$.

Proof Theorem 8A.1(a) says that any function in $\mathbf{L}^2[-\pi, \pi]$ can be approximated arbitrarily closely by a 'trigonometric polynomial' of the form $A_0 + \sum_{n=1}^{N} A_n \mathbf{C}_n + \sum_{n=1}^{N} B_n \mathbf{S}_n$. But all trigonometric polynomials are in $\mathcal{C}^\omega_{\text{per}}[-\pi, \pi]$ (because they are finite linear combinations of the functions $\mathbf{S}_n(x) := \sin(nx)$ and $\mathbf{C}_n(x) := \cos(nx)$, which are all in $\mathcal{C}^\omega_{\text{per}}[-\pi, \pi]$). Thus, any function in $\mathbf{L}^2[-\pi, \pi]$ can be approximated arbitrarily closely by an element of $\mathcal{C}^\omega_{\text{per}}[-\pi, \pi]$ – in other words, $\mathcal{C}^\omega_{\text{per}}[-\pi, \pi]$ is dense in $\mathbf{L}^2[-\pi, \pi]$. □

Remarks

(a) Proposition 17G.3, p. 415, provides a 'pointwise' version of Theorem 10D.1 for con-volutional smoothing on $\mathbb{R}^D$.

(b) For another proof of the L^2-convergence of real Fourier series, see Broman (1989), Theorems 1.5.4 and 2.3.10. For a proof of the L^2-convergence of complex Fourier series (which is very similar), see Katznelson (1976), §I.5.5.

---

[5] See Appendix H(i), p. 573.

Part IV

BVP solutions via eigenfunction expansions

A powerful and general method for solving linear PDEs is to represent the solutions using *eigenfunction expansions*. Rather than first deploying this idea in full abstract generality, we will start by illustrating it in a variety of special cases. We will gradually escalate the level of abstraction, so that the general theory is almost obvious when it is finally stated explicitly.

The orthogonal trigonometric functions $\mathbf{S_n}$ and $\mathbf{C_n}$ in a Fourier series are *eigenfunctions* of the Laplacian operator $\triangle$. Furthermore, the eigenfunctions $\mathbf{S_n}$ and $\mathbf{C_n}$ are particularly 'well adapted' to domains like the interval $[0, \pi]$, the square $[0, \pi]^2$, or the cube $[0, \pi]^3$, for two reasons:

- the functions $\mathbf{S_n}$ and $\mathbf{C_n}$ and the domain $[0, \pi]^k$ are easily expressed in a Cartesian coordinate system;
- the functions $\mathbf{S_n}$ and $\mathbf{C_n}$ satisfy desirable boundary conditions (e.g. homogeneous Dirichlet/Neumann) on the boundaries of the domain $[0, \pi]^k$.

Thus, we can use $\mathbf{S}_n$ and $\mathbf{C}_n$ as 'building blocks' to construct a solution to a given partial differential equation – a solution which also satisfies specified initial conditions and/or boundary conditions on $[0, \pi]^k$. In particular, we will use Fourier sine series to obtain homogeneous *Dirichlet* boundary conditions (by Theorems 7A.1(d), 9A.3(d), and 9B.1(d)), and Fourier cosine series to obtain homogeneous *Neumann* boundary conditions (by Theorems 7A.4(d), 9A.3(e), and 9B.1(e)). This basic strategy underlies all the solution methods developed in Chapters 11–13.

When we consider other domains (e.g. disks, annuli, balls, etc.), the functions $\mathbf{C}_n$ and $\mathbf{S}_n$ are no longer so 'well adapted'. In Chapter 14, we discover that, in polar coordinates, the 'well adapted' eigenfunctions are combinations of trigonometric functions ($\mathbf{C}_n$ and $\mathbf{S}_n$) with another class of transcendental functions called *Bessel functions*. This yields another orthogonal system of eigenfunctions. We can then represent most functions on the disks and annuli using *Fourier–Bessel expansions* (analogous to Fourier series), and we can then mimic the solution methods of Chapters 11–13.

11

Boundary value problems on a line segment

> Mathematics is the music of reason.
>
> *James Joseph Sylvester*

Prerequisites: §7A, §5C.

This chapter concerns boundary value problems on the line segment $[0, L]$, and provides solutions in the form of infinite series involving the functions $\mathbf{S}_n(x) := \sin\left(\frac{n\pi}{L}x\right)$ and $\mathbf{C}_n(x) := \cos\left(\frac{n\pi}{L}x\right)$. For simplicity, we will assume throughout the chapter that $L = \pi$. Thus $\mathbf{S}_n(x) = \sin(nx)$ and $\mathbf{C}_n(x) = \cos(nx)$. We will also assume (through an appropriate choice of time units) that the physical constants in the various equations are all equal to one. Thus, the heat equation becomes '$\partial_t u = \Delta u$', the wave equation is '$\partial_t^2 u = \Delta u$', etc.

This does not limit the generality of our results. For example, faced with a general heat equation of the form '$\partial_t u(x, t) = \kappa \cdot \Delta u$' for $x \in [0, L]$ (with $\kappa \neq 1$ and $L \neq \pi$), we can simply replace the coordinate x with a new space coordinate $y = \frac{\pi}{L}x$, and replace t with a new time coordinate $s = \kappa t$, to reformulate the problem in a way compatible with the following methods.

11A The heat equation on a line segment

Prerequisites: §7B, §5B, §5C, §1B(i), Appendix F.   **Recommended:** §7C(v).

Proposition 11A.1 Heat equation; homogeneous Dirichlet boundary

Let $\mathbb{X} = [0, \pi]$, and let $f \in \mathbf{L}^2[0, \pi]$ be some function describing an initial heat distribution. Suppose f has Fourier sine series $f(x) \underset{L2}{\approx} \sum_{n=1}^{\infty} B_n \sin(nx)$, and define the function $u : \mathbb{X} \times \mathbb{R}_{\neq} \longrightarrow \mathbb{R}$ by

$$u(x; t) \underset{L2}{\approx} \sum_{n=1}^{\infty} B_n \sin(nx) \cdot \exp\left(-n^2 \cdot t\right), \quad \text{for all } x \in [0, \pi] \text{ and } t \geq 0.$$

229

Then u is the unique solution to the one-dimensional heat equation '$\partial_t u = \partial_x^2 u$',
with homogeneous Dirichlet boundary conditions

$$u(0;t) = u(\pi;t) = 0, \quad \text{for all } t > 0,$$

and initial conditions $u(x;0) = f(x)$, for all $x \in [0, \pi]$.
Furthermore, the series defining u converges semiuniformly on $\mathbb{X} \times \mathbb{R}_+$.

Ⓔ   *Proof* **Exercise 11A.1.** *Hint:*

(a) Show that, when $t = 0$, the Fourier series of $u(x;0)$ agrees with that of $f(x)$; hence
     $u(x;0) = f(x)$.
(b) Show that, for all $t > 0$, $\sum_{n=1}^{\infty} |n^2 \cdot B_n \cdot e^{-n^2 t}| < \infty$.
(c) For any $T > 0$, apply Proposition F.1, p. 569, to conclude that

$$\partial_t u(x;t) \underset{\text{unif}}{\equiv} \sum_{n=1}^{\infty} -n^2 B_n \sin(nx) \cdot \exp\left(-n^2 \cdot t\right) \underset{\text{unif}}{\equiv} \triangle u(x;t) \quad \text{on } [T, \infty).$$

(d) Observe that, for any fixed $t > 0$, $\sum_{n=1}^{\infty} |B_n \cdot e^{-n^2 t}| < \infty$.
(e) Apply part (c) of Theorem 7A.1, p. 142, to show that the Fourier series of $u(x;t)$
     converges uniformly for all $t > 0$.
(f) Apply part (d) of Theorem 7A.1, p. 142, to conclude that $u(0;t) = 0 = u(\pi, t)$ for all
     $t > 0$.
(g) Apply Theorem 5D.8, p. 93, to show that this solution is unique.   □

Example 11A.2 Consider a metal rod of length π, with initial temperature distribution $f(x) = \tau \cdot \sinh(\alpha x)$ (where $\tau, \alpha > 0$ are constants), and homogeneous
Dirichlet boundary condition. Proposition 11A.1 tells us to find the Fourier sine
series for $f(x)$. In Example 7A.3, p. 144, we computed this to be

$$\frac{2\tau \sinh(\alpha\pi)}{\pi} \sum_{n=1}^{\infty} \frac{n(-1)^{n+1}}{\alpha^2 + n^2} \cdot \sin(nx).$$

The evolving temperature distribution is therefore given by

$$u(x;t) = \frac{2\tau \sinh(\alpha\pi)}{\pi} \sum_{n=1}^{\infty} \frac{n(-1)^{n+1}}{\alpha^2 + n^2} \cdot \sin(nx) \cdot e^{-n^2 t}.$$ ◇

Proposition 11A.3 Heat equation; homogeneous Neumann boundary
*Let $\mathbb{X} = [0, \pi]$, and let $f \in \mathbf{L}^2[0, \pi]$ be some function describing an initial heat
distribution. Suppose f has Fourier cosine series $f(x) \underset{12}{\approx} \sum_{n=0}^{\infty} A_n \cos(nx)$, and
define the function $u : \mathbb{X} \times \mathbb{R}_+ \longrightarrow \mathbb{R}$ by*

$$u(x;t) \underset{12}{\approx} \sum_{n=0}^{\infty} A_n \cos(nx) \cdot \exp\left(-n^2 \cdot t\right), \quad \text{for all } x \in [0, \pi] \text{ and } t \geq 0.$$

*Then u is the unique solution to the one-dimensional heat equation '$\partial_t u = \partial_x^2 u$',
with homogeneous* **Neumann** *boundary conditions*

$$\partial_x u(0;t) = \partial_x u(\pi;t) = 0, \quad \text{for all } t > 0,$$

*and initial conditions $u(x;0) = f(x)$, for all $x \in [0,\pi]$.
   Furthermore, the series defining u converges semiuniformly on $\mathbb{X} \times \mathbb{R}_+$.*

Proof Setting $t = 0$, we obtain

$$u(x;0) = \sum_{n=1}^{\infty} A_n \cos(nx) \cdot \exp\left(-n^2 \cdot 0\right) = \sum_{n=1}^{\infty} A_n \cos(nx) \cdot \exp(0)$$

$$= \sum_{n=1}^{\infty} A_n \cos(nx) \cdot 1 = \sum_{n=1}^{\infty} A_n \cos(nx) = f(x),$$

so we have the desired initial conditions.
   Let $M := \max_{n \in \mathbb{N}} |A_n|$. Then $M < \infty$ (because $f \in \mathbf{L}^2$).

Claim 1 *For all $t > 0$, $\sum_{n=0}^{\infty} |n^2 \cdot A_n \cdot e^{-n^2 t}| < \infty$.*

Proof Since $M = \max_{n \in \mathbb{N}} |A_n|$, we know that $|A_n| < M$ for all $n \in \mathbb{N}$. Thus,

$$\sum_{n=0}^{\infty} \left| n^2 \cdot A_n \cdot e^{-n^2 t} \right| \leq \sum_{n=0}^{\infty} |n^2| \cdot M \cdot \left| e^{-n^2 t} \right| = M \cdot \sum_{n=0}^{\infty} n^2 \cdot e^{-n^2 t}.$$

Hence, it suffices to show that $\sum_{n=0}^{\infty} n^2 \cdot e^{-n^2 t} < \infty$. To see this, let $E = e^t$.
Then $E > 1$ (because $t > 0$). Also, $n^2 \cdot e^{-n^2 t} = n^2/E^{n^2}$, for each $n \in \mathbb{N}$. Thus,

$$\sum_{n=1}^{\infty} n^2 \cdot e^{-n^2 t} = \sum_{n=1}^{\infty} \frac{n^2}{E^{n^2}} \leq \sum_{m=1}^{\infty} \frac{m}{E^m}. \tag{11A.1}$$

We must show that the right-hand series in equation (11A.1) converges. We
apply the ratio test:

$$\lim_{m \to \infty} \frac{(m+1)/E^{m+1}}{m/E^m} = \lim_{m \to \infty} \frac{m+1}{m} \frac{E^m}{E^{m+1}} = \lim_{m \to \infty} \frac{1}{E} < 1.$$

Hence the right-hand series in equation (11A.1) converges. $\diamond_{\text{Claim 1}}$

Claim 2 *For any $T > 0$, we have*

$$\partial_x u(x;t) \underset{\text{unif}}{\equiv} -\sum_{n=1}^{\infty} n A_n \sin(nx) \cdot \exp\left(-n^2 \cdot t\right)$$

on $\mathbb{X} \times [T, \infty)$, *and also*

$$\partial_x^2 u(x;t) \underset{\text{unif}}{\equiv} -\sum_{n=1}^{\infty} n^2 A_n \cos(nx) \cdot \exp\left(-n^2 \cdot t\right)$$

on $\mathbb{X} \times [T, \infty)$.

Proof   This follows from Claim 1 and two applications of Proposition F.1, p. 569.

$\diamond$ Claim 2

Claim 3 *For any* $T > 0$, *we have*

$$\partial_t u(x;t) \underset{\text{unif}}{\equiv} -\sum_{n=1}^{\infty} n^2 A_n \cos(nx) \cdot \exp\left(-n^2 \cdot t\right)$$

on $[T, \infty)$.

Proof $\partial_t u(x;t) = \partial_t \sum_{n=1}^{\infty} A_n \cos(nx) \cdot \exp\left(-n^2 \cdot t\right)$

$$\underset{(*)}{=} \sum_{n=1}^{\infty} A_n \cos(nx) \cdot \partial_t \exp\left(-n^2 \cdot t\right)$$

$$= \sum_{n=1}^{\infty} A_n \cos(nx) \cdot (-n^2) \exp\left(-n^2 \cdot t\right),$$

where $(*)$ is by Claim 1 and Proposition F.1, p. 569.

$\diamond$ Claim 3

Combining Claims 2 and 3, we conclude that $\partial_t u(x;t) = \Delta u(x;t)$. Claim 1 also implies that, for any $t > 0$,

$$\sum_{n=0}^{\infty} \left| n \cdot A_n \cdot e^{-n^2 t} \right| < \sum_{n=0}^{\infty} \left| n^2 \cdot A_n \cdot e^{-n^2 t} \right| < \infty.$$

Hence, Theorem 7A.4(d)(ii), p. 145, implies that $u(x;t)$ satisfies homogeneous Neumann boundary conditions for any $t > 0$. (This can also be seen directly via Claim 2.)

Finally, Theorem 5D.8, p. 93, implies that this solution is unique. $\square$

Example 11A.4 Consider a metal rod of length π, with initial temperature distribution $f(x) = \cosh(x)$ and homogeneous Neumann boundary condition. Proposition 11A.3 tells us to find the Fourier cosine series for $f(x)$. In Example 7A.6, p. 147, we computed this to be

$$\frac{\sinh(\pi)}{\pi} + \frac{2\sinh(\pi)}{\pi} \sum_{n=1}^{\infty} \frac{(-1)^n \cdot \cos(nx)}{n^2 + 1}.$$

The evolving temperature distribution is therefore given by

$$u(x;t) \underset{12}{\approx} \frac{\sinh(\pi)}{\pi} + \frac{2\sinh(\pi)}{\pi} \sum_{n=1}^{\infty} \frac{(-1)^n \cdot \cos(nx)}{n^2 + 1} \cdot e^{-n^2 t}. \qquad \diamond$$

Exercise 11A.2 Let $L > 0$ and let $\mathbb{X} := [0, L]$. Let $\kappa > 0$ be a diffusion constant,   Ⓔ
and consider the general one-dimensional heat equation:

$$\partial_t u = \kappa \, \partial_x^2 u. \qquad (11A.2)$$

(a) Generalize Proposition 11A.1 to find the solution to equation (11A.2) on $\mathbb{X}$ satisfying
    prescribed initial conditions and homogeneous Dirichlet boundary conditions.
(b) Generalize Proposition 11A.3 to find the solution to equation (11A.2) on $\mathbb{X}$ satisfying
    prescribed initial conditions and homogeneous Neumann boundary conditions.

In both cases, prove that your solution converges, satisfies the desired initial conditions and boundary conditions, and satisfies equation (11A.2). (*Hint:* Imitate the
strategy suggested in Exercise 11A.1.)   ◆

Exercise 11A.3 Let $\mathbb{X} = [0, \pi]$, and let $f \in \mathbf{L}^2(\mathbb{X})$ be a function whose Fourier   Ⓔ
sine series satisfies $\sum_{n=1}^{\infty} n^2 |B_n| < \infty$. Imitate Proposition 11A.1 to find a
'Fourier series' solution to the initial value problem for the one-dimensional *free
Schrödinger equation*

$$i\partial_t \omega = \frac{-1}{2} \partial_x^2 \omega \qquad (11A.3)$$

on $\mathbb{X}$, with initial conditions $\omega_0 = f$, and satisfying homogeneous Dirichlet boundary conditions. Prove that your solution converges, satisfies the desired initial conditions and boundary conditions, and satisfies equation (11A.3). (*Hint:* Imitate the
strategy suggested in Exercise 11A.1.)   ◆

11B The wave equation on a line (the vibrating string)

Prerequisites: §7B(i), §5B, §5C, §2B(i).   **Recommended:** §17D(ii).

Imagine a piano string stretched tightly between two points. At equilibrium, the
string is perfectly flat, but if we pluck or strike the string, it will vibrate, meaning
there will be a vertical displacement from equilibrium. Let $\mathbb{X} = [0, \pi]$ represent
the string, and, for any point $x \in \mathbb{X}$ on the string and time $t > 0$, let $u(x;t)$ be the
vertical displacement of the string. Then u will obey the one-dimensional wave
equation:

$$\partial_t^2 u(x;t) = \triangle u(x;t). \qquad (11B.1)$$

However, since the string is fixed at its endpoints, the function u will also exhibit homogeneous *Dirichlet* boundary conditions

$$u(0;t) = u(\pi;t) = 0, \quad \text{for all } t > 0. \tag{11B.2}$$

Proposition 11B.1 Initial position problem for vibrating string with fixed endpoints
Let $f_0 : \mathbb{X} \longrightarrow \mathbb{R}$ be a function describing the initial displacement of the string. Suppose f_0 has Fourier sine series $f_0(x) \underset{L2}{\approx} \sum_{n=1}^{\infty} B_n \sin(nx)$, and define the function $w : \mathbb{X} \times \mathbb{R}_{\not{}} \longrightarrow \mathbb{R}$ by

$$w(x;t) \underset{L2}{\approx} \sum_{n=1}^{\infty} B_n \sin(nx) \cdot \cos(nt), \quad \text{for all } x \in [0, \pi] \text{ and } t \geq 0. \tag{11B.3}$$

Then w is the unique solution to the wave equation (11B.1), satisfying the Dirichlet boundary conditions (11B.2), as well as

$$\left. \begin{array}{ll} \text{initial position:} & w(x, 0) = f_0(x) \\ \text{initial velocity:} & \partial_t w(x, 0) = 0 \end{array} \right\} \quad \text{for all } x \in [0, \pi].$$

Ⓔ *Proof* **Exercise 11B.1.** *Hint:*

(a) Prove the trigonometric identity

$$\sin(nx)\cos(nt) = \frac{1}{2} \Big(\sin\left(n(x - t) \right) + \sin\left(n(x + t) \right) \Big).$$

(b) Use this identity to show that the Fourier sine series (11B.3) converges in L^2 to the d'Alembert solution from Theorem 17D.8(a), p. 407.

(c) Apply Theorem 5D.11, p. 96, to show that this solution is unique. □

Example 11B.2 Let $f_0(x) = \sin(5x)$. Thus, $B_5 = 1$ and $B_n = 0$ for all $n \neq 5$. Proposition 11B.1 tells us that the corresponding solution to the wave equation is $w(x, t) = \cos(5t)\sin(5x)$. To see that w satisfies the wave equation, note that, for any $x \in [0, \pi]$ and $t > 0$,

$$\partial_t w(x, t) = -5\sin(5t)\sin(5x) \quad \text{and} \quad 5\cos(5t)\cos(5x) = \partial_x w(x, t);$$

Thus

$$\partial_t^2 w(x, t) = -25\cos(5t)\sin(5x) = -25\cos(5t)\sin(5x) = \partial_x^2 w(x, t).$$

Also w has the desired initial position because, for any $x \in [0, \pi]$, we have $w(x;0) = \cos(0)\sin(5x) = \sin(5x) = f_0(x)$, because $\cos(0) = 1$.

Next, w has the desired initial velocity because, for any $x \in [0, \pi]$, we have $\partial_t w(x;0) = 5\sin(0)\sin(5x) = 0$, because $\sin(0) = 0$.

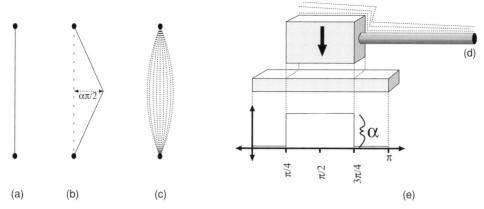

(a) (b) (c) (d) (e)

Figure 11B.1. (a) Harpstring at rest. (b) Harpstring being plucked. (c) Harpstring vibrating. (d) A big hammer striking a xylophone. (e) The initial velocity of the xylophone when struck.

Finally w satisfies homogeneous Dirichlet BC because, for any $t > 0$, we have $w(0, t) = \cos(5t)\sin(0) = 0$ and $w(\pi, t) = \cos(5t)\sin(5\pi) = 0$, because $\sin(0) = 0 = \sin(5\pi)$.   ◇

Example 11B.3 The plucked harp string
A harpist places her fingers at the midpoint of a harp string and plucks it. What is the formula describing the vibration of the string?

Solution For simplicity, we imagine the string has length π. The tight string forms a straight line when at rest (Figure 11B.1(a)); the harpist plucks the string by pulling it away from this resting position and then releasing it. At the moment she releases it, the string's *initial velocity* is zero, and its *initial position* is described by a *tent function*, like the one in Example 7C.7, p. 159:

$$f_0(x) = \begin{cases} \alpha x & \text{if } 0 \le x \le \frac{\pi}{2}; \\ \alpha(\pi - x) & \text{if } \frac{\pi}{2} < x \le \pi \end{cases}$$

(Figure 11B.1(b)), where $\alpha > 0$ is a constant describing the force with which she plucks the string (and its resulting amplitude).

The endpoints of the harp string are fixed, so it vibrates with *homogeneous Dirichlet* boundary conditions. Thus, Proposition 11B.1 tells us to find the Fourier sine series for f_0. In Example 7C.7, we computed this to be:

$$f_0 \underset{L2}{\approx} \frac{4 \cdot \alpha}{\pi} \sum_{\substack{n=1 \\ n \text{ odd;} \\ n=2k+1}}^{\infty} \frac{(-1)^k}{n^2} \sin(nx).$$

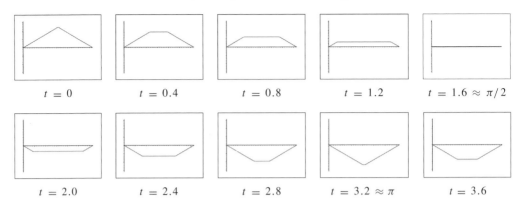

$t = 0$ $t = 0.4$ $t = 0.8$ $t = 1.2$ $t = 1.6 \approx \pi/2$

$t = 2.0$ $t = 2.4$ $t = 2.8$ $t = 3.2 \approx \pi$ $t = 3.6$

Figure 11B.2. The plucked harpstring of Example 11B.3. From $t = 0$ to $t = \pi/2$, the initially triangular shape is blunted; at $t = \pi/2$ it is totally flat. From $t = \pi/2$ to $t = \pi$, the process happens in reverse, only the triangle grows back upside down. At $t = \pi$, the original triangle reappears, upside down. Then the entire process happens in reverse, until the original triangle reappears at $t = 2\pi$.

Thus, the resulting solution is given by

$$u(x;t) \underset{L2}{\approx} \frac{4 \cdot \alpha}{\pi} \sum_{\substack{n=1 \\ n \text{odd}; \\ n=2k+1}}^{\infty} \frac{(-1)^k}{n^2} \sin(nx)\cos(nt);$$

see Figure 11B.2. This is not a very accurate model because we have not accounted for energy loss due to friction. In a real harpstring, these 'perfectly triangular' waveforms rapidly decay into the gently curving waves depicted in Figure 11B.1(c); these slowly settle down to a stationary state. $\diamond$

Proposition 11B.4 Initial velocity problem for vibrating string with fixed endpoints
Let $f_1 : \mathbb{X} \longrightarrow \mathbb{R}$ be a function describing the initial velocity of the string. Suppose f_1 has Fourier sine series $f_1(x) \underset{L2}{\approx} \sum_{n=1}^{\infty} B_n \sin(nx)$, and define the function $v : \mathbb{X} \times \mathbb{R}_+ \longrightarrow \mathbb{R}$ by

$$v(x;t) \underset{L2}{\approx} \sum_{n=1}^{\infty} \frac{B_n}{n} \sin(nx) \cdot \sin(nt), \quad \text{for all } x \in [0, \pi] \text{ and } t \geq 0. \quad (11\text{B}.4)$$

Then v is the unique solution to the wave equation (11B.1), satisfying the Dirichlet boundary conditions (11B.2), as well as

$$\left. \begin{array}{ll} \text{initial position:} & v(x, 0) = 0 \\ \text{initial velocity:} & \partial_t v(x, 0) = f_1(x) \end{array} \right\} \text{ for all } x \in [0, \pi].$$

Proof **Exercise 11B.2.** *Hint:* Ⓔ

(a) Prove the trigonometric identity

$$-\sin(nx)\sin(nt) = \frac{1}{2}\Big(\cos\left(n(x+t)\right) - \cos\left(n(x-t)\right)\Big).$$

(b) Use this identity to show that the Fourier sine series (11B.4) converges in L^2 to the d'Alembert solution from Theorem 17D.8(b), p. 407.

(c) Apply Theorem 5D.11, p. 96 to show that this solution is unique. ☐

Example 11B.5 Let $f_1(x) = 3\sin(8x)$. Thus, $B_8 = 3$ and $B_n = 0$ for all $n \neq 8$. Proposition 11B.4 tells us that the corresponding solution to the wave equation is $w(x,t) = \frac{3}{8}\sin(8t)\sin(8x)$. To see that w satisfies the wave equation, note that, for any $x \in [0, \pi]$ and $t > 0$,

$$\partial_t\, w(x,t) = 3\sin(8t)\cos(8x) \quad \text{and} \quad 3\cos(8t)\sin(8x) = \partial_x\, w(x,t);$$

Thus

$$\partial_t^2\, w(x,t) = -24\cos(8t)\cos(8x) = -24\cos(8t)\cos(8x) = \partial_x^2\, w(x,t).$$

Also w has the desired initial position because, for any $x \in [0, \pi]$, we have $w(x;0) = \frac{3}{8}\sin(0)\sin(8x) = 0$, because $\sin(0) = 0$.

Next, w has the desired initial velocity because, for any $x \in [0, \pi]$, we have $\partial_t w(x;0) = \frac{3}{8}8\cos(0)\sin(8x) = 3\sin(8x) = f_1(x)$, because $\cos(0) = 1$.

Finally w satisfies homogeneous Dirichlet BC because, for any $t > 0$, we have $w(0, t) = \frac{3}{8}\sin(8t)\sin(0) = 0$ and $w(\pi, t) = \frac{3}{8}\sin(8t)\sin(8\pi) = 0$, because $\sin(0) = 0 = \sin(8\pi)$. ◇

Example 11B.6 The xylophone

A musician strikes the midpoint of a xylophone bar with a broad, flat hammer. What is the formula describing the vibration of the string?

Solution For simplicity, we imagine the bar has length π and is fixed at its endpoints (actually most xylophones satisfy neither requirement). At the moment when the hammer strikes it, the string's *initial position* is zero, and its *initial velocity* is determined by the distribution of force imparted by the hammer head. For simplicity, we will assume the hammer head has width $\pi/2$, and hits the bar squarely at its midpoint (Figure 11B.1(d)). Thus, the initial velocity is given by the function

$$f_1(x) = \begin{cases} \alpha & \text{if } \frac{\pi}{4} \leq x \leq \frac{3\pi}{4}; \\ 0 & \text{otherwise} \end{cases}$$

(Figure 11B.1(e)), where $\alpha > 0$ is a constant describing the force of the impact. Proposition 11B.4 tells us to find the Fourier sine series for $f_1(x)$. From Example 7C.4, p. 155, we know this to be

$$f_1(x) \underset{12}{\approx} \frac{2\alpha\sqrt{2}}{\pi} \left(\sin(x) + \sum_{k=1}^{\infty} (-1)^k \frac{\sin\left((4k-1)x\right)}{4k-1} \right.$$

$$\left. + \sum_{k=1}^{\infty} (-1)^k \frac{\sin\left((4k+1)x\right)}{4k+1} \right).$$

The resulting vibrational motion is therefore described by

$$v(x,t) \underset{12}{\approx} \frac{2\alpha\sqrt{2}}{\pi} \left(\sin(x)\sin(t) + \sum_{k=1}^{\infty} (-1)^k \frac{\sin\left((4k-1)x\right)\sin\left((4k-1)t\right)}{(4k-1)^2} \right.$$

$$\left. + \sum_{k=1}^{\infty} (-1)^k \frac{\sin\left((4k+1)x\right)\sin\left((4k+1)t\right)}{(4k+1)^2} \right). \quad \diamondsuit$$

Ⓔ **Exercise 11B.3** Let $L > 0$ and let $\mathbb{X} := [0, L]$. Let $\lambda > 0$ be a parameter describing wave velocity (determined by the string's tension, elasticity, density, etc.), and consider the general one-dimensional wave equation

$$\partial_t^2 u = \lambda^2 \partial_x^2 u. \tag{11B.5}$$

(a) Generalize Proposition 11B.1 to find the solution to equation (11B.5) on $\mathbb{X}$, having zero initial velocity, a prescribed initial position, and homogeneous Dirichlet boundary conditions.

(b) Generalize Proposition 11B.4 to find the solution to equation (11B.5) on $\mathbb{X}$, having zero initial position, a prescribed initial velocity, and homogeneous Dirichlet boundary conditions.

In both cases, prove that your solution converges, satisfies the desired initial conditions and boundary conditions, and satisfies equation (11B.5) (*Hint:* Imitate the strategy suggested in Exercises 11B.1 and 11B.2.) ◆

11C The Poisson problem on a line segment

Prerequisites: §7B, §5C, §1D.   **Recommended:** §7C(v).

We can also use Fourier series to solve the one-dimensional Poisson problem on a line segment. This is not usually a practical solution method, because we already have a simple, complete solution to this problem using a double integral (see Example 1D.1, p. 14). However, we include this result anyway, as a simple illustration of Fourier techniques.

Proposition 11C.1 *Let* $\mathbb{X} = [0, \pi]$, *and let* $q : \mathbb{X} \longrightarrow \mathbb{R}$ *be some function, with semiuniformly convergent Fourier sine series:* $q(x) \underset{12}{\approx} \sum_{n=1}^{\infty} Q_n \sin(nx)$. *Define the function* $u : \mathbb{X} \longrightarrow \mathbb{R}$ *by*

$$u(x) \underset{\text{unif}}{\equiv} \sum_{n=1}^{\infty} \frac{-Q_n}{n^2} \sin(nx), \quad \textit{for all } x \in [0, \pi].$$

Then u *is the unique solution to the Poisson equation '$\triangle u(x) = q(x)$' satisfying homogeneous Dirichlet boundary conditions* $u(0) = u(\pi) = 0$.

Proof **Exercise 11C.1.** *Hint:* ⓔ

(a) Apply Proposition F.1, p. 569, twice to show that $\triangle u(x) \underset{\text{unif}}{\equiv} \sum_{n=1}^{\infty} Q_n \sin(nx) = q(x)$, for all $x \in \text{int}(\mathbb{X})$. (*Hint:* The Fourier series of q is semiuniformly convergent.)
(b) Observe that $\sum_{n=1}^{\infty} \left| \frac{Q_n}{n^2} \right| < \infty$.
(c) Apply Theorem 7A.1(c), p. 142, to show that the given Fourier sine series for $u(x)$ converges uniformly.
(d) Apply Theorem 7A.1(d)(ii), p. 142, to conclude that $u(0) = 0 = u(\pi)$.
(e) Apply Theorem 5D.5(a), p. 91, to conclude that this solution is unique. □

Proposition 11C.2 *Let* $\mathbb{X} = [0, \pi]$, *and let* $q : \mathbb{X} \longrightarrow \mathbb{R}$ *be some function with semiuniformly convergent Fourier cosine series* $q(x) \underset{12}{\approx} \sum_{n=1}^{\infty} Q_n \cos(nx)$, *and suppose that* $Q_0 = 0$. *Fix any constant* $K \in \mathbb{R}$, *and define the function* $u : \mathbb{X} \longrightarrow \mathbb{R}$ *by*

$$u(x) \underset{\text{unif}}{\equiv} \sum_{n=1}^{\infty} \frac{-Q_n}{n^2} \cos(nx) + K, \quad \textit{for all } x \in [0, \pi]. \tag{11C.1}$$

Then u *is a solution to the Poisson equation '$\triangle u(x) = q(x)$', satisfying homogeneous Neumann boundary conditions* $u'(0) = u'(\pi) = 0$.

Furthermore, all solutions to this Poisson equation with these boundary conditions have the form (11C.1) *for some choice of* K.

If $Q_0 \neq 0$, *however, the problem has no solution.*

Proof **Exercise 11C.2.** *Hint:* ⓔ

(a) Apply Proposition F.1, p. 569, twice to show that $\triangle u(x) \underset{\text{unif}}{\equiv} \sum_{n=1}^{\infty} Q_n \cos(nx) = q(x)$, for all $x \in \text{int}(\mathbb{X})$. (*Hint:* The Fourier series of q is semiuniformly convergent.)
(b) Observe that $\sum_{n=1}^{\infty} \left| \frac{Q_n}{n} \right| < \infty$.
(c) Apply Theorem 7A.4(d)(ii), p. 145, to conclude that $u'(0) = 0 = u'(\pi)$.
(d) Apply Theorem 5D.5(c), p. 91, to conclude that this solution is unique up to addition of a constant. □

Ⓔ **Exercise 11C.3** Mathematically, it is clear that the solution of Proposition 11C.2 cannot be well defined if $Q_0 \neq 0$. Provide a physical explanation for why this is to be expected. ◆

11D Practice problems

11.1 Let $g(x) = \begin{cases} 1 & \text{if } 0 \leq x < \frac{\pi}{2}; \\ 0 & \text{if } \frac{\pi}{2} \leq x \end{cases}$

(see Problem 7.5).

(a) Find the solution to the one-dimensional heat equation $\partial_t u(x, t) = \Delta u(x, t)$ on the interval $[0, \pi]$, with initial conditions $u(x, 0) = g(x)$ and homogeneous *Dirichlet* boundary conditions.

(b) Find the solution to the one-dimensional heat equation $\partial_t u(x, t) = \Delta u(x, t)$ on the interval $[0, \pi]$, with initial conditions $u(x, 0) = g(x)$ and homogeneous *Neumann* boundary conditions.

(c) Find the solution to the one-dimensional wave equation $\partial_t^2 w(x, t) = \Delta w(x, t)$ on the interval $[0, \pi]$, satisfying homogeneous *Dirichlet* boundary conditions, with initial *position* $w(x, 0) = 0$ and initial *velocity* $\partial_t w(x, 0) = g(x)$.

11.2 Let $f(x) = \sin(3x)$, for $x \in [0, \pi]$.

(a) Compute the Fourier *sine* series of $f(x)$ as an element of $\mathbf{L}^2[0, \pi]$.

(b) Compute the Fourier *cosine* series of $f(x)$ as an element of $\mathbf{L}^2[0, \pi]$.

(c) Solve the one-dimensional *heat equation* ($\partial_t u = \Delta u$) on the domain $\mathbb{X} = [0, \pi]$, with *initial conditions* $u(x; 0) = f(x)$, and the following boundary conditions:

   (i) homogeneous *Dirichlet* boundary conditions;

   (ii) homogeneous *Neumann* boundary conditions.

(d) Solve the the one-dimensional *wave equation* ($\partial_t^2 v = \Delta v$) on the domain $\mathbb{X} = [0, \pi]$, with homogeneous *Dirichlet* boundary conditions, and with

$$\begin{aligned} \text{initial position:} &\quad v(x; 0) = 0, \\ \text{initial velocity:} &\quad \partial_t v(x; 0) = f(x). \end{aligned}$$

11.3 Let $f : [0, \pi] \longrightarrow \mathbb{R}$, and suppose f has

$$\text{Fourier cosine series:} \quad f(x) = \sum_{n=0}^{\infty} \frac{1}{2^n} \cos(nx);$$

$$\text{Fourier sine series:} \quad f(x) = \sum_{n=1}^{\infty} \frac{1}{n!} \sin(nx).$$

(a) Find the solution to the one-dimensional heat equation $\partial_t u = \Delta u$, with homogeneous *Neumann* boundary conditions, and initial conditions $u(x; 0) = f(x)$ for all $x \in [0, \pi]$.

(b) *Verify* your solution in part (a). Check the heat equation, the initial conditions, and boundary conditions. (*Hint:* Use Proposition F.1, p. 569.)

(c) Find the solution to the one-dimensional *wave equation* $\partial_t^2 u(x; t) = \Delta u(x; t)$ with homogeneous *Dirichlet* boundary conditions, and

$$\begin{aligned} &\text{initial position:} \quad u(x; 0) = f(x), \quad &&\text{for all } x \in [0, \pi], \\ &\text{initial velocity:} \quad \partial_t u(x; 0) = 0, \quad &&\text{for all } x \in [0, \pi]. \end{aligned}$$

11.4 Let $f : [0, \pi] \longrightarrow \mathbb{R}$ be defined by $f(x) = x$.

(a) Compute the Fourier *sine* series for f.

(b) Does the Fourier sine series converge *pointwise* to f on $(0, \pi)$? Justify your answer.

(c) Does the Fourier sine series converge *uniformly* to f on $[0, \pi]$? Justify your answer in *two different* ways.

(d) Compute the Fourier *cosine* series for f.

(e) Solve the one-dimensional *heat equation* $(\partial_t u = \Delta u)$ on the domain $\mathbb{X} := [0, \pi]$, with initial conditions $u(x, 0) := f(x)$, and with the following boundary conditions:

(i) homogeneous *Dirichlet* boundary conditions;

(ii) homogeneous *Neumann* boundary conditions.

(f) *Verify* your solution to (e)(i). That is: check that your solution satisfies the heat equation, the desired initial conditions, and homogeneous Dirichlet BC. (You may assume that the relevent series converge uniformly, if necessary. You may differentiate Fourier series termwise, if necessary.)

(g) Find the solution to the one-dimensional *wave equation* on the domain $\mathbb{X} :=$ $[0, \pi]$, with homogeneous Dirichlet boundary conditions, and with

$$\begin{aligned} &\text{initial position:} \quad u(x; 0) = f(x), \quad &&\text{for all } x \in [0, \pi]; \\ &\text{initial velocity:} \quad \partial_t u(x; 0) = 0, \quad &&\text{for all } x \in [0, \pi]. \end{aligned}$$

12

Boundary value problems on a square

Each problem that I solved became a rule which served afterwards to
solve other problems.

René Descartes

Prerequisites: §9A, §5C. **Recommended:** §11.

Multiple Fourier series can be used to find solutions to boundary value problems on a box $[0, X] \times [0, Y]$. The key idea is that the functions $\mathbf{S}_{n,m}(x, y) :=$ $\sin\left(\frac{n\pi}{X}x\right)\sin\left(\frac{m\pi}{Y}y\right)$ and $\mathbf{C}_{n,m}(x, y) := \cos\left(\frac{n\pi}{X}x\right)\cos\left(\frac{m\pi}{Y}y\right)$ are *eigenfunctions* of the Laplacian operator. Furthermore, $\mathbf{S}_{n,m}$ satisfies *Dirichlet* boundary conditions, so any (uniformly convergent) Fourier sine series will also do so. Likewise, $\mathbf{C}_{n,m}$ satisfies *Neumann* boundary conditions, so any (sufficiently convergent) Fourier cosine series will also do so.

For simplicity, we will assume throughout that $X = Y = \pi$. Thus $\mathbf{S}_{n,m}(x) = \sin(nx)\sin(my)$ and $\mathbf{C}_{n,m}(x) = \cos(nx)\cos(my)$. We will also assume (through an appropriate choice of time units) that the physical constants in the various equations are all equal to one. Thus, the heat equation becomes '$\partial_t u = \Delta u$', the wave equation is '$\partial_t^2 u = \Delta u$', etc. This will allow us to develop the solution methods in the simplest possible scenario, without a lot of distracting technicalities.

The extension of these solution methods to equations with arbitrary physical constants on an arbitrary rectangular domain $[0, X] \times [0, Y]$ (for some $X, Y > 0$) are left as exercises. These exercises are quite straightforward, but are an effective test of your understanding of the solution techniques.

Remark on notation Throughout this chapter (and the following ones) we will often write a function $u(x, y; t)$ in the form $u_t(x, y)$. This emphasizes the distinguished role of the 'time' coordinate t, and makes it natural to think of fixing t at some value and applying the two-dimensional Laplacian $\Delta = \partial_x^2 + \partial_y^2$ to the resulting two-dimensional function u_t.

Some authors use the subscript notation 'u_t' to denote the partial derivative $\partial_t u$. We *never* use this notation. In this book, partial derivatives are always denoted by '$\partial_t u$', etc.

12A The Dirichlet problem on a square

Prerequisites: §9A, §5C(i), §1C, Appendix F.   **Recommended:** §7C(v).

In this section we will learn to solve the *Dirichlet problem* on a square domain $\mathbb{X}$: that is, to find a function which is harmonic on the interior of $\mathbb{X}$ and which satisfies specified Dirichlet boundary conditions on the boundary $\mathbb{X}$. Solutions to the Dirichlet problem have several physical interpretations.

Heat: Imagine that the boundaries of $\mathbb{X}$ are perfect heat conductors, which are in contact with external 'heat reservoirs' with fixed temperatures. For example, one boundary might be in contact with a heat source, while another is in contact with a coolant liquid. The solution to the Dirichlet problem is then the equilibrium temperature distribution on the interior of the box, given these constraints.

Electrostatic: Imagine that the boundaries of $\mathbb{X}$ are electrical conductors which are held at some fixed voltage by the application of an external electric potential (different boundaries, or different parts of the same boundary, may be held at different voltages). The solution to the Dirichlet problem is then the electric potential field on the interior of the box, given these constraints.

Minimal surface: Imagine a squarish frame of wire, which we have bent in the vertical direction to have some shape. If we dip this wire frame in a soap solution, we can form a soap bubble (i.e. minimal-energy surface) which must obey the 'boundary conditions' imposed by the shape of the wire. The differential equation describing a minimal surface is not *exactly* the same as the Laplace equation; however, when the surface is not too steeply slanted (i.e. when the wire frame is not too bent), the Laplace equation is a good approximation; hence the solution to the Dirichlet problem is a good approximation of the shape of the soap bubble.

We will begin with the simplest problem: a constant, nonzero Dirichlet boundary condition on one side of the box, and zero boundary conditions on the other three sides.

Proposition 12A.1 Dirichlet problem; one constant nonhomogeneous BC

Let $\mathbb{X} = [0, \pi] \times [0, \pi]$, and consider the Laplace equation '$\triangle u = 0$', with non-homogeneous Dirichlet boundary conditions (see Figure 12A.1(a)):

$$u(0, y) = u(\pi, y) = 0, \quad \text{for all } y \in [0, \pi), \qquad (12\text{A.1})$$

$$u(x, 0) = 0 \quad \text{and} \quad u(x, \pi) = 1, \quad \text{for all } x \in [0, \pi]. \qquad (12\text{A.2})$$

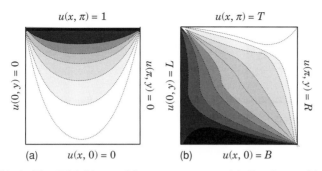

$u(x, \pi) = 1$ $u(x, \pi) = T$

(a) $u(x, 0) = 0$ (b) $u(x, 0) = B$

Figure 12A.1. The Dirichlet problem on a square. (a) See Proposition 12A.1;
(b) see Propositions 12A.2 and 12A.4.

The unique solution to this problem is the function $u : \mathbb{X} \longrightarrow \mathbb{R}$ defined:

$$u(x, y) \underset{12}{\approx} \frac{4}{\pi} \sum_{\substack{n=1 \\ n \ \text{odd}}}^{\infty} \frac{1}{n \sinh(n\pi)} \sin(nx) \cdot \sinh(ny), \qquad \text{for all } (x, y) \in \mathbb{X}$$

*(see Figures 12A.2(a) and 12A.3(a)). Furthermore, this series converges semiuni-
formly on* int $(\mathbb{X})$.

Proof **Exercise 12A.1.**   Ⓔ
(a) Check that, for all $n \in \mathbb{N}$, the function $u_n(x, y) = \sin(nx) \cdot \sinh(ny)$ satisfies the
     Laplace equation and the first boundary condition (12A.1). See Figures 12A.2(d)–
     (f) and 12A.3(d)–(f).
(b) Show that
$$\sum_{\substack{n=1 \\ n\text{odd}}}^{\infty} n^2 \left| \frac{\sinh(ny)}{n \sinh(n\pi)} \right| < \infty,$$
     for any fixed $y < \pi$. (*Hint:* If $y < \pi$, then $\sinh(ny)/\sinh(n\pi)$ decays like
     $\exp(n(y - \pi))$ as $n \to \infty$.)
(c) Apply Proposition F.1, p. 569, to conclude that $\triangle u(x, y) = 0$.
(d) Observe that
$$\sum_{\substack{n=1 \\ n \ \text{odd}}}^{\infty} \left| \frac{\sinh(ny)}{n \sinh(n\pi)} \right| < \infty,$$
     for any fixed $y < \pi$.
(e) Apply part (c) of Theorem 7A.1, p. 142, to show that the series given for $u(x, y)$
     converges uniformly for any fixed $y < \pi$.
(f) Apply part (d) of Theorem 7A.1, p. 142, to conclude that $u(0, y) = 0 = u(\pi, y)$ for all
     $y < \pi$.
(g) Observe that $\sin(nx) \cdot \sinh(n \cdot 0) = 0$ for all $n \in \mathbb{N}$ and all $x \in [0, \pi]$. Conclude that
     $u(x, 0) = 0$ for all $x \in [0, \pi]$.

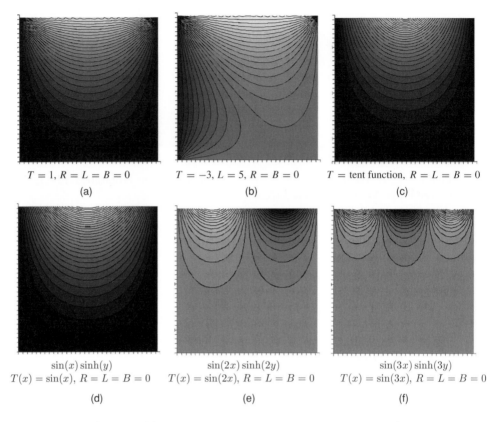

$T = 1$, $R = L = B = 0$	$T = -3$, $L = 5$, $R = B = 0$	$T = $ tent function, $R = L = B = 0$
(a)	(b)	(c)

| $\sin(x)\sinh(y)$ | $\sin(2x)\sinh(2y)$ | $\sin(3x)\sinh(3y)$ |
$T(x) = \sin(x),\ R = L = B = 0$	$T(x) = \sin(2x),\ R = L = B = 0$	$T(x) = \sin(3x),\ R = L = B = 0$
(d)	(e)	(f)

Figure 12A.2. The Dirichlet problem on a box. The curves represent isothermal contours (of a temperature distribution) or equipotential lines (of an electric voltage field).

(h) To check that the solution also satisfies the boundary condition (12A.2), subsititute $y = \pi$ to obtain

$$u(x, \pi) = \frac{4}{\pi} \sum_{\substack{n=1 \\ n \text{ odd}}}^{\infty} \frac{1}{n \sinh(n\pi)} \sin(nx) \cdot \sinh(n\pi) = \frac{4}{\pi} \sum_{\substack{n=1 \\ n \text{ odd}}}^{\infty} \frac{1}{n} \sin(nx) \underset{12}{\approx} 1$$

because $\frac{4}{\pi} \sum_{\substack{n=1 \\ n \text{ odd}}}^{\infty} \frac{1}{n} \sin(nx)$ is the (one-dimensional) Fourier sine series for the function $b(x) = 1$ (see Example 7A.2(b), p. 143).

(i) Apply Theorem 5D.5(a), p. 91, to conclude that this solution is unique.   □

Proposition 12A.2 Dirichlet problem; four constant nonhomogeneous BC
Let $\mathbb{X} = [0, \pi] \times [0, \pi]$, and consider the Laplace equation '$\triangle u = 0$', with non-homogeneous Dirichlet boundary conditions (see Figure 12A.1(b)):

$$u(0, y) = L \quad and \quad u(\pi, y) = R, \quad for\ all\ y \in (0, \pi);$$
$$u(x, \pi) = T \quad and \quad u(x, 0) = B, \quad for\ all\ x \in (0, \pi),$$

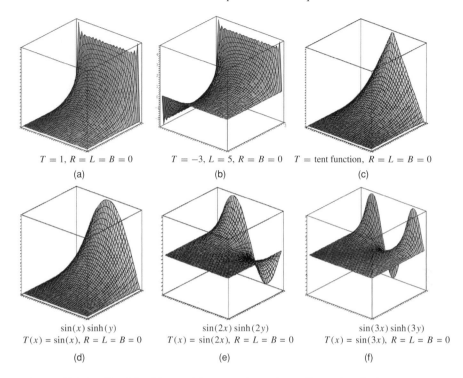

$T = 1, R = L = B = 0$

(a)

$T = -3, L = 5, R = B = 0$

(b)

T = tent function, $R = L = B = 0$

(c)

$$\frac{\sin(x)\sinh(y)}{T(x) = \sin(x), R = L = B = 0}$$

(d)

$$\frac{\sin(2x)\sinh(2y)}{T(x) = \sin(2x), R = L = B = 0}$$

(e)

$$\frac{\sin(3x)\sinh(3y)}{T(x) = \sin(3x), R = L = B = 0}$$

(f)

Figure 12A.3. The Dirichlet problem on a box: three-dimensional plots. You can imagine these as soap films.

where L, R, T, and B are four constants. The unique solution to this problem is the function $u : \mathbb{X} \longrightarrow \mathbb{R}$ *defined:*

$$u(x, y) := l(x, y) + r(x, y) + t(x, y) + b(x, y), \quad \textit{for all } (x, y) \in \mathbb{X},$$

where, for all $(x, y) \in \mathbb{X}$,

$$l(x, y) \underset{L2}{\approx} L \sum_{\substack{n=1 \\ n \ \text{odd}}}^{\infty} c_n \sinh\Big(n(\pi - x)\Big) \cdot \sin(ny),$$

$$r(x, y) \underset{L2}{\approx} R \sum_{\substack{n=1 \\ n \ \text{odd}}}^{\infty} c_n \sinh(nx) \cdot \sin(ny),$$

$$t(x, y) \underset{L2}{\approx} T \sum_{\substack{n=1 \\ n \ \text{odd}}}^{\infty} c_n \sin(nx) \cdot \sinh(ny),$$

$$b(x, y) \underset{L2}{\approx} B \sum_{\substack{n=1 \\ n \ \text{odd}}}^{\infty} c_n \sin(nx) \cdot \sinh\Big(n(\pi - y)\Big),$$

where $c_n := 4/[n\pi \ \sinh(n\pi)]$, *for all* $n \in \mathbb{N}$.

Furthermore, these four series converge semiuniformly on $\text{int}(\mathbb{X})$.

Ⓔ *Proof* **Exercise 12A.2.**
(a) Apply Proposition 12A.1 to show that each of the functions $l(x, y)$, $r(x, y)$, $t(x, y)$, and $b(x, y)$ satisfies a Dirichlet problem where one side has nonzero temperature and the other three sides have zero temperature.
(b) Add these four together to get a solution to the original problem.
(c) Apply Theorem 5D.5(a), p. 91, to conclude that this solution is unique.   □

Ⓔ **Exercise 12A.3** What happens to the solution at the four corners $(0, 0)$, $(0, \pi)$, $(\pi, 0)$, and (π, π)?   ♦

Example 12A.3 Suppose $R = 0 = B$, $T = -3$, and $L = 5$. Then the solution is given by

$$u(x, y) \underset{12}{\approx} L \sum_{\substack{n=1 \\ n \text{ odd}}}^{\infty} c_n \sinh\Big(n(\pi - x)\Big) \cdot \sin(ny) + T \sum_{\substack{n=1 \\ n \text{ odd}}}^{\infty} c_n \sin(nx) \cdot \sinh(ny)$$

$$= \frac{20}{\pi} \sum_{\substack{n=1 \\ n \text{ odd}}}^{\infty} \frac{\sinh\Big(n(\pi - x)\Big) \cdot \sin(ny)}{n \sinh(n\pi)} - \frac{12}{\pi} \sum_{\substack{n=1 \\ n \text{ odd}}}^{\infty} \frac{\sin(nx) \cdot \sinh(ny)}{n \sinh(n\pi)}$$

(see Figures 12A.2(b) and 12A.3(b)).   ◇

Proposition 12A.4 Dirichlet problem; arbitrary nonhomogeneous boundaries
Let $\mathbb{X} = [0, \pi] \times [0, \pi]$, *and consider the Laplace equation '$\Delta u = 0$', with non-homogeneous Dirichlet boundary conditions (see Figure 12A.1(b)):*

$$u(0, y) = L(y) \quad and \quad u(\pi, y) = R(y), \quad for \ all \ y \in (0, \pi);$$
$$u(x, \pi) = T(x) \quad and \quad u(x, 0) = B(x), \quad for \ all \ x \in (0, \pi),$$

where $L, R, T, B : [0, \pi] \longrightarrow \mathbb{R}$ *are four arbitrary functions. Suppose these functions have (one-dimensional) Fourier sine series:*

$$L(y) \underset{12}{\approx} \sum_{n=1}^{\infty} L_n \sin(ny), \qquad R(y) \underset{12}{\approx} \sum_{n=1}^{\infty} R_n \sin(ny), \quad for \ all \ y \in [0, \pi];$$

$$T(x) \underset{12}{\approx} \sum_{n=1}^{\infty} T_n \sin(nx), \quad and \quad B(x) \underset{12}{\approx} \sum_{n=1}^{\infty} B_n \sin(nx), \quad for \ all \ x \in [0, \pi].$$

The unique solution to this problem is the function $u : \mathbb{X} \longrightarrow \mathbb{R}$ *defined:*

$$u(x, y) := l(x, y) + r(x, y) + t(x, y) + b(x, y), \quad \text{for all } (x, y) \in \mathbb{X},$$

where, for all $(x, y) \in \mathbb{X}$,

$$l(x, y) \underset{12}{\approx} \sum_{n=1}^{\infty} \frac{L_n}{\sinh(n\pi)} \sinh\Big(n(\pi - x)\Big) \cdot \sin(ny),$$

$$r(x, y) \underset{12}{\approx} \sum_{n=1}^{\infty} \frac{R_n}{\sinh(n\pi)} \sinh(nx) \cdot \sin(ny),$$

$$t(x, y) \underset{12}{\approx} \sum_{n=1}^{\infty} \frac{T_n}{\sinh(n\pi)} \sin(nx) \cdot \sinh(ny),$$

$$and \; b(x, y) \underset{12}{\approx} \sum_{n=1}^{\infty} \frac{B_n}{\sinh(n\pi)} \sin(nx) \cdot \sinh\Big(n(\pi - y)\Big).$$

Furthermore, these four series converge semiuniformly on $\text{int}(\mathbb{X})$.

Proof **Exercise 12A.4.** First we consider the function $t(x, y)$. Ⓔ

(a) Same as Exercise 12A.1(a).

(b) For any fixed $y < \pi$, show that

$$\sum_{n=1}^{\infty} n^2 T^n \left| \frac{\sinh(ny)}{\sinh(n\pi)} \right| < \infty.$$

(*Hint:* If $y < \pi$, then $\sinh(ny)/\sinh(n\pi)$ decays like $\exp(n(y - \pi))$ as $n \to \infty$.)

(c) Combine part (b) and Proposition F.1, p. 569, to conclude that $t(x, y)$ is harmonic – i.e. $\Delta t(x, y) = 0$. Through symmetric reasoning, conclude that the functions $\ell(x, y)$, $r(x, y)$, and $b(x, y)$ are also harmonic.

(d) Same as Exercise 12A.1(d).

(e) Apply part (c) of Theorem 7A.1, p. 142, to show that the series given for $t(x, y)$ converges uniformly for any fixed $y < \pi$.

(f) Apply part (d) of Theorem 7A.1, p. 142, to conclude that $t(0, y) = 0 = t(\pi, y)$ for all $y < \pi$.

(g) Observe that $\sin(nx) \cdot \sinh(n \cdot 0) = 0$ for all $n \in \mathbb{N}$ and all $x \in [0, \pi]$. Conclude that $t(x, 0) = 0$ for all $x \in [0, \pi]$.

(h) To check that the solution also satisfies the boundary condition (12A.2), subsititute $y = \pi$ to obtain

$$t(x, \pi) = \sum_{n=1}^{\infty} \frac{T_n}{\sinh(n\pi)} \sin(nx) \cdot \sinh(n\pi) = \frac{4}{\pi} \sum_{n=1}^{\infty} T_n \sin(nx) = T(x).$$

(j) At this point, we know that $t(x, \pi) = T(x)$ for all $x \in [0, \pi]$, and $t \equiv 0$ on the other three sides of the square. Through symmetric reasoning, show that

- $\ell(0, y) = L(y)$ for all $y \in [0, \pi]$, and $\ell \equiv 0$ on the other three sides of the square;
- $r(\pi, y) = R(y)$ for all $y \in [0, \pi]$, and $r \equiv 0$ on the other three sides of the square;
- $b(x, 0) = B(x)$ for all $x \in [0, \pi]$, and $b \equiv 0$ on the other three sides of the square.

(k) Conclude that $u = t + b + r + \ell$ is harmonic and satisfies the desired boundary conditions.

(l) Apply Theorem 5D.5(a), p. 91, to conclude that this solution is unique. □

Example 12A.5 If $T(x) = \sin(3x)$, and $B \equiv L \equiv R \equiv 0$, then

$$u(x, y) = \frac{\sin(3x) \sinh(3y)}{\sinh(3\pi)}$$

(See Figures 12A.2(f) and 12A.3(f).) ◇

Example 12A.6 Let $\mathbb{X} = [0, \pi] \times [0, \pi]$. Solve the two-dimensional Laplace Equation on $\mathbb{X}$, with inhomogeneous Dirichlet boundary conditions:

$$u(0, y) = 0; \quad u(\pi, y) = 0; \quad u(x, 0) = 0;$$

$$u(x, \pi) = T(x) = \begin{cases} x & \text{if } 0 \le x \le \frac{\pi}{2}; \\ \pi - x & \text{if } \frac{\pi}{2} < x \le \pi \end{cases}$$

(see Figure 7C.4(b), p. 158).

Solution Recall from Example 7C.7, p. 159, that $T(x)$ has the following Fourier series:

$$T(x) \underset{12}{\approx} \frac{4}{\pi} \sum_{\substack{n=1 \\ n \text{ odd}; \\ n=2k+1}}^{\infty} \frac{(-1)^k}{n^2} \sin(nx).$$

Thus, the solution is

$$u(x, y) \underset{12}{\approx} \frac{4}{\pi} \sum_{\substack{n=1 \\ n \text{ odd}; \\ n=2k+1}}^{\infty} \frac{(-1)^k}{n^2 \sinh(n\pi)} \sin(nx) \sinh(ny).$$

See Figures 12A.2(c) and 12A.3(c). ◇

Ⓔ **Exercise 12A.5** Let $X, Y > 0$ and let $\mathbb{X} := [0, X] \times [0, Y]$. Generalize Proposition 12A.4 to find the solution to the Laplace equation on $\mathbb{X}$, satisfying arbitrary nonhomogeneous Dirichlet boundary conditions on the four sides of $\partial \mathbb{X}$. ♦

12B The heat equation on a square

12B(i) Homogeneous boundary conditions

Prerequisites: §9A, §5B, §5C, §1B(ii), Appendix F. **Recommended:** §11A, §7C(v).

Proposition 12B.1 Heat equation; homogeneous Dirichlet boundary
Consider the box $\mathbb{X} = [0, \pi] \times [0, \pi]$, *and let* $f : \mathbb{X} \longrightarrow \mathbb{R}$ *be some function describing an initial heat distribution. Suppose* f *has Fourier sine series*

$$f(x, y) \underset{12}{\approx} \sum_{n,m=1}^{\infty} B_{n,m} \sin(nx) \sin(my),$$

and define the function $u : \mathbb{X} \times \mathbb{R}_{\neq} \longrightarrow \mathbb{R}$ *by*

$$u_t(x, y) \underset{12}{\approx} \sum_{n,m=1}^{\infty} B_{n,m} \sin(nx) \cdot \sin(my) \cdot \exp\left(-(n^2 + m^2) \cdot t\right),$$

for all $(x, y) \in \mathbb{X}$ *and* $t \geq 0$. *Then* u *is the unique solution to the heat equation* '$\partial_t u = \Delta u$', *with homogeneous Dirichlet boundary conditions*

$$u_t(x, 0) = u_t(0, y) = u_t(\pi, y) = u_t(x, \pi) = 0, \quad \textit{for all } x, y \in [0, \pi] \textit{ and } t > 0,$$

and initial conditions $u_0(x, y) = f(x, y)$, *for all* $(x, y) \in \mathbb{X}$.
   Furthermore, the series defining u *converges semiuniformly on* $\mathbb{X} \times \mathbb{R}_+$.

Proof **Exercise 12B.1.** *Hint:*   Ⓔ

(a) Show that, when $t = 0$, the two-dimensional Fourier series of $u_0(x, y)$ agrees with that of $f(x, y)$; hence $u_0(x, y) = f(x, y)$.
(b) Show that, for all $t > 0$,

$$\sum_{n,m=1}^{\infty} \left| (n^2 + m^2) \cdot B_{n,m} \cdot e^{-(n^2+m^2)t} \right| < \infty.$$

(c) For any $T > 0$, apply Proposition F.1, p. 569, to conclude that

$$\partial_t u_t(x, y) \underset{\text{unif}}{\equiv} \sum_{n,m=1}^{\infty} -(n^2 + m^2) B_{n,m} \sin(nx) \cdot \sin(my) \cdot e^{-(n^2+m^2)t} \underset{\text{unif}}{\equiv} \Delta u_t(x, y),$$

for all $(x, y; t) \in \mathbb{X} \times [T, \infty)$.
(d) Observe that, for all $t > 0$,

$$\sum_{n,m=1}^{\infty} \left| B_{n,m} \cdot e^{-(n^2+m^2)t} \right| < \infty.$$

(e) Apply part (c)(i) of Theorem 9A.3, p. 183, to show that the two-dimensional Fourier series of u_t converges uniformly for any fixed $t > 0$.

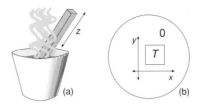

Figure 12B.1. (a) A hot metal rod quenched in a cold bucket. (b) Cross-section of the rod in the bucket.

(f) Apply part (d)(ii) of Theorem 9A.3, p. 183, to conclude that u_t satisfies homogeneous Dirichlet boundary conditions for all $t > 0$.

(g) Apply Theorem 5D.8, p. 93, to show that this solution is unique.   □

Example 12B.2 The quenched rod

On a cold January day, a blacksmith is tempering an iron rod. He pulls it out of the forge and plunges it, red-hot, into ice-cold water (Figure 12B.1(a)). The rod is very long and narrow, with a square cross section. We want to compute how the rod cooled.

Solution The rod is immersed in freezing cold water, and is a good conductor, so we can assume that its outer surface takes the surrounding water temperature of $0\,°C$. Hence, we assume homogeneous Dirichlet boundary conditions.

Endow the rod with coordinate system (x, y, z), where z runs along the length of the rod. Since the rod is extremely long relative to its cross-section, we can neglect the z coordinate, and reduce to a two-dimensional equation (Figure 12B.1(b)). Assume the rod was initially uniformly heated to a temperature of T. The initial temperature distribution is thus a constant function: $f(x, y) = T$. From Example 9A.2, p. 182, we know that the constant function 1 has a two-dimensional Fourier sine series:

$$1 \underset{12}{\approx} \frac{16}{\pi^2} \sum_{\substack{n,m=1 \\ \text{both odd}}}^{\infty} \frac{1}{n \cdot m} \sin(nx) \sin(my).$$

Thus,

$$f(x, y) \underset{12}{\approx} \frac{16T}{\pi^2} \sum_{\substack{n,m=1 \\ \text{both odd}}}^{\infty} \frac{1}{n \cdot m} \sin(nx) \sin(my).$$

Thus, the time-varying thermal profile of the rod is given by

$$u_t(x, y) \underset{12}{\approx} \frac{16T}{\pi^2} \sum_{\substack{n,m=1 \\ \text{both odd}}}^{\infty} \frac{1}{n \cdot m} \sin(nx) \sin(my) \exp\left(-(n^2 + m^2) \cdot t\right). \qquad \diamondsuit$$

Proposition 12B.3 Heat equation; homogeneous Neumann boundary

Consider the box $\mathbb{X} = [0, \pi] \times [0, \pi]$, *and let* $f : \mathbb{X} \longrightarrow \mathbb{R}$ *be some function describing an initial heat distribution. Suppose* f *has Fourier cosine series*

$$f(x, y) \underset{12}{\approx} \sum_{n,m=0}^{\infty} A_{n,m} \cos(nx) \cos(my),$$

and define the function $u : \mathbb{X} \times \mathbb{R}_+ \longrightarrow \mathbb{R}$ *by*

$$u_t(x, y) \underset{12}{\approx} \sum_{n,m=0}^{\infty} A_{n,m} \cos(nx) \cdot \cos(my) \cdot \exp\left(-(n^2 + m^2) \cdot t \right),$$

for all $(x, y) \in \mathbb{X}$ *and* $t \geq 0$. *Then* u *is the unique solution to the heat equation* '$\partial_t u = \Delta u$', *with homogeneous Neumann boundary conditions*

$$\partial_y u_t(x, 0) = \partial_y u_t(x, \pi) = \partial_x u_t(0, y)$$

$$= \partial_x u_t(\pi, y) = 0, \textit{for all } x, y \in [0, \pi] \textit{ and } t > 0,$$

and initial conditions $u_0(x, y) = f(x, y)$, *for all* $(x, y) \in \mathbb{X}$.

Furthermore, the series defining u *converges semiuniformly on* $\mathbb{X} \times \mathbb{R}_+$.

Proof **Exercise 12B.2.** *Hint:* ⓔ

(a) Show that, when $t = 0$, the two-dimensional Fourier cosine series of $u_0(x, y)$ agrees with that of $f(x, y)$; hence $u_0(x, y) = f(x, y)$.

(b) Show that, for all $t > 0$,

$$\sum_{n,m=0}^{\infty} \left| (n^2 + m^2) \cdot A_{n,m} \cdot e^{-(n^2+m^2)t} \right| < \infty.$$

(c) Apply Proposition F.1, p. 569, to conclude that

$$\partial_t u_t(x, y) \underset{\text{unif}}{\equiv} \sum_{n,m=0}^{\infty} -(n^2 + m^2) A_{n,m} \cos(nx) \cdot \cos(my) \cdot e^{-(n^2+m^2)t} \underset{\text{unif}}{\equiv} \Delta u_t(x, y),$$

for all $(x, y) \in \mathbb{X}$ and $t > 0$.

(d) Observe that, for all $t > 0$,

$$\sum_{n,m=0}^{\infty} n \cdot \left| A_{n,m} \cdot e^{-(n^2+m^2)t} \right| < \infty$$

and

$$\sum_{n,m=0}^{\infty} m \cdot \left| A_{n,m} \cdot e^{-(n^2+m^2)t} \right| < \infty.$$

(e) Apply part (e)(ii) of Theorem 9A.3, p. 183, to conclude that u_t satisfies homogeneous Neumann boundary conditions, for any fixed $t > 0$.

(f) Apply Theorem 5D.8, p. 93, to show that this solution is unique.   □

Example 12B.4 Suppose $\mathbb{X} = [0, \pi] \times [0, \pi]$.

(a) Let $f(x, y) = \cos(3x)\cos(4y) + 2\cos(5x)\cos(6y)$. Then $A_{3,4} = 1$ and $A_{5,6} = 2$, and all other Fourier coefficients are zero. Thus, $u(x, y; t) = \cos(3x)\cos(4y) \cdot e^{-25t} + \cos(5x)\cos(6y) \cdot e^{-59t}$.

(b) Suppose

$$f(x, y) = \begin{cases} 1 & \text{if } 0 \le x < \frac{\pi}{2} \text{ and } 0 \le y < \frac{\pi}{2}; \\ 0 & \text{if } \frac{\pi}{2} \le x \text{ or } \frac{\pi}{2} \le y. \end{cases}$$

We know from Example 9A.4, p. 186, that the two-dimensional Fourier cosine series of f is given by

$$f(x, y) \underset{12}{\approx} \frac{1}{4} + \frac{1}{\pi} \sum_{k=0}^{\infty} \frac{(-1)^k}{2k+1} \cos\left((2k+1)x\right) + \frac{1}{\pi} \sum_{j=0}^{\infty} \frac{(-1)^j}{2j+1} \cos\left((2j+1)y\right)$$

$$+ \frac{4}{\pi^2} \sum_{k,j=1}^{\infty} \frac{(-1)^{k+j}}{(2k+1)(2j+1)} \cos\left((2k+1)x\right) \cdot \cos\left((2j+1)y\right).$$

Thus, the solution to the heat equation, with initial conditions $u_0(x, y) = f(x, y)$ and homogeneous Neumann boundary conditions, is given by

$$u_t(x, y) \underset{12}{\approx} \frac{1}{4} + \frac{1}{\pi} \sum_{k=0}^{\infty} \frac{(-1)^k}{2k+1} \cos\left((2k+1)x\right) \cdot e^{-(2k+1)^2 t}$$

$$+ \frac{1}{\pi} \sum_{j=0}^{\infty} \frac{(-1)^j}{2j+1} \cos\left((2j+1)y\right) \cdot e^{-(2j+1)^2 t}$$

$$+ \frac{4}{\pi^2} \sum_{k,j=1}^{\infty} \frac{(-1)^{k+j}}{(2k+1)(2j+1)} \cos\left((2k+1)x\right)$$

$$\times \cos\left((2j+1)y\right) \cdot e^{-[(2k+1)^2+(2j+1)^2] \cdot t}.   \Diamond$$

Ⓔ **Exercise 12B.3** Let $X, Y > 0$, and let $\mathbb{X} := [0, X] \times [0, Y]$. Let $\kappa > 0$ be a diffusion constant, and consider the general two-dimensional heat equation

$$\partial_t u = \kappa \Delta u. \tag{12B.1}$$

(a) Generalize Proposition 12B.1 to find the solution to equation (12B.1) on $\mathbb{X}$ satisfying prescribed initial conditions and homogeneous Dirichlet boundary conditions.

(b) Generalize Proposition 12B.3 to find the solution to equation (12B.1) on $\mathbb{X}$ satisfying prescribed initial conditions and homogeneous Neumann boundary conditions.

In both cases, prove that your solution converges, satisfies the desired initial conditions and boundary conditions, and satisfies equation (12B.1) (*Hint:* Imitate the strategy suggested in Exercises 12B.1 and 12B.2.) ◆

Exercise 12B.4 Let $f : \mathbb{X} \longrightarrow \mathbb{R}$ and suppose the Fourier sine series of f satisfies ⓔ the constraint

$$\sum_{n,m=1}^{\infty} (n^2 + m^2)|B_{nm}| < \infty.$$

Imitate Proposition 12B.1 to find a Fourier series solution to the initial value problem for the two-dimensional *free Schrödinger equation*

$$\mathbf{i}\partial_t \, \omega = -\frac{1}{2}\, \triangle\, \omega \tag{12B.2}$$

on the box $\mathbb{X} = [0, \pi]^2$, with homogeneous Dirichlet boundary conditions. Prove that your solution converges, satisfies the desired initial conditions and boundary conditions, and satisfies equation (12B.2). (*Hint:* Imitate the strategy suggested in Exercise 12B.1, and also Exercise 12D.1, p. 264.) ◆

12B(ii) Nonhomogeneous boundary conditions

Prerequisites: §12B(i), §12A.　　　**Recommended:** §12C(ii).

Proposition 12B.5 *Heat equation on box; nonhomogeneous Dirichlet BC*
Let $\mathbb{X} = [0, \pi] \times [0, \pi]$. *Let* $f : \mathbb{X} \longrightarrow \mathbb{R}$, *and let* $L, R, T, B : [0, \pi] \longrightarrow \mathbb{R}$ *be functions. Consider the heat equation*

$$\partial_t u(x, y; t) = \triangle u(x, y; t),$$

with initial conditions

$$u(x, y; 0) = f(x, y), \quad \text{for all } (x, y) \in \mathbb{X}, \tag{12B.3}$$

and nonhomogeneous Dirichlet boundary conditions:

$$\left.\begin{array}{ll} u(x, \pi; t) = T(x), & u(x, 0; t) = B(x), \quad \text{for all } x \in [0, \pi] \\ u(0, y; t) = L(y), & u(\pi, y; t) = R(y), \quad \text{for all } y \in [0, \pi] \end{array}\right\} \text{ for all } t > 0.$$

$$\tag{12B.4}$$

This problem is solved as follows.

(1) *Let* $w(x, y)$ *be the solution[1] to the* Laplace equation '$\triangle w(x, y) = 0$', *with nonhomogeneous Dirichlet BC (equation (12B.4)).*

---

[1] Obtained from Proposition 12A.4, p. 248, for example.

(2) *Define $g(x, y) := f(x, y) - w(x, y)$. Let $v(x, y; t)$ be the solution[2] to the heat equation '$\partial_t v(x, y; t) = \Delta v(x, y; t)$' with initial conditions $v(x, y; 0) = g(x, y)$ and homogeneous Dirichlet BC.*

(3) *Define $u(x, y; t) := v(x, y; t) + w(x, y)$. Then $u(x, y; t)$ is a solution to the heat equation with initial conditions (12B.3) and nonhomogeneous Dirichlet BC (equation (12B.4)).*

Ⓔ *Proof* **Exercise 12B.5.** □

Interpretation In Proposition 12B.5, the function $w(x, y)$ represents the *long-term thermal equilibrium* that the system is 'trying' to attain. The function $g(x, y) = f(x, y) - w(x, y)$ thus measures the *deviation* between the current state and this equilibrium, and the function $v(x, y; t)$ thus represents how this 'transient' deviation decays to zero over time.

Example 12B.6 Suppose $T(x) = \sin(2x)$ and $R \equiv L \equiv 0$ and $B \equiv 0$. Then Proposition 12A.4, p. 248, says

$$w(x, y) = \frac{\sin(2x)\sinh(2y)}{\sinh(2\pi)}.$$

Suppose $f(x, y) := \sin(2x)\sin(y)$. Then

$$g(x, y) = f(x, y) - w(x, y) = \sin(2x)\sin(y) - \frac{\sin(2x)\sinh(2y)}{\sinh(2\pi)}$$

$$\underset{(*)}{=} \sin(2x)\sin(y) - \left(\frac{\sin(2x)}{\sinh(2\pi)}\right)\left(\frac{2\sinh(2\pi)}{\pi}\right)\sum_{m=1}^{\infty}\frac{m(-1)^{m+1}}{2^2 + m^2}\cdot\sin(my)$$

$$= \sin(2x)\sin(y) - \frac{2\sin(2x)}{\pi}\sum_{m=1}^{\infty}\frac{m(-1)^{m+1}}{4 + m^2}\cdot\sin(my).$$

Here (*) is because Example 7A.3, p. 144, says

$$\sinh(2y) = \frac{2\sinh(2\pi)}{\pi}\sum_{m=1}^{\infty}\frac{m(-1)^{m+1}}{2^2 + m^2}\cdot\sin(my).$$

Thus, Proposition 12B.1, p. 251, says that

$$v(x, y; t) = \sin(2x)\sin(y)e^{-5t} - \frac{2\sin(2x)}{\pi}\sum_{m=1}^{\infty}\frac{m(-1)^{m+1}}{4 + m^2}\cdot\sin(mx)e^{-(4+m^2)t}.$$

Finally, Proposition 12B.5 says the solution is given by

$$u(x, y; t) := v(x, y; t) + \frac{\sin(2x)\sinh(2y)}{\sinh(2\pi)}.$$ ◇

---

[2] Obtained from Proposition 12B.1, p. 251, for example.

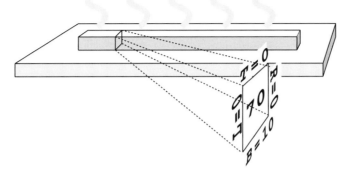

Figure 12B.2. Temperature distribution of a baguette.

Example 12B.7 The cooling baguette
A freshly baked baguette is removed from the oven and left on a wooden plank to cool near the window. The baguette is initially at a uniform temperature of 90 °C; the air temperature is 20 °C, and the temperature of the wooden plank (which was sitting in the sunlight) is 30 °C.

Mathematically model the cooling process near the center of the baguette. How long will it be before the baguette is cool enough to eat (assuming 'cool enough' is below 40 °C)?

Solution For simplicity, we will assume the baguette has a square cross-section (and dimensions $\pi \times \pi$, of course). If we confine our attention to the middle of the baguette, we are far from the endpoints, so that we can neglect the longitudinal dimension and treat this as a two-dimensional problem.

Suppose the temperature distribution along a cross-section through the center of the baguette is given by the function $u(x, y; t)$. To simplify the problem, we will subtract 20 °C off all temperatures. Thus, in the notation of Proposition 12B.5, the boundary conditions are:

$$L(y) = R(y) = T(x) = 0 \quad \text{(the air)};$$

$$B(x) = 10 \qquad\qquad\qquad \text{(the wooden plank)},$$

and our initial temperature distribution is $f(x, y) = 70$ (see Figure 12B.2).

From Proposition 12A.1, p. 244, we know that the long-term equilibrium for these boundary conditions is given by

$$w(x, y) \underset{\text{\tiny 12}}{\approx} \frac{40}{\pi} \sum_{\substack{n=1 \\ n \text{ odd}}}^{\infty} \frac{1}{n \sinh(n\pi)} \sin(nx) \cdot \sinh(n(\pi - y)).$$

We want to represent this as a two-dimensional Fourier sine series. To do this, we need the (one-dimensional) Fourier sine series for $\sinh(nx)$. We set $\alpha = n$ in Example 7A.3, p. 144, and obtain

$$\sinh(nx) \underset{L2}{\approx} \frac{2\sinh(n\pi)}{\pi} \sum_{m=1}^{\infty} \frac{m(-1)^{m+1}}{n^2 + m^2} \cdot \sin(mx). \qquad (12B.5)$$

Thus,

$$\sinh\left(n(\pi - y)\right) \underset{L2}{\approx} \frac{2\sinh(n\pi)}{\pi} \sum_{m=1}^{\infty} \frac{m(-1)^{m+1}}{n^2 + m^2} \cdot \sin(m\pi - my)$$

$$= \frac{2\sinh(n\pi)}{\pi} \sum_{m=1}^{\infty} \frac{m}{n^2 + m^2} \cdot \sin(my),$$

because $\sin(m\pi - ny) = \sin(m\pi)\cos(ny) - \cos(m\pi)\sin(ny) = (-1)^{m+1}\sin(ny)$. Substituting this into equation (12B.5) yields

$$w(x, y) \underset{L2}{\approx} \frac{80}{\pi^2} \sum_{\substack{n=1 \\ n \text{ odd}}}^{\infty} \sum_{m=1}^{\infty} \frac{m \cdot \sinh(n\pi)}{n \cdot \sinh(n\pi)(n^2 + m^2)} \sin(nx) \cdot \sin(my)$$

$$= \frac{80}{\pi^2} \sum_{\substack{n=1 \\ n \text{ odd}}}^{\infty} \sum_{m=1}^{\infty} \frac{m \cdot \sin(nx) \cdot \sin(my)}{n \cdot (n^2 + m^2)}. \qquad (12B.6)$$

Now, the initial temperature distribution is the constant function with value 70. Take the two-dimensional sine series from Example 9A.2, p. 182, and multiply it by 70, to obtain

$$f(x, y) = 70 \underset{L2}{\approx} \frac{1120}{\pi^2} \sum_{\substack{n,m=1 \\ \text{both odd}}}^{\infty} \frac{1}{n \cdot m} \sin(nx) \sin(my).$$

Thus,

$$g(x, y) = f(x, y) - w(x, y)$$

$$\underset{L2}{\approx} \frac{1120}{\pi^2} \sum_{\substack{n,m=1 \\ \text{both odd}}}^{\infty} \frac{\sin(nx) \cdot \sin(my)}{n \cdot m}$$

$$- \frac{80}{\pi^2} \sum_{\substack{n=1 \\ n \text{ odd}}}^{\infty} \sum_{m=1}^{\infty} \frac{m \cdot \sin(nx) \cdot \sin(my)}{n \cdot (n^2 + m^2)}.$$

Thus,

$$v(x, y; t) \underset{L2}{\approx} \frac{1120}{\pi^2} \sum_{\substack{n,m=1 \\ \text{both odd}}}^{\infty} \frac{\sin(nx) \cdot \sin(my)}{n \cdot m} \exp\left(-(n^2 + m^2)t\right)$$

$$-\frac{80}{\pi^2} \sum_{\substack{n=1 \\ n \text{ odd}}}^{\infty} \sum_{m=1}^{\infty} \frac{m \cdot \sin(nx) \cdot \sin(my)}{n \cdot (n^2 + m^2)} \exp\left(-(n^2 + m^2)t\right).$$

If we combine the second term in this expression with equation (12B.6), we obtain the final answer:

$$u(x, y; t) = v(x, y; t) + w(x, y)$$

$$\underset{L2}{\approx} \frac{1120}{\pi^2} \sum_{\substack{n,m=1 \\ \text{both odd}}}^{\infty} \frac{\sin(nx) \cdot \sin(my)}{n \cdot m} \exp\left(-(n^2 + m^2)t\right)$$

$$+\frac{80}{\pi^2} \sum_{\substack{n=1 \\ n \text{ odd}}}^{\infty} \sum_{m=1}^{\infty} \frac{m \cdot \sin(nx) \cdot \sin(my)}{n \cdot (n^2 + m^2)} \left[1 - e^{-(n^2+m^2)t}\right]. \qquad \diamond$$

12C The Poisson problem on a square

12C(i) Homogeneous boundary conditions

Prerequisites: §9A, §5C, §1D.   **Recommended:** §11C, §7C(v).

Proposition 12C.1 *Let* $\mathbb{X} = [0, \pi] \times [0, \pi]$, *and let* $q : \mathbb{X} \longrightarrow \mathbb{R}$ *be some function with semiuniformly convergent Fourier sine series:*

$$q(x, y) \underset{L2}{\approx} \sum_{n,m=1}^{\infty} Q_{n,m} \sin(nx) \sin(my).$$

Define the function $u : \mathbb{X} \longrightarrow \mathbb{R}$ *by*

$$u(x, y) \underset{\text{unif}}{\equiv} \sum_{n,m=1}^{\infty} \frac{-Q_{n,m}}{n^2 + m^2} \sin(nx) \sin(my),$$

for all $(x, y) \in \mathbb{X}$.

Then u *is the unique solution to the Poisson equation '*$\Delta u(x, y) = q(x, y)$*', satisfying homogeneous Dirichlet boundary conditions* $u(x, 0) = u(0, y) = u(x, \pi) = u(\pi, y) = 0$.

Proof **Exercise 12C.1.**   Ⓔ
(a) Use Proposition F.1, p. 569, to show that u satisfies the Poisson equation on int$(\mathbb{X})$.
(b) Use Proposition 9A.3(e), p. 183, to show that u satisfies homogeneous Dirichlet BC.
(c) Apply Theorem 5D.5(a), p. 91, to conclude that this solution is unique.   □

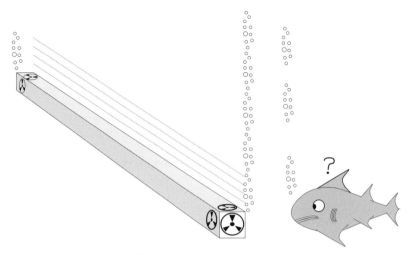

Figure 12C.1. Jettisoned fuel rod in the Arctic Ocean.

Example 12C.2 The jettisoned fuel rod

A nuclear submarine beneath the Arctic Ocean has jettisoned a fuel rod from its reactor core (Figure 12C.1). The fuel rod is a very long, narrow, enriched uranium bar with square cross-section. The radioactivity causes the fuel rod to be uniformly heated from within at a rate of Q, but the rod is immersed in freezing Arctic water. We want to compute its internal temperature distribution.

Solution The rod is immersed in freezing cold water, and is a good conductor, so we can assume that its outer surface takes the the surrounding water temperature of $0\,°C$. Hence, we assume homogeneous Dirichlet boundary conditions.

Endow the rod with coordinate system (x, y, z), where z runs along the length of the rod. Since the rod is extremely long relative to its cross-section, we can neglect the z coordinate, and reduce to a two-dimensional equation. The uniform heating is described by a constant function: $q(x, y) = Q$. From Example 9A.2, p. 182, we know that the constant function 1 has a two-dimensional Fourier sine series:

$$1 \underset{12}{\approx} \frac{16}{\pi^2} \sum_{\substack{n,m=1 \\ \text{both odd}}}^{\infty} \frac{1}{n \cdot m} \sin(nx) \sin(my).$$

Thus,

$$q(x, y) \underset{12}{\approx} \frac{16Q}{\pi^2} \sum_{\substack{n,m=1 \\ \text{both odd}}}^{\infty} \frac{1}{n \cdot m} \sin(nx) \sin(my).$$

The temperature distribution must satisfy Poisson's equation. Thus, the temperature distribution is given by

$$u(x, y) \underset{\text{unif}}{\equiv} \frac{-16Q}{\pi^2} \sum_{\substack{n,m=1 \\ \text{both odd}}}^{\infty} \frac{1}{n \cdot m \cdot (n^2 + m^2)} \sin(nx) \sin(my). \qquad \Diamond$$

Example 12C.3 Suppose $q(x, y) = x \cdot y$. Then the solution to the Poisson equation $\triangle u = q$ on the square, with homogeneous Dirichlet boundary conditions, is given by

$$u(x, y) \underset{\text{unif}}{\equiv} 4 \sum_{n,m=1}^{\infty} \frac{(-1)^{n+m+1}}{nm \cdot (n^2 + m^2)} \sin(nx) \sin(my).$$

To see this, recall from Example 9A.1, p. 181, that the two-dimensional Fourier sine series for $q(x, y)$ is given by

$$xy \underset{\text{L2}}{\approx} 4 \sum_{n,m=1}^{\infty} \frac{(-1)^{n+m}}{nm} \sin(nx) \sin(my).$$

Now apply Proposition 12C.1. $\qquad \Diamond$

Proposition 12C.4 *Let* $\mathbb{X} = [0, \pi] \times [0, \pi]$, *and let* $q : \mathbb{X} \longrightarrow \mathbb{R}$ *be some function with* semiuniformly convergent Fourier cosine series:

$$q(x, y) \underset{\text{L2}}{\approx} \sum_{n,m=0}^{\infty} Q_{n,m} \cos(nx) \cos(my).$$

Suppose that $Q_{0,0} = 0$. *Fix some constant* $K \in \mathbb{R}$, *and define the function* $u : \mathbb{X} \longrightarrow \mathbb{R}$ *by*

$$u(x, y) \underset{\text{unif}}{\equiv} \sum_{\substack{n,m=0 \\ \text{not both zero}}}^{\infty} \frac{-Q_{n,m}}{n^2 + m^2} \cos(nx) \cos(my) + K, \qquad (12\text{C}.1)$$

for all $(x, y) \in \mathbb{X}$. *Then* u *is a solution to the* Poisson equation '$\triangle u(x, y) = q(x, y)$', *satisfying homogeneous Neumann boundary conditions* $\partial_y u(x, 0) = \partial_x u(0, y) = \partial_y u(x, \pi) = \partial_x u(\pi, y) = 0$.

Furthermore, all solutions to this Poisson equation with these boundary conditions have the form of equation (12C.1).

If $Q_{0,0} \neq 0$, *however, the problem has no solution.*

Proof **Exercise 12C.2.** ⒠
(a) Use Proposition F.1, p. 569, to show that u satisfies the Poisson equation on int $(\mathbb{X})$.
(b) Use Proposition 9A.3, p. 183, to show that u satisfies homogeneous Neumann BC.

(c) Apply Theorem 5D.5(c), p. 91, to conclude that this solution is unique up to addition
    of a constant.   ☐

Ⓔ **Exercise 12C.3** Mathematically, it is clear that the solution of Proposition 12C.4
cannot be well defined if $Q_{0,0} \neq 0$. Provide a physical explanation for why this is
to be expected.   ♦

Example 12C.5 Suppose $q(x, y) = \cos(2x) \cdot \cos(3y)$. Then the solution to the
Poisson equation $\Delta u = q$ on the square, with homogeneous Neumann boundary
conditions, is given by

$$u(x, y) = \frac{-\cos(2x) \cdot \cos(3y)}{13}.$$

To see this, note that the two-dimensional Fourier cosine series of $q(x, y)$ is just
$\cos(2x) \cdot \cos(3y)$. In other words, $A_{2,3} = 1$, and $A_{n,m} = 0$ for all other n and m. In
particular, $A_{0,0} = 0$, so we can apply Proposition 12C.4 to conclude that

$$u(x, y) = \frac{-\cos(2x) \cdot \cos(3y)}{2^2 + 3^2} = \frac{-\cos(2x) \cdot \cos(3y)}{13}.$$ ◇

12C(ii) Nonhomogeneous boundary conditions

Prerequisites: §12C(i), §12A.   **Recommended:** §12B(ii).

Proposition 12C.6 Poisson equation on box; nonhomogeneous Dirichlet BC
Let $X = [0, \pi] \times [0, \pi]$. *Let* $q : X \longrightarrow \mathbb{R}$ *and* $L, R, T, B : [0, \pi] \longrightarrow \mathbb{R}$ *be functions. Consider the Poisson equation*

$$\Delta u(x, y) = q(x, y), \tag{12C.2}$$

with nonhomogeneous Dirichlet boundary conditions:

$$\begin{aligned}
u(x, \pi) = T(x) \quad and \quad u(x, 0) = B(x), &\quad for\ all\ x \in [0, \pi], \\
u(0, y) = L(y) \quad and \quad u(\pi, y) = R(y), &\quad for\ all\ y \in [0, \pi]
\end{aligned} \tag{12C.3}$$

(see Figure 12A.1(b), p. 245). This problem is solved as follows.

(1) *Let* $v(x, y)$ *be the solution*[3] *to the Poisson equation (12C.2) with homogeneous Dirichlet
    BC* $v(x, 0) = v(0, y) = v(x, \pi) = v(\pi, y) = 0$.
(2) *Let* $w(x, y)$ *be the solution*[4] *to the Laplace equation '$\Delta w(x, y) = 0$', with the nonho-
    mogeneous Dirichlet BC from equation (12C.3).*
(3) *Define* $u(x, y) := v(x, y) + w(x, y)$; *then* $u(x, y)$ *is a solution to the Poisson problem
    with the nonhomogeneous Dirichlet BC from equation (12C.3).*

---

[3] Obtained from Proposition 12C.1, p. 259, for example.
[4] Obtained from Proposition 12A.4, p. 248, for example.

Proof **Exercise 12C.4.** □ Ⓔ

Example 12C.7 Suppose $q(x, y) = x \cdot y$. Find the solution to the Poisson equation $\triangle u = q$ on the square, with *nonhomogeneous* Dirichlet boundary conditions:

$$u(0, y) = 0; \qquad u(\pi, y) = 0; \qquad u(x, 0) = 0; \tag{12C.4}$$

$$u(x, \pi) = T(x) = \begin{cases} x & \text{if } 0 \le x \le \pi/2; \\ \pi/2 - x & \text{if } \pi/2 < x \le \pi. \end{cases} \tag{12C.5}$$

(see Figure 7C.4(b), p. 158).

Solution In Example 12C.3, we found the solution to the Poisson equation $\triangle v = q$, with *homogeneous* Dirichlet boundary conditions; it was:

$$v(x, y) \underset{\text{unif}}{\equiv} 4 \sum_{n,m=1}^{\infty} \frac{(-1)^{n+m+1}}{nm \cdot (n^2 + m^2)} \sin(nx) \sin(my).$$

In Example 12A.6, p. 250, we found the solution to the Laplace Equation $\triangle w = 0$, with nonhomogeneous Dirichlet boundary conditions (12C.4) and (12C.5); it was:

$$w(x, y) \underset{12}{\approx} \frac{4}{\pi} \sum_{\substack{n=1 \\ n \text{ odd;} \\ n=2k+1}}^{\infty} \frac{(-1)^k}{n^2 \sinh(n\pi)} \sin(nx) \sinh(ny).$$

Thus, according to Proposition 12C.6, the solution to the nonhomogeneous Poisson problem is given by

$$u(x, y) = v(x, y) + w(x, y)$$

$$\underset{12}{\approx} 4 \sum_{n,m=1}^{\infty} \frac{(-1)^{n+m+1}}{nm \cdot (n^2 + m^2)} \sin(nx) \sin(my)$$

$$+ \frac{4}{\pi} \sum_{\substack{n=1 \\ n \text{ odd;} \\ n=2k+1}}^{\infty} \frac{(-1)^k}{n^2 \sinh(n\pi)} \sin(nx) \sinh(ny). \qquad \diamond$$

12D The wave equation on a square (the square drum)

Prerequisites: §9A, §5B, §5C, §2B(ii), Appendix F.　　**Recommended:** §11B, §7C(v).

Imagine a drumskin stretched tightly over a square frame. At equilibrium, the drumskin is perfectly flat, but, if we strike the skin, it will vibrate, meaning that the membrane will experience vertical displacements from equilibrium. Let

$\mathbb{X} = [0, \pi] \times [0, \pi]$ represent the square skin, and, for any point $(x, y) \in \mathbb{X}$ on the drumskin and time $t > 0$, let $u(x, y; t)$ be the vertical displacement of the drum. Then u will obey the two-dimensional wave equation:

$$\partial_t^2 u(x, y; t) = \Delta u(x, y; t). \tag{12D.1}$$

However, since the skin is held down along the edges of the box, the function u will also exhibit homogeneous *Dirichlet* boundary conditions:

$$\left.\begin{array}{ll} u(x, \pi; t) = 0, & u(x, 0; t) = 0, \quad \text{for all } x \in [0, \pi] \\ u(0, y; t) = 0, & u(\pi, y; t) = 0, \quad \text{for all } y \in [0, \pi] \end{array}\right\} \quad \text{for all } t > 0. \tag{12D.2}$$

Proposition 12D.1 *Initial position for square drumskin*

Let $\mathbb{X} = [0, \pi] \times [0, \pi]$, and let $f_0 : \mathbb{X} \longrightarrow \mathbb{R}$ be a function describing the initial displacement of the drumskin. Suppose f_0 has Fourier sine series

$$f_0(x, y) \underset{\text{unif}}{\equiv} \sum_{n,m=1}^{\infty} B_{n,m} \sin(nx) \sin(my),$$

such that

$$\sum_{n,m=1}^{\infty} (n^2 + m^2)|B_{n,m}| < \infty. \tag{12D.3}$$

Define the function $w : \mathbb{X} \times \mathbb{R}_{\not+} \longrightarrow \mathbb{R}$ by

$$w(x, y; t) \underset{\text{unif}}{\equiv} \sum_{n,m=1}^{\infty} B_{n,m} \sin(nx) \cdot \sin(my) \cdot \cos\left(\sqrt{n^2 + m^2} \cdot t\right), \tag{12D.4}$$

for all $(x, y) \in \mathbb{X}$ and $t \geq 0$. Then series (12D.4) converges uniformly, and $w(x, y; t)$ is the unique solution to the wave equation (12D.1), satisfying the Dirichlet boundary conditions (12D.2), as well as

$$\left.\begin{array}{l} \text{initial position:} \quad w(x, y, 0) = f_0(x, y) \\ \text{initial velocity:} \quad \partial_t w(x, y, 0) = 0 \end{array}\right\} \quad \text{for all } (x, y) \in \mathbb{X}.$$

Ⓔ *Proof* **Exercise 12D.1.**
(a) Use the hypothesis (12D.3) and Proposition F.1, p. 569, to conclude that

$$\partial_t^2 w(x, y; t) \underset{\text{unif}}{\equiv} - \sum_{n,m=1}^{\infty} (n^2 + m^2) \cdot B_{n,m} \sin(nx) \cdot \sin(my) \cdot \cos\left(\sqrt{n^2 + m^2} \cdot t\right)$$

$$\underset{\text{unif}}{\equiv} \Delta w(x, y; t),$$

for all $(x, y) \in \mathbb{X}$ and $t > 0$.
(b) Check that the Fourier series (12D.4) converges uniformly.

(c) Use Theorem 9A.3(d)(ii), p. 183, to conclude that w satisfies Dirichlet boundary conditions.

(d) Set $t = 0$ to check the initial position.

(e) Set $t = 0$ and use Proposition F.1, p. 569, to check the initial velocity.

(f) Apply Theorem 5D.11, p. 96, to show that this solution is unique. ☐

Example 12D.2 Suppose $f_0(x, y) = \sin(2x) \cdot \sin(3y)$. Then the solution to the wave equation on the square, with initial position f_0 and homogeneous Dirichlet boundary conditions, is given by

$$w(x, y; t) = \sin(2x) \cdot \sin(3y) \cdot \cos(\sqrt{13}\, t).$$

To see this, note that the two-dimensional Fourier sine series of $f_0(x, y)$ is just $\sin(2x) \cdot \sin(3y)$. In other words, $B_{2,3} = 1$, and $B_{n,m} = 0$ for all other n and m. Apply Proposition 12D.1 to conclude that

$$w(x, y; t) = \sin(2x) \cdot \sin(3y) \cdot \cos\left(\sqrt{2^2 + 3^2}\, t\right)$$

$$= \sin(2x) \cdot \sin(3y) \cdot \cos(\sqrt{13}\, t). \qquad \diamondsuit$$

Proposition 12D.3 Initial velocity for square drumskin

Let $X = [0, \pi] \times [0, \pi]$, and let $f_1 : X \longrightarrow \mathbb{R}$ be a function describing the initial velocity of the drumskin. Suppose f_1 has Fourier sine series

$$f_1(x, y) \underset{\text{unif}}{\equiv} \sum_{n,m=1}^{\infty} B_{n,m} \sin(nx) \sin(my),$$

such that

$$\sum_{n,m=1}^{\infty} \sqrt{n^2 + m^2} \cdot |B_{n,m}| < \infty. \qquad (12D.5)$$

Define the function $v : X \times \mathbb{R}_{\not{}} \longrightarrow \mathbb{R}$ by

$$v(x, y; t) \underset{\text{unif}}{\equiv} \sum_{n,m=1}^{\infty} \frac{B_{n,m}}{\sqrt{n^2 + m^2}} \sin(nx) \cdot \sin(my) \cdot \sin\left(\sqrt{n^2 + m^2} \cdot t\right),$$

$$(12D.6)$$

for all $(x, y) \in X$ and $t \geq 0$. Then the series (12D.6) converges uniformly, and $v(x, y; t)$ is the unique solution to the wave equation (12D.1), satisfying the Dirichlet boundary conditions (12D.2), as well as

$$\left. \begin{array}{ll} \text{initial position:} & v(x, y, 0) = 0 \\ \text{initial velocity:} & \partial_t v(x, y, 0) = f_1(x, y) \end{array} \right\} \quad \textit{for all } (x, y) \in X.$$

Ⓔ *Proof* **Exercise 12D.2.**

(a) Use the hypothesis (12D.5) and Proposition F.1, p. 569, to conclude that

$$\partial_t^2 v(x, y; t) \underset{\text{unif}}{\equiv} - \sum_{n,m=1}^{\infty} \sqrt{n^2 + m^2} \cdot B_{n,m} \sin(nx) \cdot \sin(my) \cdot \cos\left(\sqrt{n^2 + m^2} \cdot t\right)$$

$$\underset{\text{unif}}{\equiv} \Delta v(x, y; t),$$

for all $(x, y) \in \mathbb{X}$ and $t > 0$.

(b) Check that the Fourier series (12D.6) converges uniformly.

(c) Use Theorem 9A.3(d)(ii), p. 183, to conclude that $v(x, y; t)$ satisfies Dirichlet boundary conditions.

(d) Set $t = 0$ to check the initial position.

(e) Set $t = 0$ and use Proposition F.1, p. 569, to check the initial velocity.

(f) Apply Theorem 5D.11, p. 96, to show that this solution is unique.   □

Remark Note that it is important in these theorems, not only for the Fourier series (12D.4) and (12D.6) to converge uniformly, but also for their formal *second derivative* series to converge uniformly. This is not guaranteed. This is the reason for imposing the hypotheses (12D.3) and (12D.5).

Example 12D.4 Suppose

$$f_1(x, y) = \frac{16}{\pi^2} \sum_{\substack{n,m=1 \\ \text{both odd}}}^{99} \frac{1}{n \cdot m} \sin(nx) \sin(my).$$

(This is a partial sum of the two-dimensional Fourier sine series for the constant function $\tilde{f}_1(x, y) \equiv 1$, from Example 9A.2, p. 182.) Then the solution to the two-dimensional wave equation, with homogeneous Dirichlet boundary conditions and initial velocity f_1, is given by

$$w(x, y; t) \underset{12}{\approx} \frac{16}{\pi^2} \sum_{\substack{n,m=1 \\ \text{both odd}}}^{99} \frac{1}{n \cdot m \cdot \sqrt{n^2 + m^2}} \sin(nx) \sin(my) \sin\left(\sqrt{n^2 + m^2} \cdot t\right). \quad \diamondsuit$$

12E Practice problems

12.1 Let $f(y) = 4 \sin(5y)$ for all $y \in [0, \pi]$.

(a) Solve the two-dimensional Laplace equation ($\Delta u = 0$) on the square domain $\mathbb{X} = [0, \pi] \times [0, \pi]$, with nonhomogeneous Dirichlet boundary conditions:

$$u(x, 0) = 0 \quad \text{and} \quad u(x, \pi) = 0, \qquad \text{for all } x \in [0, \pi];$$
$$u(0, y) = 0 \quad \text{and} \quad u(\pi, y) = f(y), \qquad \text{for all } y \in [0, \pi].$$

(b) *Verify* your solution to part (a) (i.e. check boundary conditions, Laplacian, etc.).

12.2 Let $f_1(x, y) = \sin(3x)\sin(4y)$.

(a) Solve the two-dimensional wave equation ($\partial_t^2 u = \Delta u$) on the square domain $\mathbb{X} = [0, \pi] \times [0, \pi]$, with on the square domain $\mathbb{X} = [0, \pi] \times [0, \pi]$, with homogeneous Dirichlet boundary conditions, and initial conditions:

initial position	$u(x, y, 0) = 0$	for all $(x, y) \in \mathbb{X}$
initial velocity	$\partial_t u(x, y, 0) = f_1(x, y)$	for all $(x, y) \in \mathbb{X}$

(b) *Verify* that solution in part (a) satisfies the required initial conditions (don't worry about boundary conditions or checking the wave equation).

12.3 Solve the two-dimensional Laplace equation $\Delta h = 0$ on the square domain $\mathbb{X} = [0, \pi]^2$, with inhomogeneous Dirichlet boundary conditions:

(a) $h(\pi, y) = \sin(2y)$ and $h(0, y) = 0$,   for all $y \in [0, \pi]$;
     $h(x, 0) = 0 = h(x, \pi)$ for all $x \in [0, \pi]$;
(b) $h(\pi, y) = 0$ and $h(0, y) = \sin(4y)$,   for all $y \in [0, \pi]$;
     $h(x, \pi) = \sin(3x)$ and $h(x, 0) = 0$,   for all $x \in [0, \pi]$.

12.4 Let $\mathbb{X} = [0, \pi]^2$ and let $q(x, y) = \sin(x) \cdot \sin(3y) + 7\sin(4x) \cdot \sin(2y)$. Solve the Poisson equation $\Delta u(x, y) = q(x, y)$ with homogeneous Dirichlet boundary conditions.

12.5 Let $\mathbb{X} = [0, \pi]^2$. Solve the heat equation $\partial_t u(x, y; t) = \Delta u(x, y; t)$ on $\mathbb{X}$, with initial conditions $u(x, y; 0) = \cos(5x) \cdot \cos(y)$. and homogeneous Neumann boundary conditions.

12.6 Let $f(x, y) = \cos(2x)\cos(3y)$. Solve the following boundary value problems on the square domain $\mathbb{X} = [0, \pi]^2$ (*Hint:* See Problem 9.3, p. 195).

(a) Solve the two-dimensional heat equation $\partial_t u = \Delta u$, with homogeneous Neumann boundary conditions, and initial conditions $u(x, y; 0) = f(x, y)$.
(b) Solve the two-dimensional wave equation $\partial_t^2 u = \Delta u$, with homogeneous Dirichlet boundary conditions, initial position $w(x, y; 0) = f(x, y)$ and initial velocity $\partial_t w(x, y; 0) = 0$.
(c) Solve the two-dimensional Poisson equation $\Delta u = f$ with homogeneous Neumann boundary conditions.
(d) Solve the two-dimensional Poisson equation $\Delta u = f$ with homogeneous Dirichlet boundary conditions.
(e) Solve the two-dimensional Poisson equation $\Delta v = f$ with inhomogeneous Dirichlet boundary conditions:

$$v(\pi, y) = \sin(2y), \quad v(0, y) = 0, \quad \text{for all } y \in [0, \pi];$$

$$v(x, 0) = 0 = v(x, \pi), \quad \text{for all } x \in [0, \pi].$$

12.7 Let $\mathbb{X} = [0, \pi]^2$ be a box of sidelength π. Let $f(x, y) = \sin(3x) \cdot \sinh(3y)$. (*Hint:* See Problem 9.4, p. 195.)

(a) Solve the heat equation on $\mathbb{X}$, with initial conditions $u(x, y; 0) = f(x, y)$, and homogeneous Dirichlet boundary conditions.

(b) Let $T(x) = \sin(3x)$. Solve the Laplace equation $\triangle u(x, y) = 0$ on the box, with inhomogeneous Dirichlet boundary conditions:

$$u(x, \pi) = T(x) \quad \text{and} \quad u(x, 0) = 0 \quad \text{for} \quad x \in [0, \pi];$$

$$u(0, y) = 0 = u(\pi, y), \quad \text{for} \quad y \in [0, \pi].$$

(c) Solve the heat equation on the box with initial conditions $u(x, y; 0) = 0$, on the box $\mathbb{X}$, and the same inhomogeneous Dirichlet boundary conditions as in part (b).

12.8 In Example 12D.4, why can't we apply Theorem 12D.3 to the full Fourier series for the function $f_1 = 1$? (*Hint:* Is equation (12D.5) satisfied?)

12.9 For the solutions of the heat equation and Poisson equation, in Propositions 12B.1, 12B.3, and 12C.1, we did not need to impose explicit hypotheses guaranteeing the uniform convergence of the given series (and its derivatives). But we do need explicit hypotheses to get convergence for the wave equation in Propositions 12D.1 and 12.D3. Why is this?

13

Boundary value problems on a cube

Mathematical Analysis is as extensive as nature herself.

Jean Joseph Fourier

The Fourier series technique used to solve boundary value problems (BVPs) on a square box extends readily to three-dimensional cubes and, indeed, to rectilinear domains in any number of dimensions. As in Chapter 12, we will confine our exposition to the cube $[0, \pi]^3$, and assume that the physical constants in the various equations are all set to one. Thus, the heat equation becomes '$\partial_t u = \triangle u$', the wave equation is '$\partial_t^2 u = \triangle u$', etc. This allows us to develop the solution methods with minimum technicalities. The extension of each solution method to equations with arbitrary physical constants on an arbitrary box $[0, X] \times [0, Y] \times [0, Z]$ (for some $X, Y, Z > 0$) is left as a straightforward (but important!) exercise.

We will use the following notation.

- The cube of dimensions $\pi \times \pi \times \pi$ is denoted $\mathbb{X} = [0, \pi] \times [0, \pi] \times [0, \pi] = [0, \pi]^3$.
- A point in the cube will be indicated by a vector $\mathbf{x} = (x_1, x_2, x_3)$, where $0 \leq x_1, x_2, x_3 \leq \pi$.
- If $f : \mathbb{X} \longrightarrow \mathbb{R}$ is a function on the cube, then

$$\triangle f(\mathbf{x}) = \partial_1^2 f(\mathbf{x}) + \partial_2^2 f(\mathbf{x}) + \partial_3^2 f(\mathbf{x}).$$

- A triple of natural numbers will be denoted by $\mathbf{n} = (n_1, n_2, n_3)$, where $n_1, n_2, n_3 \in \mathbb{N} := \{0, 1, 2, 3, 4, \ldots\}$. Let $\mathbb{N}^3$ be the set of all triples $\mathbf{n} = (n_1, n_2, n_3)$, where $n_1, n_2, n_3 \in \mathbb{N}$. Thus, an expression of the form

$$\sum_{\mathbf{n} \in \mathbb{N}^3} (\text{something about } \mathbf{n})$$

269

should be read as:

$$\sum_{n_1=0}^{\infty} \sum_{n_2=0}^{\infty} \sum_{n_3=0}^{\infty} (\text{something about } (n_1, n_2, n_3)).$$

Let $\mathbb{N}_+ := \{1, 2, 3, 4, \ldots\}$ be the set of *nonzero* natural numbers, and let $\mathbb{N}_+^3$ be the set of all such triples. Thus, an expression of the form

$$\sum_{\mathbf{n} \in \mathbb{N}_+^3} (\text{something about } \mathbf{n})$$

should be read as:

$$\sum_{n_1=1}^{\infty} \sum_{n_2=1}^{\infty} \sum_{n_3=1}^{\infty} (\text{something about } (n_1, n_2, n_3)).$$

- For any $\mathbf{n} \in \mathbb{N}_+^3$, $\mathbf{S_n}(\mathbf{x}) = \sin(n_1 x_1) \cdot \sin(n_2 x_2) \cdot \sin(n_3 x_3)$. The Fourier *sine* series of a function $f(\mathbf{x})$ thus has the form:

$$f(\mathbf{x}) = \sum_{\mathbf{n} \in \mathbb{N}_+^3} B_{\mathbf{n}} \mathbf{S_n}(\mathbf{x}).$$

- For any $\mathbf{n} \in \mathbb{N}^3$, $\mathbf{C_n}(\mathbf{x}) = \cos(n_1 x_1) \cdot \cos(n_2 x_2) \cdot \cos(n_3 x_3)$. The Fourier *cosine* series of a function $f(\mathbf{x})$ thus has the form:

$$f(\mathbf{x}) = \sum_{\mathbf{n} \in \mathbb{N}^3} A_{\mathbf{n}} \mathbf{C_n}(\mathbf{x}).$$

- For any $\mathbf{n} \in \mathbb{N}^3$, let $\|\mathbf{n}\| = \sqrt{n_1^2 + n_2^2 + n_3^2}$. In particular, note that

$$\triangle \mathbf{S_n} = -\|\mathbf{n}\|^2 \cdot \mathbf{S_n} \quad \text{and} \quad \triangle \mathbf{C_n} = -\|\mathbf{n}\|^2 \cdot \mathbf{C_n}$$

Ⓔ   **(Exercise 13.1).**

13A The heat equation on a cube

Prerequisites: §9B, §5B, §5C, §1B(ii).   **Recommended:** §11A, §12B(i), §7C(v).

Proposition 13A.1 Heat equation; homogeneous Dirichlet BC
Consider the cube $\mathbb{X} = [0, \pi]^3$, and let $f : \mathbb{X} \longrightarrow \mathbb{R}$ be some function describing an initial heat distribution. Suppose f has Fourier sine series

$$f(\mathbf{x}) \underset{\text{L2}}{\approx} \sum_{\mathbf{n} \in \mathbb{N}_+^3} B_{\mathbf{n}} \mathbf{S_n}(\mathbf{x}).$$

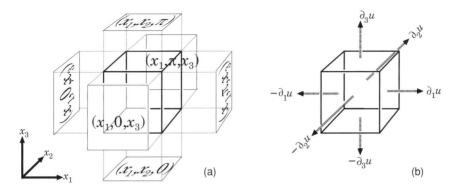

Figure 13A.1. Boundary conditions on a cube: (a) Dirichlet; (b) Neumann.

Define the function $u : \mathbb{X} \times \mathbb{R}_+ \longrightarrow \mathbb{R}$ *by:*

$$u(\mathbf{x}; t) \underset{L2}{\approx} \sum_{\mathbf{n} \in \mathbb{N}_+^3} B_{\mathbf{n}} \mathbf{S}_{\mathbf{n}}(\mathbf{x}) \cdot \exp\left(-\|\mathbf{n}\|^2 \cdot t\right).$$

Then u is the unique solution to the heat equation '$\partial_t u = \Delta u$', with homogeneous Dirichlet boundary conditions

$$u(x_1, x_2, 0; t) = u(x_1, x_2, \pi; t) = u(x_1, 0, x_3; t) \quad \textit{(see Figure 13A.1(a))}$$

$$= u(x_1, \pi, x_3; t) = u(0, x_2, x_3; t) = u(\pi, x_2, x_3, ; t) = 0,$$

and initial conditions $u(\mathbf{x}; 0) = f(\mathbf{x})$.

Furthermore, the series defining u converges semiuniformly on $\mathbb{X} \times \mathbb{R}_+$.

Proof **Exercise 13A.1.** □ Ⓔ

Example Melting ice cube
An ice cube of dimensions $\pi \times \pi \times \pi$ is removed from a freezer (ambient temperature $-10\,^\circ$C) and dropped into a pitcher of freshly brewed tea (initial temperature $+90\,^\circ$C). We want to compute how long it takes the ice cube to melt.

Solution We will assume that the cube has an initially uniform temperature of $-10\,^\circ$C and is completely immersed in the tea.[1] We will also assume that the pitcher is large enough that its temperature does not change during the experiment.

We assume that the outer surface of the cube takes the temperature of the surrounding tea. By subtracting 90 from the temperature of the cube and the water, we can set the water to have temperature 0 and the cube, to have temperature -100.

---

[1] Unrealistic, since actually the cube floats just at the surface.

Hence, we assume homogeneous Dirichlet boundary conditions; the initial temperature distribution is a constant function: $f(\mathbf{x}) = -100$. The constant function -100 has Fourier sine series:

$$-100 \underset{12}{\approx} \frac{-6400}{\pi^3} \sum_{\substack{\mathbf{n} \in \mathbb{N}_+^3 \\ n_1, n_2, n_3 \text{ all odd}}}^{\infty} \frac{1}{n_1 n_2 n_3} \mathbf{S_n(x)}.$$

Ⓔ (**Exercise 13A.2** Verify this Fourier series.) Let κ be the thermal conductivity of the ice. Thus, the time-varying thermal profile of the cube is given by[2]

$$\boxed{u(\mathbf{x}; t) \underset{12}{\approx} \frac{-6400}{\pi^3} \sum_{\substack{\mathbf{n} \in \mathbb{N}_+^3 \\ n_1, n_2, n_3 \text{ all odd}}}^{\infty} \frac{1}{n_1 n_2 n_3} \mathbf{S_n(x)} \exp\left(-\|\mathbf{n}\|^2 \cdot \kappa \cdot t \right).}$$

Thus, to determine how long it takes the cube to melt, we must solve for the minimum value of t such that $u(\mathbf{x}, t) > -90$ everywhere (recall that -90 corresponds
Ⓔ to 0 °C). The coldest point in the cube is always at its center (**Exercise 13A.3**), which has coordinates $(\pi/2, \pi/2, \pi/2)$, so we need to solve for t in the inequality $u((\pi/2, \pi/2, \pi/2); t) \geq -90$, which is equivalent to

$$\frac{90 \cdot \pi^3}{6400} \geq \sum_{\substack{\mathbf{n} \in \mathbb{N}_+^3 \\ n_1, n_2, n_3 \text{ all odd}}}^{\infty} \frac{1}{n_1 n_2 n_3} \mathbf{S_n}\left(\frac{\pi}{2}, \frac{\pi}{2}, \frac{\pi}{2}\right) \exp\left(-\|\mathbf{n}\|^2 \cdot \kappa \cdot t \right)$$

$$= \sum_{\substack{\mathbf{n} \in \mathbb{N}_+^3 \\ n_1, n_2, n_3 \text{ all odd}}}^{\infty} \frac{1}{n_1 n_2 n_3} \sin\left(\frac{n_1 \pi}{2}\right) \sin\left(\frac{n_2 \pi}{2}\right) \sin\left(\frac{n_3 \pi}{2}\right) \exp\left(-\|\mathbf{n}\|^2 \cdot \kappa \cdot t \right)$$

$$\underset{(7C.5)}{=\!=\!=} \sum_{k_1, k_2, k_3 \in \mathbb{N}_+} \frac{(-1)^{k_1 + k_2 + k_3} \exp\left(-\kappa \cdot \left[(2k_1 + 1)^2 + (2k_2 + 1)^2 + (2k_3 + 1)^2\right] \cdot t \right)}{(2k_1 + 1) \cdot (2k_2 + 1) \cdot (2k_3 + 1)},$$

where (7C.5) is by equation (7C.5), p. 151. The solution of this inequality is
Ⓔ **Exercise 13A.4**.

Ⓔ **Exercise 13A.5** Imitating Proposition 13A.1, find a Fourier series solution to the initial value problem for the *free Schrödinger equation*

$$i\partial_t \omega = \frac{-1}{2} \Delta \omega,$$

---

[2] Actually, this is physically unrealistic for two reasons. First, as the ice melts, additional thermal energy is absorbed in the phase transition from solid to liquid. Secondly, once part of the ice cube has melted, its thermal properties change; liquid water has a different thermal conductivity, and, in addition, transports heat through convection.

on the cube $\mathbb{X} = [0, \pi]^3$, with homogeneous Dirichlet boundary conditions. Prove that your solution converges, satisfies the desired initial conditions and boundary conditions, and satisfies the Schrödinger equation.   ◆

Proposition 13A.2 Heat equation; homogeneous Neumann BC
Consider the cube $\mathbb{X} = [0, \pi]^3$, and let $f : \mathbb{X} \longrightarrow \mathbb{R}$ be some function describing an initial heat distribution. Suppose f has Fourier cosine series $f(\mathbf{x}) \underset{\text{\tiny L2}}{\approx} \sum_{\mathbf{n} \in \mathbb{N}^3} A_{\mathbf{n}} C_{\mathbf{n}}(\mathbf{x})$. Define the function $u : \mathbb{X} \times \mathbb{R}_+ \longrightarrow \mathbb{R}$ by:

$$u(\mathbf{x}; t) \underset{\text{\tiny L2}}{\approx} \sum_{\mathbf{n} \in \mathbb{N}^3} A_{\mathbf{n}} C_{\mathbf{n}}(\mathbf{x}) \cdot \exp\left(-\|\mathbf{n}\|^2 \cdot t\right).$$

Then u is the unique solution to the heat equation '$\partial_t u = \Delta u$', with homogeneous Neumann boundary conditions:

$$\partial_3 u(x_1, x_2, 0; t) = \partial_3 u(x_1, x_2, \pi; t) = \partial_2 u(x_1, 0, x_3; t)$$

$$= \partial_2 u(x_1, \pi, x_3; t) = \partial_1 u(0, x_2, x_3; t) = \partial_1 u(\pi, x_2, x_3, ; t) = 0$$

(see Figure 13A.1(b)) and initial conditions $u(\mathbf{x}; 0) = f(\mathbf{x})$.
  Furthermore, the series defining u converges semiuniformly on $\mathbb{X} \times \mathbb{R}_+$.

Proof **Exercise 13A.6.**   □ Ⓔ

13B The Dirichlet problem on a cube

Prerequisites: §9B, §5C(i), §1C.   **Recommended:** §7C(v), §12A.

Proposition 13B.1 Laplace equation; one constant nonhomogeneous Dirichlet BC
Let $\mathbb{X} = [0, \pi]^3$, and consider the Laplace equation '$\Delta u = 0$', with nonhomogeneous Dirichlet boundary conditions (see Figure 13B.1(a)):

$$u(x_1, 0, x_3) = u(x_1, \pi, x_3) = u(0, x_2, x_3) = u(\pi, x_2, x_3,) = 0; \quad \text{(13B.1)}$$
$$u(x_1, x_2, 0) = 0;$$
$$u(x_1, x_2, \pi) = 1. \quad \text{(13B.2)}$$

The unique solution to this problem is the function $u : \mathbb{X} \longrightarrow \mathbb{R}$ defined by:

$$u(x_1, x_2, x_3) \underset{\text{\tiny L2}}{\approx} \sum_{\substack{n,m=1 \\ n,m \text{ both odd}}}^{\infty} \frac{16}{nm\pi \sinh(\pi\sqrt{n^2 + m^2})} \sin(nx) \sin(my)$$

$$\times \sinh(\sqrt{n^2 + m^2} \cdot x_3)$$

for all $(x_1, x_2, x_3) \in \mathbb{X}$. Furthermore, this series converges semiuniformly on int $(\mathbb{X})$.

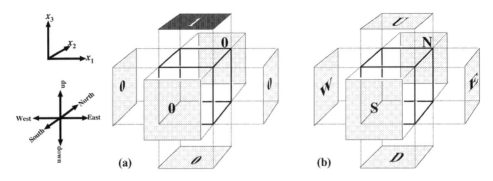

Figure 13B.1. Dirichlet boundary conditions on a cube (a) Constant; nonhomogeneous on one side only. (b) Arbitrary; nonhomogeneous on all sides. See Proposition 13B.2.

Ⓔ *Proof* **Exercise 13B.1.**

(a) Check that the series and its formal Laplacian both converge semiuniformly on int $(\mathbb{X})$.

(b) Check that each of the functions $u_{n,m}(\mathbf{x}) = \sin(nx)\sin(my) \cdot \sinh(\sqrt{n^2 + m^2}x_3)$ satisfies the Laplace equation and the first boundary condition (13B.1).

(c) To check that the solution also satisfies the boundary condition (13B.2), subsititute $x_2 = \pi$ to obtain

$$u(x_1, x_2, \pi) = \sum_{\substack{n,m=1 \\ n,m \text{ both odd}}}^{\infty} \frac{16}{nm\pi \, \sinh(\pi\sqrt{n^2 + m^2})} \sin(nx)\sin(my) \cdot \sinh(\sqrt{n^2 + m^2}\pi)$$

$$= \sum_{\substack{n,m=1 \\ n,m \text{ both odd}}}^{\infty} \frac{16}{nm\pi} \sin(nx)\sin(my) \underset{L2}{\approx} 1,$$

because this is the Fourier sine series for the function $b(x_1, x_2) = 1$, by Example 9A.2, p. 182.

(d) Apply Theorem 5D.5(a), p. 91, to conclude that this solution is unique.   □

Proposition 13B.2 Laplace equation; arbitrary nonhomogeneous Dirichlet BC
Let $\mathbb{X} = [0, \pi]^3$, and consider the Laplace equation '$\Delta h = 0$', with nonhomogeneous Dirichlet boundary conditions (see Figure 13B.1(b)):

$$h(x_1, x_2, 0) = D(x_1, x_2), \qquad h(x_1, x_2, \pi) = U(x_1, x_2),$$
$$h(x_1, 0, x_3) = S(x_1, x_3), \qquad h(x_1, \pi, x_3) = N(x_1, x_3),$$
$$h(0, x_2, x_3) = W(x_2, x_3), \qquad h(\pi, x_2, x_3) = E(x_2, x_3),$$

where $D(x_1, x_2)$, $U(x_1, x_2)$, $S(x_1, x_3)$, $N(x_1, x_3)$, $W(x_2, x_3)$, and $E(x_2, x_3)$ are six functions. Suppose that these functions have two-dimensional Fourier sine series *as follows:*

$$D(x_1, x_2) \underset{12}{\approx} \sum_{n_1,n_2=1}^{\infty} D_{n_1,n_2} \sin(n_1 x_1) \sin(n_2 x_2);$$

$$U(x_1, x_2) \underset{12}{\approx} \sum_{n_1,n_2=1}^{\infty} U_{n_1,n_2} \sin(n_1 x_1) \sin(n_2 x_2);$$

$$S(x_1, x_3) \underset{12}{\approx} \sum_{n_1,n_3=1}^{\infty} S_{n_1,n_3} \sin(n_1 x_1) \sin(n_3 x_3);$$

$$N(x_1, x_3) \underset{12}{\approx} \sum_{n_1,n_3=1}^{\infty} N_{n_1,n_3} \sin(n_1 x_1) \sin(n_3 x_3);$$

$$W(x_2, x_3) \underset{12}{\approx} \sum_{n_2,n_3=1}^{\infty} W_{n_2,n_3} \sin(n_2 x_2) \sin(n_3 x_3);$$

$$E(x_2, x_3) \underset{12}{\approx} \sum_{n_2,n_3=1}^{\infty} E_{n_2,n_3} \sin(n_2 x_2) \sin(n_3 x_3).$$

Then the unique solution to this problem is the function given by

$$h(\mathbf{x}) = d(\mathbf{x}) + u(\mathbf{x}) + s(\mathbf{x}) + n(\mathbf{x}) + w(\mathbf{x}) + e(\mathbf{x}),$$

where

$$d(x_1, x_2, x_3) \underset{12}{\approx} \sum_{n_1,n_2=1}^{\infty} \frac{D_{n_1,n_2}}{\sinh\left(\pi\sqrt{n_1^2 + n_2^2}\right)}$$

$$\times \sin(n_1 x_1) \sin(n_2 x_2) \sinh\left(\sqrt{n_1^2 + n_2^2} \cdot x_3\right);$$

$$u(x_1, x_2, x_3) \underset{12}{\approx} \sum_{n_1,n_2=1}^{\infty} \frac{U_{n_1,n_2}}{\sinh\left(\pi\sqrt{n_1^2 + n_2^2}\right)}$$

$$\times \sin(n_1 x_1) \sin(n_2 x_2) \sinh\left(\sqrt{n_1^2 + n_2^2} \cdot (\pi - x_3)\right);$$

$$s(x_1, x_2, x_3) \underset{12}{\approx} \sum_{n_1,n_3=1}^{\infty} \frac{S_{n_1,n_3}}{\sinh\left(\pi\sqrt{n_1^2 + n_3^2}\right)}$$

$$\times \sin(n_1 x_1) \sin(n_3 x_3) \sinh\left(\sqrt{n_1^2 + n_3^2} \cdot x_2\right);$$

$$n(x_1, x_2, x_3) \underset{12}{\approx} \sum_{n_1,n_3=1}^{\infty} \frac{N_{n_1,n_3}}{\sinh\left(\pi\sqrt{n_1^2 + n_3^2}\right)}$$

$$\times\ \sin(n_1 x_1)\sin(n_3 x_3)\sinh\left(\sqrt{n_1^2 + n_3^2}\cdot(\pi - x_2)\right);$$

$$w(x_1, x_2, x_3) \underset{12}{\approx} \sum_{n_2,n_3=1}^{\infty} \frac{W_{n_2,n_3}}{\sinh\left(\pi\sqrt{n_2^2 + n_3^2}\right)}$$

$$\times\ \sin(n_2 x_2)\sin(n_3 x_3)\sinh\left(\sqrt{n_2^2 + n_3^2}\cdot x_1\right);$$

$$e(x_1, x_2, x_3) \underset{12}{\approx} \sum_{n_2,n_3=1}^{\infty} \frac{E_{n_2,n_3}}{\sinh\left(\pi\sqrt{n_2^2 + n_3^2}\right)}$$

$$\times\ \sin(n_2 x_2)\sin(n_3 x_3)\sinh\left(\sqrt{n_2^2 + n_3^2}\cdot(\pi - x_1)\right).$$

Furthermore, these six series converge semiuniformly on int $(\mathbb{X})$.

Ⓔ *Proof* **Exercise 13B.2.**   □

13C The Poisson problem on a cube

Prerequisites: §9B, §5C, §1D.   **Recommended:** §11C, §12C, §7C(v).

Proposition 13C.1 Poisson problem on cube; homogeneous Dirichlet BC
Let $\mathbb{X} = [0, \pi]^3$, *and let* $q : \mathbb{X} \longrightarrow \mathbb{R}$ *be some function with semiuniformly conver-*
gent Fourier sine series:

$$q(\mathbf{x}) \underset{12}{\approx} \sum_{\mathbf{n}\in\mathbb{N}_+^3} Q_{\mathbf{n}} S_{\mathbf{n}}(\mathbf{x}).$$

Define the function $u : \mathbb{X} \longrightarrow \mathbb{R}$ *by:*

$$u(\mathbf{x}) \underset{\text{unif}}{\equiv} \sum_{\mathbf{n}\in\mathbb{N}_+^3} \frac{-Q_{\mathbf{n}}}{\|\mathbf{n}\|^2}\cdot S_{\mathbf{n}}(\mathbf{x}), \quad \text{for all } \mathbf{x} \in \mathbb{X}.$$

Then u *is the unique solution to the Poisson equation* '$\triangle u(\mathbf{x}) = q(\mathbf{x})$', *satis-*
fying homogeneous Dirichlet boundary conditions $u(x_1, x_2, 0) = u(x_1, x_2, \pi) =$
$u(x_1, 0, x_3) = u(x_1, \pi, x_3) = u(0, x_2, x_3) = u(\pi, x_2, x_3,) = 0$.

Proof **Exercise 13C.1.** □ Ⓔ

Proposition 13C.2 Poisson problem on cube; homogeneous Neumann BC

Let $\mathbb{X} = [0, \pi]^3$, *and let* $q : \mathbb{X} \longrightarrow \mathbb{R}$ *be some function with* semiuniformly *convergent Fourier cosine series:*

$$q(\mathbf{x}) \underset{\text{L2}}{\approx} \sum_{\mathbf{n} \in \mathbb{N}^3_+} Q_{\mathbf{n}} \mathbf{C}_{\mathbf{n}}(\mathbf{x}).$$

Suppose $Q_{0,0,0} = 0$. *Fix some constant* $K \in \mathbb{R}$, *and define the function* $u : \mathbb{X} \longrightarrow \mathbb{R}$ *by:*

$$u(\mathbf{x}) \underset{\text{unif}}{\equiv} \sum_{\substack{\mathbf{n} \in \mathbb{N}^3 \\ n_1, n_2, n_3 \text{ not all zero}}} \frac{-Q_{\mathbf{n}}}{\|\mathbf{n}\|^2} \cdot \mathbf{C}_{\mathbf{n}}(\mathbf{x}) + K, \quad \text{for all } \mathbf{x} \in \mathbb{X}. \quad (13\text{C}.1)$$

Then u *is a solution to the Poisson equation* '$\triangle u(\mathbf{x}) = q(\mathbf{x})$', *satisfying homogeneous Neumann boundary conditions* $\partial_3 u(x_1, x_2, 0) = \partial_3 u(x_1, x_2, \pi) = \partial_2 u(x_1, 0, x_3) = \partial_2 u(x_1, \pi, x_3) = \partial_1 u(0, x_2, x_3) = \partial_1 u(\pi, x_2, x_3) = 0$.

Furthermore, all solutions to this Poisson equation with these boundary conditions have the form (13C.1).

If $Q_{0,0,0} \neq 0$, *however, the problem has no solution.*

Proof **Exercise 13C.2.** □ Ⓔ

14

Boundary value problems in polar coordinates

> The source of all great mathematics is the special case, the concrete
> example. It is frequent in mathematics that every instance of a concept
> of seemingly great generality is in essence the same as a small and
> concrete special case.
>
> *Paul Halmos*

14A Introduction

Prerequisites: Appendix D(ii).

When solving a boundary value problem, the shape of the domain dictates the
choice of coordinate system. Seek the coordinate system yielding the simplest
description of the boundary. For rectangular domains, Cartesian coordinates are
the most convenient. For disks and annuli in the plane, *polar* coordinates are a
better choice. Recall that polar coordinates (r, θ) on $\mathbb{R}^2$ are defined by the following
transformation:

$$x = r \cdot \cos(\theta) \quad \text{and} \quad y = r \cdot \sin(\theta)$$

(see Figure 14A.1(a)), with reverse transformation:

$$r = \sqrt{x^2 + y^2} \quad \text{and} \quad \theta = \arctan\left(\frac{y}{x}\right).$$

Here, the coordinate r ranges over $\mathbb{R}_{\not{}}$, while the variable θ ranges over $[-\pi, \pi)$.
(Clearly, we could let θ range over *any* interval of length 2π; we just find $[-\pi, \pi)$
to be the most convenient.)

The three domains we will examine are as follows.

- $\mathbb{D} = \{(r, \theta); \ r \leq R\}$, the *disk* of radius R; see Figure 14A.1(b). For simplicity, we will
  usually assume $R = 1$.

279

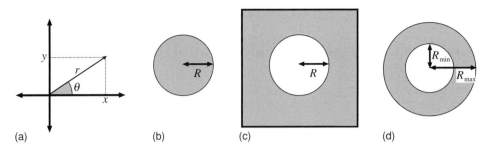

Figure 14A.1. (a) Polar coordinates; (b) the disk $\mathbb{D}$; (c) the codisk $\mathbb{D}^{\complement}$; (d) the annulus $\mathbb{A}$.

- $\mathbb{D}^{\complement} = \{(r, \theta); \ R \leq r\}$, the *codisk* or *punctured plane* of radius R; see Figure 14A.1(c). For simplicity, we will usually assume $R = 1$.
- $\mathbb{A} = \{(r, \theta); \ R_{\min} \leq r \leq R_{\max}\}$, the *annulus*, of inner radius $R_{\min}$ and outer radius $R_{\max}$; see Figure 14A.1(d).

The boundaries of these domains are circles. For example, the boundary of the disk $\mathbb{D}$ of radius R is the *circle*:

$$\partial \mathbb{D} = \mathbb{S} = \{(r, \theta); \ r = R\}.$$

The circle can be parameterized by a single angular coordinate $\theta \in [-\pi, \pi)$. Thus, the boundary conditions will be specified by a function $b : [-\pi, \pi) \longrightarrow \mathbb{R}$. Note that, if $b(\theta)$ is to be *continuous* as a function on the circle, then it must be 2π-*periodic* as a function on $[-\pi, \pi)$.

In polar coordinates, the Laplacian is written as

$$\triangle u = \partial_r^2 u + \frac{1}{r} \partial_r u + \frac{1}{r^2} \partial_\theta^2 u \qquad (14A.1)$$

Ⓔ (**Exercise 14A.1**).

14B The Laplace equation in polar coordinates

14B(i) Polar harmonic functions

Prerequisites: Appendix D(ii), §1C.

The following important harmonic functions *separate* in polar coordinates:

$$\Phi_n(r, \theta) = \cos(n\theta) \cdot r^n, \quad \Psi_n(r, \theta) = \sin(n\theta) \cdot r^n, \quad \text{for } n \in \mathbb{N}_+,$$

(see Figure 14B.1);

$$\phi_n(r, \theta) = \frac{\cos(n\theta)}{r^n}, \quad \psi_n(r, \theta) = \frac{\sin(n\theta)}{r^n}, \quad \text{for } n \in \mathbb{N}_+$$

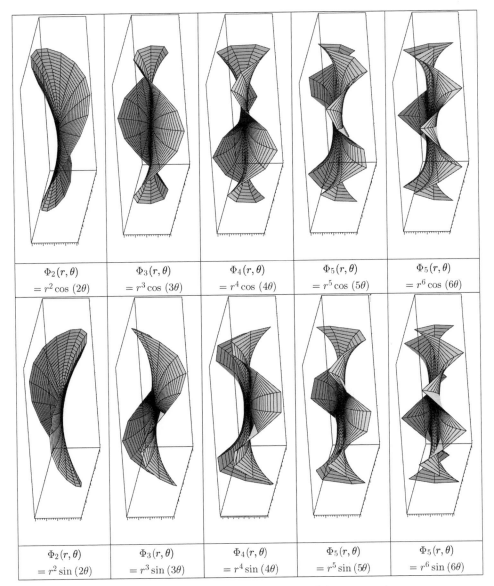

$\Phi_2(r,\theta)$	$\Phi_3(r,\theta)$	$\Phi_4(r,\theta)$	$\Phi_5(r,\theta)$	$\Phi_5(r,\theta)$
$= r^2 \cos(2\theta)$	$= r^3 \cos(3\theta)$	$= r^4 \cos(4\theta)$	$= r^5 \cos(5\theta)$	$= r^6 \cos(6\theta)$

$\Phi_2(r,\theta)$	$\Phi_3(r,\theta)$	$\Phi_4(r,\theta)$	$\Phi_5(r,\theta)$	$\Phi_5(r,\theta)$
$= r^2 \sin(2\theta)$	$= r^3 \sin(3\theta)$	$= r^4 \sin(4\theta)$	$= r^5 \sin(5\theta)$	$= r^6 \sin(6\theta)$

Figure 14B.1. Φ_n and Ψ_n for $n = 2\ldots6$ (rotate page).

(see Figure 14B.2);

$$\Phi_0(r,\theta) = 1, \ \phi_0(r,\theta) = \log(r)$$

(see Figure 14B.3).

Proposition 14B.1 *The functions* Φ_n, Ψ_n, ϕ_n, *and* ψ_n *are harmonic, for all* $n \in \mathbb{N}$.

Proof See Problems 14.1–14.5. □

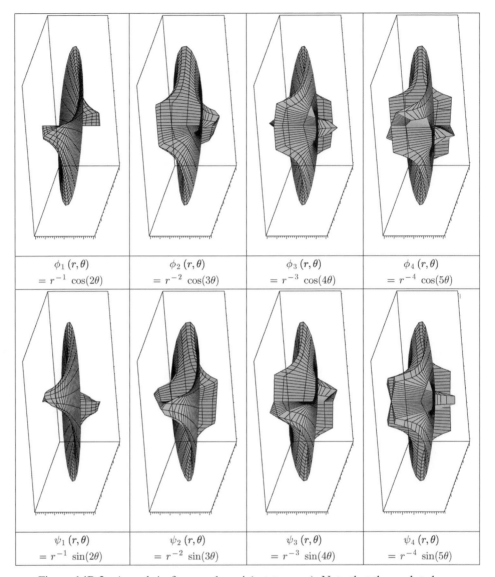

Figure 14B.2. ϕ_n and ψ_n for $n = 1 \ldots 4$ (rotate page). Note that these plots have been 'truncated' to have vertical bounds ± 3, because these functions explode to $\pm\infty$ at zero.

Ⓔ **Exercise 14B.1**

  (a) Show that $\Phi_1(r, \theta) = x$ and $\Psi_1(r, \theta) = y$ in Cartesian coordinates.

  (b) Show that $\Phi_2(r, \theta) = x^2 - y^2$ and $\Psi_2(r, \theta) = 2xy$ in Cartesian coordinates.

  (c) Define $F_n : \mathbb{C} \longrightarrow \mathbb{C}$ by $F_n(z) := z^n$. Show that $\Phi_n(x, y) = \mathrm{Re}\,[F_n(x + y\mathbf{i})]$ and $\Psi_n(x, y) = \mathrm{Im}\,[F_n(x + y\mathbf{i})]$.

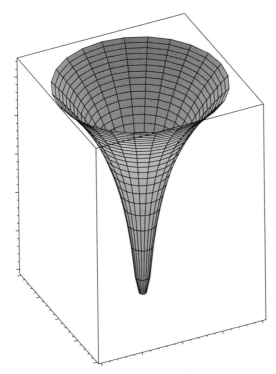

Figure 14B.3. $\phi_0(r, \theta) = \log |r|$ (vertically truncated near zero).

(d) (Hard) Show that Φ_n can be written as a homogeneous polynomial of degree n in x
and y.

(e) Show that, if $(x, y) \in \partial \mathbb{D}$ (i.e. if $x^2 + y^2 = 1$), then $\Phi_N(x, y) = \zeta_N(x)$, where

$$\zeta_N(x) := 2^{(N-1)} x^N + \sum_{n=1}^{\lfloor \frac{N}{2} \rfloor} (-1)^n 2^{(N-1-2n)} \frac{N}{n} \binom{N-n-1}{n-1} x^{(N-2n)}$$

is the Nth *Chebyshev polynomial*. (For more information, see Broman (1989),
§3.4).   ♦

We will solve the Laplace equation in polar coordinates by representing solutions
as sums of these simple functions. Note that Φ_n and Ψ_n are *bounded* at zero, but
unbounded at infinity (Figure 14B.4(a) shows the radial growth of Φ_n and Ψ_n).
Conversely, ϕ_n and ψ_n are *unbounded* at zero, but *bounded* at infinity) (Figure
14B.4(b) shows the radial decay of ϕ_n and ψ_n). Finally, Φ_0, being constant, is
bounded everywhere, while ϕ_0 is unbounded at both 0 and ∞ (see Figure 14B.4(b)).
Hence, when solving BVPs in a neighbourhood around zero (e.g. the disk), it is
preferable to use Φ_0, Φ_n, and Ψ_n. When solving BVPs on an unbounded domain

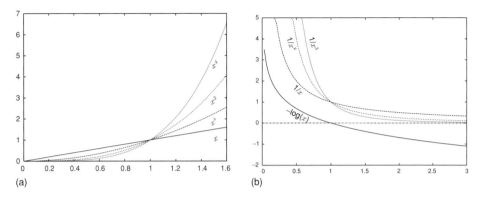

Figure 14B.4. Radial growth/decay of polar-separated harmonic functions. (a) x^n, for $n = 1, 2, 3, 4$; (b) $-\log(x)$ and $1/x^n$, for $n = 1, 2, 3$ (these plots are vertically truncated).

(i.e. one 'containing infinity') it is preferable to use Φ_0, ϕ_n, and ψ_n. When solving BVPs on a domain containing neither zero nor infinity (e.g. the annulus), we use all of Φ_n, Ψ_n, ϕ_n, ψ_n, Φ_0, and ϕ_0.

14B(ii) Boundary value problems on a disk

Prerequisites: §5C, §14A, §14B(i), §8A, Appendix F.

Proposition 14B.2 Laplace equation, unit disk, nonhomogeneous Dirichlet BC
Let $\mathbb{D} = \{(r, \theta); \ r \le 1\}$ be the unit disk, and let $b \in \mathbf{L}^2[-\pi, \pi)$ be some function. Consider the Laplace equation '$\Delta u = 0$', with nonhomogeneous Dirichlet boundary conditions:

$$u(1, \theta) = b(\theta), \qquad \text{for all } \theta \in [-\pi, \pi). \tag{14B.1}$$

Suppose b has the real Fourier series:

$$b(\theta) \underset{L^2}{\approx} A_0 + \sum_{n=1}^{\infty} A_n \cos(n\theta) + \sum_{n=1}^{\infty} B_n \sin(n\theta).$$

Then the unique solution to this problem is the function $u : \mathbb{D} \longrightarrow \mathbb{R}$ defined:

$$u(r, \theta) \underset{L^2}{\approx} A_0 + \sum_{n=1}^{\infty} A_n \Phi_n(r, \theta) + \sum_{n=1}^{\infty} B_n \Psi_n(r, \theta)$$

$$= A_0 + \sum_{n=1}^{\infty} A_n \cos(n\theta) \cdot r^n + \sum_{n=1}^{\infty} B_n \sin(n\theta) \cdot r^n. \tag{14B.2}$$

Furthermore, the series (14B.2) converges semiuniformly to u on int $(\mathbb{D})$.

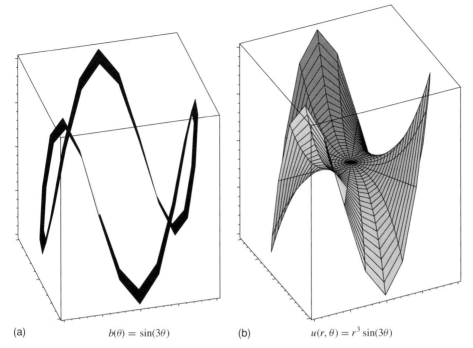

(a) $b(\theta) = \sin(3\theta)$ (b) $u(r, \theta) = r^3 \sin(3\theta)$

Figure 14B.5. (a) Bent circular wire frame. (b) Soap bubble in the frame.

Proof **Exercise 14B.2.**   Ⓔ

(a) Fix $R < 1$ and let $\mathbb{D}(R) := \{(r, \theta); \ r < R\}$. Show that on the domain $\mathbb{D}(R)$, the conditions of Proposition F.1, p. 569, are satisfied; use this to show that

$$\triangle u(r, \theta) \underset{\text{unif}}{\equiv} \sum_{n=1}^{\infty} A_n \triangle \Phi_n(r, \theta) + \sum_{n=1}^{\infty} B_n \triangle \Psi_n(r, \theta)$$

for all $(r, \theta) \in \mathbb{D}(R)$. Now use Proposition 14B.1, p. 281, to deduce that $\triangle u(r, \theta) = 0$ for all $r \leq R$. Since this works for any $R < 1$, conclude that $\triangle u \equiv 0$ on $\mathbb{D}$.

(b) To check that u also satisfies the boundary condition (14B.1), substitute $r = 1$ into equation (14B.2) to get

$$u(1, \theta) \underset{12}{\approx} A_0 + \sum_{n=1}^{\infty} A_n \cos(n\theta) + \sum_{n=1}^{\infty} B_n \sin(n\theta) = b(\theta).$$

(c) Use Proposition 5D.5(a), p. 91, to conclude that this solution is unique.   □

Example 14B.3 Soap bubble

Take a circular wire frame of radius 1, and warp it so that its vertical distortion is described by the function $b(\theta) = \sin(3\theta)$, shown in Figure 14B.5(a). Dip the frame into a soap solution to obtain a bubble with the bent wire as its boundary. What is the shape of the bubble?

Solution A soap bubble suspended from the wire is a *minimal surface*, and minimal surfaces of low curvature are well approximated by harmonic functions. Let $u(r, \theta)$ be a function describing the bubble surface. As long as the distortion $b(\theta)$ is relatively small, $u(r, \theta)$ will be a solution to Laplace's equation, with boundary conditions $u(1, \theta) = b(\theta)$. Thus, as shown in Figure 14B.5(b), $u(r, \theta) = r^3 \sin(3\theta)$.   ◇

Ⓔ **Exercise 14B.3** Let $u(x, \theta)$ be a solution to the Dirichlet problem with boundary conditions $u(1, \theta) = b(\theta)$. Let **0** be the center of the disk (i.e. the point with radius 0). Use Proposition 14B.2 to prove that

$$u(\mathbf{0}) = \frac{1}{2\pi} \int_{-\pi}^{\pi} b(\theta)d\theta.$$ ♦

Proposition 14B.4 Laplace equation, unit disk, nonhomogeneous Neumann BC
Let $\mathbb{D} = \{(r, \theta); r \leq 1\}$ be the unit disk, and let $b \in \mathbf{L}^2[-\pi, \pi)$. Consider the Laplace equation '$\triangle u = 0$', with nonhomogeneous Neumann boundary conditions:

$$\partial_r u(1, \theta) = b(\theta), \qquad \text{for all } \theta \in [-\pi, \pi).$$ (14B.3)

Suppose b has real Fourier series

$$b(\theta) \underset{L2}{\approx} A_0 + \sum_{n=1}^{\infty} A_n \cos(n\theta) + \sum_{n=1}^{\infty} B_n \sin(n\theta).$$

If $A_0 = 0$, then the solutions to this problem are all functions $u : \mathbb{D} \longrightarrow \mathbb{R}$ of the form

$$u(r, \theta) \underset{L2}{\approx} C + \sum_{n=1}^{\infty} \frac{A_n}{n} \Phi_n(r, \theta) + \sum_{n=1}^{\infty} \frac{B_n}{n} \Psi_n(r, \theta)$$

$$= C + \sum_{n=1}^{\infty} \frac{A_n}{n} \cos(n\theta) \cdot r^n + \sum_{n=1}^{\infty} \frac{B_n}{n} \sin(n\theta) \cdot r^n, \qquad (14B.4)$$

where C is any constant. Furthermore, the series (14B.4) converges semiuniformly to u on int $(\mathbb{D})$.

However, if $A_0 \neq 0$, then there is no solution.

Proof

> **Claim 1** *For any $r < 1$,*
>
> $$\sum_{n=1}^{\infty} n^2 \frac{|A_n|}{n} \cdot r^n + \sum_{n=1}^{\infty} n^2 \frac{|B_n|}{n} \cdot r^n < \infty.$$

Proof Let $M = \max\{\max\{|A_n|\}_{n=1}^{\infty}, \max\{|B_n|\}_{n=1}^{\infty}\}$. Then

$$\sum_{n=1}^{\infty} n^2 \frac{|A_n|}{n} \cdot r^n + \sum_{n=1}^{\infty} n^2 \frac{|B_n|}{n} \cdot r^n \leq \sum_{n=1}^{\infty} n^2 \frac{M}{n} \cdot r^n + \sum_{n=1}^{\infty} n^2 \frac{M}{n} \cdot r^n$$

$$= 2M \sum_{n=1}^{\infty} n r^n. \tag{14B.5}$$

Let $f(r) = 1/(1-r)$. Then $f'(r) = 1/(1-r)^2$. Recall that, for $|r| < 1$, $f(r) = \sum_{n=0}^{\infty} r^n$. Thus, $f'(r) = \sum_{n=1}^{\infty} n r^{n-1} = \frac{1}{r} \sum_{n=1}^{\infty} n r^n$. Hence, the right-hand side of equation (14B.5) is equal to

$$2M \sum_{n=1}^{\infty} n r^n = 2Mr \cdot f'(r) = 2Mr \cdot \frac{1}{(1-r)^2} < \infty,$$

for any $r < 1$. $\qquad\qquad\qquad\qquad\qquad\qquad\qquad\qquad\qquad\qquad \diamond_{\text{Claim 1}}$

Let $R < 1$ and let $\mathbb{D}(R) = \{(r, \theta); \; r \leq R\}$ be the disk of radius R. If

$$u(r, \theta) = C + \sum_{n=1}^{\infty} \frac{A_n}{n} \Phi_n(r, \theta) + \sum_{n=1}^{\infty} \frac{B_n}{n} \Psi_n(r, \theta),$$

then, for all $(r, \theta) \in \mathbb{D}(R)$,

$$\triangle u(r, \theta) \underset{\text{unif}}{\equiv} \sum_{n=1}^{\infty} \frac{A_n}{n} \triangle \Phi_n(r, \theta) + \sum_{n=1}^{\infty} \frac{B_n}{n} \triangle \Psi_n(r, \theta)$$

$$\underset{(*)}{=} \sum_{n=1}^{\infty} \frac{A_n}{n} (0) + \sum_{n=1}^{\infty} \frac{B_n}{n} (0) = 0,$$

on $\mathbb{D}(R)$. Here, $\underset{\text{unif}}{\equiv}$ is by Proposition F.1, p. 569, and Claim 1, while $(*)$ is by Proposition 14B.1, p. 281.

To check boundary conditions, observe that, for all $R < 1$ and all $(r, \theta) \in \mathbb{D}(R)$,

$$\partial_r u(r, \theta) \underset{\text{unif}}{\equiv} \sum_{n=1}^{\infty} \frac{A_n}{n} \partial_r \Phi_n(r, \theta) + \sum_{n=1}^{\infty} \frac{B_n}{n} \partial_r \Psi_n(r, \theta)$$

$$= \sum_{n=1}^{\infty} \frac{A_n}{n} n r^{n-1} \cos(n\theta) + \sum_{n=1}^{\infty} \frac{B_n}{n} n r^{n-1} \sin(n\theta)$$

$$= \sum_{n=1}^{\infty} A_n r^{n-1} \cos(n\theta) + \sum_{n=1}^{\infty} B_n r^{n-1} \sin(n\theta).$$

Here $\underset{\text{unif}}{\equiv}$ is by Proposition F.1, p. 569. Hence, letting $R \to 1$, we obtain

$$\partial_\perp u(1, \theta) = \partial_r u(1, \theta) = \sum_{n=1}^{\infty} A_n \cdot (1)^{n-1} \cos(n\theta) + \sum_{n=1}^{\infty} B_n \cdot (1)^{n-1} \sin(n\theta)$$

$$= \sum_{n=1}^{\infty} A_n \cos(n\theta) + \sum_{n=1}^{\infty} B_n \sin(n\theta) \underset{12}{\approx} b(\theta),$$

as desired. Here, $\underset{12}{\approx}$ is because this is the Fourier series for $b(\theta)$, assuming $A_0 = 0$. (If $A_0 \neq 0$, then this solution does not work.)

Finally, Proposition 5D.5(c), p. 91, implies that this solution is unique up to addition of a constant. □

Remark Physically speaking, why must $A_0 = 0$?

If $u(r, \theta)$ is an electric potential, then $\partial_r u$ is the *radial component* of the *electric field*. The requirement that $A_0 = 0$ is equivalent to requiring that the *net electric flux* entering the disk is zero, which is equivalent (via Gauss's law) to the assertion that the *net electric charge* contained in the disk is zero. If $A_0 \neq 0$, then the net electric charge within the disk must be nonzero. Thus, if $q : \mathbb{D} \longrightarrow \mathbb{R}$ is the charge density field, then we must have $q \neq 0$. However, $q = \triangle u$ (see Example 1D.2, p. 15), so this means $\triangle u \neq 0$, which means u is not harmonic.

Example 14B.5 The truth is out there

While covertly investigating mysterious electrical phenomena on a top-secret military installation in the Nevada Desert, Mulder and Scully are trapped in a cylindrical concrete silo by the Cancer Stick Man. Scully happens to have a voltmeter, and she notices an electric field in the silo. Walking around the (circular) perimeter of the silo, Scully estimates the radial component of the electric field to be the function $b(\theta) = 3 \sin(7\theta) - \cos(2\theta)$. Estimate the electric potential field inside the silo.

Solution The electric potential will be a solution to Laplace's equation, with boundary conditions $\partial_r u(1, \theta) = 3 \sin(7\theta) - \cos(2\theta)$. Thus,

$$u(r, \theta) = C + \frac{3}{7} \sin(7\theta) \cdot r^7 - \frac{1}{2} \cos(2\theta) \cdot r^2$$

(see Figure 14B.6).

Question Moments later, Mulder repeats Scully's experiment, and finds that the perimeter field has changed to $b(\theta) = 3 \sin(7\theta) - \cos(2\theta) + 6$. He immediately suspects that an alien presence has entered the silo. Why?   ◇

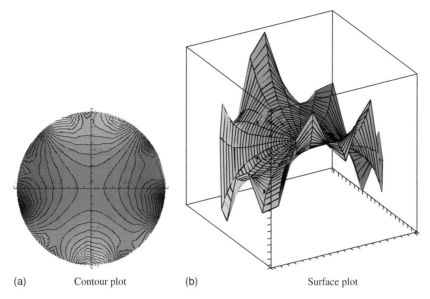

(a)   Contour plot   (b)   Surface plot

Figure 14B.6. The electric potential deduced from Scully's voltage measurements in Example 14B.5.

14B(iii) Boundary value problems on a codisk

Prerequisites: §5C, §14A, §14B(i), §8A, Appendix F.   **Recommended:** §14B(ii).

We will now solve the Dirichlet problem on an *unbounded* domain: the *codisk*

$$\mathbb{D}^{\complement} := \{(r, \theta);\ 1 \leq r, \theta \in [-\pi, \pi)\}.$$

Physical interpretations

Chemical concentration Suppose there is an unknown source of some chemical hidden inside the disk, and that this chemical diffuses into the surrounding medium. Then the solution function $u(r, \theta)$ represents the *equilibrium concentration* of the chemical. In this case, it is reasonable to expect $u(r, \theta)$ to be *bounded at infinity*, by which we mean that

$$\lim_{r \to \infty} |u(r, \theta)| \neq \infty, \quad \text{for all } \theta \in [-\pi, \pi). \tag{14B.6}$$

Otherwise the chemical concentration would become very large far away from the center, which is not realistic.

Electric potential Suppose there is an unknown charge distribution inside the disk. Then the solution function $u(r, \theta)$ represents the *electric potential field* generated by this charge. Even though we do not know the exact charge distribution, we can use the boundary conditions to extrapolate the shape of the potential field outside the disk.

If the net charge within the disk is zero, then the electric potential far away from the disk should be bounded (because, from far away, the charge distribution inside the disk 'looks' neutral); hence, the solution $u(r, \theta)$ will again satisfy the boundedness condition (14B.6).

However, if there is a *nonzero* net charge within the disk, then the electric potential will *not* be bounded (because, even from far away, the disk still 'looks' charged). Nevertheless, the electric *field* generated by this potential should still decay to zero (because the influence of the charge should be weak at large distances). This means that, while the potential is unbounded, the *gradient* of the potential must decay to zero near infinity. In other words, we must impose the *decaying gradient condition*:

$$\lim_{r \to \infty} \nabla u(r, \theta) = 0, \quad \text{for all } \theta \in [-\pi, \pi). \tag{14B.7}$$

Proposition 14B.6 Laplace equation, codisk, nonhomogeneous Dirichlet BC

Let $\mathbb{D}^{\complement} = \{(r, \theta); \ 1 \leq r\}$ *be the codisk, and let* $b \in \mathbf{L}^2[-\pi, \pi)$. *Consider the Laplace equation* '$\triangle u = 0$', *with nonhomogeneous Dirichlet boundary conditions:*

$$u(1, \theta) = b(\theta), \quad \text{for all } \theta \in [-\pi, \pi). \tag{14B.8}$$

Suppose b has real Fourier series

$$b(\theta) \underset{12}{\approx} A_0 + \sum_{n=1}^{\infty} A_n \cos(n\theta) + \sum_{n=1}^{\infty} B_n \sin(n\theta).$$

Then the unique solution to this problem, which is **bounded at infinity** *as in equation* (14B.6), *is the function* $u : \mathbb{D}^{\complement} \longrightarrow \mathbb{R}$ *defined:*

$$u(r, \theta) \underset{12}{\approx} A_0 + \sum_{n=1}^{\infty} A_n \frac{\cos(n\theta)}{r^n} + \sum_{n=1}^{\infty} B_n \frac{\sin(n\theta)}{r^n}. \tag{14B.9}$$

Furthermore, the series (14B.9) *converges semiuniformly to u on* $\left(\mathbb{D}^{\complement}\right)$.

Ⓔ *Proof* **Exercise 14B.4.**

(a) To show that u is harmonic, apply equation (14A.1) to get

$$\triangle u(r, \theta) = \partial_r^2 \left(\sum_{n=1}^{\infty} A_n \frac{\cos(n\theta)}{r^n} + \sum_{n=1}^{\infty} B_n \frac{\sin(n\theta)}{r^n} \right)$$

$$+ \frac{1}{r} \partial_r \left(\sum_{n=1}^{\infty} A_n \frac{\cos(n\theta)}{r^n} + \sum_{n=1}^{\infty} B_n \frac{\sin(n\theta)}{r^n} \right)$$

$$+ \frac{1}{r^2} \partial_\theta^2 \left(\sum_{n=1}^{\infty} A_n \frac{\cos(n\theta)}{r^n} + \sum_{n=1}^{\infty} B_n \frac{\sin(n\theta)}{r^n} \right). \tag{14B.10}$$

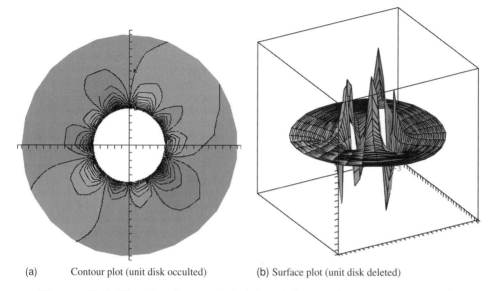

(a)   Contour plot (unit disk occulted)   (b) Surface plot (unit disk deleted)

Figure 14B.7. The electric potential deduced from voltage measurements in Example 14B.7.

Now let $R > 1$. Check that, on the domain $\mathbb{D}^{\complement}(R) = \{(r, \theta); \, r > R\}$, the conditions of Proposition F.1, p. 569, are satisfied; use this to simplify equation (14B.10). Finally, apply Proposition 14B.1, p. 281, to deduce that $\triangle u(r, \theta) = 0$ for all $r \geq R$. Since this works for any $R > 1$, conclude that $\triangle u \equiv 0$ on $\mathbb{D}^{\complement}$.

(b) To check that the solution also satisfies the boundary condition (14B.8), subsititute $r = 1$ into equation (14B.9) to get

$$u(1, \theta) \underset{12}{\approx} A_0 + \sum_{n=1}^{\infty} A_n \cos(n\theta) + \sum_{n=1}^{\infty} B_n \sin(n\theta) = b(\theta).$$

(c) Use Proposition 5D.5(a), p. 91, to conclude that this solution is unique.   □

Example 14B.7 The electric potential outside a disk

An unknown distribution of electric charges lies inside the unit disk in the plane. Using a voltmeter, the electric potential is measured along the perimeter of the circle, and is approximated by the function $b(\theta) = \sin(2\theta) + 4\cos(5\theta)$. Far away from the origin, the potential is found to be close to zero. Estimate the electric potential field.

Solution The electric potential will be a solution to Laplace's equation, with boundary conditions $u(1, \theta) = \sin(2\theta) + 4\cos(5\theta)$. Far away, the potential apparently remains bounded. Thus, as shown in Figure 14B.7,

$$u(r, \theta) = \frac{\sin(2\theta)}{r^2} + \frac{4\cos(5\theta)}{r^5}.$$ ◇

Remark Note that, for any constant $C \in \mathbb{R}$, another solution to the Dirichlet problem with boundary conditions (14B.8) is given by the function

$$u(r, \theta) = A_0 + C \log(r) + \sum_{n=1}^{\infty} A_n \frac{\cos(n\theta)}{r^n} + \sum_{n=1}^{\infty} B_n \frac{\sin(n\theta)}{r^n}.$$

Ⓔ (**Exercise 14B.5**). However, unless $C = 0$, this will *not* be bounded at infinity.

Proposition 14B.8 Laplace equation, codisk, nonhomogeneous Neumann BC
Let $\mathbb{D}^{\mathsf{C}} = \{(r, \theta); \ 1 \leq r\}$ *be the codisk, and let* $b \in \mathbf{L}^2[-\pi, \pi)$. *Consider the Laplace equation '$\Delta u = 0$', with nonhomogeneous* Neumann *boundary conditions:*

$$-\partial_{\perp} u(1, \theta) = \partial_r u(1, \theta) = b(\theta), \quad \textit{for all } \theta \in [-\pi, \pi). \qquad (14\text{B}.11)$$

Suppose b has real Fourier series

$$b(\theta) \underset{12}{\approx} A_0 + \sum_{n=1}^{\infty} A_n \cos(n\theta) + \sum_{n=1}^{\infty} B_n \sin(n\theta).$$

Fix a constant $C \in \mathbb{R}$, and define $u : \mathbb{D}^{\mathsf{C}} \longrightarrow \mathbb{R}$ by:

$$u(r, \theta) \underset{12}{\approx} C + A_0 \log(r) + \sum_{n=1}^{\infty} \frac{-A_n}{n} \frac{\cos(n\theta)}{r^n} + \sum_{n=1}^{\infty} \frac{-B_n}{n} \frac{\sin(n\theta)}{r^n} \qquad (14\text{B}.12)$$

Then u is a solution to the Laplace equation, with nonhomogeneous Neumann boundary conditions (14B.11), and furthermore it obeys the decaying gradient con- *dition (14B.7) on p. 290. Furthermore, all harmonic functions satisfying equations (14B.11) and (14B.7) must be of the form (14B.12). However, the solution (14B.12) is bounded at infinity, as in equation (14B.6), if, and only if, $A_0 = 0$.*
  Finally, the series (14B.12) converges semiuniformly to u on $\left(\mathbb{D}^{\mathsf{C}}\right)$.

Ⓔ *Proof* **Exercise 14B.6.**
  (a) To show that u is harmonic, apply equation (14A.1), p. 280, to get

$$\Delta u(r, \theta) = \partial_r^2 \left(A_0 \log(r) - \sum_{n=1}^{\infty} \frac{A_n}{n} \frac{\cos(n\theta)}{r^n} - \sum_{n=1}^{\infty} \frac{B_n}{n} \frac{\sin(n\theta)}{r^n} \right)$$

$$+ \frac{1}{r} \partial_r \left(A_0 \log(r) - \sum_{n=1}^{\infty} \frac{A_n}{n} \frac{\cos(n\theta)}{r^n} - \sum_{n=1}^{\infty} \frac{B_n}{n} \frac{\sin(n\theta)}{r^n} \right)$$

$$+ \frac{1}{r^2} \partial_\theta^2 \left(A_0 \log(r) - \sum_{n=1}^{\infty} \frac{A_n}{n} \frac{\cos(n\theta)}{r^n} - \sum_{n=1}^{\infty} \frac{B_n}{n} \frac{\sin(n\theta)}{r^n} \right). \qquad (14\text{B}.13)$$

Now let $R > 1$. Check that, on the domain $\mathbb{D}^{\mathsf{C}}(R) = \{(r, \theta); \ r > R\}$, the conditions of Proposition F.1, p. 569, are satisfied; use this to simplify equation (14B.13). Finally, apply Proposition 14B.1, p. 281, to deduce that $\triangle u(r, \theta) = 0$ for all $r \geq R$. Since this works for any $R > 1$, conclude that $\triangle u \equiv 0$ on $\mathbb{D}^{\complement}$.

(b) To check that the solution also satisfies the boundary condition (14B.11), subsititute $r = 1$ into equation (14B.12) and compute the radial derivative (using Proposition F.1, p. 569) to get

$$\partial_r u(1, \theta) = A_0 + \sum_{n=1}^{\infty} A_n \cos(n\theta) + \sum_{n=1}^{\infty} B_n \sin(n\theta) \underset{12}{\approx} b(\theta).$$

(c) Use Proposition 5D.5(c), p. 91, to show that this solution is unique up to addition of a constant.

(d) What is the physical interpretation of $A_0 = 0$? ☐

Example 14B.9 Another electric potential outside a disk
An unknown distribution of electric charges lies inside the unit disk in the plane. The *radial* component of the electric field is measured along the perimeter of the circle, and is approximated by the function $b(\theta) = 0.9 + \sin(2\theta) + 4\cos(5\theta)$. Estimate the electric potential (up to a constant).

Solution The electric potential will be a solution to the Laplace equation, with boundary conditions $\partial_r u(1, \theta) = 0.9 + \sin(2\theta) + 4\cos(5\theta)$. Thus, as shown in Figure 14B.8,

$$u(r, \theta) = C + 0.9 \log(r) + \frac{-\sin(2\theta)}{2 \cdot r^2} + \frac{-4\cos(5\theta)}{5 \cdot r^5}. \qquad \diamondsuit$$

14B(iv) Boundary value problems on an annulus

Prerequisites: §5C, §14A, §14B(i), §8A, Appendix F.   **Recommended:** §14B(ii), §14B(iii).

Proposition 14B.10 Laplace equation, annulus, nonhomogeneous Dirichlet BC
Let $\mathbb{A} = \{(r, \theta); R_{min} \leq r \leq R_{max}\}$ *be an annulus, and let* $b, B : [-\pi, \pi) \longrightarrow \mathbb{R}$ *be two functions. Consider the Laplace equation '$\triangle u = 0$', with nonhomogeneous Dirichlet boundary conditions:*

$$u(R_{min}, \theta) = b(\theta) \quad \text{and} \quad u(R_{max}, \theta) = B(\theta), \quad \text{for all } \theta \in [-\pi, \pi). \quad (14B.14)$$

Suppose b *and* B *have real Fourier series*

$$b(\theta) \underset{12}{\approx} a_0 + \sum_{n=1}^{\infty} a_n \cos(n\theta) + \sum_{n=1}^{\infty} b_n \sin(n\theta)$$

and

$$B(\theta) \underset{12}{\approx} A_0 + \sum_{n=1}^{\infty} A_n \cos(n\theta) + \sum_{n=1}^{\infty} B_n \sin(n\theta).$$

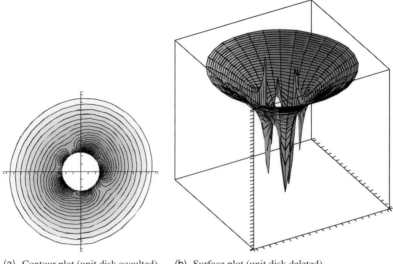

(a) Contour plot (unit disk occulted)   (b) Surface plot (unit disk deleted)

Figure 14B.8. The electric potential deduced from field measurements in Example 14B.9.

Then the unique solution to this problem is the function $u : \mathbb{A} \longrightarrow \mathbb{R}$ *defined:*

$$u(r, \theta) \underset{12}{\approx} U_0 + u_0 \log(r) + \sum_{n=1}^{\infty} \left(U_n r^n + \frac{u_n}{r^n} \right) \cos(n\theta)$$

$$+ \sum_{n=1}^{\infty} \left(V_n r^n + \frac{v_n}{r^n} \right) \sin(n\theta), \tag{14B.15}$$

where the coefficients $\{u_n, U_n, v_n, V_N\}_{n=1}^{\infty}$ *are the unique solutions to the following equations:*

$$U_0 + u_0 \log(R_{min}) = a_0; \qquad U_0 + u_0 \log(R_{max}) = A_0;$$

$$U_n R_{min}^n + \frac{u_n}{R_{min}^n} = a_n; \qquad U_n R_{max}^n + \frac{u_n}{R_{max}^n} = A_n;$$

$$V_n R_{min}^n + \frac{v_n}{R_{min}^n} = b_n; \qquad V_n R_{max}^n + \frac{v_n}{R_{max}^n} = B_n.$$

Furthermore, the series (14B.15) *converges semiuniformly to* u *on* $\mathrm{int}(\mathbb{A})$.

Ⓔ *Proof* **Exercise 14B.7.**

(a) To check that u is harmonic, generalize the strategies used to prove Proposition 14B.2, p. 284, and Proposition 14B.6, p. 290.

(b) To check that the solution also satisfies the boundary condition (14B.14), substititue $r = R_{min}$ and $r = R_{max}$ into equation (14B.15) to get the Fourier series for b and B.

(c) Use Proposition 5D.5(a), p. 91, to show that this solution is unique.   □

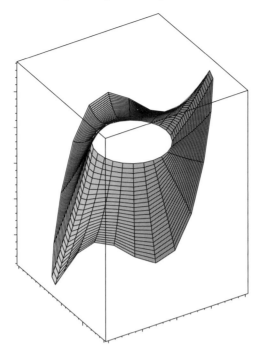

Figure 14B.9. Bubble between two concentric circular wires (see Example 14B.11).

Example 14B.11 Annular bubble

Consider an annular bubble spanning two concentric circular wire frames (see Figure 14B.9). The inner wire has radius $R_{\min} = 1$ and is unwarped, but it is elevated to a height of 4 cm, while the outer wire has radius $R_{\max} = 2$, and is twisted to have shape $B(\theta) = \cos(3\theta) - 2\sin(\theta)$. Estimate the shape of the bubble between the two wires.   ◇

Solution We have $b(\theta) = 4$, and $B(\theta) = \cos(3\theta) - 2\sin(\theta)$. Thus,

$$a_0 = 4; \qquad A_3 = 1; \qquad \text{and } B_1 = -2,$$

and all other coefficients of the boundary conditions are zero. Thus, our solution will have the form:

$$u(r, \theta) = U_0 + u_0 \log(r) + \left(U_3 r^3 + \frac{u_3}{r^3}\right) \cdot \cos(3\theta) + \left(V_1 r + \frac{v_1}{r}\right) \cdot \sin(\theta),$$

where U_0, u_0, U_3, u_3, V_1, and v_1 are chosen to solve the following equations:

$$U_0 + u_0 \log(1) = 4; \qquad U_0 + u_0 \log(2) = 0;$$

$$U_3 + u_3 = 0; \qquad 8\,U_3 + \frac{u_3}{8} = 1;$$

$$V_1 + v_1 = 0; \qquad 2V_1 + \frac{v_1}{2} = -2,$$

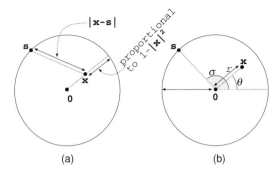

Figure 14B.10. The Poisson kernel (see also Figure 17F.1, p. 412).

which is equivalent to saying

$$U_0 = 4; \qquad\qquad u_0 = \frac{-U_0}{\log(2)} = \frac{-4}{\log(2)};$$

$$u_3 = -U_3; \quad \left(8 - \frac{1}{8}\right) U_3 = 1, \quad \text{and thus } U_3 = \frac{8}{63};$$

$$v_1 = -V_1; \quad \left(2 - \frac{1}{2}\right) V_1 = -2, \quad \text{and thus } V_1 = \frac{-4}{3}.$$

Thus

$$u(r, \theta) = 4 - \frac{4\log(r)}{\log(2)} + \frac{8}{63}\left(r^3 - \frac{1}{r^3}\right) \cdot \cos(3\theta) - \frac{4}{3}\left(r - \frac{1}{r}\right) \cdot \sin(\theta).$$

14B(v) *Poisson's solution to the Dirichlet problem on the disk*

Prerequisites: §14B(ii).   **Recommended:** §17F.[1]

Let $\mathbb{D} = \{(r, \theta); \ r \le R\}$ be the disk of radius R, and let $\partial\mathbb{D} = \mathbb{S} = \{(r, \theta); \ r = R\}$ be its boundary, the circle of radius R. Recall the *Dirichlet problem on the disk* from §14B(ii). We will now construct an 'integral representation formula' for the solution to this problem. The *Poisson kernel* is the function $\mathcal{P} : \mathbb{D} \times \mathbb{S} \longrightarrow \mathbb{R}$ defined:

$$\mathcal{P}(\mathbf{x}, \mathbf{s}) := \frac{R^2 - \|\mathbf{x}\|^2}{\|\mathbf{x} - \mathbf{s}\|^2} \quad \text{for any } \mathbf{x} \in \mathbb{D} \text{ and } \mathbf{s} \in \mathbb{S}.$$

In polar coordinates (Figure 14B.10(b)), we can parameterize $\mathbf{s} \in \mathbb{S}$ with a single angular coordinate $\sigma \in [-\pi, \pi)$, and assign $\mathbf{x}$ the coordinates (r, θ). Poisson's

---

[1] See §17F, p. 411, for a different development of the material in this section, using impulse-response functions. For yet another approach, using complex analysis, see Corollary 18C.13, p. 451.

kernel then takes the form:

$$P(\mathbf{x}, \mathbf{s}) = P(r, \theta; \sigma) = \frac{R^2 - r^2}{R^2 - 2rR\cos(\theta - \sigma) + r^2}.$$

(Exercise 14B.8).   ⒠

Proposition 14B.12 Poisson's integral formula
Let $\mathbb{D} = \{(r, \theta); r \le R\}$ be the disk of radius R, and let $b \in \mathbf{L}^2[-\pi, \pi)$. Consider the Laplace equation '$\triangle u = 0$', with nonhomogeneous Dirichlet boundary conditions $u(R, \theta) = b(\theta)$. The unique solution to this problem satisfies the following:

for any $r \in [0, R)$ and $\theta \in [-\pi, \pi)$, $u(r, \theta) = \dfrac{1}{2\pi} \displaystyle\int_{-\pi}^{\pi} P(r, \theta; \sigma) \cdot b(\sigma) d\sigma,$

$$\tag{14B.16}$$

or, more abstractly,

$$u(\mathbf{x}) = \frac{1}{2\pi} \int_{\mathbb{S}} P(\mathbf{x}, \mathbf{s}) \cdot b(\mathbf{s}) d\mathbf{s}, \quad \text{for any } \mathbf{x} \in \text{int}(\mathbb{D}).$$

Proof For simplicity, assume $R = 1$ (the general case can be obtained by rescaling). From Proposition 14B.2, p. 284, we know that

$$u(r, \theta) \underset{12}{\approx} A_0 + \sum_{n=1}^{\infty} A_n \cos(n\theta) \cdot r^n + \sum_{n=1}^{\infty} B_n \sin(n\theta) \cdot r^n,$$

where A_n and B_n are the (real) Fourier coefficients for the function b. Substituting in the definition of these coefficients (see §8A, p. 165), we get

$$u(r, \theta)$$

$$= \frac{1}{2\pi} \int_{-\pi}^{\pi} b(\sigma) d\sigma + \sum_{n=1}^{\infty} \cos(n\theta) \cdot r^n \cdot \left(\frac{1}{\pi} \int_{-\pi}^{\pi} b(\sigma) \cos(n\sigma) d\sigma \right)$$

$$+ \sum_{n=1}^{\infty} \sin(n\theta) \cdot r^n \cdot \left(\frac{1}{\pi} \int_{-\pi}^{\pi} b(\sigma) \sin(n\sigma) d\sigma \right)$$

$$= \frac{1}{2\pi} \int_{-\pi}^{\pi} b(\sigma) \left(1 + 2 \sum_{n=1}^{\infty} r^n \cdot \cos(n\theta) \cos(n\sigma) + 2 \sum_{n=1}^{\infty} r^n \cdot \sin(n\theta) \sin(n\sigma) \right) d\sigma$$

$$\underset{(*)}{=} \frac{1}{2\pi} \int_{-\pi}^{\pi} b(\sigma) \left(1 + 2 \sum_{n=1}^{\infty} r^n \cdot \cos\left(n(\theta - \sigma) \right) \right), \tag{14B.17}$$

where (*) is because $\cos(n\theta)\cos(n\sigma) + \sin(n\theta)\sin(n\sigma) = \cos(n(\theta - \sigma))$.

It now suffices to prove the following.

Claim 1

$$1 + 2 \sum_{n=1}^{\infty} r^n \cdot \cos\left(n(\theta - \sigma)\right) = P(r, \theta; \sigma).$$

Proof By Euler's formula (p. 553), $2\cos(n(\theta - \sigma)) = e^{in(\theta-\sigma)} + e^{-in(\theta-\sigma)}$. Hence,

$$1 + 2 \sum_{n=1}^{\infty} r^n \cdot \cos\left(n(\theta - \sigma)\right) = 1 + \sum_{n=1}^{\infty} r^n \cdot \left(e^{in(\theta-\sigma)} + e^{-in(\theta-\sigma)}\right). \quad (14B.18)$$

Now define the complex number $z = r \cdot e^{i(\theta-\sigma)}$; then observe that $r^n \cdot e^{in(\theta-\sigma)} = z^n$ and $r^n \cdot e^{-in(\theta-\sigma)} = \overline{z}^n$. Thus, we can rewrite the right-hand side of equation (14B.18) as follows:

$$1 + \sum_{n=1}^{\infty} r^n \cdot e^{in(\theta-\sigma)} + \sum_{n=1}^{\infty} r^n \cdot e^{-in(\theta-\sigma)}$$

$$= 1 + \sum_{n=1}^{\infty} z^n + \sum_{n=1}^{\infty} \overline{z}^n \underset{(a)}{=} 1 + \frac{z}{1-z} + \frac{\overline{z}}{1-\overline{z}}$$

$$= 1 + \frac{z - z\overline{z} + \overline{z} - z\overline{z}}{1 - z - \overline{z} + z\overline{z}} \underset{(b)}{=} 1 + \frac{2\mathrm{Re}\,[z] - 2|z|^2}{1 - 2\mathrm{Re}\,[z] + |z|^2}$$

$$= \frac{1 - 2\mathrm{Re}\,[z] + |z|^2}{1 - 2\mathrm{Re}\,[z] + |z|^2} + \frac{2\mathrm{Re}\,[z] - 2|z|^2}{1 - 2\mathrm{Re}\,[z] + |z|^2}$$

$$= \frac{1 - |z|^2}{1 - 2\mathrm{Re}\,[z] + |z|^2} \underset{(c)}{=} \frac{1 - r^2}{1 - 2r\cos(\theta - \sigma) + r^2} = P(r, \theta; \sigma).$$

(a) is because $\sum_{n=1}^{\infty} x^n = x/(1-x)$ for any $x \in \mathbb{C}$ with $|x| < 1$. (b) is because $z + \overline{z} = 2\mathrm{Re}\,[z]$ and $z\overline{z} = |z|^2$ for any $z \in \mathbb{C}$. (c) is because $|z| = r$ and $\mathrm{Re}\,[z] = \cos(\theta - \sigma)$ by definition of z. $\diamond$ $_{\text{Claim 1}}$

Now use Claim 1 to substitute $P(r, \theta; \sigma)$ into equation (14B.17); this yields the Poisson integral formula (14B.16). □

14C Bessel functions

14C(i) Bessel's equation; eigenfunctions of $\triangle$ in polar coordinates

Prerequisites: §4B, §14A.   **Recommended:** §16C.

Fix $n \in \mathbb{N}$. The (two-dimensional) *Bessel's equation* (of *order n*) is the ordinary differential equation

$$x^2 \mathcal{R}''(x) + x \mathcal{R}'(x) + (x^2 - n^2) \cdot \mathcal{R}(x) = 0, \quad (14C.1)$$

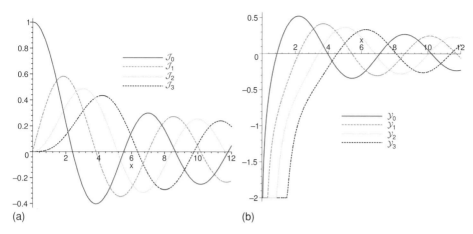

Figure 14C.1. Bessel functions near zero. (a) $\mathcal{J}_0(x)$, $\mathcal{J}_1(x)$, $\mathcal{J}_2(x)$, and $\mathcal{J}_3(x)$, for $x \in [0, 12]$; (b) $\mathcal{Y}_0(x)$, $\mathcal{Y}_1(x)$, $\mathcal{Y}_2(x)$, and $\mathcal{Y}_3(x)$ for $x \in [0, 12]$ (these four plots are vertically truncated).

where $\mathcal{R} : [0, \infty] \longrightarrow \mathbb{R}$ is an unknown function. In §16C, we will explain how this equation was first derived; in the present section, we will investigate its mathematical consequences.

The Bessel equation has two solutions:

$\mathcal{R}(x) = \mathcal{J}_n(x)$, the nth-order *Bessel function of the first kind* (see Figures 14C.1(a) and 14C.2(a));

$\mathcal{R}(x) = \mathcal{Y}_n(x)$ the nth-order *Bessel function of the second kind*, or *Neumann function* (see Figures 14C.1(b) and 14C.2(b)).

Bessel functions are like trigonometric or logarithmic functions; the 'simplest' expression for them is in terms of a power series. Hence, you should treat the functions '$\mathcal{J}_n$' and '$\mathcal{Y}_n$' the same way you treat elementary functions like 'sin', 'tan' or 'log'. In §14G we will derive an explicit power series for Bessel's functions, and in §14H we will derive some of their important properties. However, for now, we will simply take for granted that some solution functions $\mathcal{J}_n$ exist, and discuss how we can use these functions to build eigenfunctions for the Laplacian which *separate* in polar coordinates.

Proposition 14C.1 *Fix $\lambda > 0$. For any $n \in \mathbb{N}$, define the functions* $\Phi_{n,\lambda}$, $\Psi_{n,\lambda}$, $\phi_{n,\lambda}$, $\psi_{n,\lambda} : \mathbb{R}^2 \longrightarrow \mathbb{R}$ *by:*

$$\Phi_{n,\lambda}(r, \theta) = \mathcal{J}_n(\lambda \cdot r) \cdot \cos(n\theta); \quad \Psi_{n,\lambda}(r, \theta) = \mathcal{J}_n(\lambda \cdot r) \cdot \sin(n\theta);$$

$$\phi_{n,\lambda}(r, \theta) = \mathcal{Y}_n(\lambda \cdot r) \cdot \cos(n\theta); \quad \psi_{n,\lambda}(r, \theta) = \mathcal{Y}_n(\lambda \cdot r) \cdot \sin(n\theta)$$

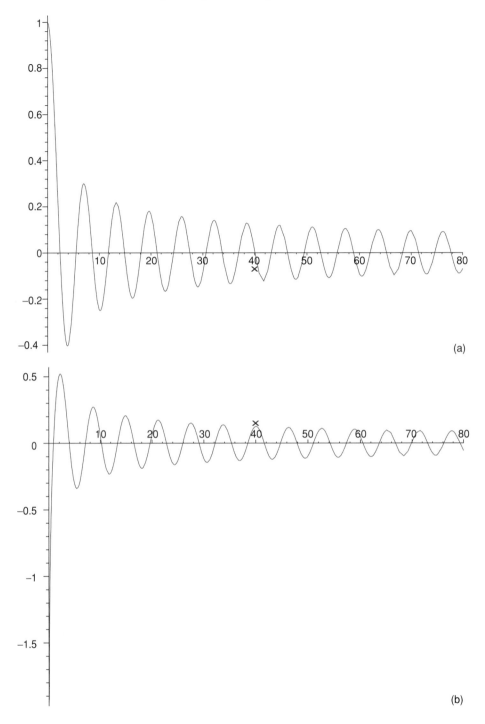

Figure 14C.2. Bessel functions are asymptotically periodic. (a) $\mathcal{J}_0(x)$, for $x \in [0, 80]$. The x intercepts of this graph are the roots $\kappa_{01}, \kappa_{02}, \kappa_{03}, \kappa_{04}, \ldots$ (b) $\mathcal{Y}_0(x)$, for $x \in [0, 80]$.

(see Figures 14C.3 and 14C.4). *Then* $\Phi_{n,\lambda}$, $\Psi_{n,\lambda}$, $\phi_{n,\lambda}$, *and* $\psi_{n,\lambda}$ *are all eigenfunctions of the Laplacian with eigenvalue* $-\lambda^2$:

$$\triangle \Phi_{n,\lambda} = -\lambda^2 \Phi_{n,\lambda}; \qquad \triangle \Psi_{n,\lambda} = -\lambda^2 \Psi_{n,\lambda};$$

$$\triangle \phi_{n,\lambda} = -\lambda^2 \phi_{n,\lambda}; \qquad \triangle \psi_{n,\lambda} = -\lambda^2 \psi_{n,\lambda}.$$

Proof See Problems 14.12–14.15. ☐

We can now use these eigenfunctions to solve PDEs in polar coordinates. Note that $\mathcal{J}_n$ – and, thus, eigenfunctions $\Phi_{n,\lambda}$ and $\Psi_{n,\lambda}$ – are *bounded* around zero (see Figure 14C.1(a)). On the other hand, $\mathcal{Y}_n$ – and thus, eigenfunctions $\phi_{n,\lambda}$ and $\psi_{n,\lambda}$ – are *unbounded* at zero (see Figure 14C.1(b)). Hence, when solving BVPs in a neighbourhood around zero (e.g. the disk), we should use $\mathcal{J}_n$, $\Phi_{n,\lambda}$, and $\Psi_{n,\lambda}$. When solving BVPs on a domain *away* from zero (e.g. the annulus), we can also use $\mathcal{Y}_n$, $\phi_{n,\lambda}$, and $\psi_{n,\lambda}$.

14C(ii) Boundary conditions; the roots of the Bessel function

Prerequisites: §5C, §14C(i).

To obtain *homogeneous Dirichlet boundary conditions* on a disk of radius R, we need an eigenfunction of the form $\Phi_{n,\lambda}$ (or $\Psi_{n,\lambda}$) such that $\Phi_{n,\lambda}(R, \theta) = 0$ for all $\theta \in [-\pi, \pi)$. Hence, we need

$$\mathcal{J}_n(\lambda \cdot R) = 0. \tag{14C.2}$$

The *roots* of the Bessel function $\mathcal{J}_n$ are the values $\kappa \in \mathbb{R}_{\not=}$ such that $\mathcal{J}_n(\kappa) = 0$. These roots form an increasing sequence:

$$0 \leq \kappa_{n1} < \kappa_{n2} < \kappa_{n3} < \kappa_{n4} < \cdots \tag{14C.3}$$

of irrational values.[2] Thus, to satisfy the homogeneous Dirichlet boundary condition (14C.2), we must set $\lambda := \kappa_{nm}/R$ for some $m \in \mathbb{N}$. This yields an increasing sequence of eigenvalues:

$$\lambda_{n1}^2 = \left(\frac{\kappa_{n1}}{R}\right)^2 < \lambda_{n2}^2 = \left(\frac{\kappa_{n2}}{R}\right)^2 < \lambda_{n3}^2 = \left(\frac{\kappa_{n3}}{R}\right)^2 < \lambda_{n4}^2 = \left(\frac{\kappa_{n4}}{R}\right)^2 < \cdots, \tag{14C.4}$$

which are the eigenvalues which we can expect to see in this problem. The corresponding eigenfunctions will then have the following form:

$$\Phi_{n,m}(r, \theta) = \mathcal{J}_n(\lambda_{n,m} \cdot r) \cdot \cos(n\theta); \qquad \Psi_{n,m}(r, \theta) = \mathcal{J}_n(\lambda_{n,m} \cdot r) \cdot \sin(n\theta) \tag{14C.5}$$

(see Figures 14C.3 and 14C.4).

---

[2] Computing these roots is difficult; tables of κ_{nm} can be found in most standard references on PDEs.

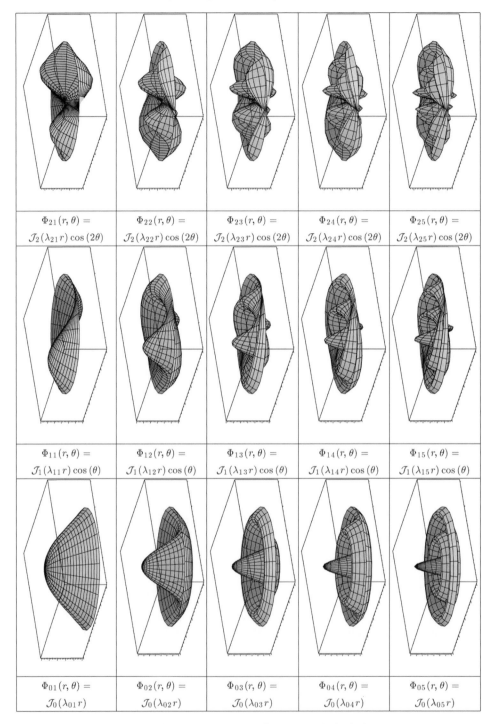

Figure 14C.3. $\Phi_{n,m}$ for $n = 0, 1, 2$ and for $m = 1, 2, 3, 4, 5$ (rotate page).

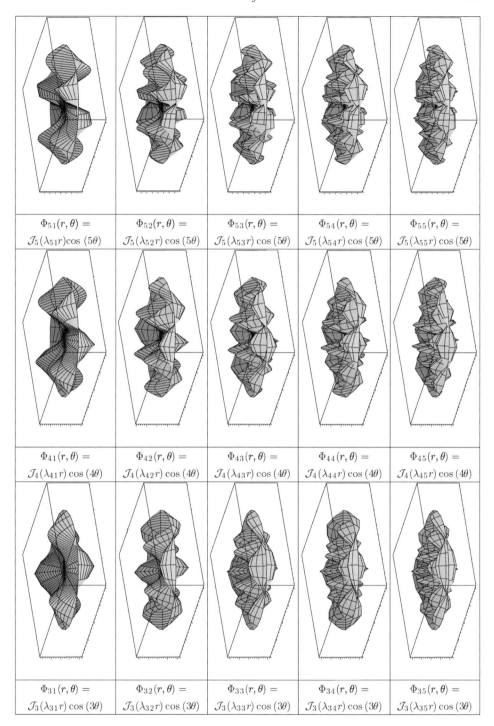

Figure 14C.4. $\Phi_{n,m}$ for $n = 3, 4, 5$ and for $m = 1, 2, 3, 4, 5$ (rotate page).

14C(iii) Initial conditions; Fourier–Bessel expansions

Prerequisites: §5B, §6F, §14C(ii).

To solve an initial value problem, while satisfying the desired boundary conditions, we express our initial conditions as a sum of the eigenfunctions from equations (14C.5). This is called a *Fourier–Bessel expansion*:

$$f(r, \theta) = \sum_{n=0}^{\infty} \sum_{m=1}^{\infty} A_{nm} \cdot \Phi_{nm}(r, \theta) + \sum_{n=1}^{\infty} \sum_{m=1}^{\infty} B_{nm} \cdot \Psi_{nm}(r, \theta)$$

$$+ \sum_{n=0}^{\infty} \sum_{m=1}^{\infty} a_{nm} \cdot \phi_{nm}(r, \theta) + \sum_{n=1}^{\infty} \sum_{m=1}^{\infty} b_{nm} \cdot \psi_{nm}(r, \theta), \quad (14C.6)$$

where A_{nm}, B_{nm}, a_{nm}, and b_{nm} are all real-valued coefficients. Suppose we are considering boundary value problems on the unit disk $\mathbb{D}$. Then we want this expansion to be bounded at 0, so we don't want the second two types of eigenfunctions. Thus, equation (14C.6) simplifies to

$$\sum_{n=0}^{\infty} \sum_{m=1}^{\infty} A_{nm} \cdot \Phi_{nm}(r, \theta) + \sum_{n=1}^{\infty} \sum_{m=1}^{\infty} B_{nm} \cdot \Psi_{nm}(r, \theta). \quad (14C.7)$$

If we substitute the explicit expressions from equations (14C.5) for $\Phi_{nm}(r, \theta)$ and $\Psi_{nm}(r, \theta)$ into equation (14C.7), we get

$$\sum_{n=0}^{\infty} \sum_{m=1}^{\infty} A_{nm} \cdot \mathcal{J}_n \left(\frac{\kappa_{nm} \cdot r}{R} \right) \cdot \cos(n\theta)$$

$$+ \sum_{n=1}^{\infty} \sum_{m=1}^{\infty} B_{nm} \cdot \mathcal{J}_n \left(\frac{\kappa_{nm} \cdot r}{R} \right) \cdot \sin(n\theta). \quad (14C.8)$$

Now, if $f : \mathbb{D} \longrightarrow \mathbb{R}$ is some function describing initial conditions, is it always possible to express f using an expansion like equation (14C.8)? If so, how do we compute the coefficients A_{nm} and B_{nm} in equation (14C.8)? The answer to these questions lies in the following result.

Theorem 14C.2   *The collection $\{\Phi_{n,m}, \Psi_{\ell,m}; n = 0 \ldots \infty, \ell \in \mathbb{N}, m \in \mathbb{N}\}$ is an orthogonal basis for $\mathbf{L}^2(\mathbb{D})$. Thus, suppose $f \in \mathbf{L}^2(\mathbb{D})$, and for all $n, m \in \mathbb{N}$, we* *define*

$$A_{nm} := \frac{\langle f, \Phi_{nm} \rangle}{\|\Phi_n m\|_2^2}$$

$$= \frac{2}{\pi R^2 \cdot \mathcal{J}_{n+1}^2(\kappa_{nm})} \cdot \int_{-\pi}^{\pi} \int_0^R f(r, \theta) \cdot \mathcal{J}_n\left(\frac{\kappa_{nm} \cdot r}{R}\right) \cdot \cos(n\theta) \cdot r \, dr \, d\theta;$$

$$B_{nm} := \frac{\langle f, \Psi_{nm} \rangle}{\|\Psi_n m\|_2^2}$$

$$= \frac{2}{\pi R^2 \cdot \mathcal{J}_{n+1}^2(\kappa_{nm})} \cdot \int_{-\pi}^{\pi} \int_0^R f(r, \theta) \cdot \mathcal{J}_n\left(\frac{\kappa_{nm} \cdot r}{R}\right) \cdot \sin(n\theta) \cdot r \, dr \, d\theta.$$

Then the Fourier–Bessel series (14C.8) *converges to* f *in* L^2-*norm.*

Proof (Sketch) The fact that the collection $\{\Phi_{n,m}, \Psi_{\ell,m}; n = 0 \ldots \infty, \ell \in \mathbb{N}, m \in \mathbb{N}\}$ is an orthogonal set will be verified in Proposition 14H.4, p. 320. The fact that this orthogonal set is actually a *basis* of $\mathbf{L}^2(\mathbb{D})$ is too complicated for us to prove here. Given that this is true, if we define $A_{nm} := \langle f, \Phi_{nm} \rangle / \|\Phi_n m\|_2^2$ and $B_{nm} := \langle f, \Psi_{nm} \rangle / \|\Psi_n m\|_2^2$, then the Fourier–Bessel series (14C.8) converges to f in L^2-norm, by definition of 'orthogonal basis' (see §6F, p. 136).

It remains to verify the integral expressions given for the two inner products. To do this, recall that

$$\langle f, \Phi_{nm} \rangle = \frac{1}{\text{area}(\mathbb{D})} \int_{\mathbb{D}} f(\mathbf{x}) \cdot \Phi_{nm}(\mathbf{x}) d\mathbf{x}$$

$$= \frac{1}{\pi R^2} \int_0^{2\pi} \int_0^R f(r, \theta) \cdot \mathcal{J}_n\left(\frac{\kappa_{nm} \cdot r}{R}\right) \cdot \cos(n\theta) \cdot r \, dr \, d\theta$$

and

$$\|\Phi_{nm}\|_2^2 = \langle \Phi_{nm}, \Phi_{nm} \rangle = \frac{1}{\pi R^2} \int_{-\pi}^{\pi} \int_0^R \mathcal{J}_n^2\left(\frac{\kappa_{nm} \cdot r}{R}\right) \cdot \cos^2(n\theta) \cdot r \, dr \, d\theta$$

$$= \left(\frac{1}{R^2} \int_0^R \mathcal{J}_n^2\left(\frac{\kappa_{nm} \cdot r}{R}\right) \cdot r \, dr\right) \cdot \left(\frac{1}{\pi} \int_{-\pi}^{\pi} \cos^2(n\theta) d\theta\right)$$

$$\underset{(\dagger)}{=\!=} \frac{1}{R^2} \int_0^R \mathcal{J}_n^2\left(\frac{\kappa_{nm} \cdot r}{R}\right) \cdot r \, dr$$

$$\underset{(\ddagger)}{=\!=} \int_0^1 \mathcal{J}_n^2(\kappa_{nm} \cdot s) \cdot s \, ds$$

$$\underset{(*)}{=\!=} \frac{1}{2} \mathcal{J}_{n+1}^2(\kappa_{nm}).$$

Here, (†) is by Proposition 6D.2, p. 117; (‡) is the change of variables $s := \frac{r}{R}$, so that $dr = R\,ds$; and (∗) is by Lemma 14H.3(b), p. 318.   □

To compute the integrals in Theorem 14C.2, one generally uses 'integration by parts' techniques similar to those used to compute trigonometic Fourier coefficients (see, e.g., §7C, p. 150). However, instead of the convenient trigonometric facts that $\sin' = \cos$ and $\cos' = -\sin$, one must make use of slightly more complicated recurrence relations of Proposition 14H.1, p. 317. See Remark 14H.2, p. 317.

We will not have space in this book to develop integration techniques for computing Fourier–Bessel coefficients. Instead, in the remaining discussion, we will simply assume that f is given to us in the form of equation (14C.8).

14D The Poisson equation in polar coordinates

Prerequisites: §1D, §14C(ii), Appendix F.   **Recommended:** §11C, §12C, §13C, §14B .

Proposition 14D.1 Poisson equation on disk; homogeneous Dirichlet BC
Let $\mathbb{D} = \{(r, \theta);\ r \leq R\}$ be a disk, and let $q \in \mathbf{L}^2(\mathbb{D})$ be some function. Consider the Poisson equation '$\triangle u(r, \theta) = q(r, \theta)$', with homogeneous Dirichlet boundary conditions. Suppose q has semiuniformly convergent Fourier–Bessel series

$$q(r, \theta) \underset{L2}{\approx} \sum_{n=0}^{\infty} \sum_{m=1}^{\infty} A_{nm} \cdot \mathcal{J}_n \left(\frac{\kappa_{nm} \cdot r}{R} \right) \cdot \cos(n\theta)$$

$$+ \sum_{n=1}^{\infty} \sum_{m=1}^{\infty} B_{nm} \cdot \mathcal{J}_n \left(\frac{\kappa_{nm} \cdot r}{R} \right) \cdot \sin(n\theta).$$

Then the unique solution to this problem is the function $u : \mathbb{D} \longrightarrow \mathbb{R}$ defined

$$u(r, \theta) \underset{\text{unif}}{\equiv} -\sum_{n=0}^{\infty} \sum_{m=1}^{\infty} \frac{R^2 \cdot A_{nm}}{\kappa_{nm}^2} \cdot \mathcal{J}_n \left(\frac{\kappa_{nm} \cdot r}{R} \right) \cdot \cos(n\theta)$$

$$- \sum_{n=1}^{\infty} \sum_{m=1}^{\infty} \frac{R^2 \cdot B_{nm}}{\kappa_{nm}^2} \cdot \mathcal{J}_n \left(\frac{\kappa_{nm} \cdot r}{R} \right) \cdot \sin(n\theta).$$

ⓔ *Proof* **Exercise 14D.1.**   □

Remark If $R = 1$, then the expression for q simplifies to

$$q(r, \theta) \underset{L2}{\approx} \sum_{n=0}^{\infty} \sum_{m=1}^{\infty} A_{nm} \cdot \mathcal{J}_n (\kappa_{nm} \cdot r) \cdot \cos(n\theta)$$

$$+ \sum_{n=1}^{\infty} \sum_{m=1}^{\infty} B_{nm} \cdot \mathcal{J}_n (\kappa_{nm} \cdot r) \cdot \sin(n\theta)$$

and the solution simplifies to

$$u(r, \theta) \underset{\text{unif}}{\equiv} - \sum_{n=0}^{\infty} \sum_{m=1}^{\infty} \frac{A_{nm}}{\kappa_{nm}^2} \cdot \mathcal{J}_n(\kappa_{nm} \cdot r) \cdot \cos(n\theta)$$

$$- \sum_{n=1}^{\infty} \sum_{m=1}^{\infty} \frac{B_{nm}}{\kappa_{nm}^2} \cdot \mathcal{J}_n(\kappa_{nm} \cdot r) \cdot \sin(n\theta).$$

Example 14D.2 Suppose $R = 1$, and $q(r, \theta) = \mathcal{J}_0(\kappa_{0,3} \cdot r) + \mathcal{J}_5(\kappa_{2,5} \cdot r) \cdot \sin(2\theta)$.
Then

$$u(r, \theta) = \frac{-\mathcal{J}_0(\kappa_{0,3} \cdot r)}{\kappa_{0,3}^2} - \frac{\mathcal{J}_5(\kappa_{2,5} \cdot r) \cdot \sin(2\theta)}{\kappa_{2,5}^2}. \qquad \diamond$$

Proposition 14D.3 Poisson equation on disk; nonhomogeneous Dirichlet BC
Let $\mathbb{D} = \{(r, \theta) ; r \leq R\}$ *be a disk. Let* $b \in \mathbf{L}^2[-\pi, \pi)$ *and* $q \in \mathbf{L}^2(\mathbb{D})$. *Consider the Poisson equation* '$\Delta u(r, \theta) = q(r, \theta)$', *with* nonhomogeneous *Dirichlet boundary conditions:*

$$u(R, \theta) = b(\theta), \qquad \text{for all } \theta \in [-\pi, \pi). \tag{14D.1}$$

(1) *Let* $w : \mathbb{D} \longrightarrow \mathbb{R}$ *be the solution[3] to the* Laplace equation '$\Delta w = 0$', *with the nonhomogeneous Dirichlet BC* (14D.1).
(2) *Let* $v : \mathbb{D} \longrightarrow \mathbb{R}$ *be the solution[4] to the* Poisson equation '$\Delta v = q$', *with the* homogeneous *Dirichlet BC.*
(3) *Define* $u(r, \theta) := v(r, \theta; t) + w(r, \theta)$. *Then* $u(r, \theta)$ *is a solution to the Poisson equation with inhomogeneous Dirichlet BC* (14D.1).

Proof **Exercise 14D.2.**   □ Ⓔ

Example 14D.4 Suppose $R = 1$, and $q(r, \theta) = \mathcal{J}_0(\kappa_{0,3} \cdot r) + \mathcal{J}_2(\kappa_{2,5} \cdot r) \cdot \sin(2\theta)$. Let $b(\theta) = \sin(3\theta)$. From Example 14B.3, p. 285, we know that the (bounded) solution to the Laplace equation with Dirichlet BC $w(1, \theta) = b(\theta)$ is given by

$$w(r, \theta) = r^3 \sin(3\theta).$$

From Example 14D.2, we know that the solution to the Poisson equation '$\Delta v = q$', with homogeneous Dirichlet BC, is given by

$$v(r, \theta) = \frac{\mathcal{J}_0(\kappa_{0,3} \cdot r)}{\kappa_{0,3}^2} + \frac{\mathcal{J}_2(\kappa_{2,5} \cdot r) \cdot \sin(2\theta)}{\kappa_{2,5}^2}.$$

---

[3] Obtained from Proposition 14B.2, p. 284, for example.   [4] Obtained from Proposition 14D.1, for example.

Thus, by Proposition 14D.3, the solution to the Poisson equation '$\Delta u = q$', with Dirichlet BC $w(1, \theta) = b(\theta)$, is given by

$$u(r, \theta) = v(r, \theta) + w(r, \theta) = \frac{\mathcal{J}_0 (\kappa_{0,3} \cdot r)}{\kappa_{0,3}^2} + \frac{\mathcal{J}_2 (\kappa_{2,5} \cdot r) \cdot \sin(2\theta)}{\kappa_{2,5}^2} + r^3 \sin(3\theta).$$

◇

14E The heat equation in polar coordinates

Prerequisites: §1B, §14C(iii), Appendix F.   **Recommended:** §11A, §12B, §13A, §14B .

Proposition 14E.1 Heat equation on disk; homogeneous Dirichlet BC
Let $\mathbb{D} = \{(r, \theta); \ r \leq R\}$ be a disk, and consider the heat equation '$\partial_t u = \Delta u$', with homogeneous Dirichlet boundary conditions, and initial conditions $u(r, \theta; 0) = f(r, \theta)$. Suppose f has Fourier–Bessel series

$$f(r, \theta) \underset{12}{\approx} \sum_{n=0}^{\infty} \sum_{m=1}^{\infty} A_{nm} \cdot \mathcal{J}_n \left(\frac{\kappa_{nm} \cdot r}{R} \right) \cdot \cos(n\theta)$$

$$+ \sum_{n=1}^{\infty} \sum_{m=1}^{\infty} B_{nm} \cdot \mathcal{J}_n \left(\frac{\kappa_{nm} \cdot r}{R} \right) \cdot \sin(n\theta).$$

Then the unique solution to this problem is the function $u : \mathbb{D} \times \mathbb{R}_+ \longrightarrow \mathbb{R}$ defined:

$$u(r, \theta; t) \underset{12}{\approx} \sum_{n=0}^{\infty} \sum_{m=1}^{\infty} A_{nm} \cdot \mathcal{J}_n \left(\frac{\kappa_{nm} \cdot r}{R} \right) \cdot \cos(n\theta) \exp \left(\frac{-\kappa_{nm}^2}{R^2} t \right)$$

$$+ \sum_{n=1}^{\infty} \sum_{m=1}^{\infty} B_{nm} \cdot \mathcal{J}_n \left(\frac{\kappa_{nm} \cdot r}{R} \right) \cdot \sin(n\theta) \exp \left(\frac{-\kappa_{nm}^2}{R^2} t \right).$$

Furthermore, the series defining u converges semiuniformly on $\mathbb{D} \times \mathbb{R}_+$.

Ⓔ   *Proof* **Exercise 14E.1.**   □

Remark If $R = 1$, then the initial conditions simplify to

$$f(r, \theta) \underset{12}{\approx} \sum_{n=0}^{\infty} \sum_{m=1}^{\infty} A_{nm} \cdot \mathcal{J}_n (\kappa_{nm} \cdot r) \cdot \cos(n\theta)$$

$$+ \sum_{n=1}^{\infty} \sum_{m=1}^{\infty} B_{nm} \cdot \mathcal{J}_n (\kappa_{nm} \cdot r) \cdot \sin(n\theta),$$

and the solution simplifies to

$$u(r, \theta; t) \underset{12}{\approx} \sum_{n=0}^{\infty} \sum_{m=1}^{\infty} A_{nm} \cdot \mathcal{J}_n\left(\kappa_{nm} \cdot r\right) \cdot \cos(n\theta) \cdot e^{-\kappa_{nm}^2 t}$$

$$+ \sum_{n=1}^{\infty} \sum_{m=1}^{\infty} B_{nm} \cdot \mathcal{J}_n\left(\kappa_{nm} \cdot r\right) \cdot \sin(n\theta) \cdot e^{-\kappa_{nm}^2 t}.$$

Example 14E.2 Suppose $R = 1$, and $f(r, \theta) = \mathcal{J}_0(\kappa_{0,7} \cdot r) - 4\mathcal{J}_3(\kappa_{3,2} \cdot r) \cdot \cos(3\theta)$. Then

$$u(r, \theta; t) = \mathcal{J}_0\left(\kappa_{0,7} \cdot r\right) \cdot e^{-\kappa_{0,7}^2 t} - 4\mathcal{J}_3\left(\kappa_{3,2} \cdot r\right) \cdot \cos(3\theta) \cdot e^{-\kappa_{3,2}^2 t}. \qquad \Diamond$$

Proposition 14E.3 Heat equation on disk; nonhomogeneous Dirichlet BC

Let $\mathbb{D} = \{(r, \theta) \, ; \, r \leq R\}$ be a disk, and let $f : \mathbb{D} \longrightarrow \mathbb{R}$ and $b : [-\pi, \pi) \longrightarrow \mathbb{R}$ be given functions. Consider the heat equation '$\partial_t u = \triangle u$', with initial conditions $u(r, \theta) = f(r, \theta)$, and nonhomogeneous Dirichlet boundary conditions:

$$u(R, \theta) = b(\theta), \qquad \text{for all } \theta \in [-\pi, \pi). \qquad (14E.1)$$

(1) *Let $w : \mathbb{D} \longrightarrow \mathbb{R}$ be the solution[5] to the Laplace equation '$\triangle w = 0$', with the nonhomogeneous Dirichlet BC (14E.1).*

(2) *Define $g(r, \theta) := f(r, \theta) - w(r, \theta)$. Let $v : \mathbb{D} \times \mathbb{R}_+ \longrightarrow \mathbb{R}$ be the solution[6] to the heat equation '$\partial_t v = \triangle v$' with initial conditions $v(r, \theta) = g(r, \theta)$, and homogeneous Dirichlet BC.*

(3) *Define $u(r, \theta; t) := v(r, \theta; t) + w(r, \theta)$. Then $u(r, \theta; t)$ is a solution to the heat equation with initial conditions $u(r, \theta) = f(r, \theta)$, and inhomogeneous Dirichlet BC (14E.1).*

Proof **Exercise 14E.2.** ⬜ Ⓔ

14F The wave equation in polar coordinates

Prerequisites: §2B, §14C(ii), §14C(iii), Appendix F.   **Recommended:** §11B, §12D, §14E.

Imagine a drumskin stretched tightly over a circular frame. At equilibrium, the drumskin is perfectly flat, but, if we strike the skin, it will vibrate, meaning that the membrane will experience vertical displacements from equilibrium. Let $\mathbb{D} = \{(r, \theta); \, r \leq R\}$ represent the round skin, and, for any point $(r, \theta) \in \mathbb{D}$ on the

---

[5] Obtained from Proposition 14B.2, p. 284, for example.   [6] Obtained from Proposition 14E.1, for example.

drumskin and time $t > 0$, let $u(r, \theta; t)$ be the vertical displacement of the drum. Then u will obey the two-dimensional wave equation

$$\partial_t^2 u(r, \theta; t) = \triangle u(r, \theta; t). \tag{14F.1}$$

However, since the skin is held down along the edges of the circle, the function u will also exhibit homogeneous *Dirichlet* boundary conditions:

$$u(R, \theta; t) = 0, \quad \text{for all } \theta \in [-\pi, \pi) \text{ and } t \geq 0. \tag{14F.2}$$

Proposition 14F.1 Wave equation on disk; homogeneous Dirichlet BC

Let $\mathbb{D} = \{(r, \theta); \ r \leq R\}$ be a disk, and consider the wave equation '$\partial_t^2 u = \triangle u$', with homogeneous Dirichlet boundary conditions, and

$$\begin{aligned}\text{initial position:} \quad &u(r, \theta; 0) = f_0(r, \theta); \\ \text{initial velocity:} \ &\partial_t\, u(r, \theta; 0) = f_1(r, \theta).\end{aligned}$$

Suppose f_0 and f_1 have Fourier–Bessel series

$$f_0(r, \theta) \underset{12}{\approx} \sum_{n=0}^{\infty} \sum_{m=1}^{\infty} A_{nm} \cdot \mathcal{J}_n \left(\frac{\kappa_{nm} \cdot r}{R} \right) \cdot \cos(n\theta)$$

$$+ \sum_{n=1}^{\infty} \sum_{m=1}^{\infty} B_{nm} \cdot \mathcal{J}_n \left(\frac{\kappa_{nm} \cdot r}{R} \right) \cdot \sin(n\theta)$$

and

$$f_1(r, \theta) \underset{12}{\approx} \sum_{n=0}^{\infty} \sum_{m=1}^{\infty} A'_{nm} \cdot \mathcal{J}_n \left(\frac{\kappa_{nm} \cdot r}{R} \right) \cdot \cos(n\theta)$$

$$+ \sum_{n=1}^{\infty} \sum_{m=1}^{\infty} B'_{nm} \cdot \mathcal{J}_n \left(\frac{\kappa_{nm} \cdot r}{R} \right) \cdot \sin(n\theta).$$

Assume that

$$\sum_{n=0}^{\infty} \sum_{m=1}^{\infty} \kappa_{nm}^2 |A_{nm}| + \sum_{n=1}^{\infty} \sum_{m=1}^{\infty} \kappa_{nm}^2 |B_{nm}| < \infty,$$

and

$$\sum_{n=0}^{\infty} \sum_{m=1}^{\infty} \kappa_{nm} |A'_{nm}| + \sum_{n=1}^{\infty} \sum_{m=1}^{\infty} \kappa_{nm} |B'_{nm}| < \infty.$$

Then the unique solution to this problem is the function $u : \mathbb{D} \times \mathbb{R}_+ \longrightarrow \mathbb{R}$ *defined:*

$$u(r, \theta; t) \underset{\text{L2}}{\approx} \sum_{n=0}^{\infty} \sum_{m=1}^{\infty} A_{nm} \cdot \mathcal{J}_n \left(\frac{\kappa_{nm} \cdot r}{R} \right) \cdot \cos(n\theta) \cdot \cos \left(\frac{\kappa_{nm}}{R} t \right)$$

$$+ \sum_{n=1}^{\infty} \sum_{m=1}^{\infty} B_{nm} \cdot \mathcal{J}_n \left(\frac{\kappa_{nm} \cdot r}{R} \right) \cdot \sin(n\theta) \cdot \cos \left(\frac{\kappa_{nm}}{R} t \right)$$

$$+ \sum_{n=0}^{\infty} \sum_{m=1}^{\infty} \frac{R \cdot A'_{nm}}{\kappa_{nm}} \cdot \mathcal{J}_n \left(\frac{\kappa_{nm} \cdot r}{R} \right) \cdot \cos(n\theta) \cdot \sin \left(\frac{\kappa_{nm}}{R} t \right)$$

$$+ \sum_{n=1}^{\infty} \sum_{m=1}^{\infty} \frac{R \cdot B'_{nm}}{\kappa_{nm}} \cdot \mathcal{J}_n \left(\frac{\kappa_{nm} \cdot r}{R} \right) \cdot \sin(n\theta) \cdot \sin \left(\frac{\kappa_{nm}}{R} t \right).$$

Proof **Exercise 14F.1.**   □   Ⓔ

Remark If $R = 1$, then the initial conditions would be

$$f_0(r, \theta) \underset{\text{L2}}{\approx} \sum_{n=0}^{\infty} \sum_{m=1}^{\infty} A_{nm} \cdot \mathcal{J}_n \left(\kappa_{nm} \cdot r \right) \cdot \cos(n\theta)$$

$$+ \sum_{n=1}^{\infty} \sum_{m=1}^{\infty} B_{nm} \cdot \mathcal{J}_n \left(\kappa_{nm} \cdot r \right) \cdot \sin(n\theta),$$

and

$$f_1(r, \theta) \underset{\text{L2}}{\approx} \sum_{n=0}^{\infty} \sum_{m=1}^{\infty} A'_{nm} \cdot \mathcal{J}_n \left(\kappa_{nm} \cdot r \right) \cdot \cos(n\theta)$$

$$+ \sum_{n=1}^{\infty} \sum_{m=1}^{\infty} B'_{nm} \cdot \mathcal{J}_n \left(\kappa_{nm} \cdot r \right) \cdot \sin(n\theta).$$

The solution then simplifies to

$$u(r, \theta; t) \underset{\text{L2}}{\approx} \sum_{n=0}^{\infty} \sum_{m=1}^{\infty} A_{nm} \cdot \mathcal{J}_n \left(\kappa_{nm} \cdot r \right) \cdot \cos(n\theta) \cdot \cos \left(\kappa_{nm} t \right)$$

$$+ \sum_{n=1}^{\infty} \sum_{m=1}^{\infty} B_{nm} \cdot \mathcal{J}_n \left(\kappa_{nm} \cdot r \right) \cdot \sin(n\theta) \cdot \cos \left(\kappa_{nm} t \right)$$

$$+ \sum_{n=0}^{\infty} \sum_{m=1}^{\infty} \frac{A'_{nm}}{\kappa_{nm}} \cdot \mathcal{J}_n \left(\kappa_{nm} \cdot r \right) \cdot \cos(n\theta) \cdot \sin \left(\kappa_{nm} t \right)$$

$$+ \sum_{n=1}^{\infty} \sum_{m=1}^{\infty} \frac{B'_{nm}}{\kappa_{nm}} \cdot \mathcal{J}_n \left(\kappa_{nm} \cdot r \right) \cdot \sin(n\theta) \cdot \sin \left(\kappa_{nm} t \right).$$

Acoustic interpretation The vibration of the drumskin is a superposition of distinct *modes* of the form

$$\Phi_{nm}(r, \theta) = \mathcal{J}_n\left(\frac{\kappa_{nm} \cdot r}{R}\right) \cdot \cos(n\theta) \quad \text{and} \quad \Psi_{nm}(r, \theta) = \mathcal{J}_n\left(\frac{\kappa_{nm} \cdot r}{R}\right) \cdot \sin(n\theta),$$

for all $m, n \in \mathbb{N}$. For fixed m and n, the modes Φ_{nm} and and Ψ_{nm} vibrate at (temporal) frequency $\lambda_{nm} = \kappa_{nm}/R$. In the case of the *vibrating string*, all the different modes vibrated at frequencies that were *integer multiples* of the fundamental frequency; musically speaking, this means that they are 'in harmony'. In the case of a *drum*, however, the frequencies are all *irrational* multiples (because the roots κ_{nm} are all irrationally related). Acoustically speaking, this means we expect a drum to sound somewhat more 'discordant' than a string.

Note also that, as the radius R gets *larger*, the frequency $\lambda_{nm} = \kappa_{nm}/R$ gets *smaller*. This means that larger drums vibrate at *lower* frequencies, which matches our experience.

Example 14F.2 Resonating loudspeaker diaphragm
A circular membrane of radius $R = 1$ is emitting a pure pitch at frequency κ_{35}. Roughly describe the space-time profile of the solution (as a pattern of distortions of the membrane).

Solution The spatial distortion of the membrane must be a combination of modes vibrating at this frequency. Thus, we expect it to be a function of the form

$$u(r, \theta; t) = \mathcal{J}_3(\kappa_{35} \cdot r)\left[\left(A \cdot \cos(3\theta) + B \cdot \sin(3\theta)\right) \cdot \cos(\kappa_{35}t)\right.$$
$$\left. + \left(\frac{A'}{\kappa_{35}} \cdot \cos(3\theta) + \frac{B'}{\kappa_{35}} \cdot \sin(3\theta)\right) \cdot \sin(\kappa_{35}t)\right].$$

By introducing some constant angular phase-shifts ϕ and ϕ', as well as new constants C and C', we can rewrite this (**Exercise 14F.2**) as follows:

Ⓔ

$$u(r, \theta; t) = \mathcal{J}_3(\kappa_{35} \cdot r)$$
$$\times \left(C \cdot \cos(3(\theta + \phi)) \cdot \cos(\kappa_{35}t) + \frac{C'}{\kappa_{35}} \cdot \cos(3(\theta + \phi')) \cdot \sin(\kappa_{35}t)\right).$$

◇

Example 14F.3 An initially silent circular drum of radius $R = 1$ is struck in its exact centre with a drumstick having a spherical head. Describe the resulting pattern of vibrations.

Solution This is a problem with *nonzero* initial *velocity* and *zero* initial *position*. Since the initial velocity (the impact of the drumstick) is rotationally symmetric (dead center, spherical head), we can write it as a Fourier–Bessel expansion with no angular dependence:

$$f_1(r, \theta) = f(r) \underset{L2}{\approx} \sum_{m=1}^{\infty} A'_m \cdot \mathcal{J}_0 (\kappa_{0m} \cdot r) \qquad (A'_1, A'_2, A'_3, \ldots \text{ some constants})$$

(all the higher-order Bessel functions disappear, since $\mathcal{J}_n$ is always associated with terms of the form $\sin(n\theta)$ and $\cos(n\theta)$, which depend on θ). Thus, the solution must have the following form:

$$u(r, \theta; t) = u(r, t) \underset{L2}{\approx} \sum_{m=1}^{\infty} \frac{A'_m}{\kappa_{0m}} \cdot \mathcal{J}_0 (\kappa_{0m} \cdot r) \cdot \sin (\kappa_{0m} t). \qquad \diamond$$

14G The power series for a Bessel function

Prerequisites: Appendix H(iii).　　　**Recommended:** §14C(i).

In §14C–§14F, we claimed that Bessel's equation had certain solutions called *Bessel functions*, and showed how to use these Bessel functions to solve differential equations in polar coordinates. Now we will derive an an explicit formula for these Bessel functions in terms of their power series.

Proposition 14G.1 *Set $\lambda := 1$. For any fixed $m \in \mathbb{N}$ there is a solution $\mathcal{J}_m :$ $\mathbb{R}_+ \longrightarrow \mathbb{R}$ to the Bessel equation*

$$x^2 \mathcal{J}''(x) + x \cdot \mathcal{J}'(x) + (x^2 - m^2) \cdot \mathcal{J}(x) = 0, \quad \text{for all } x > 0, \quad (14G.1)$$

with a power series expansion:

$$\mathcal{J}_m(x) = \left(\frac{x}{2}\right)^m \cdot \sum_{k=0}^{\infty} \frac{(-1)^k}{2^{2k} \, k! \, (m+k)!} x^{2k}. \qquad (14G.2)$$

($\mathcal{J}_m$ is called the mth-order Bessel function of the first kind.)

Proof The ODE (14G.1) satisfies the hypotheses of the Frobenius theorem (see Example H.5, p. 577). Thus, we can apply the *method of Frobenius* to solve equation (14G.1). Suppose that $\mathcal{J}$ is a solution, with an (unknown) power series $\mathcal{J}(x) = x^M \sum_{k=0}^{\infty} a_k x^k$, where $a_0, a_1, \ldots$ are unknown coefficients, and $M \geq 0$. We assume that $a_0 \neq 0$. We substitute this power series into equation (14G.1) to obtain equations relating the coefficients. The details of this computation are shown in Table 14.1.

Claim 1 $M = m$.

Proof If the Bessel equation is to be satisfied, the power series in the bottom row of Table 14.1 must be identically zero. In particular, this means that the

Table 14.1. *The method of Frobenius to solve Bessel's equation in the proof of 14G.1*

	x^M	x^{M+1}	x^{M+2}		x^{M+k}	

$$\mathcal{J}(x) = \quad a_0 x^M + \quad a_1 x^{M+1} + \quad a_2 x^{M+2} + \cdots + \quad a_k x^{M+k} + \cdots$$

Then

$$-m^2 \mathcal{J}(x) = \quad -m^2 a_0 x^M \quad - \quad m^2 a_1 x^{M+1} + \quad -m^2 a_2 x^{M+2} + \cdots + \quad -m^2 a_k x^{M+k} + \cdots$$

$$x^2 \mathcal{J}(x) = \qquad\qquad\qquad\qquad\qquad\qquad a_0 x^{M+2} + \cdots + \quad a_{k-2} x^{M+k} + \cdots$$

$$x \mathcal{J}'(x) = \quad M a_0 x^M + \quad (M+1)a_1 x^{M+1} + \quad (M+2)a_2 x^{M+2} + \cdots + \quad (M+k)a_k x^{M+k} + \cdots$$

$$x^2 \mathcal{J}''(x) = \quad M(M-1)a_0 x^M + \quad (M+1)M a_1 x^{M+1} + \quad (M+2)(M+1)a_2 x^{M+2} + \cdots + \quad (M+k)(M+k-1)a_k x^{M+k} + \cdots$$

Thus $0 = x^2 \mathcal{J}''(x) + x \cdot \mathcal{J}'(x) + (x^2 - m^2) \cdot \mathcal{J}(x)$

$$= \underbrace{(M^2 - m^2)}_{(a)} a_0 x^M + \underbrace{((M+1)^2 - m^2)a_1}_{(b)} x^{M+1} + \left(\underset{((M+2)^2 - m^2)a_2}{a_0 +} \right) x^{M+2} + \cdots + b_k x^{M+k} + \cdots$$

where

$$b_k := a_{k-2} + (M+k)a_k + (M+k)(M+k-1)a_k - m^2 a_k$$
$$= a_{k-2} + (M+k)(1 + M+k-1)a_k - m^2 a_k$$
$$= a_{k-2} + \left((M+k)^2 - m^2 \right) a_k.$$

coefficient labeled '(a)' must be zero; in other words, $a_0(M^2 - m^2) = 0$. Since we know that $a_0 \neq 0$, this means $(M^2 - m^2) = 0$, i.e. $M^2 = m^2$. But $M \geq 0$, so this means $M = m$. ◇ Claim 1

Claim 2 $a_1 = 0$.

Proof If the Bessel equation is to be satisfied, the power series in the bottom row of Table 14.1 must be identically zero. In particular, this means that the coefficient labeled '(b)' must be zero; in other words, $a_1[(M+1)^2 - m^2] = 0$. Claim 1 says that $M = m$; hence this is equivalent to $a_1[(m+1)^2 - m^2] = 0$. Clearly, $[(m+1)^2 - m^2] \neq 0$; hence we conclude that $a_1 = 0$. ◇ Claim 2

Claim 3 *For all $k \geq 2$, the coefficients $\{a_2, a_3, a_4, \ldots\}$ must satisfy the following recurrence relation:*

$$a_k = \frac{-1}{(m+k)^2 - m^2} a_{k-2}, \qquad \text{for all even } k \in \mathbb{N} \text{ with } k \geq 2. \quad (14G.3)$$

On the other hand, $a_k = 0$ for all odd $k \in \mathbb{N}$.

Proof If the Bessel equation is to be satisfied, the power series in the bottom row of Table 14.1 must be identically zero. In particular, this means that all the coefficients b_k must be zero. In other words, $a_{k-2} + ((M+k)^2 - m^2)a_k = 0$. From Claim 1, we know that $M = m$; hence this is equivalent to $a_{k-2} + ((m+k)^2 - m^2)a_k = 0$. In other words, $a_k = -a_{k-2}/((m+k)^2 - m^2)a_k$. From Claim 2, we know that $a_1 = 0$. It follows from this equation that $a_3 = 0$; hence $a_5 = 0$, etc. Inductively, $a_n = 0$ for all odd n. ◇ Claim 3

Claim 4 *Assume we have fixed a value for a_0. Define*

$$a_{2j} := \frac{(-1)^j \cdot a_0}{2^{2j} j!(m+1)(m+2)\cdots(m+j)}, \qquad \text{for all } j \in \mathbb{N}.$$

Then the sequence $\{a_0, a_2, a_4, \ldots\}$ satisfies the recurrence relation (14G.3).

Proof Set $k = 2j$ in equation (14G.3). For any $j \geq 2$, we must show that $a_{2j} = \frac{-a_{2j-2}}{(m+2j)^2 - m^2}$. Now, by definition,

$$a_{2j-2} = a_{2(j-1)} := \frac{(-1)^{j-1} \cdot a_0}{2^{2j-2}(j-1)!(m+1)(m+2)\cdots(m+j-1)}.$$

Also,

$$(m+2j)^2 - m^2 = m^2 + 4jm + 4j^2 - m^2 = 4jm + 4j^2 = 2^2 j(m+j).$$

Hence

$$\frac{-a_{2j-2}}{(m+2j)^2 - m^2} = \frac{-a_{2j-2}}{2^2 j(m+j)}$$

$$= \frac{(-1)(-1)^{j-1} \cdot a_0}{2^2 j(m+j) \cdot 2^{2j-2}(j-1)!(m+1)(m+2)\cdots(m+j-1)}$$

$$= \frac{(-1)^j \cdot a_0}{2^{2j-2+2} \cdot j(j-1)! \cdot (m+1)(m+2)\cdots(m+j-1)(m+j)}$$

$$= \frac{(-1)^j \cdot a_0}{2^{2j} j!(m+1)(m+2)\cdots(m+j-1)(m+j)} = a_{2j},$$

as desired. $\diamond$ Claim 4

By convention we define $a_0 := \frac{1}{2^m}\frac{1}{m!}$. We claim that the resulting coefficients yield the Bessel Function $\mathcal{J}_m(x)$ defined by equation (14G.2) To see this, let b_{2k} be the $2k$th coefficient of the Bessel series. By definition,

$$b_{2k} := \frac{1}{2^m} \cdot \frac{(-1)^k}{2^{2k} k! (m+k)!}$$

$$= \frac{1}{2^m} \cdot \frac{(-1)^k}{2^{2k} k! m!(m+1)(m+2)\cdots(m+k-1)(m+k)}$$

$$= \frac{1}{2^m m!} \cdot \frac{(-1)^k}{2^{2k} k! (m+1)(m+2)\cdots(m+k-1)(m+k)}$$

$$= a_0 \cdot \left(\frac{(-1)^{k+1}}{2^{2k} k!(m+1)(m+2)\cdots(m+k-1)(m+k)} \right) = a_{2k},$$

as desired. $\square$

Corollary 14G.2   *Fix $m \in \mathbb{N}$. For any $\lambda > 0$, the Bessel equation (16C.12) has solution $\mathcal{R}(r) := \mathcal{J}_m(\lambda r)$.*

Ⓔ *Proof* **Exercise 14G.1.** $\square$

Remarks

(a) We can generalize the Bessel equation by replacing m with an arbitrary real number $\mu \in \mathbb{R}$ with $\mu \geq 0$. The solution to this equation is the Bessel function

$$\mathcal{J}_\mu(x) = \left(\frac{x}{2}\right)^\mu \cdot \sum_{k=0}^{\infty} \frac{(-1)^k}{2^{2k} k! \, \Gamma(\mu+k+1)} x^{2k}.$$

Here, Γ is the *Gamma function*; if $\mu = m \in \mathbb{N}$, then $\Gamma(m+k+1) = (m+k)!$, so this expression agrees with equation (14G.2).

(b) There is a second solution to equation (14G.1); a function $\mathcal{Y}_m(x)$ which is *unbounded* at zero. This is called a *Neumann function* (or a *Bessel function of the second kind* or a

Weber–Bessel function). Its derivation is too complicated to discuss here. See Churchill and Brown (1987), §68, or Broman (1989), §6.8.

14H Properties of Bessel functions

Prerequisites: §14G.   **Recommended:** §14C(i).

Let $\mathcal{J}_n(x)$ be the Bessel function defined by equation (14G.2). In this section, we will develop some computational tools to work with these functions. First, we will define Bessel functions with *negative* order as follows: for any $n \in \mathbb{N}$, we define

$$\mathcal{J}_{-n}(x) := (-1)^n \mathcal{J}_n(x). \tag{14H.1}$$

We can now state the following useful recurrence relations.

Proposition 14H.1 *For any $m \in \mathbb{Z}$:*
(a) $\frac{2m}{x} \mathcal{J}_m(x) = \mathcal{J}_{m-1}(x) + \mathcal{J}_{m+1}(x);$
(b) $2\mathcal{J}_m'(x) = \mathcal{J}_{m-1}(x) - \mathcal{J}_{m+1}(x);$
(c) $\mathcal{J}_0'(x) = -\mathcal{J}_1(x);$
(d) $\partial_x(x^m \cdot \mathcal{J}_m(x)) = x^m \cdot \mathcal{J}_{m-1}(x);$
(e) $\partial_x(\frac{1}{x^m} \mathcal{J}_m(x)) = \frac{-1}{x^m} \cdot \mathcal{J}_{m+1}(x);$
(f) $\mathcal{J}_m'(x) = \mathcal{J}_{m-1}(x) - \frac{m}{x} \mathcal{J}_m(x);$
(g) $\mathcal{J}_m'(x) = -\mathcal{J}_{m+1}(x) + \frac{m}{x} \mathcal{J}_m(x)$

Proof **Exercise 14H.1.**   ⓔ
  (i) Prove (d) for $m \geq 1$ by substituting in the power series (14G.2) and differentiating.
  (ii) Prove (e) for $m \geq 0$ by substituting in the power series (14G.2) and differentiating.
 (iii) Use the definition (14H.1) and (i) and (ii) to prove (d) for $m \leq 0$ and (e) for $m \leq -1$.
 (iv) Set $m = 0$ in (e) to obtain (c).
  (v) Deduce (f) and (g) from (d) and (e).
 (vi) Compute the sum and difference of (f) and (g) to get (a) and (b).   □

Remark 14H.2 Integration with Bessel functions
The recurrence relations of Proposition 14H.1 can be used to simplify integrals involving Bessel functions. For example, parts (d) and (e) immediately imply that

$$\int x^m \cdot \mathcal{J}_{m-1}(x)\mathrm{d}x = x^m \cdot \mathcal{J}_m(x) + C$$

and

$$\int \frac{1}{x^m} \cdot \mathcal{J}_{m+1}(x)\mathrm{d}x = \frac{-1}{x^m} \mathcal{J}_m(x) + C.$$

The other relations are sometimes useful in an 'integration by parts' strategy.

For any $n \in \mathbb{N}$, let $0 \leq \kappa_{n,1} < \kappa_{n,2} < \kappa_{n,3} < \cdots$ be the zeros of the nth Bessel function $\mathcal{J}_n$ (i.e. $\mathcal{J}_n(\kappa_{n,m}) = 0$ for all $m \in \mathbb{N}$). Proposition 14C.1, p. 299, says we can use Bessel functions to define a sequence of polar-separated eigenfunctions of the Laplacian:

$$\Phi_{n,m}(r, \theta) := \mathcal{J}_n(\kappa_{n,m} \cdot r) \cdot \cos(n\theta); \qquad \Psi_{n,m}(r, \theta) := \mathcal{J}_n(\kappa_{n,m} \cdot r) \cdot \sin(n\theta).$$

In the proof of Theorem 14C.2, p. 304, we claimed that these eigenfunctions were *orthogonal* as elements of $\mathbf{L}^2(\mathbb{D})$. We will now verify this claim. First we must prove a technical lemma.

Lemma 14H.3 *Fix $n \in \mathbb{N}$.*

(a) *If $m \neq M$, then $\int_0^1 \mathcal{J}_n(\kappa_{n,m} \cdot r) \cdot \mathcal{J}_n(\kappa_{n,M} \cdot r)r \, dr = 0$.*

(b) *$\int_0^1 \mathcal{J}_n(\kappa_{n,m} \cdot r)^2 \cdot r \, dr = \frac{1}{2}\mathcal{J}_{n+1}(\kappa_{n,m})^2$.*

Proof (a) Let $\alpha = \kappa_{n,m}$ and $\beta = \kappa_{n,M}$. Define $f(x) := \mathcal{J}_m(\alpha x)$ and $g(x) := \mathcal{J}_m(\beta x)$. Hence we want to show that

$$\int_0^1 f(x)g(x)x \, dx = 0.$$

Define $h(x) = x \cdot (f(x)g'(x) - g(x)f'(x))$.

Claim 1 $h'(x) = (\alpha^2 - \beta^2)f(x)g(x)x$.

Proof First observe that

$$h'(x) = x \cdot \partial_x \left(f(x)g'(x) - g(x)f'(x) \right) + \left(f(x)g'(x) - g(x)f'(x) \right)$$

$$= x \cdot \left(f(x)g''(x) + f'(x)g'(x) - g'(x)f'(x) - g(x)f''(x) \right)$$

$$+ \left(f(x)g'(x) - g(x)f'(x) \right)$$

$$= x \cdot \left(f(x)g''(x) - g(x)f''(x) \right) + \left(f(x)g'(x) - g(x)f'(x) \right).$$

By setting $\mathcal{R} = f$ or $\mathcal{R} = g$ in Corollary 14G.2, we obtain

$$x^2 f''(x) + xf'(x) + (\alpha^2 x^2 - n^2)f(x) = 0,$$

and

$$x^2 g''(x) + xg'(x) + (\beta^2 x^2 - n^2)g(x) = 0.$$

We multiply the first equation by $g(x)$ and the second by $f(x)$ to get

$$x^2 f''(x)g(x) + xf'(x)g(x) + \alpha^2 x^2 f(x)g(x) - n^2 f(x)g(x) = 0,$$

and

$$x^2 g''(x) f(x) + x g'(x) f(x) + \beta^2 x^2 g(x) f(x) - n^2 g(x) f(x) = 0.$$

We then subtract these two equations to get

$$x^2 \left(f''(x) g(x) - g''(x) f(x) \right) + x \left(f'(x) g(x) - g'(x) f(x) \right)$$
$$+ \left(\alpha^2 - \beta^2 \right) f(x) g(x) x^2 = 0.$$

Divide by x to get

$$x \left(f''(x) g(x) - g''(x) f(x) \right) + \left(f'(x) g(x) - g'(x) f(x) \right)$$
$$+ \left(\alpha^2 - \beta^2 \right) f(x) g(x) x = 0.$$

Hence we conclude that

$$\left(\alpha^2 - \beta^2 \right) f(x) g(x) x = x \left(g''(x) f(x) - f''(x) g(x) \right) + \left(g'(x) f(x) - f'(x) g(x) \right)$$
$$= h'(x),$$

as desired $\Diamond$ Claim 1

It follows from Claim 1 that

$$(\alpha^2 - \beta^2) \cdot \int_0^1 f(x) g(x) x \, dx = \int_0^1 h'(x) dx = h(1) - h(0) \underset{(*)}{=} 0 - 0 = 0.$$

To see $(*)$, observe that $h(0) = 0 \cdot (f(0) g'(0) - g(0) f'(0)) = 0$. Also,

$$h(1) = (1) \cdot \left(f(1) g'(1) - g(1) f'(1) \right) = 0,$$

because $f(1) = \mathcal{J}_n(\kappa_{n,m}) = 0$ and $g(1) = \mathcal{J}_n(\kappa_{n,N}) = 0$.

(b) Let $\alpha = \kappa_{n,m}$ and $f(x) := \mathcal{J}_m(\alpha x)$. Hence we want to evaluate

$$\int_0^1 f(x)^2 x \, dx.$$

Define $h(x) := x^2 (f'(x))^2 + (\alpha^2 x^2 - n^2) f^2(x)$.

Claim 2 $h'(x) = 2\alpha^2 f(x)^2 x$.

Proof By setting $\mathcal{R} = f$ in Corollary 14G.2, we obtain

$$0 = x^2 f''(x) + x f'(x) + (\alpha^2 x^2 - n^2) f(x).$$

We multiply by $f'(x)$ to get

$$0 = x^2 f'(x) f''(x) + x(f'(x))^2 + (\alpha^2 x^2 - n^2) f(x) f'(x)$$

$$= x^2 f'(x) f''(x) + x(f'(x))^2 + (\alpha^2 x^2 - n^2) f(x) f'(x)$$

$$+ \alpha^2 x f^2(x) - \alpha^2 x f^2(x)$$

$$= \frac{1}{2} \partial_x \left[x^2 (f'(x))^2 + (\alpha^2 x^2 - n^2) f^2(x) \right] - \alpha^2 x f^2(x)$$

$$= \frac{1}{2} h'(x) - \alpha^2 x f^2(x). \qquad \diamond_{\text{Claim 2}}$$

It follows from Claim 2 that

$$2\alpha^2 \int_0^1 f(x)^2 x \, dx$$

$$= \int_0^1 h'(x) dx = h(1) - h(0)$$

$$= 1^2 (f'(1))^2 + (\alpha^2 1^2 - n^2) \cdot \underbrace{f^2(1)}_{\substack{J_n^2(\kappa_{n,m}) \\ =0}} - \underbrace{0^2 (f'(0))^2}_{0} + \underbrace{(\alpha^2 0^2 - n^2)}_{[0 \text{ if } n = 0]} \underbrace{f^2(0)}_{[0 \text{ if } n \neq 0]}$$

$$= f'(1)^2 = \left(\alpha J_n'(\alpha) \right)^2 = \alpha^2 J_n'(\alpha)^2.$$

Hence

$$\int_0^1 f(x)^2 x \, dx = \frac{1}{2} J_n'(\alpha)^2 \underset{(*)}{=} \frac{1}{2} \left(\frac{n}{\alpha} J_n(\alpha) - J_{n+1}(\alpha) \right)^2$$

$$\underset{(\dagger)}{=} \frac{1}{2} \left(\frac{n}{\kappa_{n,m}} \underbrace{J_n(\kappa_{n,m})}_{=0} - J_{n+1}(\kappa_{n,m}) \right)^2 = \frac{1}{2} J_{n+1}(\kappa_{n,m})^2,$$

where $(*)$ is by Proposition 14H.1(g) and $(\dagger)$ is because $\alpha := \kappa_{n,m}$. $\qquad \square$

Proposition 14H.4 *Let* $\mathbb{D} = \{(r, \theta); \ r \leq 1\}$ *be the unit disk. Then the collection*

$$\{\Phi_{n,m}, \Psi_{\ell,m}; \ n = 0 \ldots \infty, \ell \in \mathbb{N}, m \in \mathbb{N}\}$$

is an orthogonal set for $\mathbf{L}^2(\mathbb{D})$*. In other words, for any* $n, m, N, M \in \mathbb{N}$,

(a) $\qquad \langle \Phi_{n,m}, \Psi_{N,M} \rangle = \frac{1}{\pi} \int_0^1 \int_{-\pi}^{\pi} \Phi_{n,m}(r, \theta) \cdot \Psi_{N,M}(r, \theta) \, d\theta \, r \, dr = 0.$

Furthermore, if $(n, m) \neq (N, M)$, *then*

(b) $\qquad \langle \Phi_{n,m}, \Phi_{N,M} \rangle = \frac{1}{\pi} \int_0^1 \int_{-\pi}^{\pi} \Phi_{n,m}(r, \theta) \cdot \Phi_{N,M}(r, \theta) d\theta \, r \, dr = 0;$

(c) $\quad \langle \Psi_{n,m}, \Psi_{N,M} \rangle = \dfrac{1}{\pi} \displaystyle\int_0^1 \int_{-\pi}^{\pi} \Psi_{n,m}(r, \theta) \cdot \Psi_{N,M}(r, \theta) \mathrm{d}\theta \ r \ \mathrm{d}r = 0.$

Finally, for any (n, m),

(d) $\quad \left\| \Phi_{n,m} \right\|_2 = \dfrac{1}{\pi} \displaystyle\int_0^1 \int_{-\pi}^{\pi} \Phi_{n,m}(r, \theta)^2 \mathrm{d}\theta \ r \ \mathrm{d}r = \dfrac{1}{2} \mathcal{J}_{n+1}(\kappa_{n,m})^2;$

(e) $\quad \left\| \Psi_{n,m} \right\|_2 = \dfrac{1}{\pi} \displaystyle\int_0^1 \int_{-\pi}^{\pi} \Psi_{n,m}(r, \theta)^2 \mathrm{d}\theta \ r \ \mathrm{d}r = \dfrac{1}{2} \mathcal{J}_{n+1}(\kappa_{n,m})^2.$

Proof (a) $\Phi_{n,m}$ and $\Psi_{N,M}$ separate in the coordinates (r, θ), so the integral splits in two:

$$\int_0^1 \int_{-\pi}^{\pi} \Phi_{n,m}(r, \theta) \cdot \Psi_{N,M}(r, \theta) \mathrm{d}\theta \ r \ \mathrm{d}r$$

$$= \int_0^1 \int_{-\pi}^{\pi} \mathcal{J}_n(\kappa_{n,m} \cdot r) \cdot \cos(n\theta) \cdot \mathcal{J}_N(\kappa_{N,M} \cdot r) \cdot \sin(N\theta) \mathrm{d}\theta \ r \ \mathrm{d}r$$

$$= \int_0^1 \mathcal{J}_n(\kappa_{n,m} \cdot r) \cdot \mathcal{J}_N(\kappa_{N,M} \cdot r) \ r \ \mathrm{d}r \cdot \underbrace{\int_{-\pi}^{\pi} \cos(n\theta) \cdot \sin(N\theta) \mathrm{d}\theta}_{= \ 0 \text{ by Proposition 6D.2(c), p. 117}} = 0.$$

(b) or (c) $(n \neq N)$ Likewise, if $n \neq N$, then

$$\int_0^1 \int_{-\pi}^{\pi} \Phi_{n,m}(r, \theta) \cdot \Phi_{N,M}(r, \theta) \mathrm{d}\theta \ r \ \mathrm{d}r$$

$$= \int_0^1 \int_{-\pi}^{\pi} \mathcal{J}_n(\kappa_{n,m} \cdot r) \cdot \cos(n\theta) \cdot \mathcal{J}_N(\kappa_{N,M} \cdot r) \cdot \cos(N\theta) \mathrm{d}\theta \ r \ \mathrm{d}r$$

$$= \int_0^1 \mathcal{J}_n(\kappa_{n,m} \cdot r) \cdot \mathcal{J}_N(\kappa_{N,M} \cdot r) \ r \ \mathrm{d}r \cdot \underbrace{\int_{-\pi}^{\pi} \cos(n\theta) \cdot \cos(N\theta) \mathrm{d}\theta}_{= \ 0 \text{ by Proposition 6D.2(a), p. 117}} = 0.$$

Case (c) is proved similarly.

(b) or (c) $(n = N$ but $m \neq M)$ If $n = N$, then

$$\int_0^1 \int_{-\pi}^{\pi} \Phi_{n,m}(r, \theta) \cdot \Phi_{n,M}(r, \theta) \mathrm{d}\theta \ r \ \mathrm{d}r$$

$$= \int_0^1 \int_{-\pi}^{\pi} \mathcal{J}_n(\kappa_{n,m} \cdot r) \cdot \cos(n\theta) \cdot \mathcal{J}_n(\kappa_{n,M} \cdot r) \cdot \cos(n\theta) \mathrm{d}\theta \ r \ \mathrm{d}r$$

$$= \underbrace{\int_0^1 \mathcal{J}_n(\kappa_{n,m} \cdot r) \cdot \mathcal{J}_n(\kappa_{n,M} \cdot r) \ r \ \mathrm{d}r}_{= \ 0 \text{ by Lemma 14H.3(a)}} \cdot \underbrace{\int_{-\pi}^{\pi} \cos(n\theta)^2 \ \mathrm{d}\theta}_{\substack{= \ \pi \text{ by Proposition 6D.2(d),} \\ \text{p. 117.}}} = 0 \cdot \pi = 0.$$

(d)

$$\int_0^1 \int_{-\pi}^{\pi} \Phi_{n,m}(r, \theta)^2 d\theta \, r \, dr = \int_0^1 \int_{-\pi}^{\pi} \mathcal{J}_n(\kappa_{n,m} \cdot r)^2 \cdot \cos(n\theta)^2 d\theta \, r \, dr$$

$$= \underbrace{\int_0^1 \mathcal{J}_n(\kappa_{n,m} \cdot r)^2 \, r \, dr}_{\substack{= \frac{1}{2}\mathcal{J}_{n+1}(\kappa_{n,m})^2 \\ \text{by Lemma 14H.3(b)}}} \cdot \underbrace{\int_{-\pi}^{\pi} \cos(n\theta)^2 d\theta}_{\substack{= \pi \text{ by Proposition 6D.2(d),} \\ \text{p. 117.}}} = \frac{\pi}{2} \mathcal{J}_{n+1}(\kappa_{n,m})^2.$$

Ⓔ   The proof of (e) is **Exercise 14H.2**.   □

Ⓔ **Exercise 14H.3**

(a) Use a 'separation of variables' argument (similar to Proposition 16C.2) to prove the following Proposition.

Let $f : \mathbb{R}^2 \longrightarrow \mathbb{R}$ be a harmonic function – in other words suppose $\triangle f = 0$.

Suppose f separates in polar coordinates, meaning that there is a function Θ : $[-\pi, \pi] \longrightarrow \mathbb{R}$ (satisfying periodic boundary conditions) and a function $\mathcal{R} : \mathbb{R}_+ \longrightarrow \mathbb{R}$ such that

$$f(r, \theta) = \mathcal{R}(r) \cdot \Theta(\theta), \quad \text{for all } r \geq 0 \text{ and } \theta \in [-\pi, \pi].$$

Then there is some $m \in \mathbb{N}$ such that

$$\Theta(\theta) = A \cos(m\theta) + B \sin(m\theta), \quad \text{(for constants } A, B \in \mathbb{R}\text{),}$$

and $\mathcal{R}$ is a solution to the Cauchy–Euler equation:

$$r^2 \mathcal{R}''(r) + r \cdot \mathcal{R}'(r) - m^2 \cdot \mathcal{R}(r) = 0, \quad \text{for all } r > 0. \tag{14H.2}$$

(b) Let $\mathcal{R}(r) = r^\alpha$, where $\alpha = \pm m$. Show that $\mathcal{R}(r)$ is a solution to the Cauchy–Euler equation (14H.2).

(c) Deduce that $\Psi_m(r, \theta) = r^m \cdot \sin(m\theta)$; $\Phi_m(r, \theta) = r^m \cdot \cos(m\theta)$; $\psi_m(r, \theta) = r^{-m} \cdot \sin(m\theta)$; and $\phi_m(r, \theta) = r^{-m} \cdot \cos(m\theta)$ are harmonic functions in $\mathbb{R}^2$.   ◆

14I Practice problems

14.1 For all (r, θ), let $\Phi_n(r, \theta) = r^n \cos(n\theta)$. Show that Φ_n is harmonic.

14.2 For all (r, θ), let $\Psi_n(r, \theta) = r^n \sin(n\theta)$. Show that Ψ_n is harmonic.

14.3 For all (r, θ) with $r > 0$, let $\phi_n(r, \theta) = r^{-n} \cos(n\theta)$. Show that ϕ_n is harmonic.

14.4 For all (r, θ) with $r > 0$, let $\psi_n(r, \theta) = r^{-n} \sin(n\theta)$. Show that ψ_n is harmonic.

14.5 For all (r, θ) with $r > 0$, let $\phi_0(r, \theta) = \log |r|$. Show that ϕ_0 is harmonic.

14.6 Let $b(\theta) = \cos(3\theta) + 2 \sin(5\theta)$ for $\theta \in [-\pi, \pi)$.

(a) Find the bounded solution(s) to the Laplace equation on $\mathbb{D}$, with nonhomogeneous Dirichlet boundary conditions $u(1, \theta) = b(\theta)$. Is the solution unique?

(b) Find the bounded solution(s) to the Laplace equation on $\mathbb{D}^\mathbb{C}$, with nonhomogeneous Dirichlet boundary conditions $u(1, \theta) = b(\theta)$. Is the solution unique?

(c) Find the 'decaying gradient' solution(s) to the Laplace equation on $\mathbb{D}^\mathbb{C}$, with nonhomogeneous Neumann boundary conditions $\partial_r u(1, \theta) = b(\theta)$. Is the solution unique?

14.7 Let $b(\theta) = 2\cos(\theta) - 6\sin(2\theta)$, for $\theta \in [-\pi, \pi)$.

(a) Find the bounded solution(s) to the Laplace equation on $\mathbb{D}$, with nonhomogeneous Dirichlet boundary conditions: $u(1, \theta) = b(\theta)$ for all $\theta \in [-\pi, \pi)$. Is the solution unique?

(b) Find the bounded solution(s) to the Laplace equation on $\mathbb{D}$, with nonhomogeneous Neumann boundary conditions: $\partial_r u(1, \theta) = b(\theta)$ for all $\theta \in [-\pi, \pi)$. Is the solution unique?

14.8 Let $b(\theta) = 4\cos(5\theta)$ for $\theta \in [-\pi, \pi)$.

(a) Find the bounded solution(s) to the Laplace equation on the disk $\mathbb{D} = \{(r, \theta); r \le 1\}$, with nonhomogeneous Dirichlet boundary conditions $u(1, \theta) = b(\theta)$. Is the solution unique?

(b) *Verify* your answer in part (a) (i.e. check that the solution is harmonic and satisfies the prescribed boundary conditions.) (*Hint:* Recall that $\triangle = \partial_r^2 + \frac{1}{r}\partial_r + \frac{1}{r^2}\partial_\theta^2$.)

14.9 Let $b(\theta) = 5 + 4\sin(3\theta)$ for $\theta \in [-\pi, \pi)$.

(a) Find the 'decaying gradient' solution(s) to the Laplace equation on the codisk $\mathbb{D}^\mathbb{C} = \{(r, \theta); r \ge 1\}$, with nonhomogeneous Neumann boundary conditions $\partial_r u(1, \theta) = b(\theta)$. Is the solution unique?

(b) *Verify* that your answer in part (a) satisfies the prescribed boundary conditions. (Forget about the Laplacian.)

14.10 Let $b(\theta) = 2\cos(5\theta) + \sin(3\theta)$, for $\theta \in [-\pi, \pi)$.

(a) Find the solution(s) (if any) to the Laplace equation on the disk $\mathbb{D} = \{(r, \theta); r \le 1\}$, with nonhomogeneous Neumann boundary conditions: $\partial_\perp u(1, \theta) = b(\theta)$, for all $\theta \in [-\pi, \pi)$. Is the solution unique? Why or why not?

(b) Find the *bounded* solution(s) (if any) to the Laplace equation on the codisk $\mathbb{D}^\mathbb{C} = \{(r, \theta); r \ge 1\}$, with nonhomogeneous Dirichlet boundary conditions: $u(1, \theta) = b(\theta)$, for all $\theta \in [-\pi, \pi)$. Is the solution unique? Why or why not?

14.11 Let $\mathbb{D}$ be the unit disk. Let $b : \partial\mathbb{D} \longrightarrow \mathbb{R}$ be some function, and let $u : \mathbb{D} \longrightarrow \mathbb{R}$ be the solution to the corresponding Dirichlet problem with boundary conditions $b(\sigma)$. Prove that

$$u(0, 0) = \frac{1}{2\pi} \int_{-\pi}^{\pi} b(\sigma)d\sigma.$$

Note: This is a special case of the Mean Value Theorem for harmonic functions (Theorem, 1E.1, p. 18), but do *not* simply 'quote' Theorem 1E.1 to solve this problem. Instead, apply Proposition 14B.12, p. 297.

14.12 Let $\Phi_{n,\lambda}(r, \theta) := \mathcal{J}_n(\lambda \cdot r) \cdot \cos(n\theta)$. Show that $\triangle\Phi_{n,\lambda} = -\lambda^2\Phi_{n,\lambda}$.

14.13 Let $\Psi_{n,\lambda}(r, \theta) := \mathcal{J}_n(\lambda \cdot r) \cdot \sin(n\theta)$. Show that $\triangle\Psi_{n,\lambda} = -\lambda^2\Psi_{n,\lambda}$.

14.14 Let $\phi_{n,\lambda}(r, \theta) := \mathcal{Y}_n(\lambda \cdot r) \cdot \cos(n\theta)$. Show that $\triangle\phi_{n,\lambda} = -\lambda^2\phi_{n,\lambda}$.

14.15 $\psi_{n,\lambda}(r, \theta) := \mathcal{Y}_n(\lambda \cdot r) \cdot \sin(n\theta)$. Show that $\triangle\psi_{n,\lambda} = -\lambda^2\psi_{n,\lambda}$.

15

Eigenfunction methods on arbitrary domains

> Science is built up with facts, as a house is with stones. But a collection
> of facts is no more a science than a heap of stones is a house.
>
> *Henri Poincaré*

The methods given in Chapters 11–14 are all special cases of a single, general technique: the solution of initial/boundary value problems using *eigenfunction expansions*. The time has come to explicate this technique in full generality. The exposition in this chapter is somewhat more abstract than in previous chapters, but that is because the concepts we introduce are of such broad applicability. Technically, this chapter can be read without having read Chapters 11–14; however, this chapter will be easier to understand if you have have already read those chapters.

15A General solution to Poisson, heat, and wave equation BVPs

Prerequisites: §4B(iv), §5D, §6F, Appendix D.   **Recommended:** Chapters 11–14.

Throughout this section:

- Let $\mathbb{X} \subset \mathbb{R}^D$ be any bounded domain (e.g. a line segment, box, disk, sphere, etc. – see Appendix D). When we refer to Neumann boundary conditions, we will also assume that $\mathbb{X}$ has a piecewise smooth boundary (so the normal derivative is well defined).
- Let $\{\mathcal{S}_k\}_{k=1}^{\infty} \subset \mathbf{L}^2(\mathbb{X})$ be a *Dirichlet eigenbasis* – that is, $\{\mathcal{S}_k\}_{k=1}^{\infty}$ is an orthogonal basis of $\mathbf{L}^2(\mathbb{X})$, such that every $\mathcal{S}_k$ is an eigenfunction of the Laplacian, and satisfies homogeneous Dirichlet boundary conditions on $\mathbb{X}$ (i.e. $\mathcal{S}_k(\mathbf{x}) = 0$ for all $\mathbf{x} \in \partial \mathbb{X}$). For every $k \in \mathbb{N}$, let $-\lambda_k < 0$ be the *eigenvalue* associated with $\mathcal{S}_k$ (i.e. $\triangle \mathcal{S}_k = -\lambda_k \mathcal{S}_k$). We can assume, without loss of generality, that $\lambda_k \neq 0$ for all $k \in \mathbb{N}$ (**Exercise 15A.1** Why? *Hint:* See ⒠ Lemma 5D.3(a).)
- Let $\{\mathcal{C}_k\}_{k=0}^{\infty} \subset \mathbf{L}^2(\mathbb{X})$ be a *Neumann eigenbasis* – that is, $\{\mathcal{C}_k\}_{k=0}^{\infty}$ is an orthogonal basis $\mathbf{L}^2(\mathbb{X})$, such that every $\mathcal{C}_k$ is an eigenfunction of the Laplacian, and satisfies homogeneous

Neumann boundary conditions on $\mathbb{X}$ (i.e. $\partial_\perp \mathcal{C}_k(\mathbf{x}) = 0$ for all $\mathbf{x} \in \partial\mathbb{X}$). For every $k \in \mathbb{N}$, let $-\mu_k \leq 0$ be the *eigenvalue* associated with $\mathcal{C}_k$ (i.e. $\triangle \mathcal{C}_k = -\mu_k \mathcal{C}_k$). We can assume, without loss of generality, that $\mathcal{C}_0$ is a constant function (so that $\mu_0 = 0$), while $\mu_k \neq 0$

Ⓔ for all $k \geq 1$ (**Exercise 15A.2** Why? *Hint:* See Lemma 5D.3(b).)

Theorem 15E.12 (p. 353) will guarantee that we will be able to find a Dirichlet eigenbasis for any domain $\mathbb{X} \subset \mathbb{R}^D$, and a Neumann eigenbasis for many domains. If $f \in \mathbf{L}^2(\mathbb{X})$ is some other function (describing, for example, an initial condition), then we can express f as a combination of these basis elements, as described in §6F:

$$f \underset{L2}{\approx} \sum_{k=0}^{\infty} A_k \mathcal{C}_k, \quad \text{where } A_k := \frac{\langle f, \mathcal{C}_k \rangle}{\|\mathcal{C}_k\|_2^2}, \quad \text{for all } k \in \mathbb{N}; \tag{15A.1}$$

and

$$f \underset{L2}{\approx} \sum_{k=1}^{\infty} B_k \mathcal{S}_k, \quad \text{where } B_k := \frac{\langle f, \mathcal{S}_k \rangle}{\|\mathcal{S}_k\|_2^2}, \quad \text{for all } k \in \mathbb{N}. \tag{15A.2}$$

These expressions are called *eigenfunction expansions* for f.

Example 15A.1

(a) If $\mathbb{X} = [0, \pi] \subset \mathbb{R}$, then we could use the eigenbases $\{\mathcal{S}_k\}_{k=1}^{\infty} = \{\mathbf{S}_n\}_{n=1}^{\infty}$ and $\{\mathcal{C}_k\}_{k=0}^{\infty} = \{\mathbf{C}_n\}_{n=0}^{\infty}$, where $\mathbf{S}_n(x) := \sin(nx)$ and $\mathbf{C}_n(x) := \cos(nx)$ for all $n \in \mathbb{N}$. In this case, $\lambda_n = n^2 = \mu_n$ for all $n \in \mathbb{N}$. Also the eigenfunction expansions (15A.1) and (15A.2) are, respectively, the *Fourier cosine series* and the *Fourier sine series* for f, from §7A.

(b) If $\mathbb{X} = [0, \pi]^2 \subset \mathbb{R}^2$, then we could use the eigenbases $\{\mathcal{S}_k\}_{k=1}^{\infty} = \{\mathbf{S}_{n,m}\}_{n,m=1}^{\infty}$ and $\{\mathcal{C}_k\}_{k=0}^{\infty} = \{\mathbf{C}_{n,m}\}_{n,m=0}^{\infty}$, where $\mathbf{S}_{n,n}(x, y) := \sin(nx)\sin(my)$ and $\mathbf{C}_{n,m}(x) := \cos(nx)\cos(my)$ for all $n, m \in \mathbb{N}$. In this case, $\lambda_{n,m} = n^2 + m^2 = \mu_{n,m}$ for all $(n, m) \in \mathbb{N}$. Also, the eigenfunction expansions (15A.1) and (15A.2) are, respectively, the two-dimensional Fourier cosine series and Fourier sine series for f, from §9A.

(c) If $\mathbb{X} = \mathbb{D} \subset \mathbb{R}^2$, then we could use the Dirichlet eigenbasis $\{\mathcal{S}_n\}_{k=1}^{\infty} = \{\Phi_{n,m}\}_{n=0,m=1}^{\infty} \sqcup \{\Psi_{n,m}\}_{n,m=1}^{\infty}$, where $\Phi_{n,m}$ and $\Psi_{n,m}$ are the type-1 Fourier–Bessel eigenfunctions defined by equation (14C.5), p. 301. In this case, we have eigenvalues $\lambda_{n,m} = \kappa_{n,m}^2$, as defined in equation (14C.4), p. 301. Then the eigenfunction expansion in equation (15A.2) is the *Fourier–Bessel expansion* for f, from §14C(iii). ◇

Theorem 15A.2 General solution of the Poisson equation

Let $\mathbb{X} \subset \mathbb{R}^D$ be a bounded domain. Let $f \in \mathbf{L}^2(\mathbb{X})$, and let $b : \partial\mathbb{X} \longrightarrow \mathbb{R}$ be some other function. Let $u : \mathbb{X} \longrightarrow \mathbb{R}$ be a solution to the Poisson equation '$\triangle u = f$'.

(a) *Suppose $\{\mathcal{S}_k, \lambda_k\}_{k=1}^{\infty}$ is a Dirichlet eigenbasis, and that $\{B_n\}_{n=1}^{\infty}$ are as in equation (15A.2). Assume that $|\lambda_k| > 1$ for all but finitely many $k \in \mathbb{N}$. If u satisfies homogeneous*

*Dirichlet BC (i.e. u(**x**) = 0 for all **x** ∈ ∂X), then*

$$u \underset{12}{\approx} -\sum_{n=1}^{\infty} \frac{B_n}{\lambda_n} S_n.$$

(b) *Let h : X ⟶ ℝ be a solution to the Laplace equation 'Δh = 0' satisfying the non-homogeneous Dirichlet BC h(**x**) = b(**x**) for all **x** ∈ ∂X. If u is as in part (a), then w := u + h is a solution to the Poisson equation 'Δw = f' and also satisfies Dirichlet BC w(**x**) = b(**x**) for all **x** ∈ ∂X.*

(c) *Suppose {$\mathcal{C}_k, \mu_k$}$_{k=1}^{\infty}$ is a Neumann eigenbasis, and suppose that $|\mu_k| > 1$ for all but finitely many k ∈ ℕ. Let {A_n}$_{n=1}^{\infty}$ be as in equation (15A.1), and suppose $A_0 = 0$. For any j ∈ [1 ... D], let $\left\| \partial_j \mathcal{C}_k \right\|_{\infty}$ be the supremum of the j-derivative of $\mathcal{C}_k$ on X, and suppose that*

$$\sum_{\substack{k=1 \\ \mu_k \neq 0}}^{\infty} \frac{|A_k|}{|\mu_k|} \left\| \partial_j \mathcal{C}_k \right\|_{\infty} < \infty. \tag{15A.3}$$

*If u satisfies homogeneous Neumann BC (i.e. $\partial_{\perp} u(\mathbf{x}) = 0$ for all **x** ∈ ∂X), then*

$$u \underset{12}{\approx} C - \sum_{\substack{k=1 \\ \mu_k \neq 0}}^{\infty} \frac{A_k}{\mu_k} \mathcal{C}_k,$$

where C ∈ ℝ is an arbitrary constant. However, if $A_0 \neq 0$, then there is no solution to this problem with homogeneous Neumann BC.

(d) *Let h : X ⟶ ℝ be a solution to the Laplace equation 'Δh = 0' satisfying the non-homogeneous Neumann BC $\partial_{\perp} h(\mathbf{x}) = b(\mathbf{x})$ for all **x** ∈ ∂X. If u is as in part (c), then w := u + h is a solution to the Poisson equation 'Δw = f' and also satisfies Neumann BC $\partial_{\perp} w(\mathbf{x}) = b(\mathbf{x})$ for all **x** ∈ ∂X.*

Proof **Exercise 15A.3.** *Hint:* To show solution uniqueness, use Theorem 5D.5. (E)
For (a), imitate the proofs of Propositions 11C.1, 12C.1, 13C.1, and 14D.1.
For (b) and (d), imitate the proofs of Propositions 12C.6 and 14D.3.
For (c), imitate the proofs of Propositions 11C.2, 12C.4, and 13C.2. Note that
you need hypothesis (15A.3) to apply Proposition F.1. □

Exercise 15A.4 Show how Propositions 11C.1, 11C.2, 12C.1, 12C.4, 12C.6, (E)
13C.1, 13C.2, 14D.1, and 14D.3 are all special cases of Theorem 15A.2. For the
results involving Neumann BC, don't forget to check that equation (15A.3) is
satisfied. ♦

Theorem 15A.3 General solution of the heat equation
*Let X ⊂ ℝ^D be a bounded domain. Let f ∈ **L**^{2}(X), and let b : ∂X ⟶ ℝ be some other function. Let u : X × ℝ$_{\not{}}$ ⟶ ℝ be a solution to the heat equation '$\partial_t u = \Delta u$', with initial conditions u(**x**, 0) = f(**x**) for all **x** ∈ X.*

(a) *Suppose $\{S_k, \lambda_k\}_{k=1}^{\infty}$ is a Dirichlet eigenbasis, and $\{B_n\}_{n=1}^{\infty}$ are as in equation (15A.2). If u satisfies homogeneous Dirichlet BC (i.e. $u(\mathbf{x}, t) = 0$ for all $\mathbf{x} \in \partial X$ and $t \in \mathbb{R}_+$), then*

$$u \underset{12}{\approx} \sum_{n=1}^{\infty} B_n \exp(-\lambda_n t) S_n.$$

(b) *Let $h : \mathbb{X} \longrightarrow \mathbb{R}$ be a solution to the Laplace equation '$\triangle h = 0$' satisfying the non-homogeneous Dirichlet BC $h(\mathbf{x}) = b(\mathbf{x})$ for all $\mathbf{x} \in \partial X$. If u is as in part (a), then $w := u + h$ is a solution to the heat equation '$\partial_t w = \triangle w$', with initial conditions $w(\mathbf{x}, 0) = f(\mathbf{x}) + h(\mathbf{x})$ for all $\mathbf{x} \in \mathbb{X}$, and also satisfies Dirichlet BC $w(\mathbf{x}, t) = b(\mathbf{x})$ for all $(\mathbf{x}, t) \in \partial X \times \mathbb{R}_+$.*

(c) *Suppose $\{C_k, \mu_k\}_{k=0}^{\infty}$ is a Neumann eigenbasis, and $\{A_n\}_{n=0}^{\infty}$ are as in equation (15A.1). Suppose the sequence $\{\mu_k\}_{k=0}^{\infty}$ grows fast enough that*

$$\lim_{k\to\infty} \frac{\log(k)}{\mu_k} = 0, \text{ and, for all } j \in [1 \ldots D], \lim_{k\to\infty} \frac{\log \|\partial_j C_k\|_{\infty}}{\mu_k} = 0. \quad (15A.4)$$

If u satisfies homogeneous Neumann BC (i.e. $\partial_{\perp} u(\mathbf{x}, t) = 0$ for all $\mathbf{x} \in \partial X$ and $t \in \mathbb{R}_+$), then

$$u \underset{12}{\approx} \sum_{n=0}^{\infty} A_n \exp(-\mu_n t) C_n.$$

(d) *Let $h : \mathbb{X} \longrightarrow \mathbb{R}$ be a solution to the Laplace equation '$\triangle h = 0$' satisfying the non-homogeneous Neumann BC $\partial_{\perp} h(\mathbf{x}) = b(\mathbf{x})$ for all $\mathbf{x} \in \partial X$. If u is as in part (c), then $w := u + h$ is a solution to the heat equation '$\partial_t w = \triangle w$' with initial conditions $w(\mathbf{x}, 0) = f(\mathbf{x}) + h(\mathbf{x})$ for all $\mathbf{x} \in \mathbb{X}$, and also satisfies Neumann BC $\partial_{\perp} w(\mathbf{x}, t) = b(\mathbf{x})$ for all $(\mathbf{x}, t) \in \partial X \times \mathbb{R}_+$.*

Furthermore, in parts (a) and (c), the series defining u converges semiuniformly on $\mathbb{X} \times \mathbb{R}_+$.

Ⓔ *Proof* **Exercise 15A.5.** *Hint:* To show solution uniqueness, use Theorem 5D.8.

For part (a), imitate the proofs of Propositions 11A.1, 12B.1, 13A.1, and 14E.1.

For (b) and (d) imitate the proofs of Propositions 12B.5 and 14E.3.

For (c), imitate the proofs of Propositions 11A.3, 12B.3, and 13A.2. First use hypothesis (15A.4) to show that the sequence $\{e^{-\mu_n t} \|\partial_j C_n\|_{\infty}\}_{n=0}^{\infty}$ is square-summable for any $t > 0$. Use Parseval's equality (Theorem 6F.1) to show that the sequence $\{|A_k|\}_{k=0}^{\infty}$ is also square-summable. Use the Cauchy–Bunyakowski–Schwarz inequality to conclude that the sequence $\{e^{-\mu_n t} |A_k| \|\partial_j C_n\|_{\infty}\}_{n=0}^{\infty}$ is absolutely summable, which means the formal derivative $\partial_j u$ is absolutely convergent. Now apply Proposition F.1. □

Ⓔ **Exercise 15A.6** Show how Propositions 11A.1, 11A.3, 12B.1, 12B.3, 12B.5, 13A.1, 13A.2, 14E.1, and 14E.3 are all special cases of Theorem 15A.3. For the results involving Neumann BC, don't forget to check that equation (15A.4) is satisfied.

◆

Theorem 15A.4 General solution of the wave equation

Let $X \subset \mathbb{R}^D$ be a bounded domain and let $f \in \mathbf{L}^2(X)$. Suppose $u : X \times \mathbb{R}_+ \longrightarrow \mathbb{R}$ is a solution to the wave equation '$\partial_t^2 u = \triangle u$', and has initial position $u(\mathbf{x}; 0) = f(\mathbf{x})$ for all $\mathbf{x} \in X$.

(a) *Suppose $\{S_k, \lambda_k\}_{k=1}^{\infty}$ is a Dirichlet eigenbasis, and $\{B_n\}_{n=1}^{\infty}$ are as in equation (15A.2). Suppose $\sum_{n=1}^{\infty} |\lambda_n B_n| < \infty$. If u satisfies homogeneous Dirichlet BC (i.e. $u(\mathbf{x}, t) = 0$ for all $\mathbf{x} \in \partial X$ and $t \in \mathbb{R}_+$), then*

$$u \underset{12}{\approx} \sum_{n=1}^{\infty} B_n \, \cos(\sqrt{\lambda_n} \, t) \, S_n.$$

(b) *Suppose $\{C_k, \mu_k\}_{k=0}^{\infty}$ is a Neumann eigenbasis, and $\{A_n\}_{n=0}^{\infty}$ are as in equation (15A.1). Suppose the sequence $\{A_n\}_{n=0}^{\infty}$ decays quickly enough that*

$$\sum_{n=0}^{\infty} |\mu_n A_n| < \infty, \quad and, \text{ for all } j \in [1 \dots D], \quad \sum_{n=0}^{\infty} |A_n| \cdot \|\partial_j C_n\|_{\infty} < \infty.$$

(15A.5)

If u satisfies homogeneous Neumann BC (i.e. $\partial_\perp u(\mathbf{x}, t) = 0$ for all $\mathbf{x} \in \partial X$ and $t \in \mathbb{R}_+$), then

$$u \underset{12}{\approx} \sum_{n=0}^{\infty} A_n \, \cos(\sqrt{\mu_n} \, t) \, C_n.$$

Now suppose $u : X \times \mathbb{R}_+ \longrightarrow \mathbb{R}$ is a solution to the wave equation '$\partial_t^2 u = \triangle u$', and has initial velocity $\partial_t u(\mathbf{x}; 0) = f(\mathbf{x})$ for all $\mathbf{x} \in X$.

(c) *Suppose $\sum_{n=1}^{\infty} \sqrt{\lambda_n} |B_n| < \infty$. If u satisfies homogeneous Dirichlet BC, then $u \underset{12}{\approx} \sum_{n=1}^{\infty} \frac{B_n}{\sqrt{\lambda_n}} \sin(\sqrt{\lambda_n} \, t) \, S_n$.*

(d) *Suppose the sequence $\{A_n\}_{n=0}^{\infty}$ decays quickly enough that*

$$\sum_{n=1}^{\infty} \sqrt{\mu_n} |A_n| < \infty, \quad and, \text{ for all } j \in [1 \dots D], \quad \sum_{n=1}^{\infty} \frac{|A_n|}{\sqrt{\mu_n}} \|\partial_j C_n\|_{\infty} < \infty.$$

(15A.6)

If u satisfies homogeneous Neumann BC, then there is some constant $C \in \mathbb{R}$ such that, for all $\mathbf{x} \in X$, we have $u(\mathbf{x}; 0) = C$, and for all $t \in \mathbb{R}$, we have

$$u(\mathbf{x}; t) \underset{12}{\approx} A_0 t + \sum_{n=1}^{\infty} \frac{A_n}{\sqrt{\mu_n}} \sin(\sqrt{\mu_n} \, t) \, C_n(\mathbf{x}) + C.$$

(e) *To obtain a solution with both a specified initial position and a specified initial velocity, add the solutions from (a) and (c) for homogeneous Dirichlet BC. Add the solutions from (b) and (d) for homogeneous Neumann BC (setting $C = 0$ in part (d)).*

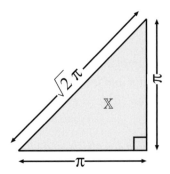

Figure 15A.1. Right-angle triangular domain of Proposition 15A.5.

ⓔ *Proof* **Exercise 15A.7.** *Hint:* To show solution uniqueness, use Theorem 5D.11.
For (a), imitate Propositions 12D.1 and 14F.1.

For (c) imitate the proof of Propositions 12D.3 and 14F.1.

For (b) and (d), use hypotheses (15A.5) and (15A.6) to apply Proposition F.1.   ☐

ⓔ **Exercise 15A.8** Show how Propositions 11B.1, 11B.4, 12D.1, 12D.3, and 14F.1
are all special cases of Theorem 15A.4(a) and (c).   ◆

ⓔ **Exercise 15A.9** What is the physical meaning of a nonzero value of A_0 in Theorem
15A.4(d)?   ◆

Theorems 15A.2, 15A.3, and 15A.4 allow us to solve I/BVPs on any domain,
once we have a suitable eigenbasis. We illustrate with a simple example.

Proposition 15A.5 Eigenbases for a triangle
Let $\mathbb{X} := \{(x, y) \in [0, \pi]^2; y \le x\}$ *be a filled right-angle triangle (Figure 15A.1).*

(a) *For any two-element subset* $\{n, m\} \subset \mathbb{N}$ *(i.e.* $n \ne m$*), let* $\mathcal{S}_{\{n,m\}} := \sin(nx)\sin(my) - \sin(mx)\sin(ny)$*, and let* $\lambda_{\{n,m\}} := n^2 + m^2$*. Then*

  (i) $\mathcal{S}_{\{n,m\}}$ *is an eigenfunction of the Laplacian:* $\triangle\mathcal{S}_{\{n,m\}} = -\lambda_{\{n,m\}}\mathcal{S}_{\{n,m\}}$;
  (ii) $\{\mathcal{S}_{\{n,m\}}\}_{\{n,m\}\subset\mathbb{N}}$ *is a Dirichlet eigenbasis for* $L^2(\mathbb{X})$.

(b) *Let* $\mathbf{C}_{0,0} = 1$*, and, for any two-element subset* $\{n, m\} \subset \mathbb{N}$*, let* $\mathcal{C}_{\{n,m\}} := \cos(nx)\cos(my) + \cos(mx)\cos(ny)$*, and let* $\lambda_{\{n,m\}} := n^2 + m^2$*. Then*

  (i) $\mathcal{C}_{\{n,m\}}$ *is an eigenfunction of the Laplacian:* $\triangle\mathcal{C}_{\{n,m\}} = -\lambda_{\{n,m\}}\mathcal{C}_{\{n,m\}}$;
  (ii) $\{\mathcal{C}_{\{n,m\}}\}_{\{n,m\}\subset\mathbb{N}}$ *is a Neumann eigenbasis for* $L^2(\mathbb{X})$.

ⓔ *Proof* **Exercise 15A.10.** *Hint:* Part (i) is a straightforward computation, as is the
verification of the homogeneous boundary conditions. (*Hint:* On the hypotenuse,
$\partial_{\perp} = \partial_2 - \partial_1$.) To verify that the specified sets are orthogonal bases, use Theorem
9A.3.   ☐

Exercise 15A.11   ⓔ

(a) Combine Proposition 15A.5 with Theorems 15A.2, 15A.3, and 15A.4 to provide a general solution method for solving the Poisson equation, heat equation, and wave equation on a right-angle triangle domain, with either Dirichlet or Neumann boundary conditions.

(b) Set up and solve some simple initial/boundary value problems using your method.   ♦

Remark 15A.6 There is nothing special about the role of the Laplacian $\triangle$ in Theorems 15A.2, 15A.3, and 15A.4. If L is any *linear* differential operator, for which we have 'solution uniqueness' results analogous to the results of §5D, then Theorems 15A.2, 15A.3, and 15A.4 are still true if you replace '$\triangle$' with 'L' everywhere (**Exercise 15A.12** Verify this). In particular, if L is an elliptic dif-   ⓔ ferential operator (see §5E), then

- Theorem 15A.2 becomes the general solution to the boundary value problem for the *nonhomogeneous elliptic PDE* '$\mathsf{L}u = f$';
- Theorem 15A.3 becomes the general solution to the initial/boundary value problem for the *homogeneous parabolic PDE* '$\partial_t u = \mathsf{L}u$';
- Theorem 15A.4 becomes the general solution to the initial value problem for the *homogeneous hyperbolic PDE* '$\partial_t^2 u = \mathsf{L}u$'.

Theorem 15E.17, p. 355, discusses the existence of Dirichlet eigenbases for other elliptic differential operators.

Exercise 15A.13 Let $\mathbb{X} \subset \mathbb{R}^3$ be a bounded domain, and consider a quantum par-   ⓔ ticle confined to the domain $\mathbb{X}$ by an 'infinite potential well' $V : \mathbb{R}^3 \longrightarrow \mathbb{R} \cup \{\infty\}$, where $V(\mathbf{x}) = 0$ for all $\mathbf{x} \in \mathbb{X}$, and $V(\mathbf{x}) = \infty$ for all $\mathbf{x} \notin \mathbb{X}$ (see Examples 3C.4 and 3C.5, pp. 51–52, for a discussion of the physical meaning of this model). Modify Theorem 15A.3 to state and prove a theorem describing the general solution to the initial value problem for the Schrödinger equation with the potential V.

    (*Hint:* If $\omega : \mathbb{R}^3 \times \mathbb{R} \longrightarrow \mathbb{C}$ is a solution to the corresponding Schrödinger equation, then we can assume $\omega_t(\mathbf{x}) = 0$ for all $\mathbf{x} \notin \mathbb{X}$. If ω is also continuous, then we can model the particle using a function $\omega : \mathbb{X} \times \mathbb{R} \longrightarrow \mathbb{C}$, which satisfies homogeneous Dirichlet boundary conditions on $\partial\mathbb{X}$.)   ♦

15B General solution to Laplace equation BVPs

Prerequisites: §4B(iv), §5B, §5C, §6F, Appendix D.   **Recommended:** §5D, §12A, §13B, §14B, §15A.

Theorems 15A.2(b) and (d) and 15A.3(b) and (d) both used the same strategy to solve a PDE with nonhomogeneous boundary conditions:

- solve the original PDE with *homogeneous* boundary conditions;
- solve the Laplace equation with the specified nonhomogeneous BC;
- add these two solutions together to get a solution to the original problem.

However, we do not yet have a general method for solving the Laplace equation. That is the goal of this section. Throughout this section, we make the following assumptions.

- Let $X \subset \mathbb{R}^D$ be a bounded domain, whose boundary ∂X is piecewise smooth. This has two consequences: (1) the normal derivative on the boundary is well defined (so we can meaningfully impose Neumann boundary conditions); and (2) we can meaningfully speak of integrating functions over ∂X. For example, if $X \subset \mathbb{R}^2$, then ∂X should be a finite union of smooth curves. If $X \subset \mathbb{R}^3$, then ∂X should be a finite union of smooth surfaces, etc. If $b, c : \partial X \longrightarrow \mathbb{R}$ are functions, then define

$$\langle b, c \rangle := \int_{\partial X} b(\mathbf{x}) \cdot c(\mathbf{x}) \mathrm{d}\mathbf{x} \quad \text{and} \quad \|b\|_2 := \sqrt{\langle b, b \rangle} := \left(\int_{\partial X} |b(\mathbf{x})|^2 \, \mathrm{d}\mathbf{x} \right)^{1/2},$$

where these are computed as contour integrals (or surface integrals, etc.) over ∂X. As usual, let $\mathbf{L}^2(\partial X)$ be the set of all integrable functions $b : \partial X \longrightarrow \mathbb{R}$ such that $\|b\|_2 < \infty$ (see §6B for further discussion).
- Let $\{\Xi_n\}_{n=1}^{\infty}$ be an orthogonal basis for $\mathbf{L}^2(\partial X)$. Thus, for any $b \in \mathbf{L}^2(\partial X)$, we can write

$$b \underset{L2}{\approx} \sum_{n=1}^{\infty} B_n \Xi_n, \quad \text{where} \quad B_n := \frac{\langle b, \Xi_n \rangle}{\|\Xi_n\|_2^2}, \quad \text{for all } n \in \mathbb{N}. \tag{15B.1}$$

- For all $n \in \mathbb{N}$, let $\mathcal{H}_n : X \longrightarrow \mathbb{R}$ be a harmonic function (i.e. $\Delta \mathcal{H}_n = 0$) satisfying the nonhomogeneous Dirichlet boundary condition $\mathcal{H}_n(\mathbf{x}) = \Xi_n(\mathbf{x})$ for all $\mathbf{x} \in \partial X$. The system $\mathfrak{H} := \{\mathcal{H}_n\}_{n=1}^{\infty}$ is called a *Dirichlet harmonic basis* for X.
- Suppose $\Xi_1 \equiv 1$ is the constant function. Then $\int_{\partial X} \Xi_n(\mathbf{x}) \mathrm{d}\mathbf{x} = \langle \Xi_n, 1 \rangle = 0$, for all $n \geq 2$ (by orthogonality). For all $n \geq 2$, let $\mathcal{G}_n : X \longrightarrow \mathbb{R}$ be a harmonic function (i.e. $\Delta \mathcal{G}_n = 0$) satisfying the nonhomogeneous Neumann boundary condition $\partial_\perp \mathcal{G}_n(\mathbf{x}) = \Xi_n(\mathbf{x})$ for all $\mathbf{x} \in \partial X$. The system $\mathfrak{G} := \{1\} \sqcup \{\mathcal{G}_n\}_{n=2}^{\infty}$ is called a *Neumann harmonic basis* for X. (Note that $\partial_\perp 1 = 0$, *not* Ξ_1.)

Note that, although they are called 'harmonic bases for X', $\mathfrak{H}$ and $\{\partial_\perp \mathcal{G}_n\}_{n=2}^{\infty}$ are actually orthogonal bases for $\mathbf{L}^2(\partial X)$, *not* for $\mathbf{L}^2(X)$.

Ⓔ **Exercise 15B.1** Show that there is no harmonic function $\mathcal{G}_1$ on X satisfying the Neumann boundary condition $\partial_\perp \mathcal{G}_1(\mathbf{x}) = 1$ for all $\mathbf{x} \in \partial X$. (*Hint:* Use Corollary 5D.4(b)(i), p. 90.) ◆

Example 15B.1 If $X = [0, \pi]^2 \subset \mathbb{R}^2$, then $\partial X = \mathbf{L} \cup \mathbf{R} \cup \mathbf{T} \cup \mathbf{B}$, where

$$\mathbf{L} := \{0\} \times [0, \pi], \quad \mathbf{R} := \{\pi\} \times [0, \pi], \quad \mathbf{B} := [0, \pi] \times \{0\}, \quad \text{and}$$
$$\mathbf{T} := [0, \pi] \times \{\pi\}.$$

(See Figure 12A.1(b), p. 245.)

(a) Let $\{\Xi_k\}_{k=1}^\infty := \{\mathcal{L}_n\}_{n=1}^\infty \sqcup \{\mathcal{R}_n\}_{n=1}^\infty \sqcup \{\mathcal{B}_n\}_{n=1}^\infty \sqcup \{\mathcal{T}_n\}_{n=1}^\infty$, where, for all $n \in \mathbb{N}$, the functions $\mathcal{L}_n, \mathcal{R}_n, \mathcal{B}_n, \mathcal{T}_n : \partial\mathbb{X} \longrightarrow \mathbb{R}$ are defined by

$$\mathcal{L}_n(x, y) := \begin{cases} \sin(ny) & \text{if } (x, y) \in \mathbf{L}; \\ 0 & \text{otherwise}; \end{cases}$$

$$\mathcal{R}_n(x, y) := \begin{cases} \sin(ny) & \text{if } (x, y) \in \mathbf{R}; \\ 0 & \text{otherwise}; \end{cases}$$

$$\mathcal{B}_n(x, y) := \begin{cases} \sin(nx) & \text{if } (x, y) \in \mathbf{B}; \\ 0 & \text{otherwise}; \end{cases}$$

$$\mathcal{T}_n(x, y) := \begin{cases} \sin(nx) & \text{if } (x, y) \in \mathbf{T}; \\ 0 & \text{otherwise}. \end{cases}$$

Now, $\{\mathcal{L}_n\}_{n=1}^\infty$ is an orthogonal basis for $\mathbf{L}^2(\mathbf{L})$ (by Theorem 7A.1). Likewise, $\{\mathcal{R}_n\}_{n=1}^\infty$ is an orthogonal basis for $\mathbf{L}^2(\mathbf{R})$, $\{\mathcal{B}_n\}_{n=1}^\infty$ is an orthogonal basis for $\mathbf{L}^2(\mathbf{B})$, and $\{\mathcal{T}_n\}_{n=1}^\infty$ is an orthogonal basis for $\mathbf{L}^2(\mathbf{T})$. Thus, $\{\Xi_k\}_{k=1}^\infty$ is an orthogonal basis for $\mathbf{L}^2(\partial\mathbb{X})$.

Let $\mathfrak{H} := \{\mathcal{H}_n^L\}_{n=1}^\infty \sqcup \{\mathcal{H}_n^R\}_{n=1}^\infty \sqcup \{\mathcal{H}_n^T\}_{n=1}^\infty \sqcup \{\mathcal{H}_n^B\}_{n=1}^\infty$, where, for all $n \in \mathbb{N}$ and all $(x, y) \in [0, \pi]^2$, we define

$$\mathcal{H}_n^L(x, y) := \frac{\sinh(n(\pi - x)) \sin(ny)}{\sinh(n\pi)};$$

$$\mathcal{H}_n^R(x, y) := \frac{\sinh(nx) \sin(ny)}{\sinh(n\pi)};$$

$$\mathcal{H}_n^B(x, y) := \frac{\sin(nx) \sinh(n(\pi - y))}{\sinh(n\pi)};$$

$$\mathcal{H}_n^T(x, y) := \frac{\sin(nx) \sinh(ny)}{\sinh(n\pi)}.$$

(See Figures 12A.2 and 12A.3, pp. 246–247.) Then $\mathfrak{H}$ is a Dirichlet harmonic basis for $\mathbb{X}$. This was the key fact employed by Proposition 12A.4, p. 248, to solve the Laplace equation on $[0, \pi]^2$ with arbitrary nonhomogeneous Dirichlet boundary conditions.

(b) Let $\{\Xi_k\}_{k=1}^\infty := \{\Xi_1, \Xi_=, \Xi_{||}, \Xi_\circ\} \sqcup \{\mathcal{L}_n\}_{n=1}^\infty \sqcup \{\mathcal{R}_n\}_{n=1}^\infty \sqcup \{\mathcal{B}_n\}_{n=1}^\infty \sqcup \{\mathcal{T}_n\}_{n=1}^\infty$. Here, for all $(x, y) \in \partial[0, \pi]^2$, we define

$$\Xi_1(x, y) := 1;$$

$$\Xi_{||}(x, y) := \begin{cases} 1 & \text{if} \quad (x, y) \in \mathbf{R}; \\ -1 & \text{if} \quad (x, y) \in \mathbf{L}; \\ 0 & \text{if} \quad (x, y) \in \mathbf{B} \sqcup \mathbf{T}; \end{cases}$$

$$\Xi_=(x, y) := \begin{cases} 1 & \text{if} \quad (x, y) \in \mathbf{T}; \\ -1 & \text{if} \quad (x, y) \in \mathbf{B}; \\ 0 & \text{if} \quad (x, y) \in \mathbf{L} \sqcup \mathbf{R}; \end{cases}$$

$$\Xi_\circ(x, y) := \begin{cases} 1 & \text{if} \quad (x, y) \in \mathbf{L} \sqcup \mathbf{R}; \\ -1 & \text{if} \quad (x, y) \in \mathbf{T} \sqcup \mathbf{B}. \end{cases}$$

Meanwhile, for all $n \in \mathbb{N}$, the functions $\mathcal{L}_n, \mathcal{R}_n, \mathcal{B}_n, \mathcal{T}_n : \partial\mathbb{X} \longrightarrow \mathbb{R}$ are defined by

$$\mathcal{L}_n(x, y) := \begin{cases} \cos(ny) & \text{if } (x, y) \in \mathbf{L}; \\ 0 & \text{otherwise}; \end{cases}$$

$$\mathcal{R}_n(x, y) := \begin{cases} \cos(ny) & \text{if } (x, y) \in \mathbf{R}; \\ 0 & \text{otherwise}; \end{cases}$$

$$\mathcal{B}_n(x, y) := \begin{cases} \cos(nx) & \text{if } (x, y) \in \mathbf{B}; \\ 0 & \text{otherwise}; \end{cases}$$

$$\mathcal{T}_n(x, y) := \begin{cases} \cos(nx) & \text{if } (x, y) \in \mathbf{T}; \\ 0 & \text{otherwise}. \end{cases}$$

Now, $\{1\} \sqcup \{\mathcal{L}_n\}_{n=1}^\infty$ is an orthogonal basis for $\mathbf{L}^2(\mathbf{L})$ (by Theorem 7A.1). Likewise, $\{1\} \sqcup \{\mathcal{R}_n\}_{n=1}^\infty$ is an orthogonal basis for $\mathbf{L}^2(\mathbf{R})$, $\{1\} \sqcup \{\mathcal{B}_n\}_{n=1}^\infty$ is an orthogonal basis for $\mathbf{L}^2(\mathbf{B})$, and $\{1\} \sqcup \{\mathcal{T}_n\}_{n=1}^\infty$ is an orthogonal basis for $\mathbf{L}^2(\mathbf{T})$. It follows that $\{\Xi_k\}_{k=1}^\infty$ is an orthogonal basis for $\mathbf{L}^2(\partial\mathbb{X})$ (**Exercise 15B.2**).

Let $\mathfrak{G} := \{1, \mathcal{G}_=, \mathcal{G}_{||}, \mathcal{G}_\circ\} \sqcup \{\mathcal{G}_n^L\}_{n=1}^\infty \sqcup \{\mathcal{G}_n^R\}_{n=1}^\infty \sqcup \{\mathcal{G}_n^B\}_{n=1}^\infty \sqcup \{\mathcal{G}_n^T\}_{n=1}^\infty$, where, for all $(x, y) \in [0, \pi]^2$,

$$\mathcal{G}_{||}(x, y) := x;$$

$$\mathcal{G}_=(x, y) := y;$$

$$\mathcal{G}_\circ(x, y) := \frac{1}{\pi}\left(\left(x - \frac{\pi}{2}\right)^2 - \left(y - \frac{\pi}{2}\right)^2\right).$$

The graphs of $\mathcal{G}_{||}(x, y)$ and $\mathcal{G}_=(x, y)$ are inclined planes at $45°$ in the x and y directions, respectively. The graph of $\mathcal{G}_\circ$ is a 'saddle' shape, very similar to Figure 1C.1(b), p. 12. Meanwhile, for all $n \geq 1$, and all $(x, y) \in [0, \pi]^2$, we define

$$\mathcal{G}_n^L(x, y) := \frac{\cosh(n(\pi - x))\cos(ny)}{n\,\sinh(n\pi)};$$

$$\mathcal{G}_n^R(x, y) := \frac{\cosh(nx)\cos(ny)}{n\,\sinh(n\pi)};$$

$$\mathcal{G}_n^B(x, y) := \frac{\cos(nx)\cosh(n(\pi - y))}{n\,\sinh(n\pi)};$$

$$\text{and } \mathcal{G}_n^T(x, y) := \frac{\cos(nx)\cosh(ny)}{n\,\sinh(n\pi)}.$$

Then $\mathfrak{G}$ is a Neumann harmonic basis for $\mathbb{X}$ (**Exercise 15B.3**).   ◇

Example 15B.2

(a) If $\mathbb{X} = \mathbb{D} = \{(r, \theta);\ r \leq 1\}$ (the unit disk in polar coordinates), then $\partial\mathbb{X} = \mathbb{S} = \{(r, \theta);\ r = 1\}$ (the unit circle). In this case, let $\{\Xi_k\}_{k=1}^\infty := \{\mathcal{C}_n\}_{n=0}^\infty \sqcup \{\mathcal{S}_n\}_{n=1}^\infty$, where, for all $n \in \mathbb{N}$ and $\theta \in [-\pi, \pi)$,

$$\mathcal{C}_n(\theta, 1) := \cos(n\theta) \quad \text{and} \quad \mathcal{S}_n(\theta, 1) := \sin(n\theta).$$

Then $\{\Xi_k\}_{k=1}^\infty$ is a basis of $\mathbf{L}^2(\mathbb{S})$, by Theorem 8A.1. Let $\mathfrak{H} := \{\Phi_n\}_{n=0}^\infty \sqcup \{\Psi_n\}_{n=1}^\infty$, where $\Phi_0 \equiv 1$, and where, for all $n \geq 1$ and $(r, \theta) \in \mathbb{D}$, we define

$$\Phi_n(r, \theta) := \cos(n\theta) \cdot r^n \quad \text{and} \quad \Psi_n(r, \theta) := \sin(n\theta) \cdot r^n.$$

(See Figure 14B.1, p. 281.) Then $\mathfrak{H}$ is a Dirichlet harmonic basis for $\mathbb{D}$; this was the key fact employed by Proposition 14B.2, p. 284, to solve the Laplace equation on $\mathbb{D}$ with arbitrary nonhomogeneous Dirichlet boundary conditions.

Suppose $\Xi_1 = \mathcal{C}_0$ (i.e. $\Xi_1 \equiv 1$). Let $\mathfrak{G} := \{1\} \sqcup \{\Phi_n/n\}_{n=1}^\infty \sqcup \{\Psi_n/n\}_{n=1}^\infty$, where, for all $n \in \mathbb{N}$ and $(r, \theta) \in \mathbb{D}$, we have

$$\Phi_n(r, \theta)/n := \frac{\cos(n\theta) \cdot r^n}{n} \quad \text{and} \quad \Psi_n(r, \theta)/n := \frac{\sin(n\theta) \cdot r^n}{n}.$$

Then $\mathfrak{G}$ is a Neumann harmonic basis for $\mathbb{D}$; this was the key fact employed by Proposition 14B.4, p. 286, to solve the Laplace equation on $\mathbb{D}$ with arbitrary nonhomogeneous Neumann boundary conditions.

(b) If $\mathbb{X} = \mathbb{D}^{\complement} = \{(r, \theta); \ r \geq 1\}$ (in polar coordinates),[1] then $\partial\mathbb{X} = \mathbb{S} = \{(r, \theta); \ r = 1\}$. In this case, let $\{\Xi_k\}_{k=1}^\infty := \{\mathcal{C}_n\}_{n=0}^\infty \sqcup \{\mathcal{S}_n\}_{n=1}^\infty$, just as in part (a). However, this time, let $\mathfrak{H} := \{\Phi_0\} \sqcup \{\phi_n\}_{n=1}^\infty \sqcup \{\psi_n\}_{n=1}^\infty$, where $\Phi_0 \equiv 1$, and where, for all $n \geq 1$ and $(r, \theta) \in \mathbb{D}$, we define

$$\phi_n(r, \theta) := \cos(n\theta)/r^n \quad \text{and} \quad \psi_n(r, \theta) := \sin(n\theta)/r^n.$$

(See Figure 14B.2, p. 282.) Then $\mathfrak{H}$ is a Dirichlet harmonic basis for $\mathbb{D}^{\complement}$; this was the key fact employed by Proposition 14B.6, p. 290, to solve the Laplace equation on $\mathbb{D}^{\complement}$ with arbitrary nonhomogeneous Neumann boundary conditions.

Recall $\Xi_1 = \mathcal{C}_0 \equiv 1$. Let $\mathfrak{G} := \{1\} \sqcup \{-\phi_n/n\}_{n=1}^\infty \sqcup \{-\psi_n/n\}_{n=1}^\infty$, where ϕ_n and ψ_n are as defined above, for all $n \geq 1$. Then $\mathfrak{G}$ is a Neumann harmonic basis for $\mathbb{D}^{\complement}$; this was the key fact employed by Proposition 14B.8, p. 292, to solve the Laplace equation on $\mathbb{D}^{\complement}$ with arbitrary nonhomogeneous Neumann boundary conditions.[2]   ◇

Theorem 15B.3 General solution to Laplace equation

Let $b \in \mathbf{L}^2(\partial\mathbb{X})$ have orthogonal expansion (15B.1). Let $\mathfrak{H} := \{\mathcal{H}_k\}_{k=1}^\infty$ be a Dirichlet harmonic basis for $\mathbb{X}$ and let $\mathfrak{G} := \{1\} \sqcup \{\mathcal{G}_k\}_{k=2}^\infty$ be a Neumann harmonic basis for $\mathbb{X}$.

---

[1] Technically, we are developing here a theory for *bounded* domains, and $\mathbb{D}^{\complement}$ is obviously not bounded. But it is interesting to note that many of our techniques still apply to $\mathbb{D}^{\complement}$. This is because $\mathbb{D}^{\complement}$ is *conformally isomorphic* to a bounded domain, once we regard $\mathbb{D}^{\complement}$ as a subset of the Riemann sphere by including the 'point at infinity'. See §18B for an introduction to conformal isomorphism. See Remark 18G.4, p. 474, for a discussion of the Riemann sphere.

[2] Note that our Neumann harmonic basis does not include the element $\phi_0(r, \theta) := \log(r)$. This is because $\partial_\perp \phi_0 = \Xi_1$. Of course, the domain $\mathbb{D}^{\complement}$ is not bounded, so Corollary 5D.4(b)(i), p. 90, does not apply, and indeed ϕ_0 is a continuous harmonic function on $\mathbb{D}^{\complement}$. However, unlike the elements of $\mathfrak{G}$, the function ϕ_0 is not bounded, and thus does *not* extend to a continuous real-valued harmonic function when we embed $\mathbb{D}^{\complement}$ in the Riemann sphere by adding the 'point at infinity'.

(a) *Let $u \underset{L2}{\approx} \sum_{k=1}^{\infty} B_k \mathcal{H}_k$. If this series converges uniformly to u on the interior of $\mathbb{X}$, then u is the unique continuous harmonic function with nonhomogeneous Dirichlet BC $u(\mathbf{x}) = b(\mathbf{x})$ for all $\mathbf{x} \in \partial \mathbb{X}$.*

(b) *Suppose $\Xi_1 \equiv 1$. If $B_1 \neq 0$, then there is no continuous harmonic function on $\mathbb{X}$ with nonhomogeneous Neumann BC $\partial_\perp u(\mathbf{x}) = b(\mathbf{x})$ for all $\mathbf{x} \in \partial \mathbb{X}$. Suppose $B_1 = 0$. Let $u \underset{L2}{\approx} \sum_{k=2}^{\infty} B_k \mathcal{G}_k + C$, where $C \in \mathbb{R}$ is any constant. If this series converges uniformly to u on the interior of $\mathbb{X}$, then it is a continuous harmonic function with nonhomogeneous Neumann BC $\partial_\perp u(\mathbf{x}) = b(\mathbf{x})$ for all $\mathbf{x} \in \partial \mathbb{X}$. Furthermore, all solutions to this BVP have this form, for some value of $C \in \mathbb{R}$.*

ⓔ **Proof** **Exercise 15B.4.** *Hint:* The boundary conditions follow from expansion (15B.1). To verify that u is harmonic, use the Mean Value Theorem (Theorem 1E.1, p. 18). (Use Proposition 6E.10(b), p. 132, to guarantee that the integral of the sum is the sum of the integrals.) Finally, use Corollary 5D.4, p. 90, to show solution uniqueness. □

ⓔ **Exercise 15B.5** Show how Propositions 12A.4, 13B.2, 14B.2, 14B.4, 14B.6, 14B.8, and 14B.10 are all special cases of Theorem 15B.3. ◆

Remark There is nothing special about the role of the Laplacian $\triangle$ in Theorem 15B.3. If L is any *linear* differential operator, then something like Theorem 15B.3 is still true if you replace '$\triangle$' with 'L' everywhere. In particular, if L is an *elliptic* differential operator (see §5E), then Theorem 15B.3 becomes the general solution to the boundary value problem for the *homogeneous elliptic PDE* 'L$u \equiv 0$'.

However, if L is an arbitrary differential operator, then there is no guarantee that you will find a 'harmonic basis' $\{\mathcal{H}_k\}_{k=1}^{\infty}$ of functions such that L$\mathcal{H}_k \equiv 0$ for all $k \in \mathbb{N}$, and such that the collection $\{\mathcal{H}_k\}_{k=1}^{\infty}$ (or $\{\partial_\perp \mathcal{H}_k\}_{k=1}^{\infty}$) provides an orthonormal basis for $\mathbf{L}^2(\partial \mathbb{X})$. (Even for the Laplacian, this is a nontrivial problem; see e.g. Corollary 15C.8, p. 340.)

Furthermore, once you define $u \underset{L2}{\approx} \sum_{k=1}^{\infty} B_k \mathcal{H}_k$ as in Theorem 15B.3, you might not be able to use something like the Mean Value Theorem to guarantee that L$u = 0$. Instead you must 'formally differentiate' the series $\sum_{k=1}^{\infty} B_k \mathcal{H}_k$ and $\sum_{k=1}^{\infty} B_k \mathcal{G}_k$ using Proposition F.1, p. 569. For this to work, you need some convergence conditions on the 'formal derivatives' of these series. For example, if L was an Nth-order differential operator, it would be sufficient to require that

$$\sum_{k=1}^{\infty} |B_k| \cdot \left\| \partial_j^N \mathcal{H}_k \right\|_{\infty} < \infty \quad \text{and} \quad \sum_{k=1}^{\infty} |B_k| \cdot \left\| \partial_j^N \mathcal{G}_k \right\|_{\infty} < \infty$$

ⓔ for all $j \in [1 \dots D]$ (**Exercise 15B.6** Verify this).

Finally, for an arbitrary differential operator, there may not be a result like Corollary 5D.4, p. 90, which guarantees a unique solution to a Dirichlet/Neumann BVP. It may be necessary to impose further constraints to get a unique solution.

15C Eigenbases on Cartesian products

Prerequisites: §4B(iv), §5B, §5C, §6F, Appendix D.

If $X_1 \subset \mathbb{R}^{D_1}$ and $X_2 \subset \mathbb{R}^{D_2}$ are two domains, then their *Cartesian product* is the set given by

$$X_1 \times X_2 := \{(\mathbf{x}_1, \mathbf{x}_2); \, \mathbf{x}_1 \in X_1 \text{ and } \mathbf{x}_2 \in X_2\} \subset \mathbb{R}^{D_1+D_2}.$$

Example 15C.1
(a) If $X_1 = [0, \pi] \subset \mathbb{R}$ and $X_2 = [0, \pi]^2 \subset \mathbb{R}^2$, then $X_1 \times X_2 = [0, \pi]^3 \subset \mathbb{R}^3$.
(b) If $X_1 = \mathbb{D} \subset \mathbb{R}^2$ and $X_2 = [0, \pi] \subset \mathbb{R}$, then $X_1 \times X_2 = \{(r, \theta, z); (r, \theta) \in \mathbb{D} \text{ and } 0 \leq z \leq \pi\} \subset \mathbb{R}^3$ is the *cylinder* of height π.

$\diamond$

To apply the solution methods from Sections 15A and 15B, we must first construct eigenbases and/or harmonic bases on the domain X; that is the goal of this section. We begin with some technical results that are useful and straightforward to prove.

Lemma 15C.2 *Let $X_1 \subset \mathbb{R}^{D_1}$ and $X_2 \subset \mathbb{R}^{D_2}$. Let $X := X_1 \times X_2 \subset \mathbb{R}^{D_1+D_2}$.*

(a) $\partial X = [(\partial X_1) \times X_2] \cup [X_1 \times (\partial X_2)]$.

Let $\Phi_1 : X_1 \longrightarrow \mathbb{R}$ and $\Phi_2 : X_2 \longrightarrow \mathbb{R}$, and define $\Phi = \Phi_1 \cdot \Phi_2 : X \longrightarrow \mathbb{R}$ by $\Phi(\mathbf{x}_1, \mathbf{x}_2) := \Phi_1(\mathbf{x}_1) \cdot \Phi_2(\mathbf{x}_2)$ for all $(\mathbf{x}_1, \mathbf{x}_2) \in X$.

(b) *If Φ_1 satisfies homogeneous Dirichlet BC on X_1 and Φ_2 satisfies homogeneous Dirichlet BC on X_2, then Φ satisfies homogeneous Dirichlet BC on X.*
(c) *If Φ_1 satisfies homogeneous Neumann BC on X_1 and Φ_2 satisfies homogeneous Neumann BC on X_2, then Φ satisfies homogeneous Neumann BC on X.*
(d) $\|\Phi\|_2 = \|\Phi_1\|_2 \cdot \|\Phi_2\|_2$. *Thus, if $\Phi_1 \in \mathbf{L}^2(X_1)$ and $\Phi_2 \in \mathbf{L}^2(X_2)$, then $\Phi \in \mathbf{L}^2(X)$.*
(e) *If $\Psi_1 \in \mathbf{L}^2(X_1)$ and $\Psi_2 \in \mathbf{L}^2(X_2)$, and $\Psi = \Psi_1 \cdot \Psi_2$, then $\langle \Phi, \Psi \rangle = \langle \Phi_1, \Psi_1 \rangle \cdot \langle \Phi_2, \Psi_2 \rangle$.*
(f) *Let $\{\Phi_n^{(1)}\}_{n=1}^{\infty}$ be an orthogonal basis for $\mathbf{L}^2(X_1)$ and let $\{\Phi_m^{(2)}\}_{m=1}^{\infty}$ be an orthogonal basis for $\mathbf{L}^2(X_2)$. For all $(n, m) \in \mathbb{N}$, let $\Phi_{n,m} := \Phi_n^{(1)} \cdot \Phi_m^{(2)}$. Then $\{\Phi_{n,m}\}_{n,m=1}^{\infty}$ is an orthogonal basis for $\mathbf{L}^2(X)$.*

Let $\triangle_1$ be the Laplacian operator on $\mathbb{R}^{D_1}$, let $\triangle_2$ be the Laplacian operator on $\mathbb{R}^{D_2}$, and let $\triangle$ be the Laplacian operator on $\mathbb{R}^{D_1+D_2}$.

(g) $\triangle\Phi(\mathbf{x}_1, \mathbf{x}_2) = (\triangle_1 \Phi_1(\mathbf{x}_1)) \cdot \Phi_2(\mathbf{x}_2) + \Phi_1(\mathbf{x}_1) \cdot (\triangle_2 \Phi_2(\mathbf{x}_2))$.
(h) *Thus, if Φ_1 is an eigenfunction of $\triangle_1$ with eigenvalue λ_1, and Φ_2 is an eigenfunction of $\triangle_2$ with eigenvalue λ_2, then Φ is an eigenfunction of $\triangle$ with eigenvalue $(\lambda_1 + \lambda_2)$.*

Proof **Exercise 15C.1.** (*Remark:* For part (f), just show that $\{\Phi_{n,m}\}_{n,m=1}^{\infty}$ is an ⓔ orthogonal collection of functions. Showing that $\{\Phi_{n,m}\}_{n,m=1}^{\infty}$ is actually a *basis* for $\mathbf{L}^2(X)$ requires methods beyond the scope of this book.) □

Corollary 15C.3 Eigenbases for Cartesian products

Let $\mathbb{X}_1 \subset \mathbb{R}^{D_1}$ and $\mathbb{X}_2 \subset \mathbb{R}^{D_2}$. Let $\mathbb{X} := \mathbb{X}_1 \times \mathbb{X}_2 \subset \mathbb{R}^{D_1+D_2}$. Let $\{\Phi_n^{(1)}\}_{n=1}^\infty$ be a Dirichlet (or Neumann) eigenbasis for $\mathbf{L}^2(\mathbb{X}_1)$, and let $\{\Phi_m^{(2)}\}_{m=1}^\infty$ be a Dirichlet (respectively Neumann) eigenbasis for $\mathbf{L}^2(\mathbb{X}_2)$. For all $(n, m) \in \mathbb{N}$, define $\Phi_{n,m} = \Phi_n^{(1)} \cdot \Phi_m^{(2)}$. Then $\{\Phi_{n,m}\}_{n,m=1}^\infty$ Dirichlet (respectively Neumann) eigenbasis for $\mathbf{L}^2(\mathbb{X})$.

Ⓔ *Proof* **Exercise 15C.2.** Just combine Lemma 15C.2(b), (c), (f), and (h). □

Example 15C.4 Let $\mathbb{X}_1 = [0, \pi]$ and $\mathbb{X}_2 = [0, \pi]^2$, so $\mathbb{X}_1 \times \mathbb{X}_2 = [0, \pi]^3$. Note that $\partial([0, \pi]^3) = (\{0, \pi\} \times [0, \pi]^2) \cup ([0, \pi] \times \partial[0, \pi]^2)$. For all $\ell \in \mathbb{N}$, define $\mathbf{C}_\ell$ and $\mathbf{S}_\ell \in \mathbf{L}^2[0, \pi]$ by $\mathbf{C}_\ell(x) := \cos(\ell x)$ and $\mathbf{S}_\ell(x) := \sin(\ell x)$. For all $m, n \in \mathbb{N}$, define $\mathbf{C}_{m,n}$ and $\mathbf{S}_{m,n} \in \mathbf{L}^2([0, \pi]^2)$ by $\mathbf{C}_{m,n}(y, z) := \cos(my)\cos(nz)$ and $\mathbf{S}_{m,n}(y, z) := \sin(my)\sin(nz)$.

For any $\ell, m, n \in \mathbb{N}$, define $\mathbf{C}_{\ell,m,n}$ and $\mathbf{S}_{\ell,m,n} \in \mathbf{L}^2(\mathbb{X})$ by $\mathbf{C}_{\ell,m,n}(x, y, z) := \mathbf{C}_\ell(x) \cdot \mathbf{C}_{m,n}(y, z) = \cos(\ell x)\cos(my)\cos(nz)$ and $\mathbf{S}_{\ell,m,n}(x, y, z) := \mathbf{S}_\ell(x) \cdot \mathbf{S}_{m,n}(y, z) = \sin(\ell x)\sin(my)\sin(nz)$.

Now, $\{\mathbf{S}_\ell\}_{\ell=1}^\infty$ is a Dirichlet eigenbasis for $[0, \pi]$ (by Theorem 7A.1, p. 142), and $\{\mathbf{S}_{m,n}\}_{m,n=1}^\infty$ is a Dirichlet eigenbasis for $[0, \pi]^2$ (by Theorem 9A.3(a), p. 183); thus, Corollary 15C.3 says that $\{\mathbf{S}_{\ell,m,n}\}_{\ell,m,n=1}^\infty$ is a Dirichlet eigenbasis for $[0, \pi]^3$ (as earlier noted by Theorem 9B.1, p. 189).

Likewise, $\{\mathbf{C}_\ell\}_{\ell=0}^\infty$ is a Neumann eigenbasis for $[0, \pi]$ (by Theorem 7A.4, p. 145), and $\{\mathbf{C}_{m,n}\}_{m,n=0}^\infty$ is a Neumann eigenbasis for $[0, \pi]^2$; (by Theorem 9A.3(b), p. 183); thus, Corollary 15C.3 says that $\{\mathbf{C}_{\ell,m,n}\}_{\ell,m,n=0}^\infty$ is a Neumann eigenbasis for $[0, \pi]^3$ (as earlier noted by Theorem 9B.1, p. 189). ◇

Example 15C.5 Let $\mathbb{X}_1 = \mathbb{D}$ and $\mathbb{X}_2 = [0, \pi]$, so that $\mathbb{X}_1 \times \mathbb{X}_2$ is the cylinder of height π and radius 1. Let $\mathbb{S} := \partial\mathbb{D}$ (the unit circle). Note that $\partial\mathbb{X} = (\mathbb{S} \times [0, \pi]) \cup (\mathbb{D} \times \{0, \pi\})$. For all $n \in \mathbb{N}$, define $\mathbf{S}_n \in \mathbf{L}^2[0, \pi]$ as in Example 15C.4. For all $\ell, m \in \mathbb{N}$, let $\Phi_{\ell,m}$ and $\Psi_{\ell,m}$ be the type-1 Fourier–Bessel eigenfunctions defined by equation (14C.5), p. 301. For any $\ell, m, n \in \mathbb{N}$, define $\Phi_{\ell,m,n}$ and $\Psi_{\ell,m,n} \in \mathbf{L}^2(\mathbb{X})$ by $\Phi_{\ell,m,n}(r, \theta, z) := \Phi_{\ell,m}(r, \theta) \cdot \mathbf{S}_n(z)$ and $\Psi_{\ell,m,n}(r, \theta, z) := \Psi_{\ell,m}(r, \theta) \cdot \mathbf{S}_n(z)$.

Now $\{\Phi_{m,n}, \Psi_{m,n}\}_{m,n=1}^\infty$ is a Dirichlet eigenbasis for the disk $\mathbb{D}$ (by Theorem 14C.2) and $\{\mathbf{S}_n\}_{n=1}^\infty$ is a Dirichlet eigenbasis for the line $[0, \pi]$ (by Theorem 7A.1, p. 142); thus, Corollary 15C.3 says that $\{\Phi_{\ell,m,n}, \Psi_{\ell,m,n}\}_{\ell,m,n=1}^\infty$ is a Dirichlet eigenbasis for the cylinder $\mathbb{X}$. ◇

Ⓔ **Exercise 15C.3**

(a) Combine Example 15C.5 with Theorems 15A.2, 15A.3, and 15A.4 to provide a general solution method for solving the Poisson equation, the heat equation, and the wave equation on a finite cylinder with Dirichlet boundary conditions.

(b) Set up and solve some simple initial/boundary value problems using your method. ◆

Exercise 15C.4 In cylindrical coordinates on $\mathbb{R}^3$, let $X = \{(r, \theta, z); \, 1 \leq r, 0 \leq$ Ⓔ
$z \leq \pi$, and $-\pi \leq \theta < \pi\}$ be the *punctured slab* of thickness π, having a cylindrical hole of radius 1.

(a) Express X as a Cartesian product of the punctured plane and a line segment.
(b) Use Corollary 15C.3 to obtain a Dirichlet eigenbasis for X.
(c) Apply Theorems 15A.2, 15A.3, and 15A.4 to provide a general solution method for
    solving the Poisson equation, the heat equation, and the wave equation on the punctured
    slab with Dirichlet boundary conditions.
(d) Set up and solve some simple initial/boundary value problems using your method. ◆

Exercise 15C.5 Let $X_1 = \{(x, y) \in [0, \pi]^2; \, y \leq x\}$ be the *right-angle triangle*   Ⓔ
from Proposition 15A.5, p. 330, and let $X_2 = [0, \pi] \subset \mathbb{R}$. Then $X = X_1 \times X_2$ is a
right-angle triangular prism.

(a) Use Proposition 15A.5 and Corollary 15C.3 to obtain Dirichlet and Neumann eigen-
    bases for the prism X.
(b) Apply Theorems 15A.2, 15A.3, and 15A.4 to provide a general solution method for
    solving the Poisson equation, the heat equation, and the wave equation on the prism
    with Dirichlet or Neumann boundary conditions.
(c) Set up and solve some simple initial/boundary value problems using your method. ◆

We now move on to the problem of constructing harmonic bases on a Cartesian
product. We will need two technical lemmas.

Lemma 15C.6 Harmonic functions on Cartesian products
Let $X_1 \subset \mathbb{R}^{D_1}$ and $X_2 \subset \mathbb{R}^{D_2}$. Let $X := X_1 \times X_2 \subset \mathbb{R}^{D_1 + D_2}$.
   *Let $\mathcal{E}_1 : X_1 \longrightarrow \mathbb{R}$ be an eigenfunction of $\triangle_1$ with eigenvalue λ, and let
$\mathcal{E}_2 : X_2 \longrightarrow \mathbb{R}$ be an eigenfunction of $\triangle_2$ with eigenvalue $-\lambda$. If we define $\mathcal{H} :=
\mathcal{E}_1 \cdot \mathcal{E}_2 : X \longrightarrow \mathbb{R}$, as in Lemma 15C.2, then $\mathcal{H}$ is a harmonic function – that is,
$\triangle \mathcal{H} = 0$.*

Proof **Exercise 15C.6.** (*Hint:* Use Lemma 15C.2(h).)   □   Ⓔ

Lemma 15C.7 Orthogonal bases on almost-disjoint unions
*Let $\mathbb{Y}_1, \mathbb{Y}_2 \subset \mathbb{R}^D$ be two $(D-1)$-dimensional subsets (e.g. two curves in $\mathbb{R}^2$,
two surfaces in $\mathbb{R}^3$, etc.). Suppose that $\mathbb{Y}_1 \cap \mathbb{Y}_2$ has dimension $(D-2)$ (e.g. it
is a discrete set of points in $\mathbb{R}^2$, or a curve in $\mathbb{R}^3$, etc.). Let $\{\Phi_n^{(1)}\}_{n=1}^\infty$ be an
orthogonal basis for $\mathbf{L}^2(\mathbb{Y}_1)$, such that $\Phi_n^{(1)}(\mathbf{y}) = 0$ for all $\mathbf{y} \in \mathbb{Y}_2$ and $n \in \mathbb{N}$.
Likewise, let $\{\Phi_n^{(2)}\}_{n=1}^\infty$ be an orthogonal basis for $\mathbf{L}^2(\mathbb{Y}_2)$, such that $\Phi_n^{(2)}(\mathbf{y}) = 0$
for all $\mathbf{y} \in \mathbb{Y}_1$ and $n \in \mathbb{N}$. Then $\{\Phi_n^{(1)}\}_{n=1}^\infty \sqcup \{\Phi_n^{(2)}\}_{n=1}^\infty$ is an orthogonal basis for
$\mathbf{L}^2(\mathbb{Y}_1 \cup \mathbb{Y}_2)$.*

Ⓔ *Proof* **Exercise 15C.7.** (*Hint:* The $(D-1)$-dimensional integral of any function on $\mathbb{Y}_1 \cap \mathbb{Y}_2$ must be zero.) $\square$

For the rest of this section we adopt the following notational convention: if $f : \mathbb{X} \longrightarrow \mathbb{R}$ is a function, then let $\tilde{f}$ denote the restriction of f to a function $\tilde{f} : \partial\mathbb{X} \longrightarrow \mathbb{R}$ (that is, $\tilde{f} := f|_{\partial\mathbb{X}}$).

Corollary 15C.8 Harmonic bases on Cartesian products

Let $\mathbb{X}_1 \subset \mathbb{R}^{D_1}$ and $\mathbb{X}_2 \subset \mathbb{R}^{D_2}$. Let $\mathbb{X} := \mathbb{X}_1 \times \mathbb{X}_2 \subset \mathbb{R}^{D_1+D_2}$.

Let $\{\Xi_m^2\}_{m\in\mathbb{M}_2}$ be an orthogonal basis for $\mathbf{L}^2(\partial\mathbb{X}_2)$ (here, $\mathbb{M}_2$ is some indexing set, either finite or infinite; e.g. $\mathbb{M}_2 = \mathbb{N}$). Let $\{\mathcal{E}_n^1\}_{n=1}^\infty$ be a Dirichlet eigenbasis for $\mathbb{X}_1$. For all $n \in \mathbb{N}$, suppose $\triangle_1\mathcal{E}_n^1 = -\lambda_n^{(1)}\mathcal{E}_n^1$, and, for all $m \in \mathbb{M}_2$, let $\mathcal{F}_{n,m}^2 \in \mathbf{L}^2(\mathbb{X}_2)$ be an eigenfunction of $\triangle_2$ with eigenvalue $+\lambda_n^{(1)}$, such that $\tilde{\mathcal{F}}_{n,m}^2 = \Xi_m^2$. Let $\mathcal{H}_{n,m}^1 := \mathcal{E}_n^1 \cdot \mathcal{F}_{n,m}^2 : \mathbb{X} \longrightarrow \mathbb{R}$, for all $n \in \mathbb{N}$ and $m \in \mathbb{M}_2$.

Likewise, let $\{\Xi_m^1\}_{m\in\mathbb{M}_1}$ be an orthogonal basis for $\mathbf{L}^2(\partial\mathbb{X}_1)$ (where $\mathbb{M}_1$ is some indexing set), and let $\{\mathcal{E}_n^2\}_{n=1}^\infty$ be a Dirichlet eigenbasis for $\mathbb{X}_2$. For all $n \in \mathbb{N}$, suppose $\triangle_2\mathcal{E}_n^2 = -\lambda_n^{(2)}\mathcal{E}_n^2$, and, for all $m \in \mathbb{M}_1$, let $\mathcal{F}_{n,m}^1 \in \mathbf{L}^2(\mathbb{X}_1)$ be an eigenfunction of $\triangle_1$ with eigenvalue $+\lambda_n^{(2)}$, such that $\tilde{\mathcal{F}}_{n,m}^1 = \Xi_m^1$. Define $\mathcal{H}_{n,m}^2 := \mathcal{F}_{n,m}^1 \cdot \mathcal{E}_n^2 : \mathbb{X} \longrightarrow \mathbb{R}$, for all $n \in \mathbb{N}$ and $m \in \mathbb{M}_1$.

Then $\mathfrak{H} := \{\mathcal{H}_{n,m}^1; \ n \in \mathbb{N}, \ m \in \mathbb{M}_2\} \sqcup \{\mathcal{H}_{n,m}^2; \ n \in \mathbb{N}, \ m \in \mathbb{M}_1\}$ is a Dirichlet harmonic basis for $\mathbf{L}^2(\partial\mathbb{X})$.

Ⓔ *Proof* **Exercise 15C.8.**

(a) Use Lemma 15C.6 to verify that all the functions $\mathcal{H}_{n,m}^1$ and $\mathcal{H}_{n,m}^2$ are harmonic on $\mathbb{X}$.

(b) Show that $\{\tilde{\mathcal{H}}_{n,m}^1\}_{n\in\mathbb{N},m\in\mathbb{M}_2}$ is an orthogonal basis for $\mathbf{L}^2(\mathbb{X}_1 \times (\partial\mathbb{X}_2))$, while $\{\tilde{\mathcal{H}}_{n,m}^2\}_{n\in\mathbb{N},m\in\mathbb{M}_1}$ is an orthogonal basis for $\mathbf{L}^2((\partial\mathbb{X}_1) \times \mathbb{X}_2)$. Use Lemma 15C.2(f).

(c) Show that $\mathfrak{H}$ is an orthogonal basis for $\mathbf{L}^2(\partial\mathbb{X})$. Use Lemma 15C.2(a) and Lemma 15C.7. $\square$

Example 15C.9 Let $\mathbb{X}_1 = [0, \pi] = \mathbb{X}_2$, so that $\mathbb{X} = [0, \pi]^2$. Observe that $\partial([0, \pi]^2) = (\{0, \pi\} \times [0, \pi]), \cup ([0, \pi] \times \{0, \pi\})$.

Observe that $\partial\mathbb{X}_1 = \{0, \pi\} = \partial\mathbb{X}_2$ (a two-element set) and that $\mathbf{L}^2\{0, \pi\}$ is a two-dimensional vector space (isomorphic to $\mathbb{R}^2$). Let $\mathbb{M}_1 := \{1, 2\} =: \mathbb{M}_2$. Let $\Xi_1^1 = \Xi_1^2 = \Xi_1$ and $\Xi_2^1 = \Xi_2^2 = \Xi_2$, where $\Xi_1, \Xi_2 : \{0, \pi\} \longrightarrow \mathbb{R}$ are defined:

$$\Xi_2(0) := 1 =: \Xi_1(\pi) \quad \text{and} \quad \Xi_2(\pi) := 0 =: \Xi_1(0).$$

Then $\{\Xi_1, \Xi_2\}$ is an orthogonal basis for $\mathbf{L}^2\{0, \pi\}$. For all $n \in \mathbb{N}$, let

$$\mathcal{E}_n^1(x) = \mathcal{E}_n^2(x) = \mathcal{E}_n(x) := \sin(nx);$$
$$\mathcal{F}_{n,1}^1(x) = \mathcal{F}_{n,1}^2(x) = \mathcal{F}_{n,1}(x) := \sinh(nx)/\sinh(n\pi);$$
$$\mathcal{F}_{n,2}^1(x) = \mathcal{F}_{n,2}^2(x) = \mathcal{F}_{n,2}(x) := \sinh(n(\pi - x))/\sinh(n\pi).$$

Then $\{\mathcal{E}_n\}_{n=1}^{\infty}$ is a Dirichlet eigenbasis for $[0, \pi]$ (by Theorem 7A.1, p. 142), while $\widetilde{\mathcal{F}}_{n,1} = \Xi_1$ and $\widetilde{\mathcal{F}}_{n,2} = \Xi_2$ for all $n \in \mathbb{N}$.

For each $n \in \mathbb{N}$, we have eigenvalue $\lambda_n := n^2$. That is, $\triangle\mathcal{E}_n(x) = -n^2\mathcal{E}_n(x)$, while $\triangle\mathcal{F}_{n,m}(x) = n^2\mathcal{F}_{n,m}(x)$. Thus, the functions $\mathcal{H}_n(x, y) := \mathcal{E}_n(x)\mathcal{F}_{n,m}(y)$ are harmonic, by Lemma 15C.6. Thus, if we define the following:

$$\mathcal{H}_{n,1}^1(x, y) := \mathcal{E}_n^1(x) \cdot \mathcal{F}_{n,1}^2(y) = \frac{\sin(nx)\sinh(ny)}{\sinh(n\pi)};$$

$$\mathcal{H}_{n,2}^1(x, y) := \mathcal{E}_n^1(x) \cdot \mathcal{F}_{n,2}^2(y) = \frac{\sin(nx)\sinh(n(\pi - y))}{\sinh(n\pi)};$$

$$\mathcal{H}_{n,1}^2(x, y) := \mathcal{F}_{n,1}^1(x) \cdot \mathcal{E}_n^2(y) = \frac{\sinh(nx)\sin(ny)}{\sinh(n\pi)};$$

$$\mathcal{H}_{n,2}^2(x, y) := \mathcal{F}_{n,2}^1(x) \cdot \mathcal{E}_n^2(y) = \frac{\sinh(n(\pi - x))\sin(ny)}{\sinh(n\pi)},$$

then Corollary 15C.8 says that the collection $\{\mathcal{H}_{n,1}^1\}_{n\in\mathbb{N}} \sqcup \{\mathcal{H}_{n,2}^1\}_{n\in\mathbb{N}} \sqcup \{\mathcal{H}_{n,1}^2\}_{n\in\mathbb{N}} \sqcup \{\mathcal{H}_{n,2}^2\}_{n\in\mathbb{N}}$ is a Dirichlet harmonic basis for $[0, \pi]^2$ – a fact we already observed in Example 15B.1(a), and exploited earlier in Proposition 12A.4, p. 248. ◇

Example 15C.10 Let $\mathbb{X}_1 = \mathbb{D} \subset \mathbb{R}^2$ and $\mathbb{X}_2 = [0, \pi]$, so that $\mathbb{X}_1 \times \mathbb{X}_2 \subset \mathbb{R}^3$ is the cylinder of height π and radius 1. Note that $\partial\mathbb{X} = (\mathbb{S} \times [0, \pi]) \cup (\mathbb{D} \times \{0, \pi\})$.

For all $n \in \mathbb{N}$ and $\ell \in \mathbb{N}$, let $\mathcal{E}_{\ell,n}^1 := \Phi_{\ell,n}$, and $\mathcal{E}_{\ell,-n}^1 := \Psi_{\ell,n}$, where $\Phi_{\ell,n}$ and $\Psi_{\ell,n}$ are the type-1 Fourier–Bessel eigenfunctions defined by equation (14C.5), p. 301. Then $\{\mathcal{E}_{\ell,n}^1; \ell \in \mathbb{N} \text{ and } n \in \mathbb{Z}\}$ is a Dirichlet eigenbasis for $\mathbb{D}$, by Theorem 14C.2, p. 297.

As in Example 15C.9, $\partial[0, \pi] = \{0, \pi\}$. Let $\mathbb{M}_2 := \{0, 1\}$ and let $\Xi_1^2 : \{0, \pi\} \longrightarrow \mathbb{R}$ and $\Xi_2^2 : \{0, \pi\} \longrightarrow \mathbb{R}$ be as in Example 15C.9. Let $\{\kappa_{\ell,n}\}_{n=1}^{\infty}$ be the roots of the Bessel function $\mathcal{J}_n$, as described in equation (14C.3), p. 301. For every $(\ell, n) \in \mathbb{N} \times \mathbb{Z}$, define $\mathcal{F}_{\ell,n;1}^2$ and $\mathcal{F}_{\ell,n;2}^2 \in \mathbf{L}^2[0, \pi]$ by

$$\mathcal{F}_{\ell,n;1}^2(z) := \frac{\sinh(\kappa_{\ell,|n|} \cdot z)}{\sinh(\kappa_{\ell,|n|} \pi)} \quad \text{and} \quad \mathcal{F}_{\ell,n;2}^2(z) := \frac{\sinh(\kappa_{\ell,|n|} \cdot (\pi - z))}{\sinh(\kappa_{\ell,|n|} \pi)},$$

for all $z \in [0, \pi]$. Then, clearly, $\widetilde{\mathcal{F}}_{\ell,n;1}^2 = \Xi_1^2$ and $\widetilde{\mathcal{F}}_{\ell,n;2}^2 = \Xi_2^2$.

For each $(\ell, n) \in \mathbb{N} \times \mathbb{Z}$, we have eigenvalue $-\kappa_{\ell,|n|}^2$ by equation (14C.4), p. 301. That is, $\triangle\Phi_{\ell,n}(r, \theta) = -\kappa_{\ell,n}^2 \Phi_{\ell,n}(r, \theta)$ and $\triangle\Psi_{\ell,n}(r, \theta) = -\kappa_{\ell,n}^2 \Psi_{\ell,n}(r, \theta)$; thus, $\triangle\mathcal{E}_{\ell,n}^1(z) = -\kappa_{\ell,|n|}^2\mathcal{E}_{\ell,n}^1(z)$ for all $(\ell, n) \in \mathbb{N} \times \mathbb{Z}$. Meanwhile, $\triangle\mathcal{F}_{\ell,n;m}^1(z) = \kappa_{\ell,|n|}^2\mathcal{F}_{\ell,n;m}^1(z)$, for all $(\ell, n; m) \in \mathbb{N} \times \mathbb{Z} \times \{1, 2\}$. Thus, the functions

$$\mathcal{H}^1_{\ell,n,1}(r, \theta, z) := \mathcal{E}^1_{\ell,n}(r, \theta) \cdot \mathcal{F}^2_{\ell,n;1}(z) = \frac{\Phi_{\ell,n}(r, \theta) \sinh(\kappa_{\ell,n} \cdot z)}{\sinh(\kappa_{\ell,n} \pi)};$$

$$\mathcal{H}^1_{\ell,n,2}(r, \theta, z) := \mathcal{E}^1_{\ell,n}(r, \theta) \cdot \mathcal{F}^2_{\ell,n;2}(z) = \frac{\Phi_{\ell,n}(r, \theta) \sinh\left(\kappa_{\ell,n} \cdot (\pi - z)\right)}{\sinh(\kappa_{\ell,n} \pi)};$$

$$\mathcal{H}^1_{\ell,-n,1}(r, \theta, z) := \mathcal{E}^1_{\ell,-n}(r, \theta) \cdot \mathcal{F}^2_{\ell,-n;1}(z) = \frac{\Psi_{\ell,n}(r, \theta) \sinh(\kappa_{\ell,n} \cdot z)}{\sinh(\kappa_{\ell,n} \pi)};$$

$$\mathcal{H}^1_{\ell,-n,2}(r, \theta, z) := \mathcal{E}^1_{\ell,-n}(r, \theta) \cdot \mathcal{F}^2_{\ell,-n;2}(z) = \frac{\Psi_{\ell,n}(r, \theta) \sinh\left(\kappa_{\ell,n} \cdot (\pi - z)\right)}{\sinh(\kappa_{\ell,n} \pi)};$$

are all harmonic, by Lemma 15C.6.

Recall that $\partial \mathbb{D} = \mathbb{S}$. Let $\mathbb{M}_1 := \mathbb{Z}$, and, for all $m \in \mathbb{Z}$, define $\Xi^1_m \in L^2(\mathbb{S})$ by $\Xi^1_m(1, \theta) := \sin(m\theta)$ (if $m > 0$) and $\Xi^1_m(1, \theta) := \cos(m\theta)$ (if $m \leq 0$), for all $\theta \in [-\pi, \pi]$; then $\{\Xi^1_m\}_{m \in \mathbb{Z}}$ is an orthogonal basis for $L^2(\mathbb{S})$, by Theorem 8A.1, p. 166. For all $n \in \mathbb{N}$ and $z \in [0, \pi]$, define $\mathcal{E}^2_n(z) := \sin(nz)$ as in Example 15C.9. Then $\{\mathcal{E}^2_n\}_{n=1}^{\infty}$ is a Dirichlet eigenbasis for $[0, \pi]$, by Theorem 7A.1, p. 142. For all $n \in \mathbb{N}$, the eigenfunction $\mathcal{E}^2_n$ has eigenvalue $\lambda^{(2)}_n := -n^2$. For all $m \in \mathbb{Z}$, let $\mathcal{F}^1_{n,m} : \mathbb{D} \longrightarrow \mathbb{R}$ be an eigenfunction of the Laplacian with eigenvalue n^2, and with boundary condition $\mathcal{F}^1_{n,m}(1, \theta) = \Xi^1_m(\theta)$ for all $\theta \in [-\pi, \pi]$ (see Exercise 15C.9(a) below). The function $\mathcal{H}^2_{n,m}(r, \theta, z) := \mathcal{F}^1_{n,m}(r, \theta) \cdot \mathcal{E}^2_n(z)$ is harmonic, by Lemma 15C.6. Thus, Corollary 15C.8 says that the collection

$$\left\{\mathcal{H}^1_{\ell,n,m} ; \ \ell \in \mathbb{N}, n \in \mathbb{Z}, m = 1, 2\right\} \sqcup \left\{\mathcal{H}^2_{n,m} ; \ n \in \mathbb{N}, m \in \mathbb{Z}\right\}$$

is a Dirichlet harmonic basis for the cylinder $\mathbb{X}$. $\diamond$

Exercise 15C.9 ⓔ

(a) Example 15C.10 posits the existence of eigenfunctions $\mathcal{F}^1_{n,m} : \mathbb{D} \longrightarrow \mathbb{R}$ of the Laplacian with eigenvalue n^2 and with boundary condition $\mathcal{F}^1_{n,m}(1, \theta) = \Xi^1_m(\theta)$ for all $\theta \in [-\pi, \pi]$. Assume $\mathcal{F}^1_{n,m}$ *separates* in polar coordinates – that is, $\mathcal{F}^1_{n,m}(r, \theta) = \mathcal{R}(r) \cdot \Xi_m(\theta)$, where $\mathcal{R} : [0, 1] \longrightarrow \mathbb{R}$ is some unknown function with $\mathcal{R}(1) = 1$. Show that $\mathcal{R}$ must satisfy the ordinary differential equation $r^2 \mathcal{R}''(r) + r\mathcal{R}'(r) - (r^2 + 1)n^2 \mathcal{R}(r) = 0$. Use the *Method of Frobenius* (Appendix H(iii)) to solve this ODE and get an expression for $\mathcal{F}^1_{n,m}$.

(b) Combine Theorem 15B.3 with Example 15C.10 to obtain a general solution to the Laplace Equation on a finite cylinder with nonhomogeneous Dirichlet boundary conditions.

(c) Set up and solve a few simple Dirichlet problems using your method.   ◆

Exercise 15C.10 ⓔ   Let $\mathbb{X} = \{(r, \theta, z); \ 1 \leq r \text{ and } 0 \leq z \leq \pi\}$ be the *punctured slab* from Exercise 15C.4.

(a) Use Corollary 15C.8 to obtain a Dirichlet harmonic basis for $\mathbb{X}$.
(b) Apply Theorem 15B.3 to obtain a general solution to the Laplace Equation on the punctured slab with nonhomogeneous Dirichlet boundary conditions.
(c) Set up and solve a few simple Dirichlet problems using your method. ♦

Exercise 15C.11 Let $\mathbb{X}$ be the *right-angle triangular prism* from Exercise 15C.5. Ⓔ

(a) Use Proposition 15A.5 and Corollary 15C.8 to obtain a Dirichlet harmonic basis for $\mathbb{X}$.
(b) Apply Theorem 15B.3 to obtain a general solution to the Laplace Equation on the prism with nonhomogeneous Dirichlet boundary conditions.
(c) Set up and solve a few simple Dirichlet problems using your method. ♦

Exercise 15C.12 State and prove a theorem analogous to Corollary 15C.8 for Ⓔ
Neumann harmonic bases. ♦

15D General method for solving I/BVPs

Prerequisites: §15A, §15B.   **Recommended:** §15C.

We now provide a general method for solving initial/boundary value problems. Throughout this section, let $\mathbb{X} \subset \mathbb{R}^D$ be a domain. Let L be a linear differential operator on $\mathbb{X}$ (e.g. $\mathsf{L} = \triangle$).

(1) Pick a suitable coordinate system Find the coordinate system in which your problem can be expressed in its simplest form. Generally, this is a coordinate system where the domain $\mathbb{X}$ can be described using a few simple inequalities. For example, if $\mathbb{X} = [0, L]^D$, then probably the Cartesian coordinate system is best. If $\mathbb{X} = \mathbb{D}$ or $\mathbb{D}^{\complement}$ or $\mathbb{A}$, then probably polar coordinates on $\mathbb{R}^2$ are the most suitable. If $\mathbb{X} = \mathbb{B}$ or $\mathbb{X} = \partial\mathbb{B}$, then probably spherical polar coordinates on $\mathbb{R}^3$ are best.

If the differential operator L has nonconstant coefficients, then you should also seek a coordinate system where these coefficients can be expressed using the simplest formulae. (If $\mathsf{L} = \triangle$, then it has constant coefficients, so this is not an issue.)

Finally, if several coordinate systems are equally suitable for describing $\mathbb{X}$ and L, then find the coordinate system where the initial conditions and/or boundary conditions can be expressed most easily. For example, if $\mathbb{X} = \mathbb{R}^2$ and $\mathsf{L} = \triangle$, then either Cartesian or polar coordinates might be appropriate. However, if the initial conditions are rotationally symmetric around zero, then polar coordinates would be more appropriate. If the initial conditions are invariant under translation in some direction, then Cartesian coordinates would be more appropriate.

Note. Don't forget to find the correct expression for L in the new coordinate system. For example, in Cartesian coordinates on $\mathbb{R}^2$, we have $\triangle u(x, y) = \partial_y^2 u(x, y) + \partial_y^2 u(x, y)$. However, in polar coordinates, $\triangle u(r, \theta) = \partial_r^2 u(r, \theta) + \frac{1}{r} \partial_r u(r, \theta) + \frac{1}{r^2} \partial_\theta^2 u(r, \theta)$. If you apply the 'Cartesian' Laplacian to a function expressed in polar coordinates, the result will be nonsense.

(2) Eliminate irrelevant coordinates A coordinate x is 'irrelevant' if:

(a) membership in the domain $\mathbb{X}$ does not depend on this coordinate; *and*
(b) the coefficients of L do not depend on this coordinate; *and*
(c) the initial and/or boundary conditions do not depend on this coordinate.

In this case, we can eliminate the x coordinate from all equations by expressing the domain $\mathbb{X}$, the operator L, and the initial/boundary conditions as functions only of the non-x coordinates. This reduces the dimension of the problem, thereby simplifying it.

To illustrate (a), suppose $\mathbb{X} = \mathbb{D}$ or $\mathbb{D}^{\complement}$ or $\mathbb{A}$, and we use the polar coordinate system (r, θ); then the angle coordinate θ is irrelevant to membership in $\mathbb{X}$. On the other hand, suppose $\mathbb{X} = \mathbb{R}^2 \times [0, L]$ is the 'slab' of thickness L in $\mathbb{R}^3$, and we use Cartesian coordinates (x, y, z). Then the coordinates x and y are irrelevant to membership in $\mathbb{X}$.

If $\mathsf{L} = \triangle$ or any other differential operator with constant coefficients, then (b) is automatically satisfied.

To illustrate (c), suppose $\mathbb{X} = \mathbb{D}$ and we use polar coordinates. Let $f : \mathbb{D} \longrightarrow \mathbb{R}$ be some initial condition. If $f(r, \theta)$ is a function only of r, and doesn't depend on θ, then θ is a redundant coordinate and can be eliminated, thereby reducing the BVP to a one-dimensional problem, as in Example 14F.3, p. 312.

On the other hand, let $b : \mathbb{S} \longrightarrow \mathbb{R}$ be a boundary condition. Then θ is only irrelevant if b is a constant function (otherwise b has nontrivial dependence on θ).

Now suppose that $\mathbb{X} = \mathbb{R}^2 \times [0, L]$ is the 'slab' of thickness L in $\mathbb{R}^3$. If the boundary condition $b : \partial\mathbb{X} \longrightarrow \mathbb{R}$ is constant on the top and bottom faces of the slab, then the x and y coordinates can be eliminated, thereby reducing the BVP to a one-dimensional problem: a BVP on the line segment $[0, L]$, which can be solved using the methods of Chapter 11.

In some cases, a certain coordinate can be eliminated if it is 'approximately' irrelevant. For example, if the domain $\mathbb{X}$ is particularly 'long' in the x dimension relative to its other dimensions, and the boundary conditions are roughly constant in the x dimension, then we can approximate 'long' with 'infinite' and 'roughly constant' with 'exactly constant', and eliminate the x dimension from the problem. This method was used in Example 12B.2, p. 252 (the 'quenched rod'), Example 12B.7, p. 257 (the 'baguette'), and Example 12C.2, p. 260 (the 'nuclear fuel rod').

(3) Find an eigenbasis for $L^2(\mathbb{X})$ If $\mathbb{X}$ is one of the 'standard' domains we have studied in this book, then use the eigenbases we have introduced in Chapters 7–9, Section 14C, or Section 15C. Otherwise, you must construct a suitable eigenbasis. Theorem 15E.12, p. 353, guarantees that such an eigenbasis exists, but it does not tell you how to construct it. The actual construction of eigenbases is usually achieved using the *separation of variables*, discussed in Chapter 16. The separation of the 'time' variable is really just a consequence of the fact that we have an eigenfunction. The separation of the 'space' variables is not *necessary* to get an eigenfunction, but it is very *convenient*, for the following two reasons.

(i) Separation of variables is a powerful strategy for finding the eigenfunctions; it reduces the problem to a set of independent ODEs which can each be solved using classical ODE methods.

(ii) If an eigenfunction $\mathcal{E}_n$ appears in 'separated' form, then it is often easier to compute the inner product $\langle \mathcal{E}_n, f \rangle$, where f is some other function. This is important when the eigenfunctions form an orthogonal basis and we want to compute the coefficients of f in this basis.

(4) Find a harmonic basis for $L^2(\partial\mathbb{X})$ (if there are nonhomogeneous boundary conditions). The same remarks apply as in Step (3).

(5) Solve the problem Express any initial conditions in terms of the eigenbasis from Step (3), as described in §15A. Express any boundary conditions in terms of the harmonic basis from Step (4), as described in §15B.

If $\mathsf{L} = \triangle$, then use Theorems 15A.2, 15A.3, 15A.4, and/or 15B.3. If L is some other linear differential operator, then use the appropriate analogues of these theorems (see Remark 15A.6, p. 331).

(6) Verify convergence Note that Theorems 15A.2, 15A.3, 15A.4, and/or 15B.3 require the eigenvalue sequences $\{\lambda_n\}_{n=1}^\infty$ and/or $\{\mu_n\}_{n=1}^\infty$ to grow at a certain speed, or require the coefficient sequences $\{A_n\}_{n=0}^\infty$ and $\{B_n\}_{n=1}^\infty$ to decay at a certain speed, so as to guarantee that the solution series and its formal derivatives are absolutely convergent. These conditions are important and must be checked. Typically, if $\mathsf{L} = \triangle$, the growth conditions on $\{\lambda_n\}_{n=1}^\infty$ and $\{\mu_n\}_{n=1}^\infty$ are easily satisfied. However, if you try to extend these theorems to some other linear differential operator, the conditions on $\{\lambda_n\}_{n=1}^\infty$ and $\{\mu_n\}_{n=1}^\infty$ must be checked.

(7) Check the uniqueness of the solution Section 5D, p. 87, describes conditions under which boundary value problems for the Poisson, Laplace, heat, and wave equations will have a *unique* solution. Check that these conditions are satisfied. If $L \neq \triangle$, then you will need to establish solution uniqueness using theorems analogous to those found in Section 5D. (General theorems for the existence/uniqueness of solutions to I/BVPs can be found in most advanced texts on PDE theory, such as Evans (1991).)

If the solution is *not* unique, then it is important to *enumerate all solutions to the problem*. Remember that your ultimate goal here is to *predict* the behaviour of some physical system in response to some initial or boundary condition. If the solution to the I/BVP is not unique, then you cannot make a precise prediction; instead, your prediction must take the form of a precisely specified *range* of possible outcomes.

15E Eigenfunctions of self-adjoint operators

Prerequisites: §4B(iv), §5C, §6F.   **Recommended:** §7A, §8A, §9B, §15A.

The solution methods of Section 15A are only relevant if we know that a suitable eigenbasis for the Laplacian exists on the domain of interest. If we want to develop similar methods for some other linear differential operator L (as described in Remark 15A.6, p. 331), then we must first know that suitable eigenbases exist for L. In this section, we will discuss the eigenfunctions and eigenvalues of an important class of linear operators: *self-adjoint* operators. This class includes the Laplacian and all other symmetric elliptic differential operators.

15E(i) Self-adjoint operators

A linear operator $F : \mathbb{R}^D \longrightarrow \mathbb{R}^D$ is *self-adjoint* if, for any vectors $\mathbf{x}, \mathbf{y} \in \mathbb{R}^D$,

$$\langle F(\mathbf{x}), \mathbf{y} \rangle = \langle \mathbf{x}, F(\mathbf{y}) \rangle.$$

Example 15E.1 The matrix $\begin{bmatrix} 1 & -2 \\ -2 & 1 \end{bmatrix}$ defines a self-adjoint operator on $\mathbb{R}^2$, because for any $\mathbf{x} = \begin{bmatrix} x_1 \\ x_2 \end{bmatrix}$ and $\mathbf{y} = \begin{bmatrix} y_1 \\ y_2 \end{bmatrix}$ in $\mathbb{R}^2$, we have

$$\langle F(\mathbf{x}), \mathbf{y} \rangle = \left\langle \begin{bmatrix} x_1 - 2x_2 \\ x_2 - 2x_1 \end{bmatrix}, \begin{bmatrix} y_1 \\ y_2 \end{bmatrix} \right\rangle = y_1 \left(x_1 - 2x_2 \right) + y_2 \left(x_2 - 2x_1 \right)$$
$$= x_1 \left(y_1 - 2y_2 \right) + x_2 \left(y_2 - 2y_1 \right) = \left\langle \begin{bmatrix} x_1 \\ x_2 \end{bmatrix}, \begin{bmatrix} y_1 - 2y_2 \\ y_2 - 2y_1 \end{bmatrix} \right\rangle$$
$$= \langle \mathbf{x}, F(\mathbf{y}) \rangle. \qquad \qquad \diamond$$

Theorem 15E.2 *Let $F : \mathbb{R}^D \longrightarrow \mathbb{R}^D$ be a linear operator with matrix $\mathbf{A}$. Then F is self-adjoint if, and only if, $\mathbf{A}$ is symmetric (i.e. $a_{ij} = a_{ji}$ for all j, i)*

Proof **Exercise 15E.1.**   □   Ⓔ

A linear operator $\mathsf{L} : \mathcal{C}^\infty \longrightarrow \mathcal{C}^\infty$ is *self-adjoint* if, for any two functions $f, g \in \mathcal{C}^\infty$,

$$\langle \mathsf{L}[f], g \rangle = \langle f, \mathsf{L}[g] \rangle$$

whenever both sides are well defined.[3]

Example 15E.3 Multiplication operators are self-adjoint
Let $\mathbb{X} \subset \mathbb{R}^D$ be any bounded domain. Let $\mathcal{C}^\infty := \mathcal{C}^\infty(\mathbb{X}; \mathbb{R})$. Fix $q \in \mathcal{C}^\infty(\mathbb{X})$, and define the operator $\mathsf{Q} : \mathcal{C}^\infty \longrightarrow \mathcal{C}^\infty$ by $\mathsf{Q}(f) := q \cdot f$ for any $f \in \mathcal{C}^\infty$. Then Q is self-adjoint. To see this, let $f, g \in \mathcal{C}^\infty$. Then

$$\langle q \cdot f, g \rangle = \int_{\mathbb{X}} (q \cdot f) \cdot g \, dx = \int_{\mathbb{X}} f \cdot (q \cdot g) dx = \langle f, q \cdot g \rangle .$$

(These integrals are all well defined because q, f, and g are all continuous and hence bounded on $\mathbb{X}$.)   ◇

Let $L > 0$, and consider the interval $[0, L]$. Recall that $\mathcal{C}^\infty[0, L]$ is the set of all smooth functions from $[0, L]$ into $\mathbb{R}$, and that:

$\mathcal{C}_0^\infty[0, L]$ is the space of all $f \in \mathcal{C}^\infty[0, L]$ satisfying homogeneous *Dirichlet* boundary conditions: $f(0) = 0 = f(L)$ (see §5C(i));

$\mathcal{C}_\perp^\infty[0, L]$ is the space of all $f \in \mathcal{C}^\infty[0, L]$ satisfying $f : [0, L] \longrightarrow \mathbb{R}$ satisfying homogeneous *Neumann* boundary conditions: $f'(0) = 0 = f'(L)$ (see §5C(ii));

$\mathcal{C}_{per}^\infty[0, L]$ is the space of all $f \in \mathcal{C}^\infty[0, L]$ satisfying $f : [0, L] \longrightarrow \mathbb{R}$ satisfying *periodic* boundary conditions: $f(0) = f(L)$ and $f'(0) = f'(L)$ (see §5C(iv));

$\mathcal{C}_{h,h_\perp}^\infty[0, L]$ is the space of all $f \in \mathcal{C}^\infty[0, L]$ satisfying homogeneous *mixed* boundary conditions, for any fixed real numbers $h(0)$, $h_\perp(0)$, $h(L)$, and $h_\perp(L)$ (see §5C(iii)).

When restricted to these function spaces, the one-dimensional Laplacian operator ∂_x^2 is self-adjoint.

Proposition 15E.4 *Let $L > 0$, and consider the operator ∂_x^2 on $\mathcal{C}^\infty[0, L]$.*

(a) *∂_x^2 is self-adjoint when restricted to $\mathcal{C}_0^\infty[0, L]$.*
(b) *∂_x^2 is self-adjoint when restricted to $\mathcal{C}_\perp^\infty[0, L]$.*
(c) *∂_x^2 is self-adjoint when restricted to $\mathcal{C}_{per}^\infty[0, L]$.*
(d) *∂_x^2 is self-adjoint when restricted to $\mathcal{C}_{h,h_\perp}^\infty[0, L]$, for any $h(0)$, $h_\perp(0)$, $h(L)$, and $h_\perp(L)$ in $\mathbb{R}$.*

---

[3] This is an important point. Often, one of these inner products (say, the left one) will *not* be well defined, because the integral $\int_{\mathbb{X}} \mathsf{L}[f] \cdot g \, dx$ does not converge, in which case 'self-adjointness' is meaningless.

Proof Let $f, g : [0, L] \longrightarrow \mathbb{R}$ be smooth functions. We apply integration by parts to get

$$\langle \partial_x^2 f, g \rangle = \int_0^L f''(x) \cdot g(x) \mathrm{d}x = f'(x) \cdot g(x) \Big|_{x=0}^{x=L} - \int_0^L f'(x) \cdot g'(x) \mathrm{d}x.$$

$$(15\mathrm{E}.1)$$

But if we apply Dirichlet, Neumann, or periodic boundary conditions, we get:

$$f'(x) \cdot g(x) \Big|_{x=0}^{x=L} = f'(L) \cdot g(L) - f'(0) \cdot g(0)$$

$$= \begin{cases} f'(L) \cdot 0 - f'(0) \cdot 0 & = 0 \quad \text{(if homogeneous Dirichlet BC);} \\ 0 \cdot g(L) - 0 \cdot g(0) & = 0 \quad \text{(if homogeneous Neumann BC);} \\ f'(0) \cdot g(0) - f'(0) \cdot g(0) = 0 & \text{(if periodic BC);} \end{cases}$$

$$= 0 \quad \text{in all cases.}$$

Thus, the first term in equation (15E.1) is zero, so

$$\langle \partial_x^2 f, g \rangle = \int_0^L f'(x) \cdot g'(x) \mathrm{d}x.$$

By the same reasoning, with f and g interchanged,

$$\int_0^L f'(x) \cdot g'(x) \mathrm{d}x = \langle f, \partial_x^2 g \rangle.$$

Thus, we have proved parts (a), (b), and (c). To prove part (d), first note that

$$f'(x) \cdot g(x) \Big|_{x=0}^{x=L} = f'(L) \cdot g(L) - f'(0) \cdot g(0)$$

$$= f(L) \cdot \frac{h(L)}{h_\perp(L)} \cdot g(L) + f(0) \cdot \frac{h(0)}{h_\perp(0)} \cdot g(0)$$

$$= f(L) \cdot g'(L) - f(0) \cdot g'(0) = f(x) \cdot g'(x) \Big|_{x=0}^{x=L}.$$

Hence, substituting $f(x) \cdot g'(x)|_{x=0}^{x=L}$ for $f'(x) \cdot g(x)|_{x=0}^{x=L}$ in equation (15E.1) yields

$$\langle \partial_x^2 f, g \rangle = \int_0^L f''(x) \cdot g(x) \mathrm{d}x = \int_0^L f(x) \cdot g''(x) \mathrm{d}x = \langle f, \partial_x^2 g \rangle. \qquad \square$$

Proposition 15E.4 generalizes to higher-dimensional Laplacians in the obvious way as follows.

Theorem 15E.5 *Let $L > 0$.*

(a) *The Laplacian operator $\triangle$ is self-adjoint on any of the spaces:* $C_0^\infty[0, L]^D$, $C_\perp^\infty[0, L]^D$, $C_{h,h_\perp}^\infty [0, L]^D$, *or* $C_{\mathrm{per}}^\infty[0, L]^D$.

(b) *More generally, if* $\mathbb{X} \subset \mathbb{R}^D$ *is any bounded domain with a smooth boundary,[4] then the Laplacian operator* $\triangle$ *is self-adjoint on any of the spaces:* $C_0^{\infty}(\mathbb{X})$, $C_{\perp}^{\infty}(\mathbb{X})$, *or* $C_{h,h_{\perp}}^{\infty}(\mathbb{X})$.

In other words, the Laplacian is self-adjoint whenever we impose homogeneous Dirichlet, Neumann, or mixed boundary conditions, or (when meaningful) periodic boundary conditions.

Proof

(a) **Exercise 15E.2.** *Hint:* The argument is similar to Proposition 15E.4. Apply integra- Ⓔ
tion by parts in each dimension and cancel the 'boundary' terms using the boundary conditions.

(b) **Exercise 15E.3.** *Hint:* Use Green's Formulae (Corollary E.5(c), p. 567) to set up an Ⓔ
'integration by parts' argument similar to that in Proposition 15E.4. □

If L_1 and L_2 are two self-adjoint operators, then their sum $L_1 + L_2$ is also self-adjoint (**Exercise 15E.4**). Ⓔ

Example 15E.6 Let $\mathbb{X} \subset \mathbb{R}^D$ be some domain (e.g. a cube), and let $V : \mathbb{X} \longrightarrow \mathbb{R}$ be a potential describing the force acting on a quantum particle (e.g. an electron) confined to the region $\mathbb{X}$ by an infinite potential barrier along $\partial\mathbb{X}$. Consider the *Hamiltonian* operator H defined in §3B, p. 43:

$$H\omega(\mathbf{x}) = \frac{-\hbar^2}{2m} \triangle \omega(\mathbf{x}) + V(\mathbf{x}) \cdot \omega(\mathbf{x}), \quad \text{for all } \mathbf{x} \in \mathbb{X}.$$

(Here, $\hbar$ is Planck's constant, m is the mass of the particle, and $\omega \in C_0^{\infty}\mathbb{X}$ is its wavefunction.) The operator H is self-adjoint on $C_0^{\infty}(\mathbb{X})$. To see this, note that $H[\omega] = \frac{-\hbar^2}{2m} \triangle \omega + V[\omega]$, where $V[\omega] = V \cdot \omega$. Now, $\triangle$ is self-adjoint by Theorem 15E.5(b), and V is self-adjoint from Example 15E.3; thus, their sum H is also self-adjoint. The *stationary Schrödinger Equation*[5] $H\omega = \lambda\omega$ simply says that ω is an eigenfunction of H with eigenvalue λ. ◇

Example 15E.7 Let $s, q : [0, L] \longrightarrow \mathbb{R}$ be differentiable. The *Sturm–Liouville operator*

$$\mathcal{S}_{s,q}[f] := s \cdot f'' + s' \cdot f' + q \cdot f$$

is self-adjoint on any of the spaces $C_0^{\infty}[0, L]$, $C_{\perp}^{\infty}[0, L]$, $C_{h,h_{\perp}}^{\infty}[0, L]$, *or* $C_{per}^{\infty}[0, L]$. To see this, note that

$$\mathcal{S}_{s,q}[f] = (s \cdot f')' + (q \cdot f) = S[f] + Q[f], \qquad (15E.2)$$

where $Q[f] = q \cdot f$ is just a multiplication operator, and $S[f] = (s \cdot f')'$. We know that Q is self-adjoint from Example 15E.3. We claim that S is also self-adjoint. To

---

[4] See p. 88 of §5D.   [5] See §3C.

see this, note that

$$\langle S[f], g \rangle = \int_0^L (s \cdot f')'(x) \cdot g(x) dx$$

$$\underset{(*)}{=} s(x) \cdot f'(x) \cdot g(x) \Big|_{x=0}^{x=L} - \int_0^L s(x) \cdot f'(x) \cdot g'(x) dx$$

$$\underset{(*)}{=} s(x) \cdot f'(x) \cdot g(x) \Big|_{x=0}^{x=L} - s(x) \cdot f(x) \cdot g'(x) \Big|_{x=0}^{x=L}$$

$$+ \int_0^L f(x) \cdot (s \cdot g')'(x) dx$$

$$\underset{(\dagger)}{=} \int_0^L f(x) \cdot (s \cdot g')'(x) dx = \langle f, S[g] \rangle.$$

Here, each (∗) is integration by parts, and (†) follows from any of the cited boundary conditions as in Proposition 15E.4, p. 347 (**Exercise 15E.5**). Thus, S is self-adjoint, so $\mathcal{L}_{s,q} = S + Q$ is self-adjoint.   ◇

If $\mathcal{L}_{s,q}$ is a Sturm–Liouville operator, then the corresponding *Sturm–Liouville equation* is the linear ordinary differential equation

$$\mathcal{L}_{s,q}[f] = \lambda \, f, \tag{15E.3}$$

where $f : [0, L] \longrightarrow \mathbb{C}$ and $\lambda \in \mathbb{C}$ are unknown. Clearly, equation (15E.3) simply asserts that f is an eigenfunction of $\mathcal{L}_{s,q}$, with eigenvalue λ. Sturm–Liouville equations appear frequently in the study of ordinary and partial differential equations.

Example 15E.8
(a) The one-dimensional *Helmholtz equation* $f''(x) = \lambda \, f(x)$ is a Sturm–Liouville equation, with $s \equiv 1$ (constant) and $q \equiv 0$.
(b) The one-dimensional *stationary Schrödinger equation*

$$\frac{-\hbar^2}{2m} f''(x) + V(x) \cdot f(x) = \lambda \, f(x), \quad \text{for all } x \in [0, L],$$

  is a Sturm–Liouville equation, with $s \equiv \frac{-\hbar^2}{2m}$ (constant) and $q(x) := V(x)$.
(c) The *Cauchy–Euler equation*[6] $x^2 f''(x) + 2x \, f'(x) - \lambda \cdot f(x) = 0$ is a Sturm–Liouville equation: let $s(x) := x^2$ and $q \equiv 0$.
(d) The *Legendre equation*[7] $(1 - x^2) f''(x) - 2x \, f'(x) + \mu f(x) = 0$ is a Sturm–Liouville equation: let $s(x) := (1 - x^2)$, $q \equiv 0$, and let $\lambda := -\mu$.

---

[6] See equation (14H.2), p. 322, and equation (16D.20), p. 366.   [7] See equation (16D.19), p. 366.

For more information about Sturm–Liouville problems, see Broman (1989), §2.6, Conway (1990), §II.6, Powers (1999), §2.7, and especially Churchill and Brown (1987), chap. 6.   ◇

15E(ii) Eigenfunctions and eigenbases

Examples 15E.6 and 15E.8, Theorem 15E.5, and the solution methods of §15A all illustrate the importance of the eigenfunctions of self-adjoint operators. One nice property of self-adjoint operators is that their eigenfunctions are orthogonal.

Proposition 15E.9 *Suppose* L *is a self-adjoint operator. If* f_1 *and* f_2 *are eigenfunctions of* L *with eigenvalues* $\lambda_1 \neq \lambda_2$, *then* f_1 *and* f_2 *are* orthogonal.

Proof By hypothesis, $L[f_k] = \lambda_k \cdot f_k$, for $k = 1, 2$. Thus,

$$\lambda_1 \cdot \langle f_1, f_2 \rangle = \langle \lambda_1 \cdot f_1, f_2 \rangle = \langle L[f_1], f_2 \rangle \underset{(*)}{=} \langle f_1, L[f_2] \rangle = \langle f_1, \lambda_2 \cdot f_2 \rangle$$
$$= \lambda_2 \cdot \langle f_1, f_2 \rangle,$$

where $(*)$ follows from self-adjointness. Since $\lambda_1 \neq \lambda_2$, this can only happen if $\langle f_1, f_2 \rangle = 0$.   □

Example 15E.10 Eigenfunctions of ∂_x^2

(a) Let ∂_x^2 act on $\mathcal{C}^\infty[0, L]$. Then *all real numbers* $\lambda \in \mathbb{R}$ are eigenvalues of ∂_x^2. For any $\mu \in \mathbb{R}$:

  - if $\lambda = \mu^2 > 0$, the eigenfunctions are of the form $\phi(x) = A \sinh(\mu \cdot x) + B \cosh(\mu \cdot x)$ for any constants $A, B \in \mathbb{R}$;
  - if $\lambda = 0$, the eigenfunctions are of the form $\phi(x) = Ax + B$ for any constants $A, B \in \mathbb{R}$;
  - if $\lambda = -\mu^2 < 0$, the eigenfunctions are of the form $\phi(x) = A \sin(\mu \cdot x) + B \cos(\mu \cdot x)$ for any constants $A, B \in \mathbb{R}$.

  Note: Because we have not imposed any boundary conditions, Proposition 15E.4 does *not* apply; indeed ∂_x^2 is *not* a self-adjoint operator on $\mathcal{C}^\infty[0, L]$.

(b) Let ∂_x^2 act on $\mathcal{C}^\infty([0, L]; \mathbb{C})$. Then *all complex numbers* $\lambda \in \mathbb{C}$ are eigenvalues of ∂_x^2. For any $\mu \in \mathbb{C}$, with $\lambda = \mu^2$, the eigenvalue λ has eigenfunctions of the form $\phi(x) = A \exp(\mu \cdot x) + B \exp(-\mu \cdot x)$ for any constants $A, B \in \mathbb{C}$. (Note that the three cases of the previous example arise by taking $\lambda \in \mathbb{R}$.) Again, Proposition 15E.4 does *not* apply in this case, because ∂_x^2 is *not* a self-adjoint operator on $\mathcal{C}^\infty([0, L]; \mathbb{C})$.

(c) Now let ∂_x^2 act on $\mathcal{C}_0^\infty[0, L]$. Then the eigenvalues of ∂_x^2 are $\lambda_n = -(n\pi/L)^2$ for every $n \in \mathbb{N}$, each of multiplicity 1; the corresponding eigenfunctions are all scalar multiples of $\mathbf{S}_n(x) := \sin(n\pi x/L)$.

(d) If ∂_x^2 acts on $\mathcal{C}_\perp^\infty[0, L]$, then the eigenvalues of ∂_x^2 are again $\lambda_n = -(n\pi/L)^2$ for every $n \in \mathbb{N}$, each of multiplicity 1, but the corresponding eigenfunctions are now all scalar multiples of $C_n(x) := \cos(n\pi x/L)$. Also, 0 is an eigenvalue, with eigenfunction $C_0 = 1$.

(e) Let $h > 0$, and let ∂_x^2 act on $C = \{f \in C^\infty[0, L]; f(0) = 0 \text{ and } h \cdot f(L) + f'(L) = 0\}$. Then the eigenfunctions of ∂_x^2 are all scalar multiples of

$$\Phi_n(x) := \sin(\mu_n \cdot x),$$

with eigenvalue $\lambda_n = -\mu_n^2$, where $\mu_n > 0$ is any real number such that

$$\tan(L \cdot \mu_n) = \frac{-\mu_n}{h}.$$

This is a *transcendental equation* in the unknown μ_n. Thus, although there is an infinite sequence of solutions $\{\mu_0 < \mu_1 < \mu_2 < \cdots\}$, there is no closed-form algebraic expression for μ_n. At best, we can estimate μ_n through numerical methods.

(f) Let $h(0)$, $h_\perp(0)$, $h(L)$, and $h_\perp(L)$ be real numbers, and let ∂_x^2 act on $C_{h,h_\perp}^\infty[0, L]$. Then the eigenfunctions of ∂_x^2 are all scalar multiples of

$$\Phi_n(x) := \sin\left(\theta_n + \mu_n \cdot x\right),$$

with eigenvalue $\lambda_n = -\mu_n^2$, where $\theta_n \in [0, 2\pi]$ and $\mu_n > 0$ are constants satisfying the transcendental equations:

$$\tan(\theta_n) = \mu_n \cdot \frac{h_\perp(0)}{h(0)} \quad \text{and} \quad \tan(\mu_n \cdot L + \theta_n) = -\mu_n \cdot \frac{h_\perp(L)}{h(L)}.$$

(E) **(Exercise 15E.6)**. In particular, if $h_\perp(0) = 0$, then we must have $\theta = 0$. If $h(L) = h$ and $h_\perp(L) = 1$, then we return to part (e).

(g) Let ∂_x^2 act on $C_{per}^\infty[-L, L]$. Then the eigenvalues of ∂_x^2 are again $\lambda_n = -(n\pi/L)^2$, for every $n \in \mathbb{N}$, each having multiplicity 2. The corresponding eigenfunctions are of the form $A \cdot S_n + B \cdot C_n$ for any $A, B \in \mathbb{R}$. In particular, 0 is an eigenvalue, with eigenfunction $C_0 = 1$.

(h) Let ∂_x^2 act on $C_{per}^\infty([-L, L]; \mathbb{C})$. Then the eigenvalues of ∂_x^2 are again $\lambda_n = -(n\pi/L)^2$, for every $n \in \mathbb{N}$, each having multiplicity 2. The corresponding eigenfunctions are of the form $A \cdot E_n + B \cdot E_{-n}$ for any $A, B \in \mathbb{R}$, where $E_n(x) := \exp(\pi i n x/L)$. In particular 0 is an eigenvalue, with eigenfunction $E_0 = 1$. $\diamond$

Example 15E.11 Eigenfunctions of $\triangle$

(a) Let $\triangle$ act on $C_0^\infty[0, L]^D$. Then the eigenvalues of $\triangle$ are the numbers $\lambda_\mathbf{m} := -(\pi/L)^2 \cdot \|\mathbf{m}\|^2$ for all $\mathbf{m} \in \mathbb{N}_+^D$. (Here, if $\mathbf{m} = (m_1, \ldots, m_D)$, then $\|\mathbf{m}\|^2 := m_1^2 + \cdots + m_d^2$). The eigenspace of $\lambda_\mathbf{m}$ is spanned by all functions

$$\mathbf{S_n}(x_1, \ldots, x_D) := \sin\left(\frac{\pi n_1 x_1}{L}\right) \sin\left(\frac{\pi n_2 x_2}{L}\right) \cdots \sin\left(\frac{\pi n_D x_D}{L}\right),$$

for all $\mathbf{n} = (n_1, \ldots, n_D) \in \mathbb{N}_+^D$, such that $\|\mathbf{n}\| = \|\mathbf{m}\|$.

(b) Now let $\triangle$ act on $C_\perp^\infty[0, L]^D$. Then the eigenvalues of $\triangle$ are $\lambda_\mathbf{m}$ for all $\mathbf{m} \in \mathbb{N}^D$. The eigenspace of $\lambda_\mathbf{m}$ is spanned by all functions

$$\mathbf{C_n}(x_1, \ldots, x_D) := \cos\left(\frac{\pi n_1 x_1}{L}\right) \cos\left(\frac{\pi n_2 x_2}{L}\right) \cdots \cos\left(\frac{\pi n_D x_D}{L}\right),$$

for all $\mathbf{n} \in \mathbb{N}^D$, such that $\|\mathbf{n}\| = \|\mathbf{m}\|$. In particular, 0 is an eigenvalue whose eigenspace is the set of *constant* functions – i.e. multiples of $\mathbf{C_0} = \mathbb{1}$.

(c) Let $\triangle$ act on $C_{\text{per}}^{\infty}[-L, L]^D$. Then the eigenvalues of $\triangle$ are again $\lambda_{\mathbf{m}}$ for all $\mathbf{m} \in \mathbb{N}^D$. The eigenspace of $\lambda_{\mathbf{m}}$ contains $\mathbf{C_n}$ and $\mathbf{S_n}$ for all $\mathbf{n} \in \mathbb{N}^D$ such that $\|\mathbf{n}\| = \|\mathbf{m}\|$.

(d) Let $\triangle$ act on $C_{\text{per}}^{\infty}([-L, L]^D; \mathbb{C})$. Then the eigenvalues of $\triangle$ are again $\lambda_{\mathbf{m}}$ for all $\mathbf{m} \in \mathbb{N}^D$. The eigenspace of $\lambda_{\mathbf{m}}$ is spanned by all functions

$$\mathbf{E_n}(x_1, \ldots, x_D) := \exp\left(\frac{\pi \mathbf{i} n_1 x_1}{L}\right) \cdots \exp\left(\frac{\pi \mathbf{i} n_D x_D}{L}\right),$$

for all $\mathbf{n} \in \mathbb{Z}^D$ such that $\|\mathbf{n}\| = \|\mathbf{m}\|$. ◇

The alert reader will notice that, in each of the above scenarios (except Examples 15E.10(a) and 15E.10(b), where ∂_x^2 is not self-adjoint), the eigenfunctions are not only orthogonal, but also form an *orthogonal basis* for the corresponding L^2-space. This is not a coincidence. If C is a subspace of $\mathbf{L}^2(\mathbb{X})$, and $\mathsf{L} : C \longrightarrow C$ is a linear operator, then a set $\{\Phi_n\}_{n=1}^{\infty} \subset C$ is an L-*eigenbasis* for $\mathbf{L}^2(\mathbb{X})$ if $\{\Phi_n\}_{n=1}^{\infty}$ is an orthogonal basis for $\mathbf{L}^2(\mathbb{X})$, and, for every $n \in \mathbb{N}$, Φ_n is an eigenfunction for L.

Theorem 15E.12 Eigenbases of the Laplacian

(a) *Let $L > 0$. Let C be any one of $C_0^{\infty}[0, L]^D$, $C_{\perp}^{\infty}[0, L]^D$, or $C_{\text{per}}^{\infty}[0, L]^D$, and treat $\triangle$ as a linear operator on C. Then there is a $\triangle$-eigenbasis for $\mathbf{L}^2[0, L]^D$ consisting of elements of C. The corresponding eigenvalues of $\triangle$ are the values $\lambda_{\mathbf{m}}$ defined in Example 15E.11(a), for all $\mathbf{m} \in \mathbb{N}^D$.*

(b) *More generally, if $\mathbb{X} \subset \mathbb{R}^D$ is any bounded open domain, then there is a $\triangle$-eigenbasis for $\mathbf{L}^2[\mathbb{X}]$ consisting of elements of $C_0^{\infty}[\mathbb{X}]$. The corresponding eigenvalues of $\triangle$ on C form a decreasing sequence $0 > \lambda_1 \geq \lambda_2 \geq \lambda_3 \geq \cdots$ with $\lim_{n \to \infty} \lambda_n = -\infty$.*

In both (a) and (b), some of the eigenspaces may be many-dimensional.

Proof (a) we have already established. The eigenfunctions of the Laplacian in these contexts are $\{\mathbf{C_n}; \mathbf{n} \in \mathbb{N}^D\}$ and/or $\{\mathbf{S_n}; \mathbf{n} \in \mathbb{N}_+^D\}$. Theorem 8A.1(a), p. 166, and Theorem 9B.1(a), p. 189, tell us that these form orthogonal bases for $\mathbf{L}^2[0, L]^D$.

(b) follows from Theorem 15E.17. Alternatively, see Warner (1983), chap. 6, Ex. 16(g), or Chavel (1993), Theorem 3.21. □

Example 15E.13

(a) Let $\mathbb{B} = \{\mathbf{x} \in \mathbb{R}^D; \|\mathbf{x}\| < R\}$ be the ball of radius R. Then there is a $\triangle$-eigenbasis for $\mathbf{L}^2(\mathbb{B})$ consisting of functions which are zero on the spherical boundary of $\mathbb{B}$.

(b) Let $\mathbb{A} = \{(x, y) \in \mathbb{R}^2; r^2 < x^2 + y^2 < R^2\}$ be the annulus of inner radius r and outer radius R in the plane. Then there is a $\triangle$-eigenbasis for $\mathbf{L}^2(\mathbb{A})$ consisting of functions which are zero on the inner and outer boundary circles of $\mathbb{A}$. ◇

Theorem 15E.14 Eigenbases for Sturm–Liouville operators

Let $L > 0$, let $s, q : [0, L] \longrightarrow \mathbb{R}$ be differentiable functions, and let $\mathcal{SL}_{s,q}$ be the Sturm–Liouville operator defined by s and q on $C_0^\infty[0, L]$. Then there exists an $\mathcal{SL}_{s,q}$-eigenbasis for $\mathbf{L}^2[0, L]$ consisting of elements of $C_0^\infty[0, L]$. The corresponding eigenvalues of $\mathcal{SL}_{s,q}$ form an infinite increasing sequence $0 < \lambda_0 < \lambda_1 < \lambda_2 < \cdots$, with $\lim_{n \to \infty} \lambda_n = \infty$. Each eigenspace is one-dimensional.

Proof See Titchmarsh (1962), Theorem 1.9. For a proof in the special case when $s \equiv 1$, see Conway (1990), Theorem 6.12. $\square$

15E(iii) Symmetric elliptic operators

The rest of this section concerns the eigenfunctions of symmetric elliptic operators. (Please see §5E for the definition of an elliptic operator.)

Lemma 15E.15 *Let $\mathbb{X} \subset \mathbb{R}^D$. If L is an elliptic differential operator on $C^\infty(\mathbb{X})$, then there are functions $\omega_{cd} : \mathbb{X} \longrightarrow \mathbb{R}$ for all $c, d \in [1 \ldots D]$, and functions $\alpha, \xi_1, \ldots, \xi_D : \mathbb{X} \longrightarrow \mathbb{R}$ such that L can be written in* divergence form:

$$\mathsf{L}[u] = \sum_{c,d=1}^{D} \partial_c(\omega_{cd} \cdot \partial_d u) + \sum_{d=1}^{D} \xi_d \cdot \partial_d u + \alpha \cdot u$$

$$= \operatorname{div} [\boldsymbol{\Omega} \cdot \nabla \phi] + \langle \Xi, \nabla \phi \rangle + \alpha \cdot u,$$

where $\Xi = \begin{bmatrix} \xi_1 \\ \vdots \\ \xi_D \end{bmatrix}$, *and*

$$\boldsymbol{\Omega} = \begin{bmatrix} \omega_{11} & \cdots & \omega_{1D} \\ \vdots & \ddots & \vdots \\ \omega_{D1} & \cdots & \omega_{DD} \end{bmatrix}$$

is a symmetric, positive-definite matrix.

Ⓔ   *Proof* **Exercise 15E.7.** (*Hint:* Use the same strategy as in equation (15E.2).) $\square$

L is called *symmetric* if, in the divergence form, $\Xi \equiv 0$. For example, in the case when $\mathsf{L} = \triangle$, we have $\Omega = \mathbf{Id}$ and $\Xi = 0$, so $\triangle$ is symmetric.

Theorem 15E.16 *If $\mathbb{X} \subset \mathbb{R}^D$ is an open, bounded domain, then any symmetric elliptic differential operator on $C_0^\infty(\mathbb{X})$ is self-adjoint.*

Proof This is a generalization of the integration by parts argument used to prove Proposition 15E.4, p. 347, and Theorem 15E.5, p. 348. See Evans (1991), §6.5. $\square$

Theorem 15E.17 *Let* $\mathbb{X} \subset \mathbb{R}^D$ *be an open, bounded domain, and let* L *be any symmetric, elliptic differential operator on* $\mathcal{C}_0^\infty(\mathbb{X})$. *Then there exists an* L*-eigenbasis for* $\mathbf{L}^2(\mathbb{X})$ *consisting of elements of* $\mathcal{C}_0^\infty(\mathbb{X})$. *The corresponding eigenvalues of* L *form an infinite decreasing series* $0 > \lambda_0 \geq \lambda_1 \geq \lambda_2 \geq \cdots$, *with* $\lim_{n\to\infty} \lambda_n = -\infty$.

Proof See Evans (1991), Theorem 1, §6.5.1. □

Remark Theorems 15E.12, 15E.14, and 15E.17 are all manifestations of a far more general result, the *spectral theorem for unbounded self-adjoint operators*. Unfortunately, it would take us too far afield to even set up the necessary background to *state* this theorem precisely. See Conway (1990), §X.4, for a good exposition.

15F Further reading

The study of eigenfunctions and eigenvalues is sometimes called *spectral theory*. For a good introduction to the spectral theory of linear operators on function spaces, see Conway (1990). An analogy of the Laplacian can be defined on any Riemannian manifold; it is often called the *Laplace–Beltrami operator*, and its eigenfunctions reveal much about the geometry of the manifold; see Warner (1983), chap. 6, or Chavel (1993), §3.9. In particular, the eigenfunctions of the Laplacian on *spheres* have been extensively studied. These are called *spherical harmonics*, and a sort of 'Fourier theory' can be developed on spheres, analogous to multivariate Fourier theory on the cube $[0, L]^D$, but with the spherical harmonics forming the orthonormal basis (see Müller (1966) and Takeuchi (1994)). Much of this theory generalizes to a broader family of manifolds called *symmetric spaces* (see Helgason (1981) and Terras (1985)). The eigenfunctions of the Laplacian on symmetric spaces are closely related to the theory of *Lie groups* and their representations (see Coifman and Wiess (1968) and Sugiura (1975)), a subject which is sometimes called *noncommutative harmonic analysis* (Taylor (1986)).

Part V

Miscellaneous solution methods

In Chapters 11–15, we saw how initial/boundary value problems for linear partial differential equations could be solved by first identifying an orthogonal basis of eigenfunctions for the relevant differential operator (usually the Laplacian), and then representing the desired initial conditions or boundary conditions as an infinite summation of these eigenfunctions. For each bounded domain, each boundary condition, and each coordinate system we considered, we found a system of eigenfunctions that was 'adapted' to that domain, boundary conditions, and coordinate system.

This method is extremely powerful, but it raises several questions as follows.

(1) What if you are confronted with a new domain or coordinate system, where none of the known eigenfunction bases is applicable? Theorem 15E.12, p. 353, says that a suitable eigenfunction basis for this domain always exists, *in principle*. But how do you go about discovering such a basis *in practice*? For that matter, how were eigenfunctions bases like the Fourier–Bessel functions discovered in the first place? Where did Bessel's equation come from?

(2) What if you are dealing with an *unbounded* domain, such as diffusion in $\mathbb{R}^3$? In this case, Theorem 15E.12 is not applicable, and in general it may not be possible (or at least, not feasible) to represent initial/boundary conditions in terms of eigenfunctions. What alternative methods are available?

(3) The eigenfunction method is difficult to connect to our physical intuition. For example, intuitively, heat 'seaps' slowly through space, and temperature distributions gradually and irreversibly decay towards uniformity. It is thus impossible to send a long-distance 'signal' using heat. On the other hand, waves maintain their shape and propagate across great distances with a constant velocity; hence they can be used to send signals through space. This familiar intuition is not explained or justified by the eigenfunction method. Is there an alternative solution method where this intuition has a clear mathematical expression?

Part V provides answers to these questions. In Chapter 16, we introduce a powerful and versatile technique called *separation of variables*, to construct eigenfunctions adapted to any coordinate system. In Chapter 17, we develop the entirely different solution technology of *impulse-response functions*, which allows you to solve differential equations on unbounded domains, and which has an an appealing intuitive interpretation. Finally, in Chapter 18, we explore some surprising and beautiful applications of complex analysis to harmonic functions and Fourier theory.

16

Separation of variables

Before creation God did just pure mathematics. Then He thought it
would be a pleasant change to do some applied.

J. E. Littlewood

16A Separation of variables in Cartesian coordinates on $\mathbb{R}^2$

Prerequisites: §1B, §1C.

A function $u : \mathbb{R}^2 \longrightarrow \mathbb{R}$ is said to *separate* if we can write $u(x, y) = X(x) \cdot Y(y)$
for some functions $X, Y : \mathbb{R} \longrightarrow \mathbb{R}$. If u is a solution to some partial differential
equation, we say u is a *separated solution*.

Example 16A.1 *The heat equation on $\mathbb{R}$*
We wish to find $u : \mathbb{R} \times \mathbb{R}_{\not{/}} \longrightarrow \mathbb{R}$ such that $\partial_t u = \partial_x^2 u$. Suppose $u(x; t) = X(x) \cdot T(t)$, where

$$X(x) = \exp(\mathbf{i}\mu x) \quad \text{and} \quad T(t) = \exp(-\mu^2 t),$$

for some constant $\mu \in \mathbb{R}$. Then $u(x; t) = \exp(\mu \mathbf{i} x - \mu^2 t)$, so that $\partial_x^2 u = -\mu^2 \cdot u = \partial_t u$. Thus, u is a separated solution to the heat equation. $\diamond$

Separation of variables is a strategy for solving partial differential equations by
specifically looking for separated solutions. At first, it seems like we are making
our lives harder by insisting on a solution in separated form. However, often, we
can use the hypothesis of separation to *simplify* the problem.

Suppose we are given some PDE for a function $u : \mathbb{R}^2 \longrightarrow \mathbb{R}$ of two variables.
Separation of variables is the following strategy.

(1) Hypothesize that u can be written as a product of two functions, $X(x)$ and $Y(y)$, each
depending on only one coordinate; in other words, assume that

$$u(x, y) = X(x) \cdot Y(y). \tag{16A.1}$$

(2) When we evaluate the PDE on a function of type (16A.1), we may find that the PDE decomposes into two separate, *ordinary* differential equations for each of the two functions X and Y. Thus, we can solve these ODEs independently, and combine the resulting solutions to get a solution for u.

Example 16A.2 *Laplace's equation in* $\mathbb{R}^2$
Suppose we want to find a function $u : \mathbb{R}^2 \longrightarrow \mathbb{R}$ such that $\triangle u \equiv 0$. If $u(x, y) = X(x) \cdot Y(y)$, then

$$\triangle u = \partial_x^2 (X \cdot Y) + \partial_y^2 (X \cdot Y) = \left(\partial_x^2 X\right) \cdot Y + X \cdot \left(\partial_y^2 Y\right) = X'' \cdot Y + X \cdot Y'',$$

where we denote $X'' = \partial_x^2 X$ and $Y'' = \partial_y^2 Y$. Thus,

$$\triangle u(x, y) = X''(x) \cdot Y(y) + X(x) \cdot Y''(y)$$

$$= \left(X''(x) \cdot Y(y) + X(x) \cdot Y''(y)\right) \frac{X(x)Y(y)}{X(x)Y(y)}$$

$$= \left(\frac{X''(x)}{X(x)} + \frac{Y''(y)}{Y(y)}\right) \cdot u(x, y).$$

Thus, dividing by $u(x, y)$, Laplace's equation is equivalent to

$$0 = \frac{\triangle u(x, y)}{u(x, y)} = \frac{X''(x)}{X(x)} + \frac{Y''(y)}{Y(y)}.$$

This is a sum of two functions which depend on *different* variables. The only way the sum can be identically zero is if each of the component functions is constant, i.e.

$$\frac{X''}{X} \equiv \lambda, \qquad \frac{Y''}{Y} \equiv -\lambda$$

So, pick some *separation constant* $\lambda \in \mathbb{R}$, and then solve the two ordinary differential equations:

$$X''(x) = \lambda \cdot X(x) \quad \text{and} \quad Y''(y) = -\lambda \cdot Y(y). \tag{16A.2}$$

The (real-valued) solutions to equations (16A.2) depend on the sign of λ. Let $\mu = \sqrt{|\lambda|}$. Then the solutions of equations (16A.2) have the following form:

$$X(x) = \begin{cases} A \sinh(\mu x) + B \cosh(\mu x) & \text{if } \lambda > 0; \\ Ax + B & \text{if } \lambda = 0; \\ A \sin(\mu x) + B \cos(\mu x) & \text{if } \lambda < 0, \end{cases}$$

where A and B are arbitrary constants. Assuming $\lambda < 0$, and $\mu = \sqrt{|\lambda|}$, we obtain:

$$X(x) = A \sin(\mu x) + B \cos(\mu x) \quad \text{and} \quad Y(y) = C \sinh(\mu x) + D \cosh(\mu x).$$

This yields the following separated solution to Laplace's equation:

$$u(x, y) = X(x) \cdot Y(y) = \Big(A \sin(\mu x) + B \cos(\mu x)\Big) \cdot \Big(C \sinh(\mu x) + D \cosh(\mu x)\Big)$$

$$(16A.3)$$

Alternatively, we could consider the general *complex* solution to equations (16A.2), given by

$$X(x) = \exp\Big(\sqrt{\lambda} \cdot x\Big),$$

where $\sqrt{\lambda} \in \mathbb{C}$ is some complex number. For example, if $\lambda < 0$ and $\mu = \sqrt{|\lambda|}$, then $\sqrt{\lambda} = \pm\mu\mathbf{i}$ are imaginary, and

$$X_1(x) = \exp(\mathbf{i}\mu x) = \cos(\mu x) + \mathbf{i} \sin(\mu x);$$
$$X_2(x) = \exp(-\mathbf{i}\mu x) = \cos(\mu x) - \mathbf{i} \sin(\mu x)$$

are two linearly independent solutions to equations (16A.2). The general solution is then given by

$$X(x) = a \cdot X_1(x) + b \cdot X_2(x) = (a + b) \cdot \cos(\mu x) + \mathbf{i} \cdot (a - b) \cdot \sin(\mu x).$$

Meanwhile, the general form for $Y(y)$ is given by

$$Y(y) = c \cdot \exp(\mu y) + d \cdot \exp(-\mu y) = (c + d) \cosh(\mu y) + (c - d) \sinh(\mu y).$$

The corresponding separated solution to Laplace's equation is given by

$$u(x, y) = X(x) \cdot Y(y)$$
$$= \Big(A \sin(\mu x) + B\mathbf{i} \cos(\mu x)\Big) \cdot \Big(C \sinh(\mu x) + D \cosh(\mu x)\Big), \quad (16A.4)$$

where $A = (a + b)$, $B = (a - b)$, $C = (c + d)$, and $D = (c - d)$. In this case, we simply recover solution (16A.3). However, we could also construct separated solutions in which $\lambda \in \mathbb{C}$ is an arbitrary complex number and $\sqrt{\lambda}$ is one of its square roots.   ◇

16B Separation of variables in Cartesian coordinates on $\mathbb{R}^D$

Recommended: §16A.

Given some PDE for a function $u : \mathbb{R}^D \longrightarrow \mathbb{R}$, we apply the strategy of *separation of variables* as follows.

(1) Hypothesize that u can be written as a product of D functions, each depending on only one coordinate; in other words, assume that

$$u(x_1, \ldots, x_D) = u_1(x_1) \cdot u_2(x_2) \cdots u_D(x_D) \tag{16B.5}$$

(2) When we evaluate the PDE on a function of type (16B.5), we may find that the PDE decomposes into D separate, *ordinary* differential equations for each of the D functions $u_1, \ldots, u_D$. Thus, we can solve these ODEs independently, and combine the resulting solutions to get a solution for u.

Example 16B.1 *Laplace's equation in $\mathbb{R}^D$*

Suppose we want to find a function $u : \mathbb{R}^D \longrightarrow \mathbb{R}$ such that $\triangle u \equiv 0$. As in the two-dimensional case (Example 16A.2), we reason as follows. If

$$u(\mathbf{x}) = X_1(x_1) \cdot X_2(x_2) \cdots X_D(x_D),$$

then

$$\triangle u = \left(\frac{X_1''}{X_1} + \frac{X_2''}{X_2} + \cdots + \frac{X_D''}{X_D} \right) \cdot u.$$

Thus, Laplace's equation is equivalent to

$$0 = \frac{\triangle u}{u}(\mathbf{x}) = \frac{X_1''}{X_1}(x_1) + \frac{X_2''}{X_2}(x_2) + \cdots + \frac{X_D''}{X_D}(x_D).$$

This is a sum of D distinct functions, each of which depends on a different variable. The only way the sum can be identically zero is if each of the component functions is constant:

$$\frac{X_1''}{X_1} \equiv \lambda_1, \quad \frac{X_2''}{X_2} \equiv \lambda_2, \quad \ldots, \quad \frac{X_D''}{X_D} \equiv \lambda_D, \tag{16B.6}$$

such that

$$\lambda_1 + \lambda_2 + \cdots + \lambda_D = 0. \tag{16B.7}$$

So, pick some *separation constant* $\lambda = (\lambda_1, \lambda_2, \ldots, \lambda_D) \in \mathbb{R}^D$ satisfying equation (16B.7), and then solve the ODEs:

$$X_d'' = \lambda_d \cdot X_d \quad \text{for } d = 1, 2, \ldots, D. \tag{16B.8}$$

The (real-valued) solutions to equation (16B.8) depend on the sign of λ (and clearly, if (16B.7) is going to be true, either all λ_d are zero, or some are negative and some are positive). Let $\mu = \sqrt{|\lambda|}$. Then the solutions of equation (16B.8) have the following form:

$$X(x) = \begin{cases} A \exp(\mu x) + B \exp(-\mu x) & \text{if } \lambda > 0; \\ Ax + B & \text{if } \lambda = 0; \\ A \sin(\mu x) + B \cos(\mu x) & \text{if } \lambda < 0, \end{cases}$$

where A and B are arbitrary constants. We then combine these as in Example 16A.2.   ◇

16C Separation of variables in polar coordinates: Bessel's equation

Prerequisites: Appendix D(ii), §1C.   **Recommended:** §14C, §16A.

In §14C–§14F, we explained how to use solutions of Bessel's equation to solve the heat equation or wave equation in polar coordinates. In this section, we will see how Bessel derived his equation in the first place: it arises naturally when one uses 'separation of variables' to find eigenfunctions of the Laplacian in polar coordinates. First, we present a technical lemma from the theory of ordinary differential equations.

Lemma 16C.1 *Let* $\Theta : [-\pi, \pi] \longrightarrow \mathbb{R}$ *be a function satisfying periodic boundary conditions (i.e.* $\Theta(-\pi) = \Theta(\pi)$ *and* $\Theta'(-\pi) = \Theta'(\pi)$*). Let* $\mu > 0$ *be some constant, and suppose* Θ *satisfies the linear ordinary differential equation*

$$\Theta''(\theta) = -\mu \cdot \Theta(\theta), \quad \text{for all } \theta \in [-\pi, \pi]. \tag{16C.9}$$

Then $\mu = m^2$ *for some* $m \in \mathbb{N}$*, and* Θ *must be a function of the following form:*

$$\Theta(\theta) = A \cos(m\theta) + B \sin(m\theta) \quad \text{(for constants } A, B \in \mathbb{C}.)$$

Proof Equation (16C.9) is a second-order linear ODE, so the set of all solutions to this equation is a two-dimensional vector space. This vector space is spanned by functions of the form $\Theta(\theta) = e^{r\theta}$, where r is any root of the characteristic polynomial $p(x) = x^2 + \mu$. The two roots of this polynomial are, of course, $r = \pm\sqrt{\mu}i$. Let $m = \sqrt{\mu}$ (it will turn out that m is an integer, although we don't know this yet). Hence the general solution to equation (16C.9) is given by

$$\Theta(\theta) = C_1 e^{mi\theta} + C_2 e^{-mi\theta},$$

where C_1 and C_2 are any two constants. The periodic boundary conditions mean that

$$\Theta(-\pi) = \Theta(\pi) \quad \text{and} \quad \Theta'(-\pi) = \Theta'(\pi),$$

which means

$$C_1 e^{-mi\pi} + C_2 e^{mi\pi} = C_1 e^{mi\pi} + C_2 e^{-mi\pi}; \tag{16C.10}$$

$$mi C_1 e^{-mi\pi} - mi C_2 e^{mi\pi} = mi C_1 e^{mi\pi} - mi C_2 e^{-mi\pi}. \tag{16C.11}$$

If we divide both sides of equation (16C.11) by mi, we get

$$C_1 e^{-mi\pi} - C_2 e^{mi\pi} = C_1 e^{mi\pi} - C_2 e^{-mi\pi}.$$

If we add this to equation (16C.10), we obtain

$$2 C_1 e^{-mi\pi} = 2 C_1 e^{mi\pi},$$

which is equivalent to $e^{2mi\pi} = 1$. Hence, m must be some integer, and $\mu = m^2$.

Now, let $A := C_1 + C_2$ and $B' := C_1 - C_2$. Then $C_1 = \frac{1}{2}(A + B')$ and $C_2 = \frac{1}{2}(A - B')$. Thus,

$$\Theta(\theta) = C_1 e^{mi\theta} + C_2 e^{-mi\theta} = (A + B')e^{mi\theta} + (A - B')e^{-mi\theta}$$

$$= \frac{A}{2}\left(e^{mi\theta} + e^{-mi\theta}\right) + \frac{B'i}{2i}\left(e^{mi\theta} - e^{-mi\theta}\right) = A\cos(m\theta) + B'i\sin(m\theta)$$

because of the Euler formulas $\cos(x) = \frac{1}{2}(e^{ix} + e^{-ix})$ and $\sin(x) = \frac{1}{2i}(e^{ix} - e^{-ix})$.
Now let $B = B'i$; then $\Theta(\theta) = A\cos(m\theta) + B\sin(m\theta)$, as desired. $\qquad\square$

Proposition 16C.2 *Let* $f : \mathbb{R}^2 \longrightarrow \mathbb{R}$ *be an eigenfunction of the Laplacian (i.e.* $\triangle f = -\lambda^2 \cdot f$ *for some constant* $\lambda \in \mathbb{R}$*). Suppose* f *separates in polar coordinates, meaning that there is a function* $\Theta : [-\pi, \pi] \longrightarrow \mathbb{R}$ *(satisfying periodic boundary conditions) and a function* $\mathcal{R} : \mathbb{R}_+ \longrightarrow \mathbb{R}$ *such that*

$$f(r, \theta) = \mathcal{R}(r) \cdot \Theta(\theta), \quad \text{for all } r \geq 0 \text{ and } \theta \in [-\pi, \pi].$$

Then there is some $m \in \mathbb{N}$ *such that*

$$\Theta(\theta) = A\cos(m\theta) + B\sin(m\theta) \qquad \text{(for constants } A, B \in \mathbb{R})$$

and $\mathcal{R}$ *is a solution to the (mth-order)* Bessel *equation:*

$$r^2 \mathcal{R}''(r) + r \cdot \mathcal{R}'(r) + (\lambda^2 r^2 - m^2) \cdot \mathcal{R}(r) = 0, \quad \text{for all } r > 0. \qquad (16C.12)$$

Proof Recall that, in polar coordinates,

$$\triangle f = \partial_r^2 f + \frac{1}{r}\partial_r f + \frac{1}{r^2}\partial_\theta^2 f.$$

Thus, if $f(r, \theta) = \mathcal{R}(r) \cdot \Theta(\theta)$, then the eigenvector equation $\triangle f = -\lambda^2 \cdot f$ becomes

$$-\lambda^2 \cdot \mathcal{R}(r) \cdot \Theta(\theta) = \triangle \mathcal{R}(r) \cdot \Theta(\theta)$$

$$= \partial_r^2 \mathcal{R}(r) \cdot \Theta(\theta) + \frac{1}{r}\partial_r \mathcal{R}(r) \cdot \Theta(\theta) + \frac{1}{r^2}\partial_\theta^2 \mathcal{R}(r) \cdot \Theta(\theta)$$

$$= \mathcal{R}''(r)\Theta(\theta) + \frac{1}{r}\mathcal{R}'(r)\Theta(\theta) + \frac{1}{r^2}\mathcal{R}(r)\Theta''(\theta),$$

which is equivalent to

$$-\lambda^2 = \frac{\mathcal{R}''(r)\Theta(\theta) + \frac{1}{r}\mathcal{R}'(r)\Theta(\theta) + \frac{1}{r^2}\mathcal{R}(r)\Theta''(\theta)}{\mathcal{R}(r) \cdot \Theta(\theta)}$$

$$= \frac{\mathcal{R}''(r)}{\mathcal{R}(r)} + \frac{\mathcal{R}'(r)}{r\mathcal{R}(r)} + \frac{\Theta''(\theta)}{r^2\Theta(\theta)}. \qquad (16C.13)$$

If we multiply both sides of equation (16C.13) by r^2 and isolate the Θ'' term, we get

$$-\lambda^2 r^2 - \frac{r^2 \mathcal{R}''(r)}{\mathcal{R}(r)} + \frac{r \mathcal{R}'(r)}{\mathcal{R}(r)} = \frac{\Theta''(\theta)}{\Theta(\theta)}. \tag{16C.14}$$

Abstractly, equation (16C.14) has the form: $F(r) = G(\theta)$, where F is a function depending only on r and G is a function depending only on θ. The only way this can be true is if there is some constant $\mu \in \mathbb{R}$ such that $F(r) = -\mu$ for all $r > 0$ and $G(\theta) = -\mu$ for all $\theta \in [-\pi, \pi)$. In other words,

$$\frac{\Theta''(\theta)}{\Theta(\theta)} = -\mu, \qquad \text{for all } \theta \in [-\pi, \pi); \tag{16C.15}$$

$$\lambda^2 r^2 + \frac{r^2 \mathcal{R}''(r)}{\mathcal{R}(r)} + \frac{r \mathcal{R}'(r)}{\mathcal{R}(r)} = \mu, \qquad \text{for all } r \geq 0. \tag{16C.16}$$

Multiply both sides of equation (16C.15) by $\Theta(\theta)$ to get

$$\Theta''(\theta) = -\mu \cdot \Theta(\theta), \qquad \text{for all } \theta \in [-\pi, \pi). \tag{16C.17}$$

Multiply both sides of equation (16C.16) by $\mathcal{R}(r)$ to get

$$r^2 \mathcal{R}''(r) + r \cdot \mathcal{R}'(r) + \lambda^2 r^2 \mathcal{R}(r) = \mu \mathcal{R}(r), \qquad \text{for all } r > 0. \tag{16C.18}$$

Apply Lemma 16C.1 to equation (16C.17) to deduce that $\mu = m^2$ for some $m \in \mathbb{N}$, and that $\Theta(\theta) = A \cos(m\theta) + B \sin(m\theta)$. Substitute $\mu = m^2$ into equation (16C.18) to get

$$r^2 \mathcal{R}''(r) + r \cdot \mathcal{R}'(r) + \lambda^2 r^2 \mathcal{R}(r) = m^2 \mathcal{R}(r).$$

Now subtract $m^2 \mathcal{R}(r)$ from both sides to get Bessel's equation (16C.12).   □

16D Separation of variables in spherical coordinates: Legendre's equation

Prerequisites: Appendix D(iv), §1C, §5C(i), §6F, Appendix H(iii).   **Recommended:** §16C.

Recall that *spherical coordinates* (r, θ, ϕ) on $\mathbb{R}^3$ are defined by the following transformations:

$$x = r \cdot \sin(\phi) \cdot \cos(\theta), \quad y = r \cdot \sin(\phi) \cdot \sin(\theta) \quad \text{and} \quad z = r \cdot \cos(\phi),$$

where $r \in \mathbb{R}_+$, $\theta \in [-\pi, \pi)$, and $\phi \in [0, \pi]$. The reverse transformations are defined:

$$r = \sqrt{x^2 + y^2 + z^2}, \quad \theta = \arctan\left(\frac{y}{x}\right) \quad \text{and} \quad \phi = \arctan\left(\frac{\sqrt{x^2 + y^2}}{z}\right).$$

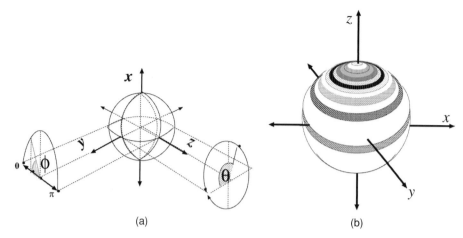

Figure 16D.1. (a) Spherical coordinates; (b) zonal functions.

See Figure 16D.1(a). Geometrically, r is the radial distance from the origin. If we fix $r = 1$, then we get a sphere of radius 1. On the surface of this sphere, θ is *longitude* and ϕ is *latitude*. In terms of these coordinates, the Laplacian is written as follows:

$$\triangle f(r, \theta, \phi) = \partial_r^2 f + \frac{2}{r} \partial_r f + \frac{1}{r^2 \sin(\phi)} \partial_\phi^2 f + \frac{\cot(\phi)}{r^2} \partial_\phi f + \frac{1}{r^2 \sin(\phi)^2} \partial_\theta^2 f.$$

ⓔ **(Exercise 16D.1).**

A function $f : \mathbb{R}^3 \longrightarrow \mathbb{R}$ is *zonal* if $f(r, \theta, \phi)$ depends only on r and ϕ; in other words, $f(r, \theta, \phi) = F(r, \phi)$, where $F : \mathbb{R}_{\not\!+} \times [0, \pi] \longrightarrow \mathbb{R}$ is some other function. If we restrict f to the aforementioned sphere of radius 1, then f is invariant under rotations around the 'north–south axis' of the sphere. Thus, f is constant along lines of equal latitude around the sphere, so it divides the sphere into 'zones' from north to south (Figure 16D.1(b)).

Proposition 16D.1 *Let $f : \mathbb{R}^3 \longrightarrow \mathbb{R}$ be zonal. Suppose f is a harmonic function (i.e. $\triangle f = 0$). Suppose f separates in spherical coordinates, meaning that there are (bounded) functions $\Phi : [0, \pi] \longrightarrow \mathbb{R}$ and $\mathcal{R} : \mathbb{R}_{\not\!+} \longrightarrow \mathbb{R}$ such that*

$$f(r, \theta, \phi) = \mathcal{R}(r) \cdot \Phi(\phi), \quad \text{for all } r \geq 0,\ \phi \in [0, \pi],\ \text{and } \theta \in [-\pi, \pi].$$

Then there is some $\mu \in \mathbb{R}$ such that $\Phi(\phi) = \mathcal{L}[\cos(\phi)]$, where $\mathcal{L} : [-1, 1] \longrightarrow \mathbb{R}$ is a (bounded) solution of the Legendre equation:

$$(1 - x^2)\mathcal{L}''(x) - 2x\mathcal{L}'(x) + \mu\mathcal{L}(x) = 0, \tag{16D.19}$$

and $\mathcal{R}$ is a (bounded) solution to the Cauchy–Euler equation:

$$r^2 \mathcal{R}''(r) + 2r \cdot \mathcal{R}'(r) - \mu \cdot \mathcal{R}(r) = 0, \quad \text{for all } r > 0. \tag{16D.20}$$

Proof By hypothesis

$$0 = \triangle f(r, \theta, \phi)$$

$$= \partial_r^2 f + \frac{2}{r} \partial_r f + \frac{1}{r^2 \sin(\phi)} \partial_\phi^2 f + \frac{\cot(\phi)}{r^2} \partial_\phi f + \frac{1}{r^2 \sin(\phi)^2} \partial_\theta^2 f$$

$$\underset{(*)}{=} \mathcal{R}''(r) \cdot \Phi(\phi) + \frac{2}{r} \mathcal{R}'(r) \cdot \Phi(\phi)$$

$$+ \frac{1}{r^2 \sin(\phi)} \mathcal{R}(r) \cdot \Phi''(\phi) + \frac{\cot(\phi)}{r^2} \mathcal{R}(r) \cdot \Phi'(\phi) + 0$$

(where (∗) is because $f(r, \theta, \phi) = \mathcal{R}(r) \cdot \Phi(\phi)$.) Hence, multiplying both sides by $r^2/(\mathcal{R}(r) \cdot \Phi(\phi))$, we obtain

$$0 = \frac{r^2 \mathcal{R}''(r)}{\mathcal{R}(r)} + \frac{2r \mathcal{R}'(r)}{\mathcal{R}(r)} + \frac{1}{\sin(\phi)} \frac{\Phi''(\phi)}{\Phi(\phi)} + \frac{\cot(\phi)\Phi'(\phi)}{\Phi(\phi)},$$

or, equivalently,

$$\frac{r^2 \mathcal{R}''(r)}{\mathcal{R}(r)} + \frac{2r \mathcal{R}'(r)}{\mathcal{R}(r)} = \frac{-1}{\sin(\phi)} \frac{\Phi''(\phi)}{\Phi(\phi)} - \frac{\cot(\phi)\Phi'(\phi)}{\Phi(\phi)}. \tag{16D.21}$$

The left-hand side of equation (16D.21) depends only on the variable r, whereas the right-hand side depends only on ϕ. The only way that these two expressions can be equal for *all* values of r and ϕ is if both expressions are constants. In other words, there is some constant $\mu \in \mathbb{R}$ (called a *separation constant*) such that

$$\frac{r^2 \mathcal{R}''(r)}{\mathcal{R}(r)} + \frac{2r \mathcal{R}'(r)}{\mathcal{R}(r)} = \mu, \qquad \text{for all } r \geq 0;$$

$$\frac{1}{\sin(\phi)} \frac{\Phi''(\phi)}{\Phi(\phi)} + \frac{\cot(\phi)\Phi'(\phi)}{\Phi(\phi)} = -\mu, \quad \text{for all } \phi \in [0, \pi].$$

Or, equivalently,

$$r^2 \mathcal{R}''(r) + 2r \mathcal{R}'(r) = \mu \mathcal{R}(r), \qquad \text{for all } r \geq 0; \tag{16D.22}$$

$$\frac{\Phi''(\phi)}{\sin(\phi)} + \cot(\phi)\Phi'(\phi) = -\mu \Phi(\phi), \quad \text{for all } \phi \in [0, \pi]. \tag{16D.23}$$

If we make the change of variables $x = \cos(\phi)$ (so that $\phi = \arccos(x)$, where $x \in [-1, 1]$), then $\Phi(\phi) = \mathcal{L}(\cos(\phi)) = \mathcal{L}(x)$, where $\mathcal{L}$ is some other (unknown) function.

Claim 1 *The function Φ satisfies the ODE (16D.23) if, and only if, $\mathcal{L}$ satisfies the Legendre equation (16D.19).*

Ⓔ *Proof* **Exercise 16D.2.** (*Hint:* This is a straightforward application of the chain
rule.) ◇ Claim 1

Finally, observe that the ODE (16D.22) is equivalent to the Cauchy–Euler equation
(16D.20). □

For all $n \in \mathbb{N}$, we define the *n*th *Legendre polynomial* by

$$P_n(x) := \frac{1}{n! \, 2^n} \partial_x^n \left[(x^2 - 1) \right]^n. \tag{16D.24}$$

For example:

$$P_0(x) = 1, \qquad P_3(x) = \tfrac{1}{2}(5x^3 - 3x),$$

$$P_1(x) = x, \qquad P_4(x) = \tfrac{1}{8}(35x^4 - 30x^2 + 3),$$

$$P_2(x) = \tfrac{1}{2}(3x^2 - 1), \quad P_5(x) = \tfrac{1}{8}(63x^5 - 70x^3 + 15x).$$

(see Figure 16D.2).

Lemma 16D.2 *Let $n \in \mathbb{N}$. Then the Legendre polynomial P_n is a solution to the
Legendre equation* (16D.19) *with $\mu = n(n+1)$.*

Ⓔ *Proof* **Exercise 16D.3.** (Direct computation.) □

Is P_n the *only* solution to the Legendre equation (16D.19)? No, because the
Legendre equation is an order-two linear ODE, so the set of solutions forms a two-
dimensional vector space $\mathcal{V}$. The scalar multiples of P_n form a one-dimensional
subspace of $\mathcal{V}$. However, to be physically meaningful, we need the solutions
to be bounded at $x = \pm 1$. So instead we ask: is P_n the only *bounded* solution to the Legendre equation (16D.19)? Also, what happens if $\mu \neq n(n+1)$ for
any $n \in \mathbb{N}$?

Lemma 16D.3
(a) *If $\mu = n(n+1)$ for some $n \in \mathbb{N}$, then (up to multiplication by a scalar), the Legendre
polynomial $P_n(x)$ is the unique solution to the Legendre equation* (16D.19), *which is
bounded on $[-1, 1]$.*
(b) *If $\mu \neq n(n+1)$ for any $n \in \mathbb{N}$, then all solutions to the Legendre equation* (16D.19) *are
infinite power series which diverge at $x = \pm 1$ (and thus are unsuitable for Proposition
16D.1).*

Proof We apply the *power series method* (see Appendix H(iii), p. 575). Sup-
pose $\mathcal{L}(x) = \sum_{n=0}^{\infty} a_n x^n$ is some analytic function defined on $[-1, 1]$ (where the
coefficients $\{a_n\}_{n=1}^{\infty}$ are as yet unknown).

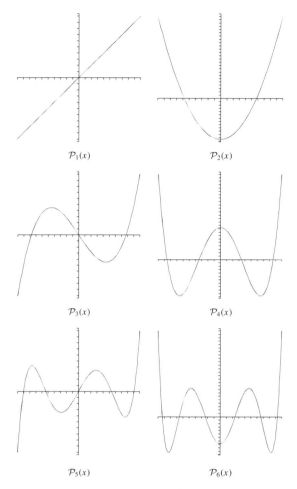

Figure 16D.2. The Legendre polynomials $\mathcal{P}_1(x)$ to $\mathcal{P}_6(x)$, plotted for $x \in [-1, 1]$.

Claim 1 *$\mathcal{L}(x)$ satisfies the Legendre equation (16D.19) if, and only if, the coefficients $\{a_0, a_1, a_2, \ldots\}$ satisfy the recurrence relation*

$$a_{k+2} = \frac{k(k+1) - \mu}{(k+2)(k+1)} a_k, \qquad \text{for all } k \in \mathbb{N}. \tag{16D.25}$$

In particular, $a_2 = \dfrac{-\mu}{2} a_0$ and $a_3 = \dfrac{2 - \mu}{6} a_1$.

Proof We will substitute the power series $\sum_{n=0}^{\infty} a_n x^n$ into the Legendre equation (16D.19). The details of the computation are shown in Table 16.1. The computation yields the equation $0 = \sum_{k=0}^{\infty} b_k x_k$, where $b_k := (k+2)(k+1)a_{k+2} + [\mu - k(k+1)]a_k$ for all $k \in \mathbb{N}$. It follows that $b_k = 0$ for all $k \in \mathbb{N}$; in other

Table 16.1. *Substitution of the power series into the proof of Claim 1 of Lemma 16D.3.*

Here $b_k := (k+2)(k+1)a_{k+2} + \left[\mu - k(k+1)\right]a_k$ for all $k \in \mathbb{N}$.

If	$\mathcal{L}(x) =$	$a_0 +$	$a_1 x +$	$a_2 x^2 + \cdots +$	$a_k x^k + \cdots$
then	$\mu\mathcal{L}(x) =$	$\mu a_0 +$	$\mu a_1 x +$	$\mu a_2 x^2 + \cdots +$	$\mu a_k x^k + \cdots$
and	$-2x\mathcal{L}'(x) =$		$-2a_1 x +$	$-4a_2 x^2 + \cdots +$	$-2k a_k x^k + \cdots$
and	$\mathcal{L}''(x) =$	$2a_2 +$	$6a_3 x +$	$12a_4 x^2 + \cdots + (k+2)(k+1)a_{k+2}x^k +$	
and	$-x^2\mathcal{L}''(x) =$			$-2a_2 x^2 + \cdots + -(k-1)k a_k x^k + \cdots$	

Thus, $0 =$
$(1 - x^2)\mathcal{L}''(x) - 2x\mathcal{L}'(x) + \mu\mathcal{L}(x) = (\mu a_0 - 2a_2) + ((\mu - 2)a_1 + 6a_3)\,x + ((\mu - 6)a_2 + 12a_4)\,x^2 + \cdots + b_k x^k + \cdots$

words, that

$$(k + 2)(k + 1)a_{k+2} + \left[\mu - k(k + 1)\right]a_k = 0, \quad \text{for all } k \in \mathbb{N}.$$

Rearranging this equation produces the desired recurrence relation (16D.25). ◇ Claim 1

The space of all solutions to the Legendre equation (16D.19) is a two-dimensional vector space, because the Legendre equation is a *linear* differential equation of order 2. We will now find a basis for this space. Recall that $\mathcal{L}$ is *even* if $\mathcal{L}(-x) = \mathcal{L}(x)$ for all $x \in [-1, 1]$, and $\mathcal{L}$ is *odd* if $\mathcal{L}(-x) = -\mathcal{L}(x)$ for all $x \in [-1, 1]$.

Claim 2 *There is a unique even analytic function $\mathcal{E}(x)$ and a unique odd analytic function $\mathcal{O}(x)$ which satisfy the Legendre equation (16D.19), such that $\mathcal{E}(1) = 1 = \mathcal{O}(1)$, and such that any other solution $\mathcal{L}(x)$ can be written as a linear combination $\mathcal{L}(x) = a\,\mathcal{E}(x) + b\,\mathcal{O}(x)$, for some constants $a, b \in \mathbb{R}$.*

Proof Claim 1 implies that the power series $\mathcal{L}(x) = \sum_{n=0}^{\infty} a_n x^n$ is entirely determined by the coefficients a_0 and a_1. To be precise, $\mathcal{L}(x) = \mathcal{E}(x) + \mathcal{O}(x)$, where

$$\mathcal{E}(x) = \sum_{n=0}^{\infty} a_{2n} x^{2n} \text{ and } \mathcal{O}(x) = \sum_{n=0}^{\infty} a_{2n+1} x^{2n+1}$$

both satisfy the recurrence relation (16D.25) and, thus, are solutions to the Legendre equation (16D.19). ◇ Claim 2

Claim 3 *Suppose $\mu = n(n + 1)$ for some $n \in \mathbb{N}$. Then the Legendre equation (16D.19) has a degree-n polynomial as one of its solutions. To be precise we state the following.*

(a) *If n is even, then $a_k = 0$ for all even $k > n$. Hence, $\mathcal{E}(x)$ is a degree-n polynomial.*
(b) *If n is odd, then $a_k = 0$ for all odd $k > n$. Hence, $\mathcal{O}(x)$ is a degree-n polynomial.*

Proof **Exercise 16D.4.** ◇ Claim 3 Ⓔ

Thus, there is a one-dimensional space of *polynomial* solutions to the Legendre equation – namely all scalar multiples of $\mathcal{E}(x)$ (if n is even) or $\mathcal{O}(x)$ (if n is odd).

Claim 4 *If $\mu \neq n(n + 1)$ for any $n \in \mathbb{N}$, the series $\mathcal{E}(x)$ and $\mathcal{O}(x)$ both diverge at $x = \pm 1$.*

Proof **Exercise 16D.5.** Ⓔ
(a) First note that an infinite number of coefficients $\{a_n\}_{n=0}^{\infty}$ are nonzero.
(b) Show that $\lim_{n \to \infty} |a_n| = 1$.
(c) Conclude that the series $\mathcal{E}(x)$ and $\mathcal{O}(x)$ diverge when $x = \pm 1$. ◇ Claim 4

So, there exist solutions to the Legendre equation (16D.19) that are bounded on $[-1, 1]$ if, *and only if*, $\mu = n(n + 1)$ for some $n \in \mathbb{N}$, and, in this case, the bounded solutions are all scalar multiples of a polynomial of degree n (either $\mathcal{E}(x)$ or $\mathcal{O}(x)$). But Lemma 16D.2 says that the Legendre polynomial $P_n(x)$ is a solution to the Legendre equation (16D.19). Thus, (up to multiplication by a constant), $P_n(x)$ must be equal to $\mathcal{E}(x)$ (if n is even) or $\mathcal{O}(x)$ (if n is odd). $\square$

Remark Sometimes the Legendre polynomials are *defined* as the (unique) polynomial solutions to Legendre's equation; the definition we have given in equation (16D.24) is then *derived* from this definition, and is called the *Rodrigues formula*.

Lemma 16D.4 *Let $\mathcal{R} : \mathbb{R}_{\not{=}} \longrightarrow \mathbb{R}$ be a solution to the Cauchy–Euler equation*

$$r^2 \mathcal{R}''(r) + 2r \cdot \mathcal{R}'(r) - n(n + 1) \cdot \mathcal{R}(r) = 0, \quad \text{for all } r > 0. \quad (16D.26)$$

Then $\mathcal{R}(r) = Ar^n + \frac{B}{r^{n+1}}$ for some constants A and B.
If $\mathcal{R}$ is bounded at zero, then $B = 0$, so $\mathcal{R}(r) = Ar^n$.

Proof Check that $f(r) = r^n$ and $g(r) = r^{-n-1}$ are solutions to equation (16D.26), but this is a second-order linear ODE, so the solutions form a two-dimensional vector space. Since f and g are linearly independent, they span this vector space. $\square$

Corollary 16D.5 *Let $f : \mathbb{R}^3 \longrightarrow \mathbb{R}$ be a zonal harmonic function that separates in spherical coordinates (as in Proposition 16D.1). Then there is some $m \in \mathbb{N}$ such that $f(r, \phi, \theta) = Cr^n \cdot P_n[\cos(\phi)]$, where P_n is the nth Legendre polynomial and $C \in \mathbb{R}$ is some constant (see Figure 16D.3).*

Proof Combine Proposition 16D.1 with Lemmas 16D.3 and 16D.4. $\square$

Thus, the Legendre polynomials are important when solving the Laplace equation on spherical domains. We now describe some of their important properties.

Proposition 16D.6 *Legendre polynomials satisfy the following recurrence relations:*

(a) $(2n + 1)P_n(x) = P'_{n+1}(x) - P'_{n-1}(x)$;
(b) $(2n + 1)xP_n(x) = (n + 1)P_{n+1}(x) + nP'_{n-1}(x)$.

Ⓔ *Proof* **Exercise 16D.6.** $\square$

Proposition 16D.7 *The Legendre polynomials form an orthogonal set for $\mathbf{L}^2[-1, 1]$. That is:*

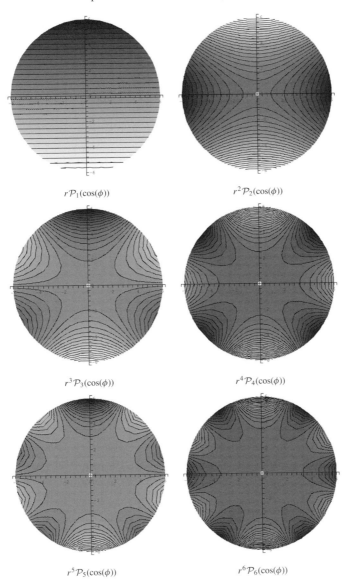

$r\mathcal{P}_1(\cos(\phi))$ $r^2\mathcal{P}_2(\cos(\phi))$

$r^3\mathcal{P}_3(\cos(\phi))$ $r^4\mathcal{P}_4(\cos(\phi))$

$r^5\mathcal{P}_5(\cos(\phi))$ $r^6\mathcal{P}_6(\cos(\phi))$

Figure 16D.3. Planar cross-sections of the zonal harmonic functions $r\mathcal{P}_1(\cos(\phi))$ to $r^6\mathcal{P}_6(\cos(\phi))$, plotted for $r \in [0...6]$; see Corollary 16D.5. Remember that these are functions in $\mathbb{R}^3$. To visualize these functions in three dimensions, take the above contour plots and mentally rotate them around the vertical axis.

(a) *for any $n \neq m$,*

$$\langle \mathcal{P}_n, \mathcal{P}_m \rangle = \frac{1}{2} \int_{-1}^{1} \mathcal{P}_n(x)\mathcal{P}_m(x)\mathrm{d}x = 0;$$

(b) *for any $n \in \mathbb{N}$,*

$$\|\mathcal{P}_n\|_2^2 = \frac{1}{2} \int_{-1}^{1} \mathcal{P}_n^2(x)\mathrm{d}x = \frac{1}{2n+1}.$$

Ⓔ *Proof* (a) **Exercise 16D.7.** (*Hint:* Start with the Rodrigues formula (16D.24). Apply integration by parts n times.)

Ⓔ   (b) **Exercise 16D.8.** (*Hint:* Use Proposition 16D.6(b).)   □

Because of Proposition 16D.7, we can try to represent an arbitrary function $f \in \mathbf{L}^2[-1, 1]$ in terms of Legendre polynomials, to obtain a *Legendre series*:

$$f(x) \underset{L2}{\approx} \sum_{n=0}^{\infty} a_n \mathcal{P}_n(x), \tag{16D.27}$$

where

$$a_n := \frac{\langle f, \mathcal{P}_n \rangle}{\|\mathcal{P}_n\|_2^2} = \frac{2n+1}{2} \int_{-1}^{1} f(x)\mathcal{P}_n(x)\mathrm{d}x$$

is the *nth Legendre coefficient* of f.

Theorem 16D.8 *The Legendre polynomials form an* orthogonal basis *for* $\mathbf{L}^2[-1, 1]$. *Thus, if $f \in \mathbf{L}^2[-1, 1]$, then the Legendre series (16D.27) converges to f in L^2.*

Proof See Broman (1989), Theorem 3.2.4.   □

Let $\mathbb{B} = \{(r, \theta, \phi); \ r \leq 1, \theta \in [-\pi, \pi], \phi \in [0, \pi]\}$ be the unit ball in spherical coordinates. Thus, $\partial\mathbb{B} = \{(1, \theta, \phi); \ \theta \in [-\pi, \pi], \phi \in [0, \pi]\}$ is the unit sphere. Recall that a *zonal* function on $\partial\mathbb{B}$ is a function which depends only on the 'latitude' coordinate ϕ, and not on the 'longitude' coordinate θ.

Theorem 16D.9 Dirichlet problem on a ball
Let $f : \partial\mathbb{B} \longrightarrow \mathbb{R}$ be some function describing a heat distribution on the surface of the ball. Suppose f is zonal, i.e. $f(1, \theta, \phi) = F(\cos(\phi))$, where $F \in \mathbf{L}^2[-1, 1]$, and F has Legendre series

$$F(x) \underset{L2}{\approx} \sum_{n=0}^{\infty} a_n \mathcal{P}_n(x).$$

Define $u : \mathbb{B} \longrightarrow \mathbb{R}$ *by*

$$u(r, \phi, \theta) = \sum_{n=0}^{\infty} a_n r^n \mathcal{P}_n (\cos(\phi)).$$

Then u *is the unique solution to the Laplace equation, satisfying the nonhomogeneous Dirichlet boundary conditions*

$$u(1, \theta, \phi) \underset{\text{L2}}{\approx} f(\theta, \phi), \qquad \text{for all } (1, \theta, \phi) \in \partial\mathbb{B}.$$

Proof **Exercise 16D.9.** □ Ⓔ

16E Separated vs. quasiseparated

Prerequisites: §16B.

If we use complex-valued functions like equation (16A.4) as the components of the separated solution (16B.5), p. 361, then we will still get mathematically valid solutions to Laplace's equation (as long as equation (16B.7) is true). However, these solutions are not physically meaningful – what does a *complex*-valued heat distribution feel like? This is not a problem, because we can extract *real*-valued solutions from the complex solution as follows.

Proposition 16E.1 *Suppose* L *is a linear differential operator with real-valued coefficients, and* $g : \mathbb{R}^D \longrightarrow \mathbb{R}$; *consider the nonhomogeneous PDE* 'L$u = g$'.

If $u : \mathbb{R}^D \longrightarrow \mathbb{C}$ *is a (complex-valued) solution to this PDE, and we define* $u_R(\mathbf{x}) = \mathrm{Re}\,[u(\mathbf{x})]$ *and* $u_I(\mathbf{x}) = \mathrm{Im}\,[u(\mathbf{x})]$, *then* L$u_R = g$ *and* L$u_I = 0$.

Proof **Exercise 16E.1.** □ Ⓔ

In this case, the solutions u_R and u_I are not themselves generally going to be in separated form. Since they arise as the real and imaginary components of a complex separated solution, we call u_R and u_I *quasiseparated* solutions.

Example 16E.2 Recall the separated solutions to the two-dimensional Laplace equation from Example 16A.2, p. 360. Here, L $= \triangle$ and $g \equiv 0$, and, for any fixed $\mu \in \mathbb{R}$, the function

$$u(x, y) = X(x) \cdot Y(y) = \exp(\mu y) \cdot \exp(\mu \mathbf{i} y)$$

is a complex solution to Laplace's equation. Thus,

$$u_R(x, y) = \exp(\mu x) \cos(\mu y) \quad \text{and} \quad u_I(x, y) = \exp(\mu x) \sin(\mu y)$$

are real-valued solutions of the form obtained earlier. ◇

16F The polynomial formalism

Prerequisites: §16B, §4B.

Separation of variables seems like a bit of a miracle. Just how generally applicable is it? To answer this, it is convenient to adopt a *polynomial formalism* for differential operators. If L is a differential operator with *constant*[1] coefficients, we will formally represent L as a 'polynomial' in the 'variables' $\partial_1, \partial_2, \ldots, \partial_D$. For example, we can write the Laplacian as follows:

$$\triangle = \partial_1^2 + \partial_2^2 + \cdots + \partial_D^2 = \mathcal{P}(\partial_1, \partial_2, \ldots, \partial_D),$$

where $\mathcal{P}(x_1, x_2, \ldots, x_D) = x_1^2 + x_2^2 + \cdots + x_D^2$.

In another example, the general second-order linear PDE

$$A\partial_x^2 u + B\partial_x \partial_y u + C\partial_y^2 u + D\partial_x u + E\partial_y u + Fu = G$$

(where $A, B, C, \ldots, F$ are constants) can be rewritten as follows:

$$\mathcal{P}(\partial_x, \partial_y)u = g,$$

where $\mathcal{P}(x, y) = Ax^2 + Bxy + Cy^2 + Dx + Ey + F$.

The polynomial $\mathcal{P}$ is called the *polynomial symbol* of L, and provides a convenient method for generating separated solutions

Proposition 16F.1 *Suppose that* L *is a linear differential operator on* $\mathbb{R}^D$ *with polynomial symbol* $\mathcal{P}$. *Regard* $\mathcal{P} : \mathbb{C}^D \longrightarrow \mathbb{C}$ *as a function.*

Fix $\mathbf{z} = (z_1, \ldots, z_D) \in \mathbb{C}^D$, *and define* $u_{\mathbf{z}} : \mathbb{R}^D \longrightarrow \mathbb{R}$ *by*

$$u_{\mathbf{z}}(x_1, \ldots, x_D) = \exp(z_1 x_1) \cdot \exp(z_2 x_2) \ldots \exp(z_D x_D) = \exp(\mathbf{z} \bullet \mathbf{x}).$$

Then $\mathsf{L}u_{\mathbf{z}}(\mathbf{x}) = \mathcal{P}(\mathbf{z}) \cdot u_{\mathbf{z}}(\mathbf{x})$ *for all* $\mathbf{x} \in \mathbb{R}^D$.

In particular, if $\mathbf{z}$ *is a root of* $\mathcal{P}$ *(that is,* $\mathcal{P}(z_1, \ldots, z_D) = 0$*), then* $\mathsf{L}u = 0$.

Ⓔ *Proof* **Exercise 16F.1.** (*Hint:* First, use equation (C.1), p. 553, to show that $\partial_d u_{\mathbf{z}} = z_d \cdot u_{\mathbf{z}}$, and, more generally, $\partial_d^n u_{\mathbf{z}} = z_d^n \cdot u_{\mathbf{z}}$.) □

Thus, many[2] separated solutions of the *differential* equation 'Lu = 0' are defined by the the complex-valued solutions of the *algebraic* equation '$\mathcal{P}(\mathbf{z}) = 0$'.

Example 16F.2 Consider again the two-dimensional Laplace equation:

$$\partial_x^2 u + \partial_y^2 u = 0.$$

---

[1] This is important.   [2] But not all.

The corresponding polynomial is $\mathcal{P}(x, y) = x^2 + y^2$. Thus, if $z_1, z_2 \in \mathbb{C}$ are any complex numbers such that $z_1^2 + z_2^2 = 0$, then

$$u(x, y) = \exp(z_1 x + z_2 y) = \exp(z_1 x) \cdot \exp(z_2 y)$$

is a solution to Laplace's Equation. In particular, if $z_1 = 1$, then we must have $z_2 = \pm \mathbf{i}$. Say we pick $z_2 = \mathbf{i}$; then the solution becomes

$$u(x, y) = \exp(x) \cdot \exp(\mathbf{i}y) = e^x \cdot \left(\cos(y) + \mathbf{i} \sin(y) \right).$$

More generally, if we choose $z_1 = \mu \in \mathbb{R}$ to be a real number, then we must choose $z_2 = \pm \mu \mathbf{i}$ to be purely imaginary, and the solution becomes

$$u(x, y) = \exp(\mu x) \cdot \exp(\pm \mu \mathbf{i} y) = e^{\mu x} \cdot \left(\cos(\pm \mu y) + \mathbf{i} \sin(\pm \mu y) \right).$$

Compare this with the separated solutions obtained from Example 16A.2, p. 360.

$\diamondsuit$

Example 16F.3 Consider the one-dimensional *telegraph equation*:

$$\partial_t^2 u + 2 \partial_t u + u = \Delta u. \qquad (16\text{F.}28)$$

We can rewrite this as follows:

$$\partial_t^2 u + 2 \partial_t u + u - \partial_x^2 u = 0,$$

which is equivalent to '$\mathsf{L}u = 0$', where L is the linear differential operator

$$\mathsf{L} = \partial_t^2 + 2 \partial_t + u - \partial_x^2,$$

with polynomial symbol

$$\mathcal{P}(x, t) = t^2 + 2t + 1 - x^2 = (t + 1 + x)(t + 1 - x).$$

Thus, the equation '$\mathcal{P}(\alpha, \beta) = 0$' has solutions:

$$\alpha = \pm(\beta + 1).$$

So, if we define $u(x, t) = \exp(\alpha \cdot x) \exp(\beta \cdot t)$, then u is a separated solution to equation (16F.28). (**Exercise 16F.2** Check this.) In particular, suppose we ⒠ choose $\alpha = -\beta - 1$. Then the separated solution is $u(x, t) = \exp(\beta(t - x) - x)$. If $\beta = \beta_R + \beta_I \mathbf{i}$ is a complex number, then the quasiseparated solutions are given by

$$u_R = \exp\left(\beta_R(x + t) - x\right) \cdot \cos\left(\beta_I(x + t)\right);$$
$$u_I = \exp\left(\beta_R(x + t) - x\right) \cdot \sin\left(\beta_I(x + t)\right).$$

$\diamondsuit$

Remark 16F.4 The polynomial formalism provides part of the motivation for the classification of PDEs as *elliptic, hyperbolic*,[3] etc. Note that, if L is an elliptic differential operator on $\mathbb{R}^2$, then the real-valued solutions to $\mathcal{P}(z_1, z_2) = 0$ (if any) form an *ellipse* in $\mathbb{R}^2$. In $\mathbb{R}^D$, the solutions form an *ellipsoid*.

Similarly, if we consider the parabolic PDE '$\partial_t u = Lu$', the corresponding differential operator $L - \partial_t$ has polynomial symbol $\mathcal{Q}(\mathbf{x}; t) = \mathcal{P}(\mathbf{x}) - t$. The real-valued solutions to $\mathcal{Q}(\mathbf{x}; t) = 0$ form a *paraboloid* in $\mathbb{R}^D \times \mathbb{R}$. For example, the one-dimensional heat equation '$\partial_x^2 u - \partial_t u = 0$' yields the classic equation '$t = x^2$' for a parabola in the (x, t) plane. Similarly, with a hyperbolic PDE, the differential operator $L - \partial_t^2$ has polynomial symbol $\mathcal{Q}(\mathbf{x}; t) = \mathcal{P}(\mathbf{x}) - t^2$, and the roots form a *hyperboloid*.

16G Constraints

Prerequisites: §16F.

Normally, we are not interested in just *any* solution to a PDE; we want a solution which satisfies certain constraints. The most common constraints are the following.

- Boundary conditions. If the PDE is defined on some bounded domain $\mathbb{X} \subset \mathbb{R}^D$, then we may want the solution function u (or its derivatives) to have certain values on the boundary of this domain.
- Boundedness. If the domain $\mathbb{X}$ is unbounded (e.g. $\mathbb{X} = \mathbb{R}^D$), then we may want the solution u to be *bounded*; in other words, we want some finite $M > 0$ such that $|u(\mathbf{x})| < M$ for all values of some coordinate x_d.

16G(i) Boundedness

The solution obtained through Proposition 16F.1 is not generally going to be bounded, because the exponential function $f(x) = \exp(\lambda x)$ is not bounded as a function of x, unless λ is a purely imaginary number. More generally, we note the following.

Proposition 16G.1 *Fix* $\mathbf{z} = (z_1, \ldots, z_D) \in \mathbb{C}^D$, *and suppose* $u_{\mathbf{z}} : \mathbb{R}^D \longrightarrow \mathbb{R}$ *is defined as in Proposition 16F.1:*

$$u_{\mathbf{z}}(x_1, \ldots, x_D) = \exp(z_1 x_1) \cdot \exp(z_2 x_2) \cdots \exp(z_D x_D) = \exp(\mathbf{z} \bullet \mathbf{x}).$$

Then

(1) $u(\mathbf{x})$ *is bounded for all values of the variable* $x_d \in \mathbb{R}$ *if, and only if,* $z_d = \lambda \mathbf{i}$ *for some* $\lambda \in \mathbb{R}$;

(2) $u(\mathbf{x})$ *is bounded for all* $x_d > 0$ *if, and only if,* $z_d = \rho + \lambda \mathbf{i}$ *for some* $\rho \leq 0$;

(3) $u(\mathbf{x})$ *is bounded for all* $x_d < 0$ *if, and only if,* $z_d = \rho + \lambda \mathbf{i}$ *for some* $\rho \geq 0$.

[3] See §5E, p. 97.

Proof **Exercise 16G.1.** □ Ⓔ

Example 16G.2 Recall the one-dimensional telegraph equation of Example 16F.3:

$$\partial_t^2 u + 2\partial_t u + u = \Delta u.$$

We constructed a separated solution of the form: $u(x, t) = \exp(\alpha x + \beta t)$, where $\alpha = \pm(\beta + 1)$. This solution will be bounded in time if, and only if, β is a purely imaginary number; i.e. $\beta = \beta_I \cdot \mathbf{i}$. Then $\alpha = \pm(\beta_I \cdot \mathbf{i} + 1)$, so that $u(x, t) = \exp(\pm x) \cdot \exp(\beta_I \cdot (t \pm x) \cdot \mathbf{i})$; thus, the quasiseparated solutions are given by

$$u_R = \exp(\pm x) \cdot \cos\left(\beta_I \cdot (t \pm x)\right) \quad \text{and} \quad u_I = \exp(\pm x) \cdot \sin\left(\beta_I \cdot (t \pm x)\right).$$

Unfortunately, this solution is *unbounded* in *space*, which is probably not what we want. An alternative is to set $\beta = \beta_I \mathbf{i} - 1$, and then set $\alpha = \beta + 1 = \beta_I \mathbf{i}$. Then the solution becomes $u(x, t) = \exp(\beta_I \mathbf{i}(x + t) - t) = e^{-t} \exp(\beta_I \mathbf{i}(x + t))$, and the quasiseparated solutions are given by

$$u_R = e^{-t} \cdot \cos(\beta_I(x + t)) \quad \text{and} \quad u_I = e^{-t} \cdot \sin(\beta_I(x + t)).$$

These solutions are exponentially decaying as $t \to \infty$, and thus they are bounded in 'forward time'. For any fixed time t, they are also bounded (and actually periodic) functions of the space variable x. ◇

16G(ii) Boundary conditions

Prerequisites: §5C.

There is no 'cure all' like Proposition 16G.1 for satisfying boundary conditions, since generally they are different in each problem. Generally, a single separated solution (say, from Proposition 16F.3) will *not* be able to satisfy the conditions; we must sum together several solutions, so that they 'cancel out' in suitable ways along the boundaries. For these purposes, the following *Euler identities* are often useful:

$$\sin(x) = \frac{e^{x\mathbf{i}} - e^{-x\mathbf{i}}}{2\mathbf{i}}; \quad \cos(x) = \frac{e^{x\mathbf{i}} + e^{-x\mathbf{i}}}{2\mathbf{i}};$$

$$\sinh(x) = \frac{e^x - e^{-x}}{2}; \quad \cosh(x) = \frac{e^x + e^{-x}}{2}.$$

We can utilize these along with the following boundary information:

$$-\cos'(n\pi) = \sin(n\pi) = 0, \quad \text{for all } n \in \mathbb{Z};$$

$$\sin'\left(\left(n + \frac{1}{2}\right)\pi\right) = \cos\left(\left(n + \frac{1}{2}\right)\pi\right) = 0, \quad \text{for all } n \in \mathbb{Z};$$

$$\cosh'(0) = \sinh(0) = 0.$$

For 'rectangular' domains, the boundaries are obtained by fixing a particular coordinate at a particular value; i.e. they are each of the form $\{\mathbf{x} \in \mathbb{R}^D; x_d = K\}$ for some constant K and some dimension d. The convenient thing about a separated solution is that it is a product of D functions, and only *one* of them is involved in satisfying this boundary condition.

For example, recall Example 16F.2, p. 376, which gave the separated solution $u(x, y) = e^{\mu x} \cdot (\cos(\pm \mu y) + \mathbf{i} \sin(\pm \mu y))$ for the two-dimensional Laplace equation, where $\mu \in \mathbb{R}$. Suppose we want the solution to satisfy the following *homogeneous Dirichlet boundary conditions*:

$$u(x, y) = 0 \quad \text{if} \quad x = 0, \quad \text{or} \quad y = 0, \quad \text{or} \quad y = \pi.$$

To satisfy these three conditions, we proceed as follows. First, let

$$u_1(x, y) = e^{\mu x} \cdot \Big(\cos(\mu y) + \mathbf{i} \sin(\mu y) \Big)$$

and

$$u_2(x, y) = e^{\mu x} \cdot \Big(\cos(-\mu y) + \mathbf{i} \sin(-\mu y) \Big) = e^{\mu x} \cdot \Big(\cos(\mu y) - \mathbf{i} \sin(\mu y) \Big).$$

If we define $v(x, y) = u_1(x, y) - u_2(x, y)$, then

$$v(x, y) = 2e^{\mu x} \cdot \mathbf{i} \sin(\mu y).$$

At this point, $v(x, y)$ already satisfies the boundary conditions for $\{y = 0\}$ and $\{y = \pi\}$. To satisfy the remaining condition, let

$$v_1(x, y) = 2e^{\mu x} \cdot \mathbf{i} \sin(\mu y)$$

and

$$v_1(x, y) = 2e^{-\mu x} \cdot \mathbf{i} \sin(\mu y).$$

If we define $w(x, y) = v_1(x, y) - v_2(x, y)$, then

$$w(x, y) = 4 \sinh(\mu x) \cdot \mathbf{i} \sin(\mu y)$$

also satisfies the boundary condition at $\{x = 0\}$.

17

Impulse-response methods

Nature laughs at the difficulties of integration.

Pierre-Simon Laplace

17A Introduction

A fundamental concept in science is *causality*: an initial event (an *impulse*) at some location $\mathbf{y}$ causes a later event (a *response*) at another location $\mathbf{x}$ (Figure 17A.1(a)). In an evolving, spatially distributed system (e.g. a temperature distribution, a rippling pond, etc.), the system state at each location results from a *combination* of the responses to the impulses from all other locations (as in Figure 17A.1(b)).

If the system is described by a linear PDE, then we expect some sort of 'superposition principle' to apply (see Theorem 4C.3, p. 69). Hence, we can replace the word 'combination' with 'sum', and state the following:

> *The state of the system at $\mathbf{x}$ is a* sum *of the responses to the impulses from all other locations.* (17A.1)

(see Figure 17A.1(b)). However, there are an infinite number – indeed, a continuum – of 'other locations', so we are 'summing' over a *continuum* of responses. But a 'sum' over a continuum is just an *integral*. Hence, statement (17A.1) becomes:

> *In a linear PDE, the solution at $\mathbf{x}$ is an* integral *of the responses to the impulses from all other locations.* (17A.2)

The relation between impulse and response (i.e. between cause and effect) is described by an *impulse-response function*, $\Gamma(\mathbf{y} \to \mathbf{x})$, which measures the degree of 'influence' which point $\mathbf{y}$ has on point $\mathbf{x}$. In other words, $\Gamma(\mathbf{y} \to \mathbf{x})$ measures the strength of the response at $\mathbf{x}$ to an impulse at $\mathbf{y}$. In a system which evolves in time, Γ may also depend on time (since it takes time for the effect from $\mathbf{y}$ to propagate to $\mathbf{x}$), so Γ also depends on time, and is written $\Gamma_t(\mathbf{y} \to \mathbf{x})$.

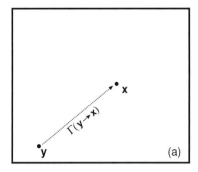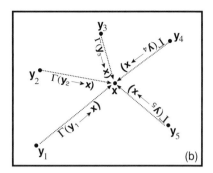

Figure 17A.1. (a) $\Gamma(\mathbf{y} \to \mathbf{x})$ describes the 'response' at $\mathbf{x}$ to an 'impulse' at $\mathbf{y}$. (b) The state at $\mathbf{x}$ is a sum of its responses to the impulses at $\mathbf{y}_1, \mathbf{y}_2, \ldots, \mathbf{y}_5$.

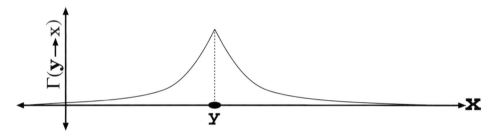

Figure 17A.2. The influence of $\mathbf{y}$ on $\mathbf{x}$ becomes small as the distance from $\mathbf{y}$ to $\mathbf{x}$ grows larger.

Intuitively, Γ should have four properties.

(i) Influence should *decay with distance*. In other words, if $\mathbf{y}$ and $\mathbf{x}$ are close together, then $\Gamma(\mathbf{y} \to \mathbf{x})$ should be large; if $\mathbf{y}$ and $\mathbf{x}$ are far apart, then $\Gamma(\mathbf{y} \to \mathbf{x})$ should be small (Figure 17A.2).

(ii) In a *spatially homogeneous* or *translation-invariant* system (Figure 17A.3(a)), Γ should only depend on the *displacement* from $\mathbf{y}$ to $\mathbf{x}$, so that we can write $\Gamma(\mathbf{y} \to \mathbf{x}) = \gamma(\mathbf{x} - \mathbf{y})$, where γ is some other function.

(iii) In an *isotropic* or *rotation-invariant* system (Figure 17A.3(b)), Γ should only depend on the *distance* between $\mathbf{y}$ and $\mathbf{x}$, so that we can write $\Gamma(\mathbf{y} \to \mathbf{x}) = \psi(|\mathbf{x} - \mathbf{y}|)$, where ψ is a function of one real variable, and $\lim_{r \to \infty} \psi(r) = 0$.

(iv) In a *time-evolving* system, the value of $\Gamma_t(\mathbf{y} \to \mathbf{x})$ should first grow as t increases (as the effect 'propagates' from $\mathbf{y}$ to $\mathbf{x}$), reach a maximum value, and then decrease to zero as t grows large (as the effect 'dissipates' through space) (see Figure 17A.4).

Thus, if there is an 'impulse' of magnitude $\mathcal{I}$ at $\mathbf{y}$, and $\mathcal{R}(\mathbf{x})$ is the 'response' at $\mathbf{x}$, then

$$\mathcal{R}(\mathbf{x}) = \mathcal{I} \cdot \Gamma(\mathbf{y} \to \mathbf{x})$$

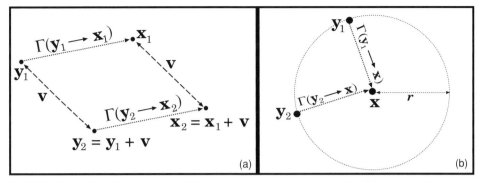

Figure 17A.3. (a) *Translation invariance.* If $\mathbf{y}_2 = \mathbf{y}_1 + \mathbf{v}$ and $\mathbf{x}_2 = \mathbf{x}_1 + \mathbf{v}$, then $\Gamma(\mathbf{y}_2 \to \mathbf{x}_2) = \Gamma(\mathbf{y}_1 \to \mathbf{x}_1)$. (b) *Rotation invariance.* If $\mathbf{y}_1$ and $\mathbf{y}_2$ are both the same distance from $\mathbf{x}$ (i.e. they lie on the circle of radius r around $\mathbf{x}$), then $\Gamma(\mathbf{y}_2 \to \mathbf{x}) = \Gamma(\mathbf{y}_1 \to \mathbf{x})$.

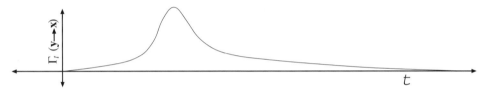

Figure 17A.4. The time-dependent impulse-response function first grows large, and then decays to zero.

(see Figure 17A.5(a)). What if there is an impulse $\mathcal{I}(\mathbf{y}_1)$ at $\mathbf{y}_1$, an impulse $\mathcal{I}(\mathbf{y}_2)$ at $\mathbf{y}_2$, and an impulse $\mathcal{I}(\mathbf{y}_3)$ at $\mathbf{y}_3$? Then statement (17A.1) implies

$$\mathcal{R}(\mathbf{x}) = \mathcal{I}(\mathbf{y}_1) \cdot \Gamma(\mathbf{y}_1 \to \mathbf{x}) + \mathcal{I}(\mathbf{y}_2) \cdot \Gamma(\mathbf{y}_2 \to \mathbf{x}) + \mathcal{I}(\mathbf{y}_3) \cdot \Gamma(\mathbf{y}_3 \to \mathbf{x})$$

(see Figure 17A.5(b)). If $\mathbb{X}$ is the domain of the PDE, then suppose, for every $\mathbf{y}$ in $\mathbb{X}$, that $\mathcal{I}(\mathbf{y})$ is the impulse at $\mathbf{y}$. Then statement (17A.1) takes the following form:

$$\mathcal{R}(\mathbf{x}) = \sum_{\mathbf{y} \in \mathbb{X}} \mathcal{I}(\mathbf{y}) \cdot \Gamma(\mathbf{y} \to \mathbf{x}). \qquad (17A.3)$$

But now we are summing over all $\mathbf{y}$ in $\mathbb{X}$, and, usually, $\mathbb{X} = \mathbb{R}^D$ or some subset, so the 'summation' in equation (17A.3) doesn't make mathematical sense. We must replace the sum with an *integral*, as in statement (17A.2), to obtain

$$\mathcal{R}(\mathbf{x}) = \int_{\mathbb{X}} \mathcal{I}(\mathbf{y}) \cdot \Gamma(\mathbf{y} \to \mathbf{x}) d\mathbf{y}. \qquad (17A.4)$$

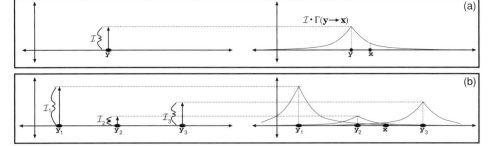

Figure 17A.5. (a) An 'impulse' of magnitude $\mathcal{I}$ at $\mathbf{y}$ triggers a 'response' of magnitude $\mathcal{I} \cdot \Gamma(\mathbf{y} \to \mathbf{x})$ at $\mathbf{x}$. (b) Multiple 'impulses' of magnitude $\mathcal{I}_1$, $\mathcal{I}_2$, and $\mathcal{I}_3$ at $\mathbf{y}_1$, $\mathbf{y}_2$, and $\mathbf{y}_3$, respectively, trigger a 'response' at $\mathbf{x}$ of magnitude $\mathcal{I}_1 \cdot \Gamma(\mathbf{y}_1 \to \mathbf{x}) + \mathcal{I}_2 \cdot \Gamma(\mathbf{y}_2 \to \mathbf{x}) + \mathcal{I}_3 \cdot \Gamma(\mathbf{y}_3 \to \mathbf{x})$.

If the system is spatially homogeneous, then, according to property (ii), this becomes

$$\mathcal{R}(\mathbf{x}) = \int \mathcal{I}(\mathbf{y}) \cdot \gamma(\mathbf{x} - \mathbf{y}) d\mathbf{y}.$$

This integral is called a *convolution*, and is usually written as $\mathcal{I} * \gamma$. In other words,

$$\mathcal{R}(\mathbf{x}) = \mathcal{I} * \gamma(\mathbf{x}), \quad \text{where} \quad \mathcal{I} * \gamma(\mathbf{x}) := \int \mathcal{I}(\mathbf{y}) \cdot \gamma(\mathbf{x} - \mathbf{y}) d\mathbf{y}. \quad (17\text{A}.5)$$

Note that $\mathcal{I} * \gamma$ is a function of $\mathbf{x}$. The variable $\mathbf{y}$ appears on the right-hand side, but as only an *integration* variable.

In a time-dependent system, equation (17A.4) becomes:

$$\mathcal{R}(\mathbf{x}; t) = \int_{\mathbb{X}} \mathcal{I}(\mathbf{y}) \cdot \Gamma_t(\mathbf{y} \to \mathbf{x}) d\mathbf{y},$$

while equation (17A.5) becomes:

$$\mathcal{R}(\mathbf{x}; t) = \mathcal{I} * \gamma_t(\mathbf{x}), \quad \text{where} \quad \mathcal{I} * \gamma_t(\mathbf{x}) = \int \mathcal{I}(\mathbf{y}) \cdot \gamma_t(\mathbf{x} - \mathbf{y}) d\mathbf{y}. \quad (17\text{A}.6)$$

The following surprising property is often useful.

Proposition 17A.1 *If* $f, g : \mathbb{R}^D \longrightarrow \mathbb{R}$ *are integrable functions, then* $g * f = f * g$.

Proof (Case $D = 1$.) Fix $x \in \mathbb{R}$. Then

$$(g * f)(x) = \int_{-\infty}^{\infty} g(y) \cdot f(x - y) dy \underset{(s)}{=} \int_{\infty}^{-\infty} g(x - z) \cdot f(z) \cdot (-1) dz$$

$$= \int_{-\infty}^{\infty} f(z) \cdot g(x - z) dz = (f * g)(x).$$

Here, step (s) was the substitution $z = x - y$, so that $y = x - z$ and $dy = -dz$. $\square$

Exercise 17A.1 Generalize this proof to the case $D \geq 2$. ◆ Ⓔ

Remarks
(a) Depending on the context, impulse-response functions are sometimes called *solution kernels*, or *Green's functions* or *impulse functions*.
(b) If f and g are analytic functions, then there is an efficient way to compute $f * g$ using complex analysis; see Corollary 18H.3, p. 479.

17B Approximations of identity

17B(i) ... in one dimension

Prerequisites: §17A.

Suppose $\gamma : \mathbb{R} \times \mathbb{R}_+ \longrightarrow \mathbb{R}$ was a one-dimensional *impulse response function*, as in equation (17A.6). Thus, if $\mathcal{I} : \mathbb{R} \longrightarrow \mathbb{R}$ is a function describing the initial 'impulse', then, for any time $t > 0$, the 'response' is given by the function $\mathcal{R}_t$ defined:

$$\mathcal{R}_t(x) := \mathcal{I} * \gamma_t(x) = \int_{-\infty}^{\infty} \mathcal{I}(y) \cdot \gamma_t(x - y) dy. \tag{17B.1}$$

Intuitively, if t is close to zero, then the response $\mathcal{R}_t$ should be concentrated near the locations where the impulse $\mathcal{I}$ is concentrated (because the energy has not yet been able to propagate very far). By inspecting equation (17B.1), we see that this means that the mass of γ_t should be 'concentrated' near zero. Formally, we say that γ is an *approximation of the identity* if it has the following properties (Figure 17B.1):

(AI1) $\gamma_t(x) \geq 0$ everywhere and $\int_{-\infty}^{\infty} \gamma_t(x) dx = 1$ for any fixed $t > 0$;
(AI2) for any $\epsilon > 0$, $\lim_{t \to 0} \int_{-\epsilon}^{\epsilon} \gamma_t(x) dx = 1$.

Property (AI1) says that γ_t is a probability density; (AI2) says that γ_t concentrates all of its 'mass' at zero as $t \to 0$. (Heuristically speaking, the function γ_t is converging to the 'Dirac delta function' δ_0 as $t \to 0$.)

Example 17B.1
(a) Let

$$\gamma_t(x) = \begin{cases} 1/t & \text{if} \quad 0 \leq x \leq t; \\ 0 & \text{if} \quad x < 0 \text{ or } t < x \end{cases}$$

(see Figure 17B.2). Thus, for any $t > 0$, the graph of γ_t is a 'box' of width t and height $1/t$. Then γ is an approximation of identity. (See Problem 17.11, p. 418.)
(b) Let

$$\gamma_t(x) = \begin{cases} 1/2t & \text{if} \quad |x| \leq t; \\ 0 & \text{if} \quad t < |x|. \end{cases}$$

Thus, for any $t > 0$, the graph of γ_t is a 'box' of width $2t$ and height $1/2t$. Then γ is an approximation of identity. (See Problem 17.12, p. 418.) ◇

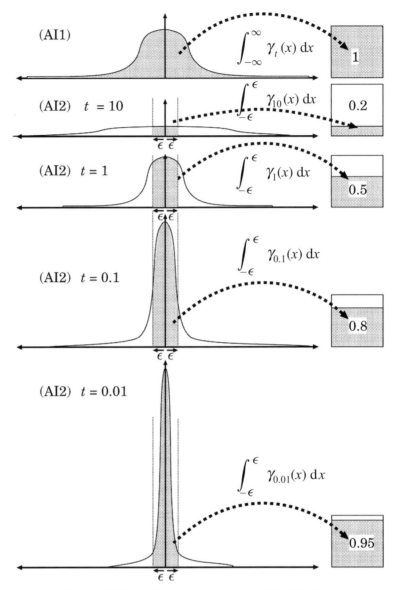

Figure 17B.1. γ is an approximation of the identity.

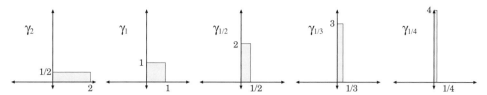

Figure 17B.2. Graph of γ_t for $t = 2, 1, 1/2, 1/3,$ and $1/4$. See Example 17B.1(a).

A function satisfying properties (AI1) and (AI2) is called an *approximation of the identity* because of the following result.

Proposition 17B.2 *Let* $\gamma : \mathbb{R} \times \mathbb{R}_+ \longrightarrow \mathbb{R}$ *be an approximation of identity.*

(a) *Let* $\mathcal{I} : \mathbb{R} \longrightarrow \mathbb{R}$ *be a bounded continuous function. Then, for all* $x \in \mathbb{R}$, *we have* $\lim_{t \to 0} \mathcal{I} * \gamma_t(x) = \mathcal{I}(x)$.

(b) *Let* $\mathcal{I} : \mathbb{R} \longrightarrow \mathbb{R}$ *be any bounded integrable function. If* $x \in \mathbb{R}$ *is any continuity point of* $\mathcal{I}$, *then* $\lim_{t \to 0} \mathcal{I} * \gamma_t(x) = \mathcal{I}(x)$.

Proof (a) Fix $x \in \mathbb{R}$. Given any $\epsilon > 0$, find $\delta > 0$ such that

$$\text{for all } y \in \mathbb{R}, \quad \left(|y - x| < \delta \right) \Longrightarrow \left(\left| \mathcal{I}(y) - \mathcal{I}(x) \right| < \epsilon/3 \right).$$

(Such an ϵ exists because $\mathcal{I}$ is continuous.) Thus,

$$\left| \mathcal{I}(x) \cdot \int_{x-\delta}^{x+\delta} \gamma_t(x - y)\mathrm{d}y - \int_{x-\delta}^{x+\delta} \mathcal{I}(y) \cdot \gamma_t(x - y)\mathrm{d}y \right|$$

$$= \left| \int_{x-\delta}^{x+\delta} \left(\mathcal{I}(x) - \mathcal{I}(y) \right) \cdot \gamma_t(x - y)\mathrm{d}y \right| \leq \int_{x-\delta}^{x+\delta} \left| \mathcal{I}(x) - \mathcal{I}(y) \right| \cdot \gamma_t(x - y)\mathrm{d}y$$

$$< \frac{\epsilon}{3} \int_{x-\delta}^{x+\delta} \gamma_t(x - y)\mathrm{d}y \underset{\text{(AI1)}}{<} \frac{\epsilon}{3}. \tag{17B.2}$$

(Here (AI1) is by property (AI1) of γ_t.)

Recall that $\mathcal{I}$ is *bounded*. Suppose $|\mathcal{I}(y)| < M$ for all $y \in \mathbb{R}$; using (AI2), find some small $\tau > 0$ such that, if $t < \tau$, then

$$\int_{x-\delta}^{x+\delta} \gamma_t(y)\mathrm{d}y > 1 - \frac{\epsilon}{3M};$$

hence,

$$\int_{-\infty}^{x-\delta} \gamma_t(y)\mathrm{d}y + \int_{x+\delta}^{\infty} \gamma_t(y)\mathrm{d}y = \int_{-\infty}^{\infty} \gamma_t(y)\mathrm{d}y - \int_{x-\delta}^{x+\delta} \gamma_t(y)\mathrm{d}y$$

$$\underset{\text{(AI1)}}{<} 1 - \left(1 - \frac{\epsilon}{3M} \right) = \frac{\epsilon}{3M}. \tag{17B.3}$$

(Here (AI1) is by property (AI1) of γ_t.) Thus,

$$\left| \mathcal{I} * \gamma_t(x) - \int_{x-\delta}^{x+\delta} \mathcal{I}(y) \cdot \gamma_t(x-y)dy \right|$$

$$\leq \left| \int_{-\infty}^{\infty} \mathcal{I}(y) \cdot \gamma_t(x-y)dy - \int_{x-\delta}^{x+\delta} \mathcal{I}(y) \cdot \gamma_t(x-y)dy \right|$$

$$= \left| \int_{-\infty}^{x-\delta} \mathcal{I}(y) \cdot \gamma_t(x-y)dy + \int_{x+\delta}^{\infty} \mathcal{I}(y) \cdot \gamma_t(x-y)dy \right|$$

$$\leq \int_{-\infty}^{x-\delta} \left| \mathcal{I}(y) \cdot \gamma_t(x-y) \right| dy + \int_{x+\delta}^{\infty} \left| \mathcal{I}(y) \cdot \gamma_t(x-y) \right| dy$$

$$\leq \int_{-\infty}^{x-\delta} M \cdot \gamma_t(x-y)dy + \int_{x+\delta}^{\infty} M \cdot \gamma_t(x-y)dy$$

$$\leq M \cdot \left(\int_{-\infty}^{x-\delta} \gamma_t(x-y)dy + \int_{x+\delta}^{\infty} \gamma_t(x-y)dy \right)$$

$$\underset{(*)}{\leq} M \cdot \frac{\epsilon}{3M} = \frac{\epsilon}{3}. \tag{17B.4}$$

(Here, (*) is by equation (17B.3).) Combining equations (17B.2) and (17B.4), we have the following:

$$\left| \mathcal{I}(x) \cdot \int_{x-\delta}^{x+\delta} \gamma_t(x-y)dy - \mathcal{I} * \gamma_t(x) \right|$$

$$\leq \left| \mathcal{I}(x) \cdot \int_{x-\delta}^{x+\delta} \gamma_t(x-y)dy - \int_{x-\delta}^{x+\delta} \mathcal{I}(y) \cdot \gamma_t(x-y)dy \right|$$

$$+ \left| \int_{x-\delta}^{x+\delta} \mathcal{I}(y) \cdot \gamma_t(x-y)dy - \mathcal{I} * \gamma_t(x) \right|$$

$$\leq \frac{\epsilon}{3} + \frac{\epsilon}{3} = \frac{2\epsilon}{3}. \tag{17B.5}$$

But if $t < \tau$, then

$$\left| 1 - \int_{x-\delta}^{x+\delta} \gamma_t(x-y)dy \right| < \frac{\epsilon}{3M}.$$

Thus,

$$\left| \mathcal{I}(x) - \mathcal{I}(x) \cdot \int_{x-\delta}^{x+\delta} \gamma_t(x-y)dy \right| \leq |\mathcal{I}(x)| \cdot \left| 1 - \int_{x-\delta}^{x+\delta} \gamma_t(x-y)dy \right|$$

$$< |\mathcal{I}(x)| \cdot \frac{\epsilon}{3M} \leq M \cdot \frac{\epsilon}{3M} = \frac{\epsilon}{3}. \tag{17B.6}$$

Combining equations (17B.5) and (17B.6), we have the following:

$$|\mathcal{I}(x) - \mathcal{I} * \gamma_t(x)|$$

$$\leq \left| \mathcal{I}(x) - \mathcal{I}(x) \cdot \int_{x-\delta}^{x+\delta} \gamma_t(x - y) \mathrm{d}y \right| + \left| \mathcal{I}(x) \cdot \int_{x-\delta}^{x+\delta} \gamma_t(x - y) \mathrm{d}y - \mathcal{I} * \gamma_t(x) \right|$$

$$\leq \frac{\epsilon}{3} + \frac{2\epsilon}{3} = \epsilon.$$

Since ϵ can be made arbitrarily small, we are done.

(b) **Exercise 17B.1.** (*Hint:* Imitate part (a).)   □   Ⓔ

In other words, as $t \to 0$, the convolution $\mathcal{I} * \gamma_t$ resembles $\mathcal{I}$ with arbitrarily high accuracy. Similar convergence results can be proved in other norms (e.g. L^2-convergence, uniform convergence).

Example 17B.3 Let

$$\gamma_t(x) = \begin{cases} 1/t & \text{if} \quad 0 \leq x \leq t; \\ 0 & \text{if} \quad x < 0 \text{ or } t < x, \end{cases}$$

as in Example 17B.1(a). Suppose $\mathcal{I} : \mathbb{R} \longrightarrow \mathbb{R}$ is a continuous function. Then, for any $x \in \mathbb{R}$,

$$\mathcal{I} * \gamma_t(x) = \int_{-\infty}^{\infty} \mathcal{I}(y) \cdot \gamma_t(x - y) \mathrm{d}y = \frac{1}{t} \int_{x-t}^{x} \mathcal{I}(y) \mathrm{d}y = \frac{1}{t} \Big(\mathcal{J}(x) - \mathcal{J}(x - t) \Big),$$

where $\mathcal{J}$ is an antiderivative of $\mathcal{I}$. Thus, as implied by Proposition 17B.2,

$$\lim_{t \to 0} \mathcal{I} * \gamma_t(x) = \lim_{t \to 0} \frac{\mathcal{J}(x) - \mathcal{J}(x - t)}{t} \underset{(*)}{=\!=} \mathcal{J}'(x) \underset{(\dagger)}{=\!=} \mathcal{I}(x).$$

(Here (∗) is just the definition of differentiation, and (†) is because $\mathcal{J}$ is an antiderivative of $\mathcal{I}$.)   ◇

17B(ii) ... in many dimensions

Prerequisites: §17B(i).   **Recommended:** §17C(i).

A non-negative function $\gamma : \mathbb{R}^D \times \mathbb{R}_+ \longrightarrow \mathbb{R}_+$ is called an *approximation of the identity* if it has the following two properties:

(AI1) $\int_{\mathbb{R}^D} \gamma_t(\mathbf{x}) \mathrm{d}\mathbf{x} = 1$ for all $t \in \mathbb{R}_+$;
(AI2) for any $\epsilon > 0$, $\lim_{t \to 0} \int_{\mathbb{B}(0;\epsilon)} \gamma_t(\mathbf{x}) \mathrm{d}\mathbf{x} = 1$.

Property (AI1) says that γ_t is a probability density; (AI2) says that γ_t concentrates all of its 'mass' at zero as $t \to 0$.

Example 17B.4 Define $\gamma : \mathbb{R}^2 \times \mathbb{R}_+ \longrightarrow \mathbb{R}$ by

$$\gamma_t(x, y) = \begin{cases} 1/4t^2 & \text{if } |x| \leq t \text{ and } |y| \leq t; \\ 0 & \text{otherwise.} \end{cases}$$

Ⓔ   Then γ is an approximation of the identity on $\mathbb{R}^2$ (**Exercise 17B.2**).   ◇

Proposition 17B.5 *Let* $\gamma : \mathbb{R}^D \times \mathbb{R}_+ \longrightarrow \mathbb{R}$ *be an approximation of the identity.*

(a) *Let* $\mathcal{I} : \mathbb{R}^D \longrightarrow \mathbb{R}$ *be a bounded continuous function. Then, for every* $\mathbf{x} \in \mathbb{R}^D$, *we have* $\lim_{t \to 0} \mathcal{I} * \gamma_t(\mathbf{x}) = \mathcal{I}(\mathbf{x})$.

(b) *Let* $\mathcal{I} : \mathbb{R}^D \longrightarrow \mathbb{R}$ *be any bounded integrable function. If* $\mathbf{x} \in \mathbb{R}^D$ *is any continuity point of* $\mathcal{I}$, *then* $\lim_{t \to 0} \mathcal{I} * \gamma_t(\mathbf{x}) = \mathcal{I}(\mathbf{x})$.

Ⓔ   *Proof*   **Exercise 17B.3.** (*Hint:* The argument is basically identical to that of Proposition 17B.2; just replace the interval $(-\epsilon, \epsilon)$ with a ball of radius ϵ.)   □

In other words, as $t \to 0$, the convolution $\mathcal{I} * \gamma_t$ resembles $\mathcal{I}$ with arbitrarily high accuracy. Similar convergence results can be proved in other norms (e.g. L^2-convergence, uniform convergence).

When solving partial differential equations, approximations of identity are invariably used in conjunction with the following result.

Proposition 17B.6 *Let* L *be a linear differential operator on* $\mathcal{C}^\infty(\mathbb{R}^D; \mathbb{R})$.

(a) *If* $\gamma : \mathbb{R}^D \longrightarrow \mathbb{R}$ *is a solution to the homogeneous equation '*$\mathsf{L}\gamma = 0$*', then, for any function* $\mathcal{I} : \mathbb{R}^D \longrightarrow \mathbb{R}$, *the function* $u = \mathcal{I} * \gamma$ *satisfies* $\mathsf{L}u = 0$.

(b) *If* $\gamma : \mathbb{R}^D \times \mathbb{R}_+ \longrightarrow \mathbb{R}$ *satisfies the evolution equation '*$\partial_t^n \gamma = \mathsf{L}\gamma$*', and we define* $\gamma_t(\mathbf{x}) := \gamma(\mathbf{x}; t)$, *then, for any function* $\mathcal{I} : \mathbb{R}^D \longrightarrow \mathbb{R}$, *the function* $u_t = \mathcal{I} * \gamma_t$ *satisfies:* $\partial_t^n u = \mathsf{L}u$.

Ⓔ   *Proof*   **Exercise 17B.4** (*Hint:* Generalize the proof of Proposition 17C.1, by replacing the one-dimensional convolution integral with a D-dimensional convolution integral, and by replacing the Laplacian with an arbitrary linear operator L.)

□

Corollary 17B.7 *Suppose* γ *is an approximation of the identity and satisfies the evolution equation '*$\partial_t^n \gamma = \mathsf{L}\gamma$*'. For any* $\mathcal{I} : \mathbb{R}^D \longrightarrow \mathbb{R}$, *define* $u : \mathbb{R}^D \times \mathbb{R}_{\not\!+} \longrightarrow \mathbb{R}$:

- $u(\mathbf{x}; 0) = \mathcal{I}(\mathbf{x})$;
- $u_t = \mathcal{I} * \gamma_t$, *for all* $t > 0$.

Then u *is a solution to the equation '*$\partial_t^n u = \mathsf{L}u$*', and* u *satisfies the initial conditions* $u(\mathbf{x}, 0) = \mathcal{I}(\mathbf{x})$ *for all* $\mathbf{x} \in \mathbb{R}^D$.

Proof Combine Propositions 17B.5 and 17B.6.   □

We say that γ is the *fundamental solution* (or *solution kernel*, or *Green's function* or *impulse function*) for the PDE. For example, the D-dimensional Gauss–Weierstrass kernel is a fundamental solution for the D-dimensional heat equation.

17C The Gaussian convolution solution (heat equation)

17C(i) ...in one dimension

Prerequisites: §1B(i), §17B(i), Appendix G.　　　Recommended: §17A, §20A(ii).

Given two functions $\mathcal{I}, \mathcal{G} : \mathbb{R} \longrightarrow \mathbb{R}$, recall (from §17A) that their *convolution* is the function $\mathcal{I} * \mathcal{G} : \mathbb{R} \longrightarrow \mathbb{R}$ defined:

$$\mathcal{I} * \mathcal{G}(x) := \int_{-\infty}^{\infty} \mathcal{I}(y) \cdot \mathcal{G}(x - y)\mathrm{d}y, \qquad \text{for all } x \in \mathbb{R}.$$

Recall the *Gauss–Weierstrass kernel* from Example 1B.1, p. 8:

$$\mathcal{G}_t(x) := \frac{1}{2\sqrt{\pi t}} \exp\left(\frac{-x^2}{4t}\right), \qquad \text{for all } x \in \mathbb{R} \text{ and } t > 0.$$

We will use $\mathcal{G}_t(x)$ as an *impulse-response function* to solve the one-dimensional heat equation.

Proposition 17C.1 *Let* $\mathcal{I} : \mathbb{R} \longrightarrow \mathbb{R}$ *be a bounded integrable function. Define* $u : \mathbb{R} \times \mathbb{R}_+ \longrightarrow \mathbb{R}$ *by* $u(x;t) := \mathcal{I} * \mathcal{G}_t(x)$ *for all* $x \in \mathbb{R}$ *and* $t > 0$. *Then* u *is a solution to the one-dimensional heat equation.*

Proof For any fixed $y \in \mathbb{R}$, define $u_y(x;t) = \mathcal{I}(y) \cdot \mathcal{G}_t(x - y)$.

Claim 1 $u_y(x;t)$ *is a solution of the one-dimensional heat equation.*

Proof First note that $\partial_t \mathcal{G}_t(x - y) = \partial_x^2 \mathcal{G}_t(x - y)$ (**Exercise 17C.1**).　　　Ⓔ
　　Now, y is a constant, so we treat $\mathcal{I}(y)$ as a constant when differentiating by x or by t. Thus,

$$\partial_t u_y(x, t) = \mathcal{I}(y) \cdot \partial_t \mathcal{G}_t(x - y) = \mathcal{I}(y) \cdot \partial_x^2 \mathcal{G}_t(x - y)$$
$$= \partial_x^2 u_y(x, t) = \Delta u_y(x, t),$$

as desired.　　　　　　　　　　　　　　　　　　　　◇ Claim 1

Now,

$$u(x, t) = \mathcal{I} * \mathcal{G}_t = \int_{-\infty}^{\infty} \mathcal{I}(y) \cdot \mathcal{G}_t(x - y)\mathrm{d}y = \int_{-\infty}^{\infty} u_y(x;t)\mathrm{d}y.$$

Thus,

$$\partial_t u(x, t) \underset{(*)}{=} \int_{-\infty}^{\infty} \partial_t u_y(x; t)\mathrm{d}y \underset{(\dagger)}{=} \int_{-\infty}^{\infty} \Delta u_y(x; t)\mathrm{d}y \underset{(*)}{=} \Delta u(x, t).$$

Here, ($\dagger$) is by Claim 1, and ($*$) is by Proposition G.1, p. 571.   □

Ⓔ **Exercise 17C.2**   Verify that the conditions of Proposition G.1 are satisfied.   ◆

Remark One way to visualize the 'Gaussian convolution' $u(x; t) = \mathcal{I} * \mathcal{G}_t(x)$ is as follows. Consider a finely spaced 'ϵ-mesh' of points on the real line,

$$\epsilon \cdot \mathbb{Z} = \{n\epsilon ; \ n \in \mathbb{Z}\}.$$

For every $n \in \mathbb{Z}$, define the function $\mathcal{G}_t^{(n)}(x) = \mathcal{G}_t(x - n\epsilon)$. For example, $\mathcal{G}_t^{(5)}(x) = \mathcal{G}_t(x - 5\epsilon)$ looks like a copy of the Gauss–Weierstrass kernel, but centred at 5ϵ (see Figure 17C.1(a)).

For each $n \in \mathbb{Z}$, let $I_n = \mathcal{I}(n \cdot \epsilon)$ (see Figure 17C.1(c)). Now consider the infinite linear combination of Gauss–Weierstrass kernels (see Figure 17C.1(d)):

$$u_\epsilon(x; t) = \epsilon \cdot \sum_{n=-\infty}^{\infty} I_n \cdot \mathcal{G}_t^{(n)}(x).$$

Now imagine that the ϵ-mesh becomes 'infinitely dense', by letting $\epsilon \to 0$. Define $u(x; t) = \lim_{\epsilon \to 0} u_\epsilon(x; t)$. I claim that $u(x; t) = \mathcal{I} * \mathcal{G}_t(x)$. To see this, note that

$$u(x; t) = \lim_{\epsilon \to 0} \epsilon \cdot \sum_{n=-\infty}^{\infty} I_n \cdot \mathcal{G}_t^{(n)}(x) = \lim_{\epsilon \to 0} \epsilon \cdot \sum_{n=-\infty}^{\infty} \mathcal{I}(n\epsilon) \cdot \mathcal{G}_t(x - n\epsilon)$$

$$\underset{(*)}{=} \int_{-\infty}^{\infty} \mathcal{I}(y) \cdot \mathcal{G}_t(x - y)\mathrm{d}y = \mathcal{I} * \mathcal{G}_t(y),$$

as shown in Figure 17C.2.

Ⓔ **Exercise 17C.3**   Rigorously justify step ($*$) in the previous computation. (*Hint:* Use a Riemann sum.)   ◆

Proposition 17C.2 *The Gauss–Weierstrass kernel is an approximation of identity (see §17B(i)), meaning that it satisfies the following two properties:*

(AI1) *$\mathcal{G}_t(x) \geq 0$ everywhere, and $\int_{-\infty}^{\infty} \mathcal{G}_t(x)\mathrm{d}x = 1$ for any fixed $t > 0$;*
(AI2) *for any $\epsilon > 0$, $\lim_{t \to 0} \int_{-\epsilon}^{\epsilon} \mathcal{G}_t(x)\mathrm{d}x = 1$.*

Ⓔ *Proof* **Exercise 17C.4.**   □

Corollary 17C.3 *Let $\mathcal{I} : \mathbb{R} \longrightarrow \mathbb{R}$ be a bounded integrable function. Define the function $u : \mathbb{R} \times \mathbb{R}_+ \longrightarrow \mathbb{R}$ by*

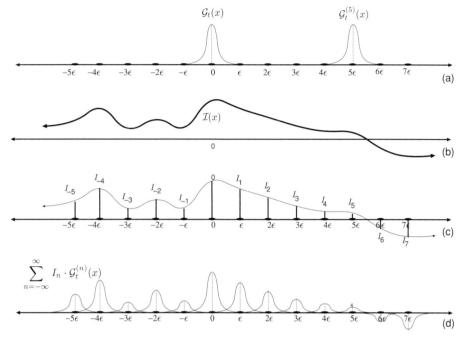

Figure 17C.1. Discrete convolution: a superposition of Gaussians.

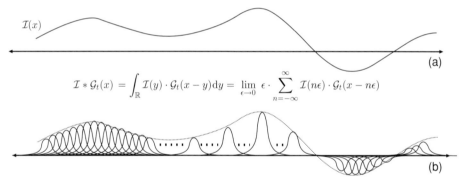

$$\mathcal{I} * \mathcal{G}_t(x) = \int_{\mathbb{R}} \mathcal{I}(y) \cdot \mathcal{G}_t(x-y)\mathrm{d}y = \lim_{\epsilon \to 0} \epsilon \cdot \sum_{n=-\infty}^{\infty} \mathcal{I}(n\epsilon) \cdot \mathcal{G}_t(x-n\epsilon)$$

Figure 17C.2. Convolution as a limit of 'discrete' convolutions.

- $u_0(x) := \mathcal{I}(x)$ *for all* $x \in \mathbb{R}$ *(initial conditions);*
- $u_t := \mathcal{I} * \mathcal{G}_t$, *for all* $t > 0$.

Then u is a solution to the one-dimensional heat equation. Furthermore:

(a) *if $\mathcal{I}$ is continuous on $\mathbb{R}$, then u is continuous on $\mathbb{R} \times \mathbb{R}_{\not=}$;*

(b) *even if $\mathcal{I}$ is not continuous, the function u is still continuous on $\mathbb{R} \times \mathbb{R}_+$, and u is also continuous at $(x, 0)$ for any $x \in \mathbb{R}$ where $\mathcal{I}$ is continuous.*

Proof Proposition 17C.1 says that u is a solution to the heat equation. Combine Proposition 17C.2 with Proposition 17B.2, p. 387, to verify the continuity assertions (a) and (b). □

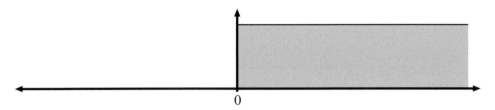

Figure 17C.3. The Heaviside step function $\mathcal{H}(x)$.

The 'continuity' part of Corollary 17C.3 means that u is the solution to the *initial value problem* for the heat equation with *initial conditions* $\mathcal{I}$. Because of Corollary 17C.3, we say that $\mathcal{G}$ is the *fundamental solution* (or *solution kernel*, or *Green's function* or *impulse function*) for the heat equation.

Example 17C.4 The Heaviside step function
Consider the *Heaviside step function*

$$\mathcal{H}(x) = \begin{cases} 1 & \text{if } x \geq 0; \\ 0 & \text{if } x < 0 \end{cases}$$

(see Figure 17C.3). The solution to the one-dimensional heat equation with initial conditions $u(x, 0) = \mathcal{H}(x)$ is given by

$$u(x, t) \underset{(*)}{=} \mathcal{H} * \mathcal{G}_t(x) \underset{(\dagger)}{=} \mathcal{G}_t * \mathcal{H}(x) = \int_{-\infty}^{\infty} \mathcal{G}_t(y) \cdot \mathcal{H}(x - y) dy$$

$$= \frac{1}{2\sqrt{\pi t}} \int_{-\infty}^{\infty} \exp\left(\frac{-y^2}{4t}\right) \mathcal{H}(x - y) dy \underset{(\ddagger)}{=} \frac{1}{2\sqrt{\pi t}} \int_{-\infty}^{x} \exp\left(\frac{-y^2}{4t}\right) dy$$

$$\underset{(\diamond)}{=} \frac{1}{\sqrt{2\pi}} \int_{-\infty}^{x/\sqrt{2t}} \exp\left(\frac{-z^2}{2}\right) dz = \Phi\left(\frac{x}{\sqrt{2t}}\right).$$

Here, $(*)$ is by Proposition 17C.1, p. 391; $(\dagger)$ is by Prop. 17A.1, p. 384; $(\ddagger)$ is because

$$\mathcal{H}(x - y) = \begin{cases} 1 & \text{if } y \leq x; \\ 0 & \text{if } y > x; \end{cases}$$

and $(\diamond)$ is where we make the substitution $z = y/\sqrt{2t}$; thus, $dy = \sqrt{2t} \, dz$.

Here, $\Phi(x)$ is the *cumulative distribution function* of the standard normal probability measure,[1] defined:

$$\boxed{\Phi(x) := \frac{1}{\sqrt{2\pi}} \int_{-\infty}^{x} \exp\left(\frac{-z^2}{2}\right) dz.}$$

---

[1] This is sometimes called the *error function* or *sigmoid function*. Unfortunately, no simple formula exists for $\Phi(x)$. It can be computed with arbitrary accuracy using a Taylor series, and tables of values for $\Phi(x)$ can be found in most statistics texts.

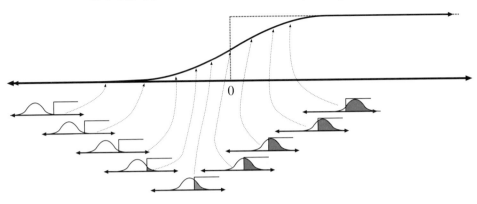

Figure 17C.4. $u_t(x) = (\mathcal{H} * \mathcal{G}_t)(x)$ evaluated at several $x \in \mathbb{R}$.

(see Figure 17C.4). At time zero, $u(x, 0) = \mathcal{H}(x)$ is a step function. For $t > 0$, $u(x, t)$ looks like a compressed version of $\Phi(x)$: a steep sigmoid function. As t increases, this sigmoid becomes broader and flatter (see Figure 17C.5). ◇

When computing convolutions, you can often avoid a lot of messy integrals by exploiting the following properties.

Proposition 17C.5 *Let $f, g : \mathbb{R} \longrightarrow \mathbb{R}$ be integrable functions. Then*

(a) *if $h : \mathbb{R} \longrightarrow \mathbb{R}$ is another integrable function, then $f * (g + h) = (f * g) + (f * h)$;*
(b) *if $r \in \mathbb{R}$ is a constant, then $f * (r \cdot g) = r \cdot (f * g)$.*
(c) *Suppose $d \in \mathbb{R}$ is some 'displacement', and we define $f_{\triangleright d}(x) = f(x - d)$. Then $(f_{\triangleright d} * g)(x) = (f * g)(x - d)$ (i.e. $(f_{\triangleright d}) * g = (f * g)_{\triangleright d}$.)*

Proof See Problems 17.2 and 17.3, p. 416 ☐

Example 17C.6 A staircase function

Suppose $\mathcal{I}(x) = \begin{cases} 0 & \text{if } x < 0; \\ 1 & \text{if } 0 \le x < 1; \\ 2 & \text{if } 1 \le x < 2; \\ 0 & \text{if } 2 \le x \end{cases}$

(see Figure 17C.6(a)). Let $\Phi(x)$ be the sigmoid function from Example 17C.4. Then

$$u(x, t) = \Phi\left(\frac{x}{\sqrt{2t}}\right) + \Phi\left(\frac{x - 1}{\sqrt{2t}}\right) - 2\Phi\left(\frac{x - 2}{\sqrt{2t}}\right)$$

(see Figure 17C.6(b)).
To see this, observe that we can write

$$\mathcal{I}(x) = \mathcal{H}(x) + \mathcal{H}(x - 1) - 2\mathcal{H}(x - 2) \qquad (17C.1)$$

$$= \mathcal{H} + \mathcal{H}_{\triangleright 1}(x) - 2\mathcal{H}_{\triangleright 2}(x), \qquad (17C.2)$$

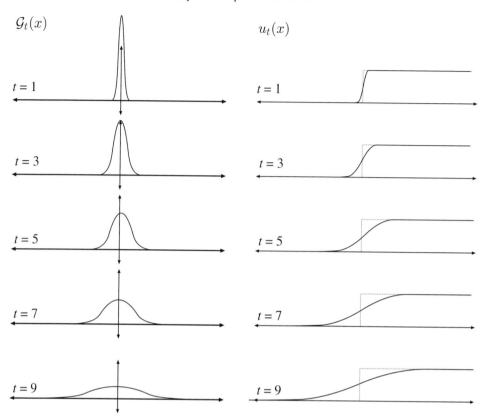

Figure 17C.5. $u_t(x) = (\mathcal{H} * \mathcal{G}_t)(x)$ for several $t > 0$.

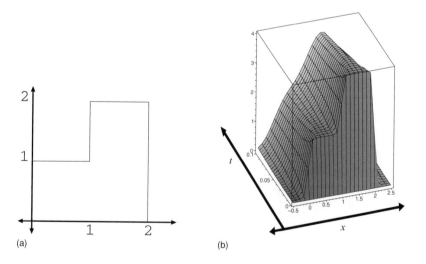

Figure 17C.6. (a) A staircase function. (b) The resulting solution to the heat equation.

where equation (17C.2) uses the notation of Proposition 17C.5(c). Thus,

$$u(x;t) \underset{(*)}{=} \mathcal{I} * \mathcal{G}_t(x) \underset{(\dagger)}{=} \left(\mathcal{H} + \mathcal{H}_{\triangleright 1} - 2\mathcal{H}_{\triangleright 2} \right) * \mathcal{G}_t(x)$$

$$\underset{(\ddagger)}{=} \mathcal{H} * \mathcal{G}_t(x) + \mathcal{H}_{\triangleright 1} * \mathcal{G}_t(x) - 2\mathcal{H}_{\triangleright 2} * \mathcal{G}_t(x)$$

$$\underset{(\diamond)}{=} \mathcal{H} * \mathcal{G}_t(x) + \mathcal{H} * \mathcal{G}_t(x - 1) - 2\mathcal{H} * \mathcal{G}_t(x - 2)$$

$$\underset{(\P)}{=} \Phi \left(\frac{x}{\sqrt{2t}} \right) + \Phi \left(\frac{x - 1}{\sqrt{2t}} \right) - 2\Phi \left(\frac{x - 2}{\sqrt{2t}} \right). \tag{17C.3}$$

Here, (*) is by Proposition 17C.1, p. 391; (†) is by equation (17C.2); (‡) is by Proposition 17C.5(a) and (b); (◇) is by Proposition 17C.5(c); and (¶) is by Example 17C.4.

Another approach Begin with equation (17C.1), and, rather than using Proposition 17C.5, use instead the linearity of the heat equation, along with Theorem 4C.3, p. 69, to deduce that the solution must have the following form:

$$u(x, t) = u_0(x, t) + u_1(x, t) - 2u_2(x, t), \tag{17C.4}$$

where

- $u_0(x, t)$ is the solution with initial conditions $u_0(x, 0) = \mathcal{H}(x)$;
- $u_1(x, t)$ is the solution with initial conditions $u_1(x, 0) = \mathcal{H}(x - 1)$;
- $u_2(x, t)$ is the solution with initial conditions $u_2(x, 0) = \mathcal{H}(x - 2)$.

But then we know, from Example 17C.4, that

$$u_0(x, t) = \Phi \left(\frac{x}{\sqrt{2t}} \right); \quad u_1(x, t) = \Phi \left(\frac{x - 1}{\sqrt{2t}} \right); \quad \text{and} \quad u_2(x, t) = \Phi \left(\frac{x - 2}{\sqrt{2t}} \right).$$

$$\tag{17C.5}$$

Now combine equation (17C.4) with equations (17C.5) to obtain again the solution (17C.3). ◇

Remark The Gaussian convolution solution to the heat equation is revisited in §20A(ii), p. 534, using the methods of Fourier transforms.

17C(ii) ... in many dimensions

Prerequisites: §1B(ii), §17B(ii). **Recommended:** §17A, §17C(i).

Given two functions $\mathcal{I}, \mathcal{G} : \mathbb{R}^D \longrightarrow \mathbb{R}$, their *convolution* is the function $\mathcal{I} * \mathcal{G} : \mathbb{R}^D \longrightarrow \mathbb{R}$ defined:

$$\mathcal{I} * \mathcal{G}(\mathbf{x}) := \int_{\mathbb{R}^D} \mathcal{I}(\mathbf{y}) \cdot \mathcal{G}(\mathbf{x} - \mathbf{y}) d\mathbf{y}.$$

Note that $\mathcal{I} * \mathcal{G}$ is a function of $\mathbf{x}$. The variable $\mathbf{y}$ appears on the right-hand side, but as an *integration* variable.

Consider the the D-dimensional Gauss–Weierstrass kernel:

$$\mathcal{G}_t(\mathbf{x}) := \frac{1}{(4\pi t)^{D/2}} \exp\left(\frac{-\|\mathbf{x}\|^2}{4t}\right), \quad \text{for all } \mathbf{x} \in \mathbb{R}^D \text{ and } t > 0.$$

(See Examples 1B.2(b) and (c), p. 10.) We will use $\mathcal{G}_t(x)$ as an *impulse-response function* to solve the D-dimensional heat equation.

Theorem 17C.7 *Suppose* $\mathcal{I} : \mathbb{R}^D \longrightarrow \mathbb{R}$ *is a bounded continuous function. Define the function* $u : \mathbb{R}^D \times \mathbb{R}_+ \longrightarrow \mathbb{R}$ *as follows:*

- $u_0(\mathbf{x}) := \mathcal{I}(\mathbf{x})$ *for all* $\mathbf{x} \in \mathbb{R}^D$ *(initial conditions);*
- $u_t := \mathcal{I} * \mathcal{G}_t$, *for all* $t > 0$.

Then u *is a continuous solution to the heat equation on* $\mathbb{R}^D$ *with initial conditions* $\mathcal{I}$.

Proof

Claim 1 $u(\mathbf{x}; t)$ *is a solution to the D-dimensional heat equation.*

Ⓔ   *Proof* **Exercise 17C.5.** (*Hint:* Combine Example 1B.2(c), p. 10, with Proposition 17B.6(b), p. 390.)   ◇ Claim 1

Claim 2 $\mathcal{G}$ *is an approximation of the identity on* $\mathbb{R}^D$.

Ⓔ   *Proof* **Exercise 17C.6.**   ◇ Claim 2

Now apply Corollary 17B.7, p. 390.   □

Because of Theorem 17C.7, we say that $\mathcal{G}$ is the *fundamental solution* for the heat equation.

Ⓔ **Exercise 17C.7** In Theorem 17C.7, suppose the initial condition $\mathcal{I}$ had some points of discontinuity in $\mathbb{R}^D$. What can you say about the continuity of the function u? In what sense is u still a solution to the initial value problem with initial conditions $u_0 = \mathcal{I}$?   ◆

17D d'Alembert's solution (one-dimensional wave equation)

> Algebra is generous; she often gives more than is asked of her.
>
>   *Jean le Rond d'Alembert*

d'Alembert's method provides a solution to the one-dimensional wave equation

$$\partial_t^2 u = \partial_x^2 u \tag{17D.1}$$

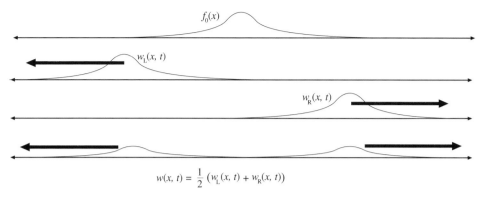

$$w(x, t) = \frac{1}{2} \left(w_{\mathrm{L}}(x, t) + w_{\mathrm{R}}(x, t) \right)$$

Figure 17D.1. The d'Alembert travelling wave solution; $f_0(x) = 1/(x^2 + 1)$ from Example 17D.2.

with any initial conditions, using combinations of *travelling waves* and *ripples*. First we will discuss this in the infinite domain $\mathbb{X} = \mathbb{R}$, then we will consider a finite domain like $\mathbb{X} = [a, b]$.

17D(i) Unbounded domain

Prerequisites: §2B(i).   **Recommended:** §17A.

Lemma 17D.1 Travelling wave solution

Let $f_0 : \mathbb{R} \longrightarrow \mathbb{R}$ be any twice-differentiable function. Define the functions w_{L}, $w_{\mathrm{R}} :$ $\mathbb{R} \times \mathbb{R}_{+} \longrightarrow \mathbb{R}$ by $w_{\mathrm{L}}(x, t) := f_0(x + t)$ and $w_{\mathrm{R}}(x, t) := f_0(x - t)$, for any $x \in \mathbb{R}$ and any $t \geq 0$ (see Figure 17D.1). Then w_{L} and w_{R} are solutions to the wave equation (17D.1), *with*

> initial position: $w_{\mathrm{L}}(x, 0) = f_0(x) = w_{\mathrm{R}}(x, 0);$
> initial velocities: $\partial_t \, w_{\mathrm{L}}(x, 0) = f_0'(x); \; \partial_t \, w_{\mathrm{R}}(x, 0) = -f_0'(x).$

Define $w : \mathbb{R} \times \mathbb{R}_{+} \longrightarrow \mathbb{R}$ by $w(x, t) := \frac{1}{2}(w_{\mathrm{L}}(x, t) + w_{\mathrm{R}}(x, t))$, for all $x \in \mathbb{R}$ and $t \geq 0$. Then w is a solution to the wave equation (17D.1), *with initial position $w(x, 0) = f_0(x)$ and initial velocity $\partial_t \, w(x, 0) = 0$.*

Proof See Problem 17.5, p. 417.   ☐

   Physically, w_{L} represents a leftwards-travelling wave: take a copy of the function f_0 and just rigidly translate it to the left. Similarly, w_{R} represents a rightwards-travelling wave. (Naïvely, it seems that $w_{\mathrm{L}}(x, t) = f_0(x + t)$ should be a *rightwards*-travelling wave, while w_{R} should be *leftwards*-travelling wave. Yet the opposite is true. Think about this until you understand it. It may be helpful to do the following: let $f_0(x) = x^2$. Plot $f_0(x)$, and then plot $w_{\mathrm{L}}(x, 5) = f(x + 5) = (x + 5)^2$. Observe the 'motion' of the parabola.)

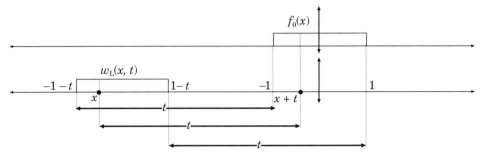

Figure 17D.2. The travelling box wave $w_L(x, t) = f_0(x + t)$ from Example 17D.2(c).

Example 17D.2

(a) If $f_0(x) = 1/(x^2 + 1)$, then

$$w(x) = \frac{1}{2} \left(\frac{1}{(x + t)^2 + 1} + \frac{1}{(x - t)^2 + 1} \right)$$

(see Figure 17D.1).

(b) If $f_0(x) = \sin(x)$, then

$$w(x; t) = \frac{1}{2} \left(\sin(x + t) + \sin(x - t) \right)$$

$$= \frac{1}{2} \left(\sin(x)\cos(t) + \cos(x)\sin(t) + \sin(x)\cos(t) - \cos(x)\sin(t) \right)$$

$$= \frac{1}{2} \left(2\sin(x)\cos(t) \right) = \cos(t)\sin(x).$$

In other words, two sinusoidal waves, travelling in opposite directions, when superposed, result in a sinusoidal *standing* wave.

(c) See Figure 17D.2. Suppose

$$f_0(x) = \begin{cases} 1 & \text{if } -1 < x < 1; \\ 0 & \text{otherwise.} \end{cases}$$

Then

$$w_L(x, t) = f_0(x + t) = \begin{cases} 1 & \text{if } -1 < x + t < 1; \\ 0 & \text{otherwise} \end{cases}$$

$$= \begin{cases} 1 & \text{if } -1 - t < x < 1 - t; \\ 0 & \text{otherwise} \end{cases}.$$

(Note that the solutions w_L and w_R are continuous (or differentiable) only when f_0 is continuous (or differentiable). But the formulae of Lemma 17D.1 make sense even when the original wave equation itself ceases to make sense, as in part (c). This is an example of a *generalized solution* of the wave equation.)   ◇

Lemma 17D.3 Ripple solution

Let $f_1 : \mathbb{R} \longrightarrow \mathbb{R}$ be a differentiable function. Define the function $v : \mathbb{R} \times \mathbb{R}_{\neq} \longrightarrow \mathbb{R}$ by

$$v(x, t) := \frac{1}{2} \int_{x-t}^{x+t} f_1(y) dy,$$

for any $x \in \mathbb{R}$ and any $t \geq 0$. Then v is a solution to the wave equation (17D.1), with

    initial position: $v(x, 0) = 0$;
    initial velocity: $\partial_t v(x, 0) = f_1(x)$.

Proof See Problem 17.6.   □

Physically, v represents a 'ripple'. You can imagine that f_1 describes the energy profile of an 'impulse' which is imparted into the vibrating medium at time zero; this energy propagates outwards, leaving a disturbance in its wake (see Figure 17D.5).

Example 17D.4
(a) If $f_1(x) = 1/(1 + x^2)$, then the d'Alembert solution to the initial velocity problem is given by

$$v(x, t) = \frac{1}{2} \int_{x-t}^{x+t} f_1(y) dy = \frac{1}{2} \int_{x-t}^{x+t} \frac{1}{1 + y^2} dy$$

$$= \frac{1}{2} \arctan(y) \Big|_{y=x-t}^{|y=x+t} = \frac{1}{2} \left(\arctan(x + t) - \arctan(x - t) \right)$$

(see Figure 17D.3).
(b) If $f_1(x) = \cos(x)$, then

$$v(x, t) = \frac{1}{2} \int_{x-t}^{x+t} \cos(y) dy = \frac{1}{2} \left(\sin(x + t) - \sin(x - t) \right)$$

$$= \frac{1}{2} \left(\sin(x) \cos(t) + \cos(x) \sin(t) - \sin(x) \cos(t) + \cos(x) \sin(t) \right)$$

$$= \frac{1}{2} \left(2 \cos(x) \sin(t) \right) = \sin(t) \cos(x).$$

(c) Let

$$f_1(x) = \begin{cases} 2 & \text{if } -1 < x < 1; \\ 0 & \text{otherwise} \end{cases}$$

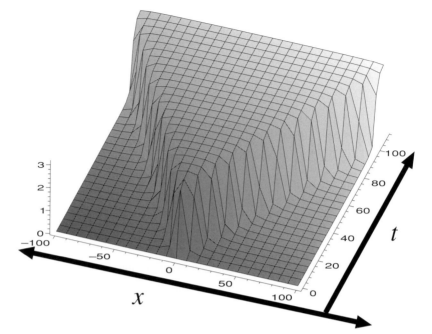

Figure 17D.3. The ripple solution with initial velocity $f_1(x) = 1/(1 + x^2)$ (see Example 17D.4(a)).

(Figures 17D.4 and 17D.5). If $t > 2$, then

$$
v(x, t) = \begin{cases}
0 & \text{if } x + t < -1; \\
x + t + 1 & \text{if } -1 \le x + t < 1; \\
2 & \text{if } x - t \le -1 < 1 \le x + t; \\
t + 1 - x & \text{if } -1 \le x - t < 1; \\
0 & \text{if } 1 \le x - t.
\end{cases}
$$

$$
= \begin{cases}
0 & \text{if } x < -1 - t; \\
x + t + 1 & \text{if } -1 - t \le x < 1 - t; \\
2 & \text{if } 1 - t \le x < t - 1; \\
t + 1 - x & \text{if } t - 1 \le x < t + 1; \\
0 & \text{if } t + 1 \le x.
\end{cases} \tag{17D.2}
$$

Note that, in this example, the wave of displacement propagates outwards through the medium, and the medium *remains displaced*. The model contains no 'restoring force' which would cause the displacement to return to zero.

(d) If $f_1(x) = -2x/(x^2 + 1)^2$, then $g(x) = 1/(x^2 + 1)$, and

$$
v(x) = \frac{1}{2} \left(\frac{1}{(x + t)^2 + 1} - \frac{1}{(x - t)^2 + 1} \right)
$$

(see Figure 17D.6).   ◇

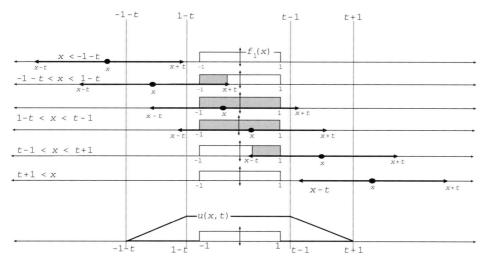

Figure 17D.4. The d'Alembert ripple solution from Example 17D.4(c), evaluated for various $x \in \mathbb{R}$, assuming $t > 2$.

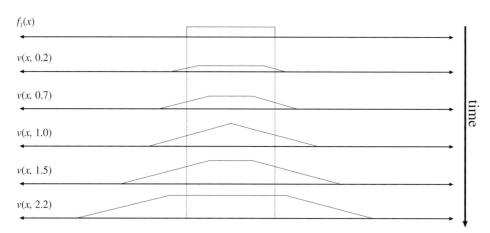

Figure 17D.5. The d'Alembert ripple solution from Example 17D.4(c), evolving in time.

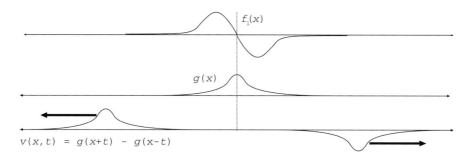

Figure 17D.6. The ripple solution with initial velocity $f_1(x) = -2x/(x^2 + 1)^2$ (see Example 17D.4(d)).

Ⓔ **Exercise 17D.1**   Verify equation (17D.2). Find a similar formula for when $t < 2$.   ◆

Remark If $g : \mathbb{R} \longrightarrow \mathbb{R}$ is an antiderivative of f_1 (i.e. $g'(x) = f_1(x)$), then $v(x, t) = g(x + t) - g(x - t)$. Thus, the d'Alembert 'ripple' solution looks like the d'Alembert 'travelling wave' solution, but with the rightward-travelling wave being vertically *inverted*.

Ⓔ **Exercise 17D.2**
 (a) Express the d'Alembert 'ripple' solution as a *convolution*, as described in §17A, p. 381. *Hint:* Find an impulse-response function $\Gamma_t(x)$, such that

$$f_1 * \Gamma_t(x) = \frac{1}{2} \int_{x-t}^{x+t} f_1(y) dy.$$

 (b) Is Γ_t an approximation of identity? Why or why not?   ◆

Proposition 17D.5 d'Alembert solution on an infinite wire
Let $f_0 : \mathbb{R} \longrightarrow \mathbb{R}$ be twice-differentiable, and let $f_1 : \mathbb{R} \longrightarrow \mathbb{R}$ be differentiable. Define the function $u : \mathbb{R} \times \mathbb{R}_{\neq} \longrightarrow \mathbb{R}$ by

$$u(x, t) := \frac{1}{2} \left(w_L(x, t) + w_R(x, t) \right) + v(x, t), \quad \text{for all } x \in \mathbb{R} \text{ and } t \geq 0,$$

where w_L, w_R, and v are as in Lemmas 17D.1 and 17D.3. Then u satisfies the wave equation, with

 initial position: $v(x, 0) = f_0(x)$;
 initial velocity: $\partial_t v(x, 0) = f_1(x)$.

Furthermore, all solutions to the wave equation with these initial conditions are of this type.

Proof   This follows from Lemmas 17D.1 and 17D.3.   □

Remark There is no nice extension of the d'Alembert solution in higher dimensions. The closest analogy is Poisson's *spherical mean solution* to the three-dimensional wave equation in free space, which is discussed in § 20B(ii), p. 537.

17D(ii) Bounded domain

Prerequisites: §17D(i), §5C(i).

The d'Alembert solution in §17D(i) works fine if $\mathbb{X} = \mathbb{R}$, but what if $\mathbb{X} = [0, L)$? We must 'extend' the initial conditions in some way. If $f : [0, L) \longrightarrow \mathbb{R}$ is any function, then an *extension* of f is any function $\overline{f} : \mathbb{R} \longrightarrow \mathbb{R}$ such that $\overline{f}(x) = f(x)$

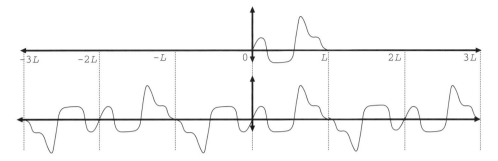

Figure 17D.7. The odd $2L$-periodic extension.

whenever $0 \le x \le L$. If f is continuous and differentiable, then we normally require its extension also to be continuous and differentiable.

The extension we want is the *odd, $2L$-periodic extension*, which is defined as the unique function $\overline{f} : \mathbb{R} \longrightarrow \mathbb{R}$ with the following three properties (see Figure 17D.7):

(1) $\overline{f}(x) = f(x)$ whenever $0 \le x \le L$;
(2) $\overline{f}$ is an *odd* function,[2] meaning that $\overline{f}(-x) = -\overline{f}(x)$ for all $x \in \mathbb{R}$;
(3) $\overline{f}$ is $2L$-*periodic*, meaning that $\overline{f}(x + 2L) = \overline{f}(x)$ for all $x \in \mathbb{R}$.

Example 17D.6
(a) Suppose $L = 1$, and $f(x) = 1$ for all $x \in [0, 1)$ (Figure 17D.8(a)). Then the odd, 2-periodic extension is defined:

$$\overline{f}(x) = \begin{cases} 1 & \text{if } x \in \cdots \cup [-2, -1) \cup [0, 1) \cup [2, 3) \cup \cdots ; \\ -1 & \text{if } x \in \cdots \cup [-1, 0) \cup [1, 2) \cup [3, 4) \cup \cdots \end{cases}$$

(Figure 17D.8(b)).
(b) Suppose $L = 1$, and

$$f(x) = \begin{cases} 1 & \text{if } x \in \left[0, \frac{1}{2}\right); \\ 0 & \text{if } x \in \left[\frac{1}{2}, 1\right) \end{cases}$$

(Figure 17D.8(c)). Then the odd, 2-periodic extension is defined:

$$\overline{f}(x) = \begin{cases} 1 & \text{if } x \in \cdots \cup \left[-2, -1\frac{1}{2}\right) \cup \left[0, \frac{1}{2}\right) \cup \left[2, 2\frac{1}{2}\right) \cup \cdots ; \\ -1 & \text{if } x \in \cdots \cup \left[-\frac{1}{2}, 0\right) \cup \left[1\frac{1}{2}, 2\right) \cup \left[3\frac{1}{2}, 4\right) \cup \cdots ; \\ 0 & \text{otherwise} \end{cases}$$

(Figure 17D.8(d)).
(c) Suppose $L = \pi$, and $f(x) = \sin(x)$ for all $x \in [0, \pi)$ (see Figure 17D.8(e)). Then the odd, 2π-periodic extension is given by $\overline{f}(x) = \sin(x)$ for all $x \in \mathbb{R}$ (Figure 17D.8(f)). $\diamond$

---

[2] See §8C, p. 172, for more information about odd functions.

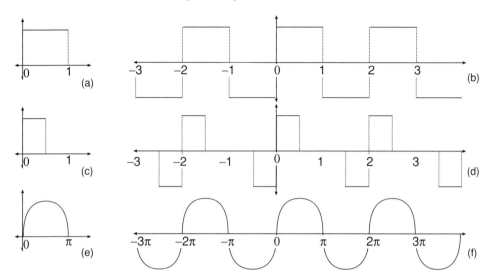

Figure 17D.8. The odd, $2L$-periodic extension.

Ⓔ **Exercise 17D.3** Verify Example 17D.6(c).   ◆

We will now provide a general formula for the odd periodic extension, and characterize its continuity and/or differentiability. First some terminology. If $f : [0, L) \longrightarrow \mathbb{R}$ is a function, then we say that f is *right-differentiable* at 0 if the *right-hand derivative* $f^{\backslash}(0)$ is well defined (see p. 203). We can usually extend f to a function $f : [0, L] \longrightarrow \mathbb{R}$ by defining $f(L^-) := \lim_{x \nearrow L} f(x)$, where this denotes the *left-hand limit* of f at L, if this limit exists (see p. 202 for definition). We then say that f is *left-differentiable* at L if the *left-hand derivative* $f^{\backslash}(L)$ exists.

Proposition 17D.7 *Let* $f : [0, L) \longrightarrow \mathbb{R}$ *be any function.*

(a) *The odd, $2L$-periodic extension of f is given by*

$$
\overline{f}(x) = \begin{cases}
f(x) & \text{if} & 0 \le x < L; \\
-f(-x) & \text{if} & -L \le x < 0; \\
f(x - 2nL) & \text{if} & 2nL \le x \le (2n+1)L, \text{ for some } n \in \mathbb{Z}; \\
-f(2nL - x) & \text{if} & (2n-1)L \le x \le 2nL, \text{ for some } n \in \mathbb{Z}.
\end{cases}
$$

(b) $\overline{f}$ *is continuous at* $0, L, 2L$, *etc. if, and only if,* $f(0) = f(L^-) = 0$.
(c) $\overline{f}$ *is differentiable at* $0, L, 2L$, *etc. if, and only if, it is continuous, f is right-differentiable at 0, and f is left-differentiable at L.*

Ⓔ *Proof* **Exercise 17D.4.**   □

Proposition 17D.8 d'Alembert solution on a finite string

Let $f_0 : [0, L) \longrightarrow \mathbb{R}$ and $f_1 : [0, L) \longrightarrow \mathbb{R}$ be differentiable functions, and let their odd periodic extensions be $\overline{f}_0 : \mathbb{R} \longrightarrow \mathbb{R}$ and $\overline{f}_1 : \mathbb{R} \longrightarrow \mathbb{R}$.

(a) *Define $w : [0, L] \times \mathbb{R}_{\not+} \longrightarrow \mathbb{R}$ by*

$$w(x, t) := \frac{1}{2} \left(\overline{f}_0(x - t) + \overline{f}_0(x + t) \right), \quad \text{for all } x \in [0, L] \text{ and } t \geq 0.$$

Then w is a solution to the wave equation (17D.1) with initial conditions:

$$w(x, 0) = f_0(x) \quad \text{and} \quad \partial_t w(x, 0) = 0, \quad \text{for all } x \in [0, L],$$

and homogeneous Dirichlet boundary conditions:

$$w(0, t) = 0 = w(L, t), \quad \text{for all } t \geq 0.$$

The function w is continuous *if, and only if, f_0 satisfies homogeneous Dirichlet boundary conditions (i.e. $f(0) = f(L^-) = 0$). In addition, w is* differentiable *if, and only if, f_0 is also right-differentiable at 0 and left-differentiable at L.*

(b) *Define $v : [0, L] \times \mathbb{R}_{\not+} \longrightarrow \mathbb{R}$ by*

$$v(x, t) := \frac{1}{2} \int_{x-t}^{x+t} \overline{f}_1(y) dy, \quad \text{for all } x \in [0, L] \text{ and } t \geq 0.$$

Then v is a solution to the wave equation (17D.1) with initial conditions:

$$v(x, 0) = 0 \quad \text{and} \quad \partial_t v(x, 0) = f_1(x), \quad \text{for all } x \in [0, L],$$

and homogeneous Dirichlet boundary conditions:

$$v(0, t) = 0 = v(L, t), \quad \text{for all } t \geq 0.$$

The function v is always continuous. However, v is differentiable if, and only if, f_1 satisfies homogeneous Dirichlet boundary conditions.

(c) *Define $u : [0, L] \times \mathbb{R}_{\not+} \longrightarrow \mathbb{R}$ by $u(x, t) := w(x, t) + v(x, t)$, for all $x \in [0, L]$ and $t \geq 0$. Then $u(x, t)$ is a solution to the wave equation (17D.1) with initial conditions:*

$$u(x, 0) = f_0(x) \quad \text{and} \quad \partial_t u(x, 0) = f_1(x), \quad \text{for all } x \in [0, L],$$

and homogeneous Dirichlet boundary conditions:

$$u(0, t) = 0 = u(L, t), \quad \text{for all } t \geq 0.$$

Clearly, u is continuous (respectively, differentiable) whenever v and w are continuous (respectively, differentiable).

Proof The fact that u, w, and v are solutions to their respective initial value problems follows from Proposition 17D.5, p. 404. The verification of homogeneous Dirichlet conditions is **Exercise 17D.5**. The conditions for continuity/differentiability are **Exercise 17D.6**.   □ Ⓔ

Ⓔ

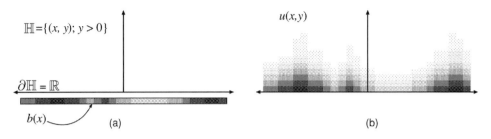

Figure 17E.1. The Dirichlet problem on a half-plane.

17E Poisson's solution (Dirichlet problem on half-plane)

Prerequisites: §1C, §5C, Appendix G, §17B(i).   **Recommended:** §17A.

Consider the *half-plane domain* $\mathbb{H} := \{(x, y) \in \mathbb{R}^2; y \geq 0\}$. The boundary of this domain is just the x axis: $\partial\mathbb{H} = \{(x, 0); x \in \mathbb{R}\}$. Thus, we impose boundary conditions by choosing some function $b : \mathbb{R} \longrightarrow \mathbb{R}$. Figure 17E.1 illustrates the corresponding *Dirichlet problem*: find a continuous function $u : \mathbb{H} \longrightarrow \mathbb{R}$ such that

(1) u is *harmonic* – i.e. u satisfies the Laplace equation: $\Delta u(x, y) = 0$ for all $x \in \mathbb{R}$ and $y > 0$;

(2) u satisfies the *nonhomogeneous Dirichlet boundary condition*: $u(x, 0) = b(x)$, for all $x \in \mathbb{R}$.

Physical interpretation   Imagine that $\mathbb{H}$ is an infinite 'ocean', so that $\partial\mathbb{H}$ is the beach. Imagine that $b(x)$ is the concentration of some chemical which has soaked into the sand of the beach. The harmonic function $u(x, y)$ on $\mathbb{H}$ describes the equilibrium concentration of this chemical, as it seeps from the sandy beach and diffuses into the water.[3] The boundary condition '$u(x, 0) = b(x)$' represents the chemical content of the sand. Note that $b(x)$ is constant in time; this represents the assumption that the chemical content of the sand is large compared to the amount seeping into the water; hence, we can assume the sand's chemical content remains effectively constant over time, as small amounts diffuse into the water.

We will solve the half-plane Dirichlet problem using the impulse-response method. For any $y > 0$, define the *Poisson kernel* $\mathcal{K}_y : \mathbb{R} \longrightarrow \mathbb{R}$ as follows:

$$\mathcal{K}_y(x) := \frac{y}{\pi(x^2 + y^2)} \tag{17E.1}$$

(Figure 17E.2). Observe the following:

----

[3] Of course this an unrealistic model: in a *real* ocean, currents, wave action, and weather transport chemicals far more quickly than mere diffusion alone.

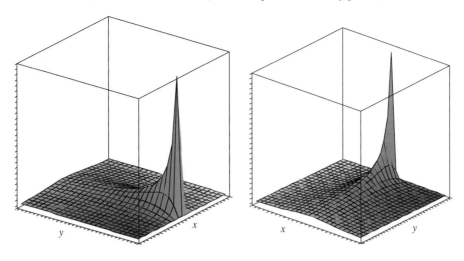

Figure 17E.2. Two views of the Poisson kernel $\mathcal{K}_y(x)$.

- $\mathcal{K}_y(x)$ is smooth for all $y > 0$ and $x \in \mathbb{R}$;
- $\mathcal{K}_y(x)$ has a singularity at $(0, 0)$; that is;

$$\lim_{(x,y)\to(0,0)} \mathcal{K}_y(x) = \infty;$$

- $\mathcal{K}_y(x)$ decays near infinity; that is, for any fixed $y > 0$,

$$\lim_{x\to\pm\infty} \mathcal{K}_y(x) = 0,$$

and also, for any fixed $x \in \mathbb{R}$,

$$\lim_{y\to\infty} \mathcal{K}_y(x) = 0.$$

Thus, $\mathcal{K}_y(x)$ has the profile of an *impulse-response function* as described in §17A, p. 381. Heuristically speaking, you can think of $\mathcal{K}_y(x)$ as the solution to the Dirichlet problem on $\mathbb{H}$, with boundary condition $b(x) = \delta_0(x)$, where δ_0 is the infamous 'Dirac delta function'. In other words, $\mathcal{K}_y(x)$ is the equilibrium concentration of a chemical diffusing into the water from an 'infinite' concentration of chemical localized at a single point on the beach (say, a leaking barrel of toxic waste).

Proposition 17E.1 Poisson kernel solution to half-plane Dirichlet problem

Let $b : \mathbb{R} \longrightarrow \mathbb{R}$ be a bounded, continuous, integrable function. Define $u : \mathbb{H} \longrightarrow \mathbb{R}$ as follows:

$$u(x, y) := b * \mathcal{K}_y(x) = \frac{y}{\pi} \int_{-\infty}^{\infty} \frac{b(z)}{(x - z)^2 + y^2} \, dz,$$

for all $x \in \mathbb{R}$ and $y > 0$, while, for all $x \in \mathbb{R}$, we define, $u(x, 0) := b(x)$. Then u is the solution to the Laplace equation ($\triangle u = 0$) which is bounded at infinity and *which satisfies the nonhomogeneous Dirichlet boundary condition* $u(x, 0) = b(x)$, *for all* $x \in \mathbb{R}$.

Proof (Sketch)

Claim 1 *Define* $\mathcal{K}(x, y) = \mathcal{K}_y(x)$ *for all* $(x, y) \in \mathbb{H}$, *except* $(0, 0)$. *Then the function* $\mathcal{K} : \mathbb{H} \longrightarrow \mathbb{R}$ *is harmonic on the interior of* $\mathbb{H}$.

Proof See Problem 17.14.   ◇Claim 1

Claim 2 *Thus, the function* $u : \mathbb{H} \longrightarrow \mathbb{R}$ *is harmonic on the interior of* $\mathbb{H}$.

Ⓔ   *Proof* **Exercise 17E.1.** (*Hint*: Combine Claim 1 with Proposition G.1, p. 571.)
                                                                    ◇Claim 2

Recall that we defined u on the boundary of $\mathbb{H}$ by $u(x, 0) = b(x)$. It remains to show that u is *continuous* when defined in this way.

Claim 3 *For any* $x \in \mathbb{R}$, $\lim_{y \to 0} u(x, y) = b(x)$.

Ⓔ   *Proof*   **Exercise 17E.2.** Show that the kernel $\mathcal{K}_y$ is an *approximation of the identity* as $y \to 0$. Then apply Proposition 17B.2, p. 387, to conclude that $\lim_{y \to 0}(b * \mathcal{K}_y)(x) = b(x)$ for all $x \in \mathbb{R}$.   ◇Claim 3

Finally, this solution is unique by Theorem 5D.5(a), p. 91.   □

Example 17E.2 Let $A < B$ be real numbers. Let

$$b(x) := \begin{cases} 1 & \text{if } A < x < B; \\ 0 & \text{otherwise.} \end{cases}$$

Then Proposition 20C.3, p. 542, yields the following solution:

$$U(x, y) \underset{(*)}{=} b * \mathcal{K}_y(x) \underset{(\dagger)}{=} \frac{y}{\pi} \int_A^B \frac{1}{(x - z)^2 + y^2} \, dz \underset{(S)}{=} \frac{y^2}{\pi} \int_{\frac{A-x}{y}}^{\frac{B-x}{y}} \frac{1}{y^2 w^2 + y^2} \, dw$$

$$= \frac{1}{\pi} \int_{\frac{A-x}{y}}^{\frac{B-x}{y}} \frac{1}{w^2 + 1} \, dw = \frac{1}{\pi} \arctan(w) \Big|_{w=\frac{A-x}{y}}^{w=\frac{B-x}{y}}$$

$$= \frac{1}{\pi} \arctan\left(\frac{B - x}{y}\right) - \arctan\left(\frac{A - x}{y}\right) \underset{(T)}{=} \frac{1}{\pi}\left(\theta_B - \theta_A\right),$$

where θ_B and θ_A are as in Figure 17E.3. Here, $(*)$ is by Proposition 20C.3; $(\dagger)$ is by equation (17E.1); (S) is the substitution $w = (z - x)/y$, so that $dw = (1/y)dz$ and $dz = y\,dw$; and (T) follows from elementary trigonometry.

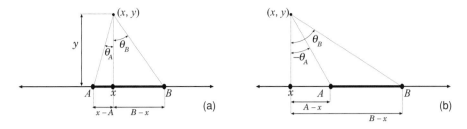

Figure 17E.3. Diagrams for Example 17E.2: (a) $A < x$; (b) $A > x$.

Note that, if $A < x$ (as in Figure 17E.3(a)), then $A - x < 0$, so θ_A is negative, so that $U(x, y) = \frac{1}{\pi}(\theta_B + |\theta_A|)$. If $A > x$, then we have the situation in Figure 17E.3(b). In either case, the interpretation is the same:

$$U(x, y) = \frac{1}{\pi}\left(\theta_B - \theta_A\right) = \frac{1}{\pi}\left(\begin{array}{l} \text{the angle subtended by interval } [A, B], \text{ as} \\ \text{seen by an observer at the point } (x, y) \end{array}\right).$$

This is reasonable, because, if this observer moves far away from the interval $[A, B]$, or views it at an acute angle, then the subtended angle $(\theta_B - \theta_A)$ will become small – hence, the value of $U(x, y)$ will also become small. ◇

Remark We will revisit the Poisson kernel solution to the half-plane Dirichlet problem in §20C(ii), p. 542, where we will prove Proposition 17E.1 using Fourier transform methods.

17F Poisson's solution (Dirichlet problem on the disk)

Prerequisites: §1C, Appendix D(ii), §5C, Appendix G.   **Recommended:** §17A, §14B(v).[4]

Let

$$\mathbb{D} := \left\{(x, y) \in \mathbb{R}^2; \sqrt{x^2 + y^2} \leq R\right\}$$

be the *disk* of radius R in $\mathbb{R}^2$. Thus, $\mathbb{D}$ has boundary

$$\partial\mathbb{D} = \mathbb{S} := \left\{(x, y) \in \mathbb{R}^2; \sqrt{x^2 + y^2} = R\right\}$$

(the circle of radius R). Suppose $b : \partial\mathbb{D} \longrightarrow \mathbb{R}$ is some function on the boundary. The *Dirichlet problem* on $\mathbb{D}$ asks for a continuous function $u : \mathbb{D} \longrightarrow \mathbb{R}$ such that:

- u is *harmonic* – i.e. u satisfies the Laplace equation $\triangle u \equiv 0$.
- u satisfies the *nonhomogeneous* Dirichlet boundary condition $u(x, y) = b(x, y)$ for all $(x, y) \in \partial\mathbb{D}$.

---

[4] See §14B(v), p. 296, for a different development of the material in this section, using the methods of polar-separated harmonic functions. For yet another approach, using complex analysis, see Corollary 18C.13, p. 451.

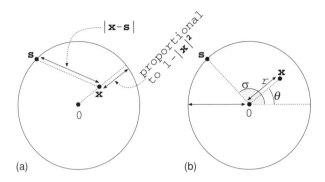

Figure 17F.1. The Poisson kernel.

If u represents the concentration of some chemical diffusing into $\mathbb{D}$ from the boundary, then the value of $u(x, y)$ at any point (x, y) in the interior of the disk should represent some sort of 'average' of the chemical reaching (x, y) from all points on the boundary. This is the inspiration of *Poisson's solution*. We define the *Poisson kernel* $\mathcal{P} : \mathbb{D} \times \mathbb{S} \longrightarrow \mathbb{R}$ as follows:

$$\mathcal{P}(\mathbf{x}, \mathbf{s}) := \frac{R^2 - \|\mathbf{x}\|^2}{\|\mathbf{x} - \mathbf{s}\|^2}, \qquad \text{for all } \mathbf{x} \in \mathbb{D} \text{ and } \mathbf{s} \in \mathbb{S}.$$

As shown in Figure 17F.1(a), the denominator, $\|\mathbf{x} - \mathbf{s}\|^2$, is just the squared-distance from $\mathbf{x}$ to $\mathbf{s}$. The numerator, $R^2 - \|\mathbf{x}\|^2$, roughly measures the distance from $\mathbf{x}$ to the boundary $\mathbb{S}$; if $\mathbf{x}$ is close to $\mathbb{S}$, then $R^2 - \|\mathbf{x}\|^2$ becomes very small. Intuitively speaking, $\mathcal{P}(\mathbf{x}, \mathbf{s})$ measures the 'influence' of the boundary condition at the point $\mathbf{s}$ on the value of u at $\mathbf{x}$; see Figure 17F.2.

In polar coordinates (Figure 17F.1(b)), we can parameterize $\mathbf{s} \in \mathbb{S}$ with a single angular coordinate $\sigma \in [-\pi, \pi)$, so that $\mathbf{s} = (R \cos(\sigma), R \sin(\sigma))$. If $\mathbf{x}$ has coordinates (x, y), then Poisson's kernel takes the following form:

$$\mathcal{P}(\mathbf{x}, \mathbf{s}) = \mathcal{P}_\sigma(x, y) = \frac{R^2 - x^2 - y^2}{(x - R \cos(\sigma))^2 + (y - R \sin(\sigma))^2}.$$

Proposition 17F.1 Poisson's integral formula

Let $\mathbb{D} = \{(x, y); x^2 + y^2 \leq R^2\}$ be the disk of radius R, and let $b : \partial\mathbb{D} \longrightarrow \mathbb{R}$ be continuous. The unique solution to the corresponding Dirichlet problem is the function $u : \mathbb{D} \longrightarrow \mathbb{R}$ defined as follows. For any (x, y) on the interior *of $\mathbb{D}$*

$$u(x, y) := \frac{1}{2\pi} \int_{-\pi}^{\pi} b(\sigma) \cdot \mathcal{P}_\sigma(x, y) d\sigma,$$

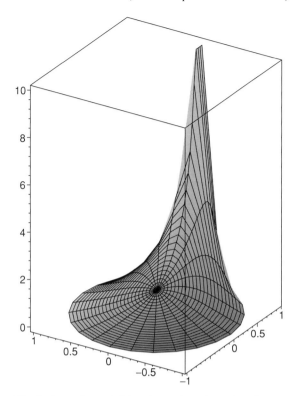

Figure 17F.2. The Poisson kernel $\mathcal{P}(\mathbf{x}, \mathbf{s})$ as a function of $\mathbf{x}$ (for some fixed value of $\mathbf{s}$). This surface illustrates the 'influence' of the boundary condition at the point $\mathbf{s}$ on the point $\mathbf{x}$. (The point $\mathbf{s}$ is located at the 'peak' of the surface.)

while, for $(x, y) \in \partial \mathbb{D}$, we define $u(x, y) := b(x, y)$. That is, for any $\mathbf{x} \in \mathbb{D}$,

$$u(\mathbf{x}) := \begin{cases} \dfrac{1}{2\pi} \displaystyle\int_{\mathbb{S}} b(\mathbf{s}) \cdot \mathcal{P}(\mathbf{x}, \mathbf{s}) \mathrm{d}s & if \quad \|x\| < R; \\ b(\mathbf{x}) & if \quad \|x\| = R. \end{cases}$$

Proof (Sketch) For simplicity, assume $R = 1$ (the proof for $R \neq 1$ is similar). Thus,

$$\mathcal{P}_\sigma(x, y) = \frac{1 - x^2 - y^2}{(x - \cos(\sigma))^2 + (y - \sin(\sigma))^2}.$$

Claim 1 *Fix $\sigma \in [-\pi, \pi)$. The function $\mathcal{P}_\sigma : \mathbb{D} \longrightarrow \mathbb{R}$ is harmonic on the interior of $\mathbb{D}$.*

Proof **Exercise 17F.1.** $\Diamond$ Claim 1   Ⓔ

Claim 2 *Thus, the function u is harmonic on the interior of $\mathbb{D}$.*

Ⓔ   *Proof* **Exercise 17F.2.** (*Hint:* Combine Claim 1 with Proposition G.1, p. 571.)

◇ Claim 2

Recall that we defined u on the boundary $\mathbb{S}$ of $\mathbb{D}$ by $u(\mathbf{s}) = b(\mathbf{s})$. It remains to show that u is *continuous* when defined in this way.

Claim 3 *For any* $\mathbf{s} \in \mathbb{S}$, $\lim_{(x,y) \to \mathbf{s}} u(x, y) = b(\mathbf{s})$.

Ⓔ   *Proof* **Exercise 17F.3** (Hard). *Hint:* Write (x, y) in polar coordinates as (r, θ). Thus, our claim becomes $\lim_{\theta \to \sigma} \lim_{r \to 1} u(r, \theta) = b(\sigma)$.

(a) Show that $\mathcal{P}_\sigma(x, y) = \mathcal{P}_r(\theta - \sigma)$, where, for any $r \in [0, 1)$, we define

$$\mathcal{P}_r(\phi) = \frac{1 - r^2}{1 - 2r \cos(\phi) + r^2}, \qquad \text{for all } \phi \in [-\pi, \pi).$$

(b) Thus,

$$u(r, \theta) = \frac{1}{2\pi} \int_{-\pi}^{\pi} b(\sigma) \cdot \mathcal{P}_r(\theta - \sigma) d\sigma$$

is a sort of 'convolution on a circle'. We can write this as $u(r, \theta) = (b \star \mathcal{P}_r)(\theta)$.

(c) Show that the function $\mathcal{P}_r$ is an 'approximation of the identity' as $r \to 1$, meaning that, for any continuous function $b : \mathbb{S} \longrightarrow \mathbb{R}$, $\lim_{r \to 1} (b \star \mathcal{P}_r)(\theta) = b(\theta)$. For your proof, borrow from the proof of Proposition 17B.2, p. 387

◇ Claim 3

Finally, this solution is unique by Theorem 5D.5(a), p. 91.   □

17G* Properties of convolution

Prerequisites: §17A.   **Recommended:** §17C.

We have introduced the convolution operator to solve the heat equation, but it is actually ubiquitous, not only in the theory of PDEs, but also in other areas of mathematics, especially probability theory, harmonic analysis, and group representation theory. We can define an *algebra* of functions using the operations of convolution and addition; this algebra is as natural as the one you would form using 'normal' multiplication and addition.[5]

Proposition 17G.1 Algebraic properties of convolution
Let $f, g, h : \mathbb{R}^D \longrightarrow \mathbb{R}$ *be integrable functions. Then the convolutions of* f, g, *and* h *have the following relations:*

---

[5] Indeed, in a sense, it is the *same* algebra, seen through the prism of the Fourier transform; see Theorem 19B.2, p. 499.

commutativity $f * g = g * f$;

associativity $f * (g * h) = (f * g) * h$;

distribution $f * (g + h) = (f * g) + (f * h)$;

linearity $f * (r \cdot g) = r \cdot (f * g)$ *for any constant* $r \in \mathbb{R}$.

Proof Commutativity is just Proposition 17A.1. In the case $D = 1$, the proofs of the other three properties are Problems 17.1 and 17.2. The proofs for $D \geq 2$ are **Exercise 17G.1**. □ Ⓔ

Remark Let $\mathbf{L}^1(\mathbb{R}^D)$ be the set of all integrable functions on $\mathbb{R}^D$. The properties of *commutativity*, *associativity*, and *distribution* mean that the set $\mathbf{L}^1(\mathbb{R}^D)$, together with the operations '+' (pointwise addition) and '∗' (convolution), is a *ring* (in the language of abstract algebra). This, together with *linearity*, makes $\mathbf{L}^1(\mathbb{R}^D)$ an *algebra* over $\mathbb{R}$.

Example 17C.4, p. 394, exemplifies the convenient 'smoothing' properties of convolution. If we convolve a 'rough' function with a 'smooth' function, then this 'smooths out' the rough function.

Proposition 17G.2 Regularity properties of convolution

Let $f, g : \mathbb{R}^D \longrightarrow \mathbb{R}$ *be integrable functions.*

(a) *If f is continuous, then so is $f * g$ (regardless of whether g is).*

(b) *If f is differentiable, then so is $f * g$. Furthermore, $\partial_d(f * g) = (\partial_d f) * g$, for all* $d \in [l \ldots D]$.

(c) *If f is N times differentiable, then so is $f * g$, and*

$$\partial_1^{n_1} \partial_2^{n_2} \cdots \partial_D^{n_D}(f * g) = \left(\partial_1^{n_1} \partial_2^{n_2} \cdots \partial_D^{n_D} f\right) * g,$$

for any $n_1, n_2, \ldots, n_D$ *such that* $n_1 + \cdots + n_D \leq N$.

(d) *More generally, if* L *is any linear differential operator of degree N or less, with constant coefficients, then* $\mathsf{L}(f * g) = (\mathsf{L}\, f) * g$.

(e) *Thus, if f is a solution to the homogeneous linear equation '*$\mathsf{L}\, f = 0$*', then so is $f * g$.*

(f) *If f is infinitely differentiable, then so is $f * g$.*

Proof **Exercise 17G.2.** □ Ⓔ

This has a convenient consequence: any function, no matter how 'rough', can be approximated arbitrarily closely by smooth functions.

Proposition 17G.3 *Suppose $f : \mathbb{R}^D \longrightarrow \mathbb{R}$ is integrable. Then there is a sequence* $f_1, f_2, f_3, \ldots$ *of infinitely differentiable functions which converges pointwise to f. In other words, for every* $\mathbf{x} \in \mathbb{R}^D$, $\lim_{n \to \infty} f_n(\mathbf{x}) = f(\mathbf{x})$.

Proof **Exercise 17G.3.** (*Hint:* Use the fact that the Gauss–Weierstrass kernel Ⓔ is infinitely differentiable, and is also an approximation of identity. Then use Proposition 17G.2(f).) □

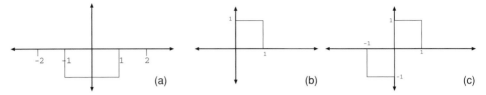

Figure 17H.1. Initial conditions $I(x)$ for: (a) Problem 17.4(a); (b) Problem 17.4(b); (c) Problem 17.4(c).

Remarks

(a) We have formulated Proposition 17G.3 in terms of *pointwise* convergence, but similar results hold for L^2-convergence, L^1-convergence, uniform convergence, etc. We are neglecting these to avoid technicalities.

(b) In §10D(ii), p. 215, we discuss the convolution of periodic functions on the interval $[-\pi, \pi]$, and develop a theory quite similar to the theory developed here. In particular, Lemma 10D.6, p. 216, is analogous to Proposition 17G.1; Lemma 10D.7, p. 217, is analogous to Proposition 17G.2; and Theorem 10D.1, p. 209, is analogous to Proposition 17G.3, except that the convergence is in L^2-norm.

17H Practice problems

17.1 Let $f, g, h : \mathbb{R} \longrightarrow \mathbb{R}$ be integrable functions. Show that $f * (g * h) = (f * g) * h$.

17.2 Let $f, g, h : \mathbb{R} \longrightarrow \mathbb{R}$ be integrable functions, and let $r \in \mathbb{R}$ be a constant. Prove that $f * (r \cdot g + h) = r \cdot (f * g) + (f * h)$.

17.3 Let $f, g : \mathbb{R} \longrightarrow \mathbb{R}$ be integrable functions. Let $d \in \mathbb{R}$ be some 'displacement' and define $f_{\triangleright d}(x) = f(x - d)$. Prove that $(f_{\triangleright d}) * g = (f * g)_{\triangleright d}$.

17.4 In each of the following, use the method of Gaussian convolutions to find the solution to the one-dimensional *heat equation* $\partial_t u(x; t) = \partial_x^2 u(x; t)$ with *initial conditions* $u(x, 0) = I(x)$.

(a) $I(x) = \begin{cases} -1 & \text{if} \quad -1 \leq x \leq 1; \\ 0 & \text{if} \quad x < -1 \text{or } 1 < x \end{cases}$

(see Figure 17H.1(a)).

(In this case, sketch your solution evolving in time.)

(b) $I(x) = \begin{cases} 1 & \text{if} \ \ 0 \leq x \leq 1; \\ 0 & \text{otherwise} \end{cases}$

(see Figure 17H.1(b));

(c) $I(x) = \begin{cases} -1 & \text{if} \ -1 \leq x \leq 0; \\ 1 & \text{if} \ \ 0 \leq x \leq 1; \\ 0 & \text{otherwise} \end{cases}$

(see Figure 17H.1(c)).

17.5 Let $f : \mathbb{R} \longrightarrow \mathbb{R}$ be some differentiable function. Define $v(x;t) = \frac{1}{2}(f(x+t) + f(x-t))$.

- (a) Show that $v(x;t)$ satisfies the one-dimensional wave equation $\partial_t^2 v(x;t) = \partial_x^2 v(x;t)$
- (b) Compute the *initial position* $v(x;0)$.
- (c) Compute the *initial velocity* $\partial_t v(x;0)$.

17.6 Let $f_1 : \mathbb{R} \longrightarrow \mathbb{R}$ be a differentiable function. For any $x \in \mathbb{R}$ and any $t \geq 0$, define

$$v(x, t) = \frac{1}{2} \int_{x-t}^{x+t} f_1(y) \, dy.$$

- (a) Show that $v(x;t)$ satisfies the one-dimensional wave equation $\partial_t^2 v(x;t) = \partial_x^2 v(x;t)$
- (b) Compute the *initial position* $v(x;0)$.
- (c) Compute the *initial velocity* $\partial_t v(x;0)$.

17.7 In each of the following, use the d'Alembert method to find the solution to the one-dimensional *wave equation* $\partial_t^2 u(x;t) = \partial_x^2 u(x;t)$ with *initial position* $u(x,0) = f_0(x)$ and *initial velocity* $\partial_t u(x, 0) = f_1(x)$. In each case, identify whether the solution satisfies homogeneous Dirichlet boundary conditions when treated as a function on the interval $[0, \pi]$. Justify your answer.

- (a) $f_0(x) = \begin{cases} 1 & \text{if } 0 \leq x < 1; \\ 0 & \text{otherwise,} \end{cases}$ and $f_1(x) = 0$

  (see Figure 17H.1(b)).
- (b) $f_0(x) = \sin(3x)$ and $f_1(x) = 0$.
- (c) $f_0(x) = 0$ and $f_1(x) = \sin(5x)$.
- (d) $f_0(x) = \cos(2x)$ and $f_1(x) = 0$.
- (e) $f_0(x) = 0$ and $f_1(x) = \cos(4x)$.
- (f) $f_0(x) = x^{1/3}$ and $f_1(x) = 0$.
- (g) $f_0(x) = 0$ and $f_1(x) = x^{1/3}$.
- (h) $f_0(x) = 0$ and $f_1(x) = \tanh(x) = \frac{\sinh(x)}{\cosh(x)}$.

17.8 Let

$$\mathcal{G}_t(x) = \frac{1}{2\sqrt{\pi t}} \exp\left(\frac{-x^2}{4t}\right)$$

be the Gauss–Weierstrass kernel. Fix $s, t > 0$; we claim that $\mathcal{G}_s * \mathcal{G}_t = \mathcal{G}_{s+t}$. (For example, if $s = 3$ and $t = 5$, this means that $\mathcal{G}_3 * \mathcal{G}_5 = \mathcal{G}_8$.)

- (a) Prove that $\mathcal{G}_s * \mathcal{G}_t = \mathcal{G}_{s+t}$ by directly computing the convolution integral.
- (b) Use Corollary 17C.3, p. 392, to find a short and elegant proof that $\mathcal{G}_s * \mathcal{G}_t = \mathcal{G}_{s+t}$ *without* computing any convolution integrals.

Remark Because of this result, probabilists say that the set $\{\mathcal{G}_t\}_{t \in \mathbb{R}_+}$ forms a *stable family of probability distributions* on $\mathbb{R}$. Analysts say that $\{\mathcal{G}_t\}_{t \in \mathbb{R}_+}$ is a *one-parameter semigroup* under convolution.

17.9 Let

$$\mathcal{G}_t(x, y) = \frac{1}{4\pi t} \exp\left(\frac{-(x^2 + y^2)}{4t}\right)$$

be the two-dimensional Gauss–Weierstrass kernel. Suppose $h : \mathbb{R}^2 \longrightarrow \mathbb{R}$ is a harmonic function. Show that $h * \mathcal{G}_t = h$ for all $t > 0$.

17.10 Let $\mathbb{D}$ be the unit disk. Let $b : \partial\mathbb{D} \longrightarrow \mathbb{R}$ be some function, and let $u : \mathbb{D} \longrightarrow \mathbb{R}$ be the solution to the corresponding Dirichlet problem with boundary conditions $b(\sigma)$. Prove that

$$u(0, 0) = \frac{1}{2\pi} \int_{-\pi}^{\pi} b(\sigma)d\sigma.$$

Remark This is a special case of the Mean Value Theorem for harmonic functions (Theorem 1E.1, p. 18), but do *not* simply 'quote' Theorem 1E.1 to solve this problem. Instead, apply Proposition 17F.1, p. 412.

17.11 Let

$$\gamma_t(x) = \begin{cases} \frac{1}{t} & \text{if} \quad 0 \leq x \leq t; \\ 0 & \text{if} \quad x < 0 \text{ or } t < x \end{cases}$$

(Figure 17B.2). Show that γ is an approximation of identity.

17.12 Let

$$\gamma_t(x) = \begin{cases} \frac{1}{2t} & \text{if} \quad |x| \leq t; \\ 0 & \text{if} \quad t < |x|. \end{cases}$$

Show that γ is an approximation of identity.

17.13 Let $\mathbb{D} = \{\mathbf{x} \in \mathbb{R}^2 ; |\mathbf{x}| \leq 1\}$ be the unit disk.

(a) Let $u : \mathbb{D} \longrightarrow \mathbb{R}$ be the unique solution to the *Laplace equation* ($\triangle u = 0$) satisfying the *nonhomogeneous* Dirichlet boundary conditions $u(\mathbf{s}) = 1$, for all $\mathbf{s} \in \mathbb{S}$. Show that u must be constant: $u(\mathbf{x}) = 1$ for all $\mathbf{x} \in \mathbb{D}$.

(b) Recall that the Poisson kernel $\mathcal{P} : \mathbb{D} \times \mathbb{S} \longrightarrow \mathbb{R}$ is defined by

$$\mathcal{P}(\mathbf{x}, \mathbf{s}) = \frac{1 - \|\mathbf{x}\|^2}{\|\mathbf{x} - \mathbf{s}\|^2},$$

for any $\mathbf{x} \in \mathbb{D}$ and $\mathbf{s} \in \mathbb{S}$. Show that, for any fixed $\mathbf{x} \in \mathbb{D}$,

$$\frac{1}{2\pi} \int_{\mathbb{S}} \mathcal{P}(\mathbf{x}, \mathbf{s})d\mathbf{s} = 1.$$

(c) Let $b : \mathbb{S} \longrightarrow \mathbb{R}$ be any function, and let $\mu : \mathbb{D} \longrightarrow \mathbb{R}$ be the unique solution to the Laplace equation ($\triangle u = 0$) satisfying the nonhomogeneous Dirichlet boundary conditions $u(\mathbf{s}) = b(\mathbf{s})$, for all $\mathbf{s} \in \mathbb{S}$.

Let $m := \min_{\mathbf{s} \in \mathbb{S}} b(\mathbf{s})$, and $M := \max_{\mathbf{s} \in \mathbb{S}} b(\mathbf{s})$. Show that, for all $\mathbf{x} \in \mathbb{D}$,

$$m \le u(\mathbf{x}) \le M.$$

Remark In other words, the harmonic function u must take its *maximal* and *minimal* values on the *boundary* of the domain $\mathbb{D}$. This is a special case of the *maximum principle* for harmonic functions; see Corollary 1E.2, p. 19.

17.14 Let $\mathbb{H} := \{(x, y) \in \mathbb{R}^2 ; \ y \ge 0\}$ be the *half-plane*. Recall that the *half-plane Poisson kernel* is the function $\mathcal{K} : \mathbb{H} \longrightarrow \mathbb{R}$ defined:

$$\mathcal{K}(x, y) := \frac{y}{\pi(x^2 + y^2)}$$

for all $(x, y) \in \mathbb{H}$ except $(0, 0)$ (where it is not defined). Show that $\mathcal{K}$ is harmonic on the interior of $\mathbb{H}$.

18

Applications of complex analysis

> The shortest path between two truths in the real domain passes through
> the complex domain.
>
> *Jacques Hadamard*

Complex analysis is one of the most surprising and beautiful areas of mathematics. It also has some unexpected applications to PDEs and Fourier theory, which we will briefly survey in this chapter. Our survey is far from comprehensive – that would require another entire book. Instead, our goal in this chapter is merely to sketch the possibilities. If you are interested in further exploring the interactions between complex analysis and PDEs, we suggest Asmar (2002) and Churchill and Brown (2003), as well as Lang (1985), chap. VIII, Fisher (1999), chaps. 4 and 5, Asmar (2005), chap. 12, or the innovative and lavishly illustrated Needham (1997), chap.12.

This chapter assumes no prior knowledge of complex analysis. However, the presentation is slightly more abstract than most of the book, and is intended for more 'theoretically inclined' students. Nevertheless, someone who only wants the computational machinery of residue calculus can skip Sections 18B, 18E, and 18F, and skim the proofs in Sections 18C, 18D, and 18G, proceeding rapidly to Section 18H.

18A Holomorphic functions

Prerequisites: Appendix C, §1C.

Let $\mathbb{U} \subset \mathbb{C}$ be a open set, and let $f : \mathbb{U} \longrightarrow \mathbb{C}$ be a complex-valued function. If $u \in \mathbb{U}$, then the (complex) *derivative* of f at u is defined:

$$f'(u) := \lim_{\substack{c \to u \\ c \in \mathbb{C}}} \frac{f(c) - f(u)}{c - u}, \tag{18A.1}$$

where all terms in this formula are understood as complex numbers. We say that f is *complex-differentiable* at u if $f'(u)$ exists.

If we identify $\mathbb{C}$ with $\mathbb{R}^2$ in the obvious way, then we might imagine f as a function from a domain $\mathbb{U} \subset \mathbb{R}^2$ into $\mathbb{R}^2$, and assume that the complex derivate f' was just another way of expressing the (real-valued) Jacobian matrix of f. But this is not the case. Not all (real-)differentiable functions on $\mathbb{R}^2$ can be regarded as complex-differentiable functions on $\mathbb{C}$. To see this, let $f_r := \text{Re}\,[f] : \mathbb{U} \longrightarrow \mathbb{R}$ and $f_i := \text{Im}\,[f] : \mathbb{U} \longrightarrow \mathbb{R}$ be the real and imaginary parts of f, so that we can write $f(u) = f_r(u) + f_i(u)\mathbf{i}$ for any $u \in \mathbb{U}$. For any $u \in \mathbb{U}$, let $u_r := \text{Re}\,[u]$ and $u_i := \text{Im}\,[u]$, so that $u = u_r + u_i\mathbf{i}$. Then the (real-valued) Jacobian matrix of f has the form

$$\begin{bmatrix} \partial_r f_r & \partial_r f_i \\ \partial_i f_r & \partial_i f_i \end{bmatrix}. \tag{18A.2}$$

The relationship between the complex derivative (18A.1) and the Jacobian (18A.2) is the subject of the following fundamental result.

Theorem 18A.1 Cauchy–Riemann

Let $f : \mathbb{U} \longrightarrow \mathbb{C}$ and let $u \in \mathbb{U}$. Then f is complex-differentiable at u if, and only if, the partial derivatives $\partial_r f_r(u)$, $\partial_r f_i(u)$, $\partial_i f_r(u)$, and $\partial_i f_i(u)$ all exist, and, furthermore, satisfy the Cauchy–Riemann differential equations (CRDEs)

$$\partial_r f_r(u) = \partial_i f_i(u) \quad and \quad \partial_i f_r(u) = -\partial_r f_i(u). \tag{18A.3}$$

In this case, $f'(u) = \partial_r f_r(u) - \mathbf{i}\partial_i f_r(u) = \partial_i f_i(u) + \mathbf{i}\partial_r f_i(u)$.

Ⓔ *Proof* **Exercise 18A.1.**
(a) Compute the limit (18A.1) along the 'real' axis – that is, let $c = u + \epsilon$, where $\epsilon \in \mathbb{R}$, and show that

$$\lim_{\mathbb{R} \ni \epsilon \to 0} \frac{f(u + \epsilon) - f(u)}{\epsilon} = \partial_r f_r(u) + \mathbf{i}\partial_r f_i(u).$$

(b) Compute the limit (18A.1) along the 'imaginary' axis – that is, let $c = u + \epsilon\mathbf{i}$, where $\epsilon \in \mathbb{R}$, and show that

$$\lim_{\mathbb{R} \ni \epsilon \to 0} \frac{f(u + \epsilon\mathbf{i}) - f(u)}{\epsilon\mathbf{i}} = \partial_i f_i(u) - \mathbf{i}\partial_i f_r(u).$$

(c) If the limit (18A.1) is well defined, then it must be the same, no matter the direction from which c approaches u. Conclude that the results of (a) and (b) must be equal. Derive equations (18A.3).   ☐

Thus, the complex-differentiable functions are actually a very special subclass of the set of all (real-) differentiable functions on the plane. The function f is called *holomorphic* on $\mathbb{U}$ if f is complex-differentiable at all $u \in \mathbb{U}$. This is actually a *much* stronger requirement than merely requiring a real-valued function to be (real-) differentiable everywhere in some open subset of $\mathbb{R}^2$. For example, later we will show that every holomorphic function is *analytic* (Theorem 18D.1, p. 455). But one immediate indication of the special nature of holomorphic functions is their close relationship to two-dimensional harmonic functions.

Proposition 18A.2 *Let* $\mathbb{U} \subset \mathbb{C}$ *be an open set, and also regard* $\mathbb{U}$ *as a subset of* $\mathbb{R}^2$ *in the obvious way. If* $f : \mathbb{U} \longrightarrow \mathbb{C}$ *is any holomorphic function, then* $f_r : \mathbb{U} \longrightarrow \mathbb{R}$ *and* $f_i : \mathbb{U} \longrightarrow \mathbb{R}$ *are both harmonic functions.*

Proof **Exercise 18A.2.** *Hint:* (Apply the Cauchy–Riemann differential equations (18A.3) twice to get Laplace's equation.) ⓔ □

So, we can convert any holomorphic map into a pair of harmonic functions. Conversely, we can convert any harmonic function into a holomorphic map. To see this, suppose $h : \mathbb{U} \longrightarrow \mathbb{R}$ is a harmonic function. A *harmonic conjugate* for h is a function $g : \mathbb{U} \longrightarrow \mathbb{R}$ which satisfies the following differential equation:

$$\partial_2 g(u) = \partial_1 h(u) \quad \text{and} \quad \partial_1 g(u) = -\partial_2 h(u), \quad \text{for all } u \in \mathbb{U}. \quad (18A.4)$$

Proposition 18A.3 *Let* $\mathbb{U} \subset \mathbb{R}^2$ *be a convex open set (e.g. a disk or a rectangle). Let* $h : \mathbb{U} \longrightarrow \mathbb{R}$ *be any harmonic function.*

(a) *There exist harmonic conjugates for* h *on* $\mathbb{U}$ – *that is, equations* (18A.4) *have solutions.*
(b) *Any two harmonic conjugates for* h *differ by a constant.*
(c) *If* g *is a harmonic conjugate to* h, *and we define* $f : \mathbb{U} \longrightarrow \mathbb{C}$ *by* $f(u) = h(u) + g(u)\mathbf{i}$, *then* f *is holomorphic.*

Proof **Exercise 18A.3.** *Hint:* (a) Define $g(0)$ arbitrarily, and then, for any $u = (u_1, u_2) \in \mathbb{U}$, define ⓔ

$$g(u) = -\int_0^{u_1} \partial_2 h(0, x)\mathrm{d}x + \int_0^{u_2} \partial_1 h(u_1, y)\mathrm{d}y.$$

Show that g is differentiable and that it satisfies equations (18A.4).

For (b), suppose g_1 and g_2 both satisfy equations (18A.4); show that $g_1 - g_2$ is a constant by showing that $\partial_1(g_1 - g_2) = 0 = \partial_2(g_1 - g_2)$.

For (c), derive the CRDEs (18A.3) from the harmonic conjugacy equation (18A.4). □

Remark

(a) If h satisfies a Dirichlet boundary condition on $\partial\mathbb{U}$, then its harmonic conjugate satisfies an associated Neumann boundary condition on $\partial\mathbb{U}$, and vice versa; see Exercise 18A.7, p. 427. Thus, harmonic conjugation can be used to convert a Dirichlet BVP into a Neumann BVP, and vice versa.

(b) The 'convexity' requirement in Proposition 18A.2 can be weakened to 'simply connected'. However, Proposition 18A.2 is *not* true if the domain $\mathbb{U}$ is not simply connected (i.e. has a 'hole'); see Exercise 18C.16(e), p. 453.

Holomorphic functions have a rich and beautiful geometric structure, with many surprising properties. The study of such functions is called *complex analysis*. Propositions 18A.2 and 18A.3 imply that *every fact about harmonic functions in $\mathbb{R}^2$ is also a fact about complex analysis, and vice versa.*

Complex analysis also has important applications to fluid dynamics and electrostatics, because any holomorphic function can be interpreted as *sourceless, irrotational flow*, as we now explain. Let $\mathbb{U} \subset \mathbb{R}^2$ and let $\vec{V} : \mathbb{U} \longrightarrow \mathbb{R}^2$ be a two-dimensional vector field. Recall that the *divergence* of $\vec{V}$ is the scalar field $\operatorname{div} \vec{V} : \mathbb{U} \longrightarrow \mathbb{R}$ defined by $\operatorname{div} \vec{V}(\mathbf{u}) := \partial_1 V_1(\mathbf{u}) + \partial_2 V_2(\mathbf{u})$ for all $\mathbf{u} \in \mathbb{U}$ (see Appendix E(ii), p. 562). We say $\vec{V}$ is *locally sourceless* if $\operatorname{div} \vec{V} \equiv 0$. If $\vec{V}$ represents the two-dimensional flow of an incompressible fluid (e.g. water) in $\mathbb{U}$, then $\operatorname{div} \vec{V} \equiv 0$ means there are no sources or sinks in $\mathbb{U}$. If $\vec{V}$ represents a two-dimensional electric (or gravitational) field, then $\operatorname{div} \vec{V} \equiv 0$ means there are no charges (or masses) inside $\mathbb{U}$.

The *curl* of $\vec{V}$ is the scalar field $\operatorname{curl} \vec{V} : \mathbb{U} \longrightarrow \mathbb{R}$ defined by $\operatorname{curl} \vec{V}(\mathbf{u}) := \partial_1 V_2(\mathbf{u}) - \partial_2 V_1(\mathbf{u})$ for all $\mathbf{u} \in \mathbb{U}$. We say $\vec{V}$ is *locally irrotational* if $\operatorname{curl} \vec{V} \equiv 0$. If $\vec{V}$ represents a force field, then $\operatorname{curl} \vec{V} \equiv 0$ means that the net energy absorbed by a particle moving around a closed path in $\vec{V}$ is zero (i.e. the field is 'conservative'). If $\vec{V}$ represents the flow of a fluid, then $\operatorname{curl} \vec{V} \equiv 0$ means there are no 'vortices' in $\mathbb{U}$. (Note that this does *not* mean that the fluid must move in straight lines without turning. It simply means that the fluid turns in a uniform manner, without turbulence.)

Regard $\mathbb{U}$ as a subset of $\mathbb{C}$, and let $f : \mathbb{U} \longrightarrow \mathbb{C}$ be some function, with real and imaginary parts $f_r : \mathbb{U} \longrightarrow \mathbb{R}$ and $f_i : \mathbb{U} \longrightarrow \mathbb{R}$. The *complex conjugate* of f is the function $\overline{f} : \mathbb{U} \longrightarrow \mathbb{C}$ defined by $\overline{f}(u) = f_r(u) - \mathbf{i} f_i(u)$. We can treat $\overline{f}$ as a vector field $\vec{V} : \mathbb{U} \longrightarrow \mathbb{R}^2$, where $V_1 \equiv f_r$ and $V_2 \equiv -f_i$.

Proposition 18A.4 Holomorphic $\Longleftrightarrow$ sourceless irrotational flow
The function f is holomorphic on $\mathbb{U}$ if, and only if, $\vec{V}$ is locally sourceless and irrotational on $\mathbb{U}$.

Ⓔ *Proof* **Exercise 18A.4.** □

In §18B, we shall see that Proposition 18A.4 yields a powerful technique for studying fluids (or electric fields) confined to a subset of the plane (see Proposition 18B.6, p. 436). In §18C, we shall see that Proposition 18A.4 is also the key to understanding complex contour integration, through its role in the proof of Cauchy's Theorem 18C.5, p. 444.

To begin our study of complex analysis, we will verify that all the standard facts about the differentiation of real-valued functions carry over to complex differentiation, pretty much verbatim.

Proposition 18A.5 Closure properties of holomorphic functions

Let $\mathbb{U} \subset \mathbb{C}$ be an open set. Let $f, g : \mathbb{U} \longrightarrow \mathbb{C}$ be holomorphic functions.

(a) *The function $h(u) := f(u) + g(u)$ is also holomorphic on $\mathbb{U}$, and $h'(u) = f'(u) + g'(u)$ for all $u \in \mathbb{U}$.*

(b) *(Leibniz rule) The function $h(u) := f(u) \cdot g(u)$ is also holomorphic on $\mathbb{U}$, and $h'(u) = f'(u)g(u) + g'(u)f(u)$ for all $u \in \mathbb{U}$.*

(c) *(Quotient rule) Let $\mathbb{U}^* := \{u \in \mathbb{U}; \ g(u) \neq 0\}$. The function $h(u) := f(u)/g(u)$ is also holomorphic on $\mathbb{U}^*$, and $h'(u) = [g(u)f'(u) - f(u)g'(u)]/g(u)^2$ for all $u \in \mathbb{U}^*$.*

(d) *For any $n \in \mathbb{N}$, the function $h(u) := f^n(u)$ is holomorphic on $\mathbb{U}$, and $h'(u) = n \, f^{n-1}(u) \cdot f'(u)$ for all $u \in \mathbb{U}$.*

(e) *Thus, for any $c_0, c_1, \ldots, c_n \in \mathbb{C}$, the polynomial function $h(z) := c_n z^n + \cdots + c_1 z + c_0$ is holomorphic on $\mathbb{C}$.*

(f) *For any $n \in \mathbb{N}$, the function $h(u) := 1/g^n(u)$ is holomorphic on $\mathbb{U}^* := \{u \in \mathbb{U}; \ g(u) \neq 0\}$ and $h'(u) = -ng'(u)/g^{n+1}(u)$ for all $u \in \mathbb{U}^*$.*

(g) *For all $n \in \mathbb{N}$, let $f_n : \mathbb{U} \longrightarrow \mathbb{C}$ be a holomorphic function. Let $f, F : \mathbb{U} \longrightarrow \mathbb{C}$ be two other functions. If $\text{unif}-\lim_{n \to \infty} f_n = f$ and $\text{unif}-\lim_{n \to \infty} f'_n = F$, then f is holomorphic on $\mathbb{U}$, and $f' = F$.*

(h) *Let $\{c_n\}_{n=0}^{\infty}$ be any sequence of complex numbers, and consider the power series*

$$\sum_{n=0}^{\infty} c_n z^n = c_0 + c_1 z + c_2 z^2 + c_3 z^3 + c_4 z^4 + \cdots.$$

Suppose this series converges on $\mathbb{U}$ to define a function $f : \mathbb{U} \longrightarrow \mathbb{C}$. Then f is holomorphic on $\mathbb{U}$. Furthermore, f' is given by the 'formal derivative' of the power series. That is:

$$f'(u) = \sum_{n=1}^{\infty} nc_n z^{n-1} = c_1 + 2c_2 z + 3c_3 z^2 + 4c_4 z^3 + \cdots.$$

(i) *Let $\mathbb{X} \subset \mathbb{R}$ be open, let $f : \mathbb{X} \longrightarrow \mathbb{R}$, and suppose f is analytic at $x \in \mathbb{X}$, with a Taylor series[1] $T_x f$ which converges in the interval $(x - R, x + R)$ for some $R > 0$. Let $\mathbb{D} := \{c \in \mathbb{C}; \ |c - x| < R\}$ be the open disk of radius R around x in the complex plane. Then the Taylor series $T_x f$ converges uniformly on $\mathbb{D}$, and defines a holomorphic function $F : \mathbb{D} \longrightarrow \mathbb{C}$ which extends f (i.e. $F(r) = f(r)$ for all $r \in (x - R, x + R) \subset \mathbb{R}$).*

(j) *(Chain rule) Let $\mathbb{U}, \mathbb{V} \subset \mathbb{C}$ be open sets. Let $g : \mathbb{U} \longrightarrow \mathbb{V}$ and $f : \mathbb{V} \longrightarrow \mathbb{C}$ be holomorphic functions. Then the function $h(u) = f \circ g(u) = f[g(u)]$ is holomorphic on $\mathbb{U}$, and $h'(u) = f'[g(u)] \cdot g'(u)$ for all $u \in \mathbb{U}$.*

---

[1] See Appendix H(ii), p. 574.

(k) (Inverse function rule) *Let* $\mathbb{U}, \mathbb{V} \subset \mathbb{C}$ *be open sets. Let* $g : \mathbb{U} \longrightarrow \mathbb{V}$ *be a holomorphic function. Let* $f : \mathbb{V} \longrightarrow \mathbb{U}$ *be an inverse for* g *– that is,* $f[g(u)] = u$ *for all* $u \in \mathbb{U}$. *Let* $u \in \mathbb{U}$ *and* $v = g(u) \in \mathbb{V}$. *If* $g'(u) \neq 0$, *then* f *is holomorphic in a neighbourhood of* v, *and* $f'(v) = 1/g'(u)$.

Ⓔ   *Proof* **Exercise 18A.5.** (*Hint:* For each part, the proof from single-variable (real) differential calculus generally translates verbatim to complex numbers.)   □

Theorem 18A.5(i) implies that all the standard real-analytic functions have natural extensions to the complex plane, obtained by evaluating their Taylor series on $\mathbb{C}$.

Example 18A.6

(a) We define $\exp : \mathbb{C} \longrightarrow \mathbb{C}$ by

$$\exp(z) = \sum_{n=0}^{\infty} \frac{z^n}{n!} \quad \text{for all} \quad z \in \mathbb{C}.$$

The function defined by this power series is the same as the exponential function defined by Euler's formula (C.1) in Appendix C, p. 553. It satisfies the same properties as the real exponential function – that is, $\exp'(z) = \exp(z)$, $\exp(z_1 + z_2) = \exp(z_1) \cdot \exp(z_2)$, etc. (See Exercise 18A.8, p. 428.)

(b) We define $\sin : \mathbb{C} \longrightarrow \mathbb{C}$ by

$$\sin(z) = \sum_{n=0}^{\infty} \frac{(-1)^n z^{2n+1}}{(2n + 1)!} \quad \text{for all} \quad z \in \mathbb{C}.$$

(c) We define $\cos : \mathbb{C} \longrightarrow \mathbb{C}$ by

$$\cos(z) = \sum_{n=0}^{\infty} \frac{(-1)^n z^{2n}}{(2n)!} \quad \text{for all} \quad z \in \mathbb{C}.$$

(d) We define $\sinh : \mathbb{C} \longrightarrow \mathbb{C}$ by

$$\sinh(z) = \sum_{n=0}^{\infty} \frac{z^{2n+1}}{(2n + 1)!} \quad \text{for all} \quad z \in \mathbb{C}.$$

(e) We define $\cosh : \mathbb{C} \longrightarrow \mathbb{C}$ by

$$\cosh(z) = \sum_{n=0}^{\infty} \frac{z^{2n}}{(2n)!} \quad \text{for all} \quad z \in \mathbb{C}. \qquad \diamond$$

The complex trigonometric functions satisfy the same algebraic relations and differentiation rules as the real trigonometric functions (see Exercise 18A.9, p. 428). We will later show that any analytic function on $\mathbb{R}$ has a *unique* extension to a holomorphic function on some open subset of $\mathbb{C}$ (see Corollary 18D.4, p. 459).

Exercise 18A.6 Proposition 18A.2 says that the real and imaginary parts of any ⒺE
holomorphic function will be harmonic functions.

(a) Let $r_0, r_1, \ldots, r_n \in \mathbb{R}$, and consider the real-valued polynomial $f(x) = r_n x^n + \cdots + r_1 x + r_0$. Proposition 18A.5(e) says that f extends to a holomorphic function $f : \mathbb{C} \longrightarrow \mathbb{C}$. Express the real and imaginary parts of f in terms of the polar harmonic functions $\{\phi_n\}_{n=0}^\infty$ and $\{\psi_n\}_{n=0}^\infty$ introduced in §14B, p. 280.
(b) Express the real and imaginary parts of each of the holomorphic functions sin, cos, sinh, and cosh (from Example 18A.6) in terms of the harmonic functions introduced in §12A, p. 244. ◆

Exercise 18A.7 Harmonic conjugation of boundary conditions ⒺE
Let $\mathbb{U} \subset \mathbb{R}^2$ be an open subset whose boundary $\partial\mathbb{U}$ is a smooth curve. Let $\gamma :$ $[0, S] \longrightarrow \partial\mathbb{U}$ be a clockwise, arc-length parameterization of $\partial\mathbb{U}$. That is: γ is a differentiable bijection from $[0, S)$ into $\partial\mathbb{U}$ with $\gamma(0) = \gamma(S)$, and $|\dot{\gamma}(s)| = 1$ for all $s \in [0, S]$. Let $b : \partial\mathbb{U} \longrightarrow \mathbb{R}$ be a continuous function describing a Dirichlet boundary condition on $\mathbb{U}$, and define $B := b \circ \gamma : [0, S] \longrightarrow \mathbb{R}$. Suppose B is differentiable; let $B' : [0, S] \longrightarrow \mathbb{R}$ be its derivative, and then define the function $b' : \partial\mathbb{U} \longrightarrow \mathbb{R}$ by $b'(\gamma(s)) = B'(s)$ for all $s \in [0, S)$ (this defines b' on $\partial\mathbb{U}$ because γ is a bijection). Thus, we can regard b' as the derivative of b 'along' the boundary of $\mathbb{U}$.

Let $h : \mathbb{U} \longrightarrow \mathbb{R}$ be a harmonic function, and let $g : \mathbb{U} \longrightarrow \mathbb{R}$ be a harmonic conjugate for h. Show that h satisfies the Dirichlet boundary condition[2] $h(x) = b(x) + C$ for all $x \in \partial\mathbb{U}$ (where C is some constant) if, and only if, g satisfies the Neumann boundary condition $\partial_\perp g(x) = b'(x)$ for all $x \in \partial\mathbb{U}$.

Hint: For all $s \in [0, S]$, let $\vec{\mathbf{N}}(s)$ denote the outward unit normal vector of $\partial\mathbb{U}$ at $\gamma(s)$. Let $\mathbf{R} = \begin{bmatrix} 0 & -1 \\ 1 & 0 \end{bmatrix}$ (thus, left-multiplying the matrix $\mathbf{R}$ rotates a vector clockwise by 90°).

(a) Show that $(\nabla g) \cdot \mathbf{R} = \nabla h$. (Here we regard ∇h and ∇g as 2×1 'row matrices'.)
(b) Show that $\mathbf{R} \cdot \dot{\gamma}(s) = \vec{\mathbf{N}}(s)$ for all $s \in [0, S]$ (Here we regard $\dot{\gamma}$ and $\vec{\mathbf{N}}$ as 1×2 'column matrices'. *Hint:* Recall that γ is a *clockwise* parameterization.)
(c) Show that $(h \circ \gamma)'(s) = \nabla h[\gamma(s)] \cdot \dot{\gamma}(s)$, for all $s \in [0, S]$. (To make sense of this, recall that ∇h is a 2×1 matrix, while $\dot{\gamma}$ is a 1×2 matrix. *Hint:* Use the chain rule.)
(d) Show that $(\partial_\perp g)[\gamma(s)] = (h \circ \gamma)'(s)$ for all $s \in [0, S]$. (*Hint:* Recall that $(\partial_\perp g)[\gamma(s)] = (\nabla g)[\gamma(s)] \cdot \vec{\mathbf{N}}(s)$.)
(e) Conclude that $\partial_\perp g[\gamma(s)] = b'[\gamma(s)]$ for all $s \in [0, S]$ if, and only if, $h[\gamma(s)] = b(s) + C$ for all $s \in [0, S]$ (where C is some constant). ◆

---

[2] See §5C(i), p. 76, and §5C(ii), p. 80.

Ⓔ **Exercise 18A.8**

(a) Show that $\exp'(z) = \exp(z)$ for all $z \in \mathbb{C}$.

(b) Fix $x \in \mathbb{R}$, and consider the smooth path $\gamma : \mathbb{R} \longrightarrow \mathbb{R}^2$ defined by

$$\gamma(t) := [\exp_r(x + \mathbf{i}t), \exp_i(x + \mathbf{i}t)],$$

where $\exp_r(z)$ and $\exp_i(z)$ denote the real and imaginary parts of $\exp(z)$. Let $R := e^x$; note that $\gamma(0) = (R, 0)$. Use (a) to show that γ satisfies the ordinary differential equation

$$\begin{bmatrix} \dot{\gamma}_1(t) \\ \dot{\gamma}_2(t) \end{bmatrix} = \begin{bmatrix} -\gamma_2(t) \\ \gamma_1(t) \end{bmatrix}.$$

Conclude that $\gamma(t) = [R\cos(t),\ R\sin(t)]$ for all $t \in \mathbb{R}$.

(c) For any $x, y \in \mathbb{R}$, use (b) to show that $\exp(x + \mathbf{i}y) = e^x(\cos(y) + \mathbf{i}\sin(y))$.

(d) Deduce that $\exp(c_1 + c_2) = \exp(c_1) \cdot \exp(c_2)$ for all $c_1, c_2 \in \mathbb{C}$.   ◆

Ⓔ **Exercise 18A.9**

(a) Show that $\sin'(z) = \cos(z)$, $\cos'(z) = -\sin(z)$, $\sinh'(z) = \cos(z)$, and $\cosh'(z) = -\sinh(z)$, for all $z \in \mathbb{C}$.

(b) For all $z \in \mathbb{C}$, verify the Euler identities:

$$\sin(z) = \frac{\exp(z\mathbf{i}) - \exp(-z\mathbf{i})}{2\mathbf{i}}; \qquad \cos(z) = \frac{\exp(-z\mathbf{i}) + \exp(z\mathbf{i})}{2};$$

$$\sinh(z) = \frac{\exp(z) - \exp(-z)}{2}; \qquad \cosh(z) = \frac{\exp(z) + \exp(-z)}{2}.$$

(c) Deduce that $\sinh(z) = \mathbf{i}\sin(\mathbf{i}z)$ and $\cosh(z) = \cos(\mathbf{i}z)$.

(d) For all $x, y \in \mathbb{R}$, prove the following identities:

$$\cos(x + y\mathbf{i}) = \cos(x)\cosh(y) - \mathbf{i}\sin(x)\sinh(y);$$
$$\sin(x + y\mathbf{i}) = \sin(x)\cosh(y) + \mathbf{i}\cos(x)\sinh(y).$$

(e) For all $z \in \mathbb{C}$, verify the Pythagorean identities:

$$\cos(z)^2 + \sin(z)^2 = 1 \quad \text{and} \quad \cosh(z)^2 - \sinh(z)^2 = 1.$$

(Later we will show that pretty much every 'trigonometric identity' that is true on $\mathbb{R}$ will also be true over all of $\mathbb{C}$; see Exercise 18D.4, p. 459.)   ◆

18B Conformal maps

Prerequisites: §1B, §5C, §18A.

A linear map $f : \mathbb{R}^D \longrightarrow \mathbb{R}^D$ is called *conformal* if it preserves the angles between vectors. Thus, for example, *rotations*, *reflections*, and *dilations* are all conformal maps.

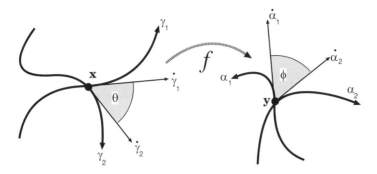

Figure 18B.1. A conformal map preserves the angle of intersection between two paths.

Let $\mathbb{U}, \mathbb{V} \subset \mathbb{R}^D$ be open subsets of $\mathbb{R}^D$. A differentiable map $f : \mathbb{U} \longrightarrow \mathbb{V}$ is called *conformal* if its derivative $\mathsf{D} f(\mathbf{x})$ is a conformal linear map, for every $\mathbf{x} \in \mathbb{U}$. One way to interpret this is depicted in Figure 18B.1. Suppose two smooth paths γ_1 and γ_2 cross at $\mathbf{x}$, and their velocity vectors $\dot{\gamma}_1$ and $\dot{\gamma}_2$ form an angle θ at $\mathbf{x}$. Let $\alpha_1 = f \circ \gamma_1$ and $\alpha_2 = f \circ \gamma_2$, and let $\mathbf{y} = f(\mathbf{x})$. Then α_1 and α_2 are smooth paths, and cross at $\mathbf{y}$, forming an angle ϕ. The map f is conformal if, for every $\mathbf{x}$, γ_1, and γ_2, the angles θ and ϕ are equal.

Complex analysis could be redefined as 'the study of two-dimensional conformal maps', because of the next result.

Proposition 18B.1 Holomorphic $\Longleftrightarrow$ conformal
Let $\mathbb{U} \subset \mathbb{R}^2$ be an open subset, and let $f : \mathbb{U} \longrightarrow \mathbb{R}^2$ be a differentiable function, with $f(\mathbf{u}) = (f_1(\mathbf{u}), f_2(\mathbf{u}))$ for all $\mathbf{u} \in \mathbb{U}$. Identify $\mathbb{U}$ with a subset $\widetilde{\mathbb{U}}$ of the plane $\mathbb{C}$ in the obvious way, and define $\widetilde{f} : \widetilde{\mathbb{U}} \longrightarrow \mathbb{C}$ by $\widetilde{f}(x + y\mathbf{i}) = f_1(x, y) + f_2(x, y)\mathbf{i}$ — that is, $\widetilde{f}$ is just the representation of f as a complex-valued function on $\mathbb{C}$. Then $(f$ is conformal$) \Longleftrightarrow (\widetilde{f}$ is holomorphic$)$.

Proof **Exercise 18B.1.** (*Hint:* The derivative $\mathsf{D} f$ is a linear map on $\mathbb{R}^2$. Show ⒠ that $\mathsf{D} f$ is conformal if, and only if, $\widetilde{f}$ satisfies the CRDEs (18A.3), p. 422.) □

If $\mathbb{U} \subset \mathbb{C}$ is open, then Proposition 18B.1 means that every holomorphic map $f : \mathbb{U} \longrightarrow \mathbb{C}$ can be treated as a conformal transformation of $\mathbb{U}$. In particular we can often conformally identify $\mathbb{U}$ with some other domain in the complex plane via a suitable holomorphic map. A function $f : \mathbb{U} \longrightarrow \mathbb{V}$ is a *conformal isomorphism* if f is conformal, invertible, and $f^{-1} : \mathbb{V} \longrightarrow \mathbb{U}$ is also conformal. Proposition 18B.1 says that this is equivalent to requiring f and f^{-1} to be holomorphic.

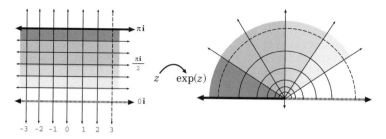

Figure 18B.2. The map $f(z) = \exp(z)$ conformally identifies a bi-infinite horizontal strip with the upper half-plane (see Example 18B.2(a)).

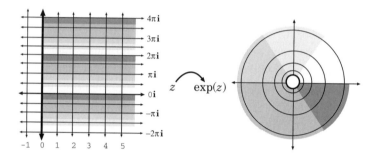

Figure 18B.3. The map $f(z) = \exp(z)$ conformally projects the right half-plane onto the complement of the unit disk (see Example 18B.2(b)).

Example 18B.2

(a) In Figure 18B.2, $\mathbb{U} = \{x + y\mathbf{i}; \ x \in \mathbb{R}, 0 < y < \pi\}$ is a bi-infinite horizontal strip, and $\mathbb{C}_+ = \{x + y\mathbf{i}; \ x \in \mathbb{R}, \ y > 0\}$ is the open upper half-plane, and $f(z) = \exp(z)$. Then $f : \mathbb{U} \longrightarrow \mathbb{C}_+$ is a conformal isomorphism from $\mathbb{U}$ to $\mathbb{C}_+$.

(b) In Figure 18B.3, $\mathbb{U} = \{x + y\mathbf{i}; \ x > 0, y \in \mathbb{R}\}$ is the open right half of the complex plane, and $\mathfrak{D}^{\complement} = \{x + y\mathbf{i}; x^2 + y^2 > 1\}$ is the complement of the closed unit disk, and $f(z) = \exp(z)$. Then $f : \mathbb{U} \longrightarrow \mathfrak{D}^{\complement}$ is not a conformal isomorphism (because it is many-to-one). However, f is a conformal *covering map*. This means that f is *locally* one-to-one: for any point $u \in \mathbb{U}$, with $v = f(u) \in \mathfrak{D}^{\complement}$, there is a neighbourhood $\mathcal{V} \subset \mathfrak{D}^{\complement}$ of v and a neighbourhood $\mathcal{U} \subset \mathbb{U}$ of u such that $f_{|} : \mathcal{U} \longrightarrow \mathcal{V}$ is one-to-one. (Note that f is *not* globally one-to-one because it is periodic in the imaginary coordinate.)

(c) In Figure 18B.4, $\mathbb{U} = \{x + y\mathbf{i}; \ x < 0, 0 < y < \pi\}$ is a left half-infinite rectangle, $\mathbb{V} = \{x + y\mathbf{i}; y > 1, x^2 + y^2 < 1\}$ is the open half-disk, and $f(z) = \exp(z)$. Then $f : \mathbb{U} \longrightarrow \mathbb{V}$ is a conformal isomorphism from $\mathbb{U}$ to $\mathbb{V}$.

(d) In Figure 18B.5, $\mathbb{U} = \{x + y\mathbf{i}; \ x > 0, 0 < y < \pi\}$ is a right half-infinite rectangle, $\mathbb{V} = \{x + y\mathbf{i}; y > 1, x^2 + y^2 > 1\}$ is the 'amphitheatre', and $f(z) = \exp(z)$. Then $f : \mathbb{U} \longrightarrow \mathbb{V}$ is a conformal isomorphism from $\mathbb{U}$ to $\mathbb{V}$.

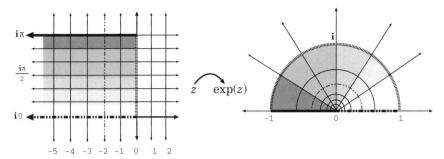

Figure 18B.4. The map $f(z) = \exp(z)$ conformally identifies a left half-infinite rectangle with the half-disk (Example 18B.2(c)).

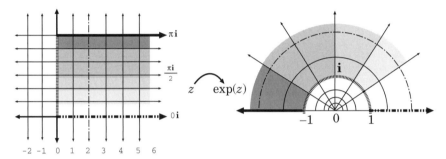

Figure 18B.5. The map $f(z) = \exp(z)$ conformally identifies a right half-infinite rectangle with the 'amphitheatre' (Example 18B.2(d)).

(e) In Figure 18B.6(a) and (b), $\mathbb{U} = \{x + y\mathbf{i} \; ; \; x, y > 0\}$ is the open upper right quarter-plane, $\mathbb{C}_+ = \{x + y\mathbf{i} \; ; \; y > 0\}$ is the open upper half-plane, and $f(z) = z^2$. Then f is a conformal isomorphism from $\mathbb{U}$ to $\mathbb{C}_+$.

(f) Let $\mathbb{C}_+ := \{x + y\mathbf{i} \; ; \; y > 0\}$ be the upper half-plane, and $\mathbb{U} := \mathbb{C}_+ \setminus \{y\mathbf{i} \; ; \; 0 < y < 1\}$; that is, $\mathbb{U}$ is the upper half-plane with a vertical line-segment of length 1 removed above the origin. Let $f(z) = (z^2 + 1)^{1/2}$; then f is a conformal isomorphism from $\mathbb{U}$ to $\mathbb{C}_+$, as shown in Figure 18B.7(a).

(g) Let $\mathbb{U} := \{x + y\mathbf{i}; \text{either } y \neq 0 \text{ or } -1 < x < 1\}$, and let $\mathbb{V} := \{x + y\mathbf{i}; \frac{-\pi}{2} < y < \frac{\pi}{2}\}$ be a bi-infinite horizontal strip of width π. Let $f(z) := \mathbf{i} \cdot \arcsin(z)$; then f is a conformal isomorphism from $\mathbb{U}$ to $\mathbb{V}$, as shown in Figure 18B.7(b). ◇

Exercise 18B.2   Verify Example 18B.2(a)–(g).   ♦   Ⓔ

Conformal maps are very useful for solving boundary value problems, because of the following result.

Proposition 18B.3 *Let* $\mathbb{X}, \mathbb{Y} \subset \mathbb{R}^2$ *be open domains with closures* $\overline{\mathbb{X}}$ *and* $\overline{\mathbb{Y}}$. *Let* $f : \overline{\mathbb{X}} \longrightarrow \overline{\mathbb{Y}}$ *be a continuous surjection which conformally maps* $\mathbb{X}$ *into* $\mathbb{Y}$. *Let* $h : \overline{\mathbb{Y}} \longrightarrow \mathbb{R}$ *be some smooth function, and define* $H = h \circ f : \overline{\mathbb{X}} \longrightarrow \mathbb{R}$.

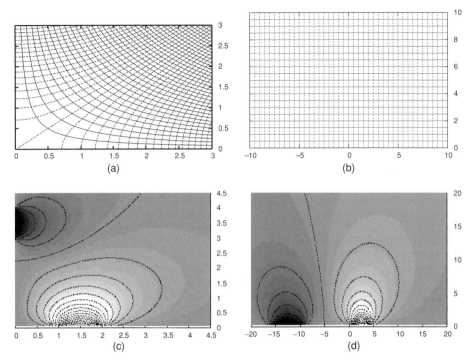

Figure 18B.6. Parts (a) and (b) refer to Example 18B.2(e). The map $f(z) = z^2$ conformally identifies the quarter-plane (a) and the half-plane (b). The mesh of curves in (a) is the preimage of the Cartesian grid in (b). Note that these curves always intersect at right angles; this is because f is a conformal map. The solid curves are the *streamlines*: the preimages of horizontal grid lines. The streamlines describe a sourceless, irrotational flow confined to the quarter-plane (see Proposition 18B.6, p. 436). The dashed curves are the *equipotential contours*: the preimages of vertical grid lines. The streamlines and equipotentials can be interpreted as the level curves of two harmonic functions (by Proposition 18A.2). They can also be interpreted as the voltage contours and field lines of an electric field in a quarter-plane bounded by perfect conductors on the x and y axes. Parts (c) and (d) refer to Example 18B.4, p. 434. The map $f(z) = z^2$ can be used to 'pull back' solutions to BVPs from the half-plane to the quarter-plane. (c) Greyscale plot of the harmonic function H defined on the quarter-plane by equation (18B.2). (d) Greyscale plot of the harmonic function h defined on the half-plane by equation (18B.1). The two functions are related by $H = h \circ f$.

(a) *h is harmonic on $\mathbb{X}$ if, and only if, H is harmonic on $\mathbb{Y}$.*

(b) *Let $b : \partial \mathbb{Y} \longrightarrow \mathbb{R}$ be some function on the boundary of $\mathbb{Y}$. Then $B = b \circ f : \partial \mathbb{X} \longrightarrow \mathbb{R}$ is a function on the boundary of $\mathbb{X}$. The function h satisfies the nonhomogeneous Dirichlet boundary condition[3] '$h(\mathbf{y}) = b(\mathbf{y})$ for all $\mathbf{y} \in \partial \mathbb{Y}$' if, and only if, H satisfies the nonhomogeneous Dirichlet boundary condition '$H(\mathbf{x}) = B(\mathbf{x})$ for all $\mathbf{x} \in \partial \mathbb{X}$'.*

---

[3] See §5C(i), p. 76.

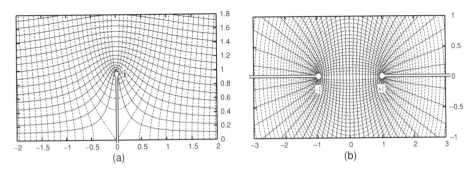

Figure 18B.7. (a) See Example 18B.2(f). The map $f(z) = (z^2 + 1)^{1/2}$ is a confor-
mal isomorphism from the set $\mathbb{C}_+ \setminus \{y\mathbf{i}; \ 0 < y < 1\}$ to the upper half-plane $\mathbb{C}_+$.
(b) See Example 18B.2(g). The map $f(z) := \mathbf{i} \cdot \arcsin(z)$ is a conformal isomorphism from the domain $\mathbb{U} := \{x + y\mathbf{i}; \text{either } y \neq 0 \text{ or } -1 < x < 1\}$ to a bi-infinite
horizontal strip. In these figures, as in Figures 18B.6(a) and (b), the mesh is the
preimage of a Cartesian grid on the image domain; the solid lines are streamlines,
and the dashed lines are equipotential contours. (a) We can interpret these streamlines as the flow of fluid over an obstacle. (b) Streamlines represent the flow of
fluid through a narrow aperature between two compartments. Alternatively, we
can interpret these curves as the voltage contours and field lines of an electric
field, where the domain boundaries are perfect conductors.

(c) *For all $\mathbf{x} \in \partial\mathbb{X}$, let $\vec{\mathbf{N}}_\mathbb{X}(\mathbf{x})$ be the outward unit normal vector to $\partial\mathbb{X}$ at $\mathbf{x}$, let $\vec{\mathbf{N}}_\mathbb{Y}(\mathbf{x})$ be
the outward unit normal vector to $\partial\mathbb{Y}$ at $f(\mathbf{x})$, and let $\mathsf{D}f(\mathbf{x})$ be the derivative of f at
$\mathbf{x}$ (a linear transformation of $\mathbb{R}^D$). Then $\mathsf{D}f(\mathbf{x})[\vec{\mathbf{N}}_\mathbb{X}(\mathbf{x})] = \phi(\mathbf{x}) \cdot \vec{\mathbf{N}}_\mathbb{Y}(\mathbf{x})$ for some scalar
$\phi(\mathbf{x}) > 0$.*
(d) *Let $b : \partial\mathbb{Y} \longrightarrow \mathbb{R}$ be some function on the boundary of $\mathbb{Y}$, and define $B : \partial\mathbb{X} \longrightarrow \mathbb{R}$ by
$B(\mathbf{x}) := \phi(\mathbf{x}) \cdot b[f(\mathbf{x})]$ for all $\mathbf{x} \in \partial\mathbb{X}$. Then h satisfies the nonhomogeneous Neumann
boundary condition[4] '$\partial_\perp h(\mathbf{y}) = b(\mathbf{y})$ for all $\mathbf{y} \in \partial\mathbb{Y}$' if, and only if, H satisfies the
nonhomogeneous Neumann boundary condition '$\partial_\perp H(\mathbf{x}) = B(\mathbf{x})$ for all $\mathbf{x} \in \partial\mathbb{X}$'.*

Proof **Exercise 18B.3.** *Hint:* For (a) combine Propositions 18A.2, 18A.3, and   Ⓔ
18B.1. For (c), use the fact that f is a conformal map, so $\mathsf{D}f(\mathbf{x})$ is a conformal
linear transformation; thus, if $\vec{\mathbf{N}}_\mathbb{X}(\mathbf{x})$ is normal to $\partial\mathbb{X}$, then $\mathsf{D}f(\mathbf{x})[\vec{\mathbf{N}}_\mathbb{X}(\mathbf{x})]$ must be
normal to $\partial\mathbb{Y}$.
   To prove (d), use (c) and the chain rule.   □

We can apply Proposition 18B.3 as follows: given a boundary value problem on
some 'nasty' domain $\mathbb{X}$, find a 'nice' domain $\mathbb{Y}$ (e.g. a box, a disk, or a half-plane),
and a conformal isomorphism $f : \mathbb{X} \longrightarrow \mathbb{Y}$. Solve the boundary value problem
in $\mathbb{Y}$ (e.g. using the methods from Chapters 12–17), to get a solution function $h :
\mathbb{Y} \longrightarrow \mathbb{R}$. Finally, 'pull back' this solution to get a solution $H = h \circ f : \mathbb{X} \longrightarrow \mathbb{R}$

---

[4] See §5C(ii), p. 80.

to the original BVP on $\mathbb{X}$. We can obtain a suitable conformal isomorphism from $\mathbb{X}$ to $\mathbb{Y}$ using holomorphic mappings, by Proposition 18B.1.

Example 18B.4 Let $\mathbb{X} = \{(x_1, x_2) \in \mathbb{R}^2; x_1, x_2 > 0\}$ be the open upper right quarter-plane. Suppose we want to find a harmonic function $H : \mathbb{X} \longrightarrow \mathbb{R}$ satisfying the nonhomogeneous Dirichlet boundary conditions $H(\mathbf{x}) = B(\mathbf{x})$ for all $\mathbf{x} \in \partial \mathbb{X}$, where $B : \partial \mathbb{X} \longrightarrow \mathbb{R}$ is defined:

$$B(x_1, 0) = \begin{cases} 3 & \text{if } 1 \leq x_1 \leq 2; \\ 0 & \text{otherwise,} \end{cases} \quad \text{and} \quad B(0, x_2) = \begin{cases} -1 & \text{if } 3 \leq x_2 \leq 4; \\ 0 & \text{otherwise.} \end{cases}$$

Identify $\mathbb{X}$ with the complex right quarter-plane $\mathbb{U}$ from Example 18B.2(e). Let $f(z) := z^2$; then f is a conformal isomorphism from $\mathbb{U}$ to the upper half-plane $\mathbb{C}^+$. If we identify $\mathbb{C}^+$ with the real upper half-plane $\mathbb{Y} := \{(y_1, y_2) \in \mathbb{R}^2; y_2 > 0\}$, then we can treat f as a function $f : \mathbb{X} \longrightarrow \mathbb{Y}$, given by the formula $f(x_1, x_2) = (x_1^2 - x_2^2, 2x_1 x_2)$.

Since f is bijective, the inverse $f^{-1} : \mathbb{Y} \longrightarrow \mathbb{X}$ is well-defined. Thus, we can define a function $b := B \circ f^{-1} : \partial \mathbb{Y} \longrightarrow \mathbb{R}$. To be concrete:

$$b(y_1, 0) = \begin{cases} 3 & \text{if } 1 \leq y_1 \leq 4; \\ -1 & \text{if } -16 \leq y_1 \leq -9; \\ 0 & \text{otherwise.} \end{cases}$$

Now we must find a harmonic function $h : \mathbb{Y} \longrightarrow \mathbb{R}$ satisfying the Dirichlet boundary conditions $h(y_1, 0) = b(y_1, 0)$ for all $y_1 \in \mathbb{R}$. By adding together two copies of the solution from Example 17E.2, p. 410, we deduce that

$$h(y_1, y_2) = \frac{3}{\pi} \left[\arcsin\left(\frac{4 - y_1}{y_2} \right) - \arcsin\left(\frac{1 - y_1}{y_2} \right) \right] \qquad (18B.1)$$
$$- \frac{1}{\pi} \left[\arcsin\left(\frac{-9 - y_1}{y_2} \right) - \arcsin\left(\frac{-16 - y_1}{y_2} \right) \right],$$

for all $(y_1, y_2) \in \mathbb{Y}$; see Figure 18B.6(d). Finally, define $H := h \circ f : \mathbb{X} \longrightarrow \mathbb{R}$. That is,

$$H(x_1, x_2) = \frac{3}{\pi} \left[\arcsin\left(\frac{4 - x_1^2 + x_2^2}{2x_1 x_2} \right) - \arcsin\left(\frac{1 - x_1^2 + x_2^2}{2x_1 x_2} \right) \right] \qquad (18B.2)$$
$$- \frac{1}{\pi} \left[\arcsin\left(\frac{-9 - x_1^2 + x_2^2}{2x_1 x_2} \right) - \arcsin\left(\frac{-16 - x_1^2 + x_2^2}{2x_1 x_2} \right) \right],$$

for all $(x_1, x_2) \in \mathbb{X}$; see Figure 18B.6(c). Proposition 18B.3(a) says that H is harmonic on $\mathbb{X}$, because h is harmonic on $\mathbb{Y}$. Finally, h satisfies the Dirichlet boundary conditions specified by b, and $B = b \circ f$; thus Proposition 18B.3(b) says that H satisfies the Dirichlet boundary conditions specified by B, as desired. ◇

For Proposition 18B.3 to be useful, we must find a conformal map from our original domain $\mathbb{X}$ to some 'nice' domain $\mathbb{Y}$ where we are able to easily solve BVPs. For example, ideally $\mathbb{Y}$ should be a disk or a half-plane, so that we can apply the Fourier techniques of Section 14B, or the Poisson kernel methods from Sections 14B(v), 17F, and 17E. If $\mathbb{X}$ is a simply connected open subset of the plane, then a deep result in complex analysis says that it is always possible to find such a conformal map. An open subset $\mathbb{U} \subset \mathbb{C}$ is *simply connected* if any closed loop in $\mathbb{U}$ can be continuously shrunk down to a point without ever leaving $\mathbb{U}$. Heuristically speaking, this means that $\mathbb{U}$ has no 'holes'. (For example, the open disk is simply connected, and so is the upper half-plane. However, the open annulus is *not* simply connected.)

Theorem 18B.5 Riemann mapping theorem

Let $\mathbb{U}, \mathbb{V} \subset \mathbb{C}$ be two open, simply connected regions of the complex plane. Then there is always a holomorphic bijection $f : \mathbb{U} \longrightarrow \mathbb{V}$.

Proof See Lang (1985), chap. XIV, pp. 340–358. □

In particular, this means that any simply connected open subset of $\mathbb{C}$ is conformally isomorphic to the disk, and also conformally isomorphic to the upper half-plane. Thus, in theory, a technique like Example 18B.4 can be applied to any such region.

Unfortunately, the Riemann mapping theorem does not tell us how to construct the conformal isomorphism – it merely tells us that such an isomorphism *exists*. This is not very useful when we want to solve a specific boundary value problem on a specific domain. If $\mathbb{V}$ is a region bounded by a polygon, and $\mathbb{U}$ is the upper half-plane, then it is possible to construct an explicit conformal isomorphism from $\mathbb{U}$ to $\mathbb{V}$ using *Schwarz–Christoffel transformations*; see Fisher (1999), §3.5, or Asmar (2005), §12.6. For further information about conformal maps in general, see Fisher (1999), §3.4, Lang (1985), chap. VII, or the innovatively visual Needham (1997), chap. 12. Older, but still highly respected, references are Bieberbach (1953), Nehari (1975) and Schwerdtfeger (1979).

Application to fluid dynamics Let $\mathbb{U} \subset \mathbb{C}$ be an open connected set, and let $\vec{\mathbf{V}} : \mathbb{U} \longrightarrow \mathbb{R}^2$ be a two-dimensional vector field (describing a flow). Define $f : \mathbb{U} \longrightarrow \mathbb{C}$ by $f(u) = V_1(u) - iV_2(u)$. Recall that Proposition 18A.4, p. 424, says that $\vec{\mathbf{V}}$ is sourceless and irrotational (e.g. describing a nonturbulent, incompressible fluid) if, and only if, f is holomorphic. Suppose F is a *complex antiderivative*[5] of f on $\mathbb{U}$ – that is $F : \mathbb{U} \longrightarrow \mathbb{C}$ is a holomorphic map such that $F' \equiv f$. Then F is called a *complex potential* for $\vec{\mathbf{V}}$. The function $\phi(u) = \mathrm{Re}\,[F(u)]$ is called the (real) *potential* of the flow. An *equipotential contour* of F is a level curve of ϕ. That is, it is a set $\mathcal{E}_x = \{u \in \mathbb{U}; \mathrm{Re}\,[F(u)] = x\}$ for some fixed $x \in \mathbb{R}$. For example, in Figures 18B.6(a) and 18B.7(a), (b), the equipotential contours are the dashed curves. A *streamline* of F is a level curve of the imaginary part of F. That is, it is a set $\mathcal{S}_y = \{u \in \mathbb{U}; \mathrm{Im}\,[F(u)] = y\}$ for some fixed $y \in \mathbb{R}$. For example, in Figures 18B.6(a) and 18B.7(a), (b), the streamlines are the solid curves.

A *trajectory* of $\vec{\mathbf{V}}$ is the path followed by a particle carried in the flow – that is, it is a smooth path $\alpha : (-T, T) \longrightarrow \mathbb{U}$ (for some $T \in (0, \infty]$) such that $\dot{\alpha}(t) = \vec{\mathbf{V}}[\alpha(t)]$ for all $t \in (-T, T)$. The flow $\vec{\mathbf{V}}$ is *confined* to $\mathbb{U}$ if no trajectories of $\vec{\mathbf{V}}$ ever pass through the boundary $\partial \mathbb{U}$. (Physically, $\partial \mathbb{U}$ represents an 'impermeable barrier'.) The equipotentials and streamlines of F are important for understanding the flow defined by $\vec{\mathbf{V}}$, because of the following result.

Proposition 18B.6 *Let* $\vec{\mathbf{V}} : \mathbb{U} \longrightarrow \mathbb{R}^2$ *be a sourceless, irrotational flow, and let* $F : \mathbb{U} \longrightarrow \mathbb{C}$ *be a complex potential for* $\vec{\mathbf{V}}$.

(a) *If* $\phi = \mathrm{Re}\,[F]$, *then* $\nabla\phi = \vec{\mathbf{V}}$. *Thus, particles in the flow can be thought of as descending the 'potential energy landscape' determined by* ϕ. *In particular, every trajectory of the flow cuts orthogonally through every equipotential contour of* F.

(b) *Every streamline of* F *also cuts orthogonally through every equipotential contour.*

(c) *Every trajectory of* $\vec{\mathbf{V}}$ *parameterizes a streamline of* F, *and every streamline can be parameterized by a trajectory. (Thus, by plotting the streamlines of* F, *we can visualize the flow* $\vec{\mathbf{V}}$.)

(d) $\vec{\mathbf{V}}$ *is confined to* $\mathbb{U}$ *if, and only if,* F *conformally maps* $\mathbb{U}$ *to a bi-infinite horizontal strip* $\mathbb{V} \subset \mathbb{C}$, *and maps each connected component of* $\partial\mathbb{U}$ *to a horizontal line in* $\mathbb{V}$.

Ⓔ *Proof* **Exercise 18B.4.** *Hint:* Part (a) follows from the definitions of F and $\vec{\mathbf{V}}$. To prove (b), use the fact that F is a conformal map. Part (c) follows by combining (a) and (b), and then (d) follows from (c). □

Thus, the set of conformal mappings from $\mathbb{U}$ onto such horizontal strips describes all possible sourceless, irrotational flows confined to $\mathbb{U}$. If $\partial\mathbb{U}$ is simply connected, then we can assume that F maps $\mathbb{U}$ to the upper half-plane and maps $\partial\mathbb{U}$ to $\mathbb{R}$ (as in Example 18B.2(e)). Or, if we are willing to allow one 'point source' (or sink) p in $\partial\mathbb{U}$, we can find a mapping from $\mathbb{U}$ to a bi-infinite horizontal strip, which maps the half of the boundary on one side of p to the top edge of strip, maps the other

---

[5] We will discuss how to construct complex antiderivatives in Exercise 18C.15, p. 453; for now, just assume that F exists.

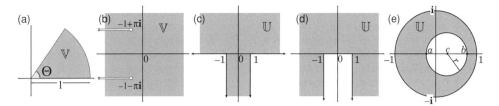

Figure 18B.8. Diagrams for Exercise 18B.5.

half of the boundary to the bottom edge, and maps p itself to ∞ (as in Example 18B.2(a); in this case, the 'point source' is at 0).

Application to electrostatics Proposition 18B.6 has another important physical interpretation. The function $\phi = \mathrm{Im}\,[f]$ is harmonic (by Proposition 18A.2, p. 423). Thus, it can be interpreted as an electrostatic potential (see Example 1D.2, p. 15). In this case, we can regard the streamlines of F as the *voltage contours* of the resulting electric field; then the 'equipotentials' F of are the *field lines* (note the reversal of roles here). If $\partial\mathbb{U}$ is a perfect conductor (e.g. a metal), then the field lines must always intersect $\partial\mathbb{U}$ orthogonally, and the voltage contours (i.e. the 'streamlines') can never intersect $\partial\mathbb{U}$ – thus, in terms of our fluid dynamical model, the 'flow' is confined to $\mathbb{U}$. Thus, the streamlines and equipotentials in Figures 18B.6(a) and 18B.7(a), (b) portray the (two-dimensional) electric field generated by charged metal plates.

For more about the applications of complex analysis to fluid dynamics and electrostatics, see Fisher (1999), §4.2, or Needham (1997), §12.V.

Exercise 18B.5   Ⓔ

(a) Let $\Theta \in (0, 2\pi]$, and consider the 'pie-wedge' domain $\mathbb{V} := \{r\,\mathrm{cis}\,\theta;\ 0 < r < 1, 0 < \theta < \Theta\}$ (in polar coordinates); see Figure 18B.8(a). Find a conformal isomorphism from $\mathbb{V}$ to the left half-infinite rectangle $\mathbb{U} = \{x + y\mathbf{i};\ x > 0, 0 < y < \pi\}$.

(b) Let $\mathbb{U} := \{x + y\mathbf{i};\ -\pi < y < \pi\}$ be a bi-infinite horizontal strip of width 2π, and let $\mathbb{V} := \{x + y\mathbf{i}\,;\ \text{either } y \neq \pm\mathbf{i} \text{ or } x > -1\}$, as shown in Figure 18B.8(b). Show that $f(z) := z + \exp(z)$ is a conformal isomorphism from $\mathbb{U}$ to $\mathbb{V}$.

(c) Let $\mathbb{C}_+ := \{x + y\mathbf{i};\ y > 0\}$ be the upper half-plane. Let $\mathbb{U} := \{x + y\mathbf{i};\ \text{either } y > 0 \text{ or } -1 < x < 1\}$. That is, $\mathbb{U}$ is the complex plane with two lower quarter-planes removed, leaving a narrow 'chasm' in between them, as shown in Figure 18B.8(c). Show that

$$f(z) = \frac{2}{\pi}\left(\sqrt{z^2 - 1} + \arcsin(1/z)\right)$$

is a conformal isomorphism from $\mathbb{C}_+$ to $\mathbb{U}$.

(d) Let $\mathbb{C}_+ := \{x + y\mathbf{i};\ y > 0\}$ be the upper half-plane. Let $\mathbb{U} := \{x + y\mathbf{i};\ \text{either } y > 0 \text{ or } x < 1 \text{ or } 1 < x\}$. That is, $\mathbb{U}$ is the complex plane with a vertical half-infinite rectangle removed, as shown in Figure 18B.8(d). Show that

$$f(z) = \frac{2}{\pi} \left(z(1 - z^2)^{1/2} + \arcsin(z) \right)$$

is a conformal isomorphism from $\mathbb{C}_+$ to $\mathbb{U}$.

(e) Let $c > 0$, let $0 < r < 1$, and let $\mathbb{U} := \{x + y\mathbf{i}; \ x^2 + y^2 < 1 \text{ and } (x - c)^2 + y^2 < r^2\}$. That is, $\mathbb{U}$ is the 'off-centre annulus', obtained by removing from the unit disk a smaller smaller disk of radius r centred at $(c, 0)$, as shown in Figure 18B.8(e). Let $a := c - r$ and $b := c + r$, and define

$$\lambda := \frac{1 + ab - \sqrt{(1 - a^2)(1 - b^2)}}{a + b} \qquad \text{and} \qquad R := \frac{1 - ab - \sqrt{(1 - a^2)(1 - b^2)}}{b - a}.$$

Let $\mathbb{A} := \{x + y\mathbf{i}; \ R < x^2 + y^2 < 1\}$ be an annulus with inner radius R and outer radius 1, and let $f(z) := (z - \lambda)/(1 - \lambda z)$. Show that f is a conformal isomorphism from $\mathbb{U}$ into $\mathbb{A}$.

(f) Let $\mathbb{U}$ be the upper half-disk shown on the right side of Figure 18B.4, and let $\mathbb{D}$ be the unit disk. Show that the function

$$f(z) = -\mathbf{i}\frac{z^2 + 2\mathbf{i}z + 1}{z^2 - 2\mathbf{i}z + 1}$$

is a conformal isomorphism from $\mathbb{U}$ into $\mathbb{D}$.

(g) Let $\mathbb{D} = \{x + y\mathbf{i}; x^2 + y^2 < 1\}$ be the open unit disk and let $\mathbb{C}_+ = \{x + y\mathbf{i}; \ y > 0\}$ be the open upper half-plane. Define $f : \mathbb{D} \longrightarrow \mathbb{C}_+$ by $f(z) = \mathbf{i}\frac{1+z}{1-z}$. Show that f is a conformal isomorphism from $\mathbb{D}$ into $\mathbb{C}_+$.   ◆

Ⓔ **Exercise 18B.6**

(a) Combine Example 18B.2(a) with Proposition 17E.1, p. 409 (or Proposition 20C.1, p. 541), to propose a general method for solving the Dirichlet problem on the bi-infinite strip $\mathbb{U} = \{x + y\mathbf{i}; \ x \in \mathbb{R}, 0 < y < \pi\}$.

(b) Now combine (a) with Exercise 18B.5(b) to propose a general method for solving the Dirichlet problem on the domain portrayed in Figure 18B.8(b). (Note that, despite the fact that the horizontal barriers are lines of zero thickness, your method allows you to assign different 'boundary conditions' to the two sides of these barriers.) Use your method to find the equilibrium heat distribution when the two barriers are each a different constant temperature. Reinterpret this solution as the electric field between two charged electrodes.

(c) Combine Exercise 18B.5(c) with Proposition 17E.1, p. 409 (or Proposition 20C.1, p. 541), to propose a general method for solving the Dirichlet problem on the 'chasm' domain portrayed in Figure 18B.8(c). Use your method to find the equilibrium heat distribution when the boundaries on either side of the chasm are two different constant temperatures. Reinterpret this solution as the electric field near the edge of a narrow gap between two large, oppositely charged parallel plates.

(d) Combine Exercise 18B.5(d) with Proposition 17E.1, p. 409 (or Proposition 20C.1, p. 541), to propose a general method for solving the Dirichlet problem on the domain portrayed in Figure 18B.8(d). Use your method to find the equilibrium heat distribution when the left side of the rectangle has temperature -1, the right side has temperature $+1$, and the top has temperature 0.

(e) Combine Exercise 18B.5(e) with Proposition 14B.10, p. 293, to propose a general method for solving the Dirichlet problem on the off-centre annulus portrayed in Figure 18B.8(e). Use your method to find the equilibrium heat distribution when the inner and outer circles are two different constant temperatures. Reinterpret this solution as an electric field between two concentric, oppositely charged cylinders.

(f) Combine Exercise 18B.5(f) with the methods of §14B, §14B(v), and/or §17F to propose a general method for solving the Dirichlet and Neumann problems on the half-disk portrayed in Figure 18B.4. Use your method to find the equilibrium temperature distribution when the semicircular top of the half-disk is one constant temperature and the base is another constant temperature.

(g) Combine Exercise 18B.5(g) with the Poisson Integral Formula on a disk (Proposition 14B.12, p. 297, or Proposition 17F.1, p. 412) to obtain another solution to the Dirichlet problem on a half-plane. Show that this is actually equivalent to the Poisson Integral Formula on a half-plane (Proposition 17E.1, p. 409).

(h) Combine Example 18B.2(f) with Proposition 17E.1, p. 409 (or Proposition 20C.1, p. 541), to propose a general method for solving the Dirichlet problem on the domain portrayed in Figure 18B.7(a). (Note that, despite the fact that the vertical obstacle is a line of zero thickness, your method allows you to assign different 'boundary conditions' to the two sides of this line.) Use your method to find the equilibrium temperature distribution when the 'obstacle' has one constant temperature and the real line has another constant temperature. Reinterpret this as the electric field generated by a charged electrode protruding, but insulated, from a horizontal, neutrally charged conducting barrier.

(i) Combine (a) with Example 18B.2(g) to propose a general method for solving the Dirichlet problem on the domain portrayed in Figure 18B.7(b). (Note that, despite the fact that the horizontal barriers are lines of zero thickness, your method allows you to assign different 'boundary conditions' to the two sides of these barriers.) Use your method to find the equilibrium temperature distribution when the two horizontal barriers have different constant temperatures. Reinterpret this as the electric field between two charged electrodes. ◆

Exercise 18B.7 Ⓔ

(a) Figure 18B.6(a) portrays the map $f(z) = z^2$ from Example 18B.2(e). Show that, in this case, the equipotential contours are all curves of the form $\{x + iy; y = \sqrt{x^2 - c}\}$ for some fixed $c > 0$. Show that the streamlines are all curves of the form $\{x + iy; y = c/x\}$ for some fixed $c > 0$.

(b) Figure 18B.7(b) portrays the map $f(z) = i \arcsin(z)$ from Example 18B.2(f). Show that, in this case, the equipotential contours are all *ellipses* of the form

$$\left\{ x + iy; \ \frac{x^2}{\cosh(r)^2} + \frac{y^2}{\sinh(r)^2} = 1 \right\},$$

for some fixed $r \in \mathbb{R}$. Likewise, show that the streamlines are all *hyperbolas*

$$\left\{ x + \mathbf{i}y; \ \frac{x^2}{\sin(r)^2} - \frac{y^2}{\cos(r)^2} = 1 \right\},$$

for some fixed $r \in \mathbb{R}$. (*Hint:* Use Exercises 18A.9(d) and (e), p. 428.)

(c) Find an equation describing all streamlines and equipotentials of the conformal map in Example 18B.2(a). Sketch the streamlines. (This describes a flow into a large body of water, from a point source on the boundary.)

(d) Fix $\Theta \in (-\pi, \pi)$, and consider the infinite wedge-shaped region $\mathbb{U} = \{r \operatorname{cis} \theta; \ r \geq 0, \ 0 < \theta < 2\pi - \Theta\}$. Find a conformal isomorphism from $\mathbb{U}$ to the upper half-plane. Sketch the streamlines of this map. (This describes the flow near the bank of a wide river, at a corner where the river bends by angle of Θ.)

(e) Suppose $\Theta = 2\pi/3$. Find an exact equation to describe the streamlines and equipotentials from (d) (analogous to the equations '$y = \sqrt{x^2 - c}$' and '$y = c/x$' from (a)).

(f) Sketch the streamlines and equipotentials defined by the conformal map in Exercise 18B.5(b). (This describes the flow out of a long pipe or channel into a large body of water.)

(g) Sketch the streamlines and equipotentials defined by the inverse of the conformal map f in Exercise 18B.5(c). (In other words, sketch the f-images of vertical and horizontal lines in $\mathbb{C}_+$.) This describes the flow over a deep 'chasm' in the streambed.

(h) Sketch the streamlines and equipotentials defined by the inverse of the conformal map f in Exercise 18B.5(d). (In other words, sketch the f-images of vertical and horizontal lines in $\mathbb{C}_+$.) This describes the flow around a long rectangular peninsula in an ocean. ◆

18C Contour integrals and Cauchy's theorem

Prerequisites: §18A.

A *contour* in $\mathbb{C}$ is a continuous function $\gamma : [0, S] \longrightarrow \mathbb{C}$ (for some $S > 0$) such that $\gamma(0) = \gamma(S)$, and such that γ does not 'self-intersect' – that is, $\gamma : [0, S) \longrightarrow \mathbb{C}$ is injective.[6] Let $\gamma_r, \gamma_i : [0, S] \longrightarrow \mathbb{R}$ be the real and imaginary parts of γ (so $\gamma(s) = \gamma_r(s) + \gamma_i(s)\mathbf{i}$, for all $s \in \mathbb{R}$). For any $s \in (0, S)$, we define the (complex) *velocity vector* of γ at s by $\dot{\gamma}(s) := \gamma_r'(s) + \gamma_i'(s)\mathbf{i}$ (if these derivatives exist). We say that γ is *smooth* if $\dot{\gamma}(s)$ exists for all $s \in (0, S)$.

Example 18C.1 Define $\gamma : [0, 2\pi] \longrightarrow \mathbb{C}$ by $\gamma(s) = \exp(\mathbf{i}s)$; then γ is a counterclockwise parameterization of the unit circle in the complex plane, as shown in Figure 18C.1(a). For any $s \in [0, 2\pi]$, we have $\gamma(s) = \cos(s) + \mathbf{i}\sin(s)$, so that $\dot{\gamma}(s) = \cos'(s) + \mathbf{i}\sin'(s) = -\sin(s) + \mathbf{i}\cos(s) = \mathbf{i}\gamma(s)$. ◇

---

[6] What we are calling a contour is sometimes called a *simple, closed curve*.

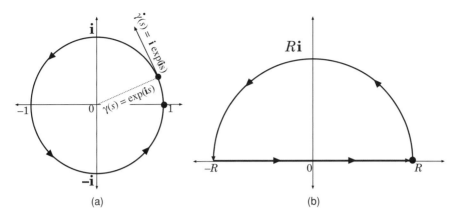

Figure 18C.1. (a) The counterclockwise unit circle contour from Example 18C.1.
(b) The 'D' contour from Example 18C.3.

Let $\mathbb{U} \subseteq \mathbb{C}$ be an open subset, let $f : \mathbb{U} \longrightarrow \mathbb{C}$ be a complex function, and
let $\gamma : [0, S] \longrightarrow \mathbb{U}$ be a smooth contour. The *contour integral* of f along γ is
defined:

$$\oint_\gamma f := \int_0^S f[\gamma(s)] \cdot \dot\gamma(s) \mathrm{d}s.$$

(Recall that $\dot\gamma(s)$ is a complex number, so $f[\gamma(s)] \cdot \dot\gamma(s)$ is a product of two complex
numbers.) Another notation we will sometimes use is $\oint_\gamma f(z)\mathrm{d}z$.

Example 18C.2 Let $\gamma : [0, 2\pi] \longrightarrow \mathbb{C}$ be the unit circle contour from Example
18C.1.

(a) Let $\mathbb{U} := \mathbb{C}$ and let $f(z) := 1$, a constant function. Then

$$\oint_\gamma f = \int_0^{2\pi} 1 \cdot \dot\gamma(s)\mathrm{d}s = \int_0^{2\pi} -\sin(s) + \mathbf{i}\cos(s)\mathrm{d}s$$

$$= -\int_0^{2\pi} \sin(s)\mathrm{d}s + \mathbf{i}\int_0^{2\pi} \cos(s)\mathrm{d}s = -0 + \mathbf{i}0 = 0.$$

(b) Let $\mathbb{U} := \mathbb{C}$ and let $f(z) := z^2$. Then

$$\oint_\gamma f = \int_0^{2\pi} \gamma(s)^2 \cdot \dot\gamma(s)\mathrm{d}s = \int_0^{2\pi} \exp(\mathbf{i}s)^2 \cdot \mathbf{i}\exp(\mathbf{i}s)\mathrm{d}s$$

$$= \mathbf{i}\int_0^{2\pi} \exp(\mathbf{i}s)^3 \,\mathrm{d}s = \mathbf{i}\int_0^{2\pi} \exp(3\mathbf{i}s)\mathrm{d}s$$

$$= \mathbf{i}\int_0^{2\pi} \cos(3s) + \mathbf{i}\sin(3s)\mathrm{d}s = \mathbf{i}\int_0^{2\pi} \cos(3s)\mathrm{d}s - \int_0^{2\pi} \sin(3s)\mathrm{d}s$$

$$= \mathbf{i}0 - 0 = 0.$$

Ⓔ (c) More generally, for any $n \in \mathbb{Z}$ *except* $n = -1$, we have $\oint_\gamma z^n dz = 0$ (**Exercise 18C.1**). (What happens if $n = -1$? See Example 18C.6 below.)

(d) It follows that, if $c_n, \ldots, c_2, c_1, c_0 \in \mathbb{C}$, and $f(z) = c_n z^n + \cdots + c_2 z^2 + c_1 z + c_0$ is a complex polynomial function, then $\oint_\gamma f = 0$. ◇

A contour $\gamma : [0, S] \longrightarrow \mathbb{U} \subseteq \mathbb{C}$ is *piecewise smooth* if $\dot{\gamma}(s)$ exists for all $s \in [0, S]$, except for perhaps finitely many points $0 = s_0 \leq s_1 \leq s_2 \leq \cdots \leq s_N = S$. If $f : \mathbb{U} \longrightarrow \mathbb{C}$ is a complex function, we define the *contour integral*:

$$\oint_\gamma f := \sum_{n=1}^{N} \int_{s_{n-1}}^{s_n} f[\gamma(s)] \cdot \dot{\gamma}(s) ds.$$

Example 18C.3 Fix $R > 0$, and define $\gamma_R : [0, \pi + 2R] \longrightarrow \mathbb{C}$ as follows:

$$\gamma_R(s) := \begin{cases} R \cdot \exp(is) & \text{if} \quad 0 \leq s \leq \pi; \\ s - \pi - R & \text{if} \quad \pi \leq s \leq \pi + 2R. \end{cases} \tag{18C.1}$$

This contour looks like a 'D' turned on its side; see Figure 18C.1(b). The first half of the contour parameterizes the upper half of the circle from R to $-R$. The second half parameterizes a straight horizontal line segment from $-R$ back to R. It follows that

$$\dot{\gamma}_R(s) := \begin{cases} R\mathbf{i} \cdot \exp(is) & \text{if} \quad 0 \leq s \leq \pi; \\ 1 & \text{if} \quad \pi \leq s \leq \pi + 2R. \end{cases} \tag{18C.2}$$

(a) Let $\mathbb{U} := \mathbb{C}$ and let $f(z) := z$. Then

$$\oint_{\gamma_R} f = \int_0^\pi \gamma(s) \cdot \dot{\gamma}(s) ds + \int_\pi^{\pi+2R} \gamma(s) \cdot \dot{\gamma}(s) ds$$

$$\underset{(*)}{=} \int_0^\pi R \exp(is) \cdot R\mathbf{i} \exp(is) ds + \int_\pi^{\pi+2R} (s - \pi - R) ds$$

$$= R^2 \mathbf{i} \int_0^\pi \exp(is)^2 ds + \int_{-R}^R t \, dt$$

$$= R^2 \mathbf{i} \int_0^\pi \cos(2s) + \mathbf{i} \sin(2s) ds + \frac{t^2}{2} \Big|_{t=-R}^{t=R}$$

$$= \frac{R^2 \mathbf{i}}{2} \Big(\sin(2s) - \mathbf{i} \cos(2s) \Big)_{s=0}^{s=\pi} + \frac{1}{2} \Big(R^2 - (-R)^2 \Big)$$

$$= \frac{R^2 \mathbf{i}}{2} \Big((0 - 0) - \mathbf{i}(1 - 1) \Big) + 0$$

$$= 0 + 0 = 0.$$

Here, $(*)$ is by equations (18C.1) and (18C.2).

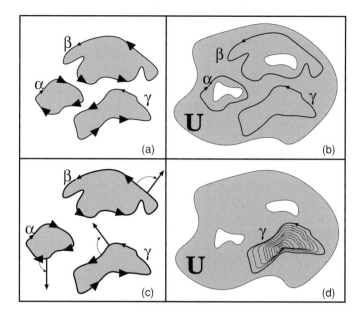

Figure 18C.2. (a) Three contours and their purviews. (b) Contour γ is nullhomo-topic in $\mathbb{U}$, but contours α and β are not nullhomotopic in $\mathbb{U}$. (c) Contour α is clockwise; contours β and γ are counterclockwise. (d) A nullhomotopy of γ.

(b) Generally, for any $n \in \mathbb{Z}$, if $n \neq -1$, then $\oint_{\gamma_R} z^n dz = 0$ (**Exercise 18C.2**). Thus, if f Ⓔ
is any complex polynomial, $\oint_{\gamma_R} f = 0$.   ◇

Any contour $\gamma : [0, S] \longrightarrow \mathbb{C}$ cuts the complex plane into exactly two pieces. Formally, the set $\mathbb{C} \setminus \gamma[0, S]$ has exactly two connected components, and exactly one of these components (the one 'inside' γ) is bounded.[7] The bounded component is called the *purview* of γ; see Figure 18C.2(a). For example, the purview of the unit circle is the unit disk. If $\mathbb{G}$ is the purview of γ, then clearly $\partial \mathbb{G} = \gamma[0, S]$. We say that γ is *counterclockwise* if the outward normal vector of $\mathbb{G}$ is always on the right-hand side of the vector $\dot{\gamma}$. We say γ is *clockwise* if the outward normal vector of $\mathbb{G}$ is always on the left-hand side of the vector $\dot{\gamma}$; see Figure 18C.2(c).

The contour γ is called *nullhomotopic* in $\mathbb{U}$ if the purview of γ is entirely contained in $\mathbb{U}$; see Figure 18C.2(b). Equivalently: it is possible to 'shrink' γ continuously down to a point without any part of the contour leaving $\mathbb{U}$; this is called a *nullhomotopy* of γ, and is portrayed in Figure 18C.2(d). Heuristically speaking, γ is nullhomotopic in $\mathbb{U}$ if, and only if, γ does not encircle any 'holes' in the domain $\mathbb{U}$.

---

[7] This seemingly innocent statement is actually the content of the *Jordan curve theorem,* which is a surprisingly difficult and deep result in planar topology.

Example 18C.4

(a) The unit circle from Examples 18C.1 and the 'D' contour from Example 18C.3 are both counterclockwise, and both are nullhomotopic in the domain $\mathbb{U} = \mathbb{C}$.

(b) The unit circle is *not* nullhomotopic on the domain $\mathbb{C}^* := \mathbb{C} \setminus \{0\}$. The purview of γ (the unit disk) is *not* entirely contained in $\mathbb{C}^*$, because the point 0 is missing. Equivalently, it is *not* possible to shrink γ down to a point without passing the curve through 0 at some moment; at this moment the curve would not be contained in $\mathbb{U}$.   ◇

The 'zero' outcomes of Examples 18C.2 and 18C.3 are not accidents; they are consequences of one of the fundamental results of complex analysis.

Theorem 18C.5 Cauchy's theorem

Let $\mathbb{U} \subseteq \mathbb{C}$ be an open subset, and let $f : \mathbb{U} \longrightarrow \mathbb{C}$ be holomorphic on $\mathbb{U}$. If $\gamma : [0, S] \longrightarrow \mathbb{U}$ is a contour which is nullhomotopic in $\mathbb{U}$, then $\oint_\gamma f = 0$.

Proof Let $\mathbb{G}$ be the purview of γ. If γ is nullhomotopic in $\mathbb{U}$, then $\mathbb{G} \subseteq \mathbb{U}$, and γ parameterizes the boundary $\partial\mathbb{G}$. Treat $\mathbb{U}$ as a subset of $\mathbb{R}^2$. Let $f_r : \mathbb{U} \longrightarrow \mathbb{R}$ and $f_i : \mathbb{U} \longrightarrow \mathbb{R}$ be the real and imaginary parts of f. The function f can be expressed as a vector field $\vec{\mathbf{V}} : \mathbb{U} \longrightarrow \mathbb{R}^2$ defined by $V_1(u) := f_r(u)$ and $V_2(u) := -f_i(u)$. For any $\mathbf{b} \in \partial\mathbb{G}$, let $\vec{\mathbf{N}}[\mathbf{b}]$ denote the outward unit normal vector to $\partial\mathbb{G}$ at $\mathbf{b}$. We define the following:

$$\text{Flux}(\vec{\mathbf{V}}, \gamma) := \int_0^S \vec{\mathbf{V}}[\gamma(s)] \bullet \vec{\mathbf{N}}[\gamma(s)]ds$$

and

$$\text{Work}(\vec{\mathbf{V}}, \gamma) := \int_0^S \vec{\mathbf{V}}[\gamma(s)] \bullet \dot{\gamma}(s)]ds.$$

The first integral is the *flux* of $\vec{\mathbf{V}}$ across the boundary of $\mathbb{G}$; this is just a reformulation of equation (E.1), p. 564 (see Figure E.1(b), p. 565). The second integral is the *work* of $\vec{\mathbf{V}}$ along the contour γ.

Claim 1

(a) $\text{Re}\left[\oint_\gamma f\right] = \text{Work}(\vec{\mathbf{V}}, \gamma)$ *and* $\text{Im}\left[\oint_\gamma f\right] = \text{Flux}(\vec{\mathbf{V}}, \gamma)$.

(b) *If* $\text{div}\,(\vec{\mathbf{V}}) \equiv 0$, *then* $\text{Flux}(\vec{\mathbf{V}}, \gamma) = 0$.

(c) *If* $\text{curl}\,(\vec{\mathbf{V}}) \equiv 0$, *then* $\text{Work}(\vec{\mathbf{V}}, \gamma) = 0$.

Ⓔ

Ⓔ

Proof (a) is **Exercise 18C.3**.

(b) is Green's Theorem (Theorem E.3, p. 565).

(c) is **Exercise 18C.4** (*Hint:* It is a variant of Green's theorem.)   ◇ Claim 1

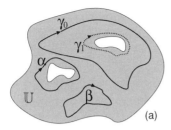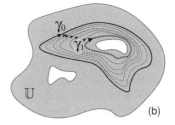

(a)   (b)

Figure 18C.3. Homotopy. (a) γ_0 is homotopic to γ_1, but *not* to α or β; γ_1 is homotopic to γ_0, but *not* to α or β; α is *not* homotopic to γ_0, γ_1, or β. Likewise, β is *not* homotopic to γ_0, γ_1, or α. (b) A homotopy from γ_0 to γ_1.

Now, if f is holomorphic on $\mathbb{U}$, then Proposition 18A.4, p. 424, says that $\mathrm{div}\,(\vec{\mathbf{V}}) \equiv 0$ and $\mathrm{curl}\,(\vec{\mathbf{V}}) \equiv 0$. Then Claim 1 implies $\oint_\gamma f = 0$. (For other proofs, see Lang (1985), §IV.3, Needham (1997), §8.X, or Fisher (1999), Theorem 1, §2.3).   □

At this point you may be wondering: what are complex contour integrals good for, if they are always equal to zero? The answer is that $\oint_\gamma f$ is only zero if the function f is holomorphic in the purview of γ. If f has a *singularity* inside this purview (i.e. a point where f is *not* complex-differentiable, or perhaps not even defined), then $\oint_\gamma f$ might be nonzero.

Example 18C.6 Let $\gamma : [0, 2\pi] \longrightarrow \mathbb{C}$ be the unit circle contour from Example 18C.1. Let $\mathbb{C}^* := \mathbb{C} \setminus \{0\}$, and define $f : \mathbb{C}^* \longrightarrow \mathbb{C}$ by $f(z) := 1/z$. Then

$$\oint_\gamma f = \int_0^{2\pi} \frac{\dot{\gamma}(s)}{\gamma(s)} \, \mathrm{d}s = \int_0^{2\pi} \frac{\mathbf{i}\exp(\mathbf{i}s)}{\exp(\mathbf{i}s)} \, \mathrm{d}s = \int_0^{2\pi} \mathbf{i}\,\mathrm{d}s = 2\pi\mathbf{i}.$$

Note that γ is *not* nullhomotopic on $\mathbb{C}^*$. Of course, we could extend f to all of $\mathbb{C}$ by defining $f(0)$ in some arbitrary way. But no matter how we do this, f will never be complex-differentiable at zero – in other words, 0 is a *singularity* of f.   ◇

If the purview of γ contains one or more singularities of f, then the value of $\oint_\gamma f$ reveals important information about these singularities. Indeed, the value of $\oint_\gamma f$ depends *only* on the singularities within the purview of γ, and *not* on the shape of γ itself. This is a consequence of the *homotopy-invariance* of contour integration.

Let $\mathbb{U} \subseteq \mathbb{C}$ be an open subset, and let $\gamma_0, \gamma_1 : [0, S] \longrightarrow \mathbb{U}$ be two contours. We say that γ_0 is *homotopic* to γ_1 *in* $\mathbb{U}$ if γ_0 can be 'continuously deformed' into γ_1 without ever moving outside of $\mathbb{U}$; see Figure 18C.3. (In particular, γ is *nullhomotopic* if γ is homotopic to a constant path in $\mathbb{U}$.) Formally, this means there is a continuous function $\Gamma : [0, 1] \times [0, S] \longrightarrow \mathbb{U}$ such that

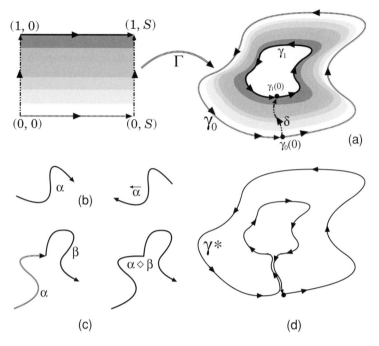

Figure 18C.4. (a) Γ is a homotopy from γ_0 to γ_1. (b) The reversal $\overleftarrow{\alpha}$ of α. (c) The linking $\alpha \diamond \beta$. (d) The contour γ^* defined by the 'boundary' of the homotopy map Γ from (a).

- for all $s \in [0, S]$, $\Gamma(0, s) = \gamma_0(s)$;
- for all $s \in [0, S]$, $\Gamma(1, s) = \gamma_1(s)$;
- for all $t \in [0, 1]$, if we fix t and define $\gamma_t : [0, S] \longrightarrow \mathbb{U}$ by $\gamma_t(s) := \Gamma(t, s)$ for all $s \in [0, S]$, then γ_t is a contour in $\mathbb{U}$.

The function Γ is called a *homotopy* of γ_0 into γ_1. See Figure 18C.4(a).

Proposition 18C.7 Homotopy invariance of contour integration

Let $\mathbb{U} \subseteq \mathbb{C}$ be an open subset, and let $f : \mathbb{U} \longrightarrow \mathbb{C}$ be a holomorphic function. Let $\gamma_0, \gamma_1 : [0, S] \longrightarrow \mathbb{U}$ be two contours. If γ_0 is homotopic to γ_1 in $\mathbb{U}$, then

$$\oint_{\gamma_0} f = \oint_{\gamma_1} f.$$

Before proving this result, it will be useful to extend somewhat our definition of contour integration. A *chain* is a piecewise continuous, piecewise differentiable function $\alpha : [0, S] \longrightarrow \mathbb{C}$ (for some $S > 0$). (Thus, a chain α is a *contour* if α is continuous, $\alpha(S) = \alpha(0)$, and α is not self-intersecting.) If $\alpha : [0, S] \longrightarrow \mathbb{U} \subseteq \mathbb{C}$ is a chain, and $f : \mathbb{U} \longrightarrow \mathbb{C}$ is a complex-valued function, then the *integral* of f along α is defined:

$$\oint_{\alpha} f = \int_0^S f[\alpha(s)] \cdot \dot{\alpha}(s) \mathrm{d}s. \qquad (18C.3)$$

Here we define $\dot{\alpha}(s) = 0$ whenever s is one of the (finitely many) points where α is nondifferentiable or discontinuous. (Thus, if α is a contour, then equation (18C.3) is just the contour integral $\oint_\alpha f$.)

The *reversal* of chain α is the chain $\overleftarrow{\alpha} : [0, S] \longrightarrow \mathbb{C}$ defined by $\overleftarrow{\alpha}(s) := \alpha(S - s)$; see Figure 18C.4(b). If $\alpha : [0, S] \longrightarrow \mathbb{C}$ and $\beta : [0, T] \longrightarrow \mathbb{C}$ are two chains, then the *linking* of α and β is the chain $\alpha \diamond \beta : [0, S + T] \longrightarrow \mathbb{C}$, defined:

$$\alpha \diamond \beta(s) := \begin{cases} \alpha(s) & \text{if } 0 \le s \le S; \\ \beta(s - S) & \text{if } S \le s \le S + T \end{cases}$$

(see Figure 18C.4(c)).

Lemma 18C.8 Let $\mathbb{U} \subseteq \mathbb{C}$ be an open set and let $f : \mathbb{U} \longrightarrow \mathbb{C}$ be a complex-valued function. Let $\alpha : [0, S] \longrightarrow \mathbb{U}$ be a chain.

(a)
$$\oint_{\overleftarrow{\alpha}} f = -\oint_\alpha f.$$

(b) If $\beta : [0, T] \longrightarrow \mathbb{C}$ is another chain, then

$$\oint_{\alpha \diamond \beta} f = \oint_\alpha f + \oint_\beta f.$$

(c) $\alpha \diamond \beta$ is continuous if, and only if, α and β are both continuous, and $\beta(0) = \alpha(S)$.

(d) The linking operation is *associative*: that is, if γ is another chain, then $(\alpha \diamond \beta) \diamond \gamma = \alpha \diamond (\beta \diamond \gamma)$. (Thus, we normally drop the brackets and just write $\alpha \diamond \beta \diamond \gamma$.)

Proof **Exercise 18C.5.**   □   Ⓔ

Proof of Proposition 18C.7   Define the continuous path $\delta : [0, 1] \longrightarrow \mathbb{C}$ by

$$\delta(t) := \Gamma(0, t) = \Gamma(S, t), \quad \text{for all } t \in [0, 1].$$

Figure 18C.4(a) shows how δ traces the path defined by the homotopy Γ from $\gamma_0(S)$ $(= \gamma_0(0))$ to $\gamma_1(S)$ $(= \gamma_1(0))$. We assert (without proof) that the homotopy Γ can always be chosen such that δ is piecewise smooth; thus we regard δ as a chain. Figure 18C.4(d) portrays the contour $\gamma^* := \gamma_0 \diamond \delta \diamond \overleftarrow{\gamma_1} \diamond \overleftarrow{\delta}$, which traces the Γ-image of the four sides of the rectangle $[0, 1] \times [0, 2\pi]$.

Claim 1 γ^* *is nullhomotopic in* $\mathbb{U}$.

Proof The purview of γ^* is simply the image of the open rectangle $(0, 1) \times (0, 2\pi)$ under the function Γ. But, by definition, Γ maps $(0, 1) \times (0, 2\pi)$ into $\mathbb{U}$; thus the purview of γ^* is contained in $\mathbb{U}$, so γ^* is nullhomotopic in $\mathbb{U}$.
                                                    ◇ Claim 1

Thus,

$$0 \underset{(C)}{=} \oint_{\gamma^*} f \underset{(*)}{=} \oint_{\overleftarrow{\gamma_0} \diamond \delta \diamond \overleftarrow{\gamma_1} \diamond \delta} f$$

$$\underset{(\dagger)}{=} \oint_{\gamma_0} f + \oint_{\delta} f - \oint_{\gamma_1} f - \oint_{\delta} f = \oint_{\gamma_0} f - \oint_{\gamma_1} f.$$

Here (C) is by Cauchy's Theorem and Claim 1, (*) is by definition of γ, and (†) is by Lemma 18C.8(a, b). Thus, we have $\oint_{\gamma_0} f - \oint_{\gamma_1} f = 0$, which means $\oint_{\gamma_0} f = \oint_{\gamma_1} f$, as claimed.   □

Example 18C.6 is a special case of the following important result.

Theorem 18C.9 Cauchy's integral formula
Let $\mathbb{U} \subseteq \mathbb{C}$ be an open subset, let $f : \mathbb{U} \longrightarrow \mathbb{C}$ be holomorphic, let $u \in \mathbb{U}$, and let $\gamma : [0, S] \longrightarrow \mathbb{U}$ be any counterclockwise contour whose purview contains u and is contained in $\mathbb{U}$. Then

$$f(u) = \frac{1}{2\pi i} \oint_{\gamma} \frac{f(z)}{z - u} \, dz.$$

In other words: if $\mathbb{U}^* := \mathbb{U} \setminus \{u\}$, and we define $F_u : \mathbb{U}^* \longrightarrow \mathbb{C}$ by $F_u(z) := f(z)/(z - u)$ for all $z \in \mathbb{U}^*$, then

$$f(u) = \frac{1}{2\pi i} \oint_{\gamma} F_u.$$

Proof For simplicity, we will prove this in the case $u = 0$. We must show that

$$\frac{1}{2\pi i} \oint_{\gamma} \frac{f(z)}{z} \, dz = f(0).$$

Let $\mathbb{G} \subset \mathbb{U}$ be the purview of γ. For any $r > 0$, let $\mathbb{D}_r$ be the disk of radius r around 0, and let β_r be a counterclockwise parameterization of $\partial \mathbb{D}_r$ (e.g. $\beta_r(s) := r e^{is}$ for all $s \in [0, 2\pi)$). Let $\mathbb{U}^* := \mathbb{U} \setminus \{0\}$.

Claim 1 *If $r > 0$ is small enough, then $\mathbb{D}_r \subset \mathbb{G}$. In this case, γ is homotopic to β_r in $\mathbb{U}^*$.*

Ⓔ   *Proof* **Exercise 18C.6.**   ◇ Claim 1

Now define $\phi : \mathbb{U} \longrightarrow \mathbb{C}$ as follows:

$$\phi(u) := \frac{f(z) - f(0)}{z} \quad \text{for all } z \in \mathbb{U}^*, \quad \text{and} \quad \phi(0) := f'(0).$$

Then ϕ is holomorphic on $\mathbb{U}^*$. Observe that

$$1 \underset{(*)}{=} \frac{1}{2\pi i} \oint_{\beta_r} \frac{1}{z} \, dz \underset{(\dagger)}{=} \frac{1}{2\pi i} \oint_\gamma \frac{1}{z} \, dz, \tag{18C.4}$$

where $(*)$ is by Example 18C.6, p. 445, and $(\dagger)$ is by Claim 1 and Proposition 18C.7. Thus,

$$f(0) \underset{(*)}{=} \frac{f(0)}{2\pi i} \oint_\gamma \frac{1}{z} \, dz = \frac{1}{2\pi i} \oint_\gamma \frac{f(0)}{z} \, dz.$$

Thus,

$$\frac{1}{2\pi i} \oint_\gamma \frac{f(z)}{z} \, dz - f(0) = \frac{1}{2\pi i} \oint_\gamma \frac{f(z)}{z} \, dz - \frac{1}{2\pi i} \int_\gamma \frac{f(0)}{z} \, dz$$

$$= \frac{1}{2\pi i} \oint_\gamma \frac{f(z)}{z} - \frac{f(0)}{z} \, dz$$

$$= \frac{1}{2\pi i} \oint_\gamma \frac{f(z) - f(0)}{z} \, dz = \frac{1}{2\pi i} \oint_\gamma \phi$$

$$\underset{(\dagger)}{=} \frac{1}{2\pi i} \oint_{\beta_r} \phi, \quad \text{for any } r > 0.$$

Here, $(*)$ is by equation (18C.4), and $(\dagger)$ is again by Claim 1 and Proposition 18C.7. Thus, we have

$$\frac{1}{2\pi i} \oint_\gamma \frac{f(z)}{z} \, dz - f(0) = \lim_{r \to 0} \frac{1}{2\pi i} \oint_{\beta_r} \phi. \tag{18C.5}$$

Thus, it suffices to show that $\lim_{r \to 0} \oint_{\beta_r} \phi = 0$. To see this, first note that ϕ is continuous at 0 (because $\lim_{z \to 0} \phi(z) = f'(0)$ by definition of the derivative), and ϕ is also continuous on the rest of $\mathbb{U}$ (where ϕ is just another holomorphic function). Thus, ϕ is bounded on $\mathbb{G}$ (because $\mathbb{G}$ is a bounded set whose closure is inside $\mathbb{U}$). Thus, if $M := \sup_{z \in \mathbb{G}} |\phi(z)|$, then $M < \infty$. But then

$$\left| \oint_{\beta_r} \phi \right| \underset{(*)}{\leq} M \cdot \text{length}(\beta_r) = M \cdot 2\pi r \xrightarrow[r \to 0]{} 0,$$

where $(*)$ is by Lemma 18C.10 (below).

Thus, $\lim_{r \to 0} \oint_{\beta_r} \phi = 0$, so equation (18C.5) implies that $\frac{1}{2\pi i} \oint_\gamma \frac{f(z)}{z} dz = f(0)$, as desired. $\square$

If $\gamma : [0, S] \longrightarrow \mathbb{C}$ is a chain, then we define $\text{length}(\gamma) := \int_0^S |\dot{\gamma}(s)| \, ds$. The proof of Theorem 18C.9 invoked the following useful lemma.

Lemma 18C.10 *Let $f : \mathbb{U} \longrightarrow \mathbb{C}$ and let γ be a chain in $\mathbb{U}$. If $M := \sup_{u \in \mathbb{U}}$*
$|f(u)|$, then

$$\left| \oint_\gamma f \right| \leq M \cdot \mathrm{length}(\gamma).$$

Ⓔ *Proof* **Exercise 18C.7.**   □

Ⓔ **Exercise 18C.8** Prove the general case of Theorem 18C.9 for an arbitrary
$u \in \mathbb{C}$.   ♦

Corollary 18C.11 Mean value theorem for holomorphic functions
Let $\mathbb{U} \subseteq \mathbb{C}$ be an open set and let $f : \mathbb{U} \longrightarrow \mathbb{C}$ be holomorphic. Let $r > 0$ be small
enough that the circle $\mathbb{S}(r)$ of radius r around u is contained in $\mathbb{U}$. Then

$$f(u) = \frac{1}{2\pi} \int_{\mathbb{S}(r)} f(s)\mathrm{d}s = \frac{1}{2\pi} \int_0^{2\pi} f(u + r\,\mathrm{e}^{\mathrm{i}\theta})\mathrm{d}\theta.$$

Proof Define $\gamma : [0, 2\pi] \longrightarrow \mathbb{U}$ by $\gamma(s) := u + r\mathrm{e}^{\mathrm{i}s}$ for all $s \in [0, 2\pi]$; thus, γ is
a counterclockwise parameterization of $\mathbb{S}(r)$, and $\dot\gamma(s) = \mathrm{i}r\mathrm{e}^{\mathrm{i}s}$ for all $s \in [0, 2\pi]$.
Then

$$f(u) \underset{(*)}{=} \frac{1}{2\pi\mathrm{i}} \oint_\gamma \frac{f(z)}{z - u} \, \mathrm{d}z = \frac{1}{2\pi\mathrm{i}} \int_0^{2\pi} \frac{f[\gamma(\theta)]}{\gamma(\theta) - u} \dot\gamma(\theta)\mathrm{d}\theta$$

$$= \frac{1}{2\pi\mathrm{i}} \int_0^{2\pi} \frac{f(u + r\mathrm{e}^{\mathrm{i}\theta})}{r\mathrm{e}^{\mathrm{i}\theta}} \, \mathrm{i}r\mathrm{e}^{\mathrm{i}\theta} \, \mathrm{d}\theta = \frac{1}{2\pi} \int_0^{2\pi} f(u + r\mathrm{e}^{\mathrm{i}\theta})\mathrm{d}\theta,$$

as desired. Here $(*)$ is by Cauchy's integral formula.   □

Ⓔ **Exercise 18C.9** Using Proposition 18C.11, derive another proof of the mean
value theorem for *harmonic* functions on $\mathbb{U}$ (Theorem 1E.1, p. 18). (*Hint:* Use
Proposition 18A.3, p. 423.)   ♦

Corollary 18C.12 Maximum modulus principle
Let $\mathbb{U} \subseteq \mathbb{C}$ be an open set and let $f : \mathbb{U} \longrightarrow \mathbb{C}$ be holomorphic. Then the function
$m(z) := |f(z)|$ has no local maxima inside $\mathbb{U}$.

Ⓔ *Proof* **Exercise 18C.10.** (*Hint:* Use the mean value theorem.)   □

Ⓔ **Exercise 18C.11** Let $f : \mathbb{U} \longrightarrow \mathbb{C}$ be holomorphic. Show that the functions
$R(z) := \mathrm{Re}\,[f(z)]$ and $I(z) := \mathrm{Im}\,[f(z)]$ have no local maxima or minima inside
$\mathbb{U}$.   ♦

   Let $\mathbb{D} := \{z \in \mathbb{C};\ |z| < 1\}$ be the open unit disk in the complex plane,
and let $\mathbb{S} := \partial\mathbb{D}$ be the unit circle. The *Poisson kernel* for $\mathbb{D}$ is the function

$\mathcal{P} : \mathbb{S} \times \mathbb{D} \longrightarrow \mathbb{R}$ defined by

$$\mathcal{P}(s, u) := \frac{1 - |u|^2}{|s - u|^2}, \quad \text{for all } s \in \mathbb{S} \text{ and } u \in \mathbb{D}.$$

Corollary 18C.13 Poisson integral formula for holomorphic functions
Let $\mathbb{U} \subseteq \mathbb{C}$ be an open subset containing the unit disk $\mathbb{D}$, and let $f : \mathbb{U} \longrightarrow \mathbb{C}$ be holomorphic. Then, for all $u \in \mathbb{D}$,

$$f(u) = \frac{1}{2\pi} \int_{\mathbb{S}} f(s) \, \mathcal{P}(s, u) \mathrm{d}s = \frac{1}{2\pi} \int_0^{2\pi} f(\mathrm{e}^{\mathrm{i}\theta}) \, \mathcal{P}(\mathrm{e}^{\mathrm{i}\theta}, u) \mathrm{d}\theta.$$

Proof If $u \in \mathbb{D}$, then $\overline{u}^{-1}$ is outside $\mathbb{D}$ (because $|\overline{u}^{-1}| = |\overline{u}|^{-1} = |u|^{-1} > 1$ if $|u| < 1$). Thus the set $\mathbb{C}_u := \mathbb{C} \setminus \{\overline{u}^{-1}\}$ contains $\mathbb{D}$. Fix $u \in \mathbb{D}$ and define the function $g_u : \mathbb{C}_u \longrightarrow \mathbb{C}$ by

$$g_u(z) := \frac{f(z) \cdot \overline{u}}{1 - \overline{u}z}.$$

Claim 1 *g_u is holomorphic on $\mathbb{C}_u$.*

Proof **Exercise 18C.12.**   ◇$_{\text{Claim 1}}$ Ⓔ

Now, define $F_u : \mathbb{U} \longrightarrow \mathbb{C}$ by $F_u(z) := f(z)/(z - u)$, and let $\gamma : [0, 2\pi] \longrightarrow \mathbb{S}$ be the unit circle contour from Example 18C.1 (i.e. $\gamma(s) = \mathrm{e}^{\mathrm{i}s}$ for all $s \in [0, 2\pi]$). Then

$$f(u) = \frac{1}{2\pi\mathrm{i}} \oint_\gamma F_u$$

by Cauchy's integral formula (Theorem 18C.9) and

$$0 = \frac{1}{2\pi\mathrm{i}} \oint_\gamma g_u$$

by Cauchy's theorem (Theorem 18C.5). Thus,

$$f(u) = \frac{1}{2\pi\mathrm{i}} \oint_\gamma (F_u + g_u) = \frac{1}{2\pi\mathrm{i}} \int_0^{2\pi} \Big(F_u[\gamma(\theta)] + g_u[\gamma(\theta)] \Big) \, \dot{\gamma}(\theta) \mathrm{d}\theta$$

$$= \frac{1}{2\pi\mathrm{i}} \int_0^{2\pi} \left(\frac{f(\mathrm{e}^{\mathrm{i}\theta})}{\mathrm{e}^{\mathrm{i}\theta} - u} + \frac{f(\mathrm{e}^{\mathrm{i}\theta}) \cdot \overline{u}}{1 - \overline{u}\mathrm{e}^{\mathrm{i}\theta}} \right) \mathrm{i}\mathrm{e}^{\mathrm{i}\theta} \, \mathrm{d}\theta$$

$$= \frac{1}{2\pi} \int_0^{2\pi} f(\mathrm{e}^{\mathrm{i}\theta}) \cdot \left(\frac{\mathrm{e}^{\mathrm{i}\theta}}{\mathrm{e}^{\mathrm{i}\theta} - u} + \frac{\mathrm{e}^{\mathrm{i}\theta}\overline{u}}{1 - \overline{u}\mathrm{e}^{\mathrm{i}\theta}} \right) \mathrm{d}\theta$$

$$\underset{(*)}{=\!=\!=} \frac{1}{2\pi} \int_0^{2\pi} f(\mathrm{e}^{\mathrm{i}\theta}) \cdot \frac{1 - |u|^2}{|\mathrm{e}^{\mathrm{i}\theta} - u|^2} \, \mathrm{d}\theta = \frac{1}{2\pi} \int_0^{2\pi} f(\mathrm{e}^{\mathrm{i}\theta}) \mathcal{P}(\mathrm{e}^{\mathrm{i}\theta}, u) \mathrm{d}\theta,$$

as desired. Here, $(*)$ uses the fact that, for any $s \in \mathbb{S}$ and $u \in \mathbb{C}$,

$$\frac{s}{s-u} + \frac{s\bar{u}}{1-\bar{u}s} = \frac{s}{s-u} + \frac{\bar{s}s\bar{u}}{\bar{s}-\bar{u}ss} = \frac{s}{s-u} + \frac{|s|^2\bar{u}}{\bar{s}-\bar{u}|s|^2}$$

$$\underset{(*)}{=} \frac{s}{s-u} + \frac{\bar{u}}{\bar{s}-\bar{u}} = \frac{s \cdot (\bar{s}-\bar{u}) + \bar{u} \cdot (s-u)}{(s-u) \cdot (\bar{s}-\bar{u})}$$

$$= \frac{|s|^2 - s\bar{u} + \bar{u}s - |u|^2}{(s-u) \cdot \overline{(s-u)}} = \frac{|s|^2 - |u|^2}{|s-u|^2} \underset{(*)}{=} \frac{1-|u|^2}{|s-u|^2},$$

where both $(*)$ are because $|s| = 1$. ☐

Ⓔ **Exercise 18C.13** Using Corollary 18C.13, derive yet another proof of the Poisson integral formula for *harmonic* functions on $\mathbb{D}$. (See Proposition 14B.12, p. 297, and also Proposition 17F.1, p. 412.) (*Hint:* Use Proposition 18A.3, p. 423.) ◆

At this point, we have proved the Poisson integral formula three entirely different ways: using Fourier series (Proposition 14B.12), using impulse-response methods (Proposition 17F.1), and now using complex analysis (Corollary 18C.13). In §18F, p. 466, we will encounter the Poisson integral formula yet again while studying the Abel mean of a Fourier series.

An equation which expresses the solution to a boundary value problem in terms of an integral over the boundary of the domain is called an *integral representation formula*. For example, the Poisson integral formula is such a formula, as is Poisson's solution to the Dirichlet problem on a half-space (Proposition 17E.1, p. 409). Cauchy's integral formula provides an integral representation formula for any holomorphic function on *any* domain in $\mathbb{C}$ which is bounded by a contour. Our proof of Corollary 18C.13 shows how this can be used to obtain integral representation formulas for harmonic functions on planar domains.

Ⓔ **Exercise 18C.14** Liouville's theorem
Suppose $f : \mathbb{C} \longrightarrow \mathbb{C}$ is holomorphic and *bounded* – i.e. there is some $M > 0$ such that $|f(z)| < M$ for all $z \in \mathbb{C}$. Show that f must be a constant function. (*Hint:* Define $g(z) := \frac{f(z)-f(0)}{z}$.)

(a) Show that g is holomorphic on $\mathbb{C}$.
(b) Show that $|g(z)| < 2M/|z|$ for all $z \in \mathbb{C}$.
(c) Let $z \in \mathbb{C}$. Let γ be a circle of radius $R > 0$ around 0, where R is large enough that z is in the purview of γ. Use Cauchy's integral formula and Lemma 18C.10, p. 450, to show that

$$|g(z)| < \frac{1}{2\pi} \frac{2M}{R} \frac{2\pi R}{R - |z|}.$$

Now let $R \to \infty$. ◆

Exercise 18C.15 Complex antiderivatives Ⓔ

Let $\mathbb{U} \subset \mathbb{C}$ be an open connected set. We say that $\mathbb{U}$ is *simply connected* if every contour in $\mathbb{U}$ is nullhomotopic. Heuristically speaking, this means $\mathbb{U}$ does not have any 'holes'. For any $u_0, u_1 \in \mathbb{U}$, a *path in* $\mathbb{U}$ from u_0 to u_1 is a continuous function $\gamma : [0, S] \longrightarrow \mathbb{U}$ such that $\gamma(0) = u_0$ and $\gamma(S) = u_1$.

Let $f : \mathbb{U} \longrightarrow \mathbb{C}$ be holomorphic. Pick a 'basepoint' $b \in \mathbb{U}$, and define a function $F : \mathbb{U} \longrightarrow \mathbb{C}$ as follows:

$$\text{for all } u \in \mathbb{U}, \; F(u) := \oint_{\gamma} f, \tag{18C.6}$$

where γ is any path in $\mathbb{U}$ from b to u.

(a) Show that $F(u)$ is well-defined by equation (18C.6), independent of the path γ you use to get from b to u. (*Hint:* If γ_1 and γ_2 are two paths from b to u, show that $\gamma_1 \diamond \overleftarrow{\gamma_2}$ is a contour. Then apply Cauchy's theorem.)

(b) For any $u_1, u_2 \in \mathbb{U}$, show that $F(u_2) - F(u_1) = \int_{\gamma} f$, where γ is any path in $\mathbb{U}$ from u_1 to u_2.

(c) Show that F is a holomorphic function, and $F'(u) = f(u)$ for all $u \in \mathbb{U}$. (*Hint:* Write $F'(u)$ as the limit (18A.1), p. 421. For any c close to u let $\gamma : [0, 1] \longrightarrow \mathbb{U}$ be the straight-line path linking u to c (i.e. $\gamma(s) = sc + (1 - s)u$). Deduce from (b) that

$$\frac{F(c) - F(u)}{c - u} = \frac{1}{c - u} \oint_{\gamma} f.$$

Now take the limit as $c \to u$.)

The function F is called a **complex antiderivative** of f, based at b. Part (c) is the complex version of the fundamental theorem of calculus.

(d) Let $\mathbb{U} = \mathbb{C}$, let $b \in \mathbb{U}$, and let $f(u) = \exp(u)$. Let F be the complex antiderivative of f based at b. Show that $F(u) = \exp(u) - \exp(b)$ for all $u \in \mathbb{C}$.

(e) Let $\mathbb{U} = \mathbb{C}$, let $b \in \mathbb{U}$, and let $f(u) = u^n$ for some $n \in \mathbb{N}$. Let F be the complex antiderivative of f based at b. Show that $F(u) = \frac{1}{n+1}(u^{n+1} - b^{n+1})$, for all $u \in \mathbb{C}$.

We have already encountered one application of complex antiderivatives in Proposition 18B.6, p. 436. Exercises 18C.16 and 18C.17 describe another important application. ◆

Exercise 18C.16 Complex logarithms (follows Exercise 18C.15) Ⓔ

(a) Let $\mathbb{U} \subset \mathbb{C}$ be an open, simply connected set which does not contain 0. Define a 'complex logarithm function' $\log : \mathbb{U} \longrightarrow \mathbb{C}$ as the complex antiderivative of $1/z$ based at 1. That is,

$$\log(u) := \oint_{\gamma} 1/z \, dz,$$

where γ is any path in $\mathbb{U}$ from 1 to z. Show that $\log$ is a *right-inverse* of the exponential function – that is, $\exp(\log(u)) = u$ for all $u \in \mathbb{U}$.

(b) What goes wrong with part (a) if $0 \in \mathbb{U}$? What goes wrong if $0 \notin \mathbb{U}$, but $\mathbb{U}$ contains an annulus which encircles 0? (*Hint:* Consider Example 18C.6.) (*Remark:* This is the reason why we required $\mathbb{U}$ to be simply connected in Exercise 18C.15.)

(c) Suppose our definition of 'complex logarithm' is 'any right-inverse of the complex exponential function' – that is, any holomorphic function $L : \mathbb{U} \longrightarrow \mathbb{C}$ such that $\exp(L(u)) = u$ for all $u \in \mathbb{U}$. Suppose $L_0 : \mathbb{U} \longrightarrow \mathbb{C}$ is one such 'logarithm' function (defined as in (a), for example). Define $L_1 : \mathbb{U} \longrightarrow \mathbb{C}$ by $L_1(u) = L_0(u) + 2\pi\mathbf{i}$. Show that L_1 is also a 'logarithm'. Relate this to the problem you found in (b).

(d) Indeed, for any $n \in \mathbb{Z}$, define $L_n : \mathbb{U} \longrightarrow \mathbb{C}$ by $L_n(u) = L_0(u) + 2n\pi\mathbf{i}$. Show that L_n is also a 'logarithm' in the sense of part (c). Make a sketch of the surface described by the functions $\mathrm{Im}\,[L_n] : \mathbb{C} \longrightarrow \mathbb{R}$, for all $n \in \mathbb{Z}$ at once.

(e) Proposition 18A.2, p. 423, asserted that any harmonic function on a convex domain $\mathbb{U} \subset \mathbb{R}^2$ can be represented as the real part of a holomorphic function on $\mathbb{U}$, treated as a subset of $\mathbb{C}$. The Remark following Proposition 18A.2 said that $\mathbb{U}$ actually does not need to be convex, but it *does* need to be simply connected. We will not prove that simple-connectedness is sufficient, but we can now show that it is necessary.

Consider the harmonic function $h(x, y) = \log(x^2 + y^2)$ defined on $\mathbb{R}^2 \setminus \{0\}$. Show that, on any simply connected subset $\mathbb{U} \subset \mathbb{C}^*$, there is a holomorphic function $L : \mathbb{U} \longrightarrow \mathbb{C}$ with $h = \mathrm{Re}\,[L]$. However, show that there is no holomorphic function $L : \mathbb{C}^* \longrightarrow \mathbb{C}$ with $h = \mathrm{Re}\,[L]$.

The functions L_n (for $n \in \mathbb{Z}$) are called the *branches of the complex logarithm*. This exercise shows that the 'complex logarithm' is a much more complicated object than the real logarithm – indeed, the complex log is best understood as a holomorphic 'multifunction' which takes countably many distinct values at each point in $\mathbb{C}^*$. The surface in part (d) is an abstract representation of the 'graph' of this multifunction – it is called a *Riemann surface*. ◆

Ⓔ **Exercise 18C.17** *Complex root functions* (follows Exercise 18C.16)

(a) Let $\mathbb{U} \subset \mathbb{C}$ be an open, simply connected set which does not contain 0, and let $\log : \mathbb{U} \longrightarrow \mathbb{C}$ be any complex logarithm function, as defined in Exercise 18C.16. Fix $n \in \mathbb{N}$. Show that $\exp(n \cdot \log(u)) = u^n$ for all $u \in \mathbb{N}$.

(b) Fix $n \in \mathbb{N}$ and now define $\sqrt[n]{\bullet} : \mathbb{U} \longrightarrow \mathbb{C}$ by $\sqrt[n]{u} = \exp(\log(u)/n)$ for all $u \in \mathbb{N}$. Show that $\sqrt[n]{\bullet}$ is a complex 'nth-root' function. That is, $\left(\sqrt[n]{u}\right)^n = u$ for all $u \in \mathbb{U}$.

Different branches of logarithm define different 'branches' of the nth-root function. However, while there are infinitely many distinct branches of logarithm, there are exactly n distinct branches of the nth-root function.

(c) Fix $n \in \mathbb{N}$, and consider the equation $z^n = 1$. Show that the set of all solutions to this equation is $\mathcal{Z}_n := \{1, e^{2\pi\mathbf{i}/n}, e^{4\pi\mathbf{i}/n}, e^{6\pi\mathbf{i}/n}, \dots, e^{2(n-1)\pi\mathbf{i}/n}\}$. (These numbers are called the *nth roots of unity*. For example, $\mathcal{Z}_2 = \{\pm 1\}$ and $\mathcal{Z}_4 = \{\pm 1, \pm\mathbf{i}\}$.)

(d) Suppose $r_1 : \mathbb{U} \longrightarrow \mathbb{C}$ and $r_2 : \mathbb{U} \longrightarrow \mathbb{C}$ are two branches of the square root function (defined by applying the definition in part (b) to different branches of the logarithm).

Show that $r_1(u) = -r_2(u)$ for all $u \in \mathbb{U}$. Sketch the Riemann surface for the complex square root function.

(e) More generally, let $n \geq 2$, and suppose $r_1 : \mathbb{U} \longrightarrow \mathbb{C}$ and $r_2 : \mathbb{U} \longrightarrow \mathbb{C}$ are two branches of the nth-root function (defined by applying the definition in part (b) to different branches of the logarithm). Show that there is some $\zeta \in \mathcal{Z}_n$ (the set of nth roots of unity from part (c)) such that $r_1(u) = \zeta \cdot r_2(u)$ for all $u \in \mathbb{U}$.

Bonus: Sketch the Riemann surface for the complex nth-root function. (Note that it is not possible to embed this surface in three dimensions without some self-intersection.) ♦

18D Analyticity of holomorphic maps

Prerequisites: §18C, Appendix H(ii).

In §18A, we said that the holomorphic functions formed a very special subclass within the set of all (real)-differentiable functions on the plane. One indication of this was Proposition 18A.2, p. 423. Another indication is the following surprising and important result.

Theorem 18D.1 Holomorphic $\Rightarrow$ analytic

Let $\mathbb{U} \subset \mathbb{C}$ be an open subset. If $f : \mathbb{U} \longrightarrow \mathbb{C}$ is holomorphic on $\mathbb{U}$, then f is infinitely (complex-)differentiable everywhere in $\mathbb{U}$. Thus, the functions $f', f'', f''', \ldots$ are also holomorphic on $\mathbb{U}$. Finally, for all $u \in \mathbb{U}$, the (complex) Taylor series of f at u converges uniformly to f in open disk around u.

Proof Since any analytic function is C^∞, it suffices to prove the final sentence, and the rest of the theorem follows. Suppose $0 \in \mathbb{U}$; we will prove that f is analytic at $u = 0$ (the general case $u \neq 0$ is similar).

Let γ be a counterclockwise circular contour in $\mathbb{U}$ centred at 0 (e.g. define $\gamma : [0, 2\pi] \longrightarrow \mathbb{U}$ by $\gamma(s) = re^{is}$ for some $r > 0$). Let $\mathbb{W} \subset \mathbb{U}$ be the purview of γ (an open disk centred at 0). For all $n \in \mathbb{N}$, let

$$c_n := \frac{1}{2\pi i} \oint_\gamma \frac{f(z)}{z^{n+1}} \, dz.$$

We will show that the power series $\sum_{n=0}^\infty c_n w^n$ converges to f for all $w \in \mathbb{W}$. For any $w \in \mathbb{W}$, we have

$$f(w) \underset{(*)}{=} \frac{1}{2\pi i} \oint_\gamma \frac{f(z)}{z - w} \, dz \underset{(\dagger)}{=} \frac{1}{2\pi i} \oint_\gamma \frac{f(z)}{z} \cdot \sum_{n=0}^\infty \left(\frac{w}{z}\right)^n \, dz$$

$$= \frac{1}{2\pi i} \oint_\gamma \sum_{n=0}^\infty \frac{f(z)}{z^{n+1}} \cdot w^n \, dz \underset{(\diamond)}{=} \sum_{n=0}^\infty \left(\frac{1}{2\pi i} \oint_\gamma \frac{f(z)}{z^{n+1}} \, dz\right) w^n \underset{(\ddagger)}{=} \sum_{n=0}^\infty c_n w^n,$$

as desired. Here, $(*)$ is Cauchy's integral formula (Theorem 18C.9, p. 448), and $(\ddagger)$ is by the definition of c_n. Step $(\dagger)$ is because

$$\frac{1}{z-w} = \left(\frac{1}{z}\right) \cdot \left(\frac{1}{1-\frac{w}{z}}\right) = \frac{1}{z} \cdot \sum_{n=0}^{\infty} \left(\frac{w}{z}\right)^n. \tag{18D.1}$$

Here, the last step in equation (18D.1) is the geometric series expansion $\frac{1}{1-x} = \sum_{n=1}^{\infty} x^n$ (with $x := w/z$), which is valid because $|w/z| < 1$, because $|w| < |z|$, because w is inside the disk $\mathbb{W}$ and z is a point on the boundary of $\mathbb{W}$.

It remains to justify step $(\diamond)$. For any $N \in \mathbb{N}$, observe that

$$\frac{1}{2\pi i} \oint_\gamma \sum_{n=0}^{\infty} \frac{f(z)}{z^{n+1}} \cdot w^n \, dz = \sum_{n=0}^{N} \left(\frac{1}{2\pi i} \oint_\gamma \frac{f(z)}{z^{n+1}} \, dz\right) w^n$$

$$+ \frac{1}{2\pi i} \oint_\gamma \sum_{n=N+1}^{\infty} \frac{f(z)}{z^{n+1}} \cdot w^n \, dz. \tag{18D.2}$$

Thus, to justify $(\diamond)$, it suffices to show that the second term on the right-hand side of equation (18D.2) tends to zero as $N \to \infty$. Let L be the length of γ (i.e. $L = 2\pi r$ if γ describes a circle of radius r). The function $z \mapsto f(z)/z$ is continuous on the boundary of $\mathbb{W}$, so it is bounded. Let $M := \sup_{z \in \partial \mathbb{W}} |\frac{f(z)}{z}|$. Fix $\epsilon > 0$ and find some $N \in \mathbb{N}$ such that $\left|\sum_{n=N}^{\infty} \left(\frac{w}{z}\right)^n\right| < \frac{\epsilon}{LM}$. (Such an N exists because the geometric series (18D.1) converges because $|w/z| < 1$.) It follows that for all $z \in \partial \mathbb{W}$,

$$\left| \frac{f(z)}{z} \cdot \sum_{n=N}^{\infty} \left(\frac{w}{z}\right)^n \right| < M \cdot \frac{\epsilon}{LM} = \frac{\epsilon}{L}. \tag{18D.3}$$

Thus,

$$\left| \oint_\gamma \sum_{n=N+1}^{\infty} \frac{f(z)}{z^{n+1}} \cdot w^n \, dz \right| = \left| \oint_\gamma \frac{f(z)}{z} \cdot \sum_{n=N+1}^{\infty} \left(\frac{w}{z}\right)^n \, dz \right| \underset{(*)}{\leq} \frac{\epsilon}{L} \cdot L = \epsilon.$$

Here, $(*)$ is by equation (18D.3) and Lemma 18C.10, p. 450. This works for any $\epsilon > 0$, so we conclude that the second term on the right-hand side of equation (18D.2) tends to zero as $N \to \infty$. This justifies step $(\diamond)$, which completes the proof. $\qquad\square$

Corollary 18D.2 Case $D = 2$ of Proposition 1E.5, p. 20

Let $\mathbb{U} \subseteq \mathbb{R}^2$ be open. If $h : \mathbb{U} \longrightarrow \mathbb{R}$ is a harmonic function, then h is analytic on $\mathbb{U}$.

Proof **Exercise 18D.1.** (*Hint:* Combine Theorem 18D.1 with Proposition 18A.3, ⓔ
p. 423. Note that this is not quite as trivial as it sounds: you must show how to
translate the (complex) Taylor series of a holomorphic function on $\mathbb{C}$ into the (real)
Taylor series of a real-valued function on $\mathbb{R}^2$.) ☐

Because of Theorem 18D.1 and Proposition 18A.5(h), p. 419, holomorphic
functions are also called *complex-analytic* functions (or even simply *analytic* functions) in some books. Analytic functions are extremely 'rigid': for any $u \in \mathbb{U}$, the
behaviour of f in a tiny neighbourhood around u determines the structure of f
everywhere on $\mathbb{U}$, as we now explain. Recall that a subset $\mathbb{U} \subset \mathbb{C}$ is *connected* if
it is not possible to write $\mathbb{U}$ as a union of two nonempty disjoint open subsets. A
subset $\mathbb{X} \subset \mathbb{C}$ is *perfect* if, for every $x \in \mathbb{X}$, every open neighbourhood around x
contains other points in $\mathbb{X}$ besides x. (Equivalently: every point in $\mathbb{X}$ is a *cluster
point* of $\mathbb{X}$.) In particular, any open subset of $\mathbb{C}$ is perfect. Also, $\mathbb{R}$ and $\mathbb{Q}$ are perfect
subsets of $\mathbb{C}$. Any disk, annulus, line segment, or unbroken curve in $\mathbb{C}$ is both
connected and perfect.

Theorem 18D.3 Identity theorem
*Let $\mathbb{U} \subset \mathbb{C}$ be a connected open set, and let $f : \mathbb{U} \longrightarrow \mathbb{C}$ and $g : \mathbb{U} \longrightarrow \mathbb{C}$ be two
holomorphic functions.*

(a) *Suppose there is some $a \in \mathbb{U}$ such that $f(a) = g(a)$, $f'(a) = g'(a)$, $f''(a) = g''(a)$,
   and, in general, $f^{(n)}(a) = g^{(n)}(a)$ for all $n \in \mathbb{N}$. Then $f(u) = g(u)$ for all $u \in \mathbb{U}$.*
(b) *Suppose there is a perfect subset $\mathbb{X} \subset \mathbb{U}$ such that $f(x) = g(x)$ for all $x \in \mathbb{X}$. Then
   $f(u) = g(u)$ for all $u \in \mathbb{U}$.*

Proof (a) Let $h := f - g$. It suffices to show that $h \equiv 0$. Let

$$\mathbb{W} := \big\{ u \in \mathbb{U}; \ h(u) = 0, \quad \text{and} \quad h^{(n)}(u) = 0 \text{ for all } n \in \mathbb{N} \big\}.$$

The set $\mathbb{W}$ is nonempty because $a \in \mathbb{W}$ by hypothesis. We will show that $\mathbb{W} = \mathbb{U}$;
it follows that $h \equiv 0$.

Claim 1 $\mathbb{W}$ *is an open subset of* $\mathbb{U}$.

Proof Let $w \in \mathbb{W}$; we must show that there is a nonempty open disk around
w that is also in $\mathbb{W}$. Now, h is analytic at w because f and g are analytic at w.
Thus, there is some nonempty open disk $\mathbb{D}$ centred at w such that the Taylor
expansion of h converges to $h(z)$ for all $z \in \mathbb{D}$. The Taylor expansion of h at w is
$c_0 + c_1(z - u) + c_2(z - u)^2 + c_3(z - u)^3 + \cdots$, where $c_n := h^{(n)}(w)/n!$, for all
$n \in \mathbb{N}$. But for all $n \in \mathbb{N}$, $c_n = 0$ because $h^{(n)}(w) = 0$, because $w \in \mathbb{W}$. Thus,
the Taylor expansion is $0 + 0(z - w) + 0(z - w)^2 + \cdots$; hence it converges to
zero. Thus, h is equal to the constant zero function on $\mathbb{D}$. Thus, $\mathbb{D} \subset \mathbb{W}$. This
holds for any $w \in \mathbb{W}$; hence $\mathbb{W}$ is an open subset of $\mathbb{C}$. ◇_{Claim 1}

Claim 2 $\mathbb{W}$ *is a closed subset of* $\mathbb{U}$.

Proof For all $n \in \mathbb{N}$, the function $h^{(n)} : \mathbb{U} \longrightarrow \mathbb{C}$ is continuous (because $f^{(n)}$ and $g^{(n)}$ are continuous, since they are differentiable). Thus, the set $\mathbb{W}_n := \{u \in \mathbb{U}; h^{(n)}(u) = 0\}$ is a closed subset of $\mathbb{U}$ (because $\{0\}$ is a closed subset of $\mathbb{C}$). But clearly $\mathbb{W} = \mathbb{W}_0 \cap \mathbb{W}_1 \cap \mathbb{W}_2 \cap \cdots$. Thus, $\mathbb{W}$ is also closed (because it is an intersection of closed sets). $\diamond$ Claim 2

Thus, $\mathbb{W}$ is nonempty, and is both open and closed in $\mathbb{U}$. Thus, the set $\mathbb{V} := \mathbb{U} \setminus \mathbb{W}$ is also open and closed in $\mathbb{U}$, and $\mathbb{U} = \mathbb{V} \sqcup \mathbb{W}$. If $\mathbb{V} \neq \emptyset$, then we have expressed $\mathbb{U}$ as a union of two nonempty disjoint open sets, which contradicts the hypothesis that $\mathbb{U}$ is connected. Thus, $\mathbb{V} = \emptyset$, which means $\mathbb{W} = \mathbb{U}$. Thus, $h \equiv 0$. Thus, $f \equiv g$.

(b) Fix $x \in \mathbb{X}$, and let $\{x_n\}_{n=1}^{\infty} \subset \mathbb{X}$ be a sequence converging to x (which exists because $\mathbb{X}$ is perfect).

Claim 3 $f'(x) = g'(x)$.

(E) *Proof* **Exercise 18D.2.** (*Hint:* Compute the limit (18A.1), p. 421, along the sequence $\{x_n\}_{n=1}^{\infty}$.) $\diamond$ Claim 3

This argument works for any $x \in \mathbb{X}$; thus $f'(x) = g'(x)$ for all $x \in \mathbb{X}$. Repeating the same argument, we get $f''(x) = g''(x)$ for all $x \in \mathbb{X}$. By induction, $f^{(n)}(x) = g^{(n)}(x)$ for all $n \in \mathbb{N}$ and all $x \in \mathbb{X}$. But then part (a) implies that $f \equiv g$. $\square$

Remark The identity theorem is true (with pretty much the same proof) for any $\mathbb{R}^M$-valued, analytic function on any connected open subset $\mathbb{U} \subset \mathbb{R}^N$, for (E) any $N, M \in \mathbb{N}$. (**Exercise 18D.3** Verify this.) In particular, the identity theorem holds for any harmonic functions defined on a connected open subset of $\mathbb{R}^N$ (for any $N \in \mathbb{N}$). This result nicely complements Corollary 5D.4, p. 90, which established the uniqueness of the harmonic function which satisfies specified boundary conditions. (Note that neither Corollary 5D.4 nor the identity theorem for harmonic functions is a special case of the other; they apply to distinct situations.)

In Proposition 18A.5(i) and Example 18A.6, pp. 425 and 426, we showed how the 'standard' real-analytic functions on $\mathbb{R}$ can be extended to holomorphic functions on $\mathbb{C}$ in a natural way. We now show that these are the *only* holomorphic extensions of these functions. In other words, there is a one-to-one relationship between real-analytic functions and their holomorphic extensions.

Corollary 18D.4 Analytic extension

Let $X \subseteq \mathbb{R}$ *be an open subset, and let* $f : X \longrightarrow \mathbb{R}$ *be an analytic function. There exists some open subset* $\mathbb{U} \subseteq \mathbb{C}$ *containing* X, *and a unique holomorphic function* $F : \mathbb{U} \longrightarrow \mathbb{C}$ *such that* $F(x) = f(x)$ *for all* $x \in X$.

Proof For any $x \in X$, Proposition 18A.5(i) says that the (real) Taylor series of f around x can be extended to define a holomorphic function $F_x : \mathbb{D}_x \longrightarrow \mathbb{C}$, where $\mathbb{D}_x \subset \mathbb{C}$ is an open disk centred at x. Let $\mathbb{U} := \bigcup_{x \in X} \mathbb{D}_x$; then $\mathbb{U}$ is an open subset of $\mathbb{C}$ containing X. We would like to define $F : \mathbb{U} \longrightarrow \mathbb{C}$ as follows:

$$F(u) := F_x(u), \quad \text{for any } x \in X \text{ and } u \in \mathbb{D}_x. \tag{18D.4}$$

But there is a problem: what if $u \in \mathbb{D}_x$ and also $u \in \mathbb{D}_y$ for some $x, y \in X$? We must make sure that $F_x(u) = F_y(u)$ – otherwise F will not be well-defined by equation (18D.4).

So, let $x, y \in X$, and suppose the disks $\mathbb{D}_x$ and $\mathbb{D}_y$ overlap. Then $\mathbb{P} := X \cap \mathbb{D}_x \cap \mathbb{D}_y$ is a nonempty open subset of $\mathbb{R}$ (hence perfect). The functions F_x and F_y both agree with f on $\mathbb{P}$; thus, they agree with each other on $\mathbb{P}$. Thus, the identity theorem says that F_x and F_y agree everywhere on $\mathbb{D}_x \cap \mathbb{D}_y$.

Thus, F is well-defined by equation (18D.4). By construction, F is a holomorphic function on $\mathbb{U}$ which extends f. Furthermore, F is the *only* holomorphic extension of f, by the identity theorem. $\qquad\square$

Exercise 18D.4 Let $I : \mathbb{R} \times \mathbb{R} \longrightarrow \mathbb{R}$ be any real-analytic function, and sup-   Ⓔ pose $I(\sin(r), \cos(r)) = 0$ for all $r \in \mathbb{R}$. Use the identity theorem to show that $I(\sin(c), \cos(c)) = 0$ for all $c \in \mathbb{C}$.

Conclude that any algebraic relation between sin and cos (i.e. any 'trigonometric identity') which is true on $\mathbb{R}$ will also be true over all of $\mathbb{C}$.   ◆

18E Fourier series as Laurent series

Prerequisites: §18D, §8D.   **Recommended:** §10D(ii).

For any $r \geq 0$, let $\mathbb{D}(r) := \{z \in \mathbb{C}; |z| < r\}$ be the open disk of radius r around 0, and let $\mathbb{D}^{\complement}(r) := \{z \in \mathbb{C}; |z| > r\}$ be the open *codisk* of 'coradius' r. Let $\mathbb{S}(r) := \{z \in \mathbb{C}; |z| = r\}$ be the circle of radius r; then $\partial\mathbb{D}(r) = \mathbb{S}(r) = \partial\mathbb{D}^{\complement}(r)$. Finally, for any $R > r \geq 0$, let $\mathbb{A}(r, R) := \{z \in \mathbb{C}; r < |z| < R\}$ be the open *annulus* of inner radius r and outer radius R.

Let $c_0, c_1, c_2, \ldots$ be complex numbers, and consider the complex-valued power series:

$$\sum_{n=0}^{\infty} c_n z^n = c_0 + c_1 z + c_2 z^2 + c_3 z^3 + c_4 z^4 + \cdots. \tag{18E.1}$$

For any coefficients $\{c_n\}_{n=0}^{\infty}$, there is some *radius of convergence* $R \geq 0$ (possibly zero) such that the power series (18E.1) converges uniformly on the open disk $\mathbb{D}(R)$ and diverges for all $z \in \mathbb{D}^{\complement}(R)$. (The series (18E.1) may or may not converge on the boundary circle $\mathbb{S}(R)$.) The series (18E.1) then defines a holomorphic function
Ⓔ on $\mathbb{D}(R)$. (**Exercise 18E.1** Prove the preceding three sentences.) Conversely, if $\mathbb{U} \subseteq \mathbb{C}$ is any open set containing 0, and $f : \mathbb{U} \longrightarrow \mathbb{C}$ is holomorphic, then Theorem 18D.1, p. 455, says that f has a power series expansion like equation (18E.1) which converges to f in a disk of nonzero radius around 0.

Next, let $c_0, c_{-1}, c_{-2}, c_{-3}, \ldots$ be complex numbers, and consider the complex-valued *inverse power series*

$$\sum_{n=-\infty}^{0} c_n z^n = c_0 + \frac{c_{-1}}{z} + \frac{c_{-2}}{z^2} + \frac{c_{-3}}{z^3} + \frac{c_{-4}}{z^4} + \cdots. \tag{18E.2}$$

For any coefficients $\{c_n\}_{n=-\infty}^{0}$, there is some *coradius of convergence* $r \geq 0$ (possibly infinity) such that the inverse power series (18E.2) converges uniformly on the open codisk $\mathbb{D}^{\complement}(r)$ and diverges for all $z \in \mathbb{D}(r)$. (The series (18E.2) may or may not converge on the boundary circle $\mathbb{S}(r)$.) The series (18E.2) then defines a holomorphic function on $\mathbb{D}^{\complement}(r)$. Conversely, if $\mathbb{U} \subseteq \mathbb{C}$ is any open set, then we say that $\mathbb{U}$ is a *neighbourhood of infinity* if $\mathbb{U}$ contains $\mathbb{D}^{\complement}(r)$ for some $r < \infty$. If $f : \mathbb{U} \longrightarrow \mathbb{C}$ is holomorphic, and $\lim_{z \to \infty} f(z)$ exists and is finite, then f has an inverse power series expansion like equation (18E.2) which converges to f in a codisk of finite coradius (i.e. a nonempty 'open disk around infinity').[8]

Ⓔ **Exercise 18E.2** Prove all statements in the preceding paragraph. (*Hint:* Consider the change of variables $w := 1/z$. Now use the results about power series from the paragraph between equations (18E.1) and (18E.2).) ♦

Finally, let $\ldots, c_{-2}, c_{-1}, c_0, c_1, c_2, \ldots$ be complex numbers, and consider the complex-valued *Laurent series*:

$$\sum_{n=-\infty}^{\infty} c_n z^n = \cdots + \frac{c_{-2}}{z^2} + \frac{c_{-1}}{z} + c_0 + c_1 z + c_2 z^2 + c_3 z^3 + \cdots. \tag{18E.3}$$

For any coefficients $\{c_n\}_{n=-\infty}^{\infty}$, there exist $0 \leq r \leq R \leq \infty$ such that the Laurent series (18E.3) converges uniformly on the open annulus[9] $^{\circ}\mathbb{A}(r, R)$ and diverges for all $z \in \mathbb{D}(r)$ and all $z \in \mathbb{D}^{\complement}(R)$. (The series (18E.3) may or may not converge on the boundary circles $\mathbb{S}(r)$ and $\mathbb{S}(R)$.) The series (18E.3) then defines a holomorphic function on $^{\circ}\mathbb{A}(r, R)$.

---

[8] This is not merely fanciful terminology; see Remark 18G.4, p. 474.
[9] Note that $^{\circ}\mathbb{A}(r, R) = \emptyset$ if $r = R$.

Exercise 18E.3 Prove all statements in the preceding paragraph. (*Hint:* Combine Ⓔ
the results about power series and inverse power series from the paragraphs between
equations (18E.1) and (18E.3).) ◆

Proposition 18E.1 *Let* $0 \leq r < R \leq \infty$, *and suppose the Laurent series* (18E.3)
converges on $°\mathbb{A}(r, R)$ *to define the function* $f : °\mathbb{A}(r, R) \longrightarrow \mathbb{C}$. *Let* γ *be a
counterclockwise contour in* $°\mathbb{A}(r, R)$ *which encircles the origin (for example,* γ
could be a counterclockwise circle of radius r_0, *where* $r < r_0 < R$). *Then for all*
$n \in \mathbb{Z}$,

$$c_n = \frac{1}{2\pi i} \oint_\gamma \frac{f(z)}{z^{n+1}} \, dz.$$

Proof For all $z \in °\mathbb{A}(r, R)$, we have $f(z) = \sum_{k=-\infty}^{\infty} c_k z^k$. Thus, for any $n \in \mathbb{Z}$,

$$\frac{f(z)}{z^{n+1}} = \frac{1}{z^{n+1}} \sum_{k=-\infty}^{\infty} c_k z^k = \sum_{k=-\infty}^{\infty} c_k z^{k-n-1} \underset{(*)}{=} \sum_{m=-\infty}^{\infty} c_{m+n+1} z^m,$$

where $(*)$ is the change of variables $m := k - n - 1$, so that $k = m + n + 1$. In
other words,

$$\frac{f(z)}{z^{n+1}} = \cdots + \frac{c_{n-1}}{z^2} + \frac{c_n}{z} + c_{n+1} + c_{n+2}z + c_{n+3}z^2 + \cdots. \tag{18E.4}$$

Thus,

$$\oint_\gamma \frac{f(z)}{z^{n+1}} \underset{(*)}{=} \cdots + \oint_\gamma \frac{c_{n-1}}{z^2} + \oint_\gamma \frac{c_n}{z} + \oint_\gamma c_{n+1} + \oint_\gamma c_{n+2}z + \oint_\gamma c_{n+3}z^2 + \cdots$$

$$\underset{(†)}{=} \cdots + 0 + 2\pi i \, c_n + 0 + 0 + 0 + \cdots$$

$$= 2\pi i \, c_n, \quad \text{as desired.}$$

Here, $(*)$ is because the series (18E.4) converges uniformly on $°\mathbb{A}(r, R)$ (because
the Laurent series (18E.3) converges uniformly on $°\mathbb{A}(r, R)$); thus, Proposition 6E.10(b), p. 132, implies we can compute the contour integral of series
(18E.4) term-by-term. Next, (†) is by Examples 18C.2(c) and 18C.6 (pp. 441
and 445). □

Laurent series are closely related to Fourier series.

Proposition 18E.2 *Suppose* $0 \leq r < 1 < R$ *and suppose the Laurent series*
(18E.3) *converges to the function* $f : °\mathbb{A}(r, R) \longrightarrow \mathbb{C}$. *Define* $g : [-\pi, \pi] \longrightarrow \mathbb{C}$
by $g(x) := f(e^{ix})$ *for all* $x \in [-\pi, \pi]$. *For all* $n \in \mathbb{Z}$, *let*

$$\widehat{g}_n := \frac{1}{2\pi} \int_{-\pi}^{\pi} g(x) \exp(-nix) dx \tag{18E.5}$$

be the nth complex Fourier coefficient of g (see §8D, p. 175). Then we have the following.

(a) $\widehat{g}_n = c_n$ *for all $n \in \mathbb{Z}$.*

(b) *For any $x \in [-\pi, \pi]$, if $z = e^{ix} \in \mathbb{S}(1)$, then, for all $N \in \mathbb{N}$, the Nth partial Fourier sum of $g(x)$ equals the Nth partial Laurent sum of $f(z)$; that is*

$$\sum_{n=-N}^{N} \widehat{g}_n \exp(nix) = \sum_{n=-N}^{N} c_n z^n.$$

Thus, the Fourier series for g converges on $[-\pi, \pi]$ in exactly the same ways (i.e. uniformly, in L^2, etc.), and at exactly the same speed, as the Laurent series for f converges on $\mathbb{S}(1)$.

Proof (a) As in Example 18C.1, p. 440, define the 'unit circle' contour γ : $[0, 2\pi] \longrightarrow \mathbb{C}$ by $\gamma(s) := \exp(is)$ for all $s \in [0, 2\pi]$. Then

$$c_n \underset{(*)}{\equiv} \frac{1}{2\pi i} \oint_{\gamma} \frac{f(z)}{z^{n+1}} = \frac{1}{2\pi i} \int_0^{2\pi} \frac{f[\gamma(s)])}{\gamma(s)^{n+1}} \cdot \dot{\gamma}(s) ds$$

$$= \frac{1}{2\pi i} \int_0^{2\pi} \frac{f(e^{is})}{e^{is(n+1)}} \cdot ie^{is} ds = \frac{1}{2\pi} \int_0^{2\pi} f(e^{is}) \cdot e^{-nis} ds$$

$$\underset{(\dagger)}{=} \frac{1}{2\pi} \int_{-\pi}^{\pi} f(e^{ix}) \cdot e^{-nix} dx = \frac{1}{2\pi} \int_{-\pi}^{\pi} g(x) \cdot e^{-nix} dx \underset{(\ddagger)}{\equiv} \widehat{g}_n.$$

Here, $(*)$ is by Proposition 18E.1, and $(\dagger)$ is because the function $s \mapsto e^{is}$ is 2π-periodic. Finally, $(\ddagger)$ is by equation (18E.5).

(b) follows immediately from (a), because if $z = e^{ix}$, then, for all $n \in \mathbb{Z}$, we have $z^n = \exp(nix)$. $\qquad\square$

We can also reverse this logic: given the Fourier series for a function g : $[-\pi, \pi] \longrightarrow \mathbb{C}$, we can interpret it as the Laurent series of some hypothetical function f defined on an open annulus in $\mathbb{C}$ (which may or may not contain $\mathbb{S}(1)$); then, by studying f and its Laurent series, we can draw conclusions about g and its Fourier series.

Let $g : [-\pi, \pi] \longrightarrow \mathbb{C}$ be some function, let $\{\widehat{g}_n\}_{n=-\infty}^{\infty}$ be its Fourier coefficients, as defined by equation (18E.5), and consider the complex Fourier series[10] $\sum_{n=-\infty}^{\infty} \widehat{g}_n \mathbf{E}_n$. If $g \in \mathbf{L}^2[-\pi, \pi]$, then the Riemann–Lebesgue lemma (Corollary 10A.3, p. 199) says that $\lim_{n \to \pm\infty} \widehat{g}_n = 0$; however, the sequence $\{\widehat{g}_n\}_{n=-\infty}^{\infty}$ might converge to zero very slowly if g is nondifferentiable and/or discontinuous. We would like the sequence $\{\widehat{g}_n\}_{n=-\infty}^{\infty}$ to converge to zero as quickly as possible, for two reasons.

---

[10] For all $n \in \mathbb{Z}$, recall that $\mathbf{E}_n(x) := \exp(nix)$ for all $x \in [-\pi, \pi]$.

(1) The faster the sequence $\{\widehat{g}_n\}_{n=-\infty}^{\infty}$ converges to zero, the easier it will be to approximate the function g using a 'partial sum' of the form $\sum_{n=-N}^{N} \widehat{g}_n \mathbf{E}_n$, for some $N \in \mathbb{N}$. (This is important for numerical computations.)

(2) The faster the sequence $\{\widehat{g}_n\}_{n=-\infty}^{\infty}$ converges to zero, the more computations we can perform with g by 'formally manipulating' its Fourier series. For example, if $\{\widehat{g}_n\}_{n=-\infty}^{\infty}$ converges to zero faster than $1/n^k$, we can compute the derivatives $g', g'', g''', \ldots, g^{(k-1)}$ by 'formally differentiating' the Fourier series for g (see §8B(iv), p. 172). This is necessary to verify the 'Fourier series' solutions to I/BVPs which we constructed in Chapters 11–14.

We say that the sequence $\{\widehat{g}_n\}_{n=-\infty}^{\infty}$ has *exponential decay* if there is some $a > 1$ such that

$$\lim_{n\to\infty} a^n |\widehat{g}_n| = 0 \quad \text{and} \quad \lim_{n\to\infty} a^n |\widehat{g}_{-n}| = 0.$$

This is an extremely rapid decay rate, which causes the partial sum $\sum_{n=-N}^{N} \widehat{g}_n \mathbf{E}_n$ to converge uniformly to g very quickly as $N \to \infty$. This means we can 'formally differentiate' the Fourier series of g as many times as we want. In particular, any 'formal' solution to an I/BVP which we obtain through such formal differentiation will converge to the correct solution.

Suppose $g : [-\pi, \pi] \longrightarrow \mathbb{C}$ has real and imaginary parts $g_r, g_i : [-\pi, \pi] \longrightarrow \mathbb{R}$ (so that $g(x) = g_r(x) + g_i(x)\mathbf{i}$ for all $x \in [-\pi, \pi]$). We say that g is *analytic and periodic* if the functions g_r and g_i are (real)-analytic everywhere on $[-\pi, \pi]$, and if we have $g(-\pi) = g(\pi)$, $g'(-\pi) = g'(\pi)$, $g''(-\pi) = g''(\pi)$, etc. (where $g'(x) = g_r'(x) + g_i'(x)\mathbf{i}$, etc.).

Proposition 18E.3 *Let* $g : [-\pi, \pi] \longrightarrow \mathbb{C}$ *have complex Fourier coefficients* $\{\widehat{g}_n\}_{n=-\infty}^{\infty}$. *Then*

$$\big(g \text{ is analytic and periodic}\big) \Longleftrightarrow \big(\text{the sequence } \{\widehat{g}_n\}_{n=-\infty}^{\infty} \text{ decays exponentially}\big).$$

Proof $\implies$ Define the function $f : \mathbb{S}(1) \longrightarrow \mathbb{C}$ by $f(e^{\mathbf{i}x}) := g(x)$ for all $x \in [-\pi, \pi]$.

Claim 1 *f can be extended to a holomorphic function $F : \mathbb{A}(r, R) \longrightarrow \mathbb{C}$, for some $0 \leq r < 1 < R \leq \infty$.*

Proof Let $\widetilde{g} : \mathbb{R} \longrightarrow \mathbb{C}$ be the 2π-periodic extension of g (i.e. $\widetilde{g}(x + 2n\pi) := g(x)$ for all $x \in [-\pi, \pi]$ and $n \in \mathbb{Z}$). Then $\widetilde{g}$ is analytic on $\mathbb{R}$, so Corollary 18D.4, p. 459, says that there is an open subset $\mathbb{U} \subset \mathbb{C}$ containing $\mathbb{R}$ and a holomorphic function $G : \mathbb{U} \longrightarrow \mathbb{C}$ which agrees with g on $\mathbb{R}$. Without loss of generality, we can assume that $\mathbb{U}$ is a horizontal strip of width $2W$ around $\mathbb{R}$, for some $W > 0$—that is, $\mathbb{U} = \{x + y\mathbf{i}; \ x \in \mathbb{R}, \ y \in (-W, W)\}$.

Claim 1.1 *G is horizontally 2π-periodic (i.e. $G(u + 2\pi) = E(u)$ for all $u \in \mathbb{U}$).*

Ⓔ *Proof* **Exercise 18E.4.** *(Hint:* Use the identity theorem 18D.3, p. 457, and the fact that g is 2π-periodic.) △$_{\text{Claim 1.1}}$

Define $E : \mathbb{C} \longrightarrow \mathbb{C}$ by $E(z) := \exp(\mathbf{i}z)$; thus, E maps $\mathbb{R}$ to the unit circle $\mathbb{S}(1)$. Let $r := \mathrm{e}^{-W}$ and $R := \mathrm{e}^{W}$; then $r < 1 < R$. Then E maps the strip $\mathbb{U}$ to the open annulus $\mathbb{A}(r, R) \subseteq \mathbb{C}$. Note that E is horizontally 2π-periodic (i.e. $E(u + 2\pi) = E(u)$ for all $u \in \mathbb{U}$). Define $F : \mathbb{A} \longrightarrow \mathbb{C}$ by $F(E(u)) := G(u)$ for all $u \in \mathbb{U}$.

Claim 1.2 *F is well-defined on $\mathbb{A}(r, R)$.*

Ⓔ *Proof* **Exercise 18E.5** *(Hint:* Use the fact that both G and E are 2π-periodic.) △$_{\text{Claim 1.2}}$

Claim 1.3 *F is holomorphic on $\mathbb{A}(r, R)$.*

Proof Let $a \in \mathbb{A}(r, R)$; we must show that F is differentiable at a. Suppose $a = G(u)$ for some $u \in \mathbb{U}$. There are open sets $\mathbb{V} \subset \mathbb{U}$ (containing u) and $\mathbb{B} \subset \mathbb{A}(r, R)$ (containing a) such that $E : \mathbb{V} \longrightarrow \mathbb{B}$ is bijective. Let $L : \mathbb{B} \longrightarrow \mathbb{V}$ be a local inverse of E. Then L is holomorphic on $\mathbb{V}$ by Proposition 18A.5(k), p. 425 (because $E'(v) \neq 0$ for all $v \in \mathbb{V}$). But, by definition, $F(b) = G(L(b))$ for all $b \in \mathbb{B}$; Thus, F is holomorphic on $\mathbb{B}$ by Proposition 18A.5(j) (the chain rule). This argument works for any $a \in \mathbb{A}(r, R)$; thus, F is holomorphic on $\mathbb{A}(r, R)$. △$_{\text{Claim 1.3}}$

It remains to show that F is an extension of f. But, by definition, $f(E(x)) = g(x)$ for all $x \in [-\pi, \pi]$. Since G is an extension of g, and $F \circ E = G$, it follows that F is an extension of f. ◇$_{\text{Claim 1}}$

Let $\{c_n\}_{n=-\infty}^{\infty}$ be the Laurent coefficients of F. Then Proposition 18E.2, p. 461, says that $c_n = \widehat{g}_n$ for all $n \in \mathbb{Z}$. However, the Laurent series (18E.3) of F (p. 460) converges on $\mathbb{A}(r, R)$. Thus, if $|z| < R$, then the power series (18E.1), p. 459, converges absolutely at z. This means that, if $1 < a < R$, then $\sum_{n=0}^{\infty} a^n |\widehat{g}_n|$ is finite. Thus, $\lim_{n\to\infty} a^n |\widehat{g}_n| = 0$.

Likewise, if $r < |z|$, then the inverse power series (18E.2), p. 460, converges absolutely at z. This means that, if $1 < a < 1/r$, then $\sum_{n=0}^{\infty} a^n |\widehat{g}_{-n}|$ is finite. Thus, $\lim_{n\to\infty} a^n |\widehat{g}_{-n}| = 0$.

⟸ Define $c_n := \widehat{g}_n$ for all $n \in \mathbb{Z}$, and consider the resulting Laurent series (18E.3). Suppose there is some $a > 1$ such that

$$\lim_{n\to\infty} a^n |c_n| = 0 \quad \text{and} \quad \lim_{n\to\infty} a^n |c_{-n}| = 0. \qquad (18\text{E}.6)$$

Claim 2 *Let $r := 1/a$ and $R := a$. For all $z \in \mathbb{A}(r, R)$, the Laurent series (18E.3) converges absolutely at z.*

Proof Let $z_+ := z/a$; then $|z_+| < 1$ because $|z| < R := a$. Also, let $z_- := 1/az$; then $|z_-| < 1$, because $|z| > r := 1/a$. Thus,

$$\sum_{n=1}^{\infty} |z_-|^n < \infty \quad \text{and} \quad \sum_{n=0}^{\infty} |z_+|^n < \infty. \tag{18E.7}$$

Evaluating the Laurent series (18E.3) at z, we see that

$$\sum_{n=-\infty}^{\infty} c_n z^n = \sum_{n=1}^{\infty} \frac{c_{-n}}{z^n} + \sum_{n=0}^{\infty} c_n z^n = \sum_{n=1}^{\infty} \frac{a^n c_{-n}}{(az)^n} + \sum_{n=0}^{\infty} a^n c_n (z/a)^n$$

$$= \sum_{n=1}^{\infty} (a^n c_{-n}) z_-^n + \sum_{n=0}^{\infty} a^n c_n z_+^n.$$

Thus,

$$\sum_{n=-\infty}^{\infty} |c_n z^n| \leq \sum_{n=1}^{\infty} a^n |c_{-n}| |z_-|^n + \sum_{n=0}^{\infty} a^n |c_n| |z_+|^n < \infty, \tag{$*$}$$

where $(*)$ is by equations (18E.6) and (18E.7). $\diamond_{\text{Claim 2}}$

Claim 2 implies that the Laurent series (18E.3) converges to some holomorphic function $f : \mathbb{A}(r, R) \longrightarrow \mathbb{C}$. But $g(x) = f(e^{ix})$ for all $x \in [-\pi, \pi]$; thus, g is (real)-analytic on $[-\pi, \pi]$, because f is (complex-)analytic on $\mathbb{A}(r, R)$, by Theorem 18D.1, p. 455. $\square$

Exercise 18E.6 Let $f : [-\pi, \pi] \longrightarrow \mathbb{R}$, and consider the real Fourier series for ⒠ f (see §8A, p. 165). Show that the real Fourier coefficients $\{A_n\}_{n=0}^{\infty}$ and $\{B_n\}_{n=0}^{\infty}$ have exponential decay if, and only if, f is analytic and periodic on $[-\pi, \pi]$. (*Hint:* Use Proposition 8D.2, p. 176.) ◆

Exercise 18E.7 Let $f : [0, \pi] \longrightarrow \mathbb{R}$, and consider the Fourier sine series and ⒠ cosine series for f (see §7A(i), p. 141, and §7A(ii), p. 145).

(a) Show that the Fourier cosine coefficients $\{A_n\}_{n=0}^{\infty}$ have exponential decay if, and only if, f is analytic on $[0, \pi]$ and $f'(0) = 0 = f'(\pi)$, and $f^{(n)}(0) = 0 = f^{(n)}(\pi)$ for all odd $n \in \mathbb{N}$.

(b) Show that the Fourier sine coefficients $\{B_n\}_{n=1}^{\infty}$ have exponential decay if, and only if, f is analytic on $[0, \pi]$ and $f(0) = 0 = f(\pi)$, and $f^{(n)}(0) = 0 = f^{(n)}(\pi)$ for all even $n \in \mathbb{N}$.

(c) Conclude that if both the sine and cosine series have exponential decay, then $f \equiv 0$.
(*Hint:* Use Exercise 18E.6 and Proposition 8C.5, p. 174.) ◆

Ⓔ **Exercise 18E.8** Let $\mathbb{X} = [0, L]$ be an interval, let $f \in \mathbf{L}^2(\mathbb{X})$ be some initial temperature distribution, and let $F : \mathbb{X} \times \mathbb{R}_{\not\!+} \longrightarrow \mathbb{R}$ be the solution to the one-dimensional heat equation ($\partial_t F = \partial_x^2 F$) with initial conditions $F(x; 0) = f(x)$ for all $x \in \mathbb{X}$, and satisfying either homogeneous Dirichlet boundary conditions, or homogeneous Neumann boundary conditions, or periodic boundary conditions on $\mathbb{X}$, for all $t > 0$. Show that, for any fixed $t > 0$, the function $F_t(x) := F(x, t)$ is analytic on $\mathbb{X}$. (*Hint:* Apply Propositions 11A.1 and 11A.3, pp. 229 and 230.)

This shows how the action of the heat equation can rapidly 'smooth' even a highly irregular initial condition. ◆

Ⓔ **Exercise 18E.9** Compute the complex Fourier series of $f : [-\pi, \pi] \longrightarrow \mathbb{C}$ when f is defined as follows:

(a) $f(x) = \sin(e^{ix})$;
(b) $f(x) = \cos(e^{-3ix})$;
(c) $f(x) = e^{2ix} \cdot \cos(e^{-3ix})$;
(d) $f(x) = (5 + e^{2ix}) \cdot \cos(e^{-3ix})$;
(e) $f(x) = \frac{1}{e^{2ix} - 4}$;
(f) $f(x) = \frac{e^{ix}}{e^{2ix} - 4}$. ◆

Ⓔ **Exercise 18E.10**
(a) Show that the Laurent series (18E.3) can be written in the form $P_+(z) + P_-(1/z)$, where P_+ and P_- are both power series.
(b) Suppose P_+ has radius of convergence R_+, and P_- has radius of convergence R_-. Let $R := R_+$ and $r := 1/R_-$, and show that the Laurent series (18E.3) converges on $\mathbb{A}(r, R)$. ◆

18F* Abel means and Poisson kernels

Prerequisites: §18E.   **Prerequisites** (for proofs): §10D(ii).

Theorem 18E.3 showed that if $g : [-\pi, \pi] \longrightarrow \mathbb{C}$ is analytic, then its Fourier series $\sum_{n=-\infty}^{\infty} \widehat{g}_n \mathbf{E}_n$ will converge uniformly and extremely quickly to g. At the opposite extreme, if g is not even differentiable, then $\sum_{n=-\infty}^{\infty} \widehat{g}_n \mathbf{E}_n$ might not converge uniformly, or even pointwise, to g. To address this problem, we introduce the Abel mean. For any $r < 1$, the rth *Abel mean* of the Fourier series $\sum_{n=-\infty}^{\infty} \widehat{g}_n \mathbf{E}_n$ is

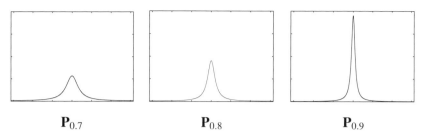

$$\mathbf{P}_{0.7} \qquad\qquad \mathbf{P}_{0.8} \qquad\qquad \mathbf{P}_{0.9}$$

Figure 18F.1. The Poisson kernels $\mathbf{P}_{0.7}$, $\mathbf{P}_{0.8}$, and $\mathbf{P}_{0.9}$, plotted on the interval $[-\pi, \pi]$. Note the increasing concentration of $\mathbf{P}_r$ near $x = 0$ as $r \nearrow 1$. (In the terminology of Section 10D(ii), the system $\{\mathbf{P}_r\}_{0 < r < 1}$ is like an *approximation of the identity*.)

defined:

$$A_r[g] := \sum_{n=-\infty}^{\infty} r^{|n|} \widehat{g}_n \mathbf{E}_n.$$

As $r \nearrow 1$, each summand $r^{|n|} \widehat{g}_n \mathbf{E}_n$ in the Abel mean converges to the corresponding summand $\widehat{g}_n \mathbf{E}_n$ in the Fourier series for g. Thus, we expect that $A_r[g]$ should converge to g as $r \nearrow 1$. The goal of this section is to verify this intuition.

For any $r \in [0, 1)$, we define the *Poisson kernel* $\mathbf{P}_r : [-2\pi, 2\pi] \longrightarrow \mathbb{R}$ by

$$\mathbf{P}_r(x) := \frac{1 - r^2}{1 - 2r \cos(x) + r^2}, \qquad \text{for all } x \in [-2\pi, 2\pi]$$

(see Figure 18F.1). Note that $\mathbf{P}_r$ is 2π-periodic (i.e. $\mathbf{P}_r(x + 2\pi) = \mathbf{P}_r(x)$ for all $x \in [-2\pi, 0]$). For any function $g : [-\pi, \pi] \longrightarrow \mathbb{C}$, the *convolution* of $\mathbf{P}_r$ and g is the function $\mathbf{P}_r * g : [-\pi, \pi] \longrightarrow \mathbb{C}$ defined by

$$\mathbf{P}_r * g(x) := \frac{1}{2\pi} \int_{-\pi}^{\pi} g(y) \, \mathbf{P}_r(x - y) dy, \qquad \text{for all } x \in [-\pi, \pi].$$

The next result tells us that $\lim_{r \nearrow 1} A_r[g](x) = g(x)$, whenever the function g is continuous at x. Furthermore, for all $r < 1$, the functions $A_r[g] : [-\pi, \pi] \longrightarrow \mathbb{C}$ are extremely smooth, and the two-variable function $G(x, r) := A_r[g](x)$ is also extremely smooth.

Proposition 18F.1 *Let $g \in \mathbf{L}^2[-\pi, \pi]$.*
(a) *For any $r \in [0, 1)$ and $x \in [-\pi, \pi]$, $\mathbf{P}_r * g(x) = A_r[g](x)$.*
(b) *For any $x \in (-\pi, \pi)$, if g is continuous at x, then $\lim_{r \nearrow 1} A_r[g](x) = g(x)$.*

(c) *Let* $\mathbb{D}$ *be the closed unit disk, and define* $f : \mathbb{D} \longrightarrow \mathbb{C}$ *by*

$$f(r\,e^{i\theta}) := \begin{cases} \mathbf{P}_r * g(\theta) & \text{if} \quad r < 1; \\ g(\theta) & \text{if} \quad r = 1, \end{cases} \quad \text{for all } \theta \in [-\pi, \pi] \text{ and } r \in [0, 1].$$

Then f *is holomorphic on* $\mathbb{D}$.

(d) *Thus, for any fixed* $r < 1$, *the function* $\mathbf{A}_r[g] : [-\pi, \pi] \longrightarrow \mathbb{C}$ *is analytic.*

(e) *Let* $\theta \in (-\pi, \pi)$, *and let* $s = e^{i\theta} \in \mathbb{S}$. *If* g *is continuous in a neighbourhood of* θ, *then* f *is continuous at* s *— i.e.*

$$\lim_{\substack{u \to s \\ u \in \mathbb{D}}} f(u) = f(s).$$

(f) *If* g *is continuous on* $[-\pi, \pi]$ *and* $g(-\pi) = g(\pi)$, *then* f *is continuous on* $\mathbb{D}$.

Ⓔ   *Proof* (a) **Exercise 18F.1.** (*Hint:* Use Lemmas 18F.2 and 18F.3 below.)
Ⓔ   (b) is **Exercise 18F.2.** (*Hint:* Use Lemma 18F.4 below and Proposition 10D.9(b), p. 220).

          (e) and (f) follow immediately from (b), while (d) follows from (c).
Ⓔ   (c) is **Exercise 18F.3.** *Hint:*

(i) Let $\mathcal{P} : \mathbb{S} \times \mathbb{D} \longrightarrow \mathbb{R}$ be the Poisson kernel defined on p. 451. For any $s \in \mathbb{S}$ and $u \in \mathbb{D}$, if $s = e^{iy}$ and $u = r \cdot e^{ix}$, show that $\mathcal{P}(s, u) = \mathbf{P}_r(x - y)$.

(ii) Use this to show that

$$\mathbf{P}_r * g(x) = \frac{1}{2\pi} \int_0^{2\pi} g(y)\, \mathcal{P}(e^{iy}, u)\mathrm{d}y.$$

(ii) Now apply the Poisson integral formula for holomorphic functions (Corollary 18C.13, p. 451).   □

To prove parts (a) and (b) of Proposition 18F.1, we require the following three lemmas.

Lemma 18F.2 *Fix* $r \in [0, 1)$. *Then*

$$\mathbf{P}_r(x) = 1 + 2\sum_{n=1}^{\infty} r^n \cos(nx) = \sum_{n=-\infty}^{\infty} r^{|n|} \exp(\mathrm{i}nx).$$

Thus, if $\{\widehat{\mathbf{P}}_r^n\}_{n=-\infty}^{\infty}$ *are the complex Fourier coefficients of the Poisson kernel* $\mathbf{P}_r$, *then* $\widehat{\mathbf{P}}_r^n = r^{|n|}$ *for all* $n \in \mathbb{Z}$.

Ⓔ   *Proof* **Exercise 18F.4.**   □

For any $f, g : [-\pi, \pi] \longrightarrow \mathbb{C}$, recall the definition of the convolution $f * g$ from §10D(ii), p. 215. The passage from a function to its Fourier coefficients converts the convolution operator into multiplication, as follows.[11]

---

[11] For the corresponding result for Fourier transforms of functions on $\mathbb{R}$, see Theorem 19B.2(b), p. 499.

Lemma 18F.3 *Let $f, g \in \mathbf{L}^2[-\pi, \pi]$, and suppose $h = f * g \in \mathbf{L}^2[-\pi, \pi]$ also. Then, for all $n \in \mathbb{Z}$, we have $\widehat{h}_n = \widehat{f}_n \cdot \widehat{g}_n$.*

Proof **Exercise 18F.5.**   □   Ⓔ

Lemma 18F.4 *The set of Poisson kernels $\{\mathbf{P}_r\}_{0 \le r < 1}$ is an approximation of identity, in the following sense:*

(AI1) $\frac{1}{2\pi} \int_{-\pi}^{\pi} \mathbf{P}_r(x)\mathrm{d}x = 1$ *for all* $r \in [0, 1)$;
(AI2) *for any* $\epsilon > 0$, $\lim_{r \nearrow 1} \frac{1}{2\pi} \int_{-\epsilon}^{\epsilon} \mathbf{P}_r(x)\mathrm{d}x = 1$. *(See Figure 10D.2, p. 220.)*

Proof **Exercise 18F.6.**   □   Ⓔ

Exercise 18F.7   For all $N \in \mathbb{N}$, the Nth *Dirichlet kernel* is the function $\mathbf{D}_N :$ Ⓔ
$[-\pi, \pi] \longrightarrow \mathbb{R}$ defined by

$$\mathbf{D}_N(x) := 1 + 2 \sum_{n=1}^{N} \cos(nx)$$

(see Figure 10B.1, p. 200).

(a) Show that $\mathbf{D}_N(x) = \sum_{n=-N}^{N} \exp(nx\mathbf{i})$.
(b) Let $g : [-\pi, \pi] \longrightarrow \mathbb{C}$ have complex Fourier series $\sum_{n=-\infty}^{\infty} \widehat{g}_n \mathbf{E}_n$. Use Lemma 18F.3
    to show that $\mathbf{D}_N * g = \sum_{n=-N}^{N} \widehat{g}_n \mathbf{E}_n$. (Compare this with Lemma 10B.1, p. 199.)   ◆

18G Poles and the residue theorem

Prerequisites: §18D.

Let $\mathbb{U} \subset \mathbb{C}$ be an open subset, let $p \in \mathbb{U}$, and let $\mathbb{U}^* := \mathbb{U} \setminus \{p\}$. Let $f : \mathbb{U}^* \longrightarrow \mathbb{C}$ be a holomorphic function. We say that p is an *isolated singularity* of f, because f is well-defined and holomorphic for all points *near* p, but not *at* p itself.

Now, it might be possible to 'extend' f to a holomorphic function $f : \mathbb{U} \longrightarrow \mathbb{C}$ by defining $f(p)$ in some suitable way. In this case, we say that p is a *removable singularity* of f; it is merely a point we 'forgot' when defining f on $\mathbb{U}^*$. However, sometimes there is no way to define $f(p)$ such that the resulting function $f : \mathbb{U} \longrightarrow \mathbb{C}$ is complex-differentiable (or even continuous) at p; in this case, we say that p is an *indelible singularity*. In this section, we will be concerned with a particularly 'nice' class of indelible singularities called *poles*.

Define $F_1 : \mathbb{U}^* \longrightarrow \mathbb{C}$ by $F_1(u) = (p - u) \cdot f(u)$. We say that p is a *simple pole* of f if p is a removable singularity of F_1 – i.e. if $F_1(p)$ can be defined such that F_1 is complex-differentiable at p. Now, F_1 is already holomorphic on $\mathbb{U}^*$ (because it is a product of two holomorphic functions f and $z \mapsto (z - u)$).

Thus, if F_1 is differentiable at p, then F_1 is holomorphic on all of $\mathbb{U}$. Then Theorem 18D.1, p. 455, says that F_1 is *analytic* at p – i.e. F_1 has a Taylor expansion near p:

$$F_1(z) = a_0 + a_1(z - p) + a_2(z - p)^2 + a_3(z - p)^3 + a_4(z - p)^4 + \cdots .$$

Thus,

$$f(z) = \frac{F_1(z)}{z - p} = \frac{a_0}{z - p} + a_1 + a_2(z - p)$$

$$+ a_3(z - p)^2 + a_4(z - p)^3 + \cdots .$$

This expression is called a *Laurent expansion* (of order 1) for f at the pole p. The coefficient a_0 is called the *residue* of f at the pole p, and is denoted $\mathrm{res}_p(f)$.

Suppose p is not a simple pole (i.e. it is not a removable singularity for F_1). Let $n \in \mathbb{N}$, and define $F_n : \mathbb{U}^* \longrightarrow \mathbb{C}$ by $F_n(u) = (p - u)^n \cdot f(u)$. We say that p is a *pole* if there is some $n \in \mathbb{N}$ such that p is a removable singularity of F_n – i.e. if $F_n(p)$ can be defined such that F_n *is* complex-differentiable at p. The smallest value of n for which this is true is called the *order* of the pole p.

Now, F_n is already holomorphic on $\mathbb{U}^*$. Thus, if F_n is differentiable at p, then F_n is holomorphic on all of $\mathbb{U}$. Again, Theorem 18D.1 says that F_n is *analytic* at p, with Taylor expansion

$$F_n(z) = a_0 + a_1(z - p) + \cdots + a_{n-1}(z - p)^{n-1} + a_n(z - p)^n$$

$$+ a_{n+1}(z - p)^{n+1} + \cdots .$$

Thus,

$$f(z) = \frac{F_n(z)}{(z - p)^n} = \frac{a_0}{(z - p)^n} + \frac{a_1}{(z - p)^{n-1}} + \cdots + \frac{a_{n-1}}{(z - p)} + a_n$$

$$+ a_{n+1}(z - p) + \cdots .$$

This expression is called a *Laurent expansion* (of order n) for f at the pole p. The coefficient a_{n-1} is called the *residue* of f at the pole p, and is denoted $\mathrm{res}_p(f)$.

Let $\widehat{\mathbb{C}} := \mathbb{C} \sqcup \{\infty\}$, where the symbol '$\infty$' represents a 'point at infinity'. If $f : \mathbb{U}^* \longrightarrow \mathbb{C}$ has a pole at p, then it is easy to check that $\lim_{z \to p} |f(z)| = \infty$ (**Exercise 18G.1**). Thus, it is natural and convenient to extend f to a function $f : \mathbb{U} \longrightarrow \widehat{\mathbb{C}}$ by defining $f(p) = \infty$. (Later, in Remark 18G.4, p. 474, we will explain why this is not merely a cute notational device, but is actually the 'correct' thing to do.) The extended function $f : \mathbb{U} \longrightarrow \widehat{\mathbb{C}}$ is called a *meromorphic* function.

Formally, if $\mathbb{U} \subset \mathbb{C}$ is an open set, then a function $f : \mathbb{U} \longrightarrow \widehat{\mathbb{C}}$ is *meromorphic* if there is a discrete subset $\mathbb{P} \subset \mathbb{U}$ (possibly empty) such that, if $\mathbb{U}^* := \mathbb{U} \setminus \mathbb{P}$,

(1) $f : \mathbb{U}^* \longrightarrow \mathbb{C}$ is holomorphic;
(2) every $p \in \mathbb{P}$ is a pole of f (hence $\lim_{z \to p} |f(z)| = \infty$);
(3) $f(p) = \infty$ for all $p \in \mathbb{P}$.

Example 18G.1

(a) Any holomorphic function is meromorphic, since it has no poles.
(b) Let $f : \mathbb{U} \longrightarrow \mathbb{C}$ be holomorphic, let $p \in \mathbb{U}$, and define $F : \mathbb{U} \longrightarrow \widehat{\mathbb{C}}$ by $F(z) = f(z)/(z - p)$. Then F is meromorphic on $\mathbb{U}$, with a single pole at p, and $\mathrm{res}_p(F) = f(p)$.
(c) Fix $y > 0$, and define $\mathcal{K}_y : \mathbb{C} \longrightarrow \widehat{\mathbb{C}}$ by

$$\mathcal{K}_y(z) = \frac{y}{\pi(z^2 + y^2)} = \frac{y}{\pi(z + y\mathbf{i})(z - y\mathbf{i})}, \qquad \text{for all } z \in \mathbb{C}.$$

Then $\mathcal{K}_y$ is meromorphic on $\mathbb{C}$, with simple poles at $z = \pm y\mathbf{i}$. Observe that

$$\mathcal{K}_y(z) = \frac{f_+(z)}{(z - y\mathbf{i})},$$

where

$$f_+(z) := \frac{y}{\pi(z + y\mathbf{i})}, \qquad \text{for all } z \in \mathbb{C}.$$

Note that f_+ is holomorphic near $y\mathbf{i}$, so (b) says that

$$\mathrm{res}_{y\mathbf{i}}(\mathcal{K}_y) = f_+(y\mathbf{i}) = \frac{y}{\pi(2y\mathbf{i})} = \frac{1}{2\pi\mathbf{i}}.$$

Likewise,

$$\mathcal{K}_y(z) = \frac{f_-(z)}{(z + y\mathbf{i})},$$

where

$$f_-(z) := \frac{y}{\pi(z - y\mathbf{i})}, \qquad \text{for all } z \in \mathbb{C}.$$

Note that f_- is holomorphic near $-y\mathbf{i}$, so (b) says that

$$\mathrm{res}_{-y\mathbf{i}}(\mathcal{K}_y) = f_-(-y\mathbf{i}) = \frac{y}{\pi(-2y\mathbf{i})} = \frac{-1}{2\pi\mathbf{i}}.$$

(d) More generally, let $P(z) = (z - p_1)^{n_1}(z - p_2)^{n_2} \cdots (z - p_J)^{n_J}$ be any complex polynomial with roots $p_1, \ldots, p_n \in \mathbb{C}$. Let $f : \mathbb{C} \longrightarrow \mathbb{C}$ be any other holomorphic function (e.g. another polynomial), and define $F : \mathbb{C} \longrightarrow \widehat{\mathbb{C}}$ by $F(z) = f(z)/P(z)$ for all $z \in \mathbb{C}$. Then F is a meromorphic function, whose poles are located

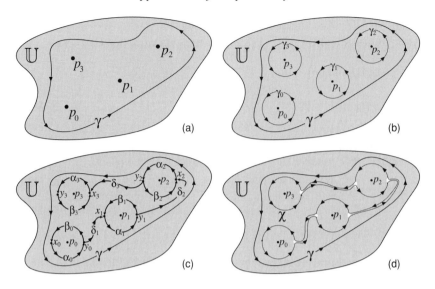

Figure 18G.1. (a) The hypotheses of the residue theorem. (b) For all $j \in [0 \dots J]$, γ_j is a small, counterclockwise circular contour around the pole p_j. (c) The paths $\alpha_0, \dots, \alpha_J$, $\beta_0, \dots, \beta_J$, and $\delta_1, \dots, \delta_J$. (d) The chain χ is a contour homotopic to γ.

at $\{p_1, p_2, \dots, p_J\}$. For any $j \in [1 \dots J]$, define $F_j(z) := f(z)/(z - p_1)^{n_1} \cdots (z - p_{j-1})^{n_{j-1}}(z - p_{j+1})^{n_{j+1}} \cdots (z - p_J)^{n_J}$. Then

$$\operatorname{res}_{p_j}(F) = \frac{F_j^{(n_j - 1)}(p_j)}{(n_j - 1)!}$$

(i.e. the $(n_j - 1)$th term in the Taylor expansion of F_j at p_j). In particular, if $n_j = 1$ (i.e. p_j is a simple pole), then $\operatorname{res}_{p_j}(F) = F_j(p_j)$.

(e) Let $g : \mathbb{U} \longrightarrow \widehat{\mathbb{C}}$ be meromorphic and let $p \in \mathbb{U}$. Suppose g has a simple pole at p. If $f : \mathbb{U} \longrightarrow \mathbb{C}$ is holomorphic, and $f(p) \neq 0$, then the function $f \cdot g$ is meromorphic, with a pole at p, and $\operatorname{res}_p(f \cdot g) = f(p) \cdot \operatorname{res}_p(g)$. ◇

(E) **Exercise 18G.2** Verify Example 18G.1(d) and (e). ◆

We now come to one of the most important results in complex analysis.

Theorem 18G.2 Residue theorem

Let $\mathbb{U} \subseteq \mathbb{C}$ be an open, simply connected subset of the plane. Let $f : \mathbb{U} \longrightarrow \widehat{\mathbb{C}}$ be meromorphic on $\mathbb{U}$. Let $\gamma : [0, S] \longrightarrow \mathbb{U}$ be a counterclockwise contour which is nullhomotopic in $\mathbb{U}$, and suppose the purview of γ contains the poles $p_0, p_1, \dots, p_J \in \mathbb{U}$, and no other poles, as shown in Figure 18G.1(a). *Then*

$$\oint_\gamma f = 2\pi \mathbf{i} \sum_{j=0}^{J} \operatorname{res}_{p_j}(f).$$

Proof For all $j \in [0 \ldots J]$, let $\gamma_j : [0, 2\pi] \longrightarrow \mathbb{U}$ be a small, counterclockwise circular contour around the pole p_j, as shown in Figure 18G.1(b).

Claim 1 *For all $j \in [0 \ldots J]$, $\oint_{\gamma_j} f = \mathrm{res}_{p_j}(f)$.*

Proof Suppose f has the following Laurent expansion around p_j:

$$f(z) = \frac{a_{-n}}{(z - p_j)^n} + \frac{a_{1-n}}{(z - p_j)^{n-1}} + \cdots + \frac{a_{-1}}{(z - p_j)} + a_0 + a_1(z - p_j) + \cdots$$

This series converges uniformly, so $\oint_{\gamma_j} f$ can integrated term-by-term to get:

$$\oint_{\gamma_j} \frac{a_{-n}}{(z - p_j)^n} + \oint_{\gamma_j} \frac{a_{1-n}}{(z - p_j)^{n-1}} + \cdots + \oint_{\gamma_j} \frac{a_{-1}}{(z - p_j)} + \oint_{\gamma_j} a_0$$

$$+ \oint_{\gamma_j} a_1(z - p_j) + \cdots \underset{(\dagger)}{=} 0 + 0 + \cdots + a_{-1} \cdot 2\pi\mathbf{i} + 0 + 0 + \cdots$$

$$= 2\pi\mathbf{i}\, a_{-1} \underset{(\ddagger)}{=} 2\pi\mathbf{i} \cdot \mathrm{res}_{p_j}(f).$$

Here, (†) is by Examples 18C.2(c) and 18C.6, pp. 441 and 445. Meanwhile, (‡) is because $a_{-1} = \mathrm{res}_{p_j}(f)$ by definition. $\diamond_{\text{Claim 1}}$

Figure 18G.1(c) portrays the smooth paths $\alpha_j : [0, \pi] \longrightarrow \mathbb{U}$ and $\beta_j : [0, \pi] \longrightarrow \mathbb{U}$ defined by

$$\alpha_j(s) := \gamma_j(s) \quad \text{and} \quad \beta_j(s) := \gamma_j(s + \pi), \qquad \text{for all } s \in [0, \pi].$$

That is: α_j and β_j parameterize the 'first half' and the 'second half' of γ_j, respectively, so that

$$\gamma_j = \alpha_j \diamond \beta_j. \tag{18G.1}$$

For all $j \in [0 \ldots J]$, let $x_j := \alpha_j(0) = \beta_j(\pi)$, and let $y_j := \alpha_j(\pi) = \beta_j(0)$. For all $j \in [1 \ldots J]$, let $\delta_j : [0, 1] \longrightarrow \mathbb{U}$ be a smooth path from y_{j-1} to x_j. For all $i \in [0 \ldots J]$, we can assume that δ_j is drawn so as not to intersect α_i or β_i, and for all $i \in [1 \ldots J]$, $i \neq j$, we can likewise assume that δ_j does not intersect δ_i. Figure 18G.1(d) portrays the following chain:

$$\chi := \alpha_0 \diamond \delta_1 \diamond \alpha_1 \diamond \delta_2 \diamond \cdots \diamond \delta_J \diamond \gamma_J \diamond \overleftarrow{\delta}_J \diamond \beta_{J-1} \diamond \overleftarrow{\delta}_{J-1} \diamond \cdots \diamond$$

$$\overleftarrow{\delta}_3 \diamond \beta_2 \diamond \overleftarrow{\delta}_2 \diamond \beta_1 \diamond \overleftarrow{\delta}_1 \diamond \beta_0. \tag{18G.2}$$

The chain χ is actually a *contour*, by Lemma 18C.8(c), p. 447.

Claim 2 γ *is homotopic to* χ *in* $\mathbb{U}$.

Ⓔ *Proof* **Exercise 18G.3.** (Not as easy as it looks.) ◇ Claim 2

Thus,

$$\oint_\gamma f \underset{(*)}{=} \oint_\chi f$$

$$\underset{(\dagger)}{=} \int_{\alpha_0} f + \int_{\delta_1} f + \int_{\alpha_1} f + \int_{\delta_2} f + \cdots + \int_{\delta_J} f + \oint_{\gamma_J} f - \int_{\delta_J} f + \int_{\beta_{J-1}} f$$

$$- \int_{\delta_{J-1}} f + \cdots - \int_{\delta_3} f + \int_{\beta_2} f - \int_{\delta_2} f + \int_{\beta_1} f - \int_{\delta_1} f + \int_{\beta_0} f.$$

$$= \int_{\alpha_0} f + \int_{\alpha_1} f + \cdots + \int_{\alpha_{J-1}} f + \oint_{\gamma_J} f + \int_{\beta_{J-1}} f + \cdots + \int_{\beta_1} f + \int_{\beta_0} f.$$

$$\underset{(@)}{=} \int_{\alpha_0 \diamond \beta_0} f + \int_{\alpha_1 \diamond \beta_1} f + \cdots + \int_{\alpha_{J-1} \diamond \beta_{J-1}} f + \oint_{\gamma_J} f$$

$$\underset{(\ddagger)}{=} \oint_{\gamma_0} f + \oint_{\gamma_1} f + \cdots + \oint_{\gamma_{J-1}} f + \oint_{\gamma_J} f$$

$$\underset{(\diamond)}{=} 2\pi\mathbf{i} \cdot \mathrm{res}_{p_0}(f) + 2\pi\mathbf{i} \cdot \mathrm{res}_{p_1}(f) + \cdots + 2\pi\mathbf{i} \cdot \mathrm{res}_{p_{J-1}}(f) + 2\pi\mathbf{i} \cdot \mathrm{res}_{p_J}(f),$$

as desired. Here, $(*)$ is by Claim 2 and Proposition 18C.7, p. 446; $(\dagger)$ is by equation (18G.2) and Lemma 18C.8(a), (b), p. 447; and $(@)$ is again by Lemma 18C.8(b). Finally, $(\ddagger)$ is by equation (18G.1), and $(\diamond)$ is by Claim 1. □

Example 18G.3

(a) Suppose f is holomorphic inside the purview of γ. Then it has no poles, so the residue sum in the residue theorem is zero. Thus, we get $\oint_\gamma f = 0$, in agreement with Cauchy's theorem (Theorem 18C.5, p. 444).

(b) Suppose $f(z) = 1/z$, and γ encircles 0. Then f has exactly one pole in the purview of γ (namely, at 0), and $\mathrm{res}_0(f) = 1$ (because the Laurent expansion of f is just $1/z$). Thus, we get $\oint_\gamma f = 2\pi\mathbf{i}$, in agreement with Example 18C.6, p. 445.

(c) Suppose f is holomorphic inside the purview of γ. Let p be in the purview of γ and define $F(z) := f(z)/(z - p)$. Then F has exactly one pole in the purview of γ (namely, at p), and $\mathrm{res}_p(F) = f(z)$, by Example 18G.1(b). Thus, we get $\oint_\gamma F = 2\pi\mathbf{i}\, f(z)$, in agreement with Cauchy's integral formula (Theorem 18C.9, p. 448). ◇

Remark 18G.4 The Riemann sphere

Earlier we introduced the notational convention of defining $f(p) = \infty$ whenever p was a pole of a holomorphic function $f : \mathbb{U} \setminus \{p\} \longrightarrow \mathbb{C}$, thereby extending f to a 'meromorphic' function $f : \mathbb{U} \longrightarrow \widehat{\mathbb{C}}$, where $\widehat{\mathbb{C}} = \mathbb{C} \sqcup \{\infty\}$. We will now explain

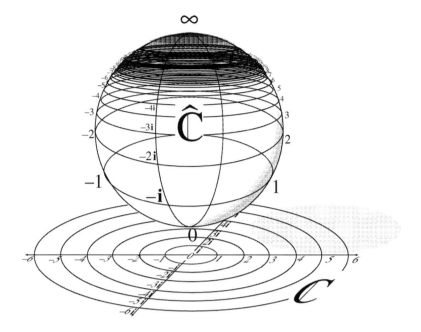

Figure 18G.2. The identification of the complex plane $\mathbb{C}$ with the Riemann sphere $\widehat{\mathbb{C}}$.

how this cute notation is actually quite sensible. The *Riemann sphere* is the topological space $\widehat{\mathbb{C}}$ constructed by taking the complex plane $\mathbb{C}$ and adding a 'point at infinity', denoted by '∞'. An open set $\mathbb{U} \subset \mathbb{C}$ is considered a 'neighbourhood of ∞' if there is some $r > 0$ such that $\mathbb{U}$ contains the codisk $\mathbb{D}^{\complement}(r) := \{c \in \mathbb{C}; \ |c| > r\}$. See Figure 18G.2.

Now, let $\mathbb{U}^* := \mathbb{U} \setminus \{p\}$ and suppose $f : \mathbb{U}^* \longrightarrow \mathbb{C}$ is a continuous function with a singularity at p. Suppose we define $f(p) = \infty$, thereby extending f to a function $f : \mathbb{U} \longrightarrow \widehat{\mathbb{C}}$. If $\lim_{z \to p} |f(p)| = \infty$ (e.g. if p is a pole of f), then this extended function will be *continuous* at p, with respect to the topology of the Riemann sphere. In particular, any meromorphic function $f : \mathbb{U} \longrightarrow \widehat{\mathbb{C}}$ is a continuous mapping from $\mathbb{U}$ into $\widehat{\mathbb{C}}$.

If $f : \mathbb{C} \longrightarrow \widehat{\mathbb{C}}$ is meromorphic, and $L := \lim_{c \to \infty} f(c)$ is well-defined, then we can extend f to a continuous function $f : \widehat{\mathbb{C}} \longrightarrow \widehat{\mathbb{C}}$ by defining $f(\infty) := L$. We can then even define the complex *derivatives* of f at ∞; f effectively becomes a complex-differentiable transformation of the entire Riemann sphere. Many of the ideas in complex analysis are best understood by regarding meromorphic functions in this way.

Remark Not all indelible singularities are poles. Suppose p is a singularity of f, and the Laurent expansion of f at p has an infinite number of negative-power terms (i.e. it looks like the Laurent series (18E.3), p. 460). Then p is called an *essential singularity* of f. The *Casorati–Weierstrass theorem* says that, if $\mathbb{B}$ is any open neighbourhood of p, however tiny, then the image $f[\mathbb{B}]$ is *dense* in $\mathbb{C}$. In other words, the value of $f(z)$ wildly oscillates all over the complex plane infinitely often as $z \to p$. This is a much more pathological behaviour than a pole, where we simply have $f(z) \to \infty$ as $z \to p$.

Ⓔ **Exercise 18G.4** Let $f(z) = \exp(1/z)$.
(a) Show that f has an essential singularity at 0.
(b) Verify the conclusion of the Casorati–Weierstrass theorem for this function. In fact, show that, for any $\epsilon > 0$, if $\mathbb{D}(\epsilon)$ is the disk of radius ϵ around 0, then $f[\mathbb{D}(\epsilon)] = \mathbb{C} \setminus \{0\}$. ◆

Ⓔ **Exercise 18G.5** For each of the following functions, find all poles and compute the residue at each pole. Then use the residue theorem to compute the contour integral along a counterclockwise circle of radius 1.8 around the origin.

(a) $f(z) = \dfrac{1}{z^4 + 1}$.   (*Hint:* $z^4 + 1 = (z - e^{\pi i/4})(z - e^{3\pi i/4})(z - e^{5\pi i/4})(z - e^{7\pi i/4})$.)

(b) $f(z) = \dfrac{z^3 - 1}{z^4 + 5z^2 + 4}$.   (*Hint:* $z^4 + 5z^2 + 4 = (z^2 + 1)(z^2 + 4)$.)

(c) $f(z) = \dfrac{z^4}{z^6 + 14z^4 + 49z^2 + 36}$.   (*Hint:* $z^6 + 14z^4 + 49z^2 + 36 = (z^2 + 1)(z^2 + 4)(z^2 + 9)$.)

(d) $f(z) = \dfrac{z + i}{z^4 + 5z^2 + 4}$.   (*Careful!*)

(e) $f(z) = \tan(z) = \sin(z)/\cos(z)$.

(f) $f(z) = \tanh(z) = \sinh(z)/\cosh(z)$. ◆

Ⓔ **Exercise 18G.6** (For algebraists.)
(a) Let $\mathfrak{H}$ be the set of all holomorphic functions $f : \mathbb{C} \longrightarrow \mathbb{C}$. (These are sometimes called *entire* functions.) Show that $\mathfrak{H}$ is an *integral domain* under the operations of pointwise addition and multiplication. That is: if $f, g \in \mathfrak{H}$, then the functions $(f + g)$ and $(f \cdot g)$ are in $\mathfrak{H}$. (*Hint:* Use Proposition 18A.5(a), (b), p. 425). Also, if $f \neq 0 \neq g$, then $f \cdot g \neq 0$. (*Hint:* Use the identity theorem 18D.3, p. 457.)
(b) Let $\mathfrak{M}$ be the set of all meromorphic functions $f : \mathbb{C} \longrightarrow \widehat{\mathbb{C}}$. Show that $\mathfrak{M}$ is a *field* under the operations of pointwise addition and multiplication. That is: if $f, g \in \mathfrak{M}$, then the functions $(f + g)$ and $(f \cdot g)$ are in $\mathfrak{M}$, and, if $g \neq 0$, then the function (f/g) is also in $\mathfrak{M}$.
(c) Suppose $f \in \mathfrak{M}$ has only a finite number of poles. Show that f can be expressed in the form $f = g/h$, where $g, h \in \mathfrak{H}$. (*Hint:* You can make h a polynomial.)
(d) (Hard) Show that *any* function $f \in \mathfrak{M}$ can be expressed in the form $f = g/h$, where $g, h \in \mathfrak{H}$. (Thus, $\mathfrak{M}$ is related to $\mathfrak{H}$ is the same way that the field of rational functions is related to the ring of polynomials, and in the same way that the field of rational numbers is related to the ring of integers. Technically, $\mathfrak{M}$ is the *field of fractions* of $\mathfrak{H}$.) ◆

18H Improper integrals and Fourier transforms

Prerequisites: §18G.   **Recommended:** §17A, §19A.

The residue theorem is a powerful tool for evaluating contour integrals in the complex plane. We shall now see that it is also useful for computing improper integrals over the real line, such as convolutions and Fourier transforms. First we introduce some notation. Let $\mathbb{C}_+ := \{c \in \mathbb{C}; \ \mathrm{Im}\,[c] > 0\}$ and $\mathbb{C}_- := \{c \in \mathbb{C}; \ \mathrm{Im}\,[c] < 0\}$ be the upper and lower halves of the complex plane. If $F : \mathbb{C} \longrightarrow \widehat{\mathbb{C}}$ is some meromorphic function, then we say that F *uniformly decays at infinity* on $\mathbb{C}_+$ with *order* $o(1/z)$ if,[12] for any $\epsilon > 0$, there is some $r > 0$ such that

$$\text{for all } z \in \mathbb{C}_+, \quad \Big(|z| > r \Big) \Longrightarrow \Big(|z| \cdot |F(z)| < \epsilon \Big). \qquad (18\text{H}.1)$$

In other words, $\lim_{\mathbb{C}_+ \ni z \to \infty} |z| \cdot |F(z)| = 0$, and this convergence is 'uniform' as $z \to \infty$ in any direction in $\mathbb{C}_+$. We define *uniform decay* on $\mathbb{C}_-$ in the same fashion.

Example 18H.1
(a) The function $f(z) = 1/z^2$ uniformly decays at infinity on both $\mathbb{C}_+$ and $\mathbb{C}_-$ with order $o(1/z)$.
(b) The function $f(z) = 1/z$, however, does *not* uniformly decay at infinity with order $o(1/z)$ (it decays just a little bit too slowly).
(c) The function $f(z) = \exp(-\mathbf{i}z)/z^2$ uniformly decays at infinity with order $o(1/z)$ on $\mathbb{C}_+$, but does *not* decay on $\mathbb{C}_-$.
(d) If $P_1, P_2 : \mathbb{C} \longrightarrow \mathbb{C}$ are two complex polynomials of degree N_1 and N_2, respectively, and $N_2 \geq N_1 + 2$, then the rational function $f(z) = P_1(z)/P_2(z)$ uniformly decay with order $o(1/z)$ on both $\mathbb{C}_+$ and $\mathbb{C}_-$.   ◇

Exercise 18H.1   Verify Example 18H.1(a)–(d).   ♦   Ⓔ

Proposition 18H.2 Improper integrals of analytic functions
Let $f : \mathbb{R} \longrightarrow \mathbb{C}$ be an analytic function, and let $F : \mathbb{C} \longrightarrow \widehat{\mathbb{C}}$ be an extension of f to a meromorphic function on $\mathbb{C}$.

(a) *Suppose that F uniformly decays with order $o(1/z)$ on $\mathbb{C}_+$. If $p_1, p_2, \ldots, p_J \in \mathbb{C}_+$ are all the poles of F in $\mathbb{C}_+$, then*

$$\int_{-\infty}^{\infty} f(x)\mathrm{d}x = 2\pi\mathbf{i} \sum_{j=1}^{J} \mathrm{res}_{p_j}(F).$$

---

[12] This is pronounced, 'small oh of $1/z$'.

(b) *Suppose that F uniformly decays with order $o(1/z)$ on $\mathbb{C}_-$. If $p_1, p_2, \ldots, p_J \in \mathbb{C}_-$ are all the poles of F in $\mathbb{C}_-$, then*

$$\int_{-\infty}^{\infty} f(x)\mathrm{d}x = -2\pi\mathbf{i} \sum_{j=1}^{J} \mathrm{res}_{p_j}(F).$$

Proof (a) Note that F has no poles on the real line $\mathbb{R}$, because f is analytic on $\mathbb{R}$. For any $R > 0$, let γ_R be the 'D'-shaped contour of radius R from Example 18C.3, p. 442. If R is made large enough, then γ_R encircles all of $p_1, p_2, \ldots, p_J$. Thus, the residue theorem, Theorem 18G.2, p. 472, says that

$$\oint_{\gamma_R} F = 2\pi\mathbf{i} \sum_{j=1}^{J} \mathrm{res}_{p_j}(F). \tag{18H.2}$$

But, by definition,

$$\oint_{\gamma_R} F = \int_{0}^{\pi+R} F[\gamma_R(s)]\dot{\gamma}_R(s) \underset{(*)}{=} \int_{0}^{\pi} F(Re^{\mathbf{i}s}) \cdot R\mathbf{i}\, e^{\mathbf{i}s}\, \mathrm{d}s + \int_{-R}^{R} f(x)\mathrm{d}x,$$

where $(*)$ is by equations (18C.1) and (18C.2), p. 442. Thus,

$$\lim_{R\to\infty} \oint_{\gamma_R} F = \lim_{R\to\infty} \int_{0}^{\pi} F(Re^{\mathbf{i}s}) \cdot R\mathbf{i}e^{\mathbf{i}s}\, \mathrm{d}s + \lim_{R\to\infty} \int_{-R}^{R} f(x)\mathrm{d}x$$

$$\underset{(*)}{=} \int_{-\infty}^{\infty} f(x)\mathrm{d}x. \tag{18H.3}$$

Now combine equations (18H.2) and (18H.3) to prove (a).

In equation (18H.3), step $(*)$ is because

$$\lim_{R\to\infty} \int_{-R}^{R} f(x)\mathrm{d}x = \int_{-\infty}^{\infty} f(x)\mathrm{d}x, \tag{18H.4}$$

while

$$\lim_{R\to\infty} \left| \int_{0}^{\pi} F(Re^{\mathbf{i}s}) \cdot R\mathbf{i}e^{\mathbf{i}s}\, \mathrm{d}s \right| = 0. \tag{18H.5}$$

Equation (18H.4) is just the definition of an improper integral. To see equation (18H.5), note that

$$\left| \int_{0}^{\pi} F(Re^{\mathbf{i}s})\, R\mathbf{i}e^{\mathbf{i}s}\, \mathrm{d}s \right| \underset{(\triangle)}{\leq} \int_{0}^{\pi} \left| F(Re^{\mathbf{i}s})\, R\mathbf{i}e^{\mathbf{i}s} \right| \mathrm{d}s = \int_{0}^{\pi} R \left| F(Re^{\mathbf{i}s}) \right| \mathrm{d}s,$$

$$\tag{18H.6}$$

where $(\triangle)$ is just the triangle inequality for integrals. But for any $\epsilon > 0$, we can find some $r > 0$ satisfying equation (18H.1). Then, for all $R > r$ and all $s \in [0, \pi]$, we have $R \cdot |F(R e^{is})| < \epsilon$, which means

$$\int_0^\pi R \cdot |F(R e^{is})| \, ds \le \int_0^\pi \epsilon = \pi \epsilon. \tag{18H.7}$$

Since $\epsilon > 0$ can be made arbitrarily small, equations (18H.6) and (18H.7) imply equation (18H.5).

Exercise 18H.2 Prove part (b) of Proposition 18H.2. □ Ⓔ

If $f, g : \mathbb{R} \longrightarrow \mathbb{C}$ are integrable functions, recall that their *convolution* is the function $f * g : \mathbb{R} \longrightarrow \mathbb{R}$ defined by $f * g(r) := \int_{-\infty}^\infty f(x) \, g(r - x) dx$, for any $r \in \mathbb{R}$. Chapter 17 illustrated how to solve I/BVPs by convolving with 'impulse-response' functions like the Poisson kernel.

Corollary 18H.3 Convolutions of analytic functions

Let $f, g : \mathbb{R} \longrightarrow \mathbb{C}$ be analytic functions, with meromorphic extensions $F, G : \mathbb{C} \longrightarrow \widehat{\mathbb{C}}$. Suppose the function $z \mapsto F(z) \cdot G(-z)$ uniformly decays with order $o(1/z)$ on $\mathbb{C}_+$. Suppose F has simple poles $p_1, p_2, \ldots, p_J \in \mathbb{C}_+$, and no other poles in $\mathbb{C}_+$. Suppose G has simple poles $q_1, q_2, \ldots, q_K \in \mathbb{C}_-$, and no other poles in $\mathbb{C}_-$. Then, for all $r \in \mathbb{R}$,

$$f * g(r) = 2\pi \mathbf{i} \sum_{j=1}^J G(r - p_j) \cdot \mathrm{res}_{p_j}(F) - 2\pi \mathbf{i} \sum_{k=1}^K F(r - q_k) \cdot \mathrm{res}_{q_k}(G).$$

Proof Fix $r \in \mathbb{R}$, and consider the function $H(z) := F(z)G(r - z)$. For all $j \in [1 \ldots J]$, Example 18G.1(e), p. 471, says that H has a simple pole at $p_j \in \mathbb{C}_+$, with residue $G(r - p_j) \cdot \mathrm{res}_{p_j}(F)$. For all $k \in [1 \ldots K]$, the function $z \mapsto G(r - z)$ has a simple pole at $r - q_k$, with residue $-\mathrm{res}_{q_k}(G)$. Thus, Example 18G.1(e) says that H has a simple pole at $r - q_k$, with residue $-F(r - q_k) \cdot \mathrm{res}_{q_k}(G)$. Note that $(r - q_k) \in \mathbb{C}_+$, because $q_k \in \mathbb{C}_-$ and $r \in \mathbb{R}$. Now apply Proposition 18H.2. □

Example 18H.4 For any $y > 0$, recall the *half-plane Poisson kernel* $\mathcal{K}_y : \mathbb{R} \longrightarrow \mathbb{R}$ from §17E, defined by

$$\mathcal{K}_y(x) := \frac{y}{\pi(x^2 + y^2)}, \quad \text{for all } x \in \mathbb{R}.$$

Let $\mathbb{H} := \{(x, y) \in \mathbb{R}^2; y \ge 0\}$ (the upper half-plane). If $b : \mathbb{R} \longrightarrow \mathbb{R}$ is bounded and continuous, then Proposition 17E.1, p. 409, says that the function $h(x, y) := \mathcal{K}_y * b(x)$ is the unique continuous harmonic function on $\mathbb{H}$ which satisfies the Dirichlet boundary condition $h(x, 0) = b(x)$ for all $x \in \mathbb{R}$. Suppose $b : \mathbb{R} \longrightarrow \mathbb{R}$ is analytic, with a meromorphic extension $B : \mathbb{C} \longrightarrow \widehat{\mathbb{C}}$ which is *asymptotically bounded* near infinity in $\mathbb{C}_-$ – that is, there exist $K, R > 0$ such that $|B(z)| < K$

for all $z \in \mathbb{C}_-$ with $|z| > R$. Then the function $\mathcal{K}_y \cdot B$ asymptotically decays near infinity with order $o(1/z)$ on $\mathbb{C}_+$, so Corollary 18H.3 is applicable.

In Example 18G.1(c), p. 471, we saw that $\mathcal{K}_y$ has a simple pole at $y\mathbf{i}$, with $\mathrm{res}_{y\mathbf{i}}(\mathcal{K}_y) = 1/2\pi\mathbf{i}$, and no other poles in $\mathbb{C}_+$. Suppose B has simple poles $q_1, q_2, \ldots, q_K \in \mathbb{C}_-$, and no other poles in $\mathbb{C}_-$. Then setting $f := \mathcal{K}_y$, $g := b$, $J := 1$, and $p_1 := y\mathbf{i}$ in Corollary 18H.3, we get, for any $(x, y) \in \mathbb{H}$,

$$h(x, y) = 2\pi\mathbf{i}\, B(x - y\mathbf{i}) \cdot \underbrace{\mathrm{res}_{y\mathbf{i}}(\mathcal{K}_y)}_{=1/2\pi\mathbf{i}} - 2\pi\mathbf{i} \sum_{k=1}^{K} \mathcal{K}_y(x - q_k) \cdot \mathrm{res}_{q_k}(B)$$

$$= B(x - y\mathbf{i}) - 2y\mathbf{i} \sum_{k=1}^{k} \frac{\mathrm{res}_{q_k}(B)}{(x - q_k)^2 + y^2}. \qquad \diamondsuit$$

Ⓔ **Exercise 18H.3**
(a) Show that, in fact, $h(x, y) = \mathrm{Re}\,[B(x - y\mathbf{i})]$. (Thus, if we could compute B, then the BVP would already be solved, and we would not need to apply Proposition 17E.1.)
(b) Deduce that

$$\mathrm{Im}\,[B(x - y\mathbf{i})] = 2y \sum_{k=1}^{K} \frac{\mathrm{res}_{q_k}(B)}{(x - q_k)^2 + y^2}. \qquad \blacklozenge$$

Ⓔ **Exercise 18H.4** Let $f : \mathbb{R} \longrightarrow \mathbb{R}$ be a bounded analytic function, with meromorphic extension $F : \mathbb{C} \longrightarrow \widehat{\mathbb{C}}$. Let $u : \mathbb{R} \times \mathbb{R}_{\not{+}} \longrightarrow \mathbb{R}$ be the unique solution to the one-dimensional heat equation ($\partial_t u = \partial_x^2 u$) with initial conditions $u(x; 0) = f(x)$ for all $x \in \mathbb{R}$. Combine Proposition 18H.3 with Proposition 17C.1, p. 391, to find a formula for $u(x; t)$ in terms of the residues of F. $\qquad \blacklozenge$

Ⓔ **Exercise 18H.5** For any $t \geq 0$, let $\Gamma_t : \mathbb{R} \longrightarrow \mathbb{R}$ be the *d'Alembert kernel*:

$$\Gamma_t(x) = \begin{cases} \frac{1}{2} & \text{if } -t < x < t; \\ 0 & \text{otherwise.} \end{cases}$$

Let $f_1 : \mathbb{R} \longrightarrow \mathbb{R}$ be an analytic function. Lemma 17D.3, p. 401, says that we can solve the initial velocity problem for the one-dimensional wave equation by defining $v(x; t) := \Gamma_t * f_1(x)$. Explain why Proposition 18H.3 is *not* suitable for computing $\Gamma_t * f_1$. $\qquad \blacklozenge$

If $f : \mathbb{R} \longrightarrow \mathbb{C}$ is an integrable function, then its *Fourier transform* is the function $\widehat{f} : \mathbb{R} \longrightarrow \mathbb{C}$ defined by

$$\widehat{f}(\mu) := \frac{1}{2\pi} \int_{-\infty}^{\infty} \exp(-\mu x\mathbf{i}) f(x)\mathrm{d}x, \quad \text{for all } \mu \in \mathbb{R}.$$

(See §19A for more information.) Proposition 18H.2 can also be used to compute Fourier transforms, but it is not quite the strongest result for this purpose. If

$F : \mathbb{C} \longrightarrow \widehat{\mathbb{C}}$ is some meromorphic function, then we say that F *uniformly decays at infinity with order* $\mathcal{O}(1/z)$ if[13] there exist some $M > 0$ and some $r > 0$ such that

$$\text{for all } z \in \mathbb{C}, \quad \Big(|z| > r \Big) \Longrightarrow \Big(|F(z)| < M/|z| \Big). \qquad (18\text{H.8})$$

In other words, the function $|z \cdot F(z)|$ is uniformly bounded (by M) as $z \to \infty$ in any direction in $\mathbb{C}$.

Example 18H.5
(a) The function $f(z) = 1/z$ uniformly decays at infinity with order $\mathcal{O}(1/z)$.
(b) If f uniformly decays at infinity on $\mathbb{C}_{\pm}$ with order $o(1/z)$, then it also uniformly decays at infinity with order $\mathcal{O}(1/z)$. (**Exercise 18H.6** Verify this). Ⓔ
(c) In particular, if $P_1, P_2 : \mathbb{C} \longrightarrow \mathbb{C}$ are two complex polynomials of degree N_1 and N_2, respectively, and $N_2 \geq N_1 + 1$, then the rational function $f(z) = P_1(z)/P_2(z)$ uniformly decays at infinity with order $\mathcal{O}(1/z)$. ◇

Thus, decay with order $\mathcal{O}(1/z)$ is a slightly weaker requirement than decay with order $o(1/z)$.

Proposition 18H.6 Fourier transforms of analytic functions
Let $f : \mathbb{R} \longrightarrow \mathbb{C}$ *be an analytic function. Let* $F : \mathbb{C} \longrightarrow \widehat{\mathbb{C}}$ *be an extension of* f *to a meromorphic function on* $\mathbb{C}$ *which uniformly decays with order* $\mathcal{O}(1/z)$. *Let* $p_{-K}, \ldots, p_{-2}, p_{-1}, p_0, p_1, \ldots, p_J$ *be all the poles of* F *in* $\mathbb{C}$, *where* $p_{-K}, \ldots, p_{-2}, p_{-1} \in \mathbb{C}_-$ *and* $p_0, p_1, \ldots, p_J \in \mathbb{C}_+$. *Then*

$$\widehat{f}(\mu) = \mathbf{i} \sum_{j=0}^{J} \text{res}_{p_j}(\mathcal{E}_{\mu} \cdot F), \quad \text{for all } \mu < 0, \qquad (18\text{H.9})$$

and

$$\widehat{f}(\mu) = -\mathbf{i} \sum_{k=-1}^{-K} \text{res}_{p_k}(\mathcal{E}_{\mu} \cdot F), \quad \text{for all } \mu > 0,$$

where $\mathcal{E}_{\mu} : \mathbb{C} \longrightarrow \mathbb{C}$ *is the holomorphic function defined by* $\mathcal{E}_{\mu}(z) := \exp(-\mu \cdot z \cdot \mathbf{i})$ *for all* $z \in \mathbb{C}$. *In particular, if all the poles of* F *are simple, then*

$$\widehat{f}(\mu) = \mathbf{i} \sum_{j=0}^{J} \exp(-\mu p_j \, \mathbf{i}) \cdot \text{res}_{p_j}(F), \quad \text{for all } \mu < 0, \qquad (18\text{H.10})$$

and

$$\widehat{f}(\mu) = -\mathbf{i} \sum_{k=-1}^{-K} \exp(-\mu p_k \, \mathbf{i}) \cdot \text{res}_{p_k}(F), \quad \text{for all } \mu > 0.$$

---

[13] This is pronounced, 'big oh of $1/z$'.

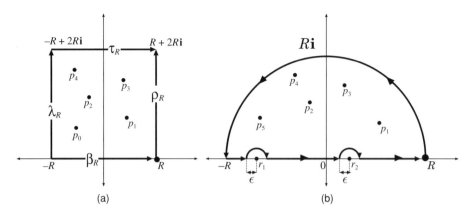

Figure 18H.1. (a) The square contour in the proof of Proposition 18H.6; (b) the contour in Exercise 18H.12, p. 485.

Proof   We will prove Proposition 18H.6 for $\mu < 0$. Fix $\mu < 0$ and define $G :$ $\mathbb{C} \longrightarrow \mathbb{C}$ by $G(z) := \exp(-\mu z \, \mathbf{i}) \cdot F(z)$. For any $R > 0$, define the chains β_R, ρ_R, τ_r, and λ_R as shown in Figure 18H.1(a):

$$
\begin{aligned}
\text{for all } s \in [-R, R], & \quad \beta_R(s) := s & \text{so that} \quad \dot{\beta}_R(s) &= 1; \\
\text{for all } s \in [0, 2R], & \quad \rho_R(s) := R + s\mathbf{i} & \text{so that} \quad \dot{\rho}_R(s) &= \mathbf{i}; \\
\text{for all } s \in [-R, R], & \quad \tau_R(s) := s + 2R\mathbf{i} & \text{so that} \quad \dot{\tau}_R(s) &= 1; \\
\text{for all } s \in [0, 2R], & \quad \lambda_R(s) := -R + s\mathbf{i} & \text{so that} \quad \dot{\lambda}_R(s) &= \mathbf{i}.
\end{aligned}
\tag{18H.11}
$$

(You can use the following mnemonic: βottom, ρight, τop, λeft.) Thus, if $\gamma = \beta \diamond \rho \diamond \overleftarrow{\tau} \diamond \overleftarrow{\lambda}$, then γ_R traces a square in $\mathbb{C}_+$ of sidelength $2R$. If R is made large enough, then γ_R encloses all of $p_0, p_2, \ldots, p_J$. Thus, for any large enough $R > 0$,

$$
2\pi \mathbf{i} \sum_{j=0}^{J} \operatorname{res}_{p_j}(G) \underset{(*)}{=} \oint_{\gamma_R} G \underset{(\dagger)}{=} \oint_{\beta_R} G + \oint_{\rho_R} G - \oint_{\tau_R} G - \oint_{\lambda_R} G, \tag{18H.12}
$$

where $(*)$ is by the residue theorem 18G.2, p. 472, and where $(\dagger)$ is by Lemma 18C.8(a), (b), p. 447.

Claim 1

$$
\lim_{R \to \infty} \oint_{\beta_R} G = 2\pi \, \widehat{f}(\mu).
$$

Ⓔ   *Proof* **Exercise 18H.7.**   ◇ Claim 1

Claim 2

(a)

$$
\lim_{R \to \infty} \oint_{\rho_R} G = 0;
$$

(b)

$$
\lim_{R \to \infty} \oint_{\lambda_R} G = 0.
$$

Proof (a) By hypothesis, f decays with order $\mathcal{O}(1/z)$. Thus, we can find some $r > 0$ and $M > 0$ satisfying equation (18H.8). If $R > r$, then, for all $s \in [0, 2R]$,

$$|G(R + si)| = |\exp[-\mu i(R + si)]| \cdot |F(R + si)|$$

$$\underset{(*)}{\leq} |\exp(\mu s - \mu Ri)| \cdot \frac{M}{|R + si|} \leq e^{\mu s} \cdot \frac{M}{R}, \qquad (18H.13)$$

where $(*)$ is by equation (18H.8). Thus,

$$\left| \oint_{\rho_R} G \right| \underset{(\diamond)}{=} \left| \int_0^{2R} G(R + si)\,i\,ds \right| \leq \int_0^{2R} |G(R + si)|\,ds \underset{(*)}{\leq} \frac{M}{R} \int_0^{2R} e^{\mu s}\,ds$$

$$= \frac{M}{\mu R} e^{\mu s} \Big|_{s=0}^{s=2R} = \frac{M}{-\mu R}(1 - e^{2\mu R}) \underset{(\dagger)}{\leq} \frac{M}{-\mu R} \xrightarrow[R\to\infty]{} 0,$$

as desired. Here, $(\diamond)$ is by equation (18H.11), $(*)$ is by equation (18H.13), and $(\dagger)$ is because $\mu < 0$. This proves (a). The proof of (b) is similar. $\diamond$ Claim 2

Claim 3
$$\lim_{R\to\infty} \oint_{\tau_R} G = 0.$$

Proof Again find $r > 0$ and $M > 0$ satisfying equation (18H.8). If $R > r$, then, for all $s \in [-R, R]$,

$$|G(s + 2Ri)| = |\exp[-\mu i(s + 2Ri)]| \cdot |F(s + 2Ri)|$$

$$\underset{(*)}{\leq} |\exp(2R\mu - s\mu i)| \cdot \frac{M}{|s + 2Ri|}$$

$$\leq e^{2R\mu} \cdot \frac{M}{2R}, \qquad (18H.14)$$

where $(*)$ is by equation (18H.8). Thus,

$$\left| \oint_{\tau_R} G \right| \underset{(*)}{\leq} e^{2R\mu} \cdot \frac{M}{2R} \cdot \text{length}(\tau_R) = e^{2R\mu} \cdot \frac{M}{2R} \cdot 2R = M e^{2R\mu} \underset{(\dagger)}{\xrightarrow[R\to\infty]{}} 0,$$

as desired. Here, $(*)$ is by equation (18H.14) and Lemma 18C.10, p. 450, while $(\dagger)$ is because $\mu < 0$. $\diamond$ Claim 3

Now we put it all together:

$$2\pi i \sum_{j=0}^{J} \text{res}_{p_j}(G) \underset{(*)}{=} \lim_{R\to\infty} \left(\oint_{\beta_R} G + \oint_{\rho_R} G - \oint_{\tau_R} G - \oint_{\lambda_R} G \right)$$

$$\underset{(\dagger)}{=} 2\pi \,\widehat{f}(\mu) + 0 + 0 + 0 = 2\pi \,\widehat{f}(\mu).$$

Now divide both sides, by 2π to get equation (18H.9). Here, $(*)$ is by equation (18H.12), and $(\dagger)$ is by Claims 1–3.

Finally, to see equation (18H.10), suppose all the poles $p_0, \ldots, p_J$ are simple. Then $\text{res}_{p_j}(G) = \exp(-i\mu p_j) \cdot \text{res}_{p_j}(F)$ for all $j \in [0 \ldots J]$, by Example 18G.1(e),

p. 471. Thus,

$$\mathbf{i}\sum_{j=0}^{J}\mathrm{res}_{p_j}(G) = \mathbf{i}\sum_{j=0}^{J}\exp(-\mu p_j\,\mathbf{i})\cdot\mathrm{res}_{p_j}(F),$$

so equation (18H.10) follows from equation (18H.9).   ☐

Ⓔ **Exercise 18H.8** Prove Proposition 18H.6 in the case $\mu > 0$.   ◆

Example 18H.7 Define $f : \mathbb{R} \longrightarrow \mathbb{R}$ by $f(x) := 1/(x^2 + 1)$ for all $x \in \mathbb{R}$. The meromorphic extension of f is simply the complex polynomial $F : \mathbb{C} \longrightarrow \hat{\mathbb{C}}$ defined:

$$F(z) := \frac{1}{z^2 + 1} = \frac{1}{(z + \mathbf{i})(z - \mathbf{i})}, \quad \text{for all } z \in \mathbb{C}.$$

Clearly, F has simple poles at $\pm\mathbf{i}$, with $\mathrm{res}_{\mathbf{i}}(F) = 1/2\mathbf{i}$ and $\mathrm{res}_{-\mathbf{i}}(F) = -1/2\mathbf{i}$. Thus, Proposition 18H.6 says

$$\text{if } \mu < 0, \text{ then } \widehat{f}(\mu) = \mathbf{i}\exp(-\mu\mathbf{i}\cdot\mathbf{i})\frac{1}{2\mathbf{i}} = \frac{e^\mu}{2} = \frac{e^{-|\mu|}}{2};$$

$$\text{if } \mu > 0, \text{ then } \widehat{f}(\mu) = -\mathbf{i}\exp(-\mu\mathbf{i}\cdot(-\mathbf{i}))\frac{-1}{2\mathbf{i}} = \frac{e^{-\mu}}{2} = \frac{e^{-|\mu|}}{2}.$$

We conclude that $\widehat{g}(\mu) = e^{-|\mu|}/2$ for all $\mu \in \mathbb{R}$.   ◇

Ⓔ **Exercise 18H.9** Compute the Fourier transforms of the following rational functions:

(a) $f(x) = \dfrac{1}{x^4 + 1}$.   (*Hint:* $x^4 + 1 = (x - e^{\pi\mathbf{i}/4})(x - e^{3\pi\mathbf{i}/4})(x - e^{5\pi\mathbf{i}/4})(x - e^{7\pi\mathbf{i}/4})$.)

(b) $f(x) = \dfrac{x^3 - 1}{x^4 + 5x^2 + 4}$.   (*Hint:* $x^4 + 5x^2 + 4 = (x^2 + 1)(x^2 + 4)$.)

(c) $f(x) = \dfrac{x^4}{x^6 + 14x^4 + 49x^2 + 36}$.   (*Hint:* $x^6 + 14x^4 + 49x^2 + 36 = (x^2 + 1)(x^2 + 4)(x^2 + 9)$.)

(d) $f(x) = \dfrac{x + \mathbf{i}}{x^4 + 5x^2 + 4}$.   (*Careful!*)   ◆

Ⓔ **Exercise 18H.10** Why is Proposition 18H.6 *not* suitable to compute the Fourier transforms of the following functions?

(a) $f(x) = \dfrac{1}{x^3 + 1}$;

(b) $f(x) = \dfrac{\sin(x)}{x^4 + 1}$;

(c) $f(x) = \dfrac{1}{|x|^3 + 1}$;

(d) $f(x) = \dfrac{\sqrt[3]{x}}{x^3 + 1}$.   ◆

Exercise 18H.11   Let $f : \mathbb{R} \longrightarrow \mathbb{R}$ be an analytic function whose meromorphic ⒠ extension $F : \mathbb{C} \longrightarrow \widehat{\mathbb{C}}$ decays with order $\mathcal{O}(1/z)$.

(a) State and prove a general formula for 'trigonometric integrals' of the form

$$\int_{-\infty}^{\infty} \cos(nx)\, f(x)\mathrm{d}x \quad \text{and} \quad \int_{-\infty}^{\infty} \sin(nx)\, f(x)\mathrm{d}x.$$

(*Hint:* Use Proposition 18H.6 and the formula $\exp(\mu x) = \cos(\mu x) + \mathbf{i}\sin(\mu x)$.)

Use your method to compute the following integrals:

(b) $\displaystyle\int_{-\infty}^{\infty} \frac{\sin(x)}{x^2 + 1}\, \mathrm{d}x.$

(c) $\displaystyle\int_{-\infty}^{\infty} \frac{\cos(x)}{x^4 + 1}\, \mathrm{d}x.$

(d) $\displaystyle\int_{-\infty}^{\infty} \frac{\sin(x)^2}{x^2 + 1}\, \mathrm{d}x.$ (*Hint:* $2\sin(x)^2 = 1 - \cos(2x)$.)   ◆

Exercise 18H.12   Proposition 18H.2 requires the function f to have no poles on ⒠ the real line $\mathbb{R}$. However, this is not really necessary.

(a) Let $\mathcal{R} := \{r_1, \ldots, r_N\} \subset \mathbb{R}$. Let $f : \mathbb{R} \setminus \mathcal{R} \longrightarrow \mathbb{R}$ be an analytic function whose mero- morphic extension $F : \mathbb{C} \longrightarrow \widehat{\mathbb{C}}$ decays with order $o(1/z)$ on $\mathbb{C}_+$, and has poles $p_1, \ldots, p_J \in \mathbb{C}_+$, and also has simple poles at $r_1, \ldots, r_N$. Show that

$$\int_{-\infty}^{\infty} f(x)\mathrm{d}x = 2\pi\mathbf{i} \sum_{j=1}^{J} \operatorname{res}_{p_j}(F) + \pi\mathbf{i} \sum_{n=1}^{N} \operatorname{res}_{r_n}(F).$$

(*Hint:* For all $\epsilon > 0$ and $R > 0$, let $\gamma_{R,\epsilon}$ be the contour shown in Figure 18H.1(b), p. 482. This is like the 'D' contour in the proof of Proposition 18H.2, except that it makes a little semicircular 'detour' of radius ϵ around each of the poles $r_1, \ldots, r_N \in \mathbb{R}$. Show that the integral along each of these ϵ-detours tends to $-\pi\mathbf{i} \cdot \operatorname{res}_{r_n}(F)$ as $\epsilon \to 0$, while the integral over the remainder of the real line tends to $\int_{-\infty}^{\infty} f(x)\mathrm{d}x$ as $\epsilon \to 0$ and $R \to \infty$.

(b) Use your method to compute

$$\int_{-\infty}^{\infty} \frac{\exp(\mathbf{i}\mu x)}{x^2}\, \mathrm{d}x.$$ ◆

Exercise 18H.13   ⒠
(a) Let $\mathcal{R} := \{r_1, \ldots, r_N\} \subset \mathbb{R}$. Let $f : \mathbb{R} \setminus \mathcal{R} \longrightarrow \mathbb{R}$ be an analytic function whose mero- morphic extension $F : \mathbb{C} \longrightarrow \widehat{\mathbb{C}}$ decays with order $\mathcal{O}(1/z)$ and has poles $p_1, \ldots, p_J \in \mathbb{C}_+$ and also has a simple poles at $r_1, \ldots, r_N$. Show that, for any $\mu < 0$,

$$\widehat{f}(\mu) = \mathbf{i} \sum_{j=1}^{J} \exp(-\mu p_j \mathbf{i}) \cdot \operatorname{res}_{p_j}(F) + \frac{\mathbf{i}}{2} \sum_{n=1}^{N} \exp(-\mu r_j \mathbf{i}) \cdot \operatorname{res}_{r_n}(F).$$

(If $\mu > 0$, it is a similar formula, only summing over the residues in $\mathbb{C}_-$ and multiplying by -1.) (*Hint:* Combine the method from Exercise 18H.12 with the proof technique from Proposition 18H.6.)

(b) Use your method to compute $\widehat{f}(\mu)$ when $f(x) = 1/x(x^2 + 1)$. ♦

Ⓔ **Exercise 18H.14** The *Laplace inversion integral* is defined by equation (19H.3), p. 522. State and prove a formula similar to Proposition 18H.6 for the computation of Laplace inversion integrals. ♦

18I* Homological extension of Cauchy's theorem

Prerequisites: §18C.

We have defined 'contours' to be non-self-intersecting curves only so as to simplify the exposition in Section 18C.[14] All of the results of Section 18C are true for any piecewise smooth closed curve in $\mathbb{C}$. Indeed, the results of Section 18C can even be extended to integrals on *chains,* as we now discuss.

Let $\mathbb{U} \subseteq \mathbb{C}$ be a connected open set, and let $\mathbb{G}_1, \mathbb{G}_2, \ldots, \mathbb{G}_N$ be the connected components of the boundary $\partial\mathbb{U}$. Suppose each $\mathbb{G}_n$ can be parameterized by a piecewise smooth contour γ_n, such that the outward normal vector field of $\mathbb{G}_n$ is always on the right-hand side of $\dot{\gamma}_n$. The chain $\gamma := \gamma_1 \diamond \gamma_2 \diamond \cdots \diamond \gamma_n$ is called the *positive boundary* of $\mathbb{U}$. Its reversal, $\overleftarrow{\gamma}$, is called the *negative boundary* of $\mathbb{U}$. Both the negative and positive boundaries of a set are called *oriented boundaries*. For example, any contour γ is the oriented boundary of the purview of γ. Theorem 18C.5, p. 444, now extends to the following theorem.

Theorem 18I.1 Cauchy's theorem on oriented boundaries
Let $\mathbb{U} \subseteq \mathbb{C}$ be any open set, and let $f : \mathbb{U} \longrightarrow \mathbb{C}$ be holomorphic on $\mathbb{U}$. If α is an oriented boundary of $\mathbb{U}$, then

$$\oint_\alpha f = 0.$$ □

Let $\alpha_1, \alpha_2, \ldots, \alpha_N : [0, 1] \longrightarrow \mathbb{C}$ be continuous, piecewise smooth curves in $\mathbb{C}$ (not necessarily closed), and consider the chain $\alpha = \alpha_1 \diamond \alpha_2 \diamond \cdots \alpha_N$ (note that any chain can be expressed in this way). Let us refer to the paths $\alpha_1, \ldots, \alpha_N$ as the 'links' of the chain α. We say that α is a *cycle* if the endpoint of each link is the starting point of exactly one other link, and the starting point of each link is the endpoint of exactly one other link. In other words, for all $m \in [1 \ldots N]$, there exists a unique $\ell, n \in [1 \ldots C]$ such that $\alpha_\ell(1) = \alpha_m(0)$ and $\alpha_m(1) = \alpha_n(0)$.

---

[14] To be precise, it made it simpler for us to define the 'purview' of the contour by invoking the Jordan curve theorem. It also made it simpler to define 'clockwise' versus 'counterclockwise' contours.

Example 18I.2
(a) Any contour is a cycle.
(b) If α and β are two cycles, then $\alpha \diamond \beta$ is also a cycle.
(c) Thus, if $\gamma_1, \ldots, \gamma_N$ are contours, then $\gamma_1 \diamond \cdots \diamond \gamma_N$ is a cycle.
(d) In particular, the oriented boundary of an open set is a cycle.
(e) If α is a cycle, then $\overleftarrow{\alpha}$ is a cycle. $\diamondsuit$

Not all cycles are oriented boundaries. For example, let γ_1 and γ_2 be two concentric counterclockwise circles around the origin; then $\gamma_1 \diamond \gamma_2$ is *not* an oriented boundary (although $\gamma_1 \diamond \overleftarrow{\gamma_2}$ is).

Let $\mathbb{U} \subseteq \mathbb{C}$ be an open set. Let α and β be two cycles in $\mathbb{U}$. We say that α is *homologous* to β in $\mathbb{U}$ if the cycle $\alpha \diamond \overleftarrow{\beta}$ is the oriented boundary of some open subset $\mathbb{V} \subseteq \mathbb{U}$. We then write '$\alpha \underset{\mathbb{U}}{\sim} \beta$'.

Example 18I.3
(a) Let α be a clockwise circle of radius 1 around the origin, and let β be a clockwise circle of radius 2 around the origin. Then α is homologous to β in $\mathbb{C}^*$, because $\alpha \diamond \overleftarrow{\beta}$ is the positive boundary of the annulus $\mathbb{A} := \{c \in \mathbb{C}; \ 1 < |c| < 2\} \subseteq \mathbb{C}^*$.
(b) If γ_0 and γ_1 are contours, then they are cycles. If γ_0 is homotopic to γ_1 in $\mathbb{U}$, then γ_0 is also *homologous* to γ_1 in $\mathbb{U}$. To see this, let $\Gamma : [0, 1] \times [0, S] \longrightarrow \mathbb{C}$ be a homotopy from γ_0 to γ_1, and let $\mathbb{V} := \Gamma((0, 1) \times [0, S])$. Then $\mathbb{V}$ is an open subset of $\mathbb{U}$, and $\gamma_1 \diamond \overleftarrow{\gamma_2}$ is an oriented boundary of $\mathbb{V}$. $\diamondsuit$

Thus, homology can be seen as a generalization of homotopy. Proposition 18C.7, p. 446, can be extended as follows.

Proposition 18I.4 Homology invariance of chain integrals
Let $\mathbb{U} \subseteq \mathbb{C}$ be any open set, and let $f : \mathbb{U} \longrightarrow \mathbb{C}$ be holomorphic on $\mathbb{U}$. If α and β are two cycles which are homologous in $\mathbb{U}$, then

$$\oint_\alpha f = \oint_\beta f.$$

□

Proof **Exercise 18I.1.** (*Hint:* Use Theorem 18I.1.) □ Ⓔ

The relation '$\underset{\mathbb{U}}{\sim}$' is an *equivalence relation*. That is, for all cycles α, β, and γ,

- $\alpha \underset{\mathbb{U}}{\sim} \alpha$;
- if $\alpha \underset{\mathbb{U}}{\sim} \beta$, then $\beta \underset{\mathbb{U}}{\sim} \alpha$;
- if $\alpha \underset{\mathbb{U}}{\sim} \beta$, and $\beta \underset{\mathbb{U}}{\sim} \gamma$, then $\alpha \underset{\mathbb{U}}{\sim} \gamma$.

Exercise 18I.2 Verify these three properties. ◆ Ⓔ

For any cycle α, let $[\alpha]_{\mathbb{U}}$ denote its equivalence class under '$\underset{\mathbb{U}}{\sim}$' (this is called a *homology class*). Let $\mathcal{H}^1(\mathbb{U})$ denote the set of all homology classes of cycles. In particular, let $[\emptyset]_{\mathbb{U}}$ denote the homology class of the empty cycle – then $[\emptyset]_{\mathbb{U}}$ contains all cycles which are oriented boundaries of subsets of $\mathbb{U}$.

Corollary 18I.5 *Let* $\mathbb{U} \subseteq \mathbb{C}$ *be any open set.*

(a) *If* $\alpha_1 \underset{\mathbb{U}}{\sim} \alpha_2$ *and* $\beta_1 \underset{\mathbb{U}}{\sim} \beta_2$, *then* $(\alpha_1 \diamond \beta_1) \underset{\mathbb{U}}{\sim} (\alpha_2 \diamond \beta_2)$. *Thus, we can define an operation* $\oplus$ *on* $\mathcal{H}^1(\mathbb{U})$ *by* $[\alpha]_{\mathbb{U}} \oplus [\beta]_{\mathbb{U}} := [\alpha \diamond \beta]_{\mathbb{U}}$.

(b) $\mathcal{H}^1(\mathbb{U})$ *is an abelian group under the operation* $\oplus$.

(c) *If* $f : \mathbb{U} \longrightarrow \mathbb{C}$ *is holomorphic, then the function*

$$[\alpha]_{\mathbb{U}} \mapsto \oint_{\alpha} f$$

is a group homomorphism from $(\mathcal{H}^1(\mathbb{U}), \oplus)$ *to the group* $(\mathbb{C}, +)$ *of complex numbers under addition.*

Ⓔ *Proof* (a) is **Exercise 18I.3.** Verify the following.

(i) The operation $\oplus$ is *commutative*. That is, for any cycles α and β, we have $\alpha \diamond \beta \underset{\mathbb{U}}{\sim} \beta \diamond \alpha$; thus, $[\alpha]_{\mathbb{U}} \oplus [\beta]_{\mathbb{U}} = [\beta]_{\mathbb{U}} \oplus [\alpha]_{\mathbb{U}}$.

(ii) The operation $\oplus$ is *associative*. That is, for any cycles α, β, and γ, we have $\alpha \diamond (\beta \diamond \gamma) \underset{\mathbb{U}}{\sim} (\alpha \diamond \beta) \diamond \gamma$; thus, $[\alpha]_{\mathbb{U}} \oplus ([\beta]_{\mathbb{U}} \oplus [\gamma]_{\mathbb{U}}) = ([\alpha]_{\mathbb{U}} \oplus [\beta]_{\mathbb{U}}) \oplus [\gamma]_{\mathbb{U}}$.

(iii) The cycle $[\emptyset]_{\mathbb{U}}$ is an *identity element*. For any cycle α, we have $[\alpha]_{\mathbb{U}} \oplus [\emptyset]_{\mathbb{U}} = [\alpha]_{\mathbb{U}}$.

(iv) For any cycle α, the class $[\overleftarrow{\alpha}]_{\mathbb{U}}$ is an *additive inverse* for $[\alpha]_{\mathbb{U}}$. That is, $[\alpha]_{\mathbb{U}} \oplus [\overleftarrow{\alpha}]_{\mathbb{U}} = [\emptyset]_{\mathbb{U}}$.

(b) follows immediately from (a), and (c) follows from Proposition 18I.4 and Lemma 18C.8(a), (b), p. 441. ☐

The group $\mathcal{H}^1(\mathbb{U})$ is called the *first homology group* of $\mathbb{U}$. In general, $\mathcal{H}^1(\mathbb{U})$ is a free abelian group of rank R, where R is the number of 'holes' in $\mathbb{U}$. One can similarly define homology groups for any subset $\mathbb{U} \subseteq \mathbb{R}^N$ for any $N \in \mathbb{N}$ (e.g. a surface or a manifold), or even for more abstract spaces. The algebraic properties of the homology groups of $\mathbb{U}$ encode the 'large-scale' topological properties of $\mathbb{U}$ (e.g. the presence of 'holes' or 'twists'). The study of homology groups is one aspect of a vast and beautiful area in mathematics called *algebraic topology*. Surprisingly, the algebraic topology of a differentiable manifold indirectly influences the behaviour of partial differential equations defined on this manifold; this is content of deep results such as the *Atiyah–Singer index theorem*. For an elementary introduction to algebraic topology, see Henle (1994). For a comprehensive text, see the beautiful book by Hatcher (2002).

Part VI

Fourier transforms on unbounded domains

In Part III, we saw that trigonometric functions such as sin and cos formed orthogonal bases of $\mathbf{L}^2(\mathbb{X})$, where $\mathbb{X}$ was one of several bounded subsets of $\mathbb{R}^D$. Thus, any function in $\mathbf{L}^2(\mathbb{X})$ could be expressed using a *Fourier series*. In Part IV we used these Fourier series to solve initial/boundary value problems (I/BVPs) on $\mathbb{X}$.

A *Fourier transform* is similar to a Fourier series, except that now $\mathbb{X}$ is an *unbounded* set (e.g. $\mathbb{X} = \mathbb{R}$ or $\mathbb{R}^D$). This introduces considerable technical complications. Nevertheless, the underlying philosophy is the same; we will construct something analogous to an orthogonal basis for $\mathbf{L}^2(\mathbb{X})$, and use this to solve partial differential equations on $\mathbb{X}$.

It is technically convenient (although not strictly necessary) to replace sin and cos with the complex exponential functions such as $\exp(x\mathbf{i}) = \cos(x) + \mathbf{i}\sin(x)$. The material on Fourier series in Part III could have also been developed using these complex exponentials, but, in that context, this would have been a needless complication. In the context of Fourier transforms, however, it is actually a simplification.

19

Fourier transforms

There is no branch of mathematics, however abstract, which may not
someday be applied to the phenomena of the real world.

Nicolai Lobachevsky

19A One-dimensional Fourier transforms

Prerequisites: Appendix C.   **Recommended:** §6C(i), §8D.

Fourier *series* help us to represent functions on a *bounded* domain, like
$\mathbb{X} = [0, 1]$ or $\mathbb{X} = [0, 1] \times [0, 1]$. But what if the domain is *unbounded*, like
$\mathbb{X} = \mathbb{R}$? Now, instead of using a *discrete* collection of Fourier coefficients like
$\{A_0, A_1, B_1, A_2, B_2, \ldots\}$ or $\{\widehat{f}_{-1}, \widehat{f}_0, \widehat{f}_1, \widehat{f}_2, \ldots\}$, we must use a continuously
parameterized family.

For every $\mu \in \mathbb{R}$, we define the function $\mathcal{E}_\mu : \mathbb{R} \longrightarrow \mathbb{C}$ by $\mathcal{E}_\mu(x) := \exp(\mu \mathbf{i} x)$.
You can visualize this function as a helix which spirals with frequency μ around the
unit circle in the complex plane (see Figure 19A.1). Indeed, using Euler's formula
(see p. 553), it is not hard to check that $\mathcal{E}_\mu(x) = \cos(\mu x) + \mathbf{i} \sin(\mu x)$ (**Exercise
19A.1**). In other words, the real and imaginary parts of $\mathcal{E}_\mu(x)$ act like a cosine wave   Ⓔ
and a sine wave, respectively, both of frequency μ.

Heuristically speaking, the (continuously parameterized) family of functions
$\{\mathcal{E}_\mu\}_{\mu \in \mathbb{R}}$ acts as a kind of 'orthogonal basis' for a certain space of functions from $\mathbb{R}$
into $\mathbb{C}$ (although making this rigorous is very complicated). This is the motivating
idea behind the Fourier transform.

Let $f : \mathbb{R} \longrightarrow \mathbb{C}$ be some function. The *Fourier transform* of f is the function
$\widehat{f} : \mathbb{R} \longrightarrow \mathbb{C}$ defined:

$$\widehat{f}(\mu) := \frac{1}{2\pi} \int_{-\infty}^{\infty} f(x)\overline{\mathcal{E}_\mu(x)}\mathrm{d}x \quad = \boxed{\frac{1}{2\pi} \int_{-\infty}^{\infty} f(x) \cdot \exp(-\mu \cdot x \cdot \mathbf{i})\mathrm{d}x,}$$

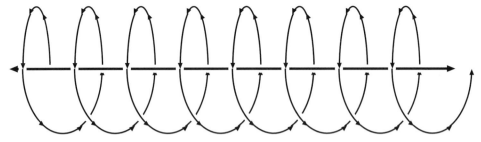

Figure 19A.1. $\mathcal{E}_\mu(x) := \exp(-\mu \cdot x \cdot \mathbf{i})$ as a function of x.

for any $\mu \in \mathbb{R}$. (In other words, $\widehat{f}(\mu) := (1/2\pi)\langle f, \mathcal{E}_\mu \rangle$, in the notation of §6C(i), p. 113.) Note that this integral may not converge in general. We need $f(x)$ to 'decay fast enough' as x goes to $\pm\infty$. To be precise, we need f to be an *absolutely integrable* function, meaning that

$$\int_{-\infty}^{\infty} |f(x)|\, \mathrm{d}x < \infty.$$

We indicate this by writing '$f \in \mathbf{L}^1(\mathbb{R})$'.

The Fourier transform $\widehat{f}(\mu)$ plays the same role that the complex Fourier coefficients $\{\ldots \widehat{f}_{-1}, \widehat{f}_0, \widehat{f}_1, \widehat{f}_2, \ldots\}$ play for a function on an interval (see §8D, p. 175). In particular, we can express $f(x)$ as a sort of generalized 'Fourier series'. We would like to write something like:

$$f(x) = \sum_{\mu \in \mathbb{R}} \widehat{f}(\mu)\mathcal{E}_\mu(x).$$

However, this expression makes no mathematical sense, because you cannot *sum* over all real numbers (there are too many). Instead of *summing* over all Fourier coefficients, we must *integrate*. For this to work, we need a technical condition. We say that f is *piecewise smooth* if there is a finite set of points $r_1 < r_2 < \cdots < r_N$ in $\mathbb{R}$ such that f is continuously differentiable on the open intervals $(-\infty, r_1)$, (r_1, r_2), (r_1, r_2), $\ldots$, (r_{N-1}, r_N), and (r_N, ∞), and, furthermore, the left-hand and right-hand limits[1] of f and f' exist at each of the points $r_1, \ldots, r_N$.

Theorem 19A.1 Fourier inversion formula

Suppose that $f \in \mathbf{L}^1(\mathbb{R})$ is piecewise smooth. For any $x \in \mathbb{R}$, if f is continuous at x, then

$$f(x) = \lim_{M \to \infty} \int_{-M}^{M} \widehat{f}(\mu) \cdot \mathcal{E}_\mu(x)\mathrm{d}\mu = \lim_{M \to \infty} \int_{-M}^{M} \widehat{f}(\mu) \cdot \exp(\mu \cdot x \cdot \mathbf{i})\mathrm{d}\mu.$$

$$(19A.1)$$

---

[1] See p. 202 for a definition.

If f is discontinuous at x, then we have

$$\lim_{M \to \infty} \int_{-M}^{M} \widehat{f}(\mu) \cdot \mathcal{E}_\mu(x) d\mu = \frac{1}{2} \left(\lim_{y \searrow x} f(y) + \lim_{y \nearrow x} f(y) \right).$$

Proof See Körner (1988), Theorem 61.1; Walker (1988), Theorem 5.17; or Fisher (1999), §5.2. □

It follows that, under mild conditions, a function can be uniquely identified from its Fourier transform.

Proposition 19A.2 *Suppose $f, g \in \mathcal{C}(\mathbb{R}) \cap \mathbf{L}^1(\mathbb{R})$ are continuous and integrable. Then $\left(\widehat{f} = \widehat{g} \right) \Longleftrightarrow (f = g)$.*

Proof $\Longleftarrow$ is obvious. The proof of $\Longrightarrow$ is **Exercise 19A.2**. (*Hints:* (a) If f and g are piecewise smooth, then show that this follows immediately from Theorem 19A.1. (b) In the general case (where f and g might not be piecewise smooth), proceed as follows. Let $h \in \mathcal{C}(\mathbb{R}) \cap \mathbf{L}^1(\mathbb{R})$. Suppose $\widehat{h} \equiv 0$; show that we must have $h \equiv 0$. Now let $h := f - g$; then $\widehat{h} = \widehat{f} - \widehat{g} \equiv 0$ (because $\widehat{f} = \widehat{g}$). Thus $h \equiv 0$; thus, $f = g$.) □

Example 19A.3 Suppose

$$f(x) = \begin{cases} 1 & \text{if } -1 < x < 1; \\ 0 & \text{otherwise} \end{cases}$$

(see Figure 19A.2(a)). Then, for all $\mu \in \mathbb{R}$,

$$\widehat{f}(\mu) = \frac{1}{2\pi} \int_{-\infty}^{\infty} f(x) \exp(-\mu \cdot x \cdot \mathbf{i}) dx = \frac{1}{2\pi} \int_{-1}^{1} \exp(-\mu \cdot x \cdot \mathbf{i}) dx$$

$$= \frac{1}{-2\pi\mu\mathbf{i}} \exp\left(-\mu \cdot x \cdot \mathbf{i} \right) \Big|_{x=-1}^{x=1} = \frac{1}{-2\pi\mu\mathbf{i}} \left(e^{-\mu\mathbf{i}} - e^{\mu\mathbf{i}} \right)$$

$$= \frac{1}{\pi\mu} \left(\frac{e^{\mu\mathbf{i}} - e^{-\mu\mathbf{i}}}{2\mathbf{i}} \right) \underset{\text{(Eu)}}{=} \frac{1}{\pi\mu} \sin(\mu)$$

(see Figure 19A.2(b)), where (Eu) is Euler's formula (p. 553). Thus, the Fourier inversion formula says that, if $-1 < x < 1$, then

$$\lim_{M \to \infty} \int_{-M}^{M} \frac{\sin(\mu)}{\pi\mu} \exp(\mu \cdot x \cdot \mathbf{i}) d\mu = 1,$$

whereas, if $x < -1$ or $x > 1$, then

$$\lim_{M \to \infty} \int_{-M}^{M} \frac{\sin(\mu)}{\pi\mu} \exp(\mu \cdot x \cdot \mathbf{i}) d\mu = 0.$$

If $x = \pm 1$, then the Fourier inversion integral will converge to $1/2$. ◇

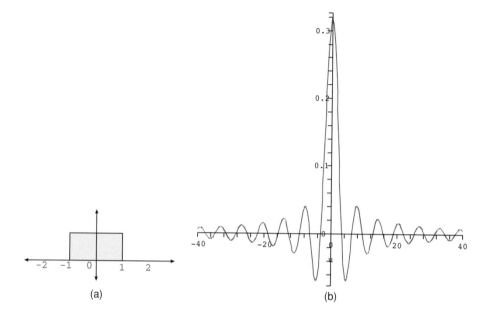

Figure 19A.2. (a) Function from Example 19A.3. (b) The Fourier transform $\widehat{f}(x) = \sin(\mu)/\pi\mu$ from Example 19A.3.

Example 19A.4 Suppose

$$f(x) = \begin{cases} 1 & \text{if } 0 < x < 1; \\ 0 & \text{otherwise} \end{cases}$$

(see Figure 19A.3(a)). Then $\widehat{f}(\mu) = \frac{1-e^{-\mu i}}{2\pi\mu i}$ (see Figure 19A.3(b)); the verification of this is Problem 19.1. Thus, the Fourier inversion formula says that, if $0 < x < 1$, then

$$\lim_{M\to\infty} \int_{-M}^{M} \frac{1-e^{-\mu i}}{2\pi\mu i} \exp(\mu \cdot x \cdot i)d\mu = 1,$$

whereas, if $x < 0$ or $x > 1$, then

$$\lim_{M\to\infty} \int_{-M}^{M} \frac{1-e^{-\mu i}}{2\pi\mu i} \exp(\mu \cdot x \cdot i)d\mu = 0.$$

If $x = 0$ or $x = 1$, then the Fourier inversion integral will converge to $1/2$. ◇

In the Fourier inversion formula, it is important that the positive and negative bounds of the integral go to infinity at the same rate in the limit (19A.1). In

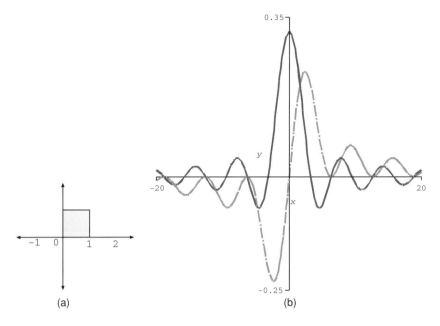

Figure 19A.3. (a) Function for Example 19A.4. (b) The real and imaginary parts of the Fourier transform $\widehat{f}(x) = (1 - e^{-\mu i})/2\pi \mu i$ from Example 19A.4.

particular, it is *not* the case that

$$f(x) = \lim_{N,M \to \infty} \int_{-N}^{M} \widehat{f}(\mu) \exp(\mu \cdot x \cdot \mathbf{i}) d\mu;$$

in general, *this* integral may not converge. The reason is this: even if f is absolutely integrable, its Fourier transform $\widehat{f}$ may *not* be. If we assume that $\widehat{f}$ is *also* absolutely integrable, then things become easier.

Theorem 19A.5 Strong Fourier inversion formula
Suppose that $f \in \mathbf{L}^1(\mathbb{R})$, and that $\widehat{f}$ is also in $\mathbf{L}^1(\mathbb{R})$. If $x \in \mathbb{R}$, and f is continuous at x, then

$$f(x) = \int_{-\infty}^{\infty} \widehat{f}(\mu) \cdot \exp(\mu \cdot x \cdot \mathbf{i}) d\mu.$$

Proof See Katznelson (1976), §VI.1.12; Folland (1984), Theorem 8.26; Körner (1988), Theorem 60.1; or Walker (1988), Theorem 4.11. $\qquad\square$

Corollary 19A.6 *Suppose $f \in \mathbf{L}^1(\mathbb{R})$, and there exists some $g \in \mathbf{L}^1(\mathbb{R})$ such that $f = \widehat{g}$. Then $\widehat{f}(\mu) = (1/2\pi)g(-\mu)$ for all $\mu \in \mathbb{R}$.*

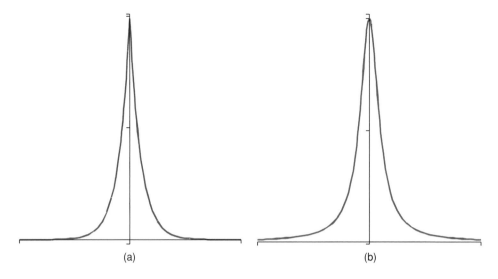

Figure 19A.4. (a) The symmetric exponential tail function $f(x) = e^{-\alpha \cdot |x|}$ from Example 19A.7. (b) The Fourier transform $\widehat{f}(x) = a/\pi(x^2 + a^2)$ of the symmetric exponential tail function from Example 19A.7.

Ⓔ   *Proof* **Exercise 19A.3.**   □

Example 19A.7   Let $\alpha > 0$ be a constant, and suppose $f(x) = e^{-\alpha \cdot |x|}$ (see Figure 19A.4(a)). Then

$$2\pi \widehat{f}(\mu) = \int_{-\infty}^{\infty} e^{-\alpha \cdot |x|} \exp(-\mu x \mathbf{i}) dx$$

$$= \int_{0}^{\infty} e^{-\alpha \cdot x} \exp(-\mu x \mathbf{i}) dx + \int_{-\infty}^{0} e^{\alpha \cdot x} \exp(-\mu x \mathbf{i}) dx$$

$$= \int_{0}^{\infty} \exp(-\alpha x - \mu x \mathbf{i}) dx + \int_{-\infty}^{0} \exp(\alpha x - \mu x \mathbf{i}) dx$$

$$= \frac{1}{-(\alpha + \mu \mathbf{i})} \exp\left(-(\alpha + \mu \mathbf{i}) \cdot x\right)\Big|_{x=0}^{x=\infty} + \frac{1}{\alpha - \mu \mathbf{i}} \exp\left((\alpha - \mu \mathbf{i}) \cdot x\right)\Big|_{x=-\infty}^{x=0}$$

$$\underset{(*)}{=} \frac{-1}{\alpha + \mu \mathbf{i}}(0 - 1) + \frac{1}{\alpha - \mu \mathbf{i}}(1 - 0) = \frac{1}{\alpha + \mu \mathbf{i}} + \frac{1}{\alpha - \mu \mathbf{i}}$$

$$= \frac{\alpha - \mu \mathbf{i} + \alpha + \mu \mathbf{i}}{(\alpha + \mu \mathbf{i})(\alpha - \mu \mathbf{i})} = \frac{2\alpha}{\alpha^2 + \mu^2}.$$

Thus, we conclude that

$$\widehat{f}(\mu) = \frac{\alpha}{\pi(\alpha^2 + \mu^2)}.$$

(see Figure 19A.4(b)). To see equality (∗), recall that $|\exp(-(\alpha + \mu\mathbf{i}) \cdot x)| = e^{-\alpha \cdot x}$. Thus,

$$\lim_{\mu \to \infty} \left|\exp\left(-(\alpha + \mu\mathbf{i}) \cdot x\right)\right| = \lim_{\mu \to \infty} e^{-\alpha \cdot x} = 0.$$

Likewise,

$$\lim_{\mu \to -\infty} \left|\exp\left((\alpha - \mu\mathbf{i}) \cdot x\right)\right| = \lim_{\mu \to -\infty} e^{\alpha \cdot x} = 0. \qquad \diamond$$

Example 19A.8 Conversely, suppose $\alpha > 0$, and $g(x) = 1/(\alpha^2 + x^2)$. Then $\widehat{g}(\mu) = (1/2\alpha)e^{-\alpha \cdot |\mu|}$; the verification of this is Problem 19.6. $\diamond$

Remark Proposition 18H.6, p. 481, provides a powerful technique for computing the Fourier transform of any analytic function $f : \mathbb{R} \longrightarrow \mathbb{C}$, using residue calculus.

19B Properties of the (one-dimensional) Fourier transform

Prerequisites: §19A, Appendix G.

Theorem 19B.1 Riemann–Lebesgue lemma
Let $f \in \mathbf{L}^1(\mathbb{R})$.

(a) *The function $\widehat{f}$ is continuous and bounded. To be precise: If $B := \int_{-\infty}^{\infty} |f(x)|dx$, then, for all $\mu \in \mathbb{R}$, we have $|\widehat{f}(\mu)| < B$.*
(b) *$\widehat{f}$ asymptotically decays near infinity. That is, $\lim_{\mu \to \pm\infty} |\widehat{f}(\mu)| = 0$.*

Proof (a) **Exercise 19B.1.** *Hint:* Boundedness follows from applying the triangle Ⓔ inequality to the integral defining $\widehat{f}(\mu)$. For continuity, fix $\mu_1, \mu_2 \in \mathbb{R}$, and define $E : \mathbb{R} \longrightarrow \mathbb{R}$ by $E(x) := \exp(-\mu_1 x\mathbf{i}) - \exp(-\mu_2 x\mathbf{i})$. For any $X > 0$, we can write

$$\widehat{f}(\mu_1) - \widehat{f}(\mu_2) = \frac{1}{2\pi} \bigg(\underbrace{\int_{-\infty}^{-X} f(x) \cdot E(x)dx}_{(A)} + \underbrace{\int_{-X}^{X} f(x) \cdot E(x)dx}_{(B)}$$

$$+ \underbrace{\int_{X}^{\infty} f(x) \cdot E(x)dx}_{(C)} \bigg).$$

(i) Show that, if X is large enough, then the integrals (A) and (C) can be made arbitrarily small, independent of the values of μ_1 and μ_2. (*Hint:* Recall that $f \in \mathbf{L}^1(\mathbb{R})$. Observe that $|E(x)| \le 2$ for all $x \in \mathbb{R}$.)

(ii) Fix $X > 0$. Show that, if μ_1 and μ_2 are close enough, then integral (B) can also be made arbitrarily small (*Hint:* If μ_1 and μ_2 are 'close', then $|E(x)|$ is 'small' for all $x \in \mathbb{R}$.)

(iii) Show that, if μ_1 and μ_2 are close enough, then $|\widehat{f}(\mu_1) - \widehat{f}(\mu_2)|$ can be made arbitrarily small. (*Hint:* Combine (i) and (ii), using the triangle inequality.) Hence, $\widehat{f}$ is continuous. ♦

Ⓔ   (b) (If f is continuous) **Exercise 19B.2.** *Hint:* For any $X > 0$, we can write

$$\widehat{f}(\mu) = \frac{1}{2\pi} \bigg(\underbrace{\int_{-\infty}^{-X} f(x) \cdot \mathcal{E}_\mu(x)dx}_{(A)} + \underbrace{\int_{-X}^{X} f(x) \cdot \mathcal{E}_\mu(x)dx}_{(B)}$$

$$+ \underbrace{\int_{X}^{\infty} f(x) \cdot \mathcal{E}_\mu(x)dx}_{(C)} \bigg).$$

(i) Show that, if X is large enough, then the integrals (A) and (C) can be made arbitrarily small, independent of the value of μ. (*Hint:* Recall that $f \in \mathbf{L}^1(\mathbb{R})$. Observe that $|\mathcal{E}_\mu(x)| = 1$ for all $x \in \mathbb{R}$.)

(ii) Fix $X > 0$. Show that, if μ is large enough, then integral (B) can also be made arbitrarily small (*Hint:* f is uniformly continuous on the interval $[-X, X]$. (Why?) Thus, for any $\epsilon > 0$, there is some M such that, for all $\mu > M$, and all $x \in [-X, X]$, we have $|f(x) - f(x + \pi/\mu)| < \epsilon$. But $\mathcal{E}_\mu(x + \pi/\mu) = -\mathcal{E}_\mu(x)$.)

(iii) Show that, if μ is large enough, then $|\widehat{f}(\mu)|$ can be made arbitrarily small. (*Hint:* Combine (i) and (ii), using the triangle inequality.) ♦

For the proof of (b) when f is an arbitrary (discontinuous) element of $\mathbf{L}^1(\mathbb{R})$, see Katznelson (1976), Theorem 1.7; Folland (1984), Theorem 8.22(f); or Fisher (1999), Ex. 15, §5.2. □

Recall that, if $f, g : \mathbb{R} \longrightarrow \mathbb{R}$ are two functions, then their *convolution* is the function $(f * g) : \mathbb{R} \longrightarrow \mathbb{R}$ defined:

$$(f * g)(x) := \int_{-\infty}^{\infty} f(y) \cdot g(x - y)dy.$$

(See §17A, p. 381, for a discussion of convolutions.) Similarly, if f has Fourier transform $\widehat{f}$ and g has Fourier transform $\widehat{g}$, we can convolve $\widehat{f}$ and $\widehat{g}$ to get a function $(\widehat{f} * \widehat{g}) : \mathbb{R} \longrightarrow \mathbb{R}$ defined:

$$(\widehat{f} * \widehat{g})(\mu) := \int_{-\infty}^{\infty} \widehat{f}(v) \cdot \widehat{g}(\mu - v)dv.$$

Theorem 19B.2 Algebraic properties of the Fourier transform
Suppose $f, g \in \mathbf{L}^1(\mathbb{R})$ are two functions.

(a) *If $h := f + g$, then, for all $\mu \in \mathbb{R}$, we have $\widehat{h}(\mu) = \widehat{f}(\mu) + \widehat{g}(\mu)$.*
(b) *If $h := f * g$, then, for all $\mu \in \mathbb{R}$, we have $\widehat{h}(\mu) = 2\pi \cdot \widehat{f}(\mu) \cdot \widehat{g}(\mu)$.*
(c) *Conversely, suppose $h := f \cdot g$. If $\widehat{f}, \widehat{g}$, and $\widehat{h}$ are in $\mathbf{L}^1(\mathbb{R})$, then, for all $\mu \in \mathbb{R}$, we have $\widehat{h}(\mu) = (\widehat{f} * \widehat{g})(\mu)$.*

Proof See Problems 19.11–19.13.   ☐

This theorem allows us to compute the Fourier transform of a complicated function by breaking it into a sum/product of simpler pieces.

Theorem 19B.3 Translation and phase shift
Suppose $f \in \mathbf{L}^1(\mathbb{R})$.

(a) *If $\tau \in \mathbb{R}$ is fixed, and $g \in \mathbf{L}^1(\mathbb{R})$ is defined by $g(x) := f(x + \tau)$, then, for all $\mu \in \mathbb{R}$, we have $\widehat{g}(\mu) = e^{\tau \mu i} \cdot \widehat{f}(\mu)$.*
(b) *Conversely, if $\nu \in \mathbb{R}$ is fixed, and $g \in \mathbf{L}^1(\mathbb{R})$ is defined by $g(x) := e^{\nu x i} f(x)$, then, for all $\mu \in \mathbb{R}$, we have $\widehat{g}(\mu) = \widehat{f}(\mu - \nu)$.*

Proof See Problems 19.14 and 19.15.   ☐

Thus, *translating* a function by τ in physical space corresponds to *phase-shifting* its Fourier transform by $e^{\tau \mu i}$, and vice versa. This means that, via a suitable translation, we can put the 'centre' of our coordinate system wherever it is most convenient to do so.

Example 19B.4 Suppose

$$g(x) = \begin{cases} 1 & \text{if } -1 - \tau < x < 1 - \tau; \\ 0 & \text{otherwise.} \end{cases}$$

Thus, $g(x) = f(x + \tau)$, where $f(x)$ is as in Example 19A.3, p. 493. We know that $\widehat{f}(\mu) = \sin(\mu)/\pi\mu$; thus, it follows from Theorem 19B.3 that

$$\widehat{g}(\mu) = e^{\tau \mu i} \cdot \frac{\sin(\mu)}{\pi\mu}. \qquad \qquad \diamondsuit$$

Theorem 19B.5 Rescaling relation
Suppose $f \in \mathbf{L}^1(\mathbb{R})$. If $\alpha > 0$ is fixed, and g is defined by: $g(x) = f(x/\alpha)$, then, for all $\mu \in \mathbb{R}$, we have $\widehat{g}(\mu) = \alpha \cdot \widehat{f}(\alpha \cdot \mu)$.

Proof See Problem 19.16.   ☐

In Theorem 19B.5, the function g is the same as function f, but expressed in a coordinate system 'rescaled' by a factor of α.

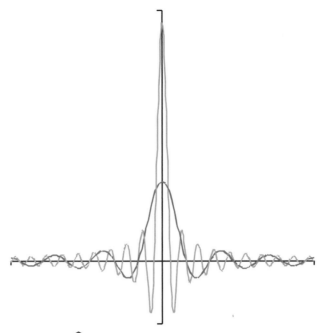

Figure 19B.1. Plot of $\widehat{f}$ (black) and $\widehat{g}$ (grey) in Example 19B.6, where $g(x) = f(x/3)$.

Example 19B.6 Suppose

$$g(x) = \begin{cases} 1 & \text{if } -3 < x < 3; \\ 0 & \text{otherwise.} \end{cases}$$

Thus, $g(x) = f(x/3)$, where $f(x)$ is as in Example 19A.3, p. 493. We know that $\widehat{f}(\mu) = \sin(\mu)/\mu\pi$; thus, it follows from Theorem 19B.5 that

$$\widehat{g}(\mu) = 3 \cdot \frac{\sin(3\mu)}{3\mu\pi} = \frac{\sin(3\mu)}{\mu\pi}.$$

See Figure 19B.1.   ◇

A function $f : \mathbb{R} \longrightarrow \mathbb{C}$ is *continuously differentiable* if $f'(x)$ exists for all $x \in \mathbb{R}$, and the function $f' : \mathbb{R} \longrightarrow \mathbb{C}$ is itself continuous. Let $\mathcal{C}^1(\mathbb{R})$ be the set of all continuously differentiable functions from $\mathbb{R}$ to $\mathbb{C}$. For any $n \in \mathbb{N}$, let $f^{(n)}(x) := \frac{d^n}{dx^n} f(x)$. The function f is *n-times continuously differentiable* if $f^{(n)}(x)$ exists for all $x \in \mathbb{R}$, and the function $f^{(n)} : \mathbb{R} \longrightarrow \mathbb{C}$ is itself continuous. Let $\mathcal{C}^n(\mathbb{R})$ be the set of all n-times continuously differentiable functions from $\mathbb{R}$ to $\mathbb{C}$.

Theorem 19B.7 Differentiation and multiplication
Suppose $f \in \mathbf{L}^1(\mathbb{R})$.

(a) *Suppose $f \in \mathcal{C}^1(\mathbb{R})$ and $\lim_{x \to \pm\infty} |f(x)| = 0$. Let $g(x) := f'(x)$. If $g \in \mathbf{L}^1(\mathbb{R})$, then, for all $\mu \in \mathbb{R}$, we have $\widehat{g}(\mu) = \mathbf{i}\mu \cdot \widehat{f}(\mu)$.*

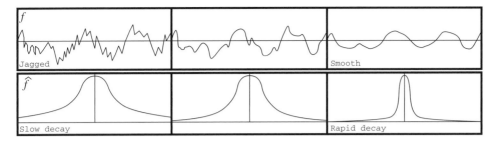

Figure 19B.2. Smoothness vs. asymptotic decay in the Fourier transform.

(b) *More generally, suppose $f \in C^n(\mathbb{R})$ and $\lim_{x \to \pm\infty} |f^{(n-1)}(x)| = 0$. Let $g(x) := f^{(n)}(x)$.*
    If $g \in \mathbf{L}^1(\mathbb{R})$, then, for all $\mu \in \mathbb{R}$, we have $\widehat{g}(\mu) = (\mathbf{i}\mu)^n \cdot \widehat{f}(\mu)$. Thus, $\widehat{f}(\mu)$ asymptot-
    ically decays faster than $1/\mu^n$ as $\mu \to \pm\infty$. That is, $\lim_{\mu \to \pm\infty} \mu^n \widehat{f}(\mu) = 0$.
(c) *Conversely, let $g(x) := x^n \cdot f(x)$, and suppose that f decays 'quickly enough' that g is*
    also in $\mathbf{L}^1(\mathbb{R})$ (this happens, for example, if $\lim_{x \to \pm\infty} x^{n+1} f(x) = 0$). Then the function
    $\widehat{f}$ is n-times differentiable, and, for all $\mu \in \mathbb{R}$,

$$\widehat{g}(\mu) = \mathbf{i}^n \cdot \frac{\mathrm{d}^n}{\mathrm{d}\mu^n} \widehat{f}(\mu).$$

Proof (a) is Problem 19.17.

   (b) is just the result of iterating (a) *n* times.

   (c) is **Exercise 19B.3.** (*Hint: Either* 'reverse' the result of (a) using the Fourier ⒺInversion Formula (Theorem 19A.1, p. 492), *or* use Proposition G.1, p. 571, to differentiate directly the integral defining $\widehat{f}(\mu)$.)   □

Theorem 19B.7 says that the Fourier transform converts *differentiation-by-x* into *multiplication-by-$\mu\mathbf{i}$*. This implies that the *smoothness* of a function f is closely related to the *asymptotic decay rate* of its Fourier transform. The 'smoother' f is (i.e. the more times we can differentiate it), the more *rapidly* $\widehat{f}(\mu)$ decays as $\mu \to \infty$ (see Figure 19B.2).

Physically, we can interpret this as follows. If we think of f as a 'signal', then $\widehat{f}(\mu)$ is the amount of 'energy' at the 'frequency' μ in the spectral decomposition of this signal. Thus, the magnitude of $\widehat{f}(\mu)$ for extremely large μ is the amount of 'very high frequency' energy in f, which corresponds to very finely featured, 'jagged' structure in the shape of f. If f is 'smooth', then we expect there will be very little of this 'jaggedness'; hence the high-frequency part of the energy spectrum will be very small.

Conversely, the asymptotic decay rate of f determines the smoothness of its Fourier transform. This makes sense because the Fourier inversion formula can be (loosely) interpreted as saying that f is itself a sort of 'backwards' Fourier transform of $\widehat{f}$.

One very important Fourier transform is the following.

Theorem 19B.8 Fourier transform of a Gaussian

(a) *If $f(x) = \exp\left(-x^2\right)$, then*

$$\widehat{f}(\mu) = \frac{1}{2\sqrt{\pi}} \cdot f\left(\frac{\mu}{2}\right) = \frac{1}{2\sqrt{\pi}} \cdot \exp\left(\frac{-\mu^2}{4}\right).$$

(b) *Fix $\sigma > 0$. If*

$$f(x) = \frac{1}{\sigma\sqrt{2\pi}} \exp\left(\frac{-x^2}{2\sigma^2}\right)$$

is a Gaussian probability distribution with mean 0 and variance σ^2, then

$$\widehat{f}(\mu) = \frac{1}{2\pi} \exp\left(\frac{-\sigma^2\mu^2}{2}\right).$$

(c) *Fix $\sigma > 0$ and $\tau \in \mathbb{R}$. If*

$$f(x) = \frac{1}{\sigma\sqrt{2\pi}} \exp\left(\frac{-|x - \tau|^2}{2\sigma^2}\right)$$

is a Gaussian probability distribution with mean τ and variance σ^2, then

$$\widehat{f}(\mu) = \frac{e^{-i\tau\mu}}{2\pi} \exp\left(\frac{-\sigma^2\mu^2}{2}\right).$$

Proof We start with part (a). Let $g(x) = f'(x)$. Then by Theorem 19B.7(a),

$$\widehat{g}(\mu) = \mathbf{i}\mu \cdot \widehat{f}(\mu). \tag{19B.1}$$

However, direct computation says $g(x) = -2x \cdot f(x)$, so $(-1/2)g(x) = x \cdot f(x)$, so Theorem 19B.7(c) implies

$$\frac{\mathbf{i}}{2}\widehat{g}(\mu) = (\widehat{f})'(\mu). \tag{19B.2}$$

Combining equation (19B.2) with equation (19B.1), we conclude the following:

$$(\widehat{f})'(\mu) \underset{\text{(19B.2)}}{=\!=\!=} \frac{\mathbf{i}}{2}\widehat{g}(\mu) \underset{\text{(19B.1)}}{=\!=\!=} \frac{\mathbf{i}}{2} \cdot \mathbf{i}\mu \cdot \widehat{f}(\mu) = \frac{-\mu}{2}\widehat{f}(\mu). \tag{19B.3}$$

Define $h(\mu) = \widehat{f}(\mu) \cdot \exp\left(\mu^2/4\right)$. If we differentiate $h(\mu)$, we obtain the following:

$$h'(\mu) \underset{\text{(dL)}}{=\!=\!=} \widehat{f}(\mu) \cdot \frac{\mu}{2}\exp\left(\frac{\mu^2}{4}\right) \underbrace{-\frac{\mu}{2}\widehat{f}(\mu) \cdot \exp\left(\frac{\mu^2}{4}\right)}_{(*)} = 0.$$

Here, (dL) is differentiating using the Leibniz rule, and $(*)$ is by equation (19B.3). In other words, $h(\mu) = H$ is a constant. Thus,

$$\widehat{f}(\mu) = \frac{h(\mu)}{\exp\left(\mu^2/4\right)} = H \cdot \exp\left(\frac{-\mu^2}{4}\right) = H \cdot f\left(\frac{\mu}{2}\right).$$

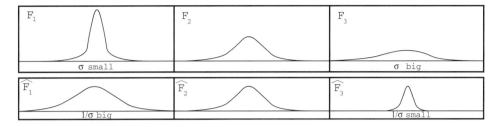

Figure 19B.3. Heisenberg uncertainty principle.

To evaluate H, set $\mu = 0$, to get

$$H = H \cdot \exp\left(\frac{-0^2}{4}\right) = \widehat{f}(0) = \frac{1}{2\pi}\int_{-\infty}^{\infty} f(x)\mathrm{d}x = \frac{1}{2\pi}\int_{-\infty}^{\infty}\exp\left(-x^2\right)$$

$$= \frac{1}{2\sqrt{\pi}}.$$

(where the last step is **Exercise 19B.4**). Thus, we conclude that

$$\widehat{f}(\mu) = \frac{1}{2\sqrt{\pi}} \cdot f\left(\frac{\mu}{2}\right).$$

Part (b) follows by setting $\alpha := \sqrt{2}\,\sigma$ in Theorem 19B.5, p. 499.
Part (c) is **Exercise 19B.5.** (*Hint:* Apply Theorem 19B.3, p. 499.) □

Loosely speaking, Theorem 19B.8 states that 'The Fourier transform of a Gaussian is another Gaussian'.[2] However, note that, in Theorem 19B.8(b), as the variance of the Gaussian (that is, σ^2) gets bigger, the 'variance' of its Fourier transform (which is effectively $1/\sigma^2$) gets *smaller* (see Figure 19B.3). If we think of the Gaussian as the probability distribution of some unknown piece of information, then the variance measures the degree of 'uncertainty'. Hence, we conclude: the greater the uncertainty embodied in the Gaussian f, the *less* the uncertainty embodied in $\widehat{f}$, and vice versa. This is a manifestation of the so-called *Heisenberg uncertainty principle* (see Theorem 19G.2, p. 518).

Proposition 19B.9 Inversion and conjugation
For any $z \in \mathbb{C}$, let $\overline{z}$ denote the complex conjugate of z. Let $f \in \mathbf{L}^1(\mathbb{R})$.

(a) *For all $\mu \in \mathbb{R}$, we have $\widehat{\overline{f}}(\mu) = \overline{\widehat{f}(-\mu)}$. In particular,*

$$\left(\, f \text{ purely real-valued}\,\right) \iff \left(\text{for all } \mu \in \mathbb{R}, \text{ we have } \widehat{f}(-\mu) = \overline{\widehat{f}(\mu)}\,\right).$$

---

[2] This is only *loosely* speaking, however, because a proper Gaussian contains the multiplier $1/\sigma\sqrt{2\pi}$ to make it a probability distribution, whereas the Fourier transform does not.

(b) *Suppose $g(x) = f(-x)$ for all $x \in \mathbb{R}$. Then, for all $\mu \in \mathbb{R}$, we have $\widehat{g}(\mu) = \widehat{f}(-\mu)$. In particular, if f is purely real-valued, then $\widehat{g}(\mu) = \overline{\widehat{f}(\mu)}$ for all $\mu \in \mathbb{R}$.*

(c) *If f is real-valued and even (i.e. $f(-x) = f(x)$), then $\widehat{f}$ is purely real-valued.*

(d) *If f is real-valued and odd (i.e. $f(-x) = -f(x)$), then $\widehat{f}$ is purely imaginary-valued.*

Ⓔ   *Proof* **Exercise 19B.6.**   □

Example 19B.10 Autocorrelation and power spectrum

If $f : \mathbb{R} \longrightarrow \mathbb{R}$, then the *autocorrelation function* of f is defined by

$$\mathbf{A}f(x) := \int_{-\infty}^{\infty} f(y) \cdot f(x+y)\mathrm{d}y.$$

Heuristically, if we think of $f(x)$ as a 'random signal', then $\mathbf{A}f(x)$ measures the degree of correlation in the signal across time intervals of length x – i.e. it provides a crude measure of how well you can predict the value of $f(y+x)$ given information about $f(x)$. In particular, if f has some sort of 'T-periodic' component, then we expect $\mathbf{A}f(x)$ to be large when $x = nT$ for any $n \in \mathbb{Z}$.

If we define $g(x) = f(-x)$, then we can see that $\mathbf{A}f(x) = (f * g)(-x)$
Ⓔ   (**Exercise 19B.7**). Thus,

$$\widehat{\mathbf{A}f}(\mu) \underset{(*)}{=} \overline{\widehat{f * g}}(\mu) \underset{(\dagger)}{=} \widehat{f}(\mu) \cdot \widehat{g}(\mu)$$

$$\underset{(*)}{=} \widehat{f}(\mu) \cdot \overline{\widehat{f}(\mu)} = \widehat{f}(\mu) \cdot \widehat{f}(\mu) = \left|\widehat{f}(\mu)\right|^2.$$

Here, both $(*)$ are by Proposition 19B.9(b), while $(\dagger)$ is by Theorem 19B.2(b). The function $|\widehat{f}(\mu)|^2$ measures the absolute magnitude of the Fourier transform of $\widehat{f}$, and is sometimes called the *power spectrum* of $\widehat{f}$.   ◇

Evil twins of the Fourier transform Unfortunately, the mathematics literature contains at least four different definitions of the Fourier transform. In this book, the Fourier transform and its inversion are defined by the integrals

$$\widehat{f}(\mu) := \frac{1}{2\pi} \int_{-\infty}^{\infty} f(x)\exp(-\mathbf{i}x\mu)\mathrm{d}\mu \quad \text{and} \quad f(x) = \int_{-\infty}^{\infty} \widehat{f}(x)\exp(\mathbf{i}x\mu)\mathrm{d}\mu.$$

Some books (for example Katznelson (1976), Haberman (1987), Körner (1988), and Fisher (1999)) instead use what we will call the *opposite Fourier transform*:

$$\check{f}(\mu) := \int_{-\infty}^{\infty} f(x)\exp(-\mathbf{i}x\mu)\mathrm{d}x,$$

with inverse transform

$$f(x) = \frac{1}{2\pi} \int_{-\infty}^{\infty} \check{f}(\mu)\exp(\mathbf{i}x\mu)\mathrm{d}\mu.$$

Other books (e.g. Asmar (2005)) instead use what we will call the *symmetric Fourier transform*:

$$\widehat{f}(\mu) := \frac{1}{\sqrt{2\pi}} \int_{-\infty}^{\infty} f(x) \exp(-i\,x\mu) dx,$$

with inverse transform

$$f(x) = \frac{1}{\sqrt{2\pi}} \int_{-\infty}^{\infty} \widehat{f}(\mu) \exp(i\,x\mu) d\mu.$$

Finally, some books (e.g. Folland (1984) and Walker (1988)) use what we will call the *canonical Fourier transform*:

$$\widetilde{f}(\mu) := \int_{-\infty}^{\infty} f(x) \exp(-2\pi i\,x\mu) dx,$$

with inverse transform

$$f(x) = \int_{-\infty}^{\infty} \widetilde{f}(\mu) \exp(2\pi i\,x\mu) d\mu.$$

All books use the symbol $\widehat{f}$ to denote the Fourier transform of f – we are using four different 'accents' simply to avoid confusing the four definitions. It is easy to translate the Fourier transform in this book into its evil twins. For any $f \in \mathbf{L}^1(\mathbb{R})$ and any $\mu \in \mathbb{R}$, we have the following:

$$\check{f}(\mu) = 2\pi\, \widehat{f}(\mu) \qquad \text{and} \quad \widehat{f}(\mu) = \frac{1}{2\pi}\, \check{f}(\mu);$$

$$\widehat{f}(\mu) = \sqrt{2\pi}\, \widehat{f}(\mu) \quad \text{and} \quad \widehat{f}(\mu) = \frac{1}{\sqrt{2\pi}}\, \widehat{f}(\mu); \qquad (19\text{B}.4)$$

$$\widetilde{f}(\mu) = 2\pi\, \widehat{f}(2\pi\,\mu) \quad \text{and} \quad \widehat{f}(\mu) = \frac{1}{2\pi}\, \widetilde{f}\left(\frac{\mu}{2\pi}\right).$$

Exercise 19B.8 Check this. ◆ Ⓔ

All of the formulae and theorems we have derived in this section are still true under these alternative definitions, except that one must multiply or divide by 2π or $\sqrt{2\pi}$ at certain key points, and replace e^i with $e^{2\pi i}$ (or vice versa) at others.

Exercise 19B.9 (Annoying!) Use the identities (19B.4) to reformulate all the Ⓔ
formulae and theorems in this chapter in terms of (a) the opposite Fourier transform $\check{f}$; (b) the symmetric Fourier transform $\widehat{f}$; or (c) the canonical Fourier transform $\widetilde{f}$. ◆

   Each of the four definitions has advantages and disadvantages; some formulae become simpler, others become more complex. Clearly, both the 'symmetric' and 'canonical' versions of the Fourier transform have some appeal because the Fourier transform and its inverse have 'symmetrical' formulae using these definitions. Furthermore, in both of these versions, the '2π' factor disappears from Parseval's and Plancherel's theorems (see §19C) – in other words, the Fourier transform becomes an *isometry* of $\mathbf{L}^2(\mathbb{R})$. The symmetric Fourier transform has the added advantage that it maps a Gaussian distribution into another Gaussian (no scalar multiplication required). The canonical Fourier transform has the added advantage that $\widehat{f * g} = \widehat{f} \cdot \widehat{g}$ (without the 2π factor required in Theorem 19B.2(a)), while simultaneously, $\widehat{f \cdot g} = \widehat{f} * \widehat{g}$ (unlike the symmetric Fourier transform).

The definition used in this book (and also in McWeeny (1972), Churchill and Brown (1987), Broman (1989), Pinsky (1998), and Powers (1999), among others) has none of these advantages. Its major advantage is that it will yield simpler expressions for the abstract solutions to partial differential equations in Chapter 20. If one uses the 'symmetric' Fourier transform, then every one of the solution formulae in Chapter 20 must be multiplied by some power of $1\sqrt{2\pi}$. If one uses the 'canonical' Fourier transform, then every spacetime variable (i.e. x, y, z, t) in every formula must by multiplied by 2π or sometimes by $4\pi^2$, which makes all the formulae look much more complicated.[3]

We end with a warning. When comparing or combining formulae from two or more books, make sure to first *compare their definitions of the Fourier transform*, and make the appropriate conversions using equations (19B.4), if necessary.

19C* Parseval and Plancherel

Prerequisites: §19A.   **Recommended:** §6C(i), §6F.

Let $\mathbf{L}^2(\mathbb{R})$ be the set of all *square-integrable* complex-valued functions on $\mathbb{R}$ – that is, all integrable functions $f : \mathbb{R} \longrightarrow \mathbb{C}$ such that $\| f \|_2 < \infty$, where we define:

$$\| f \|_2 := \left(\int_{-\infty}^{\infty} |f(x)|^2 \, \mathrm{d}x \right)^{1/2}$$

(see §6C(i) for more information). Note that $\mathbf{L}^2(\mathbb{R})$ is neither a subset nor a superset of $\mathbf{L}^1(\mathbb{R})$; however, the two spaces do overlap. If $f, g \in \mathbf{L}^2(\mathbb{R})$, then we define:

$$\langle f, g \rangle := \int_{-\infty}^{\infty} f(x) \overline{g}(x) \mathrm{d}x.$$

---

[3] Of course, when you actually apply these formulae to solve specific problems, you will end up with exactly the same solution, no matter which version of the Fourier transform you use – why?

The following identity is useful in many applications of Fourier theory, especially quantum mechanics. It can be seen as the 'continuum' analogue of Parseval's equality for an orthonormal basis (Theorem 6F.1, p. 137).

Theorem 19C.1 Parseval's equality for Fourier transforms
If $f, g \in \mathbf{L}^1(\mathbb{R}) \cap \mathbf{L}^2(\mathbb{R})$, then $\langle f, g \rangle = 2\pi \langle \widehat{f}, \widehat{g} \rangle$.

Proof Define $h : \mathbb{R} \longrightarrow \mathbb{R}$ by $h(x) := f(x)\overline{g}(x)$. Then $h \in \mathbf{L}^1(\mathbb{R})$ because $f, g \in \mathbf{L}^2(\mathbb{R})$. We have

$$\widehat{h}(0) = \frac{1}{2\pi} \int_{-\infty}^{\infty} f(x) \cdot \overline{g}(x) \cdot \exp(-\mathbf{i}0x) \mathrm{d}x$$

$$\underset{(*)}{=} \frac{1}{2\pi} \int_{-\infty}^{\infty} f(x) \cdot \overline{g}(x) \mathrm{d}x = \frac{\langle f, g \rangle}{2\pi}, \tag{19C.1}$$

where $(*)$ is because $\exp(-\mathbf{i}0x) = \exp(0) = 1$ for all $x \in \mathbb{R}$. But we also have

$$\widehat{h}(0) \underset{(*)}{=} \widehat{f} * \widehat{\overline{g}}(0) = \int_{-\infty}^{\infty} \widehat{f}(v) \cdot \widehat{\overline{g}}(-v) \mathrm{d}v$$

$$\underset{(\dagger)}{=} \int_{-\infty}^{\infty} \widehat{f}(v) \cdot \overline{\widehat{g}(v)} \mathrm{d}v = \langle \widehat{f}, \widehat{g} \rangle. \tag{19C.2}$$

Here, $(*)$ is by Theorem 19B.2(c), because $h = f \cdot \overline{g}$ so $\widehat{h} = \widehat{f} * \widehat{\overline{g}}$. Meanwhile, $(\dagger)$ is by Proposition 19B.9(a).

Combining equations (19C.1) and (19C.2) yields $\langle \widehat{f}, \widehat{g} \rangle = \widehat{h}(0) = \langle f, g \rangle / 2\pi$. The result follows. $\square$

Corollary 19C.2 Plancherel's theorem

Suppose $f \in \mathbf{L}^1(\mathbb{R}) \cap \mathbf{L}^2(\mathbb{R})$. Then $\widehat{f} \in \mathbf{L}^1(\mathbb{R}) \cap \mathbf{L}^2(\mathbb{R})$ also, and $\|f\|_2 = \sqrt{2\pi} \|\widehat{f}\|2$.

Proof Set $f = g$ in the Parseval equality. Recall that $\|f\|_2 = \sqrt{\langle f, f \rangle}$. $\square$

In fact, the Plancherel theorem says much more than this. Define the linear operator $\mathsf{F}_1 : \mathbf{L}^1(\mathbb{R}) \longrightarrow \mathbf{L}^1(\mathbb{R})$ by $\mathsf{F}_1(f) := \sqrt{2\pi} \widehat{f}$ for all $f \in \mathbf{L}^1(\mathbb{R})$; then the full Plancherel theorem says that F_1 extends uniquely to a *unitary isomorphism* $\mathsf{F}_2 : \mathbf{L}^2(\mathbb{R}) \longrightarrow \mathbf{L}^2(\mathbb{R})$ – that is, a bijective linear transformation from $\mathbf{L}^2(\mathbb{R})$ to itself such that $\|\mathsf{F}_2(f)\|_2 = \|f\|_2$ for all $f \in \mathbf{L}^2(\mathbb{R})$. For any $p \in [1, \infty)$, let $\mathbf{L}^p(\mathbb{R})$ be the set of all integrable functions $f : \mathbb{R} \longrightarrow \mathbb{C}$ such that $\|f\|_p < \infty$, where

$$\|f\|_p := \left(\int_{-\infty}^{\infty} |f(x)|^p \, \mathrm{d}x \right)^{1/p}.$$

For any $p \in [1, 2]$, let $\hat{p} \in [2, \infty]$ be the unique number such that $\frac{1}{p} + \frac{1}{\hat{p}} = 1$ (for example, if $p = 3/2$, then $\hat{p} = 3$). Then, through a process called *Riesz–Thorin interpolation*, it is possible to extend the Fourier transform even further, to get a linear transformation $\mathsf{F}_p : \mathbf{L}^p(\mathbb{R}) \longrightarrow \mathbf{L}^{\hat{p}}(\mathbb{R})$. For example, one can define a Fourier transform $\mathsf{F}_{3/2} : \mathbf{L}^{3/2}(\mathbb{R}) \longrightarrow \mathbf{L}^3(\mathbb{R})$. All these transformations agree on the overlaps of their domains, and satisfy the *Hausdorff–Young inequality*:

$$\left\| \mathsf{F}_p(f) \right\|_{\hat{p}} \leq \| f \|_p, \quad \text{for any } p \in [1, 2] \text{ and } f \in \mathbf{L}^p(\mathbb{R}).$$

However, the details are well beyond the scope of this text. For more information, see Katznelson (1976), chap. VI, or Folland (1984), chap. 8.

19D Two-dimensional Fourier transforms

Prerequisites: §19A.   **Recommended:** §9A.

Let $\mathbf{L}^1(\mathbb{R}^2)$ be the set of all functions $f : \mathbb{R}^2 \longrightarrow \mathbb{C}$ which are *absolutely integrable* on $\mathbb{R}^2$, meaning that

$$\int_{\mathbb{R}^2} |f(x, y)| \, dx \, dy < \infty.$$

If $f \in \mathbf{L}^1(\mathbb{R}^2)$, then the *Fourier transform* of f is the function $\widehat{f} : \mathbb{R}^2 \longrightarrow \mathbb{C}$ defined:

$$\boxed{\widehat{f}(\mu, \nu) := \frac{1}{4\pi^2} \int_{\mathbb{R}^2} f(x, y) \cdot \exp\left(-(\mu x + \nu y) \cdot \mathbf{i} \right) dx \, dy,}$$

for all $(\mu, \nu) \in \mathbb{R}^2$.

Theorem 19D.1 Strong Fourier inversion formula

Suppose that $f \in \mathbf{L}^1(\mathbb{R}^2)$, and that $\widehat{f}$ is also in $\mathbf{L}^1(\mathbb{R}^2)$. For any $(x, y) \in \mathbb{R}^2$, if f is continuous at (x, y), then

$$f(x, y) = \int_{\mathbb{R}^2} \widehat{f}(\mu, \nu) \cdot \exp\left((\mu x + \nu y) \cdot \mathbf{i} \right) d\mu \, d\nu.$$

Proof See Katznelson (1976), §VI.1.12, and Folland (1984), Theorem 8.26.   □

Unfortunately, not all the functions one encounters have the property that their Fourier transform is in $\mathbf{L}^1(\mathbb{R}^2)$. In particular, $\widehat{f} \in \mathbf{L}^1(\mathbb{R}^2)$ only if f agrees 'almost everywhere' with a continuous function (thus, Theorem 19D.1 is inapplicable to step functions, for example). We want a result analogous to the 'weak' Fourier inversion theorem 19A.1, p. 492. It is surprisingly difficult to find clean, simple 'inversion theorems' of this nature for multidimensional Fourier transforms. The result given here is far from the most general one in this category, but it has the advantage of being easy to state and prove. First, we must define an appropriate class of functions. Let $\widehat{\mathbf{L}}^1(\mathbb{R}^2) := \{f \in \mathbf{L}^1(\mathbb{R}^2); \ \widehat{f} \in \mathbf{L}^1(\mathbb{R}^2)\}$; this is the class considered by Theorem 19D.1. Let $\widetilde{\mathbf{L}}^1(\mathbb{R})$ be the set of all *piecewise smooth* functions in $\mathbf{L}^1(\mathbb{R})$ (the class considered by Theorem 19A.1). Let $\mathcal{F}(\mathbb{R}^2)$ be the set of all functions $f \in \mathbf{L}^1(\mathbb{R}^2)$ such that there exist $f_1, f_2 \in \widetilde{\mathbf{L}}^1(\mathbb{R})$ with $f(x_1, x_2) = f_1(x_1) \cdot f_2(x_2)$ for all $(x_1, x_2) \in \mathbb{R}^2$. Let $\mathcal{H}(\mathbb{R}^2)$ denote the set of all functions in $\mathbf{L}^1(\mathbb{R}^2)$ which can be written as a *finite sum* of elements in $\mathcal{F}(\mathbb{R}^2)$. Finally, we define:

$$\widetilde{\mathbf{L}}^1(\mathbb{R}^2) := \left\{ f \in \mathbf{L}^1(\mathbb{R}^2); \ f = g + h \text{ for some } g \in \widehat{\mathbf{L}}^1(\mathbb{R}^2) \text{ and } h \in \mathcal{H}(\mathbb{R}^2) \right\}.$$

Theorem 19D.2 Two-dimensional Fourier inversion formula
Suppose $f \in \widetilde{\mathbf{L}}^1(\mathbb{R}^2)$. If $(x, y) \in \mathbb{R}^2$ and f is is continuous at (x, y), then

$$f(x, y) = \lim_{M \to \infty} \int_{-M}^{M} \int_{-M}^{M} \widehat{f}(\mu, v) \cdot \exp\left((\mu x + vy) \cdot \mathbf{i}\right) d\mu \, dv. \quad (19D.1)$$

Proof **Exercise 19D.1.**   ⒠
(a) First show that equation (19D.1) holds for any element of $\mathcal{F}(\mathbb{R}^2)$. (*Hint:* If $f_1, f_2 \in \widetilde{\mathbf{L}}^1(\mathbb{R})$, and $f(x_1, x_2) = f_1(x_1) \cdot f_2(x_2)$ for all $(x_1, x_2) \in \mathbb{R}^2$, then show that $\widehat{f}(\mu_1, \mu_2) = \widehat{f_1}(\mu_1) \cdot \widehat{f_2}(\mu_2)$. Substitute this expression into the right-hand side of equation (19D.1); factor the integral into two one-dimensional Fourier inversion integrals, and then apply Theorem 19A.1, p. 492.)
(b) Deduce that equation (19D.1) holds for any element of $\mathcal{H}(\mathbb{R}^2)$. (*Hint:* The Fourier transform is linear.)
(c) Now combine (b) with Theorem 19D.1 to conclude that equation (19D.1) holds for any element of $\widetilde{\mathbf{L}}^1(\mathbb{R}^2)$.   □

Proposition 19D.3 *If $f, g \in \mathcal{C}(\mathbb{R}^2) \cap \mathbf{L}^1(\mathbb{R}^2)$ are continuous, integrable functions, then $(\widehat{f} = \widehat{g}) \iff (f = g)$.*   □

Example 19D.4 Let $X, Y > 0$, and let

$$f(x, y) = \begin{cases} 1 & \text{if } -X \le x \le X \text{ and } -Y \le y \le Y; \\ 0 & \text{otherwise} \end{cases}$$

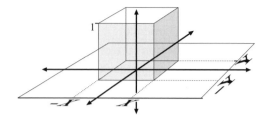

Figure 19D.1. Function for Example 19D.4.

(Figure 19D.1). Then

$$\widehat{f}(\mu, \nu) = \frac{1}{4\pi^2} \int_{-\infty}^{\infty} \int_{-\infty}^{\infty} f(x, y) \cdot \exp\left(-(\mu x + \nu y) \cdot \mathbf{i}\right) dx \, dy$$

$$= \frac{1}{4\pi^2} \int_{-X}^{X} \int_{-Y}^{Y} \exp(-\mu x \mathbf{i}) \cdot \exp(-\nu y \mathbf{i}) dx \, dy$$

$$= \frac{1}{4\pi^2} \left(\int_{-X}^{X} \exp(-\mu x \mathbf{i}) dx \right) \cdot \left(\int_{-Y}^{Y} \exp(-\nu y \mathbf{i}) dy \right)$$

$$= \frac{1}{4\pi^2} \cdot \left(\frac{-1}{\mu \mathbf{i}} \exp\left(-\mu x \mathbf{i}\right) \Big|_{x=-X}^{x=X} \right) \cdot \left(\frac{1}{\nu \mathbf{i}} \exp\left(-\nu y \mathbf{i}\right) \Big|_{y=-Y}^{y=Y} \right)$$

$$= \frac{1}{4\pi^2} \left(\frac{e^{\mu X \mathbf{i}} - e^{-\mu X \mathbf{i}}}{\mu \mathbf{i}} \right) \left(\frac{e^{\nu Y \mathbf{i}} - e^{-\nu Y \mathbf{i}}}{\nu \mathbf{i}} \right)$$

$$= \frac{1}{\pi^2 \mu \nu} \left(\frac{e^{\mu X \mathbf{i}} - e^{-\mu X \mathbf{i}}}{2 \mathbf{i}} \right) \left(\frac{e^{\nu Y \mathbf{i}} - e^{-\nu Y \mathbf{i}}}{2 \mathbf{i}} \right)$$

$$\underset{\text{(Eu)}}{=\!=} \frac{1}{\pi^2 \mu \nu} \sin(\mu X) \cdot \sin(\nu Y),$$

where (Eu) is by double application of Euler's formula (see p. 553). Note that f is in $\mathcal{F}(\mathbb{R}^2)$ (why?), and thus, in $\widetilde{\mathbf{L}}^1(\mathbb{R}^2)$. Thus, Theorem 19D.2 states that, if $-X < x < X$ and $-Y < y < Y$, then

$$\lim_{M \to \infty} \int_{-M}^{M} \int_{-M}^{M} \frac{\sin(\mu X) \cdot \sin(\nu Y)}{\pi^2 \cdot \mu \cdot \nu} \exp\left((\mu x + \nu y) \cdot \mathbf{i}\right) d\mu \, d\nu = 1,$$

while, if $(x, y) \notin [-X, X] \times [-Y, Y]$, then

$$\lim_{M \to \infty} \int_{-M}^{M} \int_{-M}^{M} \frac{\sin(\mu X) \cdot \sin(\nu Y)}{\pi^2 \cdot \mu \cdot \nu} \exp\left((\mu x + \nu y) \cdot \mathbf{i}\right) d\mu d\nu = 0.$$

At points on the *boundary* of the box $[-X, X] \times [-Y, Y]$, however, the Fourier inversion integral will converge to neither of these values.   ◇

Example 19D.5 If

$$f(x, y) = \frac{1}{2\sigma^2\pi} \exp\left(\frac{-x^2 - y^2}{2\sigma^2}\right)$$

is a two-dimensional Gaussian distribution, then

$$\widehat{f}(\mu, v) = \frac{1}{4\pi^2} \exp\left(\frac{-\sigma^2}{2}\left(\mu^2 + v^2\right)\right)$$

(**Exercise 19D.2**). ◇ Ⓔ

Exercise 19D.3 State and prove two-dimensional versions of all results in Ⓔ §19B. ♦

19E Three-dimensional Fourier transforms

Prerequisites: §19A.   **Recommended:** §9B, §19D.

In three or more dimensions, it is cumbersome to write vectors as an explicit list of coordinates. We will adopt a more compact notation. Bold-face letters will indicate vectors, and italic letters will denote their components. For example:

$$\mathbf{x} = (x_1, x_2, x_3), \quad \mathbf{y} = (y_1, y_2, y_3), \quad \boldsymbol{\mu} = (\mu_1, \mu_2, \mu_3), \quad \text{and} \quad \boldsymbol{v} = (v_1, v_2, v_3).$$

We define the *inner product* $\mathbf{x} \bullet \mathbf{y} := x_1 \cdot y_1 + x_2 \cdot y_2 + x_3 \cdot y_3$. Let $\mathbf{L}^1(\mathbb{R}^3)$ be the set of all functions $f : \mathbb{R}^3 \longrightarrow \mathbb{C}$ which are *absolutely integrable* on $\mathbb{R}^3$, meaning that

$$\int_{\mathbb{R}^3} |f(\mathbf{x})| \, d\mathbf{x} < \infty.$$

If $f \in \mathbf{L}^1(\mathbb{R}^3)$, then we can define

$$\int_{\mathbb{R}^3} f(\mathbf{x}) d\mathbf{x} := \int_{-\infty}^{\infty} \int_{-\infty}^{\infty} \int_{-\infty}^{\infty} f(x_1, x_2, x_3) dx_1 \, dx_2 \, dx_3,$$

where this integral is understood to be *absolutely convergent*. In particular, if $f \in \mathbf{L}^1(\mathbb{R}^3)$, then the *Fourier transform* of f is the function $\widehat{f} : \mathbb{R}^3 \longrightarrow \mathbb{C}$ defined:

$$\boxed{\widehat{f}(\boldsymbol{\mu}) := \frac{1}{8\pi^3} \int_{\mathbb{R}^3} f(\mathbf{x}) \cdot \exp\left(-\mathbf{x} \bullet \boldsymbol{\mu} \cdot \mathbf{i}\right) d\mathbf{x},}$$

for all $\boldsymbol{\mu} \in \mathbb{R}^3$. Define $\widetilde{\mathbf{L}}^1(\mathbb{R}^3)$ in a manner analogous to the definition of $\widetilde{\mathbf{L}}^1(\mathbb{R}^2)$, p. 509.

Theorem 19E.1 Three-dimensional Fourier inversion formula
(a) *Suppose $f \in \widetilde{\mathbf{L}}^1(\mathbb{R}^3)$. For any $\mathbf{x} \in \mathbb{R}^3$, if f is continuous at $\mathbf{x}$, then*

$$f(\mathbf{x}) = \lim_{M \to \infty} \int_{-M}^{M} \int_{-M}^{M} \int_{-M}^{M} \widehat{f}(\boldsymbol{\mu}) \cdot \exp\left(\boldsymbol{\mu} \bullet \mathbf{x} \cdot \mathbf{i}\right) d\boldsymbol{\mu}.$$

(b) *Suppose $f \in \mathbf{L}^1(\mathbb{R}^3)$, and $\widehat{f}$ is also in $\mathbf{L}^1(\mathbb{R}^3)$. For any $\mathbf{x} \in \mathbb{R}^3$, if f is continuous at $\mathbf{x}$, then*

$$f(\mathbf{x}) = \int_{\mathbb{R}^3} \widehat{f}(\boldsymbol{\mu}) \cdot \exp(\boldsymbol{\mu} \bullet \mathbf{x} \cdot \mathbf{i}) d\boldsymbol{\mu}.$$

Ⓔ   *Proof* (a) **Exercise 19E.1.** (*Hint:* Generalize the proof of Theorem 19D.2, p. 509. You may assume (b) is true.)

(b) See Katznelson (1976), §VI.1.12, or Folland (1984), Theorem 8.26.   □

Proposition 19E.2 *If $f, g \in \mathcal{C}(\mathbb{R}^3) \cap \mathbf{L}^1(\mathbb{R}^3)$ are continuous, integrable functions, then $(\widehat{f} = \widehat{g}) \Longleftrightarrow (f = g)$.*   □

Example 19E.3 A ball

For any $\mathbf{x} \in \mathbb{R}^3$, let

$$f(\mathbf{x}) = \begin{cases} 1 & \text{if } \|\mathbf{x}\| \le R; \\ 0 & \text{otherwise.} \end{cases}$$

Thus, $f(\mathbf{x})$ is nonzero on a ball of radius R around zero. Then

$$\widehat{f}(\boldsymbol{\mu}) = \frac{1}{2\pi^2} \left(\frac{\sin(\mu R)}{\mu^3} - \frac{R \cos(\mu R)}{\mu^2} \right),$$

where $\mu := \|\boldsymbol{\mu}\|$.   ◇

Ⓔ   **Exercise 19E.2** Verify Example 19E.3. *Hint:* Argue that, by spherical symmetry, we can rotate $\boldsymbol{\mu}$ without changing the integral, so we can assume that $\boldsymbol{\mu} = (\mu, 0, 0)$. Switch to the spherical coordinate system $(x_1, x_2, x_3) = (r \cdot \cos(\phi), r \cdot \sin(\phi) \sin(\theta), r \cdot \sin(\phi) \cos(\theta))$, to express the Fourier integral as follows:

$$\frac{1}{8\pi^3} \int_0^R \int_0^\pi \int_{-\pi}^\pi \exp\left(\mu \cdot r \cdot \cos(\phi) \cdot \mathbf{i}\right) \cdot r \sin(\phi) d\theta \, d\phi \, dr.$$

Use Claim 1 from Theorem 20B.6, p. 538, to simplify this to $\frac{1}{2\pi^2 \mu} \int_0^R r \cdot \sin(\mu \cdot r) dr$. Now apply integration by parts.   ◆

Ⓔ   **Exercise 19E.3** The Fourier transform of Example 19E.3 contains the terms $\sin(\mu R)/\mu^3$ and $\cos(\mu R)/\mu^2$, both of which go to infinity as $\mu \to 0$. However, these two infinities 'cancel out'. Use l'Hôpital's rule to show that

$$\lim_{\mu \to 0} \widehat{f}(\boldsymbol{\mu}) = \frac{1}{24\pi^3}.$$ ◆

Example 19E.4 A spherically symmetric function

Suppose $f : \mathbb{R}^3 \longrightarrow \mathbb{R}$ was a spherically symmetric function; in other words, $f(\mathbf{x}) = \phi\,(\|\mathbf{x}\|)$ for some function $\phi : \mathbb{R}_+ \longrightarrow \mathbb{R}$. Then, for any $\boldsymbol{\mu} \in \mathbb{R}^3$,

$$\widehat{f}(\boldsymbol{\mu}) = \frac{1}{2\pi^2} \int_0^\infty \phi(r) \cdot r \cdot \sin\left(\|\boldsymbol{\mu}\| \cdot r\right) dr$$

(**Exercise 19E.4**).   ◇

D-**dimensional Fourier transforms** Fourier transforms can be defined in an analogous way in higher dimensions. Let $\mathbf{L}^1(\mathbb{R}^D)$ be the set of all functions $f : \mathbb{R}^D \longrightarrow \mathbb{C}$ such that $\int_{\mathbb{R}^D} |f(\mathbf{x})|\, d\mathbf{x} < \infty$. If $f \in \mathbf{L}^1(\mathbb{R}^D)$, then the *Fourier transform* of f is the function $\widehat{f} : \mathbb{R}^D \longrightarrow \mathbb{C}$ defined:

$$\boxed{\widehat{f}(\boldsymbol{\mu}) := \frac{1}{(2\pi)^D} \int_{\mathbb{R}^D} f(\mathbf{x}) \cdot \exp\left(-\mathbf{x} \bullet \boldsymbol{\mu} \cdot \mathbf{i}\right) d\mathbf{x},}$$

for all $\boldsymbol{\mu} \in \mathbb{R}^D$. Define $\widetilde{\mathbf{L}}^1(\mathbb{R}^D)$ in a manner analogous to the definition of $\widetilde{\mathbf{L}}^1(\mathbb{R}^2)$, p. 509.

Theorem 19E.5 *D*-dimensional Fourier inversion formula

(a) *Suppose* $f \in \widetilde{\mathbf{L}}^1(\mathbb{R}^D)$. *For any* $\mathbf{x} \in \mathbb{R}^D$, *if* f *is continuous at* $\mathbf{x}$, *then*

$$f(\mathbf{x}) = \lim_{M \to \infty} \int_{-M}^{M} \int_{-M}^{M} \cdots \int_{-M}^{M} \widehat{f}(\boldsymbol{\mu}) \cdot \exp\left(\boldsymbol{\mu} \bullet \mathbf{x} \cdot \mathbf{i}\right) d\boldsymbol{\mu}.$$

(b) *Suppose* $f \in \mathbf{L}^1(\mathbb{R}^D)$, *and* $\widehat{f}$ *is also in* $\mathbf{L}^1(\mathbb{R}^D)$. *For any* $\mathbf{x} \in \mathbb{R}^D$, *if* f *is continuous at* $\mathbf{x}$, *then*

$$f(\mathbf{x}) = \int_{\mathbb{R}^D} \widehat{f}(\boldsymbol{\mu}) \cdot \exp(\boldsymbol{\mu} \bullet \mathbf{x} \cdot \mathbf{i}) d\boldsymbol{\mu}.$$

Proof (a) **Exercise 19E.5.**   Ⓔ

(b) See Katznelson (1976), §VI.1.12, or Folland (1984), Theorem 8.26.   □

Exercise 19E.6   State and prove *D*-dimensional versions of all results in   Ⓔ
§19B.   ♦

Evil twins of the multidimensional Fourier transform Just as with the one-dimensional Fourier transform, the mathematics literature contains at least four different definitions of the multidimensional Fourier transform. Instead of the transform we have defined here, some books use what we will call the *opposite Fourier transform*:

$$\check{f}(\boldsymbol{\mu}) := \int_{\mathbb{R}^D} f(\mathbf{x}) \exp(-\mathbf{i}\,\mathbf{x} \bullet \boldsymbol{\mu}) d\mathbf{x},$$

with inverse transform

$$f(\mathbf{x}) = \frac{1}{(2\pi)^D} \int_{\mathbb{R}^D} \check{f}(\boldsymbol{\mu}) \exp(\mathbf{i}\mathbf{x} \bullet \boldsymbol{\mu}) d\boldsymbol{\mu}.$$

Other books instead use the *symmetric Fourier transform*:

$$\widehat{f}(\boldsymbol{\mu}) := \frac{1}{(2\pi)^{D/2}} \int_{\mathbb{R}^D} f(\mathbf{x}) \exp(-\mathbf{i}\mathbf{x} \bullet \boldsymbol{\mu}) d\mathbf{x},$$

with inverse transform

$$f(\mathbf{x}) = \frac{1}{(2\pi)^{D/2}} \int_{\mathbb{R}^D} \widehat{f}(\boldsymbol{\mu}) \exp(\mathbf{i}\mathbf{x} \bullet \boldsymbol{\mu}) d\boldsymbol{\mu}.$$

Finally, some books use the *canonical Fourier transform*:

$$\widetilde{f}(\boldsymbol{\mu}) := \int_{\mathbb{R}^D} f(\mathbf{x}) \exp(-2\pi\mathbf{i}\mathbf{x} \bullet \boldsymbol{\mu}) d\mathbf{x},$$

with inverse transform

$$f(\mathbf{x}) = \int_{\mathbb{R}^D} \widetilde{f}(\boldsymbol{\mu}) \exp(2\pi\mathbf{i}\mathbf{x} \bullet \boldsymbol{\mu}) d\boldsymbol{\mu}.$$

For any $f \in \mathbf{L}^1(\mathbb{R}^D)$ and any $\boldsymbol{\mu} \in \mathbb{R}^D$, we have the following:

$$\check{f}(\boldsymbol{\mu}) = (2\pi)^D \widehat{f}(\boldsymbol{\mu}) \qquad \text{and} \quad \widehat{f}(\boldsymbol{\mu}) = \frac{1}{(2\pi)^D} \check{f}(\boldsymbol{\mu});$$

$$\widehat{f}(\boldsymbol{\mu}) = (2\pi)^{D/2} \widehat{f}(\boldsymbol{\mu}) \quad \text{and} \quad \widehat{f}(\boldsymbol{\mu}) = \frac{1}{(2\pi)^{D/2}} \widehat{f}(\boldsymbol{\mu}); \qquad (19\text{E}.1)$$

$$\widetilde{f}(\boldsymbol{\mu}) = (2\pi)^D \widehat{f}(2\pi\,\boldsymbol{\mu}) \quad \text{and} \quad \widehat{f}(\boldsymbol{\mu}) = \frac{1}{(2\pi)^D} \widetilde{f}\left(\frac{\boldsymbol{\mu}}{2\pi}\right).$$

Ⓔ (**Exercise 19E.7** Check this.) When comparing or combining formulae from two or more books, *always compare their definitions of the Fourier transform*, and make the appropriate conversions using equations (19E.1), if necessary.

19F Fourier (co)sine transforms on the half-line

Prerequisites: §19A.   Recommended: §7A, §8A.

In §8A, to represent a function on the *symmetric* interval $[-\pi, \pi]$, we used a real Fourier series (with both 'sine' and 'cosine' terms). However, to represent a function on the interval $[0, \pi]$, we found in §7A that it was only necessary to employ half as many terms, using either the Fourier sine series or the Fourier cosine series. A similar phenomenon occurs when we go from functions on the *whole* real line to functions on the positive *half*-line.

Let $\mathbb{R}_{\not{}} := \{x \in \mathbb{R};\ x \geq 0\}$ be the *half-line*: the set of all non-negative real numbers. Let

$$\mathbf{L}^1(\mathbb{R}_{\not{}}) := \left\{ f : \mathbb{R}_{\not{}} \longrightarrow \mathbb{R};\ \int_0^\infty |f(x)| \mathrm{d}x < \infty \right\}$$

be the set of *absolutely integrable* functions on the half-line.

The 'boundary' of the half-line is just the point 0. Thus, we will say that a function f satisfies homogeneous *Dirichlet* boundary conditions if $f(0) = 0$. Likewise, f satisfies homogeneous *Neumann* boundary conditions if $f'(0) = 0$.

If $f \in \mathbf{L}^1(\mathbb{R}_{\not{}})$, then the *Fourier cosine transform* of f is the function $\widehat{f}_{\cos} : \mathbb{R}_{\not{}} \longrightarrow \mathbb{R}$ defined:

$$\widehat{f}_{\cos}(\mu) := \boxed{\frac{2}{\pi} \int_0^\infty f(x) \cdot \cos(\mu x)\mathrm{d}x,} \quad \text{for all } \mu \in \mathbb{R}_{\not{}}.$$

The *Fourier sine transform* of f is the function $\widehat{f}_{\sin} : \mathbb{R}_{\not{}} \longrightarrow \mathbb{R}$ defined:

$$\widehat{f}_{\sin}(\mu) := \boxed{\frac{2}{\pi} \int_0^\infty f(x) \cdot \sin(\mu x)\mathrm{d}x,} \quad \text{for all } \mu \in \mathbb{R}_{\not{}}.$$

In both cases, for the transform to be well-defined, we require $f \in \mathbf{L}^1(\mathbb{R}_{\not{}})$.

Theorem 19F.1 Fourier (co)sine inversion formula

Suppose that $f \in \mathbf{L}^1(\mathbb{R}_{\not{}})$ is piecewise smooth. Then, for any $x \in \mathbb{R}_+$, such that f is continuous at x,

$$f(x) = \lim_{M \to \infty} \int_0^M \widehat{f}_{\cos}(\mu) \cdot \cos(\mu \cdot x)\mathrm{d}\mu,$$

and

$$f(x) = \lim_{M \to \infty} \int_0^M \widehat{f}_{\sin}(\mu) \cdot \sin(\mu \cdot x)\mathrm{d}\mu.$$

The Fourier cosine series also converges at 0. If $f(0) = 0$, then the Fourier sine series converges at 0.

Proof **Exercise 19F.1.** (*Hint:* Imitate the methods of §8C.) □ Ⓔ

19G* Momentum representation and Heisenberg uncertainty

> Anyone who is not shocked by quantum theory has not understood it.
>
> *Niels Bohr*

Prerequisites: §3B, §6B, §19C, §19E.

Let $\omega : \mathbb{R}^3 \times \mathbb{R} \longrightarrow \mathbb{C}$ be the wavefunction of a quantum particle (e.g. an electron). Fix $t \in \mathbb{R}$, and define the 'instantaneous wavefunction' $\omega_t : \mathbb{R}^3 \longrightarrow \mathbb{C}$ by

$\omega_t(\mathbf{x}) = \omega(\mathbf{x}; t)$ for all $\mathbf{x} \in \mathbb{R}^3$. Recall from §3B that ω_t encodes the probability distribution for the classical *position* of the particle at time t. However, ω_t seems to say nothing about the classical *momentum* of the particle. In Example 3B.2, p. 44, we stated (without proof) the wavefunction of a particle with a particular known velocity. Now we make a more general assertion.

Suppose a particle has instantaneous wavefunction $\omega_t : \mathbb{R}^3 \longrightarrow \mathbb{C}$. Let $\widehat{\omega}_t : \mathbb{R}^3 \longrightarrow \mathbb{C}$ be the (three-dimensional) Fourier transform of ω_t, and define $\widetilde{\omega}_t := \widehat{\omega}_t (\mathbf{p}/\hbar)$ for all $\mathbf{p} \in \mathbb{R}^3$. Then $\widetilde{\omega}_t$ is the wavefunction for the particle's classical momentum at time t. That is: if we define $\widetilde{\rho}_t(\mathbf{p}) := |\widetilde{\omega}_t|^2(\mathbf{p})/ \|\widetilde{\omega}_t\|_2^2$ for all $\mathbf{p} \in \mathbb{R}^3$, then $\widetilde{\rho}_t$ is the probability distribution for the particle's classical momentum at time t.

Recall that we can reconstruct ω_t from $\widehat{\omega}_t$ via the inverse Fourier transform. Hence, the (positional) wavefunction ω_t implicitly encodes the (momentum) wavefunction $\widetilde{\omega}_t$, and conversely the (momentum) wavefunction $\widetilde{\omega}_t$ implicitly encodes the (positional) wavefunction ω_t. This answers the question we posed on p. 40 of §3A. The same applies to multiparticle quantum systems.

Suppose an N-particle quantum system has instantaneous (position) *wavefunction $\omega_t : \mathbb{R}^{3N} \longrightarrow \mathbb{C}$. Let $\widehat{\omega}_t : \mathbb{R}^{3N} \longrightarrow \mathbb{C}$ be the* (3N-dimensional) *Fourier transform of ω_t, and define $\widetilde{\omega}_t := \widehat{\omega}_t (\mathbf{p}/\hbar)$ for all $\mathbf{p} \in \mathbb{R}^{3N}$. Then $\widetilde{\omega}_t$ is the joint wavefunction for the classical momenta of all the particles at time t. That is: if we define $\widetilde{\rho}_t(\mathbf{p}) := |\widetilde{\omega}_t|^2(\mathbf{p})/ \|\widetilde{\omega}_t\|_2^2$ for all $\mathbf{p} \in \mathbb{R}^{3N}$, then $\widetilde{\rho}_t$ is the joint probability distribution for the classical momenta of all the particles at time t.*

Because the momentum wavefunction contains exactly the same information as the positional wavefunction, we can reformulate the Schrödinger equation in momentum terms. For simplicity, we will only do this in the case of a single particle. Suppose the particle is subjected to a potential energy function $V : \mathbb{R}^3 \longrightarrow \mathbb{R}$. Let $\widehat{V}$ be the Fourier transform of V, and define

$$\widetilde{V} := \frac{1}{\hbar^3} \widehat{V} \left(\frac{\mathbf{p}}{\hbar}\right)$$

for all $\mathbf{p} \in \mathbb{R}^3$. Then the momentum wavefunction $\widetilde{\omega}$ evolves according to the *momentum Schrödinger equation*:

$$i\partial_t \, \widetilde{\omega}(\mathbf{p}; t) = \frac{\hbar^2}{2m}|\mathbf{p}|^2 \cdot \widetilde{\omega}_t(\mathbf{p}) + (\widetilde{V} * \widetilde{\omega}_t)(\mathbf{p}). \tag{19G.1}$$

(Here, if $\mathbf{p} = (p_1, p_2, p_3)$, then $|\mathbf{p}|^2 := p_1^2 + p_2^2 + p_3^2$.) In particular, if the potential field is trivial, we get the *free* momentum Schrödinger equation:

$$i\partial_t \, \widetilde{\omega}(p_1, p_2, p_3; t) = \frac{\hbar^2}{2m}(p_1^2 + p_2^2 + p_3^2) \cdot \widetilde{\omega}(p_1, p_2, p_3; t).$$

Exercise 19G.1 Verify equation (19G.1) by applying the Fourier trans- ⓔ form to the (positional) Schrödinger equation (3B.3), p. 43. (*Hint:* Use Theorem 19B.7, p. 500, to show that $\widehat{\Delta\omega_t}(\mathbf{p}) = -|\mathbf{p}|^2 \cdot \widehat{\omega}_t(\mathbf{p})$. Use Theorem 19B.2(c), p. 494, to show that $\widehat{(V \cdot \omega_t)}(\mathbf{p}/\hbar) = \widetilde{V} * \widetilde{\omega}_t(\mathbf{p})$.) ◆

Exercise 19G.2 Formulate the momentum Schrödinger equation for a single ⓔ particle confined to a one- or two-dimensional environment. Be careful how you define $\widetilde{V}$. ◆

Exercise 19G.3 Formulate the momentum Schrödinger equation for an N-particle ⓔ quantum system. Be careful how you define $\widetilde{V}$. ◆

Recall that Theorem 19B.8, p. 502, said: if f is a Gaussian distribution, then $\widehat{f}$ is also a 'Gaussian' (after multiplying by a scalar), but the variance of $\widehat{f}$ is inversely proportional to the variance of f. This is an example of a general phenomenon, called *Heisenberg's inequality*. To state this formally, we need some notation. Recall from §6B that $\mathbf{L}^2(\mathbb{R})$ is the set of all *square-integrable* complex-valued functions on $\mathbb{R}$ – that is, all integrable functions $f : \mathbb{R} \longrightarrow \mathbb{C}$ such that $\|f\|_2 < \infty$, where

$$\|f\|_2 := \left(\int_{-\infty}^{\infty} |f(x)|^2 \, dx \right)^{1/2}.$$

If $f \in \mathbf{L}^2(\mathbb{R})$, and $x \in \mathbb{R}$, then define the *uncertainty* of f around x to be

$$\blacktriangle_x (f) := \frac{1}{\|f\|_2^2} \int_{-\infty}^{\infty} |f(y)|^2 \cdot |y - x|^2 \, dy.$$

(In most physics texts, the uncertainty of f is denoted by $\triangle_x f$; however, we will not use this symbol because it looks too much like the Laplacian operator.)

Example 19G.1

(a) If $f \in \mathbf{L}^2(\mathbb{R})$, then $\rho(x) := f(x)^2 / \|f\|_2^2$ is a probability density function on $\mathbb{R}$ (why?). If $\overline{x}$ is the mean of the distribution ρ (i.e. $\overline{x} = \int_{-\infty}^{\infty} x \, \rho(x) dx$), then

$$\blacktriangle_{\overline{x}} (f) = \int_{-\infty}^{\infty} \rho(y) \cdot |y - \overline{x}|^2 \, dy$$

is the *variance* of the distribution. Thus, if ρ describes the probability density of a random variable $X \in \mathbb{R}$, then $\overline{x}$ is the expected value of X, and $\blacktriangle_{\overline{x}} (\omega)$ measures the degee of 'uncertainty' we have about the value of X. If $\blacktriangle_{\overline{x}} (\omega)$ is small, then the distribution is tightly concentrated around $\overline{x}$, so we can be fairly confident that X is close to $\overline{x}$. If $\blacktriangle_{\overline{x}} (\omega)$ is large, then the distribution is broadly dispersed around $\overline{x}$, so we really have only a vague idea where X might be.

(b) In particular, suppose $f(x) = \exp(-x^2/4\sigma^2)$. Then

$$\frac{f^2}{\|f\|_2^2} = \frac{1}{\sigma\sqrt{2\pi}} \exp\left(\frac{-x^2}{2\sigma^2}\right)$$

is a *Gaussian distribution* with mean 0 and variance σ^2. It follows that
$\blacktriangle_0 (f) = \sigma^2$.

(c) Suppose $\omega : \mathbb{R} \times \mathbb{R} \longrightarrow \mathbb{C}$ is a one-dimensional wavefunction, and fix $t \in \mathbb{R}$; thus, the
function $\rho_t(x) = |\omega_t|^2(x)/\|\omega_t\|_2$ is the probability density for the classical position
of the particle at time t in a one-dimensional environment (e.g. an electron in a thin
wire). If $\bar{x}$ is the mean of this distribution, then $\blacktriangle_{\bar{x}} (\omega_t)$ is the variance of the distribution; this reflects our degree of uncertainty about the particle's classical position at
time t.   ◇

Why is the subscript x present in $\blacktriangle_x (f)$? Why not just measure the uncertainty around the *mean* of the distribution as in Example 19G.1? Three reasons. First, because the distribution might not *have* a well-defined mean (i.e.
the integral $\int_{-\infty}^{\infty} x \rho(x)dx$ might not converge). Second, because it is sometimes
useful to measure the uncertainty around other points in $\mathbb{R}$ besides the mean
value. Third, because we do not need to use the mean value to state the next
result.

Theorem 19G.2 Heisenberg's inequality
*Let $f \in \mathbf{L}^2(\mathbb{R})$ be a nonzero function, and let $\widehat{f}$ be its Fourier transform. Then for
any $x, \mu \in \mathbb{R}$, we have $\blacktriangle_x (f) \cdot \blacktriangle_\mu (\widehat{f}) \geq 1/4$.*

Example 19G.3
(a) If $f(x) = \exp\left(-x^2/4\sigma^2\right)$, then

$$\widehat{f}(p) = \frac{\sigma}{\sqrt{\pi}} \exp\left(-\sigma^2 p^2\right)$$

(E) **(Exercise 19G.4).**[4] Thus,

$$\frac{\widehat{f}(p)^2}{\|\widehat{f}\|_2^2} = \frac{2\sigma}{\sqrt{2\pi}} \exp(-2\sigma^2 p^2)$$

is a Gaussian distribution with mean 0 and variance $1/4\sigma^2$. Thus, Example 19G.1(b)
says that $\blacktriangle_0 (f) = \sigma^2$ and $\blacktriangle_0 (\widehat{f}) = 1/4\sigma^2$. Thus,

$$\blacktriangle_0 (f) \cdot \blacktriangle_0 (\widehat{f}) = \frac{\sigma^2}{4\sigma^2} = \frac{1}{4}.$$

(b) Suppose $\omega_t \in \mathbf{L}^2(\mathbb{R})$ is the instantaneous wavefunction for the position of a particle at
time t, so that $\widetilde{\omega}_t(\mathbf{p}) = \widehat{\omega}_t (\mathbf{p}/\hbar)$ is the instantaneous wavefunction for the momentum of

----

[4] *Hint:* Set $\alpha := 2\sigma$ in Theorem 19B.5, p. 499, and then apply it to Theorem 19B.8(a), p. 502.

the particle at time t. Then Heisenberg's inequality becomes *Heisenberg's uncertainty principle.* For any $x, p \in \mathbb{R}$,

$$\blacktriangle_x(\omega_t) \geq \frac{\hbar^2}{4 \cdot \blacktriangle_p(\widetilde{\omega}_t)} \quad \text{and} \quad \blacktriangle_p(\widetilde{\omega}_t) \geq \frac{\hbar^2}{4 \cdot \blacktriangle_x(\omega_t)}$$

(**Exercise 19G.5**). In other words: if our uncertainty $\blacktriangle_\mu(\widetilde{\omega}_t)$ about the particle's momentum is small, then our uncertainty $\blacktriangle_x(\omega_t)$ about its position must be big. Conversely, if our uncertainty $\blacktriangle_x(\omega_t)$ about the particle's position is small, then our uncertainty $\blacktriangle_\mu(\widetilde{\omega}_t)$ about its momentum must be big.

In physics popularizations, the uncertainty principle is usually explained as a practical problem of measurement precision: any attempt to measure an electron's position (e.g. by deflecting photons off of it) will impart some unpredictable momentum into the particle. Conversely, any attempt to measure its momentum disturbs its position. However, as you can see, Heisenberg's uncertainty principle is actually an abstract mathematical *theorem* about Fourier transforms – it has nothing to do with the limitations of experimental equipment or the unpredictable consequences of photon bombardment. ◇

Proof of Heisenberg's inequality For simplicity, assume $\lim_{x \to \pm\infty} x \, |f(x)|^2 = 0$.

Case $x = \mu = 0$ Define $\xi : \mathbb{R} \longrightarrow \mathbb{R}$ by $\xi(x) := x$ for all $x \in \mathbb{R}$. Thus, $\xi'(x) := 1$ for all $x \in \mathbb{R}$. Observe that

$$\|f \cdot \xi\|_2^2 = \int_{-\infty}^{\infty} |f \cdot \xi|^2(x) \mathrm{d}x = \frac{\|f\|_2^2}{\|f\|_2^2} \int_{-\infty}^{\infty} |f(x)|^2 |x|^2 \, \mathrm{d}x$$

$$= \|f\|_2^2 \, \blacktriangle_0(f). \tag{19G.2}$$

Also, Theorem 19B.7, p. 500, implies that

$$\widehat{(f')} = \mathbf{i} \cdot \xi \cdot \widehat{f}. \tag{19G.3}$$

Now,

$$\|f\|_2^2 := \int_{-\infty}^{\infty} |f|^2(x) \mathrm{d}x \underset{(\P)}{=} \xi(x) \cdot |f(x)|^2 \Big|_{x=-\infty}^{x=\infty} - \int_{-\infty}^{\infty} \xi(x) \cdot (|f|^2)'(x) \mathrm{d}x$$

$$\underset{(\dagger)}{=} -\int_{-\infty}^{\infty} x \cdot (|f|^2)'(x) \mathrm{d}x \underset{(*)}{=} -\int_{-\infty}^{\infty} x \cdot 2 \, \mathrm{Re}\left[f'(x)\overline{f}(x)\right] \mathrm{d}x$$

$$= -2 \, \mathrm{Re}\left[\int_{-\infty}^{\infty} x \, \overline{f}(x) f'(x) \mathrm{d}x\right]. \tag{19G.4}$$

Here, $(\P)$ is integration by parts, because $\xi'(x) = 1$. Next, $(\dagger)$ is because $\lim_{x \to \pm\infty} x \, |f(x)|^2 = 0$. Meanwhile, $(*)$ is because $|f|^2(x) = f(x)\overline{f}(x)$, so that $(|f|^2)'(x) = f'(x)\overline{f}(x) + f(x)\overline{f}'(x) = 2 \, \mathrm{Re}\left[f'(x)\overline{f}(x)\right]$, where the last step uses the identity $z + \bar{z} = 2\,\mathrm{Re}\,[z]$, with $z = f'(x)\overline{f}(x)$. Thus,

$$\frac{1}{4}\,\|f\|_2^4 \underset{(\ddagger)}{=} \frac{2^2}{4}\,\mathrm{Re}\left[\int_{-\infty}^{\infty} x\,\overline{f}(x)f'(x)\mathrm{d}x\right]^2 \le \left|\int_{-\infty}^{\infty} x\,\overline{f}(x)f'(x)\mathrm{d}x\right|^2$$

$$= \left|\langle \xi\,\overline{f},\, f'\rangle\right|^2 \underset{(\mathrm{CBS})}{\le} \left\|\xi\,\overline{f}\right\|_2^2 \cdot \left\|f'\right\|_2^2 = \left\|\xi\,f\right\|_2^2 \cdot \left\|f'\right\|_2^2$$

$$\underset{(*)}{=} \blacktriangle_0\,(f) \cdot \|f\|_2^2 \cdot \left\|f'\right\|_2^2 \underset{(\mathrm{Pl})}{=} \blacktriangle_0(f) \cdot \|f\|_2^2 \cdot (2\pi)\left\|\widehat{(f')}\right\|_2^2$$

$$\underset{(\dagger)}{=} 2\pi\,\blacktriangle_0(f) \cdot \|f\|_2^2 \cdot \left\|\mathrm{i}\xi\,\widehat{f}\right\|_2^2 = 2\pi\,\blacktriangle_0(f) \cdot \|f\|_2^2 \cdot \left\|\xi\,\widehat{f}\right\|_2^2$$

$$\underset{(*)}{=} 2\pi\,\blacktriangle_0(f) \cdot \|f\|_2^2 \cdot \blacktriangle_0\,(\widehat{f}) \cdot \left\|\widehat{f}\right\|_2^2 \underset{(\mathrm{Pl})}{=} \blacktriangle_0(f) \cdot \|f\|_2^2 \cdot \blacktriangle_0\,(\widehat{f}) \cdot \|f\|_2^2$$

$$= \blacktriangle_0\,(f) \cdot \blacktriangle_0\,(\widehat{f}) \cdot \|f\|_2^4.$$

Cancelling $\|f\|_2^4$ from both sides of this equation, we get $1/4 \le \blacktriangle_0\,(f) \cdot \blacktriangle_0\,(\widehat{f})$, as desired.

Here, $(\ddagger)$ is by equation (19G.4), while (CBS) is the Cauchy–Bunyakowski–Schwarz inequality (Theorem 6B.5, p. 112). Both $(*)$ are by equation (19G.2). Both (Pl) are by Plancharel's theorem (Corollary 19C.2, p. 507) . Finally, $(\dagger)$ is by equation (19G.3).

Ⓔ *Case $x \ne 0$ and/or $\mu \ne 0$* **Exercise 19G.6.** (*Hint:* Combine the case $x = \mu = 0$ with Theorem 19B.3.)   □

Ⓔ **Exercise 19G.7** State and prove a form of Heisenberg's inequality for a function $f \in \mathbf{L}^1(\mathbb{R}^D)$ for $D \ge 2$. (*Hint:* You must compute the 'uncertainty' in one coordinate at a time. Integrate out all the other dimensions to reduce the D-dimensional problem to a one-dimensional problem, and then apply Theorem 19G.2.)   ◆

19H* Laplace transforms

Recommended: §19A, §19B.

The Fourier transform $\widehat{f}$ is only well-defined if $f \in \mathbf{L}^1(\mathbb{R})$, which implies that $\lim_{t \to \pm\infty} |f(t)| = 0$ relatively 'quickly'.[5] This is often inconvenient in physical models where the function $f(t)$ is bounded away from zero, or even grows without bound as $t \to \pm\infty$. In some cases, we can handle this problem using a *Laplace transform*, which can be thought of as a Fourier transform 'rotated by $90°$ in the complex plane'. The price we pay is that we must work on the half-infinite line $\mathbb{R}_{\not{+}} := [0, \infty)$, instead of the entire real line.

---

[5] Actually, we can define $\widehat{f}$ if $f \in \mathbf{L}^p(\mathbb{R})$ for any $p \in [1, 2]$, as discussed in §19C. However, this is not really that much of an improvement; we still need $\lim_{t \to \pm\infty} |f(t)| = 0$ 'quickly'.

Let $f : \mathbb{R}_{\not{}} \longrightarrow \mathbb{C}$. We say that f has *exponential growth* if there are constants $\alpha \in \mathbb{R}$ and $K > 0$ such that

$$|f(t)| \leq K \, e^{\alpha t}, \quad \text{for all } t \in \mathbb{R}_{\not{}}. \tag{19H.1}$$

If $\alpha > 0$, then inequality (19H.1) even allows $\lim_{t \to \infty} f(t) = \infty$, as long as $f(t)$ doesn't grow 'too quickly'. (However, if $\alpha < 0$, then inequality (19H.1) requires $\lim_{t \to \infty} f(t) = 0$ exponentially fast.) The *exponential order* of f is the infimum of all α satisfying inequality (19H.1). Thus, if f has exponential order α_0, then inequality (19H.1) is true for all $\alpha > \alpha_0$ (but may or may not be true for $\alpha = \alpha_0$).

Example 19H.1
(a) Fix $r \geq 0$. If $f(t) = t^r$ for all $t \in \mathbb{R}_{\not{}}$, then f has exponential order 0.
(b) Fix $r < 0$ and $t_0 > 0$. If $f(t) = (t + t_0)^r$ for all $t \in \mathbb{R}_{\not{}}$, then f has exponential order 0. (However, if $t_0 \leq 0$, then $f(t)$ does *not* have exponential growth, because, in this case, $\lim_{t \to -t_0} f(t) = \infty$, so inequality (19H.1) is always false near $-t_0$.)
(c) Fix $\alpha \in \mathbb{R}$. If $f(t) = e^{\alpha t}$ for all $t \in \mathbb{R}_{\not{}}$, then f has exponential order α.
(d) Fix $\mu \in \mathbb{R}$. If $f(t) = \sin(\mu t)$ or $f(t) = \cos(\mu t)$, then f has exponential order 0.
(e) If $f : \mathbb{R}_{\not{}} \longrightarrow \mathbb{C}$ has exponential order α, and $r \in \mathbb{R}$ is any constant, then $r \, f$ also has exponential order α. If $g : \mathbb{R}_{\not{}} \longrightarrow \mathbb{C}$ has exponential order β, then $f + g$ has exponential order at most $\max\{\alpha, \beta\}$, and $f \cdot g$ has exponential order $\alpha + \beta$.
(f) Combining (a) and (e): any polynomial $f(t) = c_n t^n + \cdots + c_2 t^2 + c_1 t + c_0$ has exponential order 0. Likewise, combining (d) and (e): any trigonometric polynomial has exponential order 0. $\diamond$

Exercise 19H.1 Verify Example 19H.1(a)–(f). ♦ Ⓔ

If $c = x + y\mathbf{i}$ is a complex number, recall that $\mathrm{Re}\,[c] := x$. Let $\mathbb{H}_\alpha := \{c \in \mathbb{C}; \, \mathrm{Re}\,[c] > \alpha\}$ be the half of the complex plane to the right of the vertical line $\{c \in \mathbb{C}; \, \mathrm{Re}\,[c] = \alpha\}$. In particular, $\mathbb{H}_0 := \{c \in \mathbb{C}; \, \mathrm{Re}\,[c] > 0\}$. If f has exponential order α, then the *Laplace transform* of f is the function $\mathcal{L}[f] : \mathbb{H}_\alpha \longrightarrow \mathbb{C}$ defined as follows:

$$\text{For all } s \in \mathbb{H}_\alpha, \quad \mathcal{L}[f](s) := \boxed{\int_0^\infty f(t) \, e^{-ts} \, dt.} \tag{19H.2}$$

Lemma 19H.2 *If f has exponential order α, then the integral* (19H.2) *converges for all $s \in \mathbb{H}_\alpha$. Thus, $\mathcal{L}[f]$ is well-defined on $\mathbb{H}_\alpha$.*

Proof **Exercise 19H.2.** □ Ⓔ

Example 19H.3
(a) If $f(t) = 1$, then f has exponential order 0. For all $s \in \mathbb{H}_0$,

$$\mathcal{L}[f](s) = \int_0^\infty e^{-ts} \, dt = \frac{-e^{-ts}}{s} \bigg|_{t=0}^{t=\infty} \underset{(*)}{=} \frac{-(0-1)}{s} = \frac{1}{s}.$$

Here $(*)$ is because $\mathrm{Re}\,[s] > 0$.

(b) If $\alpha \in \mathbb{R}$ and $f(t) = e^{\alpha t}$, then f has exponential order α. For all $s \in \mathbb{H}_\alpha$,

$$\mathcal{L}[f](s) = \int_0^\infty e^{\alpha t}\, e^{-ts}\, dt = \int_0^\infty e^{t(\alpha - s)}\, dt$$

$$= \frac{e^{t(\alpha-s)}}{\alpha - s}\bigg|_{t=0}^{t=\infty} \underset{(*)}{=} \frac{(0-1)}{\alpha - s} = \frac{1}{s - \alpha}.$$

Here $(*)$ is because $\mathrm{Re}\,[\alpha - s] < 0$, because $\mathrm{Re}\,[s] > \alpha$, because $s \in \mathbb{H}_\alpha$.

(c) If $f(t) = t$, then f has exponential order 0. For all $s \in \mathbb{H}_0$,

$$\mathcal{L}[f](s) = \int_0^\infty t\, e^{-ts}\, dt \underset{(p)}{=} \frac{-t\, e^{-ts}}{s}\bigg|_{t=0}^{t=\infty} - \int_0^\infty \frac{-e^{-ts}}{s}\, dt$$

$$= \frac{(0-0)}{s} - \frac{e^{-ts}}{s^2}\bigg|_{t=0}^{t=\infty} = \frac{-(0-1)}{s^2} = \frac{1}{s^2},$$

where (p) is integration by parts.

(d) Suppose $f : \mathbb{R}_{\not{+}} \longrightarrow \mathbb{C}$ has exponential order $\alpha < 0$. Extend f to a function $f : \mathbb{R} \longrightarrow \mathbb{C}$ by defining $f(t) = 0$ for all $t < 0$. Then the Fourier transform $\widehat{f}$ of f is well-defined, and, for all $\mu \in \mathbb{R}$, we have $2\pi\, \widehat{f}(\mu) = \mathcal{L}[f](\mu \mathbf{i})$ (**Exercise 19H.3**).

(e) Fix $\mu \in \mathbb{R}$. If $f(t) = \cos(\mu t)$, then $\mathcal{L}[f] = s/(s^2 + \mu^2)$. If $f(t) = \sin(\mu t)$, then $\mathcal{L}[f] = \mu/(s^2 + \mu^2)$.   ◇

Ⓔ **Exercise 19H.4**   Verify Example 19H.3(e). (*Hint:* Recall that $\exp(\mathbf{i}\mu t) = \cos(\mu t) + \mathbf{i}\sin(\mu t)$.)   ◆

Example 19H.3(d) suggests that most properties of the Fourier transform should translate into properties of the Laplace transform, and vice versa. First, like Fourier, the Laplace transform is invertible.

Theorem 19H.4 Laplace inversion formula
Suppose $f : \mathbb{R}_{\not{+}} \longrightarrow \mathbb{C}$ has exponential order α, and let $F := \mathcal{L}[f]$. Then for any fixed $s_r > \alpha$, and any $t \in \mathbb{R}_{\not{+}}$, we have

$$f(t) = \frac{1}{2\pi\mathbf{i}} \int_{-\infty}^\infty F(s_r + s_{\mathbf{i}}\mathbf{i}) \exp(t s_r + t s_{\mathbf{i}}\mathbf{i}) ds_{\mathbf{i}}. \qquad (19\mathrm{H}.3)$$

In particular, if $g : \mathbb{R}_{\not{+}} \longrightarrow \mathbb{C}$, and $\mathcal{L}[g] = \mathcal{L}[f]$ on any infinite vertical strip in $\mathbb{C}$, then we must have $g = f$.

Ⓔ *Proof* **Exercise 19H.5**. (*Hint:* Use an argument similar to Example 19H.3(d) to represent F as a Fourier transform. Then apply the Fourier inversion formula.)   □

The integral (19H.3) is called the *Laplace inversion integral*, and it is denoted by $\mathcal{L}^{-1}[F]$. The integral (19H.3) is sometimes written as follows:

$$f(t) = \frac{1}{2\pi\mathbf{i}} \int_{s_r - \infty\mathbf{i}}^{s_r + \infty\mathbf{i}} F(s) \exp(ts) ds.$$

The integral (19H.3) can be treated as a *contour integral* along the vertical line $\{c \in \mathbb{C}; \ \mathrm{Re}\,[c] = s_r\}$ in the complex plane, and evaluated using residue calculus.[6] However, in many situations, it is neither easy nor particularly necessary to compute equation (19H.3) explicitly; instead, we can determine the inverse Laplace transform 'by inspection', by simply writing F as a sum of Laplace transforms of functions we recognize. Most books on ordinary differential equations contain an extensive table of Laplace transforms and their inverses, which is useful for this purpose.

Example 19H.5 Suppose $F(s) = \frac{3}{s} + \frac{5}{s-2} + \frac{7}{s^2}$. Then by inspecting Example 19H.3(a)–(c), we deduce that $f(t) = \mathcal{L}^{-1}[F](t) = 3 + 5e^{2t} + 7t$.   ◇

Most of the results about Fourier transforms from Section 19B have equivalent formulations for Laplace transforms.

Theorem 19H.6 Properties of the Laplace transform

Let $f : \mathbb{R}_{+} \longrightarrow \mathbb{C}$ have exponential order α.

(a) (Linearity) *Let $g : \mathbb{R}_{+} \longrightarrow \mathbb{C}$ have exponential order β, and let $\gamma = \max\{\alpha, \beta\}$. Let $b, c \in \mathbb{C}$. Then $bf + cg$ has exponential order at most γ, and, for all $s \in \mathbb{H}_{\gamma}$, we have* $\mathcal{L}[bf + cg](s) = b\mathcal{L}[f](s) + c\mathcal{L}[g](s).$

(b) (Transform of a derivative)

   (i) *Suppose $f \in \mathcal{C}^{1}(\mathbb{R}_{+})$ (i.e. f is continuously differentiable on $\mathbb{R}_{+}$) and f' has exponential order β. Let $\gamma = \max\{\alpha, \beta\}$. Then, for all $s \in \mathbb{H}_{\gamma}$,*

$$\mathcal{L}[f'](s) = s\mathcal{L}[f](s) - f(0).$$

   (ii) *Suppose $f \in \mathcal{C}^{2}(\mathbb{R}_{+})$, f' has exponential order α_1, and f'' has exponential order α_2. Let $\gamma = \max\{\alpha, \alpha_1, \alpha_2\}$. Then, for all $s \in \mathbb{H}_{\gamma}$,*

$$\mathcal{L}[f''](s) = s^2\mathcal{L}[f](s) - f(0)s - f'(0).$$

   (iii) *Let $N \in \mathbb{N}$, and suppose $f \in \mathcal{C}^{N}(\mathbb{R}_{+})$. Suppose $f^{(n)}$ has exponential order α_n for all $n \in [1 \ldots N]$. Let $\gamma = \max\{\alpha, \alpha_1, \ldots, \alpha_N\}$. Then, for all $s \in \mathbb{H}_{\gamma}$,*

$$\mathcal{L}[f^{(N)}](s) = s^N \mathcal{L}[f](s) - f(0)s^{N-1} - f'(0)s^{N-2}$$
$$- f''(0)s^{N-3} - \cdots - f^{(N-2)}(0)\,s - f^{(N-1)}(0).$$

(c) (Derivative of a transform) *For all $n \in \mathbb{N}$, the function $g_n(t) = t^n f(t)$ also has exponential order α. If f is piecewise continuous, then the function $F = \mathcal{L}[f] : \mathbb{H}_{\alpha} \longrightarrow \mathbb{C}$ is (complex)-differentiable,[7] and, for all $s \in \mathbb{H}_{\alpha}$, we have $F'(s) = -\mathcal{L}[g_1](s)$, $F''(s) = \mathcal{L}[g_2](s)$, and, in general, $F^{(n)}(s) = (-1)^n \mathcal{L}[g_n](s).$*

---

[6] See §18H, p. 477, for a discussion of residue calculus and its application to improper integrals. See Fisher (1999), §5.3, for some examples of computing Laplace inversion integrals using this method.

[7] See §18A, p. 421, for more about complex differentiation.

(d) (Translation) *Fix $T \in \mathbb{R}_{\neq}$ and define $g : \mathbb{R}_{\neq} \longrightarrow \mathbb{C}$ by $g(t) = f(t - T)$ for $t \geq T$ and $g(t) = 0$ for $t \in [0, T)$. Then g also has exponential order α. For all $s \in \mathbb{H}_{\alpha}$,*
$$\mathcal{L}[g](s) = e^{-Ts} \mathcal{L}[f](s).$$

(e) (Dual translation) *For all $\beta \in \mathbb{R}$, the function $g(t) = e^{\beta t} f(t)$ has exponential order $\alpha + \beta$. For all $s \in \mathbb{H}_{\alpha+\beta}$, $\mathcal{L}[g](s) = \mathcal{L}[f](s - \beta)$.*

Ⓔ   *Proof*   **Exercise 19H.6.** (*Hint:* Imitate the proofs of Theorems 19B.2(a), p. 499, 19B.7, p. 500, and 19B.3, p. 499.)   □

Ⓔ   **Exercise 19H.7**   Show by a counterexample that Theorem 19H.6(d) is *false* if $T < 0$.   ◆

Corollary 19H.7   *Fix $n \in \mathbb{N}$. If $f(t) = t^n$ for all $t \in \mathbb{R}_{\neq}$, then $\mathcal{L}[f](s) = n!/s^{n+1}$ for all $s \in \mathbb{H}_0$.*

Ⓔ   *Proof*   **Exercise 19H.8.** (*Hint:* Combine Theorem 19H.6(c) with Example 19H.3(a).)   □

Ⓔ   **Exercise 19H.9**   Combine Corollary 19H.7 with Theorem 19H.6(b) to get a formula for the Laplace transform of any polynomial.   ◆

Ⓔ   **Exercise 19H.10**   Combine Theorem 19H.6(c), (d) with Example 19H.3(a) to get a formula for the Laplace transform of $f(t) = 1/(1 + t)^n$ for all $n \in \mathbb{N}$.   ◆

Corollary 19H.7 does not help us to compute the Laplace transform of $f(t) = t^r$ when r is not an integer. To do this, we must introduce the *gamma function* $\Gamma : \mathbb{R}_{\neq} \longrightarrow \mathbb{C}$, which is defined:

$$\Gamma(r) := \boxed{\int_0^{\infty} t^{r-1} e^{-t} \, dt,} \quad \text{for all } r \in \mathbb{R}_{\neq}.$$

This is regarded as a 'generalized factorial' because of the following properties.

Lemma 19H.8
(a) $\Gamma(1) = 1$.
(b) *For any $r \in \mathbb{R}_{\neq}$, we have $\Gamma(r + 1) = r \cdot \Gamma(r)$.*
(c) *Thus, for any $n \in \mathbb{N}$, we have $\Gamma(n + 1) = n!$. (For example, $\Gamma(5) = 4! = 24$.)*

Ⓔ   *Proof*   **Exercise 19H.11.**   □

Ⓔ   **Exercise 19H.12**
(a) Show that $\Gamma(1/2) = \sqrt{\pi}$.
(b) Deduce that $\Gamma(3/2) = \sqrt{\pi}/2$ and $\Gamma(5/2) = \frac{3}{4}\sqrt{\pi}$.   ◆

The gamma function is useful when computing Laplace transforms because of the following result.

Proposition 19H.9 Laplace transform of $f(t) = t^r$

Fix $r > -1$, and let $f(t) := t^r$ for all $t \in \mathbb{R}_{\not=}$. Then $\mathcal{L}[f](s) = \Gamma(r + 1)/s^{r+1}$ for all $s \in \mathbb{H}_0$.

Proof **Exercise 19H.13.** □ Ⓔ

Remark If $r < 0$, then, technically, $f(t) = t^r$ does *not* have exponential growth, as noted in Example 19H.1(b). Hence Lemma 19H.2 does not apply. However, the Laplace transform integral (19H.2) converges in this case anyway, because, although $\lim_{t \searrow 0} t^r = \infty$, it goes to infinity 'slowly', so that the integral $\int_0^1 t^r \, dt$ is still finite.

Example 19H.10
(a) If $r \in \mathbb{N}$ and $f(t) = t^r$, then Proposition 19H.9 and Lemma 19H.8(c) together imply
$\mathcal{L}[f](s) = r!/s^{r+1}$, in agreement with Corollary 19H.7.
(b) Let $r = 1/2$. Then $f(t) = \sqrt{t}$, and Proposition 19H.9 says $\mathcal{L}[f](s) = \Gamma(3/2)/s^{3/2} = \sqrt{\pi}/2s^{3/2}$, where the last step is by Exercise 19H.12(b).
(c) Let $r = -1/2$. Then $f(t) = 1/\sqrt{t}$, and Proposition 19H.9 says $\mathcal{L}[f](s) = \Gamma(1/2)/s^{1/2} = \sqrt{\pi/s}$, where the last step is by Exercise 19H.12(a). ◇

Theorem 19B.2(c) showed how the Fourier transform converts function convolution into multiplication. A similar property holds for the Laplace transform. Let $f, g : \mathbb{R}_{\not=} \longrightarrow \mathbb{C}$ be two functions. The *convolution* of f, g is the function $f * g : \mathbb{R}_{\not=} \longrightarrow \mathbb{C}$ defined:

$$f * g(T) := \int_0^T f(T - t)g(t)dt, \quad \text{for all } T \in \mathbb{R}_{\not=}.$$

Note that $f * g(T)$ is an integral over a *finite* interval $[0, T]$; thus it is well-defined, no matter how fast $f(t)$ and $g(t)$ grow as $t \to \infty$.

Theorem 19H.11 *Let $f, g : \mathbb{R}_{\not=} \longrightarrow \mathbb{C}$ have exponential order. Then $\mathcal{L}[f * g] = \mathcal{L}[f] \cdot \mathcal{L}[g]$ wherever all these functions are defined.*

Proof **Exercise 19H.14.** □ Ⓔ

Theorem 19H.6(b) makes the Laplace transform a powerful tool for solving linear ordinary differential equations.

Proposition 19H.12 Laplace solution to linear ODE

Let $f, g : \mathbb{R}_{\not=} \longrightarrow \mathbb{C}$ have exponential order α and β, respectively, and let $\gamma = \max\{\alpha, \beta\}$. Let $F := \mathcal{L}[f]$ and $G := \mathcal{L}[g]$. Let $c_0, c_1, \ldots, c_n \in \mathbb{C}$ be constants.

Then f and g satisfy the linear ODE

$$g = c_0 f + c_1 f' + c_2 f'' + \cdots + c_n f^{(n)} \tag{19H.4}$$

if, and only if, F and G satisfy the algebraic equation

$$
\begin{aligned}
G(s) = {} & \left(c_0 + c_1 s + c_2 s^2 + c_3 s^3 + \cdots + c_n s^n \right) & F(s) \\
& - \left(c_1 + c_2 s^1 + c_3 s^2 + \cdots + c_n s^{n-1} \right) & f(0) \\
& - \left(c_2 + c_3 s + \cdots + c_n s^{n-2} \right) & f'(0) \\
& \;\;\ddots \qquad\qquad \vdots \qquad\quad \vdots \;\; \vdots \\
& - (c_{n-1} + c_n s) & f^{(n-2)}(0) \\
& - c_n & f^{(n-1)}(0)
\end{aligned}
$$

for all $s \in \mathbb{H}_\gamma$. In particular, f satisfies ODE (19H.4) and homogeneous boundary conditions $f(0) = f'(0) = f''(0) = \cdots = f^{(n-1)}(0) = 0$ if, and only if,

$$F(s) = \frac{G(s)}{c_0 + c_1 s + c_2 s^2 + c_3 s^3 + \cdots + c_n s^n}$$

for all $s \in \mathbb{H}_\gamma$

Ⓔ *Proof* **Exercise 19H.15.** (*Hint:* Apply Theorem 19H.6(b), then reorder terms.) ☐

Laplace transforms can also be used to solve *partial* differential equations. Let $\mathbb{X} \subseteq \mathbb{R}$ be some one-dimensional domain, let $f : \mathbb{X} \times \mathbb{R}_{\not{}} \longrightarrow \mathbb{C}$, and write $f(x; t)$ as $f_x(t)$ for all $x \in \mathbb{X}$. Fix $\alpha \in \mathbb{R}$. For all $x \in \mathbb{X}$ suppose that f_x has exponential order α, so that $\mathcal{L}[f_x]$ is a function $\mathbb{H}_\alpha \longrightarrow \mathbb{C}$. Then we define $\mathcal{L}[f] : \mathbb{X} \times \mathbb{H}_\alpha \longrightarrow \mathbb{C}$ by $\mathcal{L}[f](x; s) = \mathcal{L}[f_x](s)$ for all $x \in \mathbb{X}$ and $s \in \mathbb{H}_\alpha$.

Proposition 19H.13 *Suppose $\partial_x f(x, t)$ is defined for all $(x, t) \in \operatorname{int}(\mathbb{X}) \times \mathbb{R}_+$. Then $\partial_x \mathcal{L}[f](x, s)$ is defined for all $(x, s) \in \operatorname{int}(\mathbb{X}) \times \mathbb{H}_\alpha$, and $\partial_x \mathcal{L}[f](x, s) = \mathcal{L}[\partial_x f](x, s)$.*

Ⓔ *Proof* **Exercise 19H.16.** (*Hint:* Apply Theorem G.1, p. 571, to the Laplace transform integral (19H.2).) ☐

By iterating Proposition 19H.13, we have $\partial_x^n \mathcal{L}[f](x, s) = \mathcal{L}[\partial_x^n f](x, s)$ for any $n \in \mathbb{N}$. Through Proposition 19H.13 and Theorem 19H.6(b), we can convert a PDE about f into an ODE involving only the x-derivatives of $\mathcal{L}[f]$.

Example 19H.14 Let $f : \mathbb{X} \times \mathbb{R}_{\not{}} \longrightarrow \mathbb{C}$ and let $F := \mathcal{L}[f] : \mathbb{X} \times \mathbb{H}_\alpha \longrightarrow \mathbb{C}$.

(a) (Heat equation) Define $f_0(x) := f(x, 0)$ for all $x \in \mathbb{X}$ (the 'initial temperature distribution'). Then f satisfies the one-dimensional heat equation $\partial_t f(x, t) = \partial_x^2 f(x, t)$ if, and only if, for every $s \in \mathbb{H}_\alpha$, the function $F_s(x) = F(x, s)$ satisfies the second-order linear ODE

$$\partial_x^2 F_s(x) = s F_s(x) - f_0(x), \quad \text{for all } x \in \mathbb{X}.$$

(b) (**Wave equation**) Define $f_0(x) := f(x, 0)$ and $f_1(x) := \partial_t f(x, 0)$ for all $x \in \mathbb{X}$ (the 'initial position' and 'initial velocity', respectively). Then f satisfies the one-dimensional wave equation $\partial_t^2 f(x, t) = \partial_x^2 f(x, t)$ if and only if, for every $s \in \mathbb{H}_\alpha$, the function $F_s(x) = F(x, s)$ satisfies the second-order linear ODE

$$\partial_x^2 F_s(x) = s^2 F_s(x) - s f_0(x) - f_1(x), \quad \text{for all } x \in \mathbb{X}. \qquad \diamondsuit$$

Exercise 19H.17   Verify Example 19H.14(a) and (b).   ◆ Ⓔ

We can then use solution methods for ordinary differential equations to solve for F_s for all $s \in \mathbb{H}_\alpha$, obtain an expression for the function F, and then apply the Laplace Inversion Theorem 19H.4 to obtain an expression for f. We will not pursue this approach further here; however, we will develop a very similar approach in Chapter 20 using Fourier transforms. For more information, see Haberman (1987), chap. 13; Broman (1989), chap. 5; Fisher (1999), §5.3–5.4; or Asmar (2005), chap. 8.

19I Further reading

Almost any book on PDEs contains a discussion of Fourier transforms, but for greater depth (and rigour) it is better to seek a text dedicated to Fourier analysis. Good introductions to Fourier transforms and their applications can be found in Körner (1988), Part IV, and Walker (1988), chaps. 6–7. (In addition to a lot of serious mathematical content, Körner's book contains interesting and wide-ranging discussions about the history of Fourier theory and its many scientific applications, and is written in a delightfully informal style.)

19J Practice problems

19.1 Suppose

$$f(x) = \begin{cases} 1 & \text{if } 0 < x < 1; \\ 0 & \text{otherwise}, \end{cases}$$

as in Example 19A.4, p. 494. Check that $\widehat{f}(\mu) = \frac{1 - e^{-\mu i}}{2\pi \mu i}$.

19.2 Compute the one-dimensional Fourier transforms of $g(x)$, when

(a) $g(x) = \begin{cases} 1 & \text{if } -\tau < x < 1 - \tau; \\ 0 & \text{otherwise}; \end{cases}$

(b) $g(x) = \begin{cases} 1 & \text{if } 0 < x < \sigma; \\ 0 & \text{otherwise}. \end{cases}$

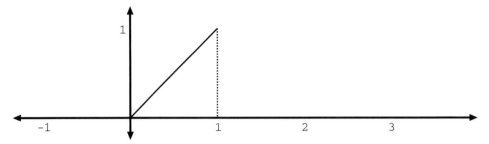

Figure 19J.1. Function for Problem 19.4.

19.3 Let $X, Y > 0$, and let

$$f(x, y) = \begin{cases} 1 & \text{if } 0 \le x \le X \text{ and } 0 \le y \le Y; \\ 0 & \text{otherwise.} \end{cases}$$

Compute the two-dimensional Fourier transform of $f(x, y)$. What does the Fourier inversion formula tell us?

19.4 Let $f : \mathbb{R} \longrightarrow \mathbb{R}$ be the function defined:

$$f(x) = \begin{cases} x & \text{if } 0 \le x \le 1; \\ 0 & \text{otherwise} \end{cases}$$

(Figure 19J.1). Compute the *Fourier transform* of f.

19.5 Let $f(x) = x \cdot \exp\left(-x^2/2\right)$. Compute the Fourier transform of f.

19.6 Let $\alpha > 0$, and let $g(x) = 1/(\alpha^2 + x^2)$. Example 19A.8, p. 497, claims that $\widehat{g}(\mu) = \frac{1}{2\alpha} e^{-\alpha|\mu|}$. Verify this. (*Hint:* use the Fourier inversion theorem.)

19.7 Fix $y > 0$, and let $\mathcal{K}_y(x) = y/\pi(x^2 + y^2)$ (this is the *half-space Poisson kernel* from §17E and §20C(ii)). Compute the one-dimensional Fourier transform

$$\widehat{\mathcal{K}}_y(\mu) = \frac{1}{2\pi} \int_{-\infty}^{\infty} \mathcal{K}_y(x) \exp\left(-\mu i x\right) d\mu.$$

19.8 Let $f(x) = 2x/(1 + x^2)^2$. Compute the Fourier transform of f.

19.9 Let

$$f(x) = \begin{cases} 1 & \text{if } -4 < x < 5; \\ 0 & \text{otherwise.} \end{cases}$$

Compute the Fourier transform $\widehat{f}(\mu)$.

19.10 Let

$$f(x) = \frac{x \cos(x) - \sin(x)}{x^2}.$$

Compute the Fourier transform $\widehat{f}(\mu)$.

19.11 Let $f, g \in \mathbf{L}^1(\mathbb{R})$, and let $h(x) = f(x) + g(x)$. Show that, for all $\mu \in \mathbb{R}$, $\widehat{h}(\mu) = \widehat{f}(\mu) + \widehat{g}(\mu)$.

19.12 Let $f, g \in \mathbf{L}^1(\mathbb{R})$, and let $h = f * g$. Show that, for all $\mu \in \mathbb{R}$, $\widehat{h}(\mu) = 2\pi \cdot \widehat{f}(\mu) \cdot \widehat{g}(\mu)$. (*Hint:* $\exp(-\mathbf{i}\mu x) = \exp(-\mathbf{i}\mu y) \cdot \exp(-\mathbf{i}\mu(x - y))$.)

19.13 Let $f, g \in \mathbf{L}^1(\mathbb{R})$, and let $h(x) = f(x) \cdot g(x)$. Suppose $\widehat{h}$ is also in $\mathbf{L}^1(\mathbb{R})$. Show that, for all $\mu \in \mathbb{R}$, $\widehat{h}(\mu) = (\widehat{f} * \widehat{g})(\mu)$. (*Hint:* Combine Problem 19.12 with the strong Fourier inversion formula (Theorem 19A.5, p. 495).)

19.14 Let $f \in \mathbf{L}^1(\mathbb{R})$. Fix $\tau \in \mathbb{R}$ and define $g : \mathbb{R} \longrightarrow \mathbb{C}$: $g(x) = f(x + \tau)$. Show that, for all $\mu \in \mathbb{R}$, $\widehat{g}(\mu) = e^{\tau\mu\mathbf{i}} \cdot \widehat{f}(\mu)$.

19.15 Let $f \in \mathbf{L}^1(\mathbb{R})$. Fix $\nu \in \mathbb{R}$ and define $g : \mathbb{R} \longrightarrow \mathbb{C}$: $g(x) = e^{\nu x \mathbf{i}} f(x)$. Show that, for all $\mu \in \mathbb{R}$, $\widehat{g}(\mu) = \widehat{f}(\mu - \nu)$.

19.16 Suppose $f \in \mathbf{L}^1(\mathbb{R})$. Fix $\sigma > 0$, and define $g : \mathbb{R} \longrightarrow \mathbb{C}$: $g(x) = f(x/\sigma)$. Show that, for all $\mu \in \mathbb{R}$, $\widehat{g}(\mu) = \sigma \cdot \widehat{f}(\sigma \cdot \mu)$.

19.17 Suppose $f : \mathbb{R} \longrightarrow \mathbb{R}$ is differentiable, and that $f \in \mathbf{L}^1(\mathbb{R})$ and $g := f' \in \mathbf{L}^1(\mathbb{R})$. Assume that $\lim_{x \to \pm\infty} f(x) = 0$. Show that $\widehat{g}(\mu) = \mathbf{i}\mu \cdot \widehat{f}(\mu)$.

19.18 Let $\mathcal{G}_t(x) = \frac{1}{2\sqrt{\pi t}} \exp(\frac{-x^2}{4t})$ be the Gauss–Weierstrass kernel. Recall that $\widehat{\mathcal{G}_t}(\mu) = \frac{1}{2\pi} e^{-\mu^2 t}$. Use this to construct a simple proof that, for any $s, t > 0$, $\mathcal{G}_t * \mathcal{G}_s = \mathcal{G}_{t+s}$.
 (*Hint:* Use Problem 19.12. Do *not* compute any convolution integrals, and do not use the 'solution to the heat equation' argument from Problem 17.8, p. 417.)

Remark Because of this result, probabilists say that the set $\{\mathcal{G}_t\}_{t \in \mathbb{R}_+}$ forms a *stable family of probability distributions* on $\mathbb{R}$. Analysts say that $\{\mathcal{G}_t\}_{t \in \mathbb{R}_+}$ is a *one-parameter semigroup* under convolution.

20

Fourier transform solutions to PDEs

Mathematics compares the most diverse phenomena and discovers the
secret analogies that unite them.

Jean Joseph Fourier

We will now see that the 'Fourier series' solutions to the PDEs on a bounded domain
(Chapters 11–14) generalize to 'Fourier transform' solutions on the unbounded
domain in a natural way.

20A The heat equation

20A(i) Fourier transform solution

Prerequisites: §1B, §19A, §5B, Appendix G.   **Recommended:** §11A, §12B, §13A, §19D, §19E.

Proposition 20A.1 Heat equation on an infinite rod

Let $F : \mathbb{R} \longrightarrow \mathbb{R}$ be a bounded function (of $\mu \in \mathbb{R}$).

(a) *Define $u : \mathbb{R} \times \mathbb{R}_+ \longrightarrow \mathbb{R}$ by*

$$u(x;t) := \int_{-\infty}^{\infty} F(\mu) \cdot \exp(\mu x \mathbf{i}) \cdot e^{-\mu^2 t} \, d\mu, \qquad (20A.1)$$

for all $t > 0$ and all $x \in \mathbb{R}$. Then u is a smooth function and satisfies the heat equation.

(b) *In particular, suppose $f \in \mathbf{L}^1(\mathbb{R})$ and $\widehat{f}(\mu) = F(\mu)$. If we define $u(x;0) := f(x)$ for all $x \in \mathbb{R}$, and define $u(x;t)$ by equation (20A.1), when $t > 0$, then u is continuous on $\mathbb{R} \times \mathbb{R}_+$, and it is a solution to the heat equation with initial conditions $u(x;0) = f(x)$.*

Proof **Exercise 20A.1.** (*Hint:* Use Proposition G.1, p. 571.)   □   Ⓔ

Example 20A.2 Suppose

$$f(x) = \begin{cases} 1 & \text{if } -1 < x < 1; \\ 0 & \text{otherwise.} \end{cases}$$

We know from Example 19A.3, p. 493, that $\widehat{f}(\mu) = \sin(\mu)/\pi\mu$. Thus,

$$u(x,t) = \int_{-\infty}^{\infty} \widehat{f}(\mu) \cdot \exp(\mu x \mathbf{i}) \cdot e^{-\mu^2 t}\, d\mu = \int_{-\infty}^{\infty} \frac{\sin(\mu)}{\pi\mu} \exp(\mu x \mathbf{i}) \cdot e^{-\mu^2 t}\, d\mu. \quad \diamond$$

Ⓔ **Exercise 20A.2** Verify that u satisfies the one-dimensional heat equation and the specified initial conditions. ◆

Example 20A.3 The Gauss–Weierstrass kernel

For all $x \in \mathbb{R}$ and $t > 0$, define the *Gauss–Weierstrass kernel*: $\mathcal{G}_t(x) :=$ $(1/2\sqrt{\pi t})\exp(-x^2/4t)$ (see Example 1B.1(c), p. 8). Fix $t > 0$; then, setting $\sigma = \sqrt{2t}$ in Theorem 19B.8(b), p. 502, we get

$$\widehat{\mathcal{G}}_t(\mu) = \frac{1}{2\pi} \exp\left(\frac{-(\sqrt{2t})^2 \mu^2}{2}\right) = \frac{1}{2\pi} \exp\left(\frac{-2t\mu^2}{2}\right) = \frac{1}{2\pi} e^{-\mu^2 t}.$$

Thus, applying the Fourier inversion formula (Theorem 19A.1, p. 492), we have

$$\mathcal{G}(x,t) = \int_{-\infty}^{\infty} \widehat{\mathcal{G}}_t(\mu) \exp(\mu x \mathbf{i}) d\mu = \frac{1}{2\pi} \int_{-\infty}^{\infty} e^{-\mu^2 t} \exp(\mu x \mathbf{i}) d\mu,$$

which, according to Proposition 20A.1, is a smooth solution of the heat equation, where we take $F(\mu)$ to be the *constant* function: $F(\mu) = 1/2\pi$. Thus, F is *not* the Fourier transform of any function f. Hence, the Gauss–Weierstrass kernel solves the heat equation, but the 'initial conditions' $\mathcal{G}_0$ do not correspond to a function, but instead a define more singular object, rather like an infinitely dense concentration of mass at a single point. Sometimes $\mathcal{G}_0$ is called the *Dirac delta function*, but this is a misnomer, since it is not really a function. Instead, $\mathcal{G}_0$ is an example of a more general class of objects called *distributions*. ◇

Proposition 20A.4 Heat equation on an infinite plane

Let $F : \mathbb{R}^2 \longrightarrow \mathbb{C}$ be some bounded function (of $(\mu, \nu) \in \mathbb{R}^2$).

(a) *Define* $u : \mathbb{R}^2 \times \mathbb{R}_+ \longrightarrow \mathbb{R}$:

$$u(x,y;t) := \int_{\mathbb{R}^2} F(\mu,\nu) \cdot \exp\left((\mu x + \nu y) \cdot \mathbf{i}\right) \cdot e^{-(\mu^2 + \nu^2)t}\, d\mu\, d\nu, \qquad (20A.2)$$

for all $t > 0$ and all $(x, y) \in \mathbb{R}^2$. Then u is continuous on $\mathbb{R}^3 \times \mathbb{R}_+$ and satisfies the two-dimensional heat equation.

(b) *In particular, suppose $f \in \mathbf{L}^1(\mathbb{R}^2)$ and $\widehat{f}(\mu,\nu) = F(\mu,\nu)$. If we define $u(x,y,0) := f(x,y)$ for all $(x,y) \in \mathbb{R}^2$, and define $u(x,y,t)$ by equation (20A.2) when $t > 0$, then u is continuous on $\mathbb{R}^2 \times \mathbb{R}_+$, and is a solution to the heat equation with initial conditions $u(x,y,0) = f(x,y)$.*

Proof **Exercise 20A.3.** (*Hint:* Use Proposition G.1, p. 571.) ☐ Ⓔ

Example 20A.5 Let $X, Y > 0$ be constants, and consider the initial conditions

$$f(x, y) = \begin{cases} 1 & \text{if } -X \le x \le X \text{ and } -Y \le y \le Y; \\ 0 & \text{otherwise.} \end{cases}$$

From Example 19D.4, p. 509, the Fourier transform of $f(x, y)$ is given by

$$\widehat{f}(\mu, \nu) = \frac{\sin(\mu X) \cdot \sin(\nu Y)}{\pi^2 \cdot \mu \cdot \nu}.$$

Thus, the corresponding solution to the two-dimensional heat equation is given by

$$u(x, y, t) = \int_{\mathbb{R}^2} \widehat{f}(\mu, \nu) \cdot \exp\left((\mu x + \nu y) \cdot \mathbf{i}\right) \cdot e^{-(\mu^2 + \nu^2)t} \, d\mu \, d\nu$$

$$= \int_{\mathbb{R}^2} \frac{\sin(\mu X) \cdot \sin(\nu Y)}{\pi^2 \cdot \mu \cdot \nu} \cdot \exp\left((\mu x + \nu y) \cdot \mathbf{i}\right) \cdot e^{-(\mu^2 + \nu^2)t} \, d\mu \, d\nu. \quad \diamond$$

Proposition 20A.6 Heat equation in infinite space

Let $F : \mathbb{R}^3 \longrightarrow \mathbb{C}$ be some bounded function (of $\boldsymbol{\mu} \in \mathbb{R}^3$).

(a) *Define $u : \mathbb{R}^3 \times \mathbb{R}_+ \longrightarrow \mathbb{R}$:*

$$u(\mathbf{x}; t) := \int_{\mathbb{R}^3} F(\boldsymbol{\mu}) \cdot \exp\left(\boldsymbol{\mu} \bullet \mathbf{x} \cdot \mathbf{i}\right) \cdot e^{-\|\boldsymbol{\mu}\|^2 t} \, d\boldsymbol{\mu}, \tag{20A.3}$$

for all $t > 0$ and all $\mathbf{x} \in \mathbb{R}^3$ (where $\|\boldsymbol{\mu}\|^2 := \mu_1^2 + \mu_2^2 + \mu_3^2$). Then u is continuous on $\mathbb{R}^3 \times \mathbb{R}_+$ and satisfies the three-dimensional heat equation.

(b) *In particular, suppose $f \in \mathbf{L}^1(\mathbb{R}^3)$ and $\widehat{f}(\boldsymbol{\mu}) = F(\boldsymbol{\mu})$. If we define $u(\mathbf{x}, 0) := f(\mathbf{x})$ for all $\mathbf{x} \in \mathbb{R}^3$, and define $u(\mathbf{x}, t)$ by equation (20A.3) when $t > 0$, then u is continuous on $\mathbb{R}^3 \times \mathbb{R}_+$, and is a solution to the heat equation with initial conditions $u(\mathbf{x}, 0) = f(\mathbf{x})$.*

Proof **Exercise 20A.4.** (*Hint:* Use Proposition G.1, p. 571.) ☐ Ⓔ

Example 20A.7 A ball of heat

Suppose the initial conditions are

$$f(\mathbf{x}) = \begin{cases} 1 & \text{if } \|\mathbf{x}\| \le 1; \\ 0 & \text{otherwise.} \end{cases}$$

Setting $R = 1$ in Example 19E.3, p. 512, yields the three-dimensional Fourier transform of f:

$$\widehat{f}(\boldsymbol{\mu}) = \frac{1}{2\pi^2} \left(\frac{\sin \|\boldsymbol{\mu}\|}{\|\boldsymbol{\mu}\|^3} - \frac{\cos \|\boldsymbol{\mu}\|}{\|\boldsymbol{\mu}\|^2} \right).$$

The resulting solution to the heat equation is given by

$$u(\mathbf{x}; t) = \int_{\mathbb{R}^3} \widehat{f}(\boldsymbol{\mu}) \cdot \exp\left(\boldsymbol{\mu} \bullet \mathbf{x} \cdot \mathbf{i}\right) \cdot e^{-\|\boldsymbol{\mu}\|^2 t} \, d\boldsymbol{\mu}$$

$$= \frac{1}{2\pi^2} \int_{\mathbb{R}^3} \left(\frac{\sin \|\boldsymbol{\mu}\|}{\|\boldsymbol{\mu}\|^3} - \frac{\cos \|\boldsymbol{\mu}\|}{\|\boldsymbol{\mu}\|^2}\right) \cdot \exp\left(\boldsymbol{\mu} \bullet \mathbf{x} \cdot \mathbf{i}\right) \cdot e^{-\|\boldsymbol{\mu}\|^2 t} \, d\boldsymbol{\mu}. \qquad \diamond$$

20A(ii) *The Gaussian convolution formula, revisited*

Prerequisites: §17C(i), §19B, §20A(i).

Recall from §17C(i), p. 391, that the Gaussian convolution formula solved the initial value problem for the heat equation by 'locally averaging' the initial conditions. Fourier methods provide another proof that this is a solution to the heat equation.

Theorem 20A.8 Gaussian convolutions and the heat equation
Let $f \in \mathbf{L}^1(\mathbb{R})$, and let $\mathcal{G}_t(x)$ be the Gauss–Weierstrass kernel from Example 20A.3 For all $t > 0$, define $U_t := f * \mathcal{G}_t$; in other words, for all $x \in \mathbb{R}$,

$$U_t(x) := \int_{-\infty}^{\infty} f(y) \cdot \mathcal{G}_t(x - y) dy.$$

Also, for all $x \in \mathbb{R}$, define $U_0(x) := f(x)$. Then U is continuous on $\mathbb{R} \times \mathbb{R}_{\not{}}$, and it is a solution to the heat equation with initial conditions $U(x, 0) = f(x)$.

Proof $U(x, 0) = f(x)$, by definition. To show that U satisfies the heat equation, we will show that it is, in fact, *equal* to the Fourier solution u described in Theorem 20A.1, p. 531. Fix $t > 0$, and let $u_t(x) = u(x, t)$; recall that, by definition,

$$u_t(x) = \int_{-\infty}^{\infty} \widehat{f}(\mu) \cdot \exp(\mu x \mathbf{i}) \cdot e^{-\mu^2 t} \, d\mu = \int_{-\infty}^{\infty} \widehat{f}(\mu) e^{-\mu^2 t} \cdot \exp(\mu x \mathbf{i}) d\mu.$$

Thus, Proposition 19A.2, p. 493, says that

$$\widehat{u_t}(\mu) = \widehat{f}(\mu) \cdot e^{-t\mu^2} \underset{(*)}{=} 2\pi \cdot \widehat{f}(\mu) \cdot \widehat{\mathcal{G}_t}(\mu). \tag{20A.4}$$

Here, $(*)$ is because Example 20A.3, p. 532, says that $e^{-t\mu^2} = 2\pi \cdot \widehat{\mathcal{G}_t}(\mu)$. But remember that $U_t = f * \mathcal{G}_t$, so Theorem 19B.2(b), p. 499, says

$$\widehat{U_t}(\mu) = 2\pi \cdot \widehat{f}(\mu) \cdot \widehat{\mathcal{G}_t}(\mu). \tag{20A.5}$$

Thus equations (20A.4) and (20A.5) mean that $\widehat{U_t} = \widehat{u_t}$. But then Proposition 19A.2, p. 493, implies that $u_t(x) = U_t(x)$. $\qquad \square$

For more discussion and examples of the Gaussian convolution approach to the heat equation, see §17C(i), p. 391.

(E) **Exercise 20A.5** State and prove a generalization of Theorem 20A.8 to solve the D-dimensional heat equation, for $D \geq 2$. ◆

20B The wave equation

20B(i) Fourier transform solution

Prerequisites: §2B, §19A, §5B, Appendix G.   **Recommended:** §11B, §12D, §19D, §19E, §20A(i).

Proposition 20B.1 Wave equation on an infinite wire

Let $f_0, f_1 \in \mathbf{L}^1(\mathbb{R})$ be twice-differentiable, and suppose f_0 and f_1 have Fourier transforms $\widehat{f_0}$ and $\widehat{f_1}$, respectively. Define $u : \mathbb{R} \times \mathbb{R}_{\not{}} \longrightarrow \mathbb{R}$:

$$u(x, t) = \int_{-\infty}^{\infty} \left(\widehat{f_0}(\mu) \cos(\mu t) + \frac{\widehat{f_1}(\mu)}{\mu} \sin(\mu t) \right) \cdot \exp(\mu x \mathbf{i}) d\mu.$$

Then u is a solution to the one-dimensional wave equation with initial position $u(x, 0) = f_0(x)$ and initial velocity $\partial_t u(x, 0) = f_1(x)$, for all $x \in \mathbb{R}$.

Proof **Exercise 20B.1.** (*Hint:* Show that this solution is equivalent to the   (E) d'Alembert solution of Proposition 17D.5, p. 404.) □

Example 20B.2 Fix $\alpha > 0$, and suppose we have initial position $f_0(x) := e^{-\alpha|x|}$, for all $x \in \mathbb{R}$, and initial velocity $f_1 \equiv 0$. From Example 19A.7, p. 496, we know that $\widehat{f_0}(\mu) = 2\alpha/(\alpha^2 + \mu^2)$. Thus, Proposition 20B.1 says

$$u(x, t) = \int_{-\infty}^{\infty} \widehat{f_0}(\mu) \cdot \exp(\mu x \mathbf{i}) \cdot \cos(\mu t) d\mu$$

$$= \int_{-\infty}^{\infty} \frac{2\alpha}{(\alpha^2 + \mu^2)} \cdot \exp(\mu x \mathbf{i}) \cdot \cos(\mu t) d\mu. \qquad \diamond$$

Exercise 20B.2 Verify that u satisfies the one-dimensional wave equation and the   (E) specified initial conditions. ◆

Proposition 20B.3 Wave equation on an infinite plane

Let $f_0, f_1 \in \mathbf{L}^1(\mathbb{R}^2)$ be twice-differentiable functions, whose Fourier transforms $\widehat{f_0}$ and $\widehat{f_1}$ decay fast enough that

$$\int_{\mathbb{R}^2} (\mu^2 + \nu^2) \cdot \left| \widehat{f_0}(\mu, \nu) \right| d\mu \, d\nu < \infty \qquad (20B.1)$$

and

$$\int_{\mathbb{R}^2} \sqrt{\mu^2 + \nu^2} \cdot \left| \widehat{f_1}(\mu, \nu) \right| d\mu \, d\nu < \infty.$$

Define $u : \mathbb{R}^2 \times \mathbb{R}_{\not{}} \longrightarrow \mathbb{R}$:

$$u(x, y, t) = \int_{\mathbb{R}^2} \widehat{f_0}(\mu, \nu) \cos\left(\sqrt{\mu^2 + \nu^2} \cdot t\right) \cdot \exp\left((\mu x + \nu y) \cdot \mathbf{i}\right) d\mu \, d\nu.$$

$$+ \int_{\mathbb{R}^2} \frac{\widehat{f_1}(\mu, \nu)}{\sqrt{\mu^2 + \nu^2}} \sin\left(\sqrt{\mu^2 + \nu^2} \cdot t\right) \cdot \exp\left((\mu x + \nu y) \cdot \mathbf{i}\right) d\mu \, d\nu.$$

Then u *is a solution to the two-dimensional wave equation with initial position* $u(x, y, 0) = f_0(x, y)$ *and initial velocity* $\partial_t u(x, y, 0) = f_1(x, y)$, *for all* $(x, y) \in \mathbb{R}^2$.

Ⓔ *Proof* **Exercise 20B.3.** (*Hint:* Equations (20B.1) make the integral absolutely convergent, and also enable us to apply Proposition G.1, p. 571, to compute the relevant derivatives of u.)   □

Example 20B.4 Let $\alpha, \beta > 0$ be constants, and suppose we have initial position $f_0 \equiv 0$ and initial velocity $f_1(x, y) = 1/(\alpha^2 + x^2)(\beta^2 + y^2)$ for all $(x, y) \in \mathbb{R}^2$. By adapting Example 19A.8, p. 497, we can check that

$$\widehat{f_1}(\mu, \nu) = \frac{1}{4\alpha\beta} \exp\left(-\alpha \cdot |\mu| - \beta \cdot |\nu|\right).$$

Thus, Proposition 20B.3 says

$$u(x, y, t) = \int_{\mathbb{R}^2} \frac{\widehat{f_1}(\mu, \nu)}{\sqrt{\mu^2 + \nu^2}} \sin\left(\sqrt{\mu^2 + \nu^2} \cdot t\right) \cdot \exp\left((\mu x + \nu y) \cdot \mathbf{i}\right) d\mu \, d\nu$$

$$= \int_{\mathbb{R}^2} \frac{\sin\left(\sqrt{\mu^2 + \nu^2} \cdot t\right) \cdot \exp\left((\mu x + \nu y) \cdot \mathbf{i} - \alpha \cdot |\mu| - \beta \cdot |\nu|\right)}{4\alpha\beta\sqrt{\mu^2 + \nu^2}} d\mu \, d\nu.$$

◇

Ⓔ **Exercise 20B.4** Verify that u satisfies the two-dimensional wave equation and the specified initial conditions.   ♦

Proposition 20B.5 Wave equation in infinite space
Let $f_0, f_1 \in \mathbf{L}^1(\mathbb{R}^3)$ *be twice-differentiable functions whose Fourier transforms* $\widehat{f_0}$ *and* $\widehat{f_1}$ *decay fast enough that*

$$\int_{\mathbb{R}^3} \|\mu\|^2 \cdot \left|\widehat{f_0}(\mu)\right| d\mu < \infty \qquad\qquad (20B.2)$$

and

$$\int_{\mathbb{R}^3} \|\mu\| \cdot \left|\widehat{f_1}(\mu)\right| d\mu < \infty.$$

Define $u : \mathbb{R}^3 \times \mathbb{R}_{\neq} \longrightarrow \mathbb{R}$:

$$u(\mathbf{x}, t) := \int_{\mathbb{R}^3} \left(\widehat{f_0}(\boldsymbol{\mu}) \cos \left(\|\boldsymbol{\mu}\| \cdot t \right) + \frac{\widehat{f_1}(\boldsymbol{\mu})}{\|\boldsymbol{\mu}\|} \sin \left(\|\boldsymbol{\mu}\| \cdot t \right) \right) \cdot \exp \left(\boldsymbol{\mu} \bullet \mathbf{xi} \right) d\boldsymbol{\mu}.$$

Then u is a solution to the three-dimensional wave equation with initial position $u(\mathbf{x}, 0) = f_0(\mathbf{x})$ and initial velocity $\partial_t u(\mathbf{x}, 0) = f_1(\mathbf{x})$, for all $\mathbf{x} \in \mathbb{R}^3$.

Proof **Exercise 20B.5.** (*Hint:* Equations (20B.2) make the integral absolutely ⓔ convergent, and also enable us to apply Proposition G.1, p. 571, to compute the relevant derivatives of u.) □

Exercise 20B.6 Show that the decay conditions (20B.1) or (20B.2) are satis- ⓔ fied if f_0 and f_1 are *asymptotically flat* in the sense that $\lim_{|\mathbf{x}| \to \infty} |f(\mathbf{x})| = 0$ and $\lim_{|\mathbf{x}| \to \infty} |\nabla f(\mathbf{x})| = 0$, while $(\partial_i \partial_j f) \in \mathbf{L}^1(\mathbb{R}^2)$ for all $i, j \in \{1, \ldots, D\}$ (where $D = 2$ or 3). (*Hint:* Apply Theorem 19B.7, p. 500, to compute the Fourier transforms of the derivative functions $\partial_i \partial_j f$; conclude that the function $\widehat{f}$ must itself decay at a certain speed.) ♦

20B(ii) Poisson's spherical mean solution; Huygens' principle

Prerequisites: §17A, §19E, §20B(i).   **Recommended:** §17D, §20A(ii).

The Gaussian convolution formula of §20A(ii) solves the initial value problem for the heat equation in terms of a kind of 'local averaging' of the initial conditions. Similarly, d'Alembert's formula (§17D) solves the initial value problem for the one-dimensional wave equation in terms of a local average.

There is an analogous result for higher-dimensional wave equations. To explain it, we must introduce the concept of spherical averages. Let $f : \mathbb{R}^3 \longrightarrow \mathbb{R}$ be some integrable function. If $\mathbf{x} \in \mathbb{R}^3$ is a point in space, and $R > 0$, then the *spherical average* of f at $\mathbf{x}$, of radius R, is defined:

$$\mathbf{M}_R f(\mathbf{x}) := \frac{1}{4\pi R^2} \int_{\mathbb{S}(R)} f(\mathbf{x} + \mathbf{s}) d\mathbf{s}.$$

Here, $\mathbb{S}(R) := \{\mathbf{s} \in \mathbb{R}^3; \|\mathbf{s}\| = R\}$ is the sphere around 0 of radius R. The total surface area of the sphere is $4\pi R^2$; we divide by this quantity to obtain an average. We adopt the notational convention that $\mathbf{M}_0 f(\mathbf{x}) := f(\mathbf{x})$. This is justified by the following exercise.

Exercise 20B.7 Suppose f is continuous at $\mathbf{x}$. Show that $\lim_{R \to 0} \mathbf{M}_R f(\mathbf{x}) =$ ⓔ $f(\mathbf{x})$. ♦

Theorem 20B.6 Poisson's spherical mean solution to the wave equation

(a) *Let $f_1 \in \mathbf{L}^1(\mathbb{R}^3)$ be twice-differentiable, such that $\widehat{f_1}$ satisfies equations* (20B.2). *Define $v : \mathbb{R}^3 \times \mathbb{R}_+ \longrightarrow \mathbb{R}$:*

$$v(\mathbf{x}; t) := t \cdot \mathbf{M}_t \, f_1(\mathbf{x}), \quad \text{for all } \mathbf{x} \in \mathbb{R}^3 \text{ and } t \geq 0.$$

Then v is a solution to the wave equation with

initial position: $v(\mathbf{x}, 0) = 0;$ initial velocity: $\partial_t v(\mathbf{x}, 0) = f_1(\mathbf{x}).$

(b) *Let $f_0 \in \mathbf{L}^1(\mathbb{R}^3)$ be twice-differentiable, such that $\widehat{f_0}$ satisfies equations* (20B.2). *536. For all $\mathbf{x} \in \mathbb{R}^3$ and $t > 0$, define $W(\mathbf{x}; t) := t \cdot \mathbf{M}_t \, f_0(\mathbf{x})$, and then define $w : \mathbb{R}^3 \times \mathbb{R}_+ \longrightarrow \mathbb{R}$:*

$$w(\mathbf{x}; t) := \partial_t \, W(\mathbf{x}; t), \quad \text{for all } \mathbf{x} \in \mathbb{R}^3 \text{ and } t \geq 0.$$

Then w is a solution to the wave equation with

initial position: $w(\mathbf{x}, 0) = f_0(\mathbf{x});$ initial velocity: $\partial_t \, w(\mathbf{x}, 0) = 0.$

(c) *Let $f_0, f_1 \in \mathbf{L}^1(\mathbb{R}^3)$ be as in (a) and (b), and define $u : \mathbb{R}^3 \times \mathbb{R}_+ \longrightarrow \mathbb{R}$:*

$$u(\mathbf{x}; t) := w(\mathbf{x}; t) + v(\mathbf{x}; t), \quad \text{for all } \mathbf{x} \in \mathbb{R}^3 \text{ and } t \geq 0,$$

where w is as in (b) and v is as in (a). Then u is the unique solution to the wave equation with

initial position: $u(\mathbf{x}, 0) = f_0(\mathbf{x});$ initial velocity: $\partial_t \, u(\mathbf{x}, 0) = f_1(\mathbf{x}).$

Proof To prove (a) we will need a certain calculation.

Claim 1 *For any $R > 0$, and any $\boldsymbol{\mu} \in \mathbb{R}^3$,*

$$\int_{\mathbb{S}(R)} \exp\left(\boldsymbol{\mu} \bullet \mathbf{si}\right) ds = \frac{4\pi R \cdot \sin(\|\boldsymbol{\mu}\| \cdot R)}{\|\boldsymbol{\mu}\|}.$$

Proof By spherical symmetry, we can rotate the vector $\boldsymbol{\mu}$ without affecting the value of the integral, so rotate $\boldsymbol{\mu}$ until it becomes $\boldsymbol{\mu} = (\mu, 0, 0)$, with $\mu > 0$. Thus, $\|\boldsymbol{\mu}\| = \mu$, and, if a point $s \in \mathbb{S}(R)$ has coordinates (s_1, s_2, s_3) in $\mathbb{R}^3$, then $\boldsymbol{\mu} \bullet \mathbf{s} = \mu \cdot s_1$. Thus, the integral simplifies to

$$\int_{\mathbb{S}(R)} \exp(\boldsymbol{\mu} \bullet \mathbf{si}) ds = \int_{\mathbb{S}(R)} \exp(\mu \cdot s_1 \cdot \mathbf{i}) ds.$$

We will integrate using a spherical coordinate system (ϕ, θ) on the sphere, where $0 < \phi < \pi$ and $-\pi < \theta < \pi$, and where

$$(s_1, s_2, s_3) = R \cdot (\cos(\phi), \sin(\phi)\sin(\theta), \sin(\phi)\cos(\theta)).$$

On the sphere of radius R, the surface area element is given by $ds = R^2 \sin(\phi)d\theta\, d\phi$. Thus,

$$\int_{S(R)} \exp(\mu \cdot s_1 \cdot \mathbf{i})\, ds = \int_0^\pi \int_{-\pi}^\pi \exp(\mu \cdot R \cdot \cos(\phi) \cdot \mathbf{i}) \cdot R^2 \sin(\phi)d\theta\, d\phi$$

$$\underset{(*)}{=\!=} 2\pi \int_0^\pi \exp(\mu \cdot R \cdot \cos(\phi) \cdot \mathbf{i}) \cdot R^2 \sin(\phi)d\phi$$

$$\underset{(\diamond)}{=\!=} 2\pi \int_{-R}^R \exp(\mu \cdot s_1 \cdot \mathbf{i}) \cdot R\, ds_1$$

$$= \frac{2\pi R}{\mu \mathbf{i}} \exp\left(\mu \cdot s_1 \cdot \mathbf{i}\right)\Big|_{s_1=-R}^{s_1=R}$$

$$= 2\frac{2\pi R}{\mu} \cdot \left(\frac{e^{\mu R \mathbf{i}} - e^{-\mu R \mathbf{i}}}{2\mathbf{i}}\right) \underset{(\dagger)}{=\!=} \frac{4\pi R}{\mu} \sin(\mu R).$$

Here, $(*)$ denotes that the integrand is constant in the θ coordinate; and $(\diamond)$ makes the substitution $s_1 = R\cos(\phi)$, so $ds_1 = -R\sin(\phi)d\phi$, and $(\dagger)$ is by Euler's formula, p. 553. $\qquad\diamond_{\text{Claim 1}}$

Now, by Proposition 20B.5, p. 536, a solution to the three-dimensional wave equation with zero initial position and initial velocity f_1 is given by

$$u(\mathbf{x}, t) = \int_{\mathbb{R}^3} \widehat{f_1}(\mu) \frac{\sin(\|\mu\| \cdot t)}{\|\mu\|} \exp(\mu \bullet \mathbf{x}\mathbf{i})d\mu. \qquad (20B.3)$$

However, if we set $R = t$ in Claim 1, we have

$$\frac{\sin(\|\mu\| \cdot t)}{\|\mu\|} = \frac{1}{4\pi t} \int_{S(t)} \exp(\mu \bullet \mathbf{s}\mathbf{i})ds.$$

Thus,

$$\frac{\sin(\|\mu\| \cdot t)}{\|\mu\|} \cdot \exp(\mu \bullet \mathbf{x}\mathbf{i}) = \exp(\mu \bullet \mathbf{x}\mathbf{i}) \cdot \frac{1}{4\pi t} \int_{S(t)} \exp(\mu \bullet \mathbf{s}\mathbf{i})ds$$

$$= \frac{1}{4\pi t} \int_{S(t)} \exp\left(\mu \bullet \mathbf{x}\mathbf{i} + \mu \bullet \mathbf{s}\mathbf{i}\right)ds$$

$$= \frac{1}{4\pi t} \int_{S(t)} \exp\left(\mu \bullet (\mathbf{x} + \mathbf{s})\mathbf{i}\right)ds.$$

Substituting this into equation (20B.3), we get

$$u(\mathbf{x}, t) = \int_{\mathbb{R}^3} \frac{\widehat{f_1}(\mu)}{4\pi t} \cdot \left(\int_{S(t)} \exp\left(\mu \bullet (\mathbf{x} + \mathbf{s})\mathbf{i}\right)ds\right)d\mu$$

$$\underset{(*)}{=\!=} \frac{1}{4\pi t} \int_{S(t)} \int_{\mathbb{R}^3} \widehat{f_1}(\mu) \cdot \exp\left(\mu \bullet (\mathbf{x} + \mathbf{s})\mathbf{i}\right)d\mu\, ds$$

$$\underset{(\diamond)}{=\!=} \frac{1}{4\pi t} \int_{S(t)} f_1(\mathbf{x} + \mathbf{s})ds = t \cdot \frac{1}{4\pi t^2} \int_{S(t)} f_1(\mathbf{x} + \mathbf{s})ds$$

$$= t \cdot \mathbf{M}_t\, f_1(\mathbf{x}).$$

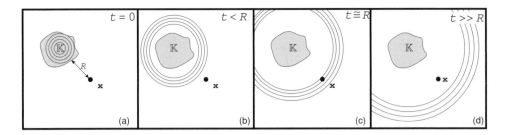

Figure 20B.1. Huygens' principle. (a) Wave originates inside K at time $t = 0$. (b) If $t < R$, wave has not yet reached x. (c) Wave reaches x around $t \cong R$. (d) For $t \gg R$, wave has completely passed by x.

In $(*)$, we simply interchange the two integrals;[1] $(\diamond)$ is just the Fourier Inversion Theorem 19E.1, p. 511.

Ⓔ   Part (b) is **Exercise 20B.8**. Part (c) follows by combining (a) and (b).   □

Corollary 20B.7 Huygens' principle

Let f_0, $f_1 \in \mathbf{L}^1(\mathbb{R}^3)$, and suppose there is some bounded region $\mathbb{K} \subset \mathbb{R}^3$ such that f_0 and f_1 are zero outside of $\mathbb{K}$ – that is, $f_0(\mathbf{y}) = 0$ and $f_1(\mathbf{y}) = 0$ for all $\mathbf{y} \notin \mathbb{K}$ (see Figure 20B.1(a)). Let $u : \mathbb{R}^3 \times \mathbb{R}_{\not{/}} \longrightarrow \mathbb{R}$ be the solution to the wave equation with initial position f_0 and initial velocity f_1, and let $\mathbf{x} \in \mathbb{R}^3$.

(a) *Let R be the distance from $\mathbb{K}$ to $\mathbf{x}$. If $t < R$ then $u(\mathbf{x}; t) = 0$ (Figure 20B.1(b)).*

(b) *If t is large enough that $\mathbb{K}$ is* entirely contained *in a ball of radius t around $\mathbf{x}$, then $u(\mathbf{x}; t) = 0$ (Figure 20B.1(d)).*

Ⓔ *Proof* **Exercise 20B.9.**   □

Part (a) of Huygens' principle says that, if a sound wave originates in the region $\mathbb{K}$ at time 0, and $\mathbf{x}$ is of distance R, then it does not reach the point $\mathbf{x}$ before time R. This is not surprising; it takes time for sound to travel through space. Part (b) says that the sound wave propagates through the point $\mathbf{x}$ in a *finite* amount of time, and leaves no wake behind it. This is somewhat more surprising, but corresponds to our experience; sounds travelling through open spaces do not 'reverberate' (except due to echo effects). It turns out, however, that Part (b) of the theorem is *not* true for waves travelling in *two* dimensions (e.g. ripples on the surface of a pond).

20C The Dirichlet problem on a half-plane

Prerequisites: §1C, §19A, §5C, Appendix G.   **Recommended:** §12A, §13B, §19D, §19E.

In §12A and §13B, we saw how to solve Laplace's equation on a bounded domain such as a rectangle or a cube, in the context of Dirichlet boundary conditions. Now

---

[1] This actually involves some subtlety, which we will gloss over.

consider the *half-plane* domain $\mathbb{H} := \{(x, y) \in \mathbb{R}^2; y \geq 0\}$. The boundary of this domain is just the x axis: $\partial\mathbb{H} = \{(x, 0); \ x \in \mathbb{R}\}$; thus, boundary conditions are imposed by choosing some function $b : \mathbb{R} \longrightarrow \mathbb{R}$. Figure 17E.1, p. 408, illustrates the corresponding *Dirichlet problem*: find a continuous function $u : \mathbb{H} \longrightarrow \mathbb{R}$ such that

(1) u satisfies the Laplace equation: $\triangle u(x, y) = 0$ for all $x \in \mathbb{R}$ and $y > 0$;
(2) u satisfies the nonhomogeneous Dirichlet boundary condition: $u(x, 0) = b(x)$ for all $x \in \mathbb{R}$.

20C(i) Fourier solution

Heuristically speaking, we will solve the problem by defining $u(x, y)$ as a continuous sequence of horizontal 'fibres', parallel to the x axis, and ranging over all values of $y > 0$. Each fibre is a function only of x, and, thus, has a one-dimensional Fourier transform. The problem then becomes one of determining these Fourier transforms from the Fourier transform of the boundary function b.

Proposition 20C.1 Fourier solution to the half-plane Dirichlet problem
Let $b \in \mathbf{L}^1(\mathbb{R})$. Suppose that b has Fourier transform $\widehat{b}$, and define $u : \mathbb{H} \longrightarrow \mathbb{R}$:

$$u(x, y) := \int_{-\infty}^{\infty} \widehat{b}(\mu) \cdot e^{-|\mu| \cdot y} \cdot \exp(\mu\mathbf{i}x)\mathrm{d}\mu, \quad \textit{for all } x \in \mathbb{R} \textit{ and } y \geq 0.$$

Then u is the solution to the Laplace equation ($\triangle u = 0$) which is bounded at infinity and which satisfies the nonhomogeneous Dirichlet boundary condition $u(x, 0) = b(x)$, for all $x \in \mathbb{R}$.

Proof For any fixed $\mu \in \mathbb{R}$, the function $f_{\mu}(x, y) = \exp(-|\mu| \cdot y)\exp(-\mu\mathbf{i}x)$ is harmonic (see Problem 20.10). Thus, Proposition G.1, p. 571, implies that the function $u(x, y)$ is also harmonic. Finally, note that, when $y = 0$, the expression for $u(x, 0)$ is just the Fourier inversion integral for $b(x)$. $\qquad\square$

Example 20C.2 Suppose

$$b(x) = \begin{cases} 1 & \text{if } -1 < x < 1; \\ 0 & \text{otherwise.} \end{cases}$$

We already know from Example 19A.3, p. 493, that $\widehat{b}(\mu) = \sin(\mu)/\pi\mu$. Thus,

$$u(x, y) = \frac{1}{\pi} \int_{-\infty}^{\infty} \frac{\sin(\mu)}{\mu} \cdot e^{-|\mu| \cdot y} \cdot \exp(\mu\mathbf{i}x)\mathrm{d}\mu. \qquad\qquad \diamondsuit$$

Ⓔ **Exercise 20C.1** Note the 'boundedness' condition in Proposition 20C.1. Find another solution to the Dirichlet problem on $\mathbb{H}$ which is *unbounded* at infinity.   ◆

20C(ii) Impulse-response solution

Prerequisites: §20C(i).   **Recommended:** §17E.

For any $y > 0$, define the *Poisson kernel* $\mathcal{K}_y : \mathbb{R} \longrightarrow \mathbb{R}$:

$$\mathcal{K}_y(x) := \frac{y}{\pi(x^2 + y^2)} \tag{20C.9}$$

(see Figure 17E.2, p. 409). In §17E, we used the Poisson kernel to solve the half-plane Dirichlet problem using *impulse-response* methods (Proposition 17E.1, p. 409). We can now use the 'Fourier' solution to provide another proof of Proposition 17E.1.

Proposition 20C.3 Poisson kernel solution to the half-plane Dirichlet problem
Let $b \in \mathbf{L}^1(\mathbb{R})$. Define $u : \mathbb{H} \longrightarrow \mathbb{R}$:

$$U(x, y) = b * \mathcal{K}_y(x) = \frac{y}{\pi} \int_{-\infty}^{\infty} \frac{b(z)}{(x - z)^2 + y^2} \, dz, \tag{20C.10}$$

for all $y > 0$ and $x \in \mathbb{R}$. Then U is the solution to the Laplace equation ($\triangle U = 0$) which is bounded at infinity and which can be continuously extended to satisfy the nonhomogeneous Dirichlet boundary condition $U(x, 0) = b(x)$ for all $x \in \mathbb{R}$.

Proof We will show that the solution U in equation (20C.10) is actually equal to the 'Fourier' solution u from Proposition 20C.1.

 Fix $y > 0$, and define $U_y(x) = U(x, y)$ for all $x \in \mathbb{R}$. Equation (20C.10) says that $U_y = b * \mathcal{K}_y$; hence Theorem 19B.2(b), p. 499, says that

$$\widehat{U}_y = 2\pi \cdot \widehat{b} \cdot \widehat{\mathcal{K}}_y. \tag{20C.11}$$

Now, by Problem 19.7, p. 528, we have the following:

$$\widehat{\mathcal{K}}_y(\mu) = \frac{e^{-y|\mu|}}{2\pi}. \tag{20C.12}$$

Combine equations (20C.11) and (20C.12) to get

$$\widehat{U}_y(\mu) = e^{-y|\mu|} \cdot \widehat{b}(\mu). \tag{20C.13}$$

Now apply the Fourier inversion formula (Theorem 19A.1, p. 492) to equation (20C.13) to obtain

$$U_y(x) = \int_{-\infty}^{\infty} \widehat{U}(\mu) \cdot \exp(\mu \cdot x \cdot \mathbf{i}) d\mu = \int_{-\infty}^{\infty} e^{-y|\mu|} \cdot \widehat{b}(\mu) \exp(\mu \cdot x \cdot \mathbf{i}) d\mu$$

$$= u(x, y),$$

where $u(x, y)$ is the solution from Proposition 20C.1. $\qquad\qquad\square$

20D PDEs on the half-line

Prerequisites: §1B(i), §19F, §5C, Appendix G.

Using the Fourier (co)sine transform, we can solve PDEs on the half-line.

Theorem 20D.1 The heat equation; Dirichlet boundary conditions
Let $f \in \mathbf{L}^1(\mathbb{R}_{\not})$ have Fourier sine transform $\widehat{f}_{\sin}$, and define $u : \mathbb{R}_{\not} \times \mathbb{R}_{\not} \longrightarrow \mathbb{R}$:

$$u(x, t) := \int_0^{\infty} \widehat{f}_{\sin}(\mu) \cdot \sin(\mu \cdot x) \cdot e^{-\mu^2 t} \, d\mu.$$

Then $u(x, t)$ is a solution to the heat equation, with initial conditions $u(x, 0) = f(x)$, and it satisfies the homogeneous Dirichlet boundary condition: $u(0, t) = 0$.

Proof **Exercise 20D.1.** (*Hint:* Use Proposition G.1, p. 571.) $\qquad\square\qquad$ Ⓔ

Theorem 20D.2 The heat equation; Neumann boundary conditions
Let $f \in \mathbf{L}^1(\mathbb{R}_{\not})$ have Fourier cosine transform $\widehat{f}_{\cos}$, and define $u : \mathbb{R}_{\not} \times \mathbb{R}_{\not} \longrightarrow \mathbb{R}$:

$$u(x, t) := \int_0^{\infty} \widehat{f}_{\cos}(\mu) \cdot \cos(\mu \cdot x) \cdot e^{-\mu^2 t} \, d\mu.$$

Then $u(x, t)$ is a solution to the heat equation, with initial conditions $u(x, 0) = f(x)$, and the homogeneous Neumann boundary condition: $\partial_x u(0, t) = 0$.

Proof **Exercise 20D.2.** (*Hint:* Use Proposition G.1, p. 571.) $\qquad\square\qquad$ Ⓔ

20E General solution to PDEs using Fourier transforms

Prerequisites: §16F, §18A, §19A, §19D, §19E.　　　**Recommended:** §20A(i), §20B(i), §20C, §20D.

Most of the results of this chapter can be subsumed into a single abstraction, which makes use of the *polynomial formalism* developed in §16F, p. 376.

Theorem 20E.1 *Fix* $D \in \mathbb{N}$, *and let* L *be a linear differential operator on* $\mathcal{C}^\infty(\mathbb{R}^D; \mathbb{R})$ *with constant coefficients. Suppose* L *has* **polynomial symbol** $\mathcal{P}$.

(a) *If* $f \in \mathbf{L}^1(\mathbb{R}^D)$ *has Fourier transform* $\widehat{f} : \mathbb{R}^D \longrightarrow \mathbb{R}$, *and* $g = \mathsf{L} f$, *then* g *has Fourier transform* $\widehat{g}(\boldsymbol{\mu}) = \mathcal{P}(\mathbf{i}\,\boldsymbol{\mu}) \cdot \widehat{f}(\boldsymbol{\mu})$, *for all* $\boldsymbol{\mu} \in \mathbb{R}^D$.

(b) *If* $q \in \mathbf{L}^1(\mathbb{R}^D)$ *has Fourier transform* $\widehat{q} \in \mathbf{L}^1(\mathbb{R}^D)$, *and* $f \in \mathbf{L}^1(\mathbb{R}^D)$ *has Fourier transform*

$$\widehat{f}(\boldsymbol{\mu}) = \frac{\widehat{q}(\boldsymbol{\mu})}{\mathcal{P}(\mathbf{i}\,\boldsymbol{\mu})}, \quad \textit{for all } \boldsymbol{\mu} \in \mathbb{R}^D,$$

then f *is a solution to the Poisson-type nonhomogeneous equation* $\mathsf{L} f = q$.

Let $u : \mathbb{R}^D \times \mathbb{R}_+ \longrightarrow \mathbb{R}$ *be another function, and, for all* $t \geq 0$, *define* $u_t :$ $\mathbb{R}^D \longrightarrow \mathbb{R}$ *by* $u_t(\mathbf{x}) := u(\mathbf{x}, t)$ *for all* $\mathbf{x} \in \mathbb{R}^D$. *Suppose* $u_t \in \mathbf{L}^1(\mathbb{R}^D)$, *and let* u_t *have Fourier transform* $\widehat{u}_t$.

(c) *Let* $f \in \mathbf{L}^1(\mathbb{R}^D)$, *and suppose* $\widehat{u}_t(\boldsymbol{\mu}) = \exp\left(\mathcal{P}(\mathbf{i}\,\boldsymbol{\mu}) \cdot t\right) \cdot \widehat{f}(\boldsymbol{\mu})$, *for all* $\boldsymbol{\mu} \in \mathbb{R}^D$ *and* $t \geq 0$. *Then* u *is a solution to the first-order evolution equation*

$$\partial_t u(\mathbf{x}, t) = \mathsf{L} u(\mathbf{x}, t), \qquad \textit{for all } \mathbf{x} \in \mathbb{R}^D \textit{ and } t > 0,$$

with initial conditions $u(\mathbf{x}, 0) = f(\mathbf{x})$, *for all* $\mathbf{x} \in \mathbb{R}^D$.

(d) *Suppose* $f \in \mathbf{L}^1(\mathbb{R}^D)$ *has Fourier transform* $\widehat{f}$ *which decays fast enough that*[2]

$$\int_{\mathbb{R}^D} \left| \mathcal{P}(\mathbf{i}\,\boldsymbol{\mu}) \cdot \cos\left(\sqrt{-\mathcal{P}(\mathbf{i}\,\boldsymbol{\mu})} \cdot t\right) \cdot \widehat{f}(\boldsymbol{\mu}) \right| d\boldsymbol{\mu} < \infty,$$

for all $t \geq 0$. *Suppose* $\widehat{u}_t(\boldsymbol{\mu}) = \cos(\sqrt{-\mathcal{P}(\mathbf{i}\,\boldsymbol{\mu})} \cdot t) \cdot \widehat{f}(\boldsymbol{\mu})$, *for all* $\boldsymbol{\mu} \in \mathbb{R}^D$ *and* $t \geq 0$. *Then* u *is a solution to the second-order evolution equation*

$$\partial_t^2 u(\mathbf{x}, t) = \mathsf{L} u(\mathbf{x}, t), \qquad \textit{for all } \mathbf{x} \in \mathbb{R}^D \textit{ and } t > 0,$$

with initial position $u(\mathbf{x}, 0) = f(\mathbf{x})$ *and initial velocity* $\partial_t u(\mathbf{x}, 0) = 0$, *for all* $\mathbf{x} \in \mathbb{R}^D$.

(e) *Suppose* $f \in \mathbf{L}^1(\mathbb{R}^D)$ *has Fourier transform* $\widehat{f}$ *which decays fast enough that*

$$\int_{\mathbb{R}^D} \left| \sqrt{\mathcal{P}(\mathbf{i}\,\boldsymbol{\mu})} \cdot \sin\left(\sqrt{-\mathcal{P}(\mathbf{i}\,\boldsymbol{\mu})} \cdot t\right) \cdot \widehat{f}(\boldsymbol{\mu}) \right| d\boldsymbol{\mu} < \infty,$$

for all $t \geq 0$. *Suppose*

$$\widehat{u}_t(\boldsymbol{\mu}) = \frac{\sin\left(\sqrt{-\mathcal{P}(\mathbf{i}\,\boldsymbol{\mu})} \cdot t\right)}{\sqrt{-\mathcal{P}(\mathbf{i}\,\boldsymbol{\mu})}} \cdot \widehat{f}(\boldsymbol{\mu}),$$

for all $\boldsymbol{\mu} \in \mathbb{R}^D$ *and* $t \geq 0$. *Then the function* $u(\mathbf{x}, t)$ *is a solution to the second-order evolution equation*

$$\partial_t^2 u(\mathbf{x}, t) = \mathsf{L} u(\mathbf{x}, t), \qquad \textit{for all } \mathbf{x} \in \mathbb{R}^D \textit{ and } t > 0,$$

with initial position $u(\mathbf{x}, 0) = 0$ *and initial velocity* $\partial_t u(\mathbf{x}, 0) = f(\mathbf{x})$, *for all* $\mathbf{x} \in \mathbb{R}^D$.

---

[2] See Example 18A.6(b), (c), p. 426, for the definitions of complex sine and cosine functions. See Exercise 18C.17, p. 454, for a discussion of complex square roots.

Proof **Exercise 20E.1.** (*Hint:* Use Proposition G.1, p. 571. In each case, be sure Ⓔ
to verify that the convergence conditions of Proposition G.1 are satisfied.)   □

Exercise 20E.2 Go back through this chapter and see how all of the different Ⓔ
solution theorems for the heat equation (§20A(i)) and the wave equation (§20B(i))
are special cases of this result. What about the solution for the Dirichlet problem
on a half-space in §20C? How does it fit into this formalism?   ◆

Exercise 20E.3 State and prove a theorem analogous to Theorem 20E.1 for solving Ⓔ
a D-dimensional Schrödinger equation using Fourier transforms.   ◆

20F Practice problems

20.1 Let

$$f(x) = \begin{cases} 1 & \text{if } 0 < x < 1; \\ 0 & \text{otherwise,} \end{cases}$$

as in Example 19A.4, p. 494.

(a) Use the Fourier method to solve the Dirichlet problem on a half-space, with
boundary condition $u(x, 0) = f(x)$.
(b) Use the Fourier method to solve the heat equation on a line, with initial condition
$u_0(x) = f(x)$.

20.2 Solve the two-dimensional heat equation, with initial conditions

$$f(x, y) = \begin{cases} 1 & \text{if } 0 \le x \le X \text{ and } 0 \le y \le Y; \\ 0 & \text{otherwise,} \end{cases}$$

where $X, Y > 0$ are constants. (*Hint:* See Problem 19.3, p. 528.)

20.3 Solve the two-dimensional wave equation, with

initial position: $u(x, y, 0) = 0$,

initial velocity: $\partial_t u(x, y, 0) = f_1(x, y) = \begin{cases} 1 & \text{if } 0 \le x \le 1 \text{ and } 0 \le y \le 1; \\ 0 & \text{otherwise.} \end{cases}$

(*Hint:* See Problem 19.3, p. 528.)

20.4 Let $f : \mathbb{R} \longrightarrow \mathbb{R}$ be the function defined:

$$f(x) = \begin{cases} x & \text{if } 0 \le x \le 1; \\ 0 & \text{otherwise} \end{cases}$$

(see Figure 19J.1, p. 528). Solve the *heat equation* on the real line, with initial
conditions $u(x; 0) = f(x)$. (*Hint:* Use the Fourier method; see Problem 19.4, p. 528.)

20.5 Let $f(x) = x \cdot \exp\left(-x^2/2\right)$. (See Problem 19.5, p. 528.)

(a) Solve the *heat equation* on the real line, with initial conditions $u(x;0) = f(x)$. (Use the Fourier method.)

(b) Solve the *wave equation* on the real line, with initial position $u(x;0) = f(x)$ and initial velocity $\partial_t u(x, 0) = 0$. (Use the Fourier method.)

20.6 Let $f(x) = 2x/(1 + x^2)^2$. (See Problem 19.8, p. 528.)

(a) Solve the *heat equation* on the real line, with initial conditions $u(x;0) = f(x)$. (Use the Fourier method.)

(b) Solve the *wave equation* on the real line, with initial position $u(x, 0) = 0$ and initial velocity $\partial_t u(x, 0) = f(x)$. (Use the Fourier method.)

20.7 Let

$$f(x) = \begin{cases} 1 & \text{if } -4 < x < 5; \\ 0 & \text{otherwise.} \end{cases}$$

(See Problem 19.9, p. 528.) Use the 'Fourier method' to solve the one-dimensional heat equation ($\partial_t u(x;t) = \Delta u(x;t)$) on the domain $\mathbb{X} = \mathbb{R}$, with initial conditions $u(x;0) = f(x)$.

20.8 Let

$$f(x) = \frac{x\cos(x) - \sin(x)}{x^2}.$$

(See Problem 19.10, p. 528.) Use the 'Fourier method' to solve the one-dimensional heat equation ($\partial_t u(x;t) = \Delta u(x;t)$) on the domain $\mathbb{X} = \mathbb{R}$, with initial conditions $u(x;0) = f(x)$.

20.9 Suppose $f : \mathbb{R} \longrightarrow \mathbb{R}$ had Fourier transform $\widehat{f}(\mu) = \mu/(\mu^4 + 1)$.

(a) Find the solution to the one-dimensional heat equation $\partial_t u = \Delta u$, with initial conditions $u(x;0) = f(x)$ for all $x \in \mathbb{R}$.

(b) Find the solution to the one-dimensional wave equation $\partial_t^2 u = \Delta u$, with

initial position: $u(x;0) = 0$, for all $x \in \mathbb{R}$;

initial velocity: $\partial_t u(x;0) = f(x)$, for all $x \in \mathbb{R}$.

(c) Find the solution to the two-dimensional Laplace equation $\Delta u(x, y) = 0$ on the half-space $\mathbb{H} = \{(x, y); \ x \in \mathbb{R}, y \geq 0\}$, with boundary condition: $u(x, 0) = f(x)$ for all $x \in \mathbb{R}$.

(d) *Verify* your solution to part (c). That is: check that your solution satisfies the Laplace equation and the desired boundary conditions.

20.10 Fix $\mu \in \mathbb{R}$, and define $f_\mu : \mathbb{R}^2 \longrightarrow \mathbb{C}$ by $f_\mu(x, y) := \exp(-|\mu| \cdot y)\exp(-\mu i x)$. Show that f is harmonic on $\mathbb{R}^2$. (This function appears in the Fourier solution to the half-plane Dirichlet problem; see Proposition 20C.1, p. 541.)

Appendix A

Sets and functions

A(i) Sets

A *set* is a collection of objects. If S is a set, then the objects in S are called *elements* of S; if s is an element of S, we write $s \in S$. A *subset* of S is a smaller set $\mathcal{R}$ such that every element of $\mathcal{R}$ is also an element of S. We indicate this by writing $\mathcal{R} \subset S$.

Sometimes we can explicitly list the elements in a set; we write $S = \{s_1, s_2, s_3, \ldots\}$.

Example A.1
(a) In Figure A.1(a), S is the set of all the cities in the world, so Toronto $\in S$. We might write $S = \{$Toronto, Beijing, London, Kuala Lumpur, Nairobi, Santiago, Pisa, Sidney, $\ldots\}$. If $\mathcal{R}$ is the set of all cities in Canada, then $\mathcal{R} \subset S$.
(b) In Figure A.1(b), the set of *natural numbers* is $\mathbb{N} := \{0, 1, 2, 3, 4, \ldots\}$. The set of *positive natural numbers* is $\mathbb{N}_+ := \{1, 2, 3, 4, \ldots\}$. Thus, $5 \in \mathbb{N}$, but $\pi \notin \mathbb{N}$ and $-2 \notin \mathbb{N}$.
(c) In Figure A.1(b), the set of *integers* is $\mathbb{Z} := \{\ldots, -3, -2, -1, 0, 1, 2, 3, 4, \ldots\}$. Thus, $5 \in \mathbb{Z}$ and $-2 \in \mathbb{Z}$, but $\pi \notin \mathbb{Z}$ and $\frac{1}{2} \notin \mathbb{Z}$. Observe that $\mathbb{N}_+ \subset \mathbb{N} \subset \mathbb{Z}$.
(d) In Figure A.1(b), the set of *real numbers* is denoted by $\mathbb{R}$. It is best visualized as an infinite line. Thus, $5 \in \mathbb{R}$, $-2 \in \mathbb{R}$, $\pi \in \mathbb{R}$ and $\frac{1}{2} \in \mathbb{R}$. Observe that $\mathbb{N} \subset \mathbb{Z} \subset \mathbb{R}$.
(e) In Figure A.1(b), the set of *non-negative real numbers* is denoted by $[0, \infty)$ or $\mathbb{R}_{\neq}$. It is best visualized as a half-infinite line, including zero. Observe that $[0, \infty) \subset \mathbb{R}$.
(f) In Figure A.1(b), the set of *positive real numbers* is denoted by $(0, \infty)$ or $\mathbb{R}_+$. It is best visualized as a half-infinite line, excluding zero. Observe that $\mathbb{R}_+ \subset \mathbb{R}_{\neq} \subset \mathbb{R}$.
(g) Figure A.1(c) depicts *two-dimensional space:* the set of all coordinate pairs (x, y), where x and y are real numbers. This set is denoted by $\mathbb{R}^2$, and is best visualised as an infinite plane.
(h) Figure A.1(d) depicts *three-dimensional space:* the set of all coordinate triples (x_1, x_2, x_3), where x_1, x_2, and x_3 are real numbers. This set is denoted by $\mathbb{R}^3$, and is best visualized as an infinite void.
(i) If D is any natural number, then *D-dimensional space* is the set of all coordinate triples $(x_1, x_2, \ldots, x_D)$, where $x_1, \ldots, x_D$ are all real numbers. This set is denoted by $\mathbb{R}^D$. It is hard to visualize this set when D is bigger than 3. $\diamond$

Cartesian products If S and T are two sets, then their *Cartesian product* is the set of all pairs (s, t), where s is an element of S and t is an element of T. We denote this set by $S \times T$.

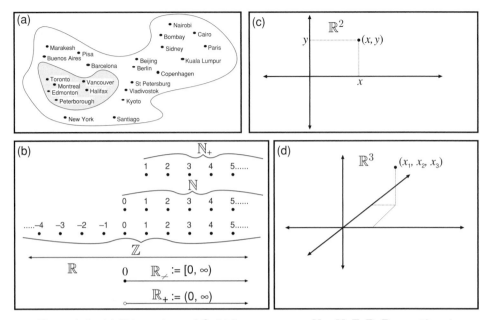

Figure A.1. (a) $\mathcal{R}$ is a subset of $\mathcal{S}$. (b) Important sets: $\mathbb{N}_+$, $\mathbb{N}$, $\mathbb{Z}$, $\mathbb{R}$, $\mathbb{R}_{\not{}} := [0, \infty)$ and $\mathbb{R}_+ := (0, \infty)$. (c) $\mathbb{R}^2$ is two-dimensional space. (d) $\mathbb{R}^3$ is three-dimensional space.

Example A.2

(a) $\mathbb{R} \times \mathbb{R}$ is the set of all pairs (x, y), where x and y are real numbers. In other words, $\mathbb{R} \times \mathbb{R} = \mathbb{R}^2$.

(b) $\mathbb{R}^2 \times \mathbb{R}$ is the set of all pairs $(\mathbf{w}, z)$, where $\mathbf{w} \in \mathbb{R}^2$ and $z \in \mathbb{R}$. But if $\mathbf{w}$ is an element of $\mathbb{R}^2$, then $\mathbf{w} = (x, y)$ for some $x \in \mathbb{R}$ and $y \in \mathbb{R}$. Thus, any element of $\mathbb{R}^2 \times \mathbb{R}$ is an object $((x, y), z)$. By suppressing the inner pair of brackets, we can write this as (x, y, z). In this way, we see that $\mathbb{R}^2 \times \mathbb{R}$ is the same as $\mathbb{R}^3$.

(c) In the same way, $\mathbb{R}^3 \times \mathbb{R}$ is the same as $\mathbb{R}^4$, once we write $((x, y, z), t)$ as (x, y, z, t). More generally, $\mathbb{R}^D \times \mathbb{R}$ is mathematically the same as $\mathbb{R}^{D+1}$.

    Often, we use the final coordinate to store a 'time' variable, so it is useful to distinguish it, by writing $((x, y, z), t)$ as $(x, y, z; t)$.   ◇

A(ii) Functions

If $\mathcal{S}$ and $\mathcal{T}$ are sets, then a *function* from $\mathcal{S}$ to $\mathcal{T}$ is a rule which assigns a specific element of $\mathcal{T}$ to every element of $\mathcal{S}$. We indicate this by writing $f : \mathcal{S} \longrightarrow \mathcal{T}$.

Example A.3

(a) In Figure A.2(a), $\mathcal{S}$ denotes the cities in the world, and $\mathcal{T} = \{A, B, C, \ldots, Z\}$ denotes the letters of the alphabet, and f is the function which is the first letter in the name of each city. Thus $f(\text{Peterborough}) = P$, $f(\text{Santiago}) = S$, etc.

(b) If $\mathbb{R}$ is the set of real numbers, then $\sin : \mathbb{R} \longrightarrow \mathbb{R}$ is a function: $\sin(0) = 0$, $\sin(\pi/2) = 1$, etc.   ◇

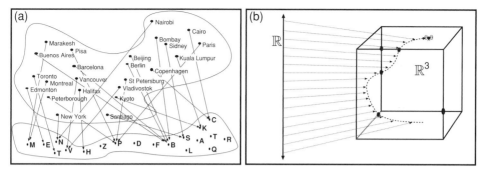

Figure A.2. (a) $f(C)$ is the first letter of city C. (b) $\mathbf{p}(t)$ is the position of the fly at time t.

Two important classes of functions are *paths* and *fields*.

Paths Imagine a fly buzzing around a room. Suppose we try to represent its trajectory as a curve through space. This defines a function $\mathbf{p}$ from $\mathbb{R}$ into $\mathbb{R}^3$, where $\mathbb{R}$ represents *time*, and $\mathbb{R}^3$ represents the (three-dimensional) room, as shown in Figure A.2(b). If $t \in \mathbb{R}$ is some moment in time, then $\mathbf{p}(t)$ is the position of the fly at time t. Since this $\mathbf{p}$ describes the path of the fly, we call $\mathbf{p}$ a *path*.

More generally, a *path* (or *trajectory* or *curve*) is a function $\mathbf{p} : \mathbb{R} \longrightarrow \mathbb{R}^D$, where D is any natural number. It describes the motion of an object through D-dimensional space. Thus, if $t \in \mathbb{R}$, then $\mathbf{p}(t)$ is the position of the object at time t.

Scalar fields Imagine a three-dimensional topographic map of Antarctica. The rugged surface of the map is obtained by assigning an altitude to every location on the continent. In other words, the map implicitly defines a function $\mathbf{h}$ from $\mathbb{R}^2$ (the Antarctic continent) to $\mathbb{R}$ (the set of altitudes, in metres above sea level). If $(x, y) \in \mathbb{R}^2$ is a location in Antarctica, then $\mathbf{h}(x, y)$ is the altitude at this location (and $\mathbf{h}(x, y) = 0$ means (x, y) is at sea level).

This is an example of a *scalar field*. A scalar field is a function $u : \mathbb{R}^D \longrightarrow \mathbb{R}$; it assigns a numerical quantity to every point in D-dimensional space.

Example A.4
(a) In Figure A.3(a), a landscape is represented by a *height function* $\mathbf{h} : \mathbb{R}^2 \longrightarrow \mathbb{R}$.
(b) Figure A.3(b) depicts a *concentration function* on a two-dimensional plane (e.g. the concentration of bacteria on a petri dish). This is a function $\rho : \mathbb{R}^2 \longrightarrow \mathbb{R}_{\not+}$ (where $\rho(x, y) = 0$ indicates zero bacteria at (x, y)).
(c) The *mass density* of a three-dimensional object is a function $\rho : \mathbb{R}^3 \longrightarrow \mathbb{R}_{\not+}$ (where $\rho(x_1, x_2, x_3) = 0$ indicates a vacuum).
(d) The *charge density* is a function $\mathbf{q} : \mathbb{R}^3 \longrightarrow \mathbb{R}$ (where $\mathbf{q}(x_1, x_2, x_3) = 0$ indicates electric neutrality)
(e) The *electric potential* (or *voltage*) is a function $\mathbf{V} : \mathbb{R}^3 \longrightarrow \mathbb{R}$.
(f) The *temperature distribution* in space is a function $u : \mathbb{R}^3 \longrightarrow \mathbb{R}$ (so $u(x_1, x_2, x_3)$ is the 'temperature at location (x_1, x_2, x_3)'). $\diamondsuit$

A *time-varying scalar field* is a function $u : \mathbb{R}^D \times \mathbb{R} \longrightarrow \mathbb{R}$, assigning a quantity to every point in space at each moment in time. Thus, for example, $u(\mathbf{x}; t)$ is the 'temperature at location $\mathbf{x}$, at time t'.

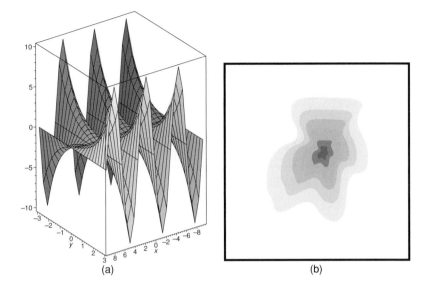

Figure A.3. (a) A height function describes a landscape. (b) A density distribution in $\mathbb{R}^2$.

Vector fields A *vector field* is a function $\vec{\mathbf{V}} : \mathbb{R}^D \longrightarrow \mathbb{R}^D$; it assigns a vector (i.e. an 'arrow') at every point in space.

Example A.5

(a) The *electric field* generated by a charge distribution is a vector field (denoted $\vec{\mathbf{E}}$).

(b) The *flux* of some material flowing through space is a vector field (often denoted by $\vec{\mathbf{F}}$).   ◇

Thus, for example, $\vec{\mathbf{F}}(\mathbf{x})$ is the 'flux' of material at location $\mathbf{x}$.

Appendix B

Derivatives – notation

If $f : \mathbb{R} \longrightarrow \mathbb{R}$, then f' is the first derivative of f, f'' is the second derivative, $f^{(n)}$ is the nth derivative, etc. If $\mathbf{x} : \mathbb{R} \longrightarrow \mathbb{R}^D$ is a path, then the *velocity* of $\mathbf{x}$ at time t is the vector

$$\dot{\mathbf{x}}(t) = \left[x_1'(t), x_2'(t), \dots, x_D'(t) \right].$$

If $u : \mathbb{R}^D \longrightarrow \mathbb{R}$ is a scalar field, then the following notations will be used interchangeably:

$$\text{for all } j \in [1 \dots D], \quad \partial_j u := \frac{\partial u}{\partial x_j}.$$

If $u : \mathbb{R}^2 \longrightarrow \mathbb{R}$ (i.e. $u(x, y)$ is a function of two variables), then we will also write

$$\partial_x u := \frac{\partial u}{\partial x} \quad \text{and} \quad \partial_y u := \frac{\partial u}{\partial y}.$$

Multiple derivatives will be indicated by iterating this procedure. For example,

$$\partial_x^3 \partial_y^2 u := \frac{\partial^3}{\partial x^3} \frac{\partial^2 u}{\partial y^2}.$$

A useful notational convention (which we rarely use) is *multiexponents*. If $\gamma_1, \dots, \gamma_D$ are positive integers, and $\gamma = (\gamma_1, \dots, \gamma_D)$, then

$$\mathbf{x}^\gamma := x_1^{\gamma_1} x_2^{\gamma_2} \cdots x_D^{\gamma_D}.$$

For example if $\gamma = (3, 4)$ and $\mathbf{z} = (x, y)$, then $\mathbf{z}^\gamma = x^3 y^4$.

This generalizes to *multi-index* notation for derivatives. If $\gamma = (\gamma_1, \dots, \gamma_D)$, then

$$\partial^\gamma u := \partial_1^{\gamma_1} \partial_2^{\gamma_2} \cdots \partial_D^{\gamma_D} u.$$

For example, if $\gamma = (1, 2)$, then

$$\partial^\gamma u = \frac{\partial}{\partial x} \frac{\partial^2 u}{\partial y^2}.$$

Remark Many authors use subscripts to indicate partial derivatives. For example, they would write

$$u_x := \partial_x u, \quad u_{xx} := \partial_x^2 u, \quad u_{xy} := \partial_x \partial_y u, \text{ etc.}$$

This notation is very compact and intuitive, but it has two major disadvantages.

(1) When dealing with an N-dimensional function $u(x_1, x_2, \ldots, x_N)$ (where N is either large or indeterminate), we have only two options. We can either either use awkward 'nested subscript' expressions such as

$$u_{x_3} := \partial_3 u, \quad u_{x_5 x_5} := \partial_5^2 u, \quad u_{x_2 x_3} := \partial_2 \partial_3 u, \text{ etc.,}$$

or we must adopt the 'numerical subscript' convention that

$$u_3 := \partial_3 u, \quad u_{55} := \partial_5^2 u, \quad u_{23} := \partial_2 \partial_3 u, \text{ etc.}$$

But once 'numerical' subscripts are reserved to indicate derivatives in this fashion, they can no longer be used for other purposes (e.g. indexing a sequence of functions, or indexing the coordinates of a vector-valued function). This can create further awkwardness.

(2) We will often be considering functions of the form $u(x, y; t)$, where (x, y) are 'space' coordinates and t is a 'time' coordinate. In this situation, it is often convenient to fix a value of t and consider the two-dimensional scalar field $u_t(x, y) := u(x, y; t)$. Normally, when we use t as a subscript, it will be indicate a 'time-frozen' scalar field of this kind.

Thus, in this book, we will *never* use subscripts to indicate partial derivatives. Partial derivatives will always be indicated by the notation $\partial_x u$ or $\frac{\partial u}{\partial x}$ (almost always the first one). However, when consulting other texts, you should be aware of the 'subscript' notation for derivatives, because it is used quite frequently.

Appendix C

Complex numbers

Complex numbers have the form $z = x + y\mathbf{i}$, where $\mathbf{i}^2 = -1$. We say that x is the *real part* of z, and y is the *imaginary part*; we write: $x = \operatorname{Re}[z]$ and $y = \operatorname{Im}[z]$.

If we imagine (x, y) as two real coordinates, then the complex numbers form a two-dimensional plane. Thus, we can also write a complex number in *polar coordinates* (see Figure C.1) If $r > 0$ and $0 \leq \theta < 2\pi$, then we define:

$$r \operatorname{cis} \theta = r \cdot [\cos(\theta) + \mathbf{i}\sin(\theta)].$$

Addition If $z_1 = x_1 + y_1\mathbf{i}$ and $z_2 = x_2 + y_2\mathbf{i}$, are two complex numbers, then $z_1 + z_2 = (x_1 + x_2) + (y_1 + y_2)\mathbf{i}$ (see Figure C.2).

Multiplication If $z_1 = x_1 + y_1\mathbf{i}$ and $z_2 = x_2 + y_2\mathbf{i}$, are two complex numbers, then $z_1 \cdot z_2 = (x_1x_2 - y_1y_2) + (x_1y_2 + x_2y_1)\mathbf{i}$.

Multiplication has a nice formulation in polar coordinates. If $z_1 = r_1 \operatorname{cis}\theta_1$ and $z_2 = r_2 \operatorname{cis}\theta_2$, then $z_1 \cdot z_2 = (r_1 \cdot r_2)\operatorname{cis}(\theta_1 + \theta_2)$. In other words, multiplication by the complex number $z = r \operatorname{cis}\theta$ is equivalent to *dilating* the complex plane by a factor of r, and *rotating* the plane by an angle of θ (see Figure C.3).

Exponential If $z = x + y\mathbf{i}$, then $\exp(z) = e^x \operatorname{cis} y = e^x \cdot [\cos(y) + \mathbf{i}\sin(y)]$ (see Figure C.4). In particular, if $x \in \mathbb{R}$, then

- $\exp(x) = e^x$ is the standard real-valued exponential function;
- $\exp(y\mathbf{i}) = \cos(y) + \sin(y)\mathbf{i}$ is a periodic function; as y moves along the real line, $\exp(y\mathbf{i})$ moves around the unit circle (this is *Euler's formula*).

The complex exponential function shares two properties with the real exponential function:

- if $z_1, z_2 \in \mathbb{C}$, then $\exp(z_1 + z_2) = \exp(z_1) \cdot \exp(z_2)$;
- if $w \in \mathbb{C}$, and we define the function $f : \mathbb{C} \longrightarrow \mathbb{C}$ by $f(z) = \exp(w \cdot z)$, then $f'(z) = w \cdot f(z)$.

Consequence If $w_1, w_2, \ldots, w_D \in \mathbb{C}$, and we define $f : \mathbb{C}^D \longrightarrow \mathbb{C}$ by

$$f(z_1, \ldots, z_D) = \exp(w_1z_1 + w_2z_2 + \cdots + w_Dz_D),$$

then $\partial_d f(\mathbf{z}) = w_d \cdot f(\mathbf{z})$. More generally,

$$\partial_1^{n_1}\partial_2^{n_2}\cdots\partial_D^{n_D} f(\mathbf{z}) = w_1^{n_1} \cdot w_2^{n_2} \cdots w_D^{n_D} \cdot f(\mathbf{z}). \tag{C.1}$$

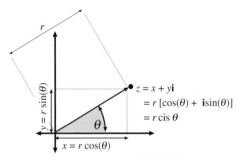

Figure C.1. $z = x + y\mathbf{i}$; $r = \sqrt{x^2 + y^2}$; $\theta = \tan(y/x)$.

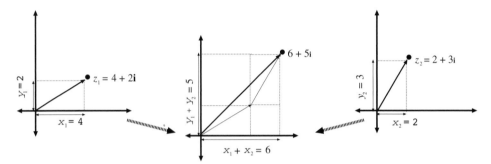

Figure C.2. The addition of complex numbers $z_1 = x_1 + y_1\mathbf{i}$ and $z_2 = x_2 + y_2\mathbf{i}$.

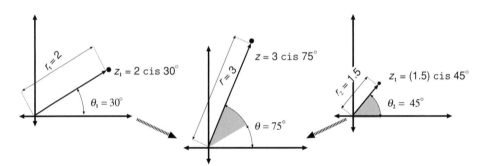

Figure C.3. The multiplication of complex numbers $z_1 = r_1 \operatorname{cis}\theta_1$ and $z_2 = r_2 \operatorname{cis}\theta_2$.

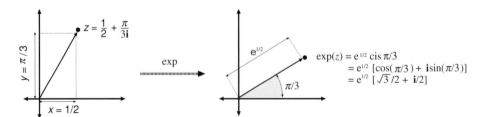

Figure C.4. The exponential of complex number $z = x + y\mathbf{i}$.

For example, if $f(x, y) = \exp(3x + 5\mathbf{i}y)$, then

$$f_{xxy}(x, y) = \partial_x^2 \, \partial_y \, f(x, y) = 45 \, \mathbf{i} \cdot \exp(3x + 5\mathbf{i}y).$$

If $\mathbf{w} = (w_1, \ldots, w_D)$ and $\mathbf{z} = (z_1, \ldots, z_D)$, then we will sometimes write

$$\exp(w_1 z_1 + w_2 z_2 + \cdots + w_D z_D) = \exp \langle \mathbf{w}, \mathbf{z} \rangle.$$

Conjugation and norm If $z = x + y\mathbf{i}$, then the *complex conjugate* of z is $\overline{z} = x - y\mathbf{i}$. In polar coordinates, if $z = r \operatorname{cis} \theta$, then $\overline{z} = r \operatorname{cis}(-\theta)$.

The *norm* of z is $|z| = \sqrt{x^2 + y^2}$. We have the following formula:

$$|z|^2 = z \cdot \overline{z}.$$

Appendix D

Coordinate systems and domains

Prerequisites: Appendix A.

Boundary value problems are usually posed on some 'domain' – some region of space. To solve the problem, it helps to have a convenient way of mathematically representing these domains, which can sometimes be simplified by adopting a suitable coordinate system. We will first give a variety of examples of 'domains' in different coordinate systems in §D(i)–§D(iv). Then in §D(v) we will give a formal definition of the word 'domain'.

D(i) Rectangular coordinates

Rectangular coordinates in $\mathbb{R}^3$ are normally denoted (x, y, z). Three common domains in rectangular coordinates are as follows.

- The *slab* $\mathbb{X} = \{(x, y, z) \in \mathbb{R}^3; \; 0 \leq z \leq L\}$, where L is the thickness of the slab (see Figure D.1(d)).
- The *unit cube* $\mathbb{X} = \{(x, y, z) \in \mathbb{R}^3; 0 \leq x \leq 1, 0 \leq y \leq 1, \text{and } 0 \leq z \leq 1\}$ (see Figure D.1(c)).
- The *box:* $\mathbb{X} = \{(x, y, z) \in \mathbb{R}^3; 0 \leq x \leq L_1, 0 \leq y \leq L_2, \text{and } 0 \leq z \leq L_3\}$, where L_1, L_2, and L_3 are the sidelengths (see Figure D.1(a)).
- The *rectangular column* $\mathbb{X} = \{(x, y, z) \in \mathbb{R}^3; \; 0 \leq x \leq L_1, \text{ and } 0 \leq y \leq L_2\}$ (see Figure D.1(e)).

D(ii) Polar coordinates on $\mathbb{R}^2$

Polar coordinates (r, θ) on $\mathbb{R}^2$ are defined by the following transformations:

$$x = r \cdot \cos(\theta) \quad \text{and} \quad y = r \cdot \sin(\theta)$$

(see Figure D.2), with reverse transformation

$$r = \sqrt{x^2 + y^2} \quad \text{and} \quad \theta = \text{Arctan}(y, x).$$

Here, the coordinate r ranges over $\mathbb{R}_{\not{}}$, while the variable θ ranges over $[-\pi, \pi)$. Finally we define:

$$\text{Arctan}(y, x) := \begin{cases} \arctan(y/x) & \text{if} \quad x > 0; \\ \arctan(y/x) + \pi & \text{if} \quad x < 0 \text{ and } y > 0; \\ \arctan(y/x) - \pi & \text{if} \quad x < 0 \text{ and } y < 0. \end{cases}$$

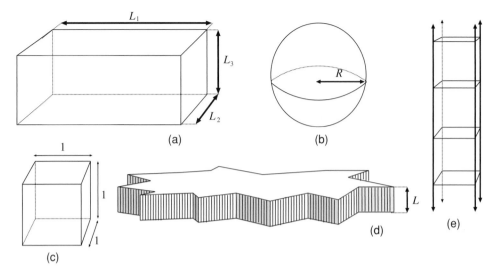

Figure D.1. Some domains in $\mathbb{R}^3$: (a) box; (b) ball; (c) cube; (d) slab; (e) rectangular column.

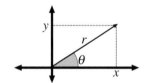

Figure D.2. Polar coordinates.

Three common domains in polar coordinates are the following.

- $\mathbb{D} = \{(r, \theta); \ r \leq R\}$ is the *disk* of radius R (see Figure D.3(a)).
- $\mathbb{D}^{\complement} = \{(r, \theta); \ R \leq r\}$ is the *codisk* of inner radius R.
- $\mathbb{A} = \{(r, \theta); \ R_{min} \leq r \leq R_{max}\}$ is the *annulus* of inner radius R_{min} and outer radius R_{max} (see Figure D.3(b)).

D(iii)   Cylindrical coordinates on $\mathbb{R}^3$

Cylindrical coordinates (r, θ, z) on $\mathbb{R}^3$ are defined by the following transformations:

$$x = r \cdot \cos(\theta), \quad y = r \cdot \sin(\theta), \quad \text{and} \quad z = z,$$

with reverse transformation

$$r = \sqrt{x^2 + y^2}, \quad \theta = \text{Arctan}\,(y, x), \quad \text{and} \quad z = z.$$

Five common domains in cylindrical coordinates are:

- $\mathbb{X} = \{(r, \theta, z); \ r \leq R\}$ is the *(infinite) cylinder* of radius R (see Figure D.3(e));
- $\mathbb{X} = \{(r, \theta, z); \ R_{min} \leq r \leq R_{max}\}$ is the *(infinite) pipe* of inner radius R_{min} and outer radius R_{max} (see Figure D.3(d));
- $\mathbb{X} = \{(r, \theta, z); \ r > R\}$ is the *wellshaft* of radius R;

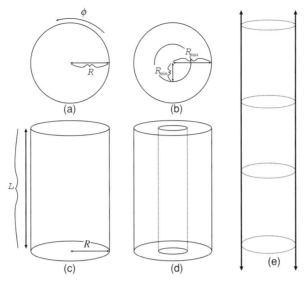

Figure D.3. Some domains in polar and cylindrical coordinates: (a) disk; (b) annulus; (c) finite cylinder; (d) pipe; (e) infinite cylinder.

- $\mathbb{X} = \{(r, \theta, z); \ r \leq R \text{ and } 0 \leq z \leq L\}$ is the *finite cylinder* of radius R and length L (see Figure D.3(c));
- in cylindrical coordinates on $\mathbb{R}^3$, we can write the *slab* as $\{(r, \theta, z); \ 0 \leq z \leq L\}$.

D(iv) Spherical coordinates on $\mathbb{R}^3$

Spherical coordinates (r, θ, ϕ) on $\mathbb{R}^3$ are defined by the following transformations:

$$x = r \cdot \sin(\phi) \cdot \cos(\theta), \quad y = r \cdot \sin(\phi) \cdot \sin(\theta),$$
$$z = r \cdot \cos(\phi)$$

(Figure D.4), with reverse transformation

$$r = \sqrt{x^2 + y^2 + z^2}, \quad \theta = \text{Arctan}\,(y, x),$$
$$\phi = \text{Arctan}\left(\sqrt{x^2 + y^2},\ z\right).$$

In spherical coordinates, the set $\mathbb{B} = \{(r, \theta, \phi); \ r \leq R\}$ is the *ball* of radius R (see Figure D.1(b)).

D(v) What is a 'domain'?

Formally, a *domain* is a subset $\mathbb{X} \subseteq \mathbb{R}^D$ which satisfies the following three conditions.

(a) $\mathbb{X}$ is *closed*. That is, $\mathbb{X}$ must contain all its boundary points.
(b) $\mathbb{X}$ has a *dense interior*. That is, every point in $\mathbb{X}$ is a limit point of a sequence $\{x_n\}_{n=1}^{\infty}$ of interior points of $\mathbb{X}$. (A point $x \in \mathbb{X}$ is an *interior point* if $\mathbb{B}(x, \epsilon) \subset \mathbb{X}$ for some $\epsilon > 0$.)
(c) $\mathbb{X}$ is *connected*. That is, we cannot find two disjoint closed subsets $\mathbb{X}_1$ and $\mathbb{X}_2$ such that $\mathbb{X} = \mathbb{X}_1 \sqcup \mathbb{X}_2$.

Observe that all of the examples in §D(i)–§D(iv) satisfy these three conditions.

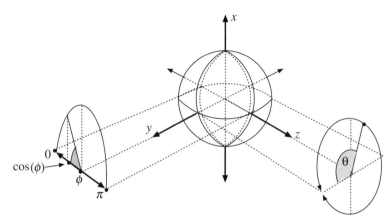

Figure D.4. Spherical coordinates.

Why do we need conditions (a), (b), and (c)? We are normally interested in finding a function $f : \mathbb{X} \longrightarrow \mathbb{R}$ which satisfies a certain partial differential equation on $\mathbb{X}$. However, such a PDE only makes sense on the interior of $\mathbb{X}$ (because the derivatives of f at x are only well-defined if x is an interior point of $\mathbb{X}$). Thus, first $\mathbb{X}$ must *have* a nonempty interior, and, second, this interior must fill 'most' of $\mathbb{X}$. This is the reason for condition (b). We often represent certain physical constraints by requiring f to satisfy certain *boundary conditions* on the boundary of $\mathbb{X}$. (That is what a 'boundary value problem' means.) But this cannot make sense unless $\mathbb{X}$ satisfies condition (a). Finally, we do not actually *need* condition (c). But if $\mathbb{X}$ is disconnected, then we could split $\mathbb{X}$ into two or more disconnected pieces and solve the equations separately on each piece. Thus, we can always assume, without loss of generality, that $\mathbb{X}$ is connected.

Appendix E

Vector calculus

Prerequisites: Appendix A, Appendix B.

E(i) Gradient

... in two dimensions

Suppose $\mathbb{X} \subset \mathbb{R}^2$ is a two-dimensional region. To define the topography of a 'landscape' on this region, it suffices[1] to specify the *height* of the land at each point. Let $u(x, y)$ be the height of the land at the point $(x, y) \in \mathbb{X}$. (Technically, we say: $u : \mathbb{X} \longrightarrow \mathbb{R}$ is a *two-dimensional scalar field*.)

The *gradient* of the landscape measures the *slope* at each point in space. To be precise, we want the gradient to be a vector pointing in the direction of *most rapid ascent*. The *length* of this vector should then measure the *rate* of ascent. Mathematically, we define the *two-dimensional gradient* of u by:

$$\nabla u(x, y) = \left[\frac{\partial u}{\partial x}(x, y), \frac{\partial u}{\partial y}(x, y) \right].$$

The gradient vector points in the direction where u is increasing the most rapidly. If $u(x, y)$ were the height of a mountain at location (x, y), and we were trying to climb the mountain, then our (naive) strategy would be always to walk in the direction $\nabla u(x, y)$. Note that, for any $(x, y) \in \mathbb{X}$, the gradient $\nabla u(x, y)$ is a two-dimensional vector – that is, $\nabla u(x, y) \in \mathbb{R}^2$. (Technically, we say '$\nabla u : \mathbb{X} \longrightarrow \mathbb{R}^2$ is a *two-dimensional vector field*'.)

... in many dimensions

This idea generalizes to any dimension. If $u : \mathbb{R}^D \longrightarrow \mathbb{R}$ is a scalar field, then the *gradient* of u is the associated vector field $\nabla u : \mathbb{R}^D \longrightarrow \mathbb{R}^D$, where, for any $\mathbf{x} \in \mathbb{R}^D$,

$$\nabla u(\mathbf{x}) = \left[\partial_1 u, \partial_2 u, \dots, \partial_D u \right] (\mathbf{x}).$$

---

[1] Assuming no overhangs!

Proposition E.1 Algebra of gradients

Let $\mathbb{X} \subset \mathbb{R}^D$ *be a domain. Let* $f, g : \mathbb{X} \longrightarrow \mathbb{R}$ *be differentiable scalar fields, and let* $(f + g) : \mathbb{X} \longrightarrow \mathbb{R}$ *and* $(f \cdot g) : \mathbb{X} \longrightarrow \mathbb{R}$ *denote the sum and product of* f *and* g, *respectively.*

(a) (Linearity) *For all* $\mathbf{x} \in \mathbb{X}$, *and any* $r \in \mathbb{R}$,

$$\nabla (rf + g)(\mathbf{x}) = r \, \nabla f(\mathbf{x}) + \nabla g(\mathbf{x}).$$

(b) (Leibniz rule) *For all* $\mathbf{x} \in \mathbb{X}$,

$$\nabla (f \cdot g)(\mathbf{x}) = f(\mathbf{x}) \cdot \left(\nabla g(\mathbf{x}) \right) + g(\mathbf{x}) \cdot \left(\nabla f(\mathbf{x}) \right).$$

Ⓔ   *Proof* **Exercise E.1.**   □

E(ii) Divergence

. . . in one dimension

Imagine a current of 'fluid' (e.g. air, water, electricity) flowing along the real line $\mathbb{R}$. For each point $x \in \mathbb{R}$, let $V(x)$ describe the rate at which fluid is flowing past this point. Now, in places where the fluid *slows down*, we expect the derivative $V'(x)$ to be negative. We also expect the fluid to *accumulate* (i.e. become 'compressed') at such locations (because fluid is entering the region more quickly than it leaves). In places where the fluid *speeds up*, we expect the derivative $V'(x)$ to be positive, and we expect the fluid to be *depleted* (i.e. to decompress) at such locations (because fluid is leaving the region more quickly than it arrives).

If the fluid is incompressible (e.g. water), then we can assume that the quantity of fluid at each point is constant. In this case, the fluid cannot 'accumulate' or 'be depleted'. In this case, a negative value of $V'(x)$ means that fluid is somehow being 'absorbed' (e.g. being destroyed or leaking out of the system) at x. Likewise, a positive value of $V'(x)$ means that fluid is somehow being 'generated' (e.g. being created, or leaking into the system) at x.

In general, positive $V'(x)$ may represent some combination of fluid depletion, decompression, or generation at x, while negative $V'(x)$ may represent some combination of local accumulation, compression, or absorption at x. Thus, if we define the *divergence* of the flow to be the *rate at which fluid is being depleted/decompressed/generated* (if positive) *or being accumulated/compressed/absorbed* (if positive), then, mathematically speaking,

$$\boxed{\operatorname{div} V(x) = V'(x).}$$

This physical model yields an interesting interpretation of the Fundamental Theorem of Calculus. Suppose $a < b \in \mathbb{R}$, and consider the interval $[a, b]$. If $V : \mathbb{R} \longrightarrow \mathbb{R}$ describes the flow of fluid, then $V(a)$ is the amount of fluid flowing into the left-hand end of the interval $[a, b]$ (or flowing *out*, if $V(a) < 0$). Likewise, $V(b)$ is the amount of fluid flowing out of the right-hand end of the interval $[a, b]$ (or flowing *in*, if $V(b) < 0$). Thus, $V(b) - V(a)$ is the net amount of fluid flowing *out* through the endpoints of $[a, b]$ (or flowing *in*, if this quantity is negative). But the *fundamental theorem of calculus* asserts that

$$V(b) - V(a) = \int_a^b V'(x)\mathrm{d}x = \int_a^b \operatorname{div} V(x)\mathrm{d}x.$$

In other words, the net amount of fluid instantaneously leaving/entering through the end-points of $[a, b]$ is equal to the integral of the divergence over the interior. But if div $V(x)$ is the amount of fluid being instantaneously 'generated' at x (or 'absorbed' if negative), this integral can be interpreted as follows.

The net amount of fluid instantaneously leaving the endpoints of $[a, b]$ is equal to the net quantity of fluid being instantaneously generated throughout the interior of $[a, b]$.

From a physical point of view, this makes perfect sense; it is simply 'conservation of mass'. This is the one-dimensional form of the *divergence theorem* (Theorem E.4, p. 566).

... in two dimensions

Let $\mathbb{X} \subset \mathbb{R}^2$ be some planar region, and consider a fluid flowing through $\mathbb{X}$. For each point $(x, y) \in \mathbb{X}$, let $\vec{\mathbf{V}}(x, y)$ be a two-dimensional vector describing the current at that point.[2]

Think of this two-dimensional current as a superposition of a *horizontal current* V_1 and a *vertical current* V_2. For each of the two currents, we can reason as in the one-dimensional case. If $\partial_x V_1(x, y) > 0$, then the horizontal current is accelerating at (x, y), so we expect it to deplete the fluid at (x, y) (or, if the fluid is incompressible, we interpret this to mean that additional fluid is being generated at (x, y)). If $\partial_x V_1(x, y) < 0$, then the horizontal current is decelerating, and we expect it to deposit fluid at (x, y) (or, if the fluid is incompressible, we interpret this to mean that fluid is being absorbed or destroyed at (x, y)).

The same reasoning applies to $\partial_y V_2(x, y)$. The *divergence* of the two-dimensional current is thus just the sum of the divergences of the horizontal and vertical currents:

$$\boxed{\text{div } \vec{\mathbf{V}}(x, y) = \partial_x V_1(x, y) + \partial_y V_2(x, y).}$$

Note that, although $\vec{\mathbf{V}}(x, y)$ is a *vector*, the divergence div $\vec{\mathbf{V}}(x, y)$ is a *scalar*.[3] Just as in the one-dimensional case, we interpret div $\vec{\mathbf{V}}(x, y)$ to be the *instantaneous rate at which fluid is being depleted/decompressed/generated at (x, y) (if positive) or being accumulated/compressed/absorbed at (x, y) (if negative)*.

For example, suppose $\mathbb{R}^2$ represents the ocean, and $\vec{\mathbf{V}} : \mathbb{R}^2 \longrightarrow \mathbb{R}^2$ is a vector field representing ocean currents. If div $\vec{\mathbf{V}}(x, y) > 0$, this means that there is a net injection of water into the ocean at the point (x, y) – e.g. due to rainfall or a river outflow. If div $\vec{\mathbf{V}}(x, y) < 0$, this means that there is a net removal of water from the ocean at the point (x, y) – e.g. due to evaporation or a hole in the bottom of the sea.

... in many dimensions

We can generalize this idea to any number of dimensions. If $\vec{\mathbf{V}} : \mathbb{R}^D \longrightarrow \mathbb{R}^D$ is a vector field, then the *divergence* of $\vec{\mathbf{V}}$ is the associated scalar field div $\vec{\mathbf{V}} : \mathbb{R}^D \longrightarrow \mathbb{R}$, where, for any $\mathbf{x} \in \mathbb{R}^D$,

$$\boxed{\text{div } \vec{\mathbf{V}}(\mathbf{x}) = \partial_1 V_1(\mathbf{x}) + \partial_2 V_2(\mathbf{x}) + \cdots + \partial_D V_D(\mathbf{x}).}$$

---

[2] Technically, we say $\vec{\mathbf{V}} : \mathbb{X} \longrightarrow \mathbb{R}^2$ is a *two-dimensional vector field*.

[3] Technically, we say div $\vec{\mathbf{V}} : \mathbb{X} \longrightarrow \mathbb{R}^2$ is a *two-dimensional scalar field*.

If $\vec{V}$ represents the flow of some fluid through $\mathbb{R}^D$, then div $\vec{V}(\mathbf{x})$ represents the *instantaneous rate at which fluid is being depleted/decompressed/generated at* $\mathbf{x}$ (if positive) *or being accumulated/compressed/absorbed at* $\mathbf{x}$ (if negative). For example, if $\vec{E}$ is the electric field, then div $\vec{E}(\mathbf{x})$ is the amount of electric field being 'generated' at $\mathbf{x}$ – that is, div $\vec{E}(\mathbf{x}) = \mathbf{q}(\mathbf{x})$ is the *charge density* at $\mathbf{x}$.

Proposition E.2 Algebra of divergences

Let $\mathbb{X} \subset \mathbb{R}^D$ *be a domain. Let* $\vec{V}, \vec{W} : \mathbb{X} \longrightarrow \mathbb{R}^D$ *be differentiable vector fields, and let* $f : \mathbb{X} \longrightarrow \mathbb{R}$ *be a differentiable scalar field, and let* $(f \cdot \vec{V}) : \mathbb{X} \longrightarrow \mathbb{R}^D$ *represent the product of* f *and* $\vec{V}$.

(a) (Linearity) *For all* $\mathbf{x} \in \mathbb{X}$, *and any* $r \in \mathbb{R}$,

$$\text{div}\,(r\vec{V} + \vec{W})(\mathbf{x}) = r\,\text{div}\,\vec{V}(\mathbf{x}) + \text{div}\,\vec{W}g(\mathbf{x}).$$

(b) (Leibniz rule) *For all* $\mathbf{x} \in \mathbb{X}$,

$$\text{div}\,(f \cdot \vec{V})(\mathbf{x}) = f(\mathbf{x}) \cdot \left(\text{div}\,\vec{V}(\mathbf{x})\right) + \left(\nabla f(\mathbf{x})\right) \bullet \vec{V}(\mathbf{x}).$$

Ⓔ *Proof* **Exercise E.2.** □

Ⓔ **Exercise E.3** Let $\vec{V}, \vec{W} : \mathbb{X} \longrightarrow \mathbb{R}^D$ be differentiable vector fields, and consider their dot product $(\vec{V} \bullet \vec{W}) : \mathbb{X} \longrightarrow \mathbb{R}^D$ (a differentiable scalar field). State and prove a Leibniz-like rule for $\nabla(\vec{V} \bullet \vec{W})$ (a) in the case $D = 3$; (b) in the case $D \geq 4$. ♦

E(iii) The Divergence Theorem

. . . in two dimensions

Let $\mathbb{X} \subset \mathbb{R}^2$ be some domain in the plane, and let $\partial\mathbb{X}$ be the *boundary* of $\mathbb{X}$. (For example, if $\mathbb{X}$ is the unit disk, then $\partial\mathbb{X}$ is the unit circle. If $\mathbb{X}$ is a square domain, then $\partial\mathbb{X}$ is the four sides of the square, etc.) Let $\mathbf{x} \in \partial\mathbb{X}$. A line segment $\mathbb{L}$ through $\mathbf{x}$ is *tangent* to $\partial\mathbb{X}$ if $\mathbb{L}$ touches $\partial\mathbb{X}$ only at $\mathbf{x}$; that is, $\mathbb{L} \cap \partial\mathbb{X} = \{\mathbf{x}\}$ (see Figure E.1(a)). A unit vector $\vec{N}$ is *normal* to $\partial\mathbb{X}$ at $\mathbf{x}$ if there is a line segment through $\mathbf{x}$ which is orthogonal to $\vec{N}$ and which is tangent to $\partial\mathbb{X}$. We say $\partial\mathbb{X}$ is *piecewise smooth* if there is a unique unit normal vector $\vec{N}(\mathbf{x})$ at every $\mathbf{x} \in \partial\mathbb{X}$, except perhaps at finitely many points (the 'corners' of the boundary). For example, the disk, the square, and any other polygonal domain have piecewise smooth boundaries. The function $\vec{N} : \partial\mathbb{X} \longrightarrow \mathbb{R}^2$ is then called the *normal vector field* for $\partial\mathbb{X}$.

If $\vec{V} = (V_1, V_2)$ and $\vec{N} = (N_1, N_2)$ are two vectors, then define $\vec{V} \bullet \vec{N} := V_1 N_1 + V_2 N_2$. If $\vec{V} : \mathbb{R}^2 \longrightarrow \mathbb{R}^2$ is a vector field, and $\mathbb{X} \subset \mathbb{R}^2$ is a domain with a smooth boundary $\partial\mathbb{X}$, then we can define the *flux* of $\vec{V}$ across $\partial\mathbb{X}$ as the integral:

$$\int_{\partial\mathbb{X}} \vec{V}(\mathbf{s}) \bullet \vec{N}(\mathbf{s}) d\mathbf{s} \tag{E.1}$$

(see Figure E.1(b)). Here, by 'integrating over $\partial\mathbb{X}$', we are assuming that $\partial\mathbb{X}$ can be parameterized as a smooth curve or a union of smooth curves; this integral can then be computed (via this parameterization) as one or more one-dimensional integrals over intervals in $\mathbb{R}$. The value of integral (E.1) is independent of the choice of parameterization used. If $\vec{V}$ describes the flow of some fluid, then the flux given by equation (E.1) represents the net quantity of fluid flowing across the boundary of $\partial\mathbb{X}$.

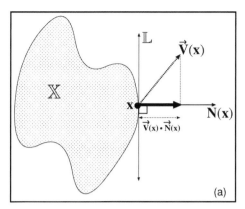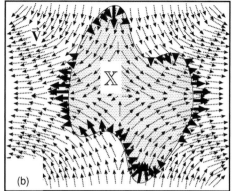

Figure E.1. (a) Line segment $\mathbb{L}$ is tangent to $\partial\mathbb{X}$ at **x**. Vector $\vec{\mathbf{N}}(\mathbf{x})$ is normal to $\partial\mathbf{X}$ at **x**. If $\vec{\mathbf{V}}(\mathbf{x})$ is another vector based at **x**, then the dot product $\vec{\mathbf{V}}(\mathbf{x}) \bullet \vec{\mathbf{N}}(\mathbf{x})$ measures the orthogonal projection of $\vec{\mathbf{V}}(\mathbf{x})$ onto $\vec{\mathbf{N}}(\mathbf{x})$ – that is, the 'part of $\vec{\mathbf{V}}(\mathbf{x})$ which is normal to $\partial\mathbb{X}$'. (b) Here $\vec{\mathbf{V}} : \mathbb{R}^2 \longrightarrow \mathbb{R}^2$ is a differentiable vector field, and we portray the scalar field $\vec{\mathbf{V}} \bullet \vec{\mathbf{N}}$ along the curve $\partial\mathbb{X}$ (although we have visualized it as a 'vector field', to help your intuition). The *flux* of $\vec{\mathbf{V}}$ across the boundary of $\mathbb{X}$ is obtained by integrating $\vec{\mathbf{V}} \bullet \vec{\mathbf{N}}$ along $\partial\mathbb{X}$.

On the other hand, if div $\vec{\mathbf{V}}(\mathbf{x})$ represents the instantaneous rate at which fluid is being generated/destroyed at the point **x**, then the two-dimensional integral

$$\int_{\mathbb{X}} \text{div } \vec{\mathbf{V}}(\mathbf{x})d\mathbf{x}$$

is the net rate at which fluid is being generated/destroyed throughout the interior of the region $\mathbb{X}$. The following result then simply says that the total 'mass' of the fluid must be conserved when we combine these two processes.

Theorem E.3 Green's theorem
If $\mathbb{X} \subset \mathbb{R}^2$ is an bounded domain with a piecewise smooth boundary, and $\vec{\mathbf{V}} : \mathbb{X} \longrightarrow \mathbb{R}^2$ is a continuously differentiable vector field, then

$$\int_{\partial\mathbb{X}} \vec{\mathbf{V}}(\mathbf{s}) \bullet \vec{\mathbf{N}}(\mathbf{s})d\mathbf{s} = \int_{\mathbb{X}} \text{div } \vec{\mathbf{V}}(\mathbf{x})d\mathbf{x}.$$

Proof See any introduction to vector calculus, for example Stewart (2008), §16.5. ☐

... *in many dimensions*

Let $\mathbb{X} \subset \mathbb{R}^D$ be some domain, and let $\partial\mathbb{X}$ be the *boundary* of $\mathbb{X}$. (For example, if $\mathbb{X}$ is the unit ball, then $\partial\mathbb{X}$ is the unit sphere.) If $D = 2$, then $\partial\mathbb{X}$ will be a one-dimensional *curve*. If $D = 3$, then $\partial\mathbb{X}$ will be a two-dimensional *surface*. In general, if $D \geq 4$, then $\partial\mathbb{X}$ will be a $(D-1)$-dimensional *hypersurface*.

Let $\mathbf{x} \in \partial\mathbb{X}$. A (hyper)plane segment $\mathbb{P}$ through **x** is *tangent* to $\partial\mathbb{X}$ if $\mathbb{P}$ touches $\partial\mathbb{X}$ only at **x**; that is, $\mathbb{P} \cap \partial\mathbb{X} = \{\mathbf{x}\}$. A unit vector $\vec{\mathbf{N}}$ is *normal* to $\partial\mathbb{X}$ at **x** if there is a (hyper)plane segment through **x** which is orthogonal to $\vec{\mathbf{N}}$ and which is tangent to $\partial\mathbb{X}$. We say $\partial\mathbb{X}$ is *smooth* if there is a unique unit normal vector $\vec{\mathbf{N}}(\mathbf{x})$ at each $\mathbf{x} \in \partial\mathbb{X}$.[4] The function $\vec{\mathbf{N}} : \partial\mathbb{X} \longrightarrow \mathbb{R}^D$ is then called the *normal vector field* for $\partial\mathbb{X}$.

If $\vec{\mathbf{V}} = (V_1, \ldots, V_D)$ and $\vec{\mathbf{N}} = (N_1, \ldots, N_D)$ are two vectors, then define $\vec{\mathbf{V}} \bullet \vec{\mathbf{N}} := V_1 N_1 + \cdots + V_D N_D$. If $\vec{\mathbf{V}} : \mathbb{R}^D \longrightarrow \mathbb{R}^D$ is a vector field, and $\mathbb{X} \subset \mathbb{R}^D$ is a domain with a smooth boundary $\partial\mathbb{X}$, then we can define the *flux* of $\vec{\mathbf{V}}$ across $\partial\mathbb{X}$ as the integral:

$$\int_{\partial\mathbb{X}} \vec{\mathbf{V}}(\mathbf{s}) \bullet \vec{\mathbf{N}}(\mathbf{s}) \mathrm{d}\mathbf{s}. \tag{E.2}$$

Here, by 'integrating over $\partial\mathbb{X}$', we are assuming that $\partial\mathbb{X}$ can be parameterized as a smooth (hyper)surface or a union of smooth (hyper)surfaces; this integral can then be computed (via this parameterization) as one or more $(D-1)$-dimensional integrals over open subsets of $\mathbb{R}^{D-1}$. The value of integral (E.2) is independent of the choice of parameterization used. If $\vec{\mathbf{V}}$ describes the flow of some fluid, then the flux given by equation (E.2) represents the net quantity of fluid flowing across the boundary of $\partial\mathbb{X}$.

On the other hand, if $\operatorname{div} \vec{\mathbf{V}}(\mathbf{x})$ represents the instantaneous rate at which fluid is being generated/destroyed at the point $\mathbf{x}$, then the D-dimensional integral

$$\int_{\mathbb{X}} \operatorname{div} \vec{\mathbf{V}}(\mathbf{x}) \mathrm{d}\mathbf{x}$$

is the net rate at which fluid is being generated/destroyed throughout the interior of the region $\mathbb{X}$. The following result then simply says that the total 'mass' of the fluid must be conserved when we combine these two processes.

Theorem E.4 Divergence theorem

If $\mathbb{X} \subset \mathbb{R}^D$ is a bounded domain with a piecewise smooth boundary, and $\vec{\mathbf{V}} : \mathbb{X} \longrightarrow \mathbb{R}^D$ is a continuously differentiable vector field, then

$$\int_{\partial\mathbb{X}} \vec{\mathbf{V}}(\mathbf{s}) \bullet \vec{\mathbf{N}}(\mathbf{s}) \mathrm{d}\mathbf{s} = \int_{\mathbb{X}} \operatorname{div} \vec{\mathbf{V}}(\mathbf{x}) \mathrm{d}\mathbf{x}.$$

Proof If $D = 1$, this just restates the Fundamental theorem of calculus. If $D = 2$, this just restates of Green's theorem E.3. For the case $D = 3$, this result can be found in any introduction to vector calculus; see, for example, Stewart (2008), §16.9. This theorem is often called *Gauss's theorem* (after C. F. Gauss) or *Ostrogradsky's theorem* (after Mikhail Ostrogradsky).

For the case $D \geq 4$, this is a special case of the *generalized Stokes theorem*, one of the fundamental results of modern differential geometry, which unifies the classic (two-dimensional) Stokes theorem, Green's theorem, Gauss's theorem, and the fundamental theorem of calculus. A statement and proof can be found in any introduction to differential geometry or tensor calculus. See, for example, Bishop and Goldberg (1980), Theorem 4.9.2.

Some texts on partial differential equations also review the divergence theorem, usually in an appendix. See, for example, Evans (1991), Appendix C.2. □

---

[4] More generally, $\partial\mathbb{X}$ is *piecewise smooth* if there is a unique unit normal vector $\vec{\mathbf{N}}(\mathbf{x})$ at 'almost every' $\mathbf{x} \in \partial\mathbb{X}$, except perhaps for a subset of dimension $(D-2)$. For example, a surface in $\mathbb{R}^3$ is piecewise smooth if it has a normal vector field everywhere except at some union of curves which represent the 'edges' between the smooth 'faces' the surface. In particular, a cube, a cylinder, or any other polyhedron has a piecewise smooth boundary.

Green's formulas Let $u : \mathbb{R}^D \longrightarrow \mathbb{R}$ be a scalar field. If $\mathbb{X}$ is a domain, and $\mathbf{s} \in \partial\mathbb{X}$, then the *outward normal derivative* of u at $\mathbf{s}$ is defined:

$$\partial_\perp u(\mathbf{s}) := \nabla u(\mathbf{s}) \bullet \vec{\mathbf{N}}(\mathbf{s})$$

(see §5C(ii), p. 80, for more information). Meanwhile, the *Laplacian* of u is defined by

$$\triangle u = \operatorname{div}(\nabla(u))$$

(see §1B(ii), p. 8, for more information). The divergence theorem then has the following useful consequences.

Corollary E.5 Green's formulae
Let $\mathbb{X} \subset \mathbb{R}^D$ be a bounded domain, and let $u : \mathbb{X} \longrightarrow \mathbb{R}$ be a scalar field which is C^2 (i.e. twice continuously differentiable). Then

(a) $\int_{\partial\mathbb{X}} \partial_\perp u(\mathbf{s})\mathrm{d}\mathbf{s} = \int_{\mathbb{X}} \triangle u(\mathbf{x})\mathrm{d}\mathbf{x}$*;*
(b) $\int_{\partial\mathbb{X}} u(\mathbf{s}) \cdot \partial_\perp u(\mathbf{s})\mathrm{d}\mathbf{s} = \int_{\mathbb{X}} u(\mathbf{x}) \triangle u(\mathbf{x}) + |\nabla u(\mathbf{x})|^2 \, \mathrm{d}\mathbf{x}$*;*
(c) *for any other C^2 function $w : \mathbb{X} \longrightarrow \mathbb{R}$,*

$$\int_{\mathbb{X}} \Big(u(\mathbf{x}) \triangle w(\mathbf{x}) - w(\mathbf{x}) \triangle u(\mathbf{x})\Big)\mathrm{d}\mathbf{x} = \int_{\partial\mathbb{X}} \Big(u(\mathbf{s}) \cdot \partial_\perp w(\mathbf{s}) - w(\mathbf{s}) \cdot \partial_\perp u(\mathbf{s})\Big)\mathrm{d}\mathbf{s}.$$

Proof (a) is **Exercise E.4**. To prove (b), note that ⓔ

$$2 \int_{\partial\mathbb{X}} u(\mathbf{s}) \cdot \partial_\perp u(\mathbf{s})\mathrm{d}\mathbf{s} \underset{(\dagger)}{=} \int_{\partial\mathbb{X}} \partial_\perp(u^2)(\mathbf{s})\mathrm{d}\mathbf{s} \underset{(*)}{=} \int_{\mathbb{X}} \triangle(u^2)(\mathbf{x})\mathrm{d}\mathbf{x}$$

$$\underset{(\diamond)}{=} 2 \int_{\mathbb{X}} u(\mathbf{x}) \triangle u(\mathbf{x}) + |\nabla u(\mathbf{x})|^2 \, \mathrm{d}\mathbf{x}.$$

Here, ($\dagger$) is because $\partial_\perp(u^2)(\mathbf{s}) = 2u(\mathbf{s}) \cdot \partial_\perp u(\mathbf{s})$, by the Leibniz rule for normal derivatives (see Exercise E.6), while ($\diamond$) is because $\triangle(u^2)(\mathbf{x}) = 2|\nabla u(\mathbf{x})|^2 + 2u(\mathbf{x}) \cdot \triangle u(\mathbf{x})$ by the Leibniz rule for Laplacians (Exercise 1B.4, p. 11). Finally, ($*$) is by part (a). The result follows.
  (c) is **Exercise E.5**. □ ⓔ

Exercise E.6 Prove the *Leibniz rule* for normal derivatives: if $f, g : \mathbb{X} \longrightarrow \mathbb{R}$ are two ⓔ scalar fields, and $(f \cdot g) : \mathbb{X} \longrightarrow \mathbb{R}$ is their product, then, for all $\mathbf{s} \in \partial\mathbb{X}$,

$$\partial_\perp(f \cdot g)(\mathbf{s}) = \Big(\partial_\perp f(\mathbf{s})\Big) \cdot g(\mathbf{s}) + f(\mathbf{s}) \cdot \Big(\partial_\perp g(\mathbf{s})\Big).$$

(*Hint:* Use the Leibniz rules for gradients (Proposition E.1(b), p. 562) and the linearity of the dot product.) ◆

Appendix F

Differentiation of function series

Recommended: §6E(iii), §6E(iv).

Many of our methods for solving partial differential equations will involve expressing the solution function as an infinite *series* of functions (e.g. Taylor series, Fourier series, etc.). To make sense of such solutions, we must be able to differentiate them.

Proposition F.1 Differentiation of series

Let $-\infty \leq a < b \leq \infty$. For all $n \in \mathbb{N}$, let $f_n : (a, b) \longrightarrow \mathbb{R}$ be a differentiable function, and define $F : (a, b) \longrightarrow \mathbb{R}$ by

$$F(x) = \sum_{n=0}^{\infty} f_n(x), \quad \text{for all } x \in (a, b).$$

(a) *Suppose that $\sum_{n=0}^{\infty} f_n$ converges uniformly[1] to F on (a, b), and that $\sum_{n=0}^{\infty} f_n'$ also converges uniformly on (a, b). Then F is differentiable, and $F'(x) = \sum_{n=0}^{\infty} f_n'(x)$ for all $x \in (a, b)$.*

(b) *Suppose there is a sequence $\{B_n\}_{n=1}^{\infty}$ of positive real numbers such that*

- *$\sum_{n=1}^{\infty} B_n < \infty$;*
- *for all $x \in (a, b)$, and all $n \in \mathbb{N}$, $|f_n(x)| \leq B_n$ and $\left| f_n'(x) \right| \leq B_n$.*

Then F is differentiable, and $F'(x) = \sum_{n=0}^{\infty} f_n'(x)$ for all $x \in (a, b)$.

Proof (a) follows immediately from Proposition 6E.10(c), p. 132.

(b) follows from (a) and the Weierstras M-test (Proposition, 6E.13, p. 135).

For a direct proof, see Folland (1984), Theorem 2.27(b), or Asmar (2005), §2.9, Theorems 1 and 5. □

Example F.2 Let $a = 0$ and $b = 1$. For all $n \in \mathbb{N}$, let $f_n(x) = x^n/n!$. Thus,

$$F(x) = \sum_{n=0}^{\infty} f_n(x) = \sum_{n=0}^{\infty} \frac{x^n}{n!} = \exp(x)$$

---

[1] See §6E(iii) and §6E(iv) for the definition of 'uniform convergence' of a function series.

(because this is the Taylor series for the exponential function). Now let $B_0 = 1$ and let $B_n = 1/(n-1)!$ for $n \geq 1$. Then, for all $x \in (0, 1)$ and all $n \in \mathbb{N}$,

$$|f_n(x)| = \frac{1}{n!}x^n < \frac{1}{n!} < \frac{1}{(n-1)!} = B_n$$

and

$$|f_n'(x)| = \frac{n}{n!}x^{n-1} = \frac{1}{(n-1)!}x^{n-1} < \frac{1}{(n-1)!} = B_n.$$

Also,

$$\sum_{n=1}^{\infty} B_n = \sum_{n=1}^{\infty} \frac{1}{(n-1)!} < \infty.$$

Hence the conditions of Proposition F.1(b) are satisfied, so we conclude that

$$F'(x) = \sum_{n=0}^{\infty} f_n'(x) = \sum_{n=0}^{\infty} \frac{n}{n!}x^{n-1} = \sum_{n=1}^{\infty} \frac{x^{n-1}}{(n-1)!} \underset{(c)}{=} \sum_{m=0}^{\infty} \frac{x^m}{m!} = \exp(x),$$

where (c) is the change of variables $m = n - 1$. In this case, the conclusion is a well-known fact. But the same technique can be applied to more mysterious functions. $\diamond$

Remarks

(a) The series $\sum_{n=0}^{\infty} f_n'(x)$ is sometimes called the *formal derivative* of the series $\sum_{n=0}^{\infty} f_n(x)$. It is 'formal' because it is obtained through a purely symbolic operation; it is not true in general that the 'formal' derivative is *really* the derivative of the series, or indeed, that the formal derivative series even converges. Proposition F.1 essentially says that, under certain conditions, the 'formal' derivative equals the *true* derivative of the series.

(b) Proposition F.1 is also true if the functions f_n involve more than one variable and/or more than one index. For example, if $f_{n,m}(x, y, z)$ is a function of three variables and two indices, and

$$F(x, y, z) = \sum_{n=0}^{\infty} \sum_{m=0}^{\infty} f_{n,m}(x, y, z), \quad \text{for all } (x, y, z) \in (a, b)^3,$$

then, under a similar hypothesis, we can conclude that

$$\partial_y F(x, y, z) = \sum_{n=0}^{\infty} \sum_{m=0}^{\infty} \partial_y f_{n,m}(x, y, z), \quad \text{for all } (x, y, z) \in (a, b)^3.$$

Appendix G

Differentiation of integrals

Recommended: Appendix F.

Many of our methods for solving partial differential equations will involve expressing the solution function $F(x)$ as an *integral* of functions; i.e. $F(x) = \int_{-\infty}^{\infty} f_y(x) \mathrm{d}y$, where, for each $y \in \mathbb{R}$, $f_y(x)$ is a differentiable function of the variable x. This is a natural generalization of the 'solution series' mentioned in Appendix F. Instead of beginning with a *discretely* parameterized family of functions $\{f_n\}_{n=1}^{\infty}$, we begin with a *continuously* parameterized family, $\{f_y\}_{y \in \mathbb{R}}$. Instead of combining these functions through a *summation* to get $F(x) = \sum_{n=1}^{\infty} f_n(x)$, we combine them through *integration*, to get $F(x) = \int_{-\infty}^{\infty} f_y(x) \mathrm{d}y$. However, to make sense of such integrals as the solutions of differential equations, we must be able to differentiate them.

Proposition G.1 Differentiation of integrals

Let $-\infty \leq a < b \leq \infty$. For all $y \in \mathbb{R}$, let $f_y : (a, b) \longrightarrow \mathbb{R}$ be a differentiable function, and define $F : (a, b) \longrightarrow \mathbb{R}$ by

$$F(x) = \int_{-\infty}^{\infty} f_y(x) \mathrm{d}y, \quad \text{for all } x \in (a, b).$$

Suppose there is a function $\beta : \mathbb{R} \longrightarrow \mathbb{R}_+$ such that

(a) $\int_{-\infty}^{\infty} \beta(y) \mathrm{d}y < \infty$;
(b) *for all $y \in \mathbb{R}$ and for all $x \in (a, b)$, $|f_y(x)| \leq \beta(y)$ and $|f_y'(x)| \leq \beta(y)$.*

Then F is differentiable, and, for all $x \in (a, b)$, $F'(x) = \int_{-\infty}^{\infty} f_y'(x) \mathrm{d}y$.

Proof See Folland (1984), Theorem 2.27(b). □

Example G.2 Let $a = 0$ and $b = 1$. For all $y \in \mathbb{R}$ and $x \in (0, 1)$, let $f_y(x) = x^{|y|+1}/(1 + y^4)$. Thus,

$$F(x) = \int_{-\infty}^{\infty} f_y(x) \mathrm{d}y = \int_{-\infty}^{\infty} \frac{x^{|y|+1}}{1 + y^4} \mathrm{d}y.$$

571

Now, let $\beta(y) = (1 + |y|)/(1 + y^4)$. Then

(a) $\int_{-\infty}^{\infty} \beta(y)dy = \int_{-\infty}^{\infty} \frac{1+|y|}{1+y^4}dy < \infty$ (check this);
(b) for all $y \in \mathbb{R}$ and all $x \in (0, 1)$,

$$|f_y(x)| = \frac{x^{|y|+1}}{1 + y^4} < \frac{1}{1 + y^4} < \frac{1 + |y|}{1 + y^4} = \beta(y),$$

and

$$|f_n'(x)| = \frac{(|y| + 1) \cdot x^{|y|}}{1 + y^4} < \frac{1 + |y|}{1 + y^4} = \beta(y).$$

Hence the conditions of Proposition G.1 are satisfied, so we conclude that

$$F'(x) = \int_{-\infty}^{\infty} f_n'(x)dy = \int_{-\infty}^{\infty} \frac{(|y| + 1) \cdot x^{|y|}}{1 + y^4}dy. \qquad \diamond$$

Remarks Proposition G.1 is also true if the functions f_y involve more than one variable. For example, if $f_{v,w}(x, y, z)$ is a function of five variables, and

$$F(x, y, z) = \int_{-\infty}^{\infty} \int_{-\infty}^{\infty} f_{u,v}(x, y, z)du\, dv \quad \text{for all } (x, y, z) \in (a, b)^3,$$

then, under a similar hypothesis, we can conclude that

$$\partial_y^2 F(x, y, z) = \int_{-\infty}^{\infty} \int_{-\infty}^{\infty} \partial_y^2 f_{u,v}(x, y, z)du\, dv, \quad \text{for all } (x, y, z) \in (a, b)^3.$$

Appendix H

Taylor polynomials

H(i) Taylor polynomials in one dimension

Let $\mathbb{X} \subset \mathbb{R}$ be an open set and let $f : \mathbb{X} \longrightarrow \mathbb{R}$ be an N-times differentiable function. Fix $a \in \mathbb{X}$. The *Taylor polynomial of order N* for f around a is the function

$$T_a^N f(x) := f(a) + f'(a) \cdot (x - a) + \frac{f''(a)}{2}(x - a)^2 + \frac{f'''(a)}{6}(x - a)^3$$

$$+ \frac{f^{(4)}(a)}{4!}(x - a)^4 + \cdots + \frac{f^{(N)}(a)}{N!}(x - a)^N. \tag{H.1}$$

Here, $f^{(n)}(a)$ denotes the nth derivative of f at a (e.g. $f^{(3)}(a) = f'''(a)$), and $n!$ (pronounced 'n factorial') is the product $n \cdot (n - 1) \cdots 4 \cdot 3 \cdot 2 \cdot 1$. For example,

$$
\begin{array}{ll}
T_a^0 f(x) = f(a) & \text{(a constant);} \\
T_a^1 f(x) = f(a) + f'(a) \cdot (x - a) & \text{(a linear function);} \\
T_a^2 f(x) = f(a) + f'(a) \cdot (x - a) + \frac{f''(a)}{2}(x - a)^2 & \text{(a quadratic function).}
\end{array}
$$

Note that $T_a^1 f(x)$ parameterizes the *tangent line* to the graph of $f(x)$ at the point $(a, f(a))$ – that is, the best *linear approximation* of f in a neighbourhood of a. Likewise, $T_a^2 f(x)$ is the best *quadratic* approximation of f in a neighbourhood of a. In general, $T_a^N f(x)$ is the polynomial of degree N which provides the best approximation of $f(x)$ if x is reasonably close to N. The formal statement of this is *Taylor's theorem*, which states that

$$f(x) = T_a^N f(x) + \mathcal{O}(|x - a|^{N+1}).$$

Here, $\mathcal{O}(|x - a|^{N+1})$ means some function which is smaller than a constant multiple of $|x - a|^{N+1}$. In other words, there is a constant $K > 0$ such that

$$\left| f(x) - T_a^N f(x) \right| \leq K \cdot |x - a|^{N+1}.$$

If $|x - a|$ is large, then $|x - a|^{N+1}$ is huge, so this inequality is not particularly useful. However, as $|x - a|$ becomes small, $|x - a|^{N+1}$ becomes really, really small. For example, if $|x - a| < 0.1$, then $|x - a|^{N+1} < 10^{-N-1}$. In this sense, $T_a^N f(x)$ is a very good approximation of $f(x)$ if x is close enough to a.

More information about Taylor polynomials can be found in any introduction to single-variable calculus; see, for example, Stewart (2008), pp. 253–254.

H(ii) Taylor series and analytic functions

Prerequisites: Appendix H(i), Appendix F.

Let $X \subset \mathbb{R}$ be an open set, let $f : X \longrightarrow \mathbb{R}$, let $a \in X$, and suppose f is infinitely differentiable at a. By letting $N \to \infty$ in equation (H.1), we obtain the *Taylor series* (or *power series*) for f at a:

$$T_a^\infty f(x) := f(a) + f'(a) \cdot (x - a) + \frac{f''(a)}{2}(x - a)^2 + \cdots = \sum_{n=0}^\infty \frac{f^{(n)}(a)}{n!} (x - a)^n.$$

$$(H.2)$$

Taylor's Theorem suggests that $T_a^\infty f(x) = f(x)$ if x is close enough to a. Unfortunately, this is not true for all infinitely differentiable functions; indeed, the series (H.2) might not even converge for any $x \neq a$. However, we have the following result.

Proposition H.1 *Suppose the series* (H.2) *converges for some* $x \neq a$. *In that case, there is some* $R > 0$ *such that the series* (H.2) *converges* uniformly *to* $f(x)$ *on the interval* $(a - R, a + R)$. *Thus,* $T_a^\infty f(x) = f(x)$ *for all* $x \in (a - R, a + R)$. *On the other hand, series* (H.2) *diverges for all* $x \in (-\infty, a - R)$ *and all* $x \in (a + R, \infty)$. $\square$

The $R > 0$ in Proposition H.1 is called the *radius of convergence* of the power series (H.2), and the interval $(a - R, a + R)$ is the *interval of convergence*. (Note that Proposition H.1 says nothing about the convergence of (H.2) at $a \pm R$; this varies from case to case.) When the conclusion of Proposition H.1 is true, we say that f is *analytic* at a.

Example H.2
(a) All the 'basic' functions of calculus are analytic everywhere on their domain: all polynomials, all rational functions, all trigonometric functions, the exponential function, the logarithm, and any sum, product, or quotient of these functions.
(b) More generally, if f and g are analytic at a, then $(f + g)$ and $(f \cdot g)$ are analytic at a. If $g(a) \neq 0$, then f/g is analytic at a.
(c) If g is analytic at a, and $g(a) = b$, and f is analytic at b, then $f \circ g$ is analytic at a. $\diamond$

If f is infinitely differentiable at $a = 0$, then we can compute the Taylor series:

$$T_0^\infty f(x) := c_0 + c_1 x + c_2 x^2 + c_3 x^3 + \cdots = \sum_{n=0}^\infty c_n x^n, \qquad (H.3)$$

where $c_n := f^{(n)}(0)/n!$, for all $n \in \mathbb{N}$. This special case of the Taylor series (with $a = 0$) is sometimes called a *Maclaurin series*.

Differentiating a Maclaurin series If f is analytic at $a = 0$, then there is some $R > 0$ such that $f(x) = T_0^\infty f(x)$ for all $x \in (-R, R)$. It follows that $f'(x) = (T_0^\infty f)'(x)$, and $f''(x) = (T_0^\infty f)''(x)$, and so on, for all $x \in (-R, R)$. Proposition F.1, p. 569, says that we can compute $(T_0^\infty f)'(x)$, $(T_0^\infty f)''(x)$, etc. by 'formally differentiating' the Maclaurin series (H.3). Thus, we get the following:

$$f(x) = c_0 + c_1 x + c_2 x^2 + c_3 x^3 + c_4 x^4 + \cdots = \sum_{n=0}^{\infty} c_n x^n;$$

$$f'(x) = c_1 + 2c_2 x + 3c_3 x^2 + 4c_4 x^3 + \cdots = \sum_{n=1}^{\infty} n c_n x^{n-1};$$

$$f''(x) = 2c_2 + 6c_3 x + 12c_4 x^2 + \cdots = \sum_{n=1}^{\infty} n(n-1) c_n x^{n-2};$$

$$f'''(x) = 6c_3 + 24c_4 x + \cdots = \sum_{n=1}^{\infty} n(n-1)(n-2) c_n x^{n-3};$$

$$(\text{H.4})$$

etc.

More information about Taylor series can be found in any introduction to single-variable calculus; see, for example, Stewart (2008), §11.10.

H(iii) Using the Taylor series to solve ordinary differential equations

Prerequisites: Appendix H(ii).

Suppose f is an unknown analytic function (so the coefficients $\{c_0, c_1, c_2, \ldots\}$ are unknown). An ordinary differential equation in f can be reformulated in terms of the Maclaurin series in equations (H.4); this yields a set of equations involving the coefficients $\{c_0, c_1, c_2, \ldots\}$. For example, let $A, B, C \in \mathbb{R}$ be constants. The second-order linear ODE

$$A f(x) + B f'(x) + C f''(x) = 0 \tag{H.5}$$

can be reformulated as a power-series equation:

$$\begin{aligned}
0 = {} & A c_0 + A c_1 x + A c_2 x^2 + A c_3 x^3 + A c_4 x^4 + \cdots \\
& + B c_1 + 2 B c_2 x + 3 B c_3 x^2 + 4 B c_4 x^3 + \cdots \\
& + 2 C c_2 + 6 C c_3 x + 12 C c_4 x^2 + \cdots .
\end{aligned} \tag{H.6}$$

When we collect like terms in the x variable, this becomes:

$$0 = (A c_0 + B c_1 + 2 C c_2) + (A c_1 + 2 B c_2 + 6 C c_3) x + (A c_2 + 3 B c_3 + 12 C c_4) x^2 + \cdots . \tag{H.7}$$

This yields an (infinite) system of linear equations

$$\begin{aligned}
0 &= A c_0 + B c_1 + 2 C c_2; \\
0 &= A c_1 + 2 B c_2 + 6 C c_3; \\
0 &= A c_2 + 3 B c_3 + 12 C c_4; \\
&\;\; \vdots
\end{aligned} \tag{H.8}$$

If we define $\widetilde{c}_n := n!\, c_n$ for all $n \in \mathbb{N}$, then the system (H.8) reduces to the simple linear recurrence relation:

$$A \widetilde{c}_n + B \widetilde{c}_{n+1} + C \widetilde{c}_{n+2} = 0, \quad \text{for all } n \in \mathbb{N}. \tag{H.9}$$

(Note the relationship between equations (H.9) and (H.5); this is because, if f is analytic and has Maclaurin series (H.3), then $\widetilde{c}_n = f^{(n)}(0)$ for all $n \in \mathbb{N}$.)

We can then solve the linear recurrence relation (H.9) using standard methods (e.g. characteristic polynomials), and obtain the coefficients $\{c_0, c_1, c_2, \ldots\}$. If the resulting power series converges, then it is a solution of the ODE (H.5), which is analytic in a neighbourhood of zero.

This technique for solving an ordinary differential equation is called the *power series method*. It is not necessary to work in a neighbourhood of zero to apply this method; we assumed $a = 0$ only to simplify the exposition. The power series method can be applied to a Taylor expansion around any point in $\mathbb{R}$.

We used the *constant-coefficient* linear ODE (H.5) inorder to provide a simple example. In fact, there are much easier ways to solve these sorts of ODEs (e.g. characteristic polynomials, matrix exponentials). However, the Power Series Method is also applicable to linear ODEs with *nonconstant* coefficients. For example, if the coefficients A, B, and C in equation (H.5) were themselves analytic functions in x, then we would simply substitute the Taylor series expansions of $A(x)$, $B(x)$, and $C(x)$ into the power series equation (H.6). This would make the simplification into equation (H.7) much more complicated, but we would still end up with a system of linear equations in $\{c_n\}_{n=0}^{\infty}$, like equation (H.8). In general, this will not simplify into a neat linear recurrence relation like equation (H.9), but it can still be solved one term at a time.

Indeed, the power series method is also applicable to *nonlinear* ODEs. In this case, we may end up with a system of *nonlinear* equations in $\{c_n\}_{n=0}^{\infty}$ instead of the linear system (H.8). For example, if the ODE (H.5) contained a term like $f(x) \cdot f''(x)$, then the system of equations (H.8) would contain quadratic terms like $c_0 c_2$, $c_1 c_3$, $c_2 c_4$, etc.

Our analysis is actually incomplete, because we did not check that the power series (H.3) had a nonzero radius of convergence when we obtained the sequence $\{\widetilde{c}_n\}_{n=0}^{\infty}$ as solutions to equation (H.9). If equation (H.9) is a linear recurrence relation (as in the example here), then the sequence $\{c_n\}_{n=0}^{\infty}$ will grow subexponentially, and it is easy to show that the radius of convergence for equation (H.3) will always be nonzero. However, in the case of *non*constant coefficients or a *non*linear ODE, the power series (H.3) may not converge; this needs to be checked. For most of the second-order linear ODEs we will encounter in this book, convergence is assured by the following result.

Theorem H.3 Fuchs's theorem
Let $a \in \mathbb{R}$, let $R > 0$, and let $\mathbb{I} := (a - R, a + R)$. Let $p, q, r : \mathbb{I} \longrightarrow \mathbb{R}$ be analytic functions whose Taylor series at a all converges everywhere in $\mathbb{I}$. Then every solution of the ODE

$$f''(x) + p(x)f'(x) + q(x)f(x) = r(x) \tag{H.10}$$

is an analytic function, whose Taylor series at a converges on $\mathbb{I}$. The coefficients of this Taylor series can be found using the power series method.

Proof See Rota and Birkhoff (1969), chap. 3.   □

If the conditions for Fuchs's theorem are satisfied (i.e. if p, q, and r are all analytic at a), then a is called an *ordinary point* for the ODE (H.10). Otherwise, if one of p, q, or r is *not* analytic at a, then a is called a *singular point* for ODE (H.10). In this case, we can sometimes use a modification of the power series method: the *method of Frobenius*. For simplicity, we will discuss this method in the case $a = 0$. Consider the homogeneous linear ODE

$$f''(x) + p(x)f'(x) + q(x)f(x) = 0. \tag{H.11}$$

Suppose that $a = 0$ is a singular point – i.e. either p or q is not analytic at 0. Indeed, perhaps p and/or q are not even defined at zero (e.g. $p(x) = 1/x$). We say that 0 is a *regular singular point* if there are functions $P(x)$ and $Q(x)$ which *are* analytic at 0, such that $p(x) = P(x)/x$ and $q(x) = Q(x)/x^2$ for all $x \neq 0$. Let $p_0 := P(0)$ and $q_0 := Q(0)$ (the zeroth terms in the Maclaurin series of P and Q), and consider the *indicial polynomial*

$$x(x-1) + p_0 x + q_0.$$

The roots $r_1 \geq r_2$ of the indicial polynomial are called the *indicial roots* of the ODE (H.11).

Theorem H.4 Frobenius's theorem
Suppose $x = 0$ is a regular singular point of the ODE (H.11), and let $\mathbb{I}$ be the largest open interval of 0 where the Taylor series of both $P(x)$ and $Q(x)$ converge. Let $\mathbb{I}^* := \mathbb{I} \setminus \{0\}$. Then there are two linearly independent functions $f_1, f_2 : \mathbb{I}^* \longrightarrow \mathbb{R}$ which satisfy the ODE (H.11) and which depend on the indicial roots $r_1 \geq r_2$ as follows.

(a) *If $r_1 - r_2$ is not an integer, then $f_1(x) = |x|^{r_1} \sum_{n=0}^{\infty} b_n x^n$ and*
    $f_2(x) = |x|^{r_2} \sum_{n=0}^{\infty} c_n x^n$.
(b) *If $r_1 = r_2 = r$, then $f_1(x) = |x|^r \sum_{n=0}^{\infty} b_n x^n$ and*
    $f_2(x) = f_1(x) \ln|x| + |x|^r \sum_{n=0}^{\infty} c_n x^n$.
(c) *If $r_1 - r_2 \in \mathbb{N}$, then $f_1(x) = |x|^r \sum_{n=0}^{\infty} b_n x^n$ and*
    $f_2(x) = k \cdot f_1(x) \ln|x| + |x|^{r_2} \sum_{n=0}^{\infty} c_n x^n$, *for some $k \in \mathbb{R}$.*

In all three cases, to obtain explicit solutions, substitute the expansions for f_1 and f_2 into the ODE (H.11), along with the power series for $P(x)$ and $Q(x)$, to obtain recurrence relations characterizing the coefficients $\{b_n\}_{n=0}^{\infty}, \{c_n\}_{n=0}^{\infty}$.

Proof See Asmar (2005), Appendix A.6. □

Example H.5 Bessel's equation
For any $n \in \mathbb{N}$, the (two-dimensional) *Bessel equation of order n* is the ordinary differential equation

$$x^2 f''(x) + x f'(x) + (x^2 - n^2) f(x) = 0. \tag{H.12}$$

To put this in the form of ODE (H.11), we divide by x^2, to get

$$f''(x) + \frac{1}{x} f'(x) + \left(1 - \frac{n^2}{x^2}\right) f(x) = 0.$$

Thus, we have $p(x) = 1/x$ and $q(x) = (1 - \frac{n^2}{x^2})$; hence 0 is a singular point of ODE (H.12), because p and q are not defined (and hence not analytic) at zero. However, clearly $p(x) = P(x)/x$ and $q(x) = Q(x)/x^2$, where $P(x) = 1$ and $Q(x) = (x^2 - n^2)$ *are* analytic at zero; thus 0 is a *regular* singular point of ODE (H.12). We have $p_0 = 1$ and $q_0 = -n^2$, so the indicial polynomial is $x(x-1) + 1x - n^2 = x^2 - n^2$, which has roots $r_1 = n$ and $r_2 = -n$. Since $r_1 - r_2 = 2n \in \mathbb{N}$, we apply Theorem H.4(c), and look for solutions of the form

$$f_1(x) = |x|^n \sum_{n=0}^{\infty} b_n x^n \quad \text{and} \quad f_2(x) = k \cdot f_1(x) \ln|x| + |x|^{-n} \sum_{n=0}^{\infty} c_n x^n.$$

$$\tag{H.13}$$

To identify the coefficients $\{b_n\}_{n=0}^{\infty}$ and $\{c_n\}_{n=0}^{\infty}$, we substitute the power series (H.13) into ODE (H.12) and simplify. The resulting solutions are called *Bessel functions* of types 1 and 2, respectively. The details can be found in the proof of Proposition, 14G.1, p. 313. ◇

Finally, we remark that a multivariate version of the power series method can be applied to a multivariate Taylor series, to obtain solutions to *partial* differential equations. (However, this book provides many other, much nicer, methods for solving linear PDEs with constant coefficients.)

More information about the power series method and the method of Frobenius can be found in any introduction to ordinary differential equations. See, for example, Coddington (1989), §3.9, and §4.6. Some books on partial differential equations also contain this information (usually in an appendix); see, for example, Asmar (2005), Appendix A.5–A.6.

H(iv) Taylor polynomials in two dimensions

Prerequisites: Appendix B. **Recommended:** Appendix H(i).

Let $\mathbb{X} \subset \mathbb{R}^2$ be an open set and let $f : \mathbb{X} \longrightarrow \mathbb{R}$ be an N-times differentiable function. Fix $\mathbf{a} = (a_1, a_2) \in \mathbb{X}$. The *Taylor polynomial of order N* for f around $\mathbf{a}$ is the function given by

$$T_{\mathbf{a}}^N f(x_1, x_2) := f(\mathbf{a}) + \partial_1 f(\mathbf{a}) \cdot (x_1 - a_1) + \partial_2 f(\mathbf{a}) \cdot (x_2 - a_2)$$

$$+ \frac{1}{2} \left(\partial_1^2 f(\mathbf{a}) \cdot (x_1 - a_1)^2 + 2 \partial_1 \partial_2 f(\mathbf{a}) \cdot (x_1 - a_1)(x_2 - a_2) + \partial_2^2 f(\mathbf{a}) \cdot (x_2 - a_2)^2 \right)$$

$$+ \frac{1}{6} \left(\partial_1^3 f(\mathbf{a}) \cdot (x_1 - a_1)^3 + 3 \partial_1^2 \partial_2 f(\mathbf{a}) \cdot (x_1 - a_1)^2 (x_2 - a_2) \right.$$

$$\left. + 3 \partial_1 \partial_2^2 f(\mathbf{a}) \cdot (x_1 - a_1)(x_2 - a_2)^2 + \partial_2^3 f(\mathbf{a}) \cdot (x_1 - a_1)^3 \right)$$

$$+ \frac{1}{4!} \left(\partial_1^4 f(\mathbf{a}) \cdot (x_1 - a_1)^4 + 4 \partial_1^3 \partial_2 f(\mathbf{a}) \cdot (x_1 - a_1)^3 (x_2 - a_2) \right.$$

$$+ 6 \partial_1^2 \partial_2^2 f(\mathbf{a}) \cdot (x_1 - a_1)^2 (x_2 - a_2)^2$$

$$\left. + 4 \partial_1 \partial_2^3 f(\mathbf{a}) \cdot (x_1 - a_1)(x_2 - a_2)^2 + \partial_2^4 f(\mathbf{a}) \cdot (x_1 - a_1)^4 \right) + \cdots$$

$$+ \frac{1}{N!} \sum_{n=0}^{N} \binom{N}{n} \partial_1^{(N-n)} \partial_2^n f(\mathbf{a}) \cdot (x_1 - a_1)^{N-n} (x_2 - a_2)^n.$$

For example, $T_{\mathbf{a}}^0 f(x_1, x_2) = f(\mathbf{a})$ is just a constant, whereas

$$T_{\mathbf{a}}^1 f(x_1, x_2) = f(\mathbf{a}) + \partial_1 f(\mathbf{a}) \cdot (x_1 - a_1) + \partial_2 f(\mathbf{a}) \cdot (x_2 - a_2)$$

is an affine function which parameterizes the *tangent plane* to the surface graph of $f(x)$ at the point $(\mathbf{a}, f(\mathbf{a}))$ – that is, the best *linear approximation* of f in a neighbourhood of $\mathbf{a}$. In general, $T_a^N f(\mathbf{x})$ is the two-variable polynomial of degree N which provides the best approximation of $f(\mathbf{x})$ if $\mathbf{x}$ is reasonably close to N. The formal statement of this is *multivariate Taylor's theorem*, which states that

$$f(\mathbf{x}) = T_{\mathbf{a}}^N f(\mathbf{x}) + \mathcal{O}(|\mathbf{x} - \mathbf{a}|^{N+1}).$$

For example, if we set $N = 1$, we get

$$f(x_1, x_2) = f(\mathbf{a}) + \partial_1 f(\mathbf{a}) \cdot (x_1 - a_1) + \partial_2 f(\mathbf{a}) \cdot (x_2 - a_2) + \mathcal{O}(|\mathbf{x} - \mathbf{a}|^2).$$

H(v) Taylor polynomials in many dimensions

Prerequisites: Appendix E(i).   **Recommended:** Appendix H(iv).

Let $\mathbb{X} \subset \mathbb{R}^D$ be an open set and let $f : \mathbb{X} \longrightarrow \mathbb{R}$ be an N-times differentiable function. Fix $\mathbf{a} = (a_1, \ldots, a_D) \in \mathbb{X}$. The *Taylor polynomial of order N* for f around $\mathbf{a}$ is the function

$$T_\mathbf{a}^N f(\mathbf{x}) := f(\mathbf{a}) + \nabla f(\mathbf{a})^\dagger \cdot (\mathbf{x} - \mathbf{a}) + \frac{1}{2}(\mathbf{x} - \mathbf{a})^\dagger \cdot \mathsf{D}^2 f(\mathbf{a}) \cdot (\mathbf{x} - \mathbf{a}) + \cdots$$

$$+ \frac{1}{N!} \sum_{n_1 + \cdots + n_D = N} \binom{N}{n_1 \ldots n_D} \partial_1^{n_1} \partial_2^{n_2} \cdots \partial_D^{n_D} f(\mathbf{a})$$

$$\times (x_1 - a_1)^{n_1} (x_2 - a_2)^{n_2} \cdots (x_D - a_D)^{n_D}.$$

Here, we regard $\mathbf{x}$ and $\mathbf{a}$ as *column* vectors, and the transposes $\mathbf{x}^\dagger$, $\mathbf{a}^\dagger$, etc. as *row* vectors. $\nabla f(\mathbf{a})^\dagger := [\partial_1 f(\mathbf{a}), \partial_2 f(\mathbf{a}), \ldots, \partial_D f(\mathbf{a})]$ is the (transposed) *gradient vector* of f at $\mathbf{a}$, and

$$\mathsf{D}^2 f := \begin{bmatrix} \partial_1^2 f & \partial_1 \partial_2 f & \cdots & \partial_1 \partial_D f \\ \partial_2 \partial_1 f & \partial_2^2 f & \cdots & \partial_2 \partial_D f \\ \vdots & \vdots & \ddots & \vdots \\ \partial_D \partial_1 f & \partial_D \partial_2 f & \cdots & \partial_D^2 f \end{bmatrix}$$

is the *Hessian derivative matrix* of f. For example, $T_\mathbf{a}^0 f(\mathbf{x}) = f(\mathbf{a})$ is just a constant, whereas

$$T_\mathbf{a}^1 f(x) = f(\mathbf{a}) + \nabla f(\mathbf{a})^\dagger \cdot (\mathbf{x} - \mathbf{a})$$

is an affine function which paramaterizes the *tangent hyperplane* to the hypersurface graph of $f(x)$ at the point $(\mathbf{a}, f(\mathbf{a}))$ – that is, the best *linear approximation* of f in a neighbourhood of $\mathbf{a}$. In general $T_a^N f(\mathbf{x})$ is the multivariate polynomial of degree N which provides the best approximation of $f(\mathbf{x})$ if $\mathbf{x}$ is reasonably close to N. The formal statement of this is *multivariate Taylor's theorem*, which states that

$$f(\mathbf{x}) = T_\mathbf{a}^N f(\mathbf{x}) + \mathcal{O}(|\mathbf{x} - \mathbf{a}|^{N+1}).$$

For example, if we set $N = 2$, we get

$$f(\mathbf{x}) = f(\mathbf{a}) + \nabla f(\mathbf{a})^\dagger \cdot (\mathbf{x} - \mathbf{a}) + \frac{1}{2}(\mathbf{x} - \mathbf{a})^\dagger \cdot \mathsf{D}^2 f(\mathbf{a}) \cdot (\mathbf{x} - \mathbf{a}) + \mathcal{O}\left(|\mathbf{x} - \mathbf{a}|^3\right).$$

References

Asmar, N. H. (2002). *Applied Complex Analysis with Partial Differential Equations*, 1st edn (Upper Saddle River, NJ: Prentice-Hall).

Asmar, N. H. (2005). *Partial Differential Equations, with Fourier Series and Boundary Value Problems*, 2nd edn (Upper Saddle River, NJ: Prentice-Hall).

Bieberbach, L. (1953). *Conformal Mapping*, trans. F. Steinhardt (New York: Chelsea Publishing Co.).

Bishop, R. L. and Goldberg, S. I. (1980). *Tensor Analysis on Manifolds* (Mineola, NY: Dover).

Blank, J., Exner, P. and Havlicek, M. (1994). *Hilbert Space Operators in Quantum Physics*, AIP series in Computational and Applied Mathematical Science (New York: American Institute of Physics).

Bohm, D. (1979). *Quantum Theory* (Mineola, NY: Dover).

Broman, A. (1989). *Introduction to Partial Differential Equations*, Dover Books on Advanced Mathematics, 2nd edn (New York: Dover Publications Inc.)

Chavel, I. (1993). *Riemannian Geometry: A Modern Introduction* (Cambridge: Cambridge University Press).

Churchill, R. V. and Brown, J. W. (1987). *Fourier Series and Boundary Value Problems*, 4th edn (New York: McGraw-Hill).

Churchill, R. V. and Brown, J. W. (2003). *Complex Variables and Applications*, 7th edn (New York: McGraw-Hill).

Coddington, E. A. (1989). *An Introduction to Ordinary Differential Equations* (Mineola, NY: Dover).

Coifman, R. R. and Wiess, G. (1968). Representations of compact groups and spherical harmonics, *L'enseignment Math.* **14**, 123–173.

Conway, J. B. (1990). *A Course in Functional Analysis*, 2nd edn (New York: Springer-Verlag).

duChateau, P. and Zachmann, D. W. (1986). *Partial Differential Equations, Schaum's Outlines* (New York: McGraw-Hill).

Evans, L. C. (1991). *Partial Differential Equations*, vol. 19 of Graduate Studies in Mathematics (Providence, RI: American Mathematical Society).

Fisher, S. D. (1999). *Complex Variables* (Mineola, NY: Dover Publications Inc.). Corrected reprint of the 2nd (1990) edition.

Folland, G. B. (1984). *Real Analysis* (New York: John Wiley and Sons).

Haberman, R. (1987). *Elementary Applied Partial Differential Equations*, 2nd edn (Englewood Cliff, NJ: Prentice Hall Inc.).

Hatcher, A. (2002). *Algebraic Topology* (Cambridge: Cambridge University Press).

Helgason, S. (1981). *Topics in Harmonic Analysis on Homogeneous Spaces* (Boston, MA: Birkhäuser).

Henle, M. (1994). *A Combinatorial Introduction to Topology* (Mineola, NY: Dover).

Katznelson, Y. (1976). *An Introduction to Harmonic Analysis*, 2nd edn (New York: Dover).

Kolmogorov, A. N. and Fomīn, S. V. (1975). *Introductory Real Analysis* (New York: Dover Publications Inc.), translated from the second Russian edition and edited by Richard A. Silverman; corrected reprinting.

Körner, T. W. (1988). *Fourier Analysis* (Cambridge: Cambridge University Press).

Lang, S. (1985). *Complex Analysis*, 2nd edn (New York: Springer-Verlag).

McWeeny, R. (1972). *Quantum Mechanics: Principles and Formalism* (Mineola, NY: Dover).

Müller, C. (1966). *Spherical Harmonics*, Lecture Notes in Mathematics 17 (New York: Springer-Verlag).

Murray, J. D. (1993). *Mathematical Biology*, 2nd edn, vol. 19 of Biomathematics (New York: Springer-Verlag).

Needham, T. (1997). *Visual Complex Analysis* (New York: The Clarendon Press, Oxford University Press).

Nehari, Z. (1975). *Conformal Mapping* (New York: Dover Publications Inc.).

Pinsky, M. A. (1998). *Partial Differential Equations and Boundary-Value Problems with Applications*, International Series in Pure and Applied Mathematics, 3rd edn (Boston, MA: McGraw-Hill).

Powers, D. L. (1999). *Boundary Value Problems*, 4th edn (San Diego, CA: Harcourt/Academic Press).

Prugovecki, E. (1981). *Quantum Mechanics in Hilbert Space*, 2nd edn (New York: Academic Press).

Rota, G.-C. and Birkhoff, G. (1969). *Ordinary Differential Equations*, 2nd edn (New York: Wiley).

Royden, H. L. (1988). *Real Analysis*, 3rd edn (New York: Macmillan Publishing Company).

Rudin, W. (1987). *Real and Complex Analysis*, 3rd edn (New York: McGraw-Hill).

Schwerdtfeger, H. (1979). *Geometry of Complex Numbers* (New York: Dover Publications Inc.).

Stevens, C. F. (1995). *The Six Core Theories of Modern Physics*, A Bradford Book (Cambridge, MA: MIT Press).

Stewart, J. (2008). *Calculus, Early Transcendentals*, 6th edn (Belmont, CA: Thomson Brooks/Cole).

Strook, D. W. (1993). *Probability Theory: An Analytic View*, rev. edn (Cambridge: Cambridge University Press).

Sugiura, M. (1975). *Unitary Representations and Harmonic Analysis* (New York: Wiley).

Takeuchi, M. (1994). *Modern Spherical Functions*, Translations of Mathematical Monographs 135 (Providence, RI: American Mathematical).

Taylor, M. E. (1986). *Noncommutative Harmonic Analysis* (Providence, RI: American Mathematical Society).

Terras, A. (1985). *Harmonic Analysis on Symmetric Spaces and Applications*, vol. I (New York: Springer-Verlag).

Titchmarsh, E. (1962). *Eigenfunction Expansions Associated with Second-Order Differential Equations, Part I*, 2nd edn (London: Oxford University Press).

Walker, J. S. (1988). *Fourier Analysis* (Oxford: Oxford University Press).

Warner, F. M. (1983). *Foundations of Differentiable Manifolds and Lie Groups* (New York: Springer-Verlag).

Wheeden, R. L. and Zygmund, A. (1977). *Measure and Integral*: *An Introduction to Real Analysis*, Pure and Applied Mathematics, vol. 43 (New York: Morcel Dekker Inc.).

Subject index

Abel mean, 466
Abel sum, 179
Abel's test, 135
abelian group, 179
absolute convergence of Fourier series, 177
absolutely integrable function
  on the half-line $\mathbb{R}_{\neq} = [0, \infty)$, 515
  on the real line $\mathbb{R}$, 492
  on the two-dimensional plane $\mathbb{R}^2$, 508
  on the two-dimensional plane $\mathbb{R}^3$, 511
Airy equation, 71, 101
algebraic topology, 488
analytic extension, 459
analytic functions
  convolutions of, 479
  definition, 574
  Fourier coefficient decay rate, 209
  Fourier transform of, 481
  improper integral of, 477
analytic harmonic functions, 20
annulus, 280
antiderivative, complex, 453
approximation of identity
  definition (on $\mathbb{R}$), 385
  definition (on $\mathbb{R}^D$), 389
  Gauss–Weierstrass kernel
    many-dimensional, 398
    one-dimensional, 392
  on $[-\pi, \pi]$, 201, 219
  Poisson kernel (on disk), 414
  Poisson kernel (on half-plane), 410
  use for smooth approximation, 415
Atiyah–Singer index theorem, 488
autocorrelation function, 504

baguette example, 257
Balmer, J. J., 55
BC, *see* boundary conditions
Beam equation, 71, 101
Bernstein's theorem, 177
Bessel functions, 578
  and eigenfunctions of Laplacian, 299

definition, 299
  roots, 301
Bessel equation, 298, 577
Bessel's inequality, 197
big 'O' notation, *see* order $\mathcal{O}(1/z)$
bilinearity, 112
binary expansion, 120
Borel-measurable set, 115
Borel-measurable subset, 213
boundary
  definition of, 75
  examples (for various domains), 75
boundary conditions
  definition, 76
  and harmonic conjugacy, 427
  homogeneous Dirichlet, *see* Dirichlet boundary
    conditions, homogeneous
  homogeneous mixed, *see* mixed boundary
    conditions, homogeneous
  homogeneous Neumann, *see* Neumann boundary
    conditions, homogeneous
  homogeneous Robin, *see* mixed boundary
    conditions, homogeneous
  nonhomogeneous Dirichlet, *see* Dirichlet boundary
    conditions, nonhomogeneous
  nonhomogeneous mixed, *see* mixed boundary
    conditions, nonhomogeneous
  nonhomogeneous Neumann, *see* Neumann
    boundary conditions, nonhomogeneous
  nonhomogeneous Robin, *see* mixed boundary
    conditions, nonhomogeneous
  periodic, *see* periodic boundary conditions
boundary value problem, 76
bounded variation, 177
branches
  of complex logarithm, 454
  of complex roots, 454
Brownian motion, 23
Burger equation, 71, 101
BVP, *see* boundary value problem

$\mathcal{C}^1[0, L]$, 148
$\mathcal{C}^1[0, \pi]$, 141

$\mathcal{C}^1$ interval, 142, 148
Casorati–Weierstrass theorem, 475
Cauchy problem, *see* initial value problem
Cauchy residue theorem, *see* residue theorem
Cauchy's criterion, 135
Cauchy's integral formula, 448
Cauchy's theorem
  on contours, 444
  on oriented boundaries, 486
Cauchy–Bunyakowski–Schwarz inequality
  for complex functions, 114
  in L^2, 112
  for sequences in $l^2(\mathbb{N})$, 206
Cauchy–Euler equation
  polar eigenfunctions of $\triangle$ (two dimensions), 322
  as Sturm–Liouville equation, 350
  zonal eigenfunctions of $\triangle$ (three dimensions), 366
Cauchy–Riemann differential equations, 422
Cauchy–Schwarz inequality, *see*
        Cauchy–Bunyakowski–Schwarz inequality
CBS inequality, *see* Cauchy–Bunyakowski–Schwarz
        inequality
Cesáro sum, 179
character (of a group), 179
chasm in streambed (flow), 440
Chebyshev polynomial, 283
codisk, 279
compact abelian topological group, 179
complex antiderivative, 453
  as complex potential, 436
complex derivative, 421
complex logarithm, 454
complex nth root, 454
complex numbers
  addition, 553
  conjugate, 555
  exponential, 553
    derivative of, 553
  multiplication, 553
  norm, 555
  polar coordinates, 553
complex potentials, 436
complex-analytic function, *see* holomorphic function
complex-differentiable, 422
componentwise addition, 61
conformal, $\Leftrightarrow$ holomorphic, 429
conformal isomorphism, 429
conformal map
  definition, 429
  Riemann mapping theorem, 435
connected, definition, 457
conservation of energy, in wave equation, 94
continuously differentiable, 141, 148
contour, 440
  piecewise smooth, 442
  purview of, 443
  smooth, 440
contour integral, 441, 442
  is homotopy-invariant, 446
convergence
  as 'approximation', 122

of complex Fourier series, 176
of Fourier cosine series, 145, 150
of Fourier series; Bernstein's theorem, 177
of Fourier sine series, 142, 148
of function series, 134
in L^2, 122
of multidimensional Fourier series, 189
of real Fourier series, 166
of two-dimensional Fourier (co)sine series, 183
of two-dimensional mixed Fourier series, 187
pointwise, 126
pointwise $\Longrightarrow L^2$, 126
semiuniform, 133
uniform, 130
uniform $\Longrightarrow$ pointwise, 133
uniform $\Longrightarrow L^2$, 133
convolution
  of analytic functions, 479
  is associative ($f * (g * h) = (f * g) * h$), 415
  is commutative ($f * g = g * f$), 384, 415
  continuity of, 415
  definition of ($f * g$), 20, 384
  differentiation of, 415
  with the Dirichlet kernel, 199, 469
  is distributive ($f * (g + h) = (f * g) + (f * h)$), 415
  Fourier transform of, 499
  $\Longrightarrow$ multiplication of Fourier coefficients, 469
  with the Poisson kernel, 467
  use for smooth approximation, 415
  of 2π-periodic functions, 215
convolution ring, 219
coordinates
  cylindrical, 558
  polar, 557
  rectangular, 557
  spherical, 559
cosine series, *see* Fourier series, cosine
Coulomb potential (electrostatics), 16
Coulomb's law (electrostatics), 17
curl, 424
cycle (in homology), 486
cylindrical coordinates, 558

$\triangle$, *see* Laplacian
d'Alembert ripple solution (initial velocity), 401
  d'Alembert solution to wave equation, 404, 407, 535
  d'Alembert travelling wave solution (initial position), 399
Davisson, C. J., 39
de Broglie, Louis
  'matter wave' hypothesis, 39
  de Broglie wavelength, 45
decay at ∞ of order $\mathcal{O}(1/z)$, *see* order $\mathcal{O}(1/z)$
decay at ∞ of order $o(1/z)$, *see* order $o(1/z)$
decaying gradient condition, 290
$\partial_\perp u$, *see* outward normal derivative
$\partial\mathbb{X}$, *see* boundary
dense subspace of $\mathbf{L}^2[-\pi, \pi]$, 209
difference operator, 64

differentiation as linear operator, 65
diffraction of 'matter waves', 39
Dirac delta function δ_0, 385, 409, 532
Dirichlet boundary conditions
  homogeneous
    definition, 76
    Fourier sine series, 142
    multidimensional Fourier sine series, 189
    physical interpretation, 76
    two-dimensional Fourier sine series, 183
  nonhomogeneous
    definition, 79
Dirichlet kernel, 199, 469
Dirichlet problem
  on annulus
    Fourier solution, 293
  on bi-infinite strip, 438
  around chasm, 438
  on codisk
    Fourier solution, 290
  on cube
    nonconstant, nonhomogeneous Dirichlet BC,
      274
    one constant nonhomogeneous Dirichlet BC,
      273
  definition, 79
  on disk
    definition, 411
    Fourier solution, 284
    Poisson (impulse-response) solution, 297, 412
  on half-disk, 439
  on half-plane
    definition, 408, 541
    Fourier solution, 541
    physical interpretation, 408
    Poisson (impulse-response) solution, 409, 542
  with vertical obstacle, 439
  on interval $[0, L]$, 79
  on off-centre annulus, 439
  on quarter-plane
    conformal mapping solution, 434
  on square
    four constant nonhomogeneous Dirichlet BC,
      246
    nonconstant nonhomogeneous Dirichlet BC,
      248
    one constant nonhomogeneous Dirichlet BC,
      244
  two compartments separated by aperture, 439
  unique solution of, 90
Dirichlet test, 135
distance
  L^2, 122
  L^∞, 130
  uniform, 130
divergence (div V)
  in many dimensions, 563
  in one dimension, 562
  in two dimensions, 563
divergence theorem, 566
dot product, 107

drumskin
  round, 310
  square, 264, 265

eigenfunctions
  of ∂_τ^2, 351
  definition, 67
  of differentiation operator, 163, 172
  of Laplacian, 67, 71, 352
    polar-separated, 299
    polar-separated; homogeneous Dirichlet BC, 301
  of self-adjoint operators, 351
eigenvalue
  definition, 67
  of Hamiltonian as energy levels, 48
eigenvector
  definition, 67
  of Hamiltonian as stationary quantum states, 48
Eikonal equation, 71, 101
electric field, 17
electric field lines, 437
electrostatic potential, 15, 437
elliptic differential equation, 100
  motivation: polynomial formalism, 377
  two-dimensional, 98
elliptic differential operator
  definition, 98, 100
  divergence form, 354
  self-adjoint
    eigenvalues of, 355
    if symmetric, 354
    symmetric, 354
entire function, 476
ϵ-tube, 130
equipotential contour, 436
error function Φ, 394
essential singularity, 475
Euler's formula, 553
even extension, 174
even function, 172
even–odd decomposition, 173
evolution equation, 73
extension
  even, *see* even extension
  odd, *see* odd extension
  odd periodic, *see* odd periodic extension

factorial, 573
  gamma function, 524
field of fractions, 476
flow
  along river bank, 440
  around peninsula, 440
  confined to domain, 436
  irrotational, 424
  out of pipe, 440
  over chasm, 440
  sourceless, 424
  sourceless and irrotational, 424, 436
fluid, incompressible and nonturbulent, 436
fluid dynamics, 435

flux across boundary
  in $\mathbb{R}^2$, 564
  in $\mathbb{R}^D$, 566
Fokker–Planck equation, 21
  is homogeneous linear, 68
  is parabolic PDE, 100
forced heat equation, unique solution of, 93
forced wave equation, unique solution of, 96
Fourier cosine series, *see* Fourier series, cosine
Fourier (co)sine transform
  definition, 515
  inversion, 515
Fourier series
  absolute convergence, 177
  convergence; Bernstein's theorem, 177
  failure to converge pointwise, 178
Fourier series, complex
  coefficients, 175
  convergence, 176
  definition, 176
  relation to real Fourier series, 176
Fourier series, (co)sine
  of derivative, 162
  of piecewise linear function, 160
  of polynomials, 151
  relation to real Fourier series, 174
  of step function, 158
Fourier series, cosine
  coefficents
    on $[0, \pi]$, 145
    on $[0, L]$, 149
  convergence, 145, 150
  definition
    on $[0, \pi]$, 145
    on $[0, L]$, 150
  is even function, 173
  of $f(x) = \cosh(\alpha x)$, 147
  of $f(x) = \sin(m\pi x/L)$, 150
  of $f(x) = \sin(mx)$, 146
  of $f(x) = x$, 152
  of $f(x) = x^2$, 152
  of $f(x) = x^3$, 152
  of $f(x) \equiv 1$, 146, 150
  of half-interval, 158
Fourier series, multidimensional
  complex, 193
  convergence, 189
  cosine
    coefficients, 188
    series, 188
  derivatives of, 194
  mixed
    coefficients, 189
    series, 189
  sine
    coefficients, 188
    series, 188
Fourier series, real
  coefficients, 165
  convergence, 166
  definition, 165

of derivative, 172
of $f(x) = x$, 168
of $f(x) = x^2$, 168
of piecewise linear function, 171
of polynomials, 167
of step function, 169
relation to complex Fourier series, 176
relation to Fourier (co)sine series, 174
Fourier series, sine
  coefficents
    on $[0, \pi]$, 141
    on $[0, L]$, 148
  convergence, 142, 148
  definition
    on $[0, L]$, 148
    on $[0, \pi]$, 141
  of $f(x) = \cos(m\pi x/L)$, 149
  of $f(x) = \cos(mx)$, 144
  of $f(x) = \sinh(\alpha\pi x/L)$, 149
  of $f(x) = \sinh(\alpha x)$, 144
  of $f(x) = x$, 152
  of $f(x) = x^2$, 152
  of $f(x) = x^3$, 152
  of $f(x) \equiv 1$, 143, 149
  is odd function, 173
  of tent function, 159, 163
Fourier series, two-dimensional
  convergence, 183
  cosine
    coefficients, 182
    definition, 183
  sine
    coefficients, 181
    definition, 181
    of $f(x, y) = x \cdot y$, 181
    of $f(x, y) \equiv 1$, 182
Fourier series, two-dimensional, mixed
  coefficients, 187
  convergence, 187
  definition, 187
Fourier sine series, *see* Fourier series, sine
Fourier transform
  of analytic function, 481
  asymptotic decay, 497
  is continuous, 497
  convolution, 499
  D-dimensional
    definition, 513
    inversion, 513
  derivative of, 500
  evil twins of, 504
  one-dimensional
    of box function, 493
    of Gaussian, 502
    definition, 491
    inversion, 492, 495
    of Poisson kernel (on half-plane), 542
    of rational functions, 484
    of symmetric exponential tail function, 496
  rescaling, 499
  smoothness vs. asymptotic decay, 501

three-dimensional
  of ball, 512
  definition, 511
  inversion, 511
translation vs. phase shift, 499
two-dimensional
  of box function, 509
  of Gaussian, 511
  definition, 508
  inversion, 508, 509
Fourier's law of heat flow
  many dimensions, 6
  one-dimension, 5
Fourier–Bessel series, 304
frequency spectrum, 55
Frobenius, method of, 576
  to solve Bessel equation, 313
Fuchs's power series solution to ODE, 576
fuel rod example, 260
functions as vectors, 61
fundamental solution, 391
  heat equation (many-dimensional), 398
  heat equation (one-dimensional), 394
fundamental theorem of calculus, 562
  as special case of divergence theorem, 566

gamma function, 524
Gauss's theorem, *see* divergence theorem
Gauss's law (electrostatics), 17
Gauss–Weierstrass kernel
  convolution with, *see* Gaussian convolution
  many-dimensional
    is approximation of identity, 398
    definition, 11
  one-dimensional, 391, 532
    is approximation of identity, 392
    definition, 8
  two-dimensional, 10
Gaussian
  one-dimensional
    cumulative distribution function of, 394
    Fourier transform of, 502
    integral of, 394
    stochastic process, 23
  two-dimensional, Fourier transform of, 511
Gaussian convolution, 392, 398, 534
general boundary conditions, 84
generation equation, 14
  equilibrium of, 14
generation-diffusion equation, 14
Germer, L. H, 39
Gibbs phenomenon, 144, 149, 156
gradient ∇u
  many-dimensional, 561
  two-dimensional, 561
gradient vector field
  many-dimensional, 561
  two-dimensional, 561
gravitational potential, 15
Green's function, 385
Green's theorem, 565

as special case of divergence theorem, 566
guitar string, 75

Hölder continuous function, 177
Haar basis, 120
Hamiltonian operator
  eigenfunctions of, 48
  in Schrödinger equation, 43
  is self-adjoint, 349
harmonic $\Rightarrow$ locally holomorphic, 423
harmonic analysis, 179
  noncommutative, 355
harmonic conjugate, 423
  swaps Neumann and Dirichlet BC, 427
harmonic function
  'saddle' shape, 12
  analyticity, 20
  convolution against Gauss–Weierstrass, 418
  definition, 11
  maximum principle, 19
  mean value theorem, 18, 324, 418
  separated (Cartesian), 362
  smoothness properties, 20
  two-dimensional, separated (Cartesian), 360
  two-dimensional, separated (polar coordinates), 280
harp string, 235
Hausdorff–Young inequality, 508
HDBC, *see* Dirichlet boundary conditions,
    homogeneous
heat equation
  on cube
    homogeneous Dirichlet BC, Fourier solution, 270
    homogeneous Neumann BC, Fourier solution, 533
  definition, 9
  derivation and physical interpretation
  on disk
    homogeneous Dirichlet BC, Fourier–Bessel solution, 308
    nonhomogeneous Dirichlet BC, Fourier–Bessel solution, 309
  equilibrium of, 11
  is evolution equation, 73
  fundamental solution of, 394, 398
  is homogeneous linear, 68
  initial conditions: Heaviside step function, 394
  on interval
    homogeneous Dirichlet BC, Fourier solution, 229
    homogeneous Neumann BC, Fourier solution, 230
  L^2-norm decay, 92
  is parabolic PDE, 98, 100
  on real line
    Fourier transform solution, 531
    Gaussian convolution solution, 392, 534
  on square
    homogeneous Dirichlet BC, Fourier solution, 251
    homogeneous Neumann BC, Fourier solution, 253
    nonhomogeneous Dirichlet BC, Fourier solution, 255

heat equation (*cont.*)
   on three-dimensional, space, Fourier transform
      solution, 273
   on two-dimensional plane, Fourier transform
      solution, 532
   on unbounded domain, Gaussian convolution
      solution, 398
   unique solution of, 93
     many dimensions, 9
     one-dimension, 7
Heaviside step function, 394
Heisenberg uncertainty principle, *see* uncertainty
      principle
Heisenberg, Werner, 518
Helmholtz equation, 70, 101
   is not evolution equation, 74
   as Sturm–Liouville equation, 350
Hermitian, 114
Hessian derivative, 28
Hessian derivative matrix, 579
HNBC, *see* Neumann boundary conditions,
      homogeneous
holomorphic
   $\Rightarrow$ complex-analytic, 455
   $\Leftrightarrow$ conformal, 429
   function, 422
   $\Rightarrow$ harmonic, 423
   $\Leftrightarrow$ sourceless irrotational flow, 424
homogeneous boundary conditions
   Dirichlet, *see* Dirichlet boundary conditions,
      homogeneous
   mixed, *see* mixed boundary conditions,
      homogeneous
   Neumann, *see* Neumann boundary conditions,
      homogeneous
   Robin, *see* mixed boundary conditions,
      homogeneous
homogeneous linear differential equation
   definition, 68
   superposition principle, 69
homologous (cycles), 487
homology group, 488
homology invariance (of chain integrals), 487
homotopic contours, 445
homotopy invariance, of contour integration, 446
Huygens' principle, 540
hydrogen atom
   Balmer lines, 55
   Bohr radius, 54
   energy spectrum, 55
   frequency spectrum, 55
   ionization potential, 54
   Schrödinger equation, 44
   stationary Schrödinger equation, 53
hyperbolic differential equation, 100
   motivation: polynomial formalism, 377
   one-dimensional, 98

I/BVP, *see* initial/boundary value problem
ice cube example, 271
identity theorem, 457

imperfect conductor (Robin BC), 84
impermeable barrier (homogeneous Neumann BC), 81
improper integral, of analytic functions, 477
impulse function, 385
impulse-response function
   four properties, 382
   interpretation, 381
impulse-response solution
   to Dirichlet problem on disk, 297, 412
   to half-plane Dirichlet problem, 409, 542
   to heat equation, 398
     one-dimensional, 392
   to wave equation (one-dimensional), 404
indelible singularity, 469
indicial polynomial, 577
indicial roots, 577
∞, *see* point at infinity
initial conditions, 74
initial position problem, 234, 264, 310, 399
initial value problem, 74
initial velocity problem, 236, 265, 310, 401
initial/boundary value problem, 76
inner product
   of complex functions, 113
   of functions, 109, 111
    complex-valued, 175
   of vectors, 107
int $(\mathbb{X})$, *see* interior
integral domain, 476
integral representation formula, 452
integration as linear operator, 66
integration by parts, 150
interior (of a domain), 75
irrotational flow, 424
IVP, *see* initial value problem

Jordan curve theorem, 443

kernel
   convolution, *see* impulse-response function
   Gauss–Weierstrass, *see* Gauss–Weierstrass kernel
   of linear operator, 67
   Poisson
     on disk, *see* Poisson kernel (on disk)
     on half-plane, *see* Poisson kernel (on half-plane)

L^2-convergence, *see* convergence in L^2
L^2-distance, 122
L^2-norm, 122
   ($\|f\|_2$), *see* norm, L^2
L^2-space, 42, 109, 112
L^∞-convergence, *see* convergence, uniform
L^∞-distance, 130
L^∞-norm ($\|f\|_\infty$), 129
$\mathbf{L}^1(\mathbb{R}_{\neq})$, 515
$\mathbf{L}^1(\mathbb{R})$, 492
$\mathbf{L}^1(\mathbb{R}^2)$, 508
$\mathbf{L}^1(\mathbb{R}^3)$, 511
$\mathbf{L}^1(\mathbb{R}^D)$, 513
$\mathbf{L}^2(\mathbb{X})$, 42, 109, 112
$\mathbf{L}^2_{\text{even}}[-\pi, \pi]$, 173

$\mathbf{L}^2_{odd}[-\pi, \pi]$, 173
$\mathbf{L}^p$-norm
   on $[-\pi, \pi]$, 178
   on $\mathbb{R}$, 507
$\mathbf{L}^p(\mathbb{R})$, 507
$\mathbf{L}^p[-\pi, \pi]$, 178
Landau big 'O' notation, *see* order $\mathcal{O}(1/z)$
Landau small 'o' notation, *see* order $o(1/z)$
Laplace equation
   on codisk
     homogeneous Neumann BC, Fourier solution, 292
     physical interpretation, 289
   definition, 11
   on disk, homogeneous Neumann BC Fourier solution, 286
   is elliptic PDE, 98, 100
   is homogeneous linear, 68
   is not evolution equation, 74
   nonhomogeneous Dirichlet BC, *see* Dirichlet problem
   one-dimensional, 11
   polynomial formalism, 376
   quasiseparated solution, 375
   separated solution (Cartesian), 13, 362
   three-dimensional, 12
   two-dimensional, 11
     separated solution (Cartesian), 360, 376
     separated solution (polar coordinates), 280
   unique solution of, 89
Laplace transform, 520
Laplace–Beltrami operator, 23, 355
Laplacian, 9
   eigenfunctions (polar-separated), 299
   eigenfunctions (polar-separated) homogeneous Dirichlet BC, 301
   eigenfunctions of, 352
   is linear operator, 66
   is self-adjoint, 348
   in polar coordinates, 280
   spherical mean formula, 18, 26
Laurent expansion, 470
Lebesgue integral, 114, 213
Lebesgue's dominated convergence theorem, 127
left-hand derivative ($f^{\rangle}(x)$), 203
left-hand limit ($\lim_{y \nearrow x} f(y)$), 202
Legendre equation, 366
   as Sturm–Liouville equation, 350
Legendre polynomial, 368
Legendre series, 374
Leibniz rule
   for divergence, 564
   for gradients, 562
   for Laplacians, 11
   for normal derivatives, 567
$\lim_{y \nearrow x} f(y)$, *see* left-hand limit
$\lim_{y \searrow x} f(y)$, *see* right-hand limit
linear differential operator, 66
linear function, *see* linear operator
linear operator
   definition, 64

kernel of, 67
linear transformation, *see* linear operator
Liouville equation, 21
Liouville's theorem, 452
logarithm, complex, 454

Maclaurin series, 574
   derivatives of, 574
maximum principle, 19
mean value theorem, 324, 418
   for harmonic functions, 18, 450
   for holomorphic functions, 450
meromorphic function, 471
method of Frobenius, *see* Frobenius, method of
Minkowski's inequality, 215
mixed boundary conditions
   homogeneous, definition, 84
   nonhomogeneous
     definition, 84
     as Dirichlet, 84
     as Neumann, 84
mollifier, 222
Monge–Ampère equation, 70, 101
multiplication operator
   continuous, 66
   discrete, 65

negative definite matrix, 98, 100
Neumann boundary conditions
   homogeneous
     definition, 80
     Fourier cosine series, 146
     multidimensional Fourier cosine series, 189
     physical interpretation, 81
     two-dimensional Fourier cosine series, 183
   nonhomogeneous
     definition, 83
     physical interpretation, 83
Neumann problem
   definition, 83
   unique solution of, 90
Newton's law of cooling, 84
nonhomogeneous boundary conditions
   Dirichlet, *see* Dirichlet boundary conditions, nonhomogeneous
   mixed, *see* mixed boundary conditions, nonhomogeneous
   Neumann, *see* Neumann boundary conditions, nonhomogeneous
   Robin, *see* mixed boundary conditions, nonhomogeneous
nonhomogeneous linear differential equation
   definition, 69
   subtraction principle, 70
norm
   L^2-($\|f\|_2$), 42, 109, 112, 113
   uniform ($\|f\|_\infty$), 129
   of a vector, 108
norm decay, in heat equation, 92
normal derivative, *see* outward normal derivative
normal vector, in $\mathbb{R}^D$, 565

normal vector field, in $\mathbb{R}^2$, 564
nullhomotopic function, 443

$o(1/z)$, *see* order $o(1/z)$
$\mathcal{O}(1/z)$, *see* order $\mathcal{O}(1/z)$
ocean pollution, 408
odd extension, 174
odd function, 172
odd periodic extension, 405
one-parameter semigroup, 418, 529
open source, xvii
order
    of differential equation, 74
    of differential operator, 74
order $o(1/z)$, 477
order $\mathcal{O}(1/z)$, 481
ordinary point for ODE, 576
oriented boundary (of a subset of $\mathbb{C}$), 486
orthogonal basis
    eigenfunctions of Laplacian, 353
    for even functions $L_{\text{even}}[-\pi, \pi]$, 173
    of functions, 136
    for $L^2([0, X] \times [0, Y])$, 183, 187
    for $L^2([0, X_1] \times \cdots \times [0, X_D])$, 189
    for $L^2(D)$, using Fourier–Bessel functions, 304
    for $L^2[-\pi, \pi]$, using (co)sine functions, 166
    for $L^2[0, \pi]$
    for odd functions $L_{\text{odd}}[-\pi, \pi]$, 173
        using cosine functions, 146
        using sine functions, 142
orthogonal eigenfunctions of self-adjoint operators, 351
orthogonal functions, 116
orthogonal set, of functions, 117, 136
orthogonal trigonometric functions, 117, 119
orthogonal vectors, 107
orthonormal basis
    of functions, 136
    for $L^2[-L, L]$, using $\exp(inx)$ functions, 176
    of vectors, 108
orthonormal set of functions, 117
Ostrogradsky's theorem, *see* divergence theorem
outward normal derivative, $(\partial_\perp u)$
    abstract definition, 567
    definition in special cases, 80
    examples (various domains), 80
    physical interpretation, 80

parabolic differential equation, 100
    motivation: polynomial formalism, 377
    one-dimensional, 98
Parseval's equality
    for Fourier transforms, 507
    for orthonormal bases, 137
    for vectors, 108
peninsula (flow), 440
perfect conductor (homogeneous Dirichlet BC), 76
perfect insulator (homogeneous Neumann BC), 81
perfect set, definiton, 457
periodic boundary conditions
    complex Fourier series, 176

definition
    on cube, 86
    on interval, 85
    on square, 85
interpretation
    on interval, 85
    on square, 86
    real Fourier series, 166
Φ ('error function' or 'sigmoid function'), 394
piano string, 75
piecewise $\mathcal{C}^1$, 142, 148
piecewise continuously differentiable function, 142, 148
piecewise linear function, 159, 170
piecewise smooth boundary, 88
    in $\mathbb{R}^2$, 564
    in $\mathbb{R}^D$, 566
pipe into lake (flow), 440
Plancherel's theorem, 507
plucked string problem, 234
point at infinity, 470
pointwise convergence, *see* convergence, pointwise
Poisson equation
    on cube
        homogeneous Dirichlet BC, Fourier solution, 276
        homogeneous Neumann BC, Fourier solution, 277
    definition, 14
    on disk
        homogeneous Dirichlet BC, Fourier–Bessel solution, 306
        nonhomogeneous Dirichlet BC, Fourier–Bessel solution, 307
    electrostatic potential, 15
    on interval
        homogeneous Dirichlet BC, Fourier solution, 239
        homogeneous Neumann BC, Fourier solution, 239
    is elliptic PDE, 100
    is nonhomogeneous, 69
    is not evolution equation, 74
    one-dimensional, 14
    on square
        homogeneous Dirichlet BC, Fourier solution, 259
        homogeneous Neumann BC, Fourier solution, 261
        nonhomogeneous Dirichlet BC, Fourier solution, 262
    unique solution of, 91
Poisson integral formula
    for harmonic functions on disk, 297
    for holomorphic functions on disk, 451
Poisson kernel
    and Abel mean of Fourier series, 467
    Fourier series of, 468
Poisson kernel (on disk)
    definition, 296, 412
    in complex plane, 450
    is approximation of identity, 414, 469
    picture, 412
    in polar coordinates, 297, 412

Poisson kernel (on half-plane)
  is approximation of identity, 410
  definition, 408, 542
  Fourier transform of, 542
  picture, 409
Poisson solution
  to Dirichlet problem on disk, 297, 412
  to half-plane Dirichlet problem, 409, 439, 542
  to three-dimensional wave equation, 538
polar coordinates, 557
pole, 470
  simple, 469
pollution, oceanic, 408
polynomial formalism
  definition, 376
  elliptic, parabolic, and hyperbolic, 377
  Laplace equation, 376
  telegraph equation, 377, 379
polynomial symbol, 376
positive-definite function, inner product is, 112, 114
positive definite matrix, 98, 99
potential, 15
  complex, 436
  Coulomb, 16
  electrostatic, 15, 437
  of a flow, 436
  gravitational, 15
potential fields and Poisson equation, 15
power series, 574
power series method, 575
  to solve Legendre equation, 368
power spectrum, 504
punctured plane, 279
purview (of a contour), 443
Pythagorean formula
  in L^2, 136
  in $\mathbb{R}^N$, 108

quantization of energy
  in finite potential well, 50
  hydrogen atom, 55
  in infinite potential well, 52
quantum numbers, 52
quasiseparated solution, 375
  of Laplace equation, 375

reaction kinetic equation, 22
reaction-diffusion equation, 23, 70, 101
  is nonlinear, 70
rectangular coordinates, 557
regular singular point for ODE, 577
removable singularity, 469
residue, 470
residue theorem, 472
Riemann integrable function, 211
Riemann integral
  of bounded function on $[-\pi, \pi]$, 211
  of step function on $[-\pi, \pi]$, 210
  of unbounded function on $[-\pi, \pi]$, 212
Riemann mapping theorem, 435
Riemann sphere, 474

Riemann surface, 454
Riemann–Lebesgue lemma
  for Fourier series, 199
  for Fourier transforms, 497
Riesz–Thorin interpolation, 508
right-hand derivative ($f^{\langle}(x)$), 203
right-hand limit ($\lim_{y \searrow x} f(y)$), 202
river bank (flow), 440
Robin boundary conditions
  homogeneous, *see* mixed boundary conditions,
    homogeneous
  nonhomogeneous, *see* mixed boundary conditions,
    nonhomogeneous
Rodrigues formula, 372
root, complex, 454
roots of unity, 454
Rydberg, J. R, 55

scalar conservation law, 70, 101
Schrödinger equation
  abstract, 43, 48
  of electron in Coulomb field, 43
  is evolution equation, 73, 101
  of free electron, 43
    solution, 44
  of hydrogen atom, 44
  is linear, 71
  momentum representation, 516
  positional, 43
Schrödinger equation, stationary, 48, 74
  of free electron, 48
  hydrogen atom, 53
  potential well (one-dimensional)
    finite voltage, 48
    infinite voltage, 51
  potential well (three-dimensional), 52
  as Sturm–Liouville equation, 350
sectionally smooth boundary, *see* piecewise smooth
    boundary 148
self-adjoint operator
  ∂_x^2, 347
  definition, 346
  eigenfunctions are orthogonal, 351
  Laplacian, 348
  multiplication operators, 347
  Sturm–Liouville operator, 349
semidifferentiable function, 203
separation constant, 360, 362
separation of variables
  boundary conditions, 379
  bounded solutions, 378
  description
    many dimensions, 361
    two dimensions, 359
  Laplace equation
    many-dimensional, 362
    two-dimensional, 360, 376
  telegraph equation, 377, 379
sesquilinearity, 114
sigmoid function Φ, 394
simple closed curve, *see* contour

simple function, 213
simple pole, 469
simply connected function, 435, 453
sine series, *see* Fourier series, sine
singular point for ODE, 576
singularity
　essential, 475
　of holomorphic function, 445
　indelible, 469
　pole, 469
　removable, 469
small 'o' notation, *see* order $o(1/z)$
smooth approximation (of function), 415
smooth boundary, 88
　in $\mathbb{R}^D$, 566
smooth graph, 87
smooth hypersurface, 88
smoothness vs. asymptotic decay
　of Fourier coefficients, 208
　of Fourier transform, 501
soap bubble example, 285
solution kernel, 385
sourceless flow, 424
spectral signature, 56
spectral theory, 355
spherical coordinates, 559
spherical harmonics, 355
spherical mean
　definition, 26
　formula for Laplacian, 18, 26
　Poisson solution to wave equation, 537
　solution to three-dimensional wave equation, 538
spherically symmetric function, 20
stable family of probability distributions,
　418, 529
standing wave
　one-dimensional, 32
　two-dimensional, 34
stationary Schrödinger equation, *see* Schrödinger
　equation, stationary
$\mathsf{Step}[-\pi, \pi]$, *see* step function
step function, 156, 168, 210
Stokes theorem, *see* divergence theorem
streamline, 436
struck string problem, 236
Sturm–Liouville equation, 350
Sturm–Liouville operator
　is self-adjoint, 349
　self-adjoint, eigenvalues of, 354
subtraction principle for nonhomogeneous linear
　PDE, 70
summation operator, 64
superposition principle for homogeneous linear PDE,
　69

tangent (hyper)plane, 565
tangent line, 564
Taylor polynomial
　many-dimensional, 579
　one-dimensional, 573
　two-dimensional, 578

Taylor series, 574
Taylor's theorem
　many-dimensional, 579
　one-dimensional, 573
　two-dimensional, 578
telegraph equation
　definition, 36
　is evolution equation., 73
　polynomial formalism, 377, 379
　separated solution, 377, 379
tent function, 159, 163
Thompson, G. P, 39
topological group, 179
torus, 86
total variation, 177
trajectory, of flow, 436
transport equation, 21
travelling wave
　one-dimensional, 32
　two-dimensional, 35
trigonometric orthogonality, 117, 119

uncertainty principle, examples
　electron with known velocity, 46
　normal (Gaussian) distribution, 503,
　518
uniform convergence, *see* convergence, uniform
　Abel's test, 135
　Cauchy's criterion, 135
　of continuous functions, 133
　of derivatives, 133
　Dirichlet test, 135
　of integrals, 133
　Weierstrass M-test, 135
uniform distance, 130
uniform norm ($\|f\|_\infty$), 129
unique solution
　to Dirichlet problem, 90
　of forced heat equation, 93
　of forced wave equation, 96
　of heat equation, 93
　of Laplace equation, 89
　to Neumann problem, 90
　of Poisson equation, 91
　of wave equation, 96

vector addition, 61
velocity vector, of a contour in $\mathbb{C}$, 440
vibrating string
　initial position, 234
　initial velocity, 236
violin string, 97
voltage contour, 437

wave equation
　conservation of energy, 94
　definition, 35
　derivation and physical interpretation
　　one dimension, 32
　　two dimensions, 34
　on disk, Fourier–Bessel solution, 310

is evolution equation, 73
is homogeneous linear, 68
is hyperbolic PDE, 98, 100
on interval
    d'Alembert solution, 407
    initial position, Fourier solution, 234
    initial velocity, Fourier solution, 236
on real line
    d'Alembert solution, 404, 535
    Fourier transform solution, 535
    initial position, d'Alembert (travelling wave)
      solution, 399
    initial velocity, d'Alembert (ripple) solution,
      401
on square
    initial position, Fourier solution, 264
    initial velocity, Fourier solution, 265
on three-dimensional space
    Fourier transform solution, 536

Huygens' principle, 540
    Poisson's (spherical mean) solution,
      538
    on two-dimensional plane, Fourier transform
      solution, 535
    unique solution of, 96
wave vector
    many dimensions, 36
    two dimensions, 35
wavefunction
    phase, 46
    probabilistic interpretation, 42
wavelet basis, 121
    convergence in L^2, 124
    pointwise convergence, 129
Weierstrass M-test, 135
wind instrument, 97

xylophone, 237

Notation index

Sets and domains

$\mathbb{A}(r, R)$ The two-dimensional *closed annulus* of inner radius r and outer radius R: the set of all $(x, y) \in \mathbb{R}^2$ such that $r \leq x^2 + y^2 \leq R$.

$°\mathbb{A}(r, R)$ The two-dimensional *open annulus* of inner radius r and outer radius R: the set of all $(x, y) \in \mathbb{R}^2$ such that $r < x^2 + y^2 < R$.

$\mathbb{B}$ A D-dimensional *closed ball* (often the unit ball centred at the origin).

$\mathbb{B}(\mathbf{x}, \epsilon)$ The D-dimensional *closed ball* of radius ϵ around the point $\mathbf{x}$: the set of all $\mathbf{y} \in \mathbb{R}^D$ such that $\|\mathbf{x} - \mathbf{y}\| < \epsilon$.

$\mathbb{C}$ The set of *complex numbers* of the form $x + y\mathbf{i}$ where $x, y \in \mathbb{R}$ and $\mathbf{i}$ is the square root of -1.

$\mathbb{C}_+$ The set of complex numbers $x + y\mathbf{i}$ with $y > 0$.

$\mathbb{C}_-$ The set of complex numbers $x + y\mathbf{i}$ with $y < 0$.

$\widehat{\mathbb{C}} = \mathbb{C} \sqcup \{\infty\}$ the *Riemann sphere* (the range of a meromorphic function).

$\mathbb{D}$ A two-dimensional *closed disk* (usually the unit disk centred at the origin).

$\mathbb{D}(R)$ A two-dimensional *closed disk of radius* R, centred at the origin: the set of all $(x, y) \in \mathbb{R}^2$ such that $x^2 + y^2 \leq R$.

$°\mathbb{D}(R)$ A two-dimensional *open disk of radius* R, centred at the origin: the set of all $(x, y) \in \mathbb{R}^2$ such that $x^2 + y^2 < R$.

$\mathbb{D}^{\mathbb{C}}(R)$ A two-dimensional *closed codisk of coradius* R, centred at the origin: the set of all $(x, y) \in \mathbb{R}^2$ such that $x^2 + y^2 \geq R$.

$°\mathbb{D}^{\mathbb{C}}(R)$ A two-dimensional *open codisk of coradius* R, centred at the origin: the set of all $(x, y) \in \mathbb{R}^2$ such that $x^2 + y^2 > R$.

$\mathbb{H}$ A *half-plane*. Usually $\mathbb{H} = \{(x, y) \in \mathbb{R}^2; \ y \geq 0\}$ (the upper half-plane).

$\mathbb{N} = \{0, 1, 2, 3, \ldots\}$ The set of *natural numbers*.

$\mathbb{N}_+ = \{1, 2, 3, \ldots\}$ The set of *positive natural numbers*.

$\mathbb{N}^D$ The set of all $\mathbf{n} = (n_1, n_2, \ldots, n_D)$, where $n_1, \ldots, n_D$ are natural numbers.

$\emptyset$ The empty set, also denoted $\{\}$.

$\mathbb{Q}$ The *rational numbers*: the set of all fractions n/m, where $n, m \in \mathbb{Z}$ and $m \neq 0$.

$\mathbb{R}$ The set of *real numbers* (e.g. $2, -3, \sqrt{7} + \pi$, etc.)

$\mathbb{R}_+ := (0, \infty) = \{r \in \mathbb{R}; \ r \geq 0\}$.

$\mathbb{R}_{\not+} := [0, \infty) = \{r \in \mathbb{R}; \ r \geq 0\}$.

$\mathbb{R}^2$ The two-dimensional infinite plane: the set of all ordered pairs (x, y), where $x, y \in \mathbb{R}$.

$\mathbb{R}^D$ D-dimensional space: the set of all D-*tuples* $(x_1, x_2, \dots, x_D)$, where $x_1, x_2, \dots, x_D \in \mathbb{R}$. Sometimes we will treat these D-tuples as *points* (representing locations in physical space); normally points will be indicated in bold face, e.g. $\mathbf{x} = (x_1, \dots, x_D)$. Sometimes we will treat the D-tuples as *vectors* (pointing in a particular direction); then they will be indicated with arrows, $\vec{\mathbf{V}} = (V_1, V_2, \dots, V_D)$.

$\mathbb{R}^D \times \mathbb{R}$ The set of all pairs $(\mathbf{x}; t)$, where $\mathbf{x} \in \mathbb{R}^D$ is a vector and $t \in \mathbb{R}$ is a number. (Of course, mathematically, this is the same as $\mathbb{R}^{D+1}$, but sometimes it is useful to regard the last dimension as 'time'.)

$\mathbb{R} \times \mathbb{R}_+$ The *half-space* of all points $(x, y) \in \mathbb{R}^2$, where $y \geq 0$.

$\mathbb{S}$ The two-dimensional *unit circle*: the set of all $(x, y) \in \mathbb{R}^2$ such that $x^2 + y^2 = 1$.

$\mathbb{S}^{D-1}(\mathbf{x}; R)$ The D-dimensional *sphere*; of radius R around the point $\mathbf{x}$: the set of all $\mathbf{y} \in \mathbb{R}^D$ such that $\|\mathbf{x} - \mathbf{y}\| = R$

$\mathbb{U}, \mathbb{V}, \mathbb{W}$ Usually denote open subsets of $\mathbb{R}^D$ or $\mathbb{C}$.

$\mathbb{X}, \mathbb{Y}$ Usually denote *domains* – closed connected subsets of $\mathbb{R}^D$ with dense interiors.

$\mathbb{Z}$ The *integers* $\{\dots, -2, -1, 0, 1, 2, 3, \dots\}$.

$\mathbb{Z}^D$ The set of all $\mathbf{n} = (n_1, n_2, \dots, n_D)$, where $n_1, \dots, n_D$ are integers.

$[1...D]$ $= \{1, 2, 3, \dots, D\}$.

$[0, \pi]$ The *closed interval* of length π: the set of all real numbers x, where $0 \leq x \leq \pi$.

$(0, \pi)$ The *open interval* of length π: the set of all real numbers x, where $0 < x < \pi$.

$[0, \pi]^2$ The (closed) $\pi \times \pi$ *square*: the set of all points $(x, y) \in \mathbb{R}^2$, where $0 \leq x, y \leq \pi$.

$[0, \pi]^D$ The D-dimensional *unit cube*: the set of all points $(x_1, \dots, x_D) \in \mathbb{R}^D$, where $0 \leq x_d \leq 1$ for all $d \in [1...D]$.

$[-L, L]$ The *interval* of all real numbers x with $-L \leq X \leq L$.

$[-L, L]^D$ The D-dimensional *cube* of all points $(x_1, \dots, x_D) \in \mathbb{R}^D$, where $-L \leq x_d \leq L$ for all $d \in [1...D]$.

Set operations

$\text{int}(\mathbb{X})$ The *interior* of the set $\mathbb{X}$ (i.e. all points in $\mathbb{X}$ *not* on the boundary of $\mathbb{X}$).

$\bigcap$ *Intersection*   If $\mathbb{X}$ and $\mathbb{Y}$ are sets, then $\mathbb{X} \cap \mathbb{Y} := \{z \; ; \; z \in \mathbb{X} \text{ and } z \in \mathbb{Y}\}$. If $\mathbb{X}_1, \dots, \mathbb{X}_N$ are sets, then $\bigcap_{n=1}^N \mathbb{X}_n := \mathbb{X}_1 \cap \mathbb{X}_2 \cap \cdots \cap \mathbb{X}_N$.

$\bigcup$ *Union*   If $\mathbb{X}$ and $\mathbb{Y}$ are sets, then $\mathbb{X} \cup \mathbb{Y} := \{z \; ; \; z \in \mathbb{X} \text{ or } z \in \mathbb{Y}\}$. If $\mathbb{X}_1, \dots, \mathbb{X}_N$ are sets, then $\bigcup_{n=1}^N \mathbb{X}_n := \mathbb{X}_1 \cup \mathbb{X}_2 \cup \cdots \cup \mathbb{X}_N$.

$\bigsqcup$ *Disjoint union*   If $\mathbb{X}$ and $\mathbb{Y}$ are sets, then $\mathbb{X} \sqcup \mathbb{Y}$ means the same as $\mathbb{X} \cup \mathbb{Y}$, but conveys the added information that $\mathbb{X}$ and $\mathbb{Y}$ are *disjoint* – i.e. $\mathbb{X} \cap \mathbb{Y} = \emptyset$. Likewise, $\bigsqcup_{n=1}^N \mathbb{X}_n := \mathbb{X}_1 \sqcup \mathbb{X}_2 \sqcup \cdots \sqcup \mathbb{X}_N$.

$\backslash$ *Difference*   If $\mathbb{X}$ and $\mathbb{Y}$ are sets, then $\mathbb{X} \setminus \mathbb{Y} = \{x \in \mathbb{X} \; ; \; x \notin \mathbb{Y}\}$.

Spaces of functions

$\mathcal{C}^\infty$ A vector space of (infinitely) differentiable functions. Some examples are as follows:

- $\mathcal{C}^\infty[\mathbb{R}^2; \mathbb{R}]$: the space of differentiable scalar fields on the plane.
- $\mathcal{C}^\infty[\mathbb{R}^D; \mathbb{R}]$: the space of differentiable scalar fields on D-dimensional space.
- $\mathcal{C}^\infty[\mathbb{R}^2; \mathbb{R}^2]$: the space of differentiable vector fields on the plane.

$\mathcal{C}_0^\infty[0,1]^D$ The space of differentiable scalar fields on the cube $[0,1]^D$ satisfying *Dirichlet boundary conditions*: $f(\mathbf{x}) = 0$ for all $\mathbf{x} \in \partial[0,1]^D$.

$\mathcal{C}_\perp^\infty[0,1]^D$ The space of differentiable scalar fields on the cube $[0,1]^D$ satisfying *Neumann boundary conditions*: $\partial_\perp f(\mathbf{x}) = 0$ for all $\mathbf{x} \in \partial[0,1]^D$.

$\mathcal{C}_h^\infty[0,1]^D$ The space of differentiable scalar fields on the cube $[0,1]^D$ satisfying *mixed boundary conditions*: $\frac{\partial_\perp f}{f}(\mathbf{x}) = h(\mathbf{x})$ for all $\mathbf{x} \in \partial[0,1]^D$.

$\mathcal{C}_{\text{per}}^\infty[-\pi,\pi]$ The space of differentiable scalar fields on the interval $[-\pi,\pi]$ satisfying *periodic boundary conditions*.

$\mathbf{L}^1(\mathbb{R})$ The set of all functions $f : \mathbb{R} \longrightarrow \mathbb{R}$ such that $\int_{-\infty}^{\infty} |f(x)|\, dx < \infty$.

$\mathbf{L}^1(\mathbb{R}^2)$ The set of all functions $f : \mathbb{R}^2 \longrightarrow \mathbb{R}$ such that $\int_{-\infty}^{\infty} \int_{-\infty}^{\infty} |f(x,y)|\, dx\, dy < \infty$.

$\mathbf{L}^1(\mathbb{R}^3)$ The set of all functions $f : \mathbb{R}^3 \longrightarrow \mathbb{R}$ such that $\int_{\mathbb{R}^3} |f(\mathbf{x})|\, d\mathbf{x} < \infty$.

$\mathbf{L}^2(\mathbb{X})$ The set of all functions $f : \mathbb{X} \longrightarrow \mathbb{R}$ such that $\|f\|_2 = (\int_{\mathbb{X}} |f(\mathbf{x})|^2\, d\mathbf{x})^{1/2} < \infty$.

$\mathbf{L}^2(\mathbb{X}; \mathbb{C})$: The set of all functions $f : \mathbb{X} \longrightarrow \mathbb{C}$ such that $\|f\|_2 = (\int_{\mathbb{X}} |f(\mathbf{x})|^2\, d\mathbf{x})^{1/2} < \infty$.

Derivatives and boundaries

$\partial_k f = \frac{df}{dx_k}$.

$\nabla f = (\partial_1 f, \partial_2 f, \ldots, \partial_D f)$ The *gradient* of scalar field f.

$\mathrm{div}\, f = \partial_1 f_1 + \partial_2 f_2 + \cdots + \partial_D f_D$ The *divergence* of vector field f.

$\partial_\perp f$ The derivative of f *normal* to the boundary of some region. Sometimes this is written as $\frac{\partial f}{\partial \mathbf{n}}$ or $\frac{\partial f}{\partial \nu}$, or as $\nabla f \cdot \mathbf{n}$.

$\triangle f = \partial_1^2 f + \partial_2^2 f + \cdots + \partial_D^2 f$ Sometimes this is written as $\nabla^2 f$.

$\mathsf{L}\, f$ Sometimes means a general *linear differential operator* L being applied to the function f.

$\mathsf{SL}_{s,q}(\phi) = s \cdot \partial^2 \phi + s' \cdot \partial \phi + q \cdot \phi$ Here, $s, q : [0, L] \longrightarrow \mathbb{R}$ are predetermined functions, and $\phi : [0, L] \longrightarrow \mathbb{R}$ is the function we are operating on by the *Sturm–Liouville operator* $\mathsf{SL}_{s,q}$.

$\dot{\gamma} = (\gamma_1', \ldots, \gamma_D')$ The *velocity vector* of the path $\gamma : \mathbb{R} \longrightarrow \mathbb{R}^D$.

$\partial \mathbb{X}$ If $\mathbb{X} \subset \mathbb{R}^D$ is some region in space, then $\partial \mathbb{X}$ is the *boundary* of that region. For example,

- $\partial[0,1] = \{0, 1\}$;
- $\partial \mathbb{B}^2(0; 1) = \mathbb{S}^2(0; 1)$;
- $\partial \mathbb{B}^D(\mathbf{x}; R) = \mathbb{S}^D(\mathbf{x}; R)$;
- $\partial (\mathbb{R} \times \mathbb{R}_{\neq}) = \mathbb{R} \times \{0\}$.

Norms and inner products

$\|\mathbf{x}\|$ If $\mathbf{x} \in \mathbb{R}^D$ is a vector, then $\|\mathbf{x}\| = \sqrt{x_1^2 + x_2^2 + \cdots + x_D^2}$ is the *norm* (or *length*) of $\mathbf{x}$.

$\|f\|_2$ Let $\mathbb{X} \subset \mathbb{R}^D$ be a bounded domain, with volume $M = \int_{\mathbb{X}} 1\, d\mathbf{x}$. If $f : \mathbb{X} \longrightarrow \mathbb{R}$ is an integrable function, then $\|f\|_2 = \frac{1}{M} \left(\int_{\mathbb{X}} |f(\mathbf{x})|\, d\mathbf{x} \right)^{1/2}$ is the *L^2-norm* of f.

$\langle f, g \rangle$ If $f, g : \mathbb{X} \longrightarrow \mathbb{R}$ are integrable functions, then their *inner product* is given by $\langle f, g \rangle = \frac{1}{M} \int_{\mathbb{X}} f(\mathbf{x}) \cdot g(\mathbf{x}) d\mathbf{x}$.

$\|f\|_1$ Let $\mathbb{X} \subseteq \mathbb{R}^D$ be any domain. If $f : \mathbb{X} \longrightarrow \mathbb{R}$ is an integrable function, then $\|f\|_\infty = \int_{\mathbb{X}} |f(\mathbf{x})|\, d\mathbf{x}$ is the *L^1-norm* of f.

$\|f\|_\infty$ Let $\mathbb{X} \subseteq \mathbb{R}^D$ be any domain. If $f : \mathbb{X} \longrightarrow \mathbb{R}$ is a bounded function, then $\|f\|_\infty = \sup_{\mathbf{x} \in \mathbb{X}} |f(\mathbf{x})|$ is the L^∞-*norm* of f.

Other operations on functions

A_n Normally denotes the nth *Fourier cosine coefficient* of a function f on $[0, \pi]$ or $[-\pi, \pi]$. That is, $A_n := \frac{2}{\pi} \int_0^\pi f(x) \cos(nx) \mathrm{d}x$ or $A_n := \frac{1}{\pi} \int_{-\pi}^\pi f(x) \cos(nx) \mathrm{d}x$.

$A_{n,m}$ Normally denotes a two-dimensional Fourier cosine coefficient, and $A_\mathbf{n}$ normally denotes a D-dimensional Fourier cosine coefficient.

B_n Normally denotes the nth *Fourier cosine coefficient* of a function f on $[0, \pi]$ or $[-\pi, \pi]$. That is, $B_n := \frac{2}{\pi} \int_0^\pi f(x) \sin(nx) \mathrm{d}x$ or $A_n := \frac{1}{\pi} \int_{-\pi}^\pi f(x) \sin(nx) \mathrm{d}x$.

$B_{n,m}$ Normally denotes a two-dimensional Fourier sine coefficient, and $B_\mathbf{n}$ normally denotes a D-dimensional Fourier sine coefficient.

$f * g$ If $f, g : \mathbb{R}^D \longrightarrow \mathbb{R}$, then their *convolution* is the function $f * g : \mathbb{R}^D \longrightarrow \mathbb{R}$ defined by $(f * g)(\mathbf{x}) = \int_{\mathbb{R}^D} f(\mathbf{y}) \cdot g(\mathbf{x} - \mathbf{y}) \mathrm{d}\mathbf{y}$.

$\widehat{f}(\boldsymbol{\mu}) = \frac{1}{(2\pi)^D} \int_{\mathbb{R}^D} f(\mathbf{x}) \exp(\mathbf{i}\,\boldsymbol{\mu} \bullet \mathbf{x}) \mathrm{d}\mathbf{x}$ The *Fourier transform* of the function $f : \mathbb{R}^D \longrightarrow \mathbb{C}$. It is defined for all $\boldsymbol{\mu} \in \mathbb{R}^D$.

$\widehat{f}_n = \frac{1}{2\pi} \int_{-\pi}^\pi f(x) \exp(\mathbf{i}nx) \mathrm{d}x$ The nth *complex Fourier coefficient* of a function $f : [-\pi, \pi] \longrightarrow \mathbb{C}$ (here $n \in \mathbb{Z}$).

$\displaystyle\oint_\gamma f = \int_0^S f[\gamma(s)] \cdot \dot{\gamma}(s) \mathrm{d}s$ A *chain integral*. Here, $f : \mathbb{U} \longrightarrow \mathbb{C}$ is some complex-valued function and $\gamma : [0, S] \longrightarrow \mathbb{U}$ is a *chain* (a piecewise continuous, piecewise differentiable path).

$\displaystyle\oint_\gamma f$ A *contour integral*. The same definition as the chain integral $\oint_\gamma f$, but γ is a contour.

$\alpha \diamond \beta$ If α and β are two chains, then $\alpha \diamond \beta$ represents the *linking* of the two chains.

$\mathcal{L}[f] = \int_0^\infty f(t) \mathrm{e}^{-ts} \, \mathrm{d}t$ The *Laplace transform* of the function $f : \mathbb{R}_+ \longrightarrow \mathbb{C}$; it is defined for all $s \in \mathbb{C}$ with $\mathrm{Re}\,[s] > \alpha$, where α is the exponential order of f.

$\mathbf{M}_R u(\mathbf{x}) = \frac{1}{4\pi R^2} \int_{\mathbb{S}(R)} f(\mathbf{x} + \mathbf{s}) \mathrm{d}\mathbf{s}$ The *spherical average* of f at $\mathbf{x}$, of radius R. Here, $\mathbf{x} \in \mathbb{R}^3$ is a point in space, $R > 0$, and $\mathbb{S}(R) = \left\{ \mathbf{s} \in \mathbb{R}^3 ; \|\mathbf{s}\| = R \right\}$.

Special functions

$\mathbf{C}_n(x) = \cos(nx)$, for all $n \in \mathbb{N}$ and $x \in [-\pi, \pi]$.

$\mathbf{C}_{n,m}(\mathbf{x}) = \cos(nx) \cdot \cos(my)$, for all $n, m \in \mathbb{N}$ and $(x, y) \in [\pi, \pi]^2$.

$\mathbf{C}_\mathbf{n}(\mathbf{x}) = \cos(n_1 x_1) \cdots \cos(n_D x_D)$, for all $\mathbf{n} \in \mathbb{N}^D$ and $\mathbf{x} \in [\pi, \pi]^D$.

$\mathbf{D}_N(x) = 1 + 2 \sum_{n=1}^N \cos(nx)$ The nth *Dirichlet kernel*, for all $n \in \mathbb{N}$ and $x \in [-2\pi, 2\pi]$.

$\mathbf{E}_n(x) = \exp(\mathbf{i}nx)$, for all $n \in \mathbb{Z}$ and $x \in [-\pi, \pi]$.

$\mathcal{E}_\mu(x) = \exp(\mathbf{i}\mu x)$, for all $\mu \in \mathbb{R}$ and $x \in [-\pi, \pi]$.

$\mathcal{G}(x; t) = \frac{1}{2\sqrt{\pi t}} \exp(\frac{-x^2}{4t})$ The (one-dimensional) *Gauss–Weierstrass kernel*.

$\mathcal{G}(x, y; t) = \frac{1}{4\pi t} \exp(\frac{-x^2 - y^2}{4t})$ The (two-dimensional) *Gauss–Weierstrass kernel*.

$\mathcal{G}(\mathbf{x}; t) = \frac{1}{(4\pi t)^{D/2}} \exp(\frac{-\|\mathbf{x}\|^2}{4t})$ The (D-dimensional) *Gauss–Weierstrass kernel*.

$\mathcal{J}_n$ The nth *Bessel function of the first kind*.

$\mathcal{K}_y(x) = \frac{y}{\pi(x^2 + y^2)}$ The *half-plane Poisson kernel*, for all $x \in \mathbb{R}$ and $y > 0$.

$\vec{\mathbf{N}}(\mathbf{x})$ The *outward unit normal vector* to a domain $\mathbb{X}$ at a point $\mathbf{x} \in \partial\mathbb{X}$.

$\mathcal{P}(\mathbf{x}, \mathbf{s}) = \frac{R^2 - \|\mathbf{x}\|^2}{\|\mathbf{x} - \mathbf{s}\|^2}$ The *Poisson kernel* on the disk, for all $\mathbf{x} \in \mathbb{D}$ and $\mathbf{s} \in \mathbb{S}$.

$\mathbf{P}_r(x) = \frac{1-r^2}{1-2r\cos(x)+r^2}$ The *Poisson kernel* in polar coordinates, for all $x \in [-2\pi, 2\pi]$ and $r < 1$.

Φ_n, ϕ_n, Ψ_n **and** ψ_n Refer to the harmonic functions on the unit disk which separate in polar coordinates: $\Phi_0(r, \theta) = 1$ and $\phi_0(r, \theta) = \log(r)$, while, for all $n \geq 1$, we have
$$\Phi_n(r, \theta) = \cos(n\theta) \cdot r^n, \quad \Psi_n(r, \theta) = \sin(n\theta) \cdot r^n, \quad \phi_n(r, \theta) = \frac{\cos(n\theta)}{r^n}, \text{ and }$$
$\psi_n(r, \theta) = \frac{\sin(n\theta)}{r^n}$.

$\Phi_{n,\lambda}, \Psi_{n,\lambda}, \phi_{n,\lambda},$ **and** $\psi_{n,\lambda}$ Refer to eigenfunctions of the Laplacian on the unit disk which separate in polar coordinates. For all $n \in \mathbb{N}$ and $\lambda > 0$,
$$\Phi_{n,\lambda}(r, \theta) = \mathcal{J}_n(\lambda \cdot r) \cdot \cos(n\theta), \quad \Psi_{n,\lambda}(r, \theta) = \mathcal{J}_n(\lambda \cdot r) \cdot \sin(n\theta),$$
$$\phi_{n,\lambda}(r, \theta) = \mathcal{Y}_n(\lambda \cdot r) \cdot \cos(n\theta), \text{ and } \psi_{n,\lambda}(r, \theta) = \mathcal{Y}_n(\lambda \cdot r) \cdot \sin(n\theta)$$

$\mathbf{S}_n(x) = \sin(nx)$, for all $n \in \mathbb{N}$ and $x \in [\pi, \pi]$.

$\mathbf{S}_{n,m}(\mathbf{x}) = \sin(nx) \cdot \sin(my)$, for all $n, m \in \mathbb{N}$ and $(x, y) \in [\pi, \pi]^2$.

$\mathbf{S}_\mathbf{n}(\mathbf{x}) = \sin(n_1 x_1) \cdots \sin(n_D x_D)$, for all $\mathbf{n} \in \mathbb{N}^D$ and $\mathbf{x} \in [\pi, \pi]^D$.

$\mathcal{Y}_n$ The nth *Bessel function of the second kind*.